AF669210

AGRICULTURAL INCENTIVES IN INDIA

Centre de Sciences Humaines (Centre for Social Sciences and Humanities): Created in New Delhi in 1989, the CSH is part of the network of research centres of the French Ministry of Foreign Affairs. The Centre's research work is primarily oriented towards the study of issues concerning the contemporary dynamics of development in India and South Asia. The activities of the Centre are focused on four main themes, namely: Economic transition and sustainable development, political dynamics, institutional set-up and social transformations, regional dynamics in South Asia and international relations and urban dynamics.

Centre de Sciences Humaines, 2, Aurangzeb Road, New Delhi 110011, India,
Tel: (91) 11 2301 62 59, Fax: (91) 11 2301 84 80
E-mail: public@csh-delhi.com
Website: http://www.csh.delhi.com

Institut Français de Pondichéry (French Institute of Pondicherry): Created in 1955, the IFP is a multidisciplinary research and advanced educational institute. Major research works are focusing on Sanskrit and Tamil languages and literatures in close collaboration with the Ecole Française d'Extrême-Orient-ecosystems, biodiversity and sustainability, dynamics of population and socio-economic development.

Institut Français de Pondichéry, 11 Saint Louis Street, PB 33, Pondicherry 605001,
Tel: (91) 413 2334 168/2334 170, Fax: (91) 413 2339 534
E-mail: ifpdir@ifpindia.org
Website: http://www.ifpindia.org

AGRICULTURAL INCENTIVES IN INDIA

Past Trends and Prospective Paths towards Sustainable Development

edited by

BRUNO DORIN AND THOMAS JULLIEN

CENTRE DE SCIENCES HUMAINES
2004

First published November 2004

© Individual Contributors, 2004

All right reserved. No part of this publication may be reproduced of transmitted, in any form or by any means, without prior permission of editors and the publishers

ISBN 81-7304-589-5

Published by
Ajay Kumar Jain for
Manohar Publishers & Distributors
4753/23 Ansari Road, Daryaganj
New Delhi 110002

Typeset by
A J Software Publishing Co. Pvt. Ltd.
New Delhi 110005

Printed at
Lordson Publishers Pvt. Ltd.
Delhi 110007

Contents

PART III

FUTURE SCENARIOS AND PROSPECTIVE MODELING

Tables

Illustrations

Abbreviations

AMS	Aggregate Measure of Support
CACP	Commission for Agricultural Costs and Prices
CAD	Command Area Development
CERC	Central Electricity Regulatory Commission
GCA	Gross Cropped Area
GIA	Gross Irrigated Area
GIS	Geographical Information System
HYVs	High Yielding Varieties
IPPs	Independent Power Producers
K	Potassium
MSP	Minimum Support Price
N	Nitrogen
NIA	Net Irrigated Area
O&M	Operation and Maintenance
P	Phosphate
PDS	Public Distribution System
RPDS	Revemped Public Distribution System
RPS	Retention Price Scheme
SEB	State Electricity Board
SERC	State Electricity Regulatory Commission
T&D	Transmission and Distribution
TPDS	Targeted Public Distribution System

Introduction

BRUNO DORIN, THOMAS JULLIEN

This book gathers twelve papers which sustained the discussions and conclusions of an Indo-French seminar organized by the Centre de Sciences Humaines (CSH, New Delhi) on 3 and 4 April 2000 at the India International Centre (IIC, New Delhi).

The objective of this meeting was rather ambitious and sensitive: to debate the relevance and sustainability of a nearly forty-years old system of public incentives to Indian agriculture, mainly subsidies to water, electricity and fertilizers.

The sensitivity of the subject, as also its pertinence, is rooted in the difficult challenge that India has had to take up since the early 1990s: to liberalize and open to the world its domestic market in order to bypass some inefficiencies or failures of its mixed economy, without selling of in the process its decision-making independence, as well as some social and environmental objectives peculiar to the subcontinent or to the world community.

This economical upheaval pushes agriculture towards the front of the stage for many good reasons. First, it has to feed one-sixth of the world population, among which there are still many families in dire need, but also now a growing middle-class which aspires to diversify its food basket. Second, it is the first employer in India, giving jobs to millions and means of subsistence not only to cultivators and agricultural labourers, but also to the multitudes working at the pre- or post-production stages. Third, it supports inestimable cultural and natural heritages, which of course must not be destroyed, but preserved and transmitted to the future generations. Finally, and since the GATT agreements on agriculture signed in 1994 at Marrakech, Indian agriculture is also involved in a process which challenges our human ability to transcend various frontiers, which is a very disturbing exercise even if it should lead us to better share and enhance in the long run the fabulous richness of the world.

The above problematics were not either easy to scan within a two-day seminar, since agricultural incentives play or can play a crucial role in all of them. However, from nutritional or equity issues to the international scene, without forgetting budgetary or environmental considerations, we had not only a rather large panorama of the past trends and lessons, but also some prospective scenarios and tools to better shape the future. Each of these presentations were followed by a discussion, which were fruitful thanks to the quality of the participants[1] as also to six eminent chairpersons[2] who have skillfully conducted the debates. An output of this seminar is the following report,[3] beside the papers presented and included in this book which, the authors hope, will serve as a basic reference to policy-makers as well as to academics.

CONSTRAINTS AND BOTTLENECKS OF INPUT SUBSIDIES

Ever since the Green Revolution, and more significantly since the 1980s, Union and states governments have been devoting an increasing share of their budget to subsidize the production and the consumption of fertilizers (S.S. Acharya), water (A. Vaidynathan) and electricity for pumping the latter (A. Sen), which obviously jeopardizes the fixed capital investments in agriculture as well as the good working of governmental administrations. However, it is difficult to assess the benefits that actually accrue to agricultural producers given the fact that there is a general system of cross-subsidies for electricity, a sharing of subsidies with industrialists for fertilizers, and some capital costs for irrigation which are difficult to evaluate. Nonetheless, it seems that at least US$ 6.6 billion have been transferred to agriculture through such subsidies in 1995 (B. Dorin and T. Jullien). This amount, which is 1 billion higher than the one declared by India to the WTO for the same year, accounts for 8.6 per cent of agricultural GDP whereas it was only 2.7 per cent fifteen years earlier. It represents also an all-India average of $35 per ha of gross cropped area in 1995, or an annual income transfer of $188 per cultivator in a rich state like Punjab, while it is only 14 in a poor state like Orissa. Similarly, only three crops absorb more than 60 per cent of a so-called 'non-product-specific' support to agriculture: rice, wheat and sugarcane whose market prices are consequently more attractive and competitive than those of other crops. At the same time, low-priced inputs lead to overuse or waste, causing alarming environmental problems like the dramatic depletion of groundwater tables (A. Vaidyanathan). They also limit the recourse to traditional inputs, such as organic manure which has become rare and expensive whereas it conditions the long-term efficiency of chemical fertilizers (B.C. Barah). All in all, they certainly made India self-sufficient in wheat and rice, but the country is now highly deficient in other food products such as oilseeds and pulses which have been relegated to dry lands or dry seasons (B. Dorin); widespread famines are over but under-nutrition and malnutrition (proteins, lipids, iron, iodine, vitamins, etc.) prevail at a large scale (A. Shariff). The country is thus not yet food secured whereas this industrialization of agriculture has placed Indian farmers under the control of few multinational suppliers of modern inputs (V. Shiva).

MERITS AND CHALLENGES OF AGRICULTURAL INCENTIVES

On the other hand, the growing consumption of subsidized inputs led to an important increase in agricultural production which would have been nearly impossible otherwise since the growing number of Indian cultivators (+ 29 per cent between 1980 and 1995) hampers technical progress which usually takes place when labour can be replaced by capital (B. Dorin, N. Pingault and J.M. Boussard). By reducing the cost of production of some essential food items, input subsidies are also a way of controlling the price of staples in a low-income economy (S.S. Acharya). Moreover, this policy of 'low inputs low outputs prices', coupled with the Public Distribution System (PDS) providing cheap rice, wheat and sugar from surplus production areas, should free deprived farmers from producing such commodities so that they can diversify and earn more from other ones (A. Sen). In any case, Indian agriculture appears to be globally highly taxed according to GATT's criteria: in 1997, the Aggregate Measurement of specific and non-specific Support (AMS) represented minus 62 per cent of the value of agricultural production (A. Gulati). This is why India has been released from any international commitment to reduce its domestic support to agriculture, and also why it considers that its farmers cannot fairly compete with those from Europe or the USA who enjoy much larger subsidies now hidden in the Blue and Green Boxes (G.S. Bhalla). Besides, supporting one's agriculture for ensuring one's food self-sufficiency is a fundamental right, just as it is for the farmer to receive water on time and in sufficient quantities (A. Vaidyanathan). In this respect, India must be able to meet its future domestic demand. By the year 2020, it should not exceed 240 million tons in foodgrains (cereals and pulses), i.e. + 69 million tons in comparison to 1993, including 24 million tons for cattle (P.C. Bansil). But the increase in other foods should be much more, especially that of meat, eggs, fish, milk, vegetables and fruits since their demand by 2030 could go up threefold as compared to 1995 with a per capita GDP growth rate of 5.5 per cent (P. Kumar and R.S. Paroda). As far as the future domestic production is concerned, two sophisticated models developed for the state of Haryana demonstrate how water will be the major limiting factor (P.K. Aggarwal, N. Kalra, S. Kumar and A.K. Vasisht), and how a change in inputs prices (higher charges) would primarily affect rice and cotton to the benefit of other products like milk (V. Alary and D. Deybe).

SCOPE FOR REFORMS TOWARDS A SECOND GREEN REVOLUTION

India must therefore initiate a second Green Revolution, with a markedly different recipe than for the first one since the limits or constraints of the present input subsidies are well acknowledged, from the financial, social, nutritional and environmental point of view. They first and foremost led to a distortion in resources use, and it is time to phase them out to better encourage, for instance, modern water saving devices (P.C. Bansil). However, despite many criticisms, the present system appears to be hard to replace by any other mechanism of incentives and redistribution since input subsidies are still considered to be the best for numerous small and socially heterogeneous farm-holdings. The shift towards de-coupled support (direct payments de-linked from production, as is being increasingly implemented in the European Union for example) would lead to unbearable transaction costs as also to uncontrollable escalation of claims. On the other hand, a higher direct support of production prices would attract strong condemnation from the WTO (procurement prices for wheat and rice are already above world prices). For want of better solutions, or for fear of the agitation radical reforms could provoke, the general feeling is therefore to move towards a more rational management of the existing system. There seems to be a consensus that agricultural inputs must not only be billed (unlike groundwater and electricity in some states), but billed at a higher level (electricity, surface water, fertilizers, etc.), just as it is agreed that these bills should be on the basis of quantities consumed (installation of water and electricity meters). But since attempts in these directions have failed so far, many are now convinced of the following three points: (1) any fiscal approach to this problem will fail as one cannot politically or reasonably ask a sector that is poorer than others to make big sacrifices; (2) the aim of any reforms should not be to reduce subsidies, but to better manage them for a greater benefit of farming communities; and (3) this calls for an active involvement of the concerned communities. In the case of water and electricity, the decentralization at the village level of both distribution and collection of resources is thus strongly recommended, along with a reward system for those who play the game by its rules. This suggestion calls in itself for a revolutionary change in behaviour, in rural areas as well as in the Union or state administrations. But it is undoubtedly changes of this kind that will bring about the second Green Revolution in India, along with revolutionary changes in terms of statistics[4] and studies.[5]

NOTES

1. More than forty, among which seventeen were invited to present a paper. However, six of the latter could not deliver their talk in a written form, or attend the meeting at the last moment: B. Debroy (Rajiv Gandhi Institute for Contemporary Studies, New Delhi), A. Gulati (Delhi University, Institute of Economic Growth, New Delhi), R. Radhakrishna (Andhra University, Visakhapatnam), A. Sen (Ministry of Agriculture, Commission of Agricultural Cost and Prices, New Delhi), R.P. Singh (National Institute of Rural Development, Hyderabad), M.H. Suryanarayana (Indira Gandhi Institute of Development Research, Mumbai).
2. In order of the sessions: J.M. Boussard, Institut National de la Recherche Agronomique, Paris (session 'Agricultural Incentives and Public Finances'), B. Barua, Ministry of Agriculture, New Delhi (session 'Agricultural Incentives and Income Distribution'), G.S. Bhalla, Jawaharlal Nehru University, New Delhi (session 'Agricultural Incentives and Food security'), A. Vaidyanathan, Madras Institute of Development Studies, Chennai (session 'Agricultural Incentives and Environment'), S. Baru, Indian Centre for Research in International Economic Relations, New Delhi (session 'Agricultural Incentives and Foreign Trade'), P. Sen, Planning Commission, New Delhi (session 'Agricultural Incentives and Prospective Scenarios').
3. This tentative report of two-day discussions mentions some names within brackets just as an indication which may unreasonably shorten, extend or distort some talks.
4. Establishment, updating and large spreading of consistent databases at multiple levels, like on livestock feeding in peculiar Indian conditions.
5. Integrated studies of economical, technical, cultural, biological (...) systems inter-linked on different scales, and taking into account risk and related aspects.

PART I

PROMOTING FOOD SECURITY: THE INDIAN CHALLENGE

1

Food and Nutrition Security in India

THE MICRO-LEVEL DIMENSION

ABUSALEH SHARIFF

INTRODUCTION

The food situation in India during the early 1940s was affected by the war that became worse after the partition: India inherited 82 per cent of the total population but only 75 per cent of the area under cereals and 69 per cent of the total irrigated area. Before Independence, 24 per cent of the cropped area was irrigated, and it declined to 19 after partition. The 'grow more food' campaign was launched in 1943 due to impending food crises as a result of cessation of supplies of rice from Burma during the Second World War. A further deterioration in the situation occurred after 1947, which led to the continuation of the campaign up to 1951. When the first Five-Year Plan started in 1950–1, the country was in the grip of a serious food shortage with high relative prices of foodgrains. To bridge the gap between demand and supply, foodgrains imports—including those from PL-480 Agreement—helped food rationing and several other food assistance programmes that become indispensable till the country became self-sufficient by about the mid-1970s. The growth rate in foodgrains production then averaged out to a healthy 3.2 per cent over the 1980 to 1998 period. Foodgrain output in 1997 reached a high of 174 million tons followed by 169 million tons in 1998.

However, in spite of a decline in macro-level food insecurity as a result of an increase in per capita per day availability of foodgrains, it has not translated into reducing the degree of household level food insecurity among the poor. The findings of the National Sample Survey (NSS) household consumer expenditure survey of 1993–4 indicate that the percentage of those below poverty line actually increased during the economic reforms decade of the 1990s. Rural poverty rose by 3.42 per cent and urban poverty declined marginally by only about 1 per cent. Both in rural and urban India, about 80 per cent of the population remained below the recommended consumption level of 2,300 kcal of energy per day. The deficits of energy, proteins, fat and iron are dangerously low among the populations of India all over. In fact, more than one-half of the population seems to face chronic food insecurity in the country.

The existence of large deficits in food intakes and consequent levels of nutrition among a substantial proportion of the population during regimes of macro (national and state level) food adequacy seems not contradictory but a reality. In the case of nutrition, the 'average food deficit' of about 13 per cent in rural and 10 per cent in urban areas underestimates the gravity of the situation. It is important to note that averages conceal more important and often embarrassing facts of distribution below the mean that should be a matter of deep concern to the policy makers both at the national and state levels. Complacency on the part of regional and national governments, civil society organizations and multilateral agencies in understanding the nature of food security and innovative mechanisms to alleviate this deficiency among the masses would be counter productive.

DIETARY ENERGY SUPPLY BY EXPENDITURE CLASS

The Dietary Energy Supply (DES) per capita measured calories in India has improved over the years although it varied considerably during the 1970s and 1980s. The DES during the mid-1970s was less than 2,000 kcal, which improved to about 2,300 kcal by 1990, and to 2,500 kcal by 1997. In spite of recent lower growth rates in agricultural production, the DES has been increasing largely because of a fall in the rate of population growth as well. Besides, India has been maintaining about 10-12 per cent of the total production of cereals as buffer stocks of wheat and rice. The share of cereals in household food consumption is declining across all expenditure classes. For example, the top 20 per cent of the population spends about 50 per cent of their income on cereals while this percentage still is about 70 to 80 amongst the bottom 20 per cent. The share of expenditures on items such as milk and milk products and meat, fish and poultry has been increasing considerably. The expectation is that while there will be some rise in consumption of cereals as incomes increase during the next 20 years, the per capita increase in milk and milk products will accelerate considerably followed by meat and eggs. The projections are that the cereal consumption per capita is likely to increase from 458 grams in 1993 to 509 g in 2020. Milk and milk products will increase from 162 g to 597 g in 2020. Meat and eggs may increase from 16 g per day to 41 g per day per capita (Bhalla et al., 1999).

CALORIES

The nutritional pattern based on NSS consumer expenditure survey data for 1987–8 and 1993–4 suggests that consumption of energy was found to be lower compared to the recommended levels in both the rural and urban parts of India (Tables 1.1 and 1.2). In rural India, the calorie deficits—compared with the recommended level of 2,400 kcal—rather increased from 10.6 per cent in 1987–8 to 12.5 per cent in 1993–4. In urban India, the deficits narrowed down from 10 per cent to 8.4 per cent during the same period. The calorie deficits are found to be far higher amongst the bottom 20 per cent and subsequent 20 per cent of the population both in rural and urban areas. By 1993–4, however, the calorie deficits amongst the bottom 20 per cent increased from 34 to 44 per cent and, for the next 20 per cent, from 20 to 31 per cent respectively in rural India. In urban areas, the respective deficits increased only marginally.

PROTEINS

In the case of proteins, the deficits amongst the bottom 20 per cent increased from 25 to 37 per cent during this period, but protein supply increased from 42 to 62 per cent amongst the top 20 per cent in rural areas. The overall deficits remain low only at about 2 per cent level. But protein deficits are much lower in Urban India at 10 per cent less than the recommended level. Protein deficits were found to have increased to 30 per cent for the bottom 20 per cent in urban areas.

FATS

Overall deficits in fat consumption continue to be very high at 16 per cent in rural areas. In the urban parts, the deficit is marginal at 3 per cent of the recommended level. In both rural and urban areas, fat consumption remained low among all population groups excepting the top 20 per cent. The bottom 20 per cent had a deficit of 26 per cent in rural and 22 per cent in urban areas. The top 20 per cent, however, enjoyed an excess of 20 per cent in rural and 22 per cent in urban India.

IRON

Iron deficiencies were 3.6 per cent in rural and 14.3 per cent in urban India. The bottom 20 per cent of the population has recorded a very high level of iron deficiency at levels 27.9 per cent in rural and 34 per cent in urban areas. Only the top 20 per cent enjoyed an excess of 27 and 5 per cent respectively in rural and urban parts.

Table 1.1: Per Capita per Day Consumption of Energy, Protein, Fat, Carbohydrate, Iron and Calcium in India (1987–8 and 1993–4)

Population	Energy (2400 kcal)		Protein (60 g)		Fat (40 g)	Carbo-hydrate (399 g)	Iron (28 mg)	Calcium (400 mg)
	1987–8	1993–4	1987–8	1993–4	1993–4	1993–4	1993–4	1993–4
Rural India								
Below 20%	1,577	1,351	45	38	14	269	20.2	326.8
Next 20%	1,930	1,661	54	46	19	327	23.9	463.5
Next 40%	2,263	2,291	64	64	32	437	28.0	689.6
Top 20%	2,957	3,325	85	97	61	598	35.5	1,119.1
All	2,146	2,100	61	59	24	399	27.0	674.0
Urban India								
Below 20%	1,498	1,483	42	41	18	287	18.45	340.36
Next 20%	1,751	1,736	49	48	27	325	21.55	502.96
Next 40%	2,047	1,992	58	56	40	351	24.50	762.08
Top 20%	2,510	2,392	73	69	62	386	29.47	1,292.00
All	1,889	1,924	54	54	37	341	24.00	735.00

Source: Computed from Shariff, Abusaleh, Mallick Ananta C., Dynamics of Food Intake and Nutrition by Expenditure Class in India, *Economic and Political Weekly*, Vol. 34, No. 27, 3–9 July 1999, pp. 1790–800.

Table 1.2: Percentage Deviation from the Recommended Level in the per Capita per Day Consumption of Energy and Nutrients in India (1987–8 and 1993–4)

Population	Energy (2400 kcal)		Protein (60g)		Fat (40g)	Carbo-hydrate (399g)	Iron (28mg)	Calcium (400mg)
	1987–8	1993–4	1987–8	1993–4	1993–4	1993–4	1993–4	1993–4
Rural India								
Below 20%	–34.29	–43.71	–25.00	–36.67	–26.0	–32.58	–27.86	–18.30
Next 20%	–19.58	–30.79	–10.00	–23.33	–21.0	–18.05	–14.64	15.75
Next 40%	–5.70	–4.54	6.67	6.67	–8.0	9.52	0.00	72.25
Top 20%	23.21	38.54	41.67	61.67	20.0	49.87	26.78	194.92
All	–10.58	–12.50	1.67	1.67	–16.0	0.00	–3.57	68.50
Urban India								
Below 20%	–28.67	–29.38	–30.00	–31.67	–22.0	–15.84	–34.11	–14.90
Next 20%	–16.62	–17.33	–18.33	–20.00	–13.0	–4.69	–23.03	25.74
Next 40%	–2.52	–5.14	–3.33	–6.67	0.0	2.93	–12.50	90.52
Top 20%	19.52	13.90	21.67	13.04	22.0	11.66	5.25	223.00
All	–10.05	–8.38	–10.00	–10.00	–3.0	0.00	14.28	83.75

Source: Computed from Shariff, Abusaleh and Mallick Ananta C., 'Dynamics of Food Intake and Nutrition by Expenditure Class in India', *Economic and Political Weekly*, Vol. 34, No. 27, 3–9 July 1999, pp. 1790–800.

MICRO-LEVEL FOOD INSECURITY IN RURAL INDIA

THE VULNERABLE

India predominantly being a rural economy, the extent of vulnerability amongst the subsistence farmers and daily wageworkers determines the state of the economy. In this context, it has been highlighted that '*outstanding vulnerability of agricultural laborers and artisans has been widely noted*' (Dreze et al., 1995).

The landless and other asset-less, those living in drought prone areas, people belonging to tribal and forest areas and those living on high altitudes have a higher degree of food insecurity. Female-headed households are highly vulnerable and food insecure. Infants, children, the elderly and women have higher levels as well as higher degrees of under-nourishment. Besides, intra-household food distribution seems to disfavour particularly the women and children. The infirm, chronically sick and disabled are others who have a high degree of food insecurity and risk of under-nutrition. Malnutrition is persistent amongst the above groups across India and it varies only marginally between states.

Access to food is also affected by food prices. There is more than normal increase in food price especially in the case of rice, wheat and pulses during the 1990s in India (NCAER, 1998). Secondly, while the share of coarse cereals has declined substantially, the price differentials between coarse cereals and wheat and rice has also declined. It would be most appropriate to ensure that a certain proportion (about one-third) of the total food production is of indigenous varieties of rice and other traditional cereals. Besides it is of utmost importance that the prices of these cereals be maintained at about half the cost price of wheat in India.

There are a number of situations when the food insecurity may become more apparent; such as those during a bad agricultural season, droughts, floods, extreme cold conditions and so on. There has to be high alert and preparedness on the part of the respective state governments to take care of such situations.

The crises is manifest in the 'malnutrition-infection nexus': deficiencies in two entitlements together—one for food and the other for hygiene or health (including lack of potable water)—is the most potent mix giving rise to higher incidence and depth of under-nutrition. Disease incidence may peak during the rainy season, which is also a period of food shortage. Period of high labour demand is also a period of reduced allocation of time on health care due to high opportunity costs.

SUBSISTENCE FARMERS

On an average, the size of operational holdings that were 105.3 million in number across India covering 165.6 mil lion ha. of land, was only 1.57 ha. in 1991. This size in fact was 2.3 ha. in 1970–1. A size class distribution suggests that the share of marginal farmers owning less than 1 ha. has increased from about 51 per cent in 1970–1 to 59 pcr cent in 1991, cultivating only 15 per cent of the total land area (Figure 1.1). The mean size of land for this group is about 0.4 ha. per household. There, small farmers are just

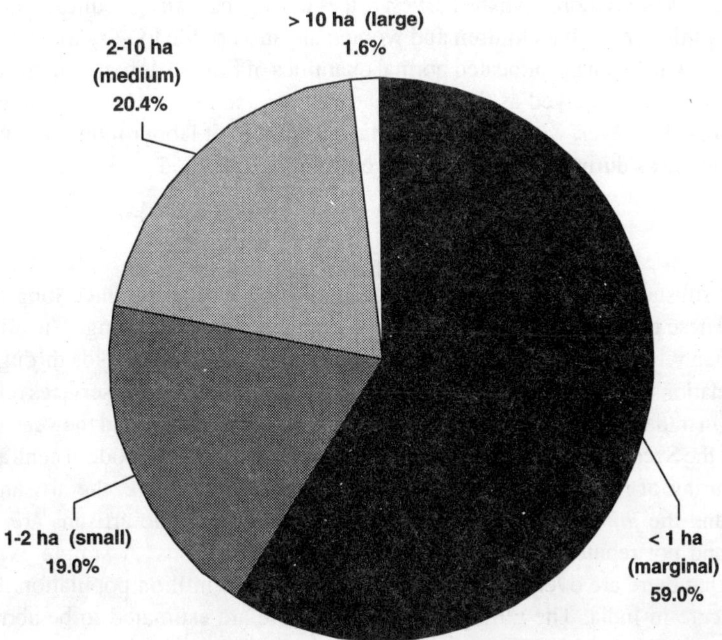

Figure 1.1: Number of Operational Holdings in Hectares by Size Class (1990–1).

a shade better: they constitute about 19 per cent in number owing about 17 per cent of land area. The number of large farmers, on the other hand, has declined from 3.9 per cent to only 1.6 per cent, but owning 17.4 per cent of land area during 1991 with an average farm size of 17.3 ha. The marginal farmers are found to have per capita income that is only 33 per cent, and small farmers have 40 per cent the level of large farmers. Thus a total of about 80 per cent of all farmers, that is, about 82 million are marginal and small farming households that can be classified as vulnerable and food insecure. Such a situation may arise because during the crisis this group of people are much less able to find jobs and also they may be less willing to migrate to other areas to mitigate stress caused by food shortages.

WAGE-WORKERS

Nearly 35 per cent or about 45 million rural households undertake wage-work both on agricultural and non-agricultural activities to eke out a living on a daily basis. This number is growing each year as a number of landed households become landless. They do not own any cultivable land nor do they posses other assets, but entirely depend upon others to provide them daily work and wage. This group of people is nutrition insecure in more than one way. First, they do not have the capacity to store food grains, say during the harvest seasons when the prices are low. The only mechanism to undertake this is by selling their labour at relatively cheaper or discounted rates in advance so as to secure a loan to buy stocks of grains. But often the trade off is such that the labourers may find it more practical to fend wage work on a daily basis and also buy food products on a daily basis. It has been found that the market price of cereals is 2 to 3 times more than the harvest prices during the lean agricultural seasons. It is during this period that children and women are susceptible to very low levels of food intakes while men are allocated normal quantities of food so as to maintain calorie levels as they are conceived as the real bread earners and as they have to undertake manual work. However, women also undertake a number of labour intensive but low productivity tasks during this period while consuming less food.

ARTISANS

The third substantial group that has been facing food insecurity since long is the artisans. These traditional occupations are no longer income generating. The artisans in the Indian villages were closely associated with the landed households through the *jajmani* relationships. This enabled the rural artisans to provide regular services relating to their own trade for a specific number of landed households all around the year. Since the end of the Second World War, however, due to the onslaught of modern centralized manufacturing of rural implements and instruments of agriculture, the artisans got affected and the *jajmani* system of relationships collapsed. The artisans are often landless and not reputed to undertake agricultural occupations. A simple exercise suggests that there are over 99 million households or 546 million population that is food insecure in India. The ultra poor and vulnerable are estimated to be about 60 million households or 327 million people.

MIGRANTS TO URBAN SLUMS

Besides, a large number of manufacturing workers such as in *bidi* (indigenous rolled leaf cigarettes) and *agarbatti* (incense sticks) industry, brick clines, construction works, the rickshaw pullers, head loaders, petty traders, garment factory workers and so on, just to name a few, become food insecure due to fluctuations in demand for their labour and irregularity of work. To this, one needs to add the slum dwellers that constitute about 30 per cent of the metropolitan urban areas since they also face extreme and constant threat of food insecurity. Moreover, there is an increase in rural to urban migration from exclusive areas in the states of Bihar, eastern parts of Uttar Pradesh, Orissa, Rajasthan and Gujarat. Often these migrants are in family units. Among all such identifiable groups, the women and children are often the worst affected due to irregular food intake, lack of adequate food during pregnancy and lactation. In terms of caste and religious identities, the most food insecure are those belonging to the Schedule Castes (SCs) and the Schedule Tribes (STs) all over India, and the minorities in selected states.

FOOD INSECURITY AT THE HOUSEHOLD LEVEL

WOMEN

Gender differentials in a number of household level parameters—such as in food and nutrition, access to health care and primary education—are paramount in the South Asian societies with the exception of Sri Lanka and the Maldives. Often, these differentials emerge from cultural stereotypes prevalent over centuries that have barely changed with the advent of modern education, information and media technologies useful in narrowing such differentials. Women at large—cutting across the economic class, caste, religion and regions in India—have relatively lower access to food within the household context. Even if they are the producers and managers of food products at home, women restrain from consuming adequate food so as to enable higher allocations to men and male children who are considered the breadwinners for the entire households. Almost always women eat last after having served adult men and children first.

Often women are left with no adequate food to eat. Men all over India, both in rural and urban areas, are, on the other hand, advantaged in that they are habitual consumers of food and beverages from outside homes almost on a daily basis. There is a common practice across India that men frequently visit local restaurants to enjoy the snacks and eateries along with tea and coffee. Women and children do not have access to such sources of nutrition at all.

In addition to the above types of nutritional disadvantages, women are also hard pressed due to nutritional denials during pregnancy, childbirth and lactation (Shariff, 1993). A number of cultural practices restrain women in maternity from eating adequate and often locally available cheap food. This practice impairs nutrition supply not only to women but also to the foetus and to the newly born child who is on full breast-feeding. For example, pregnant women are made to eat less than normal food during pregnancy for fear that normal foetal growth may cause natal difficulty and may even cause the death of the mother. This is one single reason as to why birth weights are low and also that a larger percentage of births are lower in weights. Similarly, nutritious food such as consumption of milk is restrained due to the belief that this will cause the foetus to get stuck to the mother's womb. On the other hand, women are encouraged to consume relatively larger amounts of edible oil which is believed to lubricate the internal parts of the mothers womb. Similarly, a large number of food items, vegetable and fruits are avoided by lactating mothers in anticipation of protecting the infant from the harmful effects of such food items.

Although there are a number of biological and physiological mechanisms to withstand dietary energy stress, by and large women and children in the Third World, especially, India are at risk. In this connection it is appropriate to recall '*that inadequate iodine intake during pregnancy, iron deficiency anemia in infancy, and early protein-energy malnutrition in infancy and early childhood only contributes to increased morbidity and mortality in childhood, but also resulting stunting means lower physical work capacity of adults*' (Scrimshaw, 1997). The gender difference in relative deprivation of women in the intra-family distribution—especially of food and health care—is a relatively higher morbidity and mortality over all age groups in women. One of the dominant manifestations of this is found in lower sex ratios for India. Besides, one could find relatively lower life expectancies for women in India until recently. Such differences are caused by a systematic difference between the command over resources and commodities that men and women respectively enjoy (Dreze and Sen, 1995). Often the vulnerable groups of people are characterized by lack of access to clean and potable water and primary health care. The children, for example, are worst affected since they are also deprived of nutritional supports that often are made available through the Integrated Child Development Services (ICDS) and mid-day meals programmes because these groups are less likely to be enrolled in schools and ICDS centres.

The intensity of food insecurity at the household level and the incidence of stress caused by food shortages can be gauged by the prevalence of anemia among women and low birth weights. The prevalence of anemia measured as below 11 grams per deciliter (gpdl) was found to be extremely high in states for which data are available (Figure 1.2). Incidence of anemia was about 50 per cent in Andhra Pradesh and Tamil

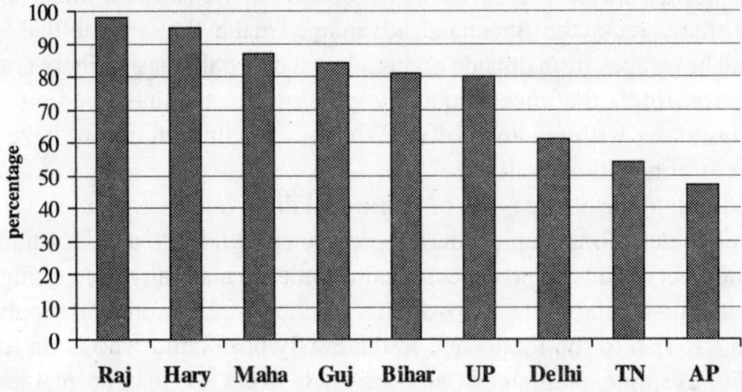

Figure 1.2: Prevalence of Anemia among Pregnant Women (1992–3).

Nadu, but was as high as 95 and 98 per cent in Haryana and Rajasthan, closer to 90 per cent in Maharashtra and Gujarat, and over 80 per cent in Bihar and Uttar Pradesh. A level of anemia below 8 gpdl was as high as 41 per cent in Maharashtra, 21 per cent in Gujarat and 14 per cent in Andhra Pradesh (Measham and Chatterjee, 1999).

CHILDREN

Low birth weight is the most important indicator of under-nutrition both at the household and national level. Its high level in India points to extremely precarious health and nutrition conditions among pregnant women. Under-nutrition among the infants and children, besides being a manifestation of low birth weights, is also associated with poor post-natal maternal and child practices relating to food intake and health care. However, data on birth weights is hard to come by. There is only scanty data, that too for selected urban centers largely accumulated from hospital records. Those who seek hospital births are likely to be relatively well off as well. Yet in Bombay (Mumbai) city, about 35 per cent of all births were found to be less than 2,500 g (Figure 1.3). This proportion was as high as 46 per cent in Baroda, 31 per cent in Varanasi and 25 per cent in Delhi. Given this scenario, in urban areas, it is expected that underweight births in the rural parts of India can be as high 60–70 per cent. On the whole, over one half of newly born children can be regarded as grossly underweight.

Malnutrition is directly or indirectly responsible for more than half of the deaths of children under four years of age worldwide. While India has successfully brought down the Infant Mortality Rate (IMR) from 146 per thousand live births in 1951 to 72 in 1996, most of the children who survived are malnourished (Government of India, 1998). Infant mortality declined from 110 per thousand live births in 1981 to 72 in 1996 overall (Figure 1.4). In rural India, it declined from 119 to 78 and, in urban India, from 62 to 46 in the respective years. In the year 1996, the IMR in India was 73 as compared to 65 in all developing countries, 13 in industrialized countries and 60 in the world average (UNDP, 1998).

India accounts for over 20 per cent of the world's child population, but it has 40 per cent of the malnourished children. The anthropometric measurements are

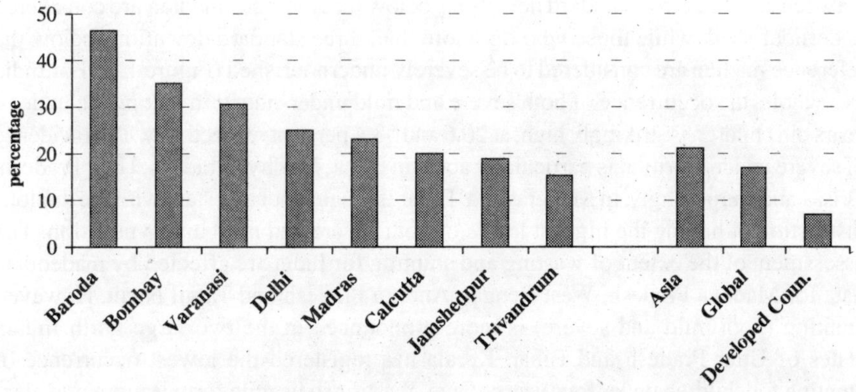

Figure 1.3: Birth Weight below 2,500 g (1990).

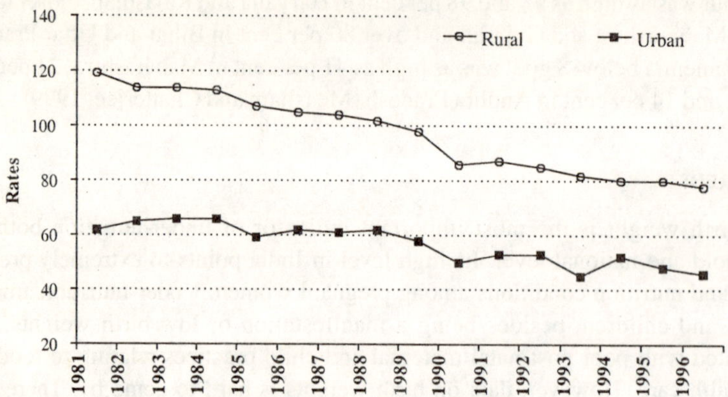

Figure 1.4: Infant Mortality Rates by Residents in India (1981–96).

dependable indicators of child nutrition and cumulative health both in the long (height-for-age) and short (weight-for-height) term perspective (Waterlow et al., 1997). These measures also highlight children's susceptibility to disease and their chances of survival. Child anthropometric levels are primarily determined right from the time the child grows in the mother's womb. Thus, it is mother's nutritional intake during pregnancy which determines birth weights and lengths. Subsequently, however, the breast-feeding practices and the timing of supplementation determine the growth in anthropometric measures. Often fully breast-fed children up to the age of about 4 months and then long duration partially breast-fed children with adequate supplementation grow normally. Another reason that causes serious damage to anthropometric growth of the young is the very high levels of diarrhoea often caused by poor quality water and feeding practices. For example, the prevalence of diarrhoea is as high as 108 per thousand children between 0–4 years of age (Shariff, 1999).

The height-for-age, weight-for-age and weight-for-height data for children under 4 years of age drawn from the National Family Health Survey for 1992–3 are discussed below (Table 1.3). The three indices are expressed in standard deviation units (z-scores) from the median of the international reference population. Children who are between two and three standard deviations below the reference median are considered undernourished, while those who fall more than three standard deviations below the reference median are considered to be severely undernourished (Figure 1.5). For India as a whole, the occurrence of both severe and mild under-nutrition among the under 4 years old children is strikingly high, at 20.6 and 53.4 per cent respectively. The incidence of severe under-nutrition is particularly acute in Bihar, Madhya Pradesh, Uttar Pradesh, Orissa and, surprisingly, in Maharashtra. Bihar is singled out as a state with the dubious distinction of having the highest levels of both severe and mild under-nutrition. The assessment of the extent of wasting and stunting for India are affected by inadequate data for Madhya Pradesh, West Bengal, Andhra Pradesh and Tamil Nadu. However, stunting (both mild and severe) is more pronounced in the two large north Indian states of Uttar Pradesh and Bihar. Kerala has registered the lowest occurrence of stunting for children under four years of age. Wasting in its mild form is more prevalent in Punjab, Rajasthan, Bihar, Orissa, Gujarat and Maharashtra. In its severe form, it is most pronounced in Rajasthan, Bihar and Maharashtra. Although the differentials are

Table 1.3: Percentage of Children under 4 years Under-nourished, Wasted and Stunted in India States (1992–3)

	Underweight (Weight for Age)		Stunted (Height for Age)		Wasted (Weight for Height)	
	<-2SD	<-3SD	<-2SD	<-3SD	<-2SD	<-3SD
Andhra Pradesh	49.1	15.6	NA	NA	NA	NA
Assam	50.4	18.7	52.2	26.3	10.8	1.7
Bihar	62.6	31.1	60.9	39.5	21.8	4.1
Gujarat	50.1	17.6	48.2	25.3	18.9	3.7
Haryana	37.9	9.0	46.7	19.3	5.9	0.6
Himachal Pradesh	47.0	12.9	NA	NA	NA	NA
Karnataka	54.3	19.4	47.6	22.7	17.4	2.6
Kerala	28.5	6.1	27.4	9.0	11.6	1.3
Madhya Pradesh	57.4	22.3	NA	NA	NA	NA
Maharashtra	54.2	21.3	48.5	23.5	20.2	4.2
Orissa	53.3	22.7	48.2	25.2	21.3	3.6
Punjab	45.9	14.2	40.0	15.7	19.9	2.8
Rajasthan	41.6	19.2	43.1	26.6	19.5	5.2
Tamil Nadu	48.2	13.3	NA	NA	NA	NA
Uttar Pradesh	59.0	24.6	59.5	35.6	16.1	2.7
West Bengal	48.8	18.6	NA	NA	NA	NA
All India	53.4	20.6	52.0	28.9	17.5	3.2

Notes: NA = Not Available, SD = Standard Deviation.

Source: *National Family Health Survey, 1992–1993, All India*, Indian Institute for Population Sciences (IIPS), August 1995, Mumbai, p. 286.

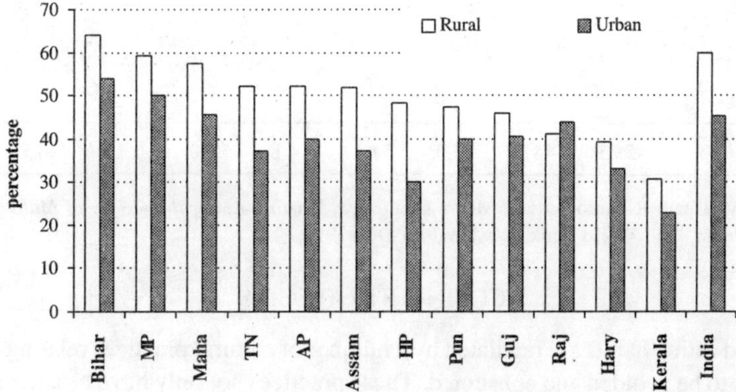

Figure 1.5: Prevalence of Malnutrition among Children Aged 1–4 Years in India and Selected States (1992–3).

not large, female children are found to suffer from under-nutrition in relatively larger percentage terms. Incidence of under-nutrition in rural areas was found to be 60 per cent compared to 45 in urban parts. Similarly, stunting and wasting proportions in the rural parts of India are found to be 54 per cent and 18 per cent respectively, significantly higher than in urban areas (Table 1.4).

Table 1.4: Prevalence of Malnutrition among Children Aged 1–4 Years saccording to Place of Residence in India and Selected States (1992–3)

(Percentage)

	Underweight (Weight for age below) 2 SD of median			Stunted (Weight for age below) 2 SD of median			Wasted (Weight for age below) 2 SD of median		
	Rural	Urban	Total	Rural	Urban	Total	Rural	Urban	Total
Andhra Pradesh	52.1	40.0	49.1	-	-	-	-	-	-
Assam	51.8	37.3	50.4	53.5	39.6	52.5	11.4	5.6	10.8
Bihar	64.1	53.8	62.6	61.8	55.2	60.9	22.7	16.3	21.8
Gujarat	45.8	40.5	44.1	44.6	41.6	43.6	20.3	16.1	18.9
Haryana	39.4	33.0	37.9	48.0	42.4	46.7	5.7	6.4	5.9
Himachal Pradesh	48.3	30.2	47.0	-	-	-	-	-	-
Jammu-Kashmir			44.5	-	-	40.8	-	-	14.8
Karnataka			54.3	-	-	47.6	-	-	17.4
Kerala	30.6	22.9	28.5	29.6	21.5	27.4	11.5	12.0	11.6
Madhya Pradesh	59.4	50.1	57.4	-	-	-	-	-	-
Maharashtra	57.5	45.5	52.6	50.8	39.1	46.0	21.5	18.3	20.2
Orissa			53.3	-	-	48.2	-	-	21.3
Punjab	47.4	40.0	45.9	40.4	38.4	40.0	21.4	14.3	19.9
Rajasthan	41.1	43.9	41.6	43.0	43.5	43.1	17.7	29.1	19.5
Tamil Nadu	52.1	37.3	46.6	-	-	-	-	-	-
Uttar Pradesh			49.8	-	-	49.2	-	-	16.2
West Bengal			56.8	-	-	43.2	-	-	11.9
All India	59.9	45.2	53.1	54.1	14.8	52.0	18.0	15.8	17.5

Source: Measham R. Anthony and Meera Chatterjee, *Wasting Away, the Crises of Malnutrition in India*, The World bank, Washington, DC, 1999.

COPING STRATEGIES

Food and eating habits are regulated by a number of cultural practices relating to types of foods to be avoided and consumed. These practices not only have broader patterns in terms of vegetarianism and non-vegetarianism, but are also closely related to caste and religious identities. The type of food eaten has a strong association with the class identities as well. However, most often the locally available food-stuffs such as coarse cereals, local varieties of vegetables and fruits and so on are the least preferred although often they have relatively higher nutritive value including the micro-nutrients. Besides, food avoidance through fasting and ban of eating specific types of food during certain days and months of the year is also practiced. On the other hand, during pregnancy and lactation, women may drain their own energy to provide nourishment to the child. During the time of acute food shortages, it is also expected that women draw upon their body fat to sustain themselves a bit more than the men.

One of the ways to distinguish between population groups is by identifying unique characteristics that regulate maintaining typical life styles, and through strategies and coping mechanisms adopted to sustain food supplies and nutrition levels among the community and households. In India about 100 million households or about 570 million people that are classified as food insecure, and 60 per cent that are ultra-vulnerable are either the landless or those who have meagre lands. They often toil day and night to earn a wage or produce an agricultural output barely enough to provide sustenance to the family round the year. Thus, the coping strategies start with the most difficult task of ensuring assured access to food round the year. There are a number of time-tested community and household level practices which harvest and preserve rain water both over- and underground through micro shed water conservative strategies. Such practices are not only efficient in conserving water for domestic and cultivation needs, but are also labour intensive and employment generating. The modern day large irrigation projects have obscured the utility of traditional practices and have shifted the direction of the so-called development irrigation investment benefits to people who are outside the affected area. This is only one of the innumerable examples of the modern planning strategies that seem to have contributed to the increasing vulnerability of local residents forcing them to migrate and often add to the teeming millions in the urban slums.

Wage workers have been identified as the most vulnerable as there is hardly scope for any long term prospects. Marginal and small farmers would have some production not often enough to pull though 365 days. In fact a substantial number of these farmers live closer to the standards of the wage workers. Thus, the first priority among these households is to ensure access to food in the lean periods and to save adequate cereals for sowing during the next season. Most of the above-identified households are found in rain-fed agricultural areas having the least amount of irrigation facilities. Hence, the coping mechanisms and behaviours emerge from these harsh facts and are attuned to the scheme of agricultural cycles and expectations.

There is a fairly vibrant private grain market in India. One rarely comes across shortages in availability of rice, wheat and pulses at the local and district level markets. However, one would find them to be priced high, often beyond the capacity of most of those living in that area. This situation arises when there are food shortages both in the local economy and at household level. However, it has been noticed that often the price shocks, although of shorter duration, are felt more in rural than urban areas in India. People even after resorting to a number of coping behaviours such as rationing of food stocks—through storage, exchange, shifting to cheaper varieties, reducing daily intake and so on may be forced to migrate, a step that is taken only under extreme vulnerability.

The coping strategies to mitigate food risks are large and complex and differ substantially between rural and urban areas, and within the rural areas among agro-climatic regions and between different population groups. The strategies may be as simple as borrowing grains from kin and neighbours, or withdrawing children from school to make them work long hours as child workers. Others in the household may work long hours in the hope of increasing income, but this possibility is low as the demand for labour in the local areas would have disappeared. Since food shortages certainly mean low or no cash reserves, there certainly will be denial of health and medical care which will affect women and children more than the others in the household. Families who own properties may resort to liquidating utility assets followed by productive assets as well. Often it is seen that households sell their highly valued livestock to ward of temporary shocks and food shortages. Large-scale indebtedness and sale or mortgaging of agricultural lands and property are the extreme response to such crisis, next only to permanent migration to other parts of the country in search of wage labour and employment.

PROGRAMS ADDRESSING FULLY OR PARTIALLY THE FOOD INSECURITY

Despite food self-sufficiency and the achievements of the Green Revolution at the national level, about 350 million people are classified as poor in India. A modest increase in per capita availability of cereals does not seem to have benefited the poor through an increase in access to food. As a result, several direct group oriented efforts were made to protect the interests of the poorer categories of the population. A time line presented below (Figure 1.6) summarizes the policy efforts in building up food security for India since Independence.

```
1950            1960            1970            1980            1990            2000

Food Shortages ------------------------ *    Food Sufficiency ---------------------------- Food Export*
          PL480 ----------------- *          SFDA/MFALA
*----------Import of Food ----------------------- * DPAP                          Nutrition Policy(93)
      Planned Schemes of food production
    Food Rationing -------------------          PDS ---------------------------------- TPDS -------*
        Buffer Stocks ------------ FCI Created -------- Price Support (76-77) ----------------- *
                               APC --------------------------------- (now) CACP --------
                        *Green Revolution*   Dairy Development   ----------------- *
                                   ICDS   ------------------------------------------------ *
                                    *Takeover              Wheat based ICDS
                                  of wheat trade           linked Prog. -------------- *
                Applied Nutrition Program    Food for Work (78) ----------- *------
                       Mid-day Meals (TN) ---------------------------- All India ------*
                                             IRDP --------------------------- *SGSY -*
                                             JRY------------------------------ *JGS --*
```

Notes:

PL480	Public Law (of the USA) number 480
SFDA	Small Farmers Development Agency
MFALA	Marginal Farmers ALA
DPAP	Drought Prone Area Programme
PDS	Public Distribution (of food) System
TPDS	Targeted Public Distribution (of food) System
FCI	Food Corporation of India
APC	Agricultural Prices Commission

(contd.)

CACP	Commission for Agricultural Costs and Prices
IRDP / SGSY	Integrated Rural Development Programme, now *Swarnajayanti Gram Swarozgar Yojna*
JRY / JGSY	*Jawahar Rozgar Yojna*, now *Jawahar Gram Samridhi Yojna*
ICDS	Integrated Child Development Services
TN	Tamil Nadu

Figure 1.6: Major Policy Initiatives over the Years that have a Bearing on Food Security in India (1950 onwards).

EMPLOYMENT ORIENTED PROGRAMS

In 1970, a number of targeted programmes aiming to improve food output were initiated: the Small Farmers Development Agency (SFDA), the Marginal Farmers and Agricultural Labourers Agency (MFALA) and the Drought Prone Area Programme (DPAP). However, subsequent evaluations suggested inefficiency due to poor implementation, weaker management and high incidents of leakages. Subsequently, during the early 1980s, programmes such as Integrated Rural Development Project (IRDP), the Employment Guarantee Scheme and Old Age Pension were launched so as to improve the purchasing power of the poor. These programmes did have some impact that was not sufficient to reduce the number and extent of malnutrition among the poor. A number of employment oriented programs have been recently restructured in an effort to improve efficiency, such as the Swarnajayanti Gram Swarozgar Yojna (SGSY) and Jawahar Gram Samridhi Yojna (JGSY), and so on.

PUBLIC DISTRIBUTION SYSTEM

The Public Distribution System (PDS), in contrast, was initiated during the early period after independence largely to distribute the PL-480 aid. PDS is considered relatively more effective because it distributes cereals at subsidized prices enabling a direct improvement of levels of nutrition. This is an income transfer strategy for the poor. On the other hand, other programmes were aimed to improve income generation instead. PDS therefore continued as a welfare measure with the advent of planned development in the 1950s. Under the First Five Year Plan, the PDS was extended to those rural areas where foodgrains deficit was most prominent. These were defined as 'statutory rationing areas' where supply from the ration shop or Fair Price Shop (FPS) was obligatory, while other areas were to be covered by non-statutory rationing. Thus, its scope was widened to include both urban and rural areas. As early as in 1957, a Food Grains Enquiry Committee recommended the creation of 'buffer stock' and control over cereal trade and 'progressive socialization of trade in foodgrains'. In view of the statutory prices, national trading in foodgrains was attempted by fixing prices in 1959. At this stage, other essential commodities such as sugar was also included in the PDS.

In 1965, two important organizations, viz., Agricultural Prices Commission now called Commission for Agricultural Costs and Prices (CACP) and the Food Corporation of India (FCI) were set up. The Procurement Prices offered by the FCI are fixed by the

CACP, and are generally lower compared to the open market, that do not justify their being labeled as incentive price. The FCI was made to implement Government's policy on procurement, storage, transportation, distribution of foodgrains and other commodities through Fair Price Shops (FPS). The CACP is an institution that is to give advice on support price and procurement.

INTEGRATED CHILD DEVELOPMENT SERVICES

The National Policy for children (1974) acknowledged that since a majority of India's children live in wretched living conditions including poverty, poor environmental sanitation, disease/infections, inadequate access to primary health care, inappropriate child caring and feeding, special interventions were required to provide equality of opportunity to these children. Founded on this conviction, the Integrated Child Development Services (ICDS) programme was launched on the second of October 1975 in 33 blocks, nearly a quarter of a century ago. Today, ICDS represents one of the world's largest and most unique programmes for early childhood development—an initiative unparalleled in history. Poised for nearly universal coverage at the turn of the century, ICDS today reaches out to 4.8 million expectant and nursing mothers and 22.9 million children (under 6 years of age) belonging to disadvantaged groups. Of them, 12.5 million children (3–6 years of age) participate in pre-school education activities as well. The network consists of 4,200 projects, covering nearly 75 per cent of the country's community development blocks and 273 urban slum pockets.

Currently, ICDS has many a meaning for the community (a homely child care centre, a play/learning centre, a peripheral health centre, a meeting place for women/mothers, a source of support during calamity), a means to fulfil aspirations for millions of young children. This is the reality of today on which India's vision for the new millennium is founded. All the ICDS services are provided through an *anganwadi* (child care centre) that is used to house all the children collected on a daily basis so as to provide all the listed inputs that includes daily nutrition supplement, some teaching and schedule of immunization services. Each *anganwadi* is run by an *anganwadi* worker (AWW) who is always a woman, and supported by an *anganwadi* helper in service delivery and improved linkages between the health system and the community, especially women (mother-for-childcare). A number of services provisioning activities through the ICDS seem to be better than those that are to be provide by the public health infrastructure.

The ICDS facilities have been extend to all the districts in the country. However this alone has not ensured that all children below six years from poorer sections of the society get covered by the scheme. There are still large areas left uncovered, especially less densely populated areas or where population is more scattered and in undeclared slums, and so on. A basic lacuna in the scheme is poor coverage of children under three years who are more likely to suffer from maximum malnutrition. Since these children are too young to be taken care of at the *anganwadi*, they must be either brought to the centre especially for feedings, or their rations must be taken home. This scheme has also sidestepped the issue of inequality within the families. A family, which is officially above the poverty line, may have some members, especially female children, who are individually at risk but remain uncovered by the scheme. Besides, more than 80 per cent of AWW's have no toilet facilities and only 21.2 per cent have a satisfactory water/sewage disposal drainage system.

There has not yet been a comprehensive national level evaluation of the ICDS. However, a few studies seem to highlight the fact that the nutritional levels of children in the ICDS areas have been marginally better than the non-ICDS areas. A few other studies have pointed out that so far as the coverage of women is concerned, the programme has been fairly successful in enrolling expectant mothers, but have rather failed to enroll expectant mothers which is one of the priority target group of the programme. It may be noted at any given point in time the number of expectant mothers would be about the double the number of lactating mothers in a given area. The lactating mothers therefore seem to be deprived of supplementary nutrition despite being eligible, while nursing mothers, once registered, are reported to continue utilizing the services beyond the recommended six months of lactation.

CONCLUSION

The micro-level food insecurity is multidimensional. It can be observed at the level of household and village on the one side, and community, caste, religion, economic status, age-sex/gender level on the other. To overcome micro-level food insecurity, one should be concerned about people below the minimum level of consumption irrespective of caste, creed, religion, community, village, and so on. However, for the purpose of targeting, identification of the poor by such classification seems a necessity. The causes for micro-level food insecurity under normal conditions could be due to lack of purchasing power, high food prices and shortage and poorly developed markets. In abnormal situations, it could be due to natural calamities like drought, flood and cyclone. Food insecurity under normal conditions requires a long term strategy of improving the lot of the poor through generation of employment opportunities, scope and assistance for employment, making available essential food items at reasonable rates and even on credit, and so on. Food insecurity in abnormal situations needs to be tackled on an emergency footing through aid and assistance, imports and schemes like 'food for work' and so on.

Suffices it to say that policies and programmes need to be divided into those which are complementary to successful adaptive strategies and those which provide relief for those individuals/households/groups which can not longer cope. Nutrition supplementation, mid-day meals, ICDS, dietary supply in hospitals, vitamin A supplementation, iodination of salt, iron and folic acid prophylaxis and so on are only some of the most important programmes to be put in place as an enduring strategy. Strategies for assistance in market interventions to contain prices and arrest reduction in supplies due to lack of demand of essential items are needed. Understanding the local grain market and distribution networks, local future markets, agricultural credit and so on needs to be explored.

Strategies to augment just the levels of income are not adequate to eliminate under-nutrition. However, employment generation as an empowering strategy needs to be strengthened. Evaluating the public programmes intended for income transfer for the poor and supply of subsidised food and nutrition should ideally be targeted and implemented in due earnest. Using modern technology for providing essential items of assured quality and in small pre-packed form may go a long way in alleviating the problem of adulteration and lower measures which is pervasive all over India.

At last, a point of caution: any amount of safety net will be inadequate not only because such programmes are difficult to implement but also because the sheer size of the poor are so large in absolute numbers that resources will be a constraint. Reasons as to why state policies and programmes relating to nutrition and health are too weak to make an impact, and why they need urgent study.

REFERENCES

ANDERSEN, Per Pinsturp and Rajul PANDYA–LORCH (eds.), *World Food Prospects: Critical Issues For The Early Twenty-First Century*, International Food Policy Research Institute, Washington, DC, 1999.

BHALLA, G.S., Peter HAZELL and John KERR, *Prospects for India's Cereal Supply and Demand to 2020*, Food, Agriculture, and the Environment Discussion Paper 29, International Food Policy Research Institute, Washington, DC, 1999.

DREZE, Jean and Amartiya SEN, *The Political Economy of Hunger. Selected Essays*, Clarendon Press, Oxford, 1995.

GOVERNMENT OF INDIA, *Sample Registration Bulletin*, New Delhi, 1998.

KHULLAR, Vandana, 'Integrated Child Development Services: A Critiques of Evaluation Techniques', *Economic and Political Weekly,* Vol. 33, 7 March 1998, pp. 537–45.

MEASHAM, Anthony R. and Meera CHATTEJEE, *Wasting Away: The Crisis of Malnutrition in India*, The World Bank, Washington, DC, 1999.

MOOIJ, Jos, 'Dilemmas in Food Policy. About Institutional Contradiction and Vested Interest', *Economic and Political Weekly,* Vol. 34, No. 52, 1999, pp. A114–20.

NCAER, *Economic Reforms and the Poor,* mimeographed, National Council of Applied Economic Research, New Delhi, 1999, pp. 1–125.

PAYNE, Philip and Michael LIPTON, *How Third World Rural Households Adapt to Dietary Energy Stress: The Evidence and the Issues*, International Food Policy Research Institute, Washington, DC, 1994.

SCRIMSHAW, Navin S., 'The Lasting Damage of Early Malnutrition', in *Ending the Inheritance of Hunger*, Lectures given by Robert William FOGEL, Navin S. SCRIMSHAW, Amartya SEN, World Food Program / United Nation University (WFP/UNU) Seminar, Rome, 31 May 1997.

SEN, Amartya, 'The Lasting Damage of Early Malnutrition', in *Ending the Inheritance of Hunger,* Lectures given by Robert William FOGEL, Navin S. SCRIMSHAW, Amartya SEN, World Food Program / United Nation University (WFP/UNU) Seminar, Rome, 31 May 1997.

SHARIFF, Abusaleh, *Factors Affecting Child Health. Search for Maternal Education Effects in Rural Gujarat,* Working Paper, No. 47, Gujarat Institute of Development Research, Gota (Ahmedabad), 1993.

———, *India: Human Development Report: A profile of the 1990s*, Oxford University Press, New Delhi, 1999.

SHARIFF, Abusaleh and C. Mallick ANANTA, 'Dynamics of Food Intake and Nutrition by Expenditure Class in India', *Economic and Political Weekly*, Vol. 34, No. 27, 3–9 July 1999, pp. 1790–800.

UNDP, *Human Development Report,* United Nations Development Programme, Oxford University Press, New York, 1998.

VYAS, V.S. and Pradeep BHARGAVA, *Poverty Reduction in Developing Countries: Experience from Asia and Africa,* Rawat Publications, Jaipur, 1999.

WATRLOW, J., R. BUZINA, W. KELLER, J. LANE, M. NICHMAN and J. TANNER, *The Presentation and Use of Height and Weight Data for Comparing the Nutritional Status of Groups of Children under the Age of Ten Years*, Bulletin of World Health Organisation, No. 55, 1997, pp. 489–98.

2

Agricultural Subsidies

A GLOBAL VIEW

P.C. BANSIL

INTRODUCTION

It may very well be appreciated that subsidies in general are intended to introduce a new product and encourage its consumption. But once the people concerned are able to realize the benefit of the policy, subsidies are to be withdrawn slowly. In India, however, food subsidies—both consumer and input—have become a part of the system and taken the role of welfare activity. The basic objective of these subsidies should actually be to provide help to the target group of poorer sections by supplying foodgrains at subsidized prices and encouraging the farming community in general to increase the production of foodgrains and other cash crops badly needed by the economy.

However, subsidies, both hidden and explicit, are today as ubiquitous as they are omnipresent and arguably, as arbitrary in their operation as any act of providence. This is not to say that subsidies are, per se, a bad thing. A well-crafted and precisely targeted subsidy can work wonders in correcting an imbalance or inequity that the free market by itself might not remedy. This is especially true for public goods like defence and policing but also for primary education, public health, sanitation and public transportation. But when subsidies are used to cushion the affluent or appease politically important sectors, they cease to have any welfare-enhancing effect. If anything, such subsidies have a negative impact, first because they introduce distortions and, second, because other categories of public expenditure that might be more deserving must be scaled down to keep the budget deficit within acceptable bounds. In India, unfortunately, the subsidy bill is so huge that even after pruning allocations under crucial budgetary heads such as public investment, the fiscal deficit has reached unmanageable proportions. To bring about some semblance of sanity, therefore, the government will have to close the deficit from both ends: mobilizing additional tax revenues and cutting unnecessary expenditure, especially on non-merit subsidies.

The subsidy trap is suicidal because it makes governments believe they are doing enough for the farmers. It fools the farmers, keeps the entire agricultural input industry still within the licence-quota-subsidy vortex and is in the process destroying our farm economy and the environment. Just how big a disaster this is causing is evident the moment one visits a farm input shop. Half-literate shop-owners push the most dangerous pesticides and unnecessary fertilizers down the farmer's throat and the farmer is tempted to buy since it all looks so cheap because of the subsidies. He soaks his soil with poison which eventually reaches our kitchens. How many of us realize that the Indian farmer uses three or four times more pesticides than the international norm? Some of the strong pesticides last at least 45 days once absorbed by a plant. Yet, our farmers continue to soak their cauliflowers, brinjal and tomato with it even in the week of plucking. We eat some of the most poisonous vegetables and fruits in the world. Because the farmer gets these inputs cheap, he overuses them. Because the Government invests everything in subsidies, it has nothing left for research or development—many of our premier agriculture universities can hardly pay salaries now. If the Government only gets out of the way completely and leaves the farmer to face the market challenges, he would perhaps be better off. And our foods less poisoned.

It works exactly the same way for electricity. Punjab claims to provide free power to the farmer. Consequently one can see the quality of the basic power infrastructure—tilted poles, hanging wires, exposed fuses, etc. Most farmers spend more by way of bribes and repair to get free power than they would pay for reliable, quality power. Also, since power is free, and they don't know which four or six hours of the day it will be available, they tend to keep the tube-well motor on round the clock. The result is too much power consumed, too much water pumped out, too much subsidy and, therefore, too much deficit in the budget.

Subsidies were needed in the 1950s when the nation-building process had just started; but now there is a strong view that they should be done away with in a phased manner and they run counter to our free-market orientation. They are now being used as political bait to lure voters. The basic purpose of subsidies has been lost. The poor have not benefited from them. Instead, the exchequer has ended up subsidizing middlemen and the rich. The system has to be made more transparent and efficient so that subsidies benefit the intended targets, and are not misused. There is now a demand for rationalizing these subsidies so that the burden of adjustment is borne equally by the recipients—manufacturers and farmers.

Agriculture is said to have preferential treatment in the form of subsidies: free power for the pumps, cheaper credit, lower prices of fertilizers, minimal water rates, low rates of land revenue, subsidized supply of agricultural machinery, constant upward revision of procurement prices and to cap them all, no income tax on agricultural income. Of these various subsidies we have examined here in detail fertilizers, power and irrigation. Agriculture credit, seeds and machinery also carry subsidies, but they have been left out of the scope of this paper primarily because their quantification was difficult. Because of the present system of Minimum Support Pricing (MSP) and Public Distribution System (PDS) commitments, food subsidies are affected to a large extent by the changes in various input subsidies.

This paper has, therefore, been divided into five parts: part one gives an overview of various subsidies, parts two to four examine in detail the present ground situation with regard to the major three subsidies (fertilizers, irrigation, and power), and the last part provides a perspective view of these subsidies.

AN OVERVIEW

SUBSIDY ON FOODGRAINS: A RATIONALE

With the introduction of new programmes of structural adjustment and fiscal tightening in India in 1991, reduction in subsidies, including food subsidies, has been the buzzword for policy makers. Bearing in mind the improved supply situation, a section of the economists are of the view that it is appropriate to phase out current procurement operation and supply to the PDS. The emphasis, in fact, is on *laissez-faire* economy—sans public interference in any form. On the other hand, those who oppose withdrawal of both production and consumption subsidies are of the view that food subsidies in India affect millions of consumers and producers and are of immense importance in a country where a majority of the population suffers from chronic hunger and malnutrition. In other words, their withdrawal would cause a sharp decline in foodgrains production and further pauperization of millions of people. It would be the negation of the broad objective of food policy in India which is to make available food to the people at reasonable prices and provide remunerative prices to cultivators. Other objectives may also be jeopardized, like supplying food at subsidized prices to the undernourished, controlling inflationary pressures, stabilizing prices for consumers and producers, reducing fluctuations in food availability, and achieving self-sufficiency in foodgrains production.

FOOD SUBSIDY

The food subsidy as defined in the Government budget includes the entire operational deficit of the state-owned Food Corporation of India (FCI) for procuring and distributing four major items: rice (and paddy), wheat, imported edible oils and sugar. The study conducted by Swaminathan[1] on food subsidy reveals that the nominal food subsidy as a proportion of Gross Domestic Product (GDP) over the 30 year period from 1966–7 to 1997–8 was rising rapidly, particularly after the 1980s, and it took a big jump in 1993–4. However, at constant prices, the increase is more subdued. The expenditure on the food subsidy rose in the mid-1980s and then remained unchanged until about 1989–90. It dipped between 1990–1 and 1992–3 and rose in 1993–4. It fell slightly in the following years, but was higher than in the 1980s. All in all, at constant prices, the food subsidy as a share of GDP appears to have remained more or less unchanged over the last 20 years, peaking at about 0.63 per cent in 1985–6 and 0.64 per cent in 1993–4, indicating that the burden of food subsidy has not risen. Besides, the subsidy for foodgrains alone is lower than the budgeted food subsidy since the latter includes now the subsidy in sugar and other items.

PRODUCTION SUBSIDY

Production subsidies fall into two categories: (a) 'product specific subsidies' given for specific crops, which in the Indian case is the minimum support price or procurement price given for about twenty agricultural commodities and, (b) 'non-product specific subsidies' given for all crops which in the Indian case are subsidies given for all the major inputs: fertilizers, water, pesticides, power for irrigation, credit and even agricultural machineries. During 1996–7, per ha. consumption of fertilizer was 76.75 kg. (estimated) which included all crops. Assuming that 75 per cent of the consumption of fertilizers was for foodgrains production, the subsidy provided for fertilizers may be apportioned accordingly. During the same year, total power consumption for agriculture was 84,331 million kWh. It is difficult to assume how much power was consumed for foodgrains production. Similar problem arises with pesticides consumption. The cost of production of each of the foodgrains is calculated on the basis of the actual price paid for the purchased inputs without deducting the subsidy element. For fixing the procurement price, the cost of production per quintal is the major consideration.

As per GATT agreement (1994), a developing country whose product specific and non-product specific Aggregate Measure of Support (AMS) does not exceed 10 per cent of the total value of agricultural product (5 per cent for developed countries) is not subject to any reduction commitments. Hence, India has no obligation to reduce agricultural subsidies as, according to the Ministry of Commerce, it is less than 10 per cent of AMS. However, to reduce the fiscal deficit, reduction or withdrawal of subsidy has become essential.

TOTAL MAGNITUDE

Not bothering immediately about the definitional and procedural issues concerning various subsidies and issues discussed above, just let us say for the time being that the various types of production subsidies amount to around Rs. 500 billion[2] at present, or say over 13 per cent of agricultural GDP. Add to this another 100 billion as food subsidies and it is the sheer magnitude of these subsidies which has raised alarm bells. In addition, serious concerns have been voiced regarding the impact of these subsidies on the efficiency in resource-use and environmental degradation in general. Despite various attempts made from time to time to reduce these subsidies, populist policies proved to be overbearing and subsidies have grown at about 12 per cent compound rate over the last two decades.

However, today, there is a general agreement that such growth cannot be sustained without a reallocation of funds towards public infrastructure. The political leadership also appears to be concerned over the increase in expenditure, particularly the untargeted flow of funds. As for the World Bank's report (1999) on poverty in India, it has noted that high allocation of funds for implicit and explicit subsidies are a constraint on the development of not only infrastructure, but also research in agriculture.

IRRIGATION

PRICING WATER

Pricing of water well below its economic value is prevalent throughout the world, resulting in misallocation and inefficient use. Farmers in both industrial and developing countries often pay little for their publicly supplied irrigation water, and so have few incentives to refrain from growing water intensive crops or to conserve water. This has invariably led to gross misallocations, whereby agriculture, which absorbs the lions' share, often includes low-value uses compared with high-value domestic and industrial uses.

Water charges are very low on an average in India, typically amounting to 2–3 per cent of the harvests value in most states where they have no influence on farmers' management decision. The cost of irrigation water, on the basis of investment, works out on an average to 91 paise per thousand litres, against which, at prevailing water charges, the highest water rates for rice and sugarcane are only 1.2 p. and 11.6 p. per thousand litres respectively. Most of the states have not revised their irrigation water rates for the last 15–20 years. On an average, gross revenue collection per ha. of irrigated command is found to be about Rs. 68 against the estimated expenses of Rs. 217 per ha., thereby resulting in financial loss of Rs. 149 per ha. per year.

Irrigation is one of the key inputs for crop production in as much as the productivity impact of better seeds, fertilizers and other inputs is critically dependent on the way water is used. It is therefore both legitimate and necessary to address the pricing of this input as one of the first steps and an integral component in the process of rationalizing the totality of the price structure, and raising the efficiency of water use.

The Vaidhyanathan Committee on Pricing of Irrigation Water (1992) calculated that in order to cover the cost of operations and maintenance (O&M) charges, the increase in revenue in various states should range from 1.5 times in Punjab and Orissa to over 14 times in West Bengal. The under-pricing of water adversely affects the availability of resources for the management of irrigation systems. Inadequate allocation for maintenance and repairs is a direct consequence of the poor financial position of the states, and is responsible for the low, possibly deteriorating, quality of services. This means that the potential increases in productivity which new technology makes possible cannot be realized in full.

A revision in the level and structure of water rates is thus necessary in the interest of both efficiency and equity. The said committee was of the view that the revision should be such as to achieve full cost recovery in due course and, in the process, promote saving, create disincentives for waste and, thereby, enable the service area to be expanded and a more reliable service assumed. The committee recommended that rates should be based on O&M norms and capital charges (interest and depreciation), and that the rates should be revised once in five years.

The World Bank has also called for hiking water and power tariffs in the agriculture sector to optimize returns on projects funded by it. The states which are reluctant to revise water rates will sooner or later realize the adverse impact of under-pricing on availability of water. Under-pricing has not fulfilled the socialist goal of ensuring equity in resource use and distorted the distribution pattern in the agricultural sector. The relatively well off farmers merrily lift water in excess of their requirement by using powerful pumpsets. This depletes the water table faster and makes it difficult for small farmers who cannot afford to hire or own pumpsets. If the trend continues, agricultural growth will be severely affected.

Since there is hardly any justification that water supplies should continue as at present, a few points merit attention in terms of a realistic strategy. One, there can be different bases for pricing water: it could be charged on the basis of crops and area irrigated, or on the basis of volume of water used. In most states in India, water is charged on the basis of crops and area. This does not always imply a link between the volume of water used and what you pay for it. Raising the price of water, keeping the rate structure intact, will not have a major impact on securing efficiency objectives. For pricing to have a significant effect on water use decisions, it has to be linked to the volume of water used.

Volumetric pricing is, therefore, a more viable option. However, it is laden with what economists call high transaction costs: the costs of information, contracting and enforcement. It is difficult to monitor how much water each of several individual farmers uses. It is difficult to collect this information and to ensure that every farmer pays his due. Further, volumetric pricing also has certain technical requirements. It requires the installation of measuring devices such as the V-notch, the parhsall flume and the standing wave flume. Thus, the introduction of volumetric pricing to individual farmers is constrained by high transaction costs, the large spread of irrigation systems and technical requirements that such a system imposes.

Another important distinction is between cost recovery and irrigation financing. Under a system of cost recovery, which is currently in vogue in most states, what is realized by way of irrigation fees goes into the state exchequer. With no link between the resources and the expenditure, there is no incentive to improve either the recovery, or the quality of service. Under a system of irrigation financing, on the other hand, these accruals are used for improving the quality of irrigation services. We therefore, need to move from the current system of cost recovery to a system of irrigation financing.

The other myth that is now wearing out is that the farmer is unable to or reluctant to pay more for water. In places with access to both surface and ground water, farmers pay as much as ten times more for ground water. This is largely because ground water enables them greater control over-irrigation. Thus, farmers would be willing to pay more for water if it is matched by productivity increases brought about through improvements in the quality of irrigation services.

This brings us to the political acceptability of water price hikes. In Maharashtra and Andhra Pradesh, recent proposals to raise irrigation prices found acceptance. In these states, there have been efforts at forming water users associations. These are groups of farmers usually below the outlet that take water from the irrigation department on a volumetric basis. It is then distributed among farmers. There develops a mutual contract wherein the irrigation department agrees to provide certain supply of water to farmers who agree to pay for it collectively. The formation of water users associations, thus, provides a forum for dual accountability from the two sides. It also makes volumetric pricing of water possible and acceptable and serves as the base for further price revisions upwards. The current process of reform in Andhra Pradesh establishes that upward price revisions are possible, but only when part of an overall reform strategy aimed at creating accountability of the irrigation departments, empowering users, putting in place improvements in the infrastructure and improving the quality of irrigation services, is accompanied.

IMPROVING WATER USE EFFICIENCY

Improvement in water use efficiency is also important because of the need for its conservation. Despite whatever we do, water problems are likely to become acute in the country in the coming two decades. If so, water use efficiency has to be the corner stone of future irrigation policy. Water use efficiency is presently estimated to be only 38 to 40 per cent for canal irrigation and about 60 per cent for ground water irrigation schemes.

On the basis of 1991 Census, per capita water availability per year in India was estimated at 2,214 cubic meters against the global average of 9,231 cubic metres, and 3,020, 3,962 and 4,792 cubic meters per year respectively for countries like Afganistan, Pakistan and Sudan. In 1990, India was ranked at the 42nd position among 100 countries by per capita water availability. In the total water use in 1990, the share of agriculture was 83 per cent, followed by domestic use (4.5 per cent), industrial use (2.5 per cent) and energy (1.8 per cent). The remaining 8 per cent was for other uses including environmental requirements. The projected total water demand by the year 2025 is around 1,050 cubic kilometres against the country's utilizable water resources of 1,140 cubic kilometres. The share of agriculture in total water demand by the year 2025 would be about 74 to 75 per cent. Thus, almost the entire utilizable water resources of the country would be required to be put to use by the year 2025. Irrigation being the major water user, its share in the total demand is bound to decrease from the present 83 per cent to 74 per cent due to more pressing and competing demands from other sectors by 2025 and, as such, the question of improving the present level of water use efficiency in general and for irrigation in particular assumes a great significance in prospective water resource planning.

It is estimated that a 10 per cent increase in the present level of water use efficiency in irrigation projects can bring under irrigation an additional 14 million ha. area from the existing irrigation capacities, which would involve a very moderate investment as compared to the investment that would be required for creating equivalent potential through new schemes. Thus, there is a need to improve the water use efficiency in most of the existing irrigation projects through modernization, renovation and upgradation, in order to realize optimum benefits on the one hand, and mitigate the consequential side effects like water-logging on the other.

A corollary of the issue allied is the water use management, which means optimization of available water for maximizing crop production in a given area. In most of the northern states, for example, it is observed from the data collected under a survey on methodological investigation into high yielding variety programme taken up by the Indian Agricultural Statistics Research Institute, that the frequency of irrigation for the wheat crop is quite high, ranging from five to seven. The situation is similar for paddy and maize in some of the states. A recent study by the Water Technology Centre of the Indian Agricultural Research Institute (IARI) says that mismanagement of irrigation leads to water logging and saline conditions not conducive to crop growth. Excess water interferes with field preparation, sowing, plant growth, yield and quality and harvesting operations. About six million ha. or 15 per cent of the area under paddy is reported to suffer due to water-logged conditions in the different parts of the country. Serious problems of water-logging have been reported in some irrigation projects like Chambal in Rajasthan and Madhya Pradesh, Tawa in Madhya Pradesh, Pune and Jayakwadi in Maharashtra, Tungabhadra and Malaprabha in Karnataka, and Gandak and Kosi in Bihar. At a number of places, the problem of excess water has led to salinity problems aggravating the situation. It may be mentioned that, although paddy requires large quantities of water for good growth, prolonged water-logged conditions are not good for it. Certain stages of growth like early flowering have been reported to be more susceptible to waterlogged conditions than other stages of crop growth. For instance, flooding in the third month of transplanting, which coincides with the flowering stage, reduces the yield by about 42 per cent. Similarly, yields get reduced by 20 to 24 per cent when flooding takes place at too early or too late stages.

Thus, there appears to be a considerable wastage of water by way of over-irrigating crops without taking into account the actual requirements of each crop. To study the scope and extent of saving water by reducing the number of irrigations without seriously affecting the crop yield, results of an experiment of production potential at different research stations were examined and are presented in Table 2.1. It will be observed that in Rajasthan the number of irrigations on wheat could be reduced from 6 to 3 without affecting the crop yield significantly. Similar results were also observed in Uttar Pradesh and Punjab. Obviously, the water so saved by reducing the number of irrigations on wheat crop can be profitably utilized for cultivation of other crops like gram and jowar. It is thus clear that for a given amount of water, there is a need for formulation of a proper irrigation schedule so as to maximize the area irrigated under different crops which in turn will bring in the highest total production.

We have now to go much beyond the simple engineering concept of water utilization which merely represents the area of surface to which water has been applied or spread and does not take into account the needs of the crops grown in respect of the quantity and timing of the irrigation. We should not really be content even with 100 per cent utilization of potential in this sense, but should aim at making optimum use of water by adopting cropping patterns and irrigation practices based on a careful study of water requirements of crops in the particular soil and climatic conditions of any given area.

Table 2.1: Production Levels for Wheat and Gram for Alternative Use of Irrigation

(Q./ha.)

State	Crop	Number of Irrigation	Yield (Q./ha.)	Enhanced or Reduced Number of Iriigation	Yield (Q./ha.)
Rajastan	Wheat	6	33.5	3	31.7
	Gram	0	11.3	2	22.4
Uttar Pradesh	Wheat	4	36.6	3	32.2
	Gram	0	15.9	2	18.7
Maharashtra	Wheat	4	28.0	2	17.1
	Gram	0	8.5	2	20.9
Punjab	Wheat	4	63.0	2	56.4
	Gram	0	9.9	2	17.5

Source: D. Singh, Presidential Address, Indian Agricultural Statistics Research Institute, 1981.

SAVING WATER WITH MODERN TECHNIQUES

There are basically three irrigation methods (each with many variations), gravity, sprinkler or drip. Gravity irrigation depends on topography. The water source must be higher than the surface, and the field contours determine the irrigation possibilities. A sprinkler system operates by water being expelled through a narrow nozzle at high velocity, resulting in spray. The system is widely used in Israel, Australia and USA. Coffee growers in India and some states like Haryana and Rajasthan have also adopted it to some extent. The aim of the sprinkler system is to distribute the drops equally over a given area. In a drip system, water is distributed to designated points on the soil without passing through the atmosphere.

A key to evaluating all systems is the Co-efficient of Uniformity, or the CU. This is a calculated figure that represents the uniformity of water distribution over an area. According to some experts, the higher the CU, the greater the uniformity and the more effective the irrigation. Drip methods typically have CUs of about 90, sprinklers around 80 and gravity methods from 50 to 75. The impact of different CU levels on crop yields can be significant, as indicated in Table 2.2. It is obvious that high CUs are important for irrigation to result in high crop yields. In the case of maize, for example, increasing the CU from 70 to 80 results in an 18 per cent increase in yield.

In the conventional method of furrow irrigation, for example, only a quarter of each 100 cubic metres of moisture reaches the roots of cotton. Sub-soil irrigation represents tubes buried in the soil to a definite depth. Through the openings in the tubes, water moistens the root layer. One variety of this method is drop-let irrigation, where a special dropping bottle is installed near each plant from the polythene irrigating main line. The whole of the water coming in small doses is assimilated by the roots. Nine years of experimentation in the USSR have shown that the yield of raw cotton was 41.3 metric centners per ha. in the case of furrow irrigation against 50.5 metric centners in the case of sub-soil irrigation with the distance of 60 cm. between the tubes. The loss of irrigation water per metric centner of raw cotton was thus reduced there by 33 per cent.

Some new irrigation techniques economizing the usage of water direct the quantity of optimum water right at the root zone of plants through an elaborate network of pipes. The benefits are: (1) reduced labour costs, (2) economy in water usage, (3) minimum land leveling, (4) decreased tillage, (5) reduced week growth, (6) low fertilizer requirement, and (7) high yield of crops. The only problem with the system is its cost.

Table 2.2: Crop Sensitivity to Water Distribution

	Yield (kg./ha.)		
CU	Banana	Potatoes	Maize
60	23.4	26.6	4.2
62	25.4	28.8	4.5
64	27.2	31.0	4.8
66	28.9	32.9	5.2
68	30.6	34.9	5.4
70	32.1	36.6	5.7
72	33.6	38.2	6.0
74	34.9	39.7	6.2
76	36.1	41.2	6.4
78	37.3	42.5	6.6
80	38.3	43.7	6.8

Source: Rain Bird International.

This can be divided into two categories, the pump and control head, which cost around 25 per cent of the total cost and the distribution system and drippers, which account for the rest. The total cost of drip irrigation system depends on the spacing adopted for planting. For a closely spaced (1 x 1 m.), the cost is placed around Rs. 50,000 per ha. For horticultural crops with wider spacing (6 x 6 m.), the cost is estimated to be around Rs. 12,000 per ha. The amount of duties and levies, particularly on plastics, is an important element in the cost. If these are reduced by the Government, costs can possibly be slashed to half. The government of Rajasthan has introduced a subsidy on the drip irrigation system, restricted to small farmers whose total installation cost is within Rs. 20,000.

All in all, present returns from irrigated area being not more than half of their potential, agricultural scientists and water management experts have come to the conclusion that the most important issue which should now receive the highest attention is that the next twenty years should be devoted primarily to a reorganization of the management of irrigation infrastructure. According to them, the overall efficiency of the irrigation command areas can be doubled during this period if only the Government takes the important policy decision with regard to future directions of irrigation development and management in the country. The foremost priority at this stage should be the scientific management of the resources created with so much investment and dedication. The watchword from now on has to be scientific management of irrigation rather than additions to it. Scientific management of irrigation water is a relatively new concept not only in this country but in the world as a whole. The developed countries of the West do not have the kind of vast irrigation networks based on harnessing of major rivers, such as those of the Himalayas, as in this country. There are no ready models available, which India can emulate. Much of the knowledge of scientific management of irrigation networks must therefore be generated within India and must accrue from our own experience of the past thirty years.

FERTILIZER

THE RETENTION PRICE SCHEME

There is no doubt that India's strides in both agriculture and fertilizer consumption were fuelled by Government policies. From a fertilizer production base of almost nil in 1951–2, it has grown to 13 million tons of nutrients now. This has become possible on account of Government policies of intervention.

Based on Marathe Formula, the Retention Price Scheme (RPS) for fertilizers was introduced in November 1977. The idea was to provide a fair price to the manufacturers based on prescribed efficiency norms *vis-à-vis* the net realization being controlled by the government, and at the same time making it affordable to the farmers with the government subsidizing the difference.

In this method, the normal *modus operandi* is to take into consideration the variable costs borne by the urea producers, which include feedstock in principal, and the capital costs incurred by any urea plant: annual interest costs, provision for depreciation, marketing and selling expenses, etc. The annual output for any plant is fixed at 85 per cent capacity utilization for the urea plants using naptha and furnace oil as feedstock, and 90 per cent for the new plants using natural gas. Over and above that, besides taking into account these costs, the current specification (which have not been modified since long) allows for a 12 per cent rate of return of the net worth. But the case study of 17 fertilizer units for a period between 1991–2 and 1996–7, which together account for more than 50 per cent of market share of urea, suggests that almost half of the subsidies is owing to the rise in feedstock prices, which actually reaches the Government only, through petroleum and gas distributing Public Sector Undertakings (PSUs).

BALLOONING SUBSIDIES

Fertilizer subsidy during 1990–1 was Rs. 4,400 crore (Table 2.3). This was perceived to be too high and the Government vowed to prune it drastically. In 1991–2, it even committed to the International Monetary Fund (IMF) that fertilizer subsidies would be removed in three years. However, by 1998–9, the pay out had already increased by 2.6 times to Rs. 11,388 crore. Against the budgetary allocation for 1999–2000 of Rs. 13,250 crore, revised estimates amount to Rs. 13,877.4 crore, which is over three times the 1990–1 level. This clearly suggests the need for an in-depth study of the fundamental causes leading to growth in subsidy.

Subsidy expressed in rupees per ton represents the amount in excess of reasonable cost of production (including a fair margin of profit) based on prescribed efficiency norms in regard to capacity utilization and consumption of raw materials and utilities, over net return from sales at the controlled price. This is reimbursed to producers under the Retention Price Scheme. The changes in subsidy amount depend on the relative movements in reasonable cost *vis-à-vis* selling price. During the 1990s, the ex-refinery price of naphtha increased from Rs. 1,982 per ton to about Rs. 9,100 per ton (for supplies to port based units), i.e. by Rs. 7,118 per ton. The increase in production cost due to this alone was thus about Rs. 4,300 per ton (based on 0.6 ton naphtha for one ton urea). The overall increase in cost would be much higher due to increase in conversion cost, viz. wages/salaries, other overheads, chemicals and catalysts, repairs and maintenance, cost of bags, marketing and selling expense, etc. In sharp contrast, the selling price of urea increased from Rs. 2,350 per ton to Rs. 4,000 per ton. Thus the hike was only Rs. 1,650 per ton, leaving an uncovered gap of Rs. 2,650 per ton towards increase in feedstock cost alone (4,300–1,650). Necessarily, this was plugged by a corresponding increase in subsidy. The scenario for units based on other feedstocks, i.e. gas, fuel oil/LSHS, etc., would be broadly similar.

In view of the above, ballooning fertilizer subsidies is sustained despite steep increase in prices of feedstocks since there is a reluctance to increase the selling price of urea even by a small amount. Interestingly, the former represents a corresponding additional income in the hands of public sector units in the oil/gas sector—Indian Oil, ONGC, GAIL, Hindustan Petroleum, Oil India, etc. This is nothing but intra-economy transfer, i.e. from budget head to PSUs.

Besides, the subsidy is available on both domestic and imported fertilizers. Import subsidy is the difference between the sale price and the cost of importing fertilizers including transportation costs (inclusive of port handling) and dealer's margin. No wonder, along with the massive jump in nutrient production and consumption over the years, the subsidy bill has also bloated. The subsidy bill which was Rs. 371 crore in

Table 2.3: Central Subsidy on Food and Fertilizers (1976–7 to 1999–2000)

(Rs. million)

No.	Year	Food	Fertilizers		
			Imported	Indigenous	Total
1	1976–7	4,773	N.A	N.A	600
2	1977–8	4,801	2,410	250	2,660
3	1978–9	5,694	1,710	1,722	3,432
4	1979–80	6,000	2,830	3,208	6,038
5	1980–1	6,500	3,350	1,700	5,050
6	1981–2	7,000	1,000	2,750	3,750
7	1982–3	7,110	550	5,500	6,050
8	1983–4	8,350	1,420	9,000	10,420
9	1984–5	11,010	7,273	12,000	19,273
10	1985–6	16,500	3,237	16,000	19,237
11	1986–7	20,000	1,971	17,000	18,971
12	1987–8	20,000	1,140	20,500	21,640
13	1988–9	22,000	2,007	30,000	32,007
14	1989–90	24,760	7,711	37,710	45,421
15	1990–1	24,500	6,593	37,297	43,890
16	1991–2	28,500	12,996	35,000	47,996
17	1992–3	28,000	9,961	48,000	57,961
18	1993–4	55,370	7,620	38,000	45,620
19	1994–5	51,000	11,660	40,750	52,410
20	1995–6	53,770	19,350	43,000	62,350
21	1996–7*	60,660	13,500	47,430	60,930
22	1997–8*	75,000	8,260	66,000	74,260
23	1998–9*	87,000	2,380	73,600	75,980
24	1999–2000@	82,000	7,500	80,000	87,500

Notes: @ = Budget Estimates, * = Revised Budget Estimates

Source: The Fertilizer Association of India, *Fertilizer Statistics, 1998–1999*, FAI, New Delhi.

1980–1 jumped up 17 times to Rs. 6,217 crore in 1991–2. With the Government embarking on reforms, it decided to go for decontrol of phosphatic (P) and potassic (K) fertilizers and allowing subsidy on urea. As was expected, it proved to be a turning point. This created an imbalance in favour of consumption of cheaper (subsidized) urea. This is against the expert advice, which advocates balanced use of plant nutrients in order to sustain crop productivity and soil-fertility. The present price/subsidy structure of different fertilizer nutrients has thus led to a lopsided use of the same. An ideal N/P/K consumption ratio is 4/2/1. But as against this, 1996–7 pattern was 10/8/1. This was slightly corrected to 7.9/2.8/1 during 1997–8. This is primarily due to the decontrol of all the fertilizers other than urea since 1992. Since urea became comparatively cheaper, farmers used more of it and disturbed the balance.

With regard to the question of subsidies, there are three basic questions. One, who gets this subsidy: the farmer, industry, or consumers of grain? Two, what happens to the fertilizer industry if the RPS is abolished? And three, what happens to fertilizer consumption and agricultural production, as one moves from administered to market pricing?

WHO GETS THE SUBSIDY?

To see who gets the subsidy, we must know what the price of urea would be without controls and if imports were de-canalized at zero import duty. Then, domestic urea price would be linked to import parity price. In recent years, world urea prices have fluctuated between $90 per metric ton to $210 per metric ton. The Long Run Marginal Cost (LRMC) can be taken between $150 per metric ton to $160 per metric ton. Adding another $15 to $20 per metric ton for freight and domestic distribution costs, we get a much higher figure than what the Indian farmer has been paying. Farmers are apparently being subsidized because urea prices are lower than world LRMC.

However, if we combine urea pricing with the pricing of output like wheat and rice, then, according to import-price parity calculations, it was found years back that farmers were getting a raw deal. This was because the prices of their output were below corresponding import parity prices. Netting out the effect of output prices on input prices, one found that the Indian farmer is not really net subsidized. In fact, he is being 'taxed' through trade and pricing policies, and the Aggregate Measure of Support (AMS) for Indian agriculture is actually minus 20 per cent of the value of agricultural produce. If so, it is either the consumer of grain or the fertilizer industry who got the fertilizer subsidy.

In fact, even if the agricultural price position has recently completely changed (India today is priced out in the international market for all the major agricultural commodities), and whatever be its shape and form, the subsidy has not much to do with the farmer/farming. Under the Procurement policy, total produce of the farmer (whatever comes to the market) is to be purchased by the Government at fixed Minimum Support Prices (MSP). And the basic criteria to determine MSP is cost-plus. This means that any increase in the cost of production of the farmer will call for an immediate increase in MSP. What really happens is that, due to the farmer lobby being quite strong, increase in the MSP is higher than the actual impact of increases in the input costs. This is what happened in June 1994, when the prices of urea were raised from Rs. 6.0 to Rs. 7.22 (20.3 per cent) and those of other fertilizers decontrolled: MSP for wheat was raised from Rs. 380 to Rs. 475 (including central bonus of Rs. 60), which meant an increase of 25 per cent although the share of fertilizers in the total cost of production would workout to be much less (say 3–4 per cent on cost A2). From 1997, the PDS was restructured when, under TPDS, prices for people below the poverty line were drastically reduced both for rice and wheat. This led to substantial increases in the food subsidy bill. The moral of the story is that subsidies on fertilizers or food are the two sides of the same coin.

But all this doesn't prevent the industry from getting the subsidy because many plants have higher costs than the import price of urea. The current landed cost of concessional naphtha is in the range of Rs. 3,200–3,500 per ton, and the price to fertilizer units works out between Rs. 5,500–5,800 after including freight and sales tax elements. The landed cost on a non-concessional basis is Rs. 6,684 ton and, after adding a sales tax of 10 per cent, freight charges and sales tax, the final price comes to around Rs. 9,000 per ton. Thus, the differential in price between subsidized and non-subsidized urea comes to Rs. 3,200–3,500 per ton. Since 0.6 ton of naphtha is utilized to manufacture one ton of urea, the cost differential per ton of urea to the manufacturer comes to Rs. 2,000. As four million tons of urea is produced using naphtha as the feedstock, the extra subsidy works out to Rs. 800 crore a year.

A similar calculation will show that the differential for units that use fuel oil comes to Rs. 2,500 per ton. Using the standard formula of 0.6 tons of oil requirement for a ton of output, the extra cost to the producer was Rs. 1,500 per ton. On an output of 2.26 million tons in 1997–8, the extra cost was a whopping Rs. 600 crore. The production figures taken into account in the above calculations were based on data available from the year 1996–7. In 1997–8, the targeted production was substantially higher and it was safe to assume that the extra burden on fertilizer units was higher by at least Rs. 100 crore. The total yearly burden, taking fuel oil and naphtha into account, would come to Rs. 1,500 crore. The Finance Ministry, however, did not bear the entire subsidy as half of it was borne by the oil pool account (for the period April–September, 1997). The incidence on the budget was calculated to be roughly Rs. 750 crore in 1997–8.

THE HANUMANTHA RAO COMMITTEE

If one were to price feedstocks on the basis of their opportunity cost—gas on the basis of Sankar Committee's recommendations (Rs. 3,050 per thousand cubic metres at land fall and Rs. 4,200 per thousand cubic metres on HBJ pipeline), and naphtha and fuel oil on import parity basis—, about one-third of domestic urea production becomes economically non-viable at an import parity price of $180 per metric ton. If the import-parity price comes down to $160 per metric ton, 45 per cent of domestic production becomes non-viable. Thus, roughly half to one-third of the subsidy on domestic urea production is going to support high-cost local urea producers. This remained largely hidden in the past as feedstock prices for urea industry carried high concession. But as distortions in feedstock pricing are corrected, the high costs of area industry become glaring.

The Hanumantha Rao Committee has questioned these costs, and feels that there is lot of gold plating. It suggested a method under which the RPS was to be replaced by a dual system. This system called for fixing the ex-factory Normative Referral Prices (NRP) of urea at Rs. 6,050 for gas based plants. For naptha based plants, the committee recommended Rs. 1,700 per ton extra, and Rs. 1,300 per ton extra for the FO/LSHS based plants. In fact, these recommendations meant raising the farm-gate price of urea by almost 65 per cent (over Rs. 3,660 a ton when the report was tabled), a real bold step. The report also recommended going for a Low Range Marginal Cost (LRMC) method, according to which subsidies on every additional ton of urea would be lower.

Because of the vested interests of those who were enjoying the benefits of RPS, as also the inability of the Government to face the aftermaths of raising urea prices by 65 per cent, the report has been put on the back burner. However, the Finance Minister, in his 2000–1 budget, has no doubt initiated a movement away from a system which has operated as a premium on inefficiency on the one hand, and, on the other hand, as an undeserved bonus for private sector companies which have even used subsidies for dividend distribution as some kind of 'other income'. This is a healthy beginning to get out of the subsidy syndrome.

FUTURE POLICY

While giving a clear indication that a subsidy free regime will prevail in the coming 10–15 years, an objective of the future fertilizer pricing policy has to lure investment to keep enlarging the fertilizer production base, imports playing only a marginal and supplementary role. The present, retention-price based subsidy scheme which stems from the Marathe Formula has achieved two broad objectives: (a) ensured that the farmers got fertilizers at an affordable price, and (b) ensured India became a major fertilizer producer, especially of nitrogen and phosphorous. This has made India immune to the blackmail of global producers. It must be realized that China and India are major players in world market and variations in their import demand has significant impact on world prices. Once the demand of these countries goes up, prices in the international market will shoot.

The focus should be on bringing down feedstock/fuel prices to reasonable levels which will automatically make indigenous production cheaper than imports. It means that fertilizer production has to be encouraged, but cost of production must be reduced by an appropriate pricing policy for feedstock/fuel prices. In order to encourage investment in fertilizer plants, the following concessions may be considered:

(i) exemption from customs, excise and other duties on capital goods (plant machinery) for setting up new plants or modernization or expansion of plants;
(ii) a ten year tax holiday for new plants;
(iii) permission to raise money from the parallel economy (as in the case of Indira Vikas Patra) since it is necessary that India continuously increased its production base for not becoming subject to the price blackmail of foreign producers;
(iv) present restrictions on the movement of fertilizer may be removed so that all plants realize maximum profits. In fact, for the new plants being set up with these concessions, there need be no restriction about their pricing. They should sell freely at market determined prices.

POWER

BURGEONING SUBSIDIES ON ELECTRICITY

Pump-sets are essential for farming and the development of rural areas. They are the heart of the irrigation system since they provide ground water and surface water resources. But they need power to operate, sometimes in huge quantities: in India, end-1990s, they are estimated to consume 30 per cent of total power generation in the country, with agricultural sector power tariff fixed on the basis of BHP (British Horse Power) rating of sets which vitiates the incentive to procure the somewhat more costly but more energy efficient pumping sets (the country can save more than one-third of its agricultural power consumption by having efficient pumps and pumping system). Nor is there any optimum use of the water for irrigation purposes, and there is also friction loss through foot valves and piping. But above all this is the present electricity pricing policy of the State Electricity Boards (SEBs) which regulate and base electricity prices on historical cost of the utility. In fact, for most of the utilities, average revenue realized per unit is actually less than the average cost of supply. The extent of subsidies for the agriculture sector is now at 80 to 100 per cent, while the subsidies for the domestic sector range between 40 and 60 per cent.

Yet, in 1976, agriculture accounted for 14 per cent of electricity consumption and was not subsidized at all. The average cost of production of electricity was then Rs. 0.18 per unit and the average tariff paid by the farm sector was Rs. 0.19. By 1978, the cost had risen to Rs. 0.20 but tariff remained at Rs. 0.18. In the next 18 years, the cost of power generation rose to Rs. 1.328 per unit, but the average tariff on agriculture increased to only Rs. 0.245. The share of the farm sector in electricity consumption, however, rose to 29 per cent of the total. As a result, in 1995–6, the subsidy to agriculture on power supply had risen to over Rs. 13,000 crore. Latest estimates indicate that gross subsidies to the power sector in 1999–2000 have been around Rs. 33,820 crore: while agriculture accounts for a bulk of the subsidy at Rs. 25,577 crore, domestic consumption accounts for Rs. 7,893 crore, the balance going to inter-state sales (Delhi excluded). The annual plan projections for 2000–01 calculate that agriculture will still account for the bulk of the subsidy at Rs. 28,217 crore, while domestic consumption will take up 9,387 crore and inter-state sales 358, the grand total reaching Rs. 37,692 crore (Economic Survey 1999–2000, Ministry of Finance).

Thus, along with uneconomic water rates, continued policy of states governments to provide subsidies to the farm sector on power has made the problem further complicated. However, some of the states governments like Karnataka, Maharashtra and Haryana have already decided to increase the power tariff. SEBs, which have now electricity regulatory commissions, are hoping to rationalize the tariff structure in the power industry. While subsidies on agriculture and domestic consumption would not be totally withdrawn, industry experts feel that rationalization would reduce the burden on industry which is cross-subsidizing the other sectors. Besides, the Regulatory Commission Act provides for subsidies only if the state government compensates the SEBs, and the new Electricity Bill 2000 wants to make mandatory for all SEBs to unbundle and rationalize the tariff structures.

NEED FOR DIFFERENT TARIFF

All over the world, the domestic tariffs are the highest and industrial tariffs the lowest: the off-peak tariffs are still lower. In India, till the end of the Fifth Five-Year Plan (ending March 1979), domestic tariff was the highest at about Rs. 0.55 per unit, whereas the industrial tariff was only 30 per cent of this, and agricultural tariff as one-third, i.e. Rs. 0.18 per unit. But this norm of international tariff formulation was completely changed after the Fifth Plan: while the domestic tariff has merely doubled till the end of the Eighth Plan, the industrial tariff has been raised by about 25 times and the agricultural tariff about three times in several states.

In fact, all over the world, there are different tariffs for peak, normal and off-peak consumers. Normally, the tariff for off-peak consumers is less than 50 per cent of that for peak consumers. Irrigation being not only off-peak during the day but also of seasonal character since the demand goes up only in certain months, the agricultural consumers should thus not be burdened with more than 50 per cent of the power tariff applicable to the domestic peak consumers. Moreover, even during the off-peak time of about 18 hours a day, the agricultural sector works in shifts and hardly gets power for 8–10 hours a day during peak seasons. The power tariff for the farm sector, therefore, should be further reduced by another 50 per cent, i.e. at only 25 per cent of the domestic tariff. At present, the average domestic rate is estimated at Rs. 1.14 per unit, and the average farm rate is around Rs. 0.35 per unit. The farm sector is thus not being subsidized, and all SEBs which have increased the average farm rate to Rs. 0.50 per unit are in fact subsidizing the domestic sector by Rs. 0.15 per unit.

Besides, it may be revealing to compare the average rates for the agricultural sector with that of Low Transmission (LT) industries. It is well known that LT industries normally get power round the clock except for some restrictions during peak time. The all-India average consumption of LT industries per kWh. of connected load in 1994–5 was less than 400 units. On the other hand, the average consumption of the agricultural sector—which hardly gets few hours of supply even during peak months—was about 2,100 units per kWh. of connected load. As there is no metering arrangement for pumping sets, it would be fair to compare the converted BHP rate of LT industries with that of the farm sector's. The average tariff of LT industries is about Rs. 2.50 per unit which, on a consumption of 700 units per kWh. corresponds to a fixed tariff of only Rs. 1,750. On the other hand, the average BHP tariff of the agricultural sector is about Rs. 1,000 per kWh. of connected load and, therefore, even if for academic purposes the average tariff of the agricultural sector is raised to match the LT industries, the average tariff in majority of the boards will still work out to only Rs. 0.75 per unit as against Rs. 0.50 at present. Therefore, it is quite obvious that there is no justification to hike the power tariff for the agricultural sector as it happens to be an off-peak consumer with limited hours of supply.

REAL AGRICULTURAL CONSUMPTION

Till the end of the Fifth Plan (March 1979), the consumption in the farm sector and the LT industries was practically equal at about 800 units per kWh. of connected load. But SEBs cleverly manipulated figures to convince the Government that heavy subsidy was needed for the farm sector. Indeed, in the absence of metering in rural areas, the excessive Transmission & Distribution (T&D) losses and pilferage has been transferred and shown as agricultural consumption. Thus, according to official sources, the share of agriculture in the total power sold has increased from 3.9 per cent in 1950 to 6.0 per cent in 1960–1, 10.2 per cent in 1970–1, 17,6 per cent in 1980–1, 26.4 per cent in 1990–1, and 31.8 per cent in 1998–9.

The fact of the matter is that most of the electricity being subsidized does not go to the farmer at all, but rather to the millions of small, ramshackle factories and machine sshops that are mushrooming in the rural areas all over the country and that are diverting the cheap agricultural electricity. Few micro-level studies have attempted to estimate the power consumption in agriculture by looking at how many hours the farmer has really used his irrigation pump, its horse power, and so to get to the actual consumption. The estimates so derived are less than even one-fifth of the figures being reported by official statistics in some cases. For example, in a report of the Centre for Advanced Development Research (1995) sponsored by the Uttar Pradesh State Planning Commission, it was found that total power consumption by pumps in UP was only 1,496 million units during 1993–4 as against 8,924 million units reported by the Uttar Pradesh SEB. In other words the actual consumption was just 17 per cent of what was being claimed officially.

Moreover, if one were to use the figures of power consumption as estimated in these survey studies to work out the actual revenue tariff being paid by agriculture, one would literally get a shock. As per this approach, for irrigation purpose only, in western UP, the farmers are already paying Rs. 2.09 to Rs. 2.53 per kWh.; in central UP the revenue tariff is even higher: between Rs. 2.47 to Rs. 3.09 per kWh.; and still higher in eastern UP: between Rs. 2.82 to Rs. 3.53 per kWh. For all agricultural operations, and for entire UP, the revenue tariff is Rs. 1.17 to Rs. 1.46 per kWh. In Karnal (Haryana), it is 0.94 per kWh. for 7.5 HP pump, on an average. Similar situation exists in Gujarat, where the revenue tariff for different capacities of irrigation pumps turns out to be between Rs. 1 to Rs. 2 per kWh. (World Bank, 1997). These estimates are 5 to 10 times higher than what the official statistics would provide. This wide variation in numbers speaks of an urgent need to do further research in this area before blindly recommending drastic increases in flat rate charges for power in rural areas. This also speaks of the possibility of enormous losses in transmission or simply theft, which is being shown as power supplied to agriculture in the official statistics.

Second, if one looks at the quality of power being supplied to rural areas, one would hear frequent complaints that wide fluctuations in power supplies often burn motors of irrigation pumps. Besides the extra cost that it imposes on the farmer to get his motors rewound, it leads to disruption in his irrigation schedule, as it is not always easy to get a mechanic that would repair the motor the same day. This factor compels many farmers to have stand-by diesel operated pump sets, which leads to much higher operational costs, besides the extra (avoidable) capital that he has to put in business to have some certainty in irrigating his fields. If one takes into account all these extra costs which are imposed by unreliable and erratic power supplies in the rural sector, the cost of irrigation to farmers may not be as low as is generally perceived.

If state governments want to keep giving cheap power to their farmers for their pumpsets, all they have to do is to license the 6.5 million pumpsets as they license cars today. Each farmer can then pay the market rate for power and get a rebate on the imputed consumption for irrigation per ha. per crop. That no one has even seen the discrepancy described above, let alone acted upon it, reflects the power of the small business class in the country and the abysmal incompetence of the state bureaucracies.

TRANSMISSION AND DISTRIBUTION PROBLEMS

'A plan is nothing, planning is everything': this observation attributed to Winston Churchill applies perfectly well to the power sector in India where the Ministry of

Power (MoP) has prepared massive plan to add 40,000 megawatts (MW) of new installed capacity during the Ninth Plan without matching provisions for associated Transmission and Distribution (T&D) facilities. And yet, it is obvious that power from new generation projects cannot be utilized efficiently due to inadequate T&D system and losses will mount at the cost of quality of power. Moreover, in earlier plans, India had recognized the need for a balanced programme within the maximum available resources, instead of focusing alone on generation at the cost of T&D as is being done at present. Apart from this, a number of mega-power projects and state trading of power have also been proposed, but no study has been done from generation to utilization of power even as the country boasts of being the leader in software. This, when all Indian Institutes of Technology (IITs) and even the Central Electricity Authority (CEA) have comprehensive software programmes for system and financial studies. It is unfortunate that even simple information of simultaneous peak demand being met in the country is not known.

Rough estimates indicate that the country is meeting a gross peak demand of only 60,000 MW against installed capacity of 95,000 MW—or 63 per cent against 85–90 per cent elsewhere in the world. The performance of our thermal machines is, therefore, deplorable for meeting peak demands. The real ailment of the power sector is serious shortage of peak power in the evening time, particularly during the summer months, whereas during the low demand period, the country is surplus in energy to an extent of about 30 per cent. All in all, the average technical loss exceeds 30 per cent and peak time loss 50 per cent corrected to voltage and frequency due to poor T&D system. The United States Assistance for International Development (USAID) also estimates that the current losses due to poor T&D system account for up to 40 to 50 per cent of the total power produced in India, whereas official figures are shown as only 17 to 23 per cent—figures which are already very high compared to other countries (Table 2.4).

Apart from this inefficiency, the theft of electricity is increasing, resulting in heavy loss of revenues to the SEBs. Over 40 per cent of the power generated is not brought into the billing process, and it is estimated that electricity worth about Rs. 1,500 crores is 'stolen' every year in the states: average losses due to thefts and pilferage in the country as a whole amounts to over Rs. 30,000 crore annually.

Consequently, losses of the SEBs could reach a whopping Rs. 1,70,000 crore by 2012 from Rs. 12,000 crore per annum now, and their outstanding to central power utilities could amount to a staggering Rs. 5,00,000 crore by 2012 from the present Rs. 23,000 crore. In fact, eighteen of the twenty SEBs have today a negative rate of return, and commercial losses of SEBs had mounted four times in the last ten years. As a result, the states could not generate funds to invest in new power projects. Most SEBs could not even pay the central power agencies such as the National Thermal Power Corporation (NTPC) for the electricity bought (these agencies have threatened to even cut off supply to defaulting states, creating further problems). And while the Electricity Act envisaged a minimum 3 per cent return on investment in power projects, loans and advances from the state governments get converted into equity with no chance of any returns.

Furthermore, there is a faster deterioration in the quality of power supply as the frequency even exceeds 52 hertz, and the voltage profile throughout is extremely poor resulting in enormous damage to the consumer equipment. The public is not aware of the magnitude of this menace in the absence of awareness. The central planners are

Table 2.4: Transmission and Distribution Losses of Power in Selected Countries (1989 to 1993)

(%)

Country	1989	1990	1991	1992	1993
Bangladesh	30.10	15.07	16.32	13.31	17.10
Canada	9.45	8.44	8.74	8.46	8.25
China	7.54	7.44	7.68	7.75	7.16
Denmark	5.65	5.20	6.92	6.70	6.93
Finland	5.64	5.10	5.04	5.48	5.01
France	8.25	8.19	8.28	7.64	8.02
Germany	-	5.45	5.17	4.87	5.08
Hungary	11.11	11.09	11.06	9.43	13.66
India*	23.28	22.89	22.83	21.80	21.41
Indonesia	21.93	22.47	20.95	17.97	17.18
Israel	5.06	5.22	5.43	5.06	4.65
Italy	8.54	7.83	8.15	7.84	8.31
Malaysia	11.05	8.71	12.13	7.89	-
Pakistan	20.64	21.35	20.55	22.94	22.89
Poland	10.58	9.15	12.43	13.85	15.19
Rep of Korea	5.93	5.49	5.50	5.66	5.48
Russia	-	8.84	8.98	9.53	9.26
Sri Lanka	17.26	16.78	18.25	16.89	17.28
Switzerland	8.03	7.97	7.97	8.03	7.98
Thailand	10.20	10.85	11.08	10.67	8.41
U.K.	8.53	8.39	8.80	7.49	7.85
U.S.A.	5.70	9.93	7.93	8.40	8.66

Note: * = For financial year.
Source: Lok Sabha Unstarred Question #2693, Ministry of Power.

also unable to specify the objectives of power reforms and hesitate to quantify the same in terms of improved availability, financial gains and likely hike in power tariff, as they are apprehensive of public backlash. They have achieved little by the reforms in the power sector and are in fact responsible for worsening the situation. They are swayed by reform bills to mislead the public.

POWER REFORMS

Irrational tariff structure, poor operational performance, over-staffing and huge transmission and distribution losses, which have become a euphemism for power pilferage, account for severe erosion in the financial viability of the SEBs. The situation has come to such a pass that the SEBs just cannot meet their day-to-day expenditure, leave alone muster the resources required for capacity addition. Given their debt-based capital structure and the poor resource position of the state governments, the SEBs have been increasingly resorting to borrowing from the market or the financial institutions, thereby adding to their interest burden, and the resultant increase in the cost of power generation and supply. The culture of cross-subsidization across the so-called 'haves' and 'have nots', the primacy given to socio-political considerations over everything else, coupled with the unwillingness and inability of the powers that be to come to grips with the simmering crises, have almost disabled the reform process initiated in the early 1990s.

In 1991, the government allowed private sector participation in power generation. But the Independent Power Producers (IPPs) were allowed to sell power only to the SEBs. Given the perpetual sickness of the latter, the IPPs were apprehensive of their bills being paid by the sole off-taker of power, and they then wanted the state government to give guarantee, and the central Government to give counter-guarantee. The counter-guarantee route, being fraught with grave financial risks, was later abandoned in favour of certain workable alternatives, like linking power generation with distribution, permitting direct supply by IPPs to high tension consumers, opening of an escrow, etc.. All said and done, for want of clarity in Government policy and a transparent mechanism, less than 1,400 MW capacity has so far been operated in the private sector. It has been reported that a foreign company which entered the power market of China and India at the same time has so far commissioned two projects and constructed another three in China, whereas it is yet to get financial closure of its first project in India.

Reforms are undertaken to improve efficiency and ensure accountability in the overall operation of an industry or a sector for the good of both the industry and the consumer. But far from achieving these goals, the power sector reforms in India have only done the reverse. They have neither toned up the performance of the SEBs, nor brought any relief to the people. The net result is that:

— the official growth rate of energy consumption in 1996–7 declined to 1 per cent against the 8 per cent projected by the Ministry of Power (MoP);
— even though energy availability in 1996–7 increased by 4.2 per cent, around 75 per cent of this additional power was lost in either transmission or theft;
— technical loss incrseased to 24.53 per cent in 1996–7 against 22.27 per cent in previous year;
— planners have not provided any funds for the new distribution system needed for utilization of power from IPPs projects: according to estimates, the investment needed for sub-transmission and distribution is to the order of Rs. 2,00,000 crore;
— there are apprehensions that all big size IPPs projects may gradually withdraw due to the inability of states/SEBs to provide escrow cover.

This state of affairs has focused attention on the urgency of new reforms in the energy sector. A number of states have initiated these reforms and are in the process of restructuring their electricity boards (SEBs): Orissa set the tone, and other states have caught up: Rajasthan, Haryana, Andhra Pradesh, Maharashtra and Karnataka. But the outcome is still wide open. Claims and counter-claims by employees and the governments tend to come out with only half-truths. Because of the politics involved in power management and free or subsidized supply to vote banks, Boards faces great difficulties in generating enough funds to make ends meet. It is indeed politically opportune for state governments to provide free power to the agricultural sector, irrespective of the capacity of the farmer to pay for the energy consumed, but knowing that most of this free supply is siphoned off to industrial and commercial consumers thanks to a collusion of interests. As a matter of fact, in this context, the easy way out is to tax industrial or commercial consumers to cross-subsidize supply to the weaker sections, instead of passing on the costs to the consumers.

It was in this dismal scenario that the Centre embarked on a thorough reform of the power sector. A Central Electricity Regulatory Commission (CERC) was set up, and the states were asked to set up their own Regulatory Commissions (SERCs). Their role is to promote efficiency in generation, transmission and distribution of power. They are to establish norms of production and productivity, study the cost structure, establish a framework for dispute resolution, promote competition, advise the state governments. Simultaneously, the states and their SERCs were asked to appoint a tariff regulator who could fix and revise the tariff depending on the costs and the policies of the governments. The aim was to de-politicize the SEBs and ensure that if the states want to subsidize power, they must provide budgetary support for it every year.

More recently, Prime Minister Atal Behari Vajpayee has also outlined a seven-point agenda to overcome problems plaguing the power sector. Increasing power through renovation of existing plants, strengthening T&D systems to improve quality of power, unburdening T&D through corporatization, breaking down SEBs into generation, transmission and distribution entities, energy audit at all levels, elimination of power theft within a specified time frame, metering all power consumption by December 2001, changing hydro-thermal ratios: here are some of the main objectives the Centre and states would have to follow to meet the '*power for all*' objective by 2012.

PERSPECTIVE

The above discussion and the empirical facts indicate succinctly why inputs are subsidized and who actually get the benefit of subsidies. The position today is rather complicated. The commitment of the Government to PDS, procurement of foodgrains and remunerative prices to the farmer are all interrelated.

We shall, perhaps, not be very much wrong to assume that during the coming two decades, per capita incomes will at least double at the constant 2000 prices. With the establishment of an egalitarian society, present disparities between rural/urban societies and different regions—say BIMARU states, Punjab, Haryana, Maharashtra, etc.—will be reduced to some extent. Some 50 million people, primarily concentrated in the eastern region, who are not getting two square meals a day, will no longer go hungry, and the food basket of the average citizen will get reasonably diversified.

In a scenario like this, our PDS commitments will be reduced drastically. It will not possibly be obligatory to continue the procurement of various agricultural commodities, particularly foodgrains, at the present scale. If so, the present regime of MSP will also have to undergo corresponding changes, and our buffer-stock policy will have to be accordingly refashioned. There may be two types of buffer-stocks, one, a normal buffer-stock and, two, a contingency buffer-stock, which shall be the sole instrument of market intervention for price stabilization. For the operations of buffer-stock, we may follow the formula developed by the International Rubber Agreement, 1979 (see Appendix). This would mean setting up a price range consisting of a reference price and three price levels above it and three price levels below it. The upper and lower intervention prices may be calculated as plus or minus certain percentage of the reference price. Again, the upper and lower trigger action prices may be calculated as plus and minus certain percentages of reference price which would be slightly higher than the intervention prices. The upper limit may be termed as upper indicative prices, and the lower limit may be termed as lower indicative prices.

As far as input subsidies are concerned, the wrong pricing policies of fertilizers, irrigation water and electricity have brought inefficiencies in the production system and are eating up the resources which could better be used for creative production potential. In fact, continued subsidies are sending wrong signals to the farmers, distorting the age-old cropping pattern and leading to environmental degradation. They are also acting against promotion of regional and interpersonal equity. However, it is not possible to come out of the subsidy regime in one go. They have to be phased out in such a manner that it helps in achieving a balanced growth.

As for power, reforms already suggested in transmission and distribution of electricity, as well as in the organizational structure of the SEBs, will help immensely—if carried out earnestly—in reducing the cost of generation and increasing availability. Simultaneously, the need for a proper pricing policy—keeping in view the load factors—can hardly be overstated, and, here, it can be safely assumed that power meters will be installed for agriculture, as already decided. All this will take care of the abysmal condition of the SEBs with no subsidies involved.

Similarly, there is an urgent need for ensuring rational use of irrigation water, not only to achieve the required productivity growth, but also for retarding environment degradation. A judicious pricing of irrigation water will help in reversing the tendency of overuse of this scarce input. Besides introducing a regular system of revising water rates at reasonable intervals, a serious attempt will have to be made to encourage the application of modern techniques of water management and conservation. This would mean providing incentives and giving training to the farmers to adopt on a large scale sprinkle, drip and mulch irrigation systems, since large parts of north and south India are already facing acute problems of water scarcity. True, it is difficult to have water a meter for the individual field, but Water Users Associations or Pani Panchayats, already existing in a small way, will have to be established throughout the country, and metered bulk supply of water will have to be given to them. These Associations will then be responsible for making payments to the Electricity Boards, while the responsibility to collect the charges from individual farmers will rest with them.

So far as subsidy on fertilizers is concerned, it is necessary to formulate a coherent pricing policy so that there is a balanced use of N, P and K. In order to phase out fertilizer subsidy, it is imperative to withstand the pressure of the big farmers lobbies and the industries. There must be a political will to bring reforms in agricultural subsidy. Our aim should be to achieve a balanced growth between the supply of and demand for various agricultural products, thereby helping to alleviate the serious difficulties arising from surpluses and shortages. At the same time, there is need to encourage the use of organic and green manures, bio-fertilizers and vermiculture. If major part of dung is being burnt today, it will be necessary to encourage village forestry and also set up 'Gobar Gas Plants'. These issues will have to be carefully studied and every possible incentive provided to make these innovative programmes more popular.

Research results have already shown that such practices are cost effective and, besides improving soil structure, lead to higher productivity levels. Subsidies given in the initial stages can certainly be withdrawn slowly, and finally removed in the coming decades.

To conclude, the subsidy policy cannot be viewed in isolation as it has its ramifications on the supply of various agricultural inputs and their judicious uses, on PDS and the output prices, besides the goal of achieving fiscal balance. In fact, withdrawal of subsidies—food, agriculture and others—is simply not an arithmetical jugglery. Above all, they need to be target- and time-oriented, and should not be a source of just additional gains for the big farmers, industries and the suppliers. They should also not distort our limited resource allocations and become a cause for environmental degradation. All future policies should actually aim at creating a healthy and robust agriculture and producing a prosperous farming community.

APPENDIX

THE INTERNATIONAL NATURAL RUBBER AGREEMENT

Starting from the first Preparatory Meeting on natural rubber in January 1977 in Geneva, at the end of its Fourth Session, on 6 October 1979, the conference established the text of the 'International Natural Rubber Agreement', 1979. Thus this agreement became the first new International Commodity Agreement successfully negotiated under the 'Integrated Programme for Commodities'. It became operational on 23 October 1980.

Main Features of the agreement

(1) The objectives of the agreement are spelt out in Article 1 as follows:

(a) to achieve a balanced growth between the supply of and demand for natural rubber, thereby helping to alleviate the serious difficulties arising from surpluses or shortages of natural rubber;

(b) to achieve stable conditions in natural rubber trade through avoiding excessive natural rubber price fluctuations, which adversely affect the long-term interests of both producers and consumers, and stabilizing these prices without distorting long-term market trends, in the interests of producers and consumers;

(c) to help to stabilize the export earnings from natural rubber of exporting members, and to increase their earnings based on expanding natural rubber export volumes at fair and remunerative prices, thereby helping to provide the necessary incentives for a dynamic and rising rate of production and the resources for accelerated economic growth and social development;

(d) to seek to ensure adequate supplies of natural rubber to meet the requirements of importing members at fair and reasonable prices and to improve the reliability and continuity of these supplies;

(e) to take feasible steps in the event of a surplus or shortage of natural rubber to mitigate the economic difficulties that members might encounter;

(f) to seek to expand international trade and to improve market access for natural rubber and processed products thereof;

(g) to improve the competitiveness of natural rubber by encouraging research and development on the problems of natural rubber;

(h) to encourage the efficient development of the natural rubber economy by seeking to facilitate and promote improvements in the processing, marketing and distribution of raw natural rubber; and

(i) to further international co-operation in and consultations on natural rubber matters affecting supply and demand, and to facilitate promotion and co-ordination of natural rubber research, assistance and other programme.

(2) In order to achieve the objectives of the agreement, an international natural rubber buffer-stock of 550,000 tons was established. This buffer-stock consists of a normal buffer-stock of 400,000 tons and a contingency buffer-stock of 150,000 tons and shall be the sole instrument of market intervention for price stabilization in the agreement.

(3) For the operations of the buffer-stock, the agreement established a price range consisting of a reference price and three price levels above and three price levels below it. The reference

price that applied on the entry into force of the Agreement was set at 210 Malaysian/Singapore cents per kg. The upper and lower intervention prices are calculated as plus or minus 15 per cent of the reference price; the upper intervention price was therefore 242 Malaysian/Singapore cents per kg. and the lower intervention price was 179 Malaysian/Singapore cents per kg. The upper and lower trigger action prices are calculated as plus and minus 20 per cent of the reference price; these prices were therefore set at 252 and 168 Malaysian/Singapore cents per kg. respectively. For the first 30 months after the entry into force of the agreement, the upper and lower limits of the price range have been set at 270 and 150 Malaysian/Singapore cents per kg. respectively. The upper limit is called 'the upper indicative prices' and the lower limit is called 'the lower indicative'.

(4) A Daily Market Indicator Price (DMIP) was established which is a composite weighted average of daily official current-month prices on the Kuala Lumpur, London, New York and Singapore markets in respects of three grades, namely, RSS I, RSS 3 and TSR 20. When the five-day average of DMIP is at the upper or lower intervention price or between them, the buffer-stock manager shall not intervene in the market. When the five-day average of DMIP is at or below the lower trigger action price, the buffer-stock manager must buy rubber until the market indicator price exceeds the lower trigger action price. Similarly, when the five-day average of DMIP is above the upper intervention price, the buffer-stock manager may sell rubber to defend the upper trigger action price, but at or above the upper trigger action price he must sell rubber until the market indicator price falls below the upper trigger action price.

(5) When sales or purchases for the buffer-stock reach the 400,000 tons level, the Council shall, by special vote, decide whether to bring the contingency buffer-stock into operation at the lower or upper trigger action price; or any price between the lower trigger action price and the lower indicative price, or the upper trigger action price and the upper indicative price.

Unless the Council, by special vote, decides otherwise, the contingency buffer-stock shall be brought into action at a point midway between the trigger action prices and the indicative prices.

(6) Review and possibly revision of the reference price shall take place every 18 months after the entry into force of the agreement and shall be based on market trends and/or net changes in the buffer-stock. In addition, the Council shall review and may, by special vote, revise the upper and lower indicative prices under the agreement every 30 months after the entry into force of the agreement or, in exceptional circumstances, at the request of a member or members holding 200 or more votes in the council.

(7) Under article 28 of the agreement, members commit themselves to financing the total costs of the international buffer-stock of 550,000 tons. The financing of both the normal and contingency stock shall be shared equally between exporting and importing categories of members. The total costs of the normal buffer-stock of 400,000 tons are to be paid in cash.

The total costs of the contingency buffer-stock of 150,000 tons shall be financed by cash contributions and/or by cash borrowed from commercial sources by the Council upon the security of both stock warrants and government guarantees/undertakings. The agreement contains a provision which allows the International Natural Rubber Council to take full advantage of the financing facilities that will be available when the Common Fund under the Integrated Programme for Commodities becomes operational.

(8) The agreement established a 'Committee on Other Measures' to promote the development of the natural rubber economy by producing members through expanded and improved production, productivity and marketing. For this purpose, the Committee on Other Measures shall undertake economic and technical analyses in order to identify:

   (a) natural rubber research and development programmes and projects of benefits to exporting and importing members, including scientific research in specific areas;
   (b) programmes and projects to improve the productivity of the natural rubber industry;
   (c) ways and means to upgrade natural rubber supplies and achieve uniformity in quality specification and presentation of natural rubber; and
   (d) methods of improving the processing, marketing and distribution of raw natural rubber.

(9) The agreement is for a period of five years, unless extended or terminated in accordance with the provisions of the Agreement.

(10) The agreement has a provision for establishment of a panel of experts from the rubber industry and trade of exporting and importing members. The panel is to provide advice and assistance to the Council and its Committees, particularly on buffer-stock operations and on other measures.

The Agreement had been opened for signature from 2 January to 13 June 1980 at the Headquarters of United Nations in New York and governments of 28 countries signed it. The agreement entered into force provisionally on 23 October 1980 and definitively on 15 April 1982.

NOTES

1. Madhura SWAMINATHAN, 'Understand the Costs of the Food Corporation', *Economic and Political Weekly*, Vol. 34, No. 52, 25-31 December 1999.
2. This consists of fertilizers (Rs. 138.7 billion), power (Rs. 255.8 billion), irrigation (Rs. 26.5 billion) and credit (Rs. 34.8 billion) plus an estimated amount of Rs. 45 billion for the remaining heads—seeds, machinery, pesticides and diesel for agriculture, etc.

3

Fertilizer Subsidy in Indian Agriculture

SOME ISSUES

S.S. ACHARYA

INTRODUCTION

Fertilizer subsidy is one of the important subsidies related to the agricultural sector in India. The other major subsidies are food subsidy, electricity subsidy and canal water subsidy. Fertilizer and food subsidies are in the central sector whereas power and canal water subsidies are in the state sector. While fertilizer and food subsidies explicitly appear in the union budget, the power and canal water subsidies are implicit subsidies arising on account of lower user charges recovered from the consumers than the average cost associated with these inputs.

The objective of this paper is to examine the trend in the fertilizer subsidies; review the system of pricing of fertilizers; and analyze the options available to reduce the fertilizer subsidies. The paper is divided into six sections. The first section deals with the trend in the volume of fertilizer subsidies in India. In the second section, we present a brief review of fertilizer pricing policy in India. The decomposition of fertilizer subsidies is presented in the third section. The role of fertilizers and place of fertilizer subsidies in the farm economy are discussed in the fourth section. Some options to reduce fertilizer subsidies are presented in the fifth section. Some concluding observations are given in the last section.

TRENDS IN FERTILIZER SUBSIDY IN THE UNION BUDGET

Since 1977, the fertilizer subsidy in the Union budget mainly represents the difference between Farm Gate Price (FGP, excluding dealer's margin and sales tax) and Retention Price (RP, inclusive of equated freight) for indigenous fertilizers, and, for imported fertilizers, between FGP and c.i.f. plus pool handling charges. The fertilizer subsidy in the Union budget increased from a meager Rs. 60 crore during 1976–7 to Rs. 505 crore during 1980–1, Rs. 4,389 crore during 1990–1 and further to Rs. 11,596 crore during 1998–9. During 1999-2000, the subsidy is estimated at Rs. 13,250 crore (Table 3.1). The subsidy on fertilizer as a percentage of GDP from agriculture had gone up from 1.2 per cent during 1980–1 to 3.2 per cent during 1990–1, but declined to 2.5 per cent during 1997–8. It may be mentioned here that during the 1990s, the share of fertilizer subsidies

Table 3.1: Total Subsidy on Indigenous and Imported Fertilizers (1976–7 to 1999–2000)

(Rs. in crore)

Year	Indigenous	Imported	Total
1976–7	-	-	60
1977–8	25	241	266
1978–9	172	171	343
1979–80	321	283	604
1980–1	170	335	505
1981–2	275	100	375
1982–3	550	55	605
1983–4	900	142	1042
1984–5	1200	727	1927
1985–6	1600	324	1924
1986–7	1700	197	1897
1987–8	2050	114	2164
1988–9	3000	201	3201
1989–90	3771	771	4542
1990–1	3730	659	4389
1991–2	3500	1300	4800
1992–3	5140	996	6136
1993–4	4327	600	4927
1994–5	4603	1166	5769
1995–6	4800	1935	6735
1996–7	6417	1163	7580
1997–8	8350	722	9072
1998–9	11263	333	11596
1999–2000*	12500	750	13250
Average			
TE 1979–80	221	263	484
TE 1989–90	2991	362	3353
TE 1999–2000	10653	602	11255

Note: * Estimation.
Source: Ministry of Finance, *Economic Survey*, Government of India, New Delhi (various issues).

in the total revenue expenditure of the Union Government had declined from around 6 per cent during 1990–1 to 5 per cent during 1997–8. The increase in subsidy during the last two decades has been mainly on domestically produced fertilizers. Of the total incremental fertilizer subsidy of Rs. 2,869 crore between triennium ending (TE) 1979–80 and TE 1989–90 (from Rs. 484 crore to Rs. 3,353 crore), as much as 96.5 per cent (Rs. 2,770 crore) was on indigenous fertilizers. Again, between TE 1989–90 and TE 1999–2000, out of the total increase of Rs. 7,902 crore in fertilizer subsidy, as much as 97.0 per cent (Rs. 7,662 crore) was on indigenous fertilizers.

In the context of bulk of the fertilizer subsidy being accounted for by the indigenous fertilizers, it may be relevant to look at the expansion in fertilizer production capacity created in the country and break-up of subsidy into different components. It may be recalled that traditionally India imported fertilizer in small quantities, mostly for use in plantations. The first major fertilizer plant was set up in Kerala in 1947 to produce 150 tons of Ammonium Sulphate (AS) per day. However, the first major public sector unit (of Fertilizer Corporation of India) was commissioned (at Sindri) in 1951 to produce AS. In 1956, fertilizer was recognized as a core sector for public sector investment and several units were commissioned for production of urea as well as AS, Calcium Ammonium Nitrate (CAN), Di-ammonium Phosphate (DAP) and complex NPK fertilizers. In 1966, distribution controls were relaxed to encourage private investment in this sector. Between 1966 and 1982, 26 urea plants were commissioned, which included seven in the private sector and two in the cooperative sector. With the discovery of gas in Bombay High, the production of fertilizers in the country got further momentum during the late 1980s and 1990s.

The production of fertilizers in terms of nutrients which had increased from a meager 0.15 million tons during 1960–1 to 1.06 million tons during 1970–1 and 3.00 million tons during 1980–1, went up to 9.04 million tons during 1990–1 and further to 13.62 million tons during 1998–9 (Table 3.2). Between 1980–1 and 1998–9, while the production of N increased from 2.16 million tons to 10.48 million tons, that of P went up from 0.84 million tons to 3.14 million tons. It may be mentioned here that despite the increase in domestic production, the country continued to depend on imported fertilizers for meeting the domestic demand. Though the volume of imports have fluctuated from year to year, the dependence on imports as observed during 1998–9 was 100 per cent for potassic fertilizers, 24 per cent for phosphatic fertilizers and about 6 per cent for nitrogenous fertilizers.

Table 3.2: Production, Imports and Total Availability of Fertilizer (1960–1 to 1998–9)*

(million tons of nutrients)

Fertilizers	1960–1	1970–1	1980–1	1990–1	1998–9
Nitrogenous					
Production	0.10	0.83	2.16	6.99	10.48
Imports	0.40	0.48	1.51	0.41	0.64
Total	0.50	1.31	3.67	7.40	11.14
Phosphatic					
Production	0.05	0.23	0.84	2.05	3.14
Imports	-	0.03	0.45	1.31	0.97
Total	0.05	0.26	1.29	3.36	4.11
Potassic					
Production	-	-	-	-	-
Imports	0.02	0.12	0.80	1.33	1.54
Total	0.02	0.12	0.80	1.33	1.54
All India					
Production	0.15	1.06	3.00	9.04	13.62
Imports	0.42	0.63	2.76	3.05	3.15
Total	0.57	1.69	5.76	12.09	16.77

Note: * Total availability here does not include carry over from previous year.

Source: Government of India, *Economic Survey 1999–2000*, Ministry of Finance, New Delhi, 2000, p. S-27.

The subsidy on indigenous fertilizers includes freight subsidy, special incentives and ad-hoc concessions. While the freight subsidy is given to certain manufacturers for carrying fertilizers to difficult areas, the ad-hoc concessions were introduced after the decontrol of distribution and pricing of phosphatic and potassic fertilizers in early 1990s with a view to encouraging the balanced use of N, P and K. While the ad-hoc concessions on phosphatic and potassic fertilizers were increased during the recent years, the subsidy on these two nutrients almost became negligible (Table 3.3). The subsidy on phosphatic fertilizers which aggregated to Rs. 1,956 crore during 1992–3, was reduced to Rs. 52 crore during 1997–8. During 1997–8, out of a total subsidy of Rs. 8,350 crore on indigenous fertilizers, 67.3 per cent was on nitrogenous fertilizers, 24.0 per cent was ad-hoc concessions on decontrolled fertilizers, and 7.9 per cent was freight subsidy.

Table 3.3: Components of Subsidy on Indigenous Fertilizers (1980–1 to 1997–8)

(Rs. in crore)

Year	Nitrogenous Fertilizer	Phosphatic Fertilizer	Single Super-phosphate	Freight Subsidy	Ad-hoc Concessions	Others	Total
1980–1	36	62	23	49	-	-	170
1981–2	78	79	25	93	-	-	275
1982–3	333	68	27	122	-	-	550
1983–4	588	108	38	166	-	-	900
1984–5	673	257	74	196	-	-	1200
1985–6	841	382	134	243	-	-	1600
1986–7	1088	227	112	273	-	-	1700
1987–8	1312	291	148	298	-	-	2050
1988–9	1673	588	223	516	-	-	3000
1989–90	2028	929	341	473	-	-	3771
1990–1	1736	1057	431	505	-	-	3730
1991–2	1572	995	331	603	-	-	3500
1992–3	2428	1765	191	416	340	-	5140
1993–4	3149	88	101	435	527	26	4327
1994–5	3288	299	36	390	528	62	4603
1995–6	3601	11	1	644	500	43	4800
1996–7	4075	25	1	610	1674	32	6417
1997–8	5617	50	2	661	2000	20	8350

Source: Ministry of Chemicals and Fertilizers, *Fertilizer Pricing Policy – Report of High Powered Review Committee*, Government of India, Department of Fertilizers, New Delhi, March 1998, Table II-1, p. 8

FERTILIZER PRICING POLICY

In order to understand the genesis of fertilizer subsidies in the Union budget, it is important to look at the main objectives and contents of fertilizer price policy. At the outset, it is necessary to distinguish between the policy to control fertilizer prices at farm gate level and the policy to subsidize fertilizer use.

The policy to control fertilizer prices dates back to the mid-1940s when the 'Grow More Food' campaign was launched in India. Its primary objective was not subsidizing fertilizer use but that of ensuring its use on food crops by controlling the distribution of limited fertilizer supply. With a view to insulating the farm gate prices from fluctuations in the world market and for tackling the problem of different cost of imported fertilizers and that of domestic production from plants with different costs of production, a Central Fertilizer Pool was created and the Fertilizer Control Order (introduced in 1944) was made effective in 1957. Prices of fertilizers for different manufacturing units were fixed individually and, based on the pooled cost of obtaining fertilizers from imports and from domestic producers, uniform prices were fixed for fertilizer consumers. There was no policy change to subsidize fertilizers even after the HYVs were introduced in the mid-1960s. For three decades, i.e. until the early 1970s, budgetary statistics on fertilizer transactions revealed surplus in all but few years. During these three decades, subsidy on fertilizers was confined to (a) off-season rebates, (b) transportation to inaccessible areas, and (c) occasionally to a few states to promote new products like CAN in place of AS and high analysis phosphatic fertilizers like Triple Super Phosphate (TSP) and DAP. However, the Central Fertilizer Pool was operated on a 'no-loss no-profit' basis.

The situation of fertilizer subsidy changed in 1973–4 due to the dramatic impact of oil crisis on world market fertilizer prices. Even after nearly doubling the farm gate prices, imported fertilizers had to be subsidized to avoid severe setbacks to farmers' demand for fertilizers. However, at one stage, farm gate prices exceeded the domestic cost of fertilizer production. At this stage, a Fertilizer Pool Equalization charge was introduced under which the difference between the farm gate prices and the cost of production had to be refunded to the pool. The pool was used to build up a good distribution system even in difficult and distant areas by absorbing the additional cost of transportation to these places.

In 1977, the Government of India introduced a Retention Price Scheme (RPS) covering urea, AS and CAN. Later, phosphatic fertilizers were also included under RPS. The spurt in fertilizer subsidies occurred after the introduction of this scheme. Under RPS, the retention prices are fixed for all the fertilizer manufacturing units individually, and each fertilizer plant is allowed a 12 per cent post tax return on net worth as a component of cost. In 1978, an equated freight scheme was introduced under which a single rate for all despatches, determined on the basis of average distance covered, was allowed to each unit. The retention prices are fixed, taking into account the cost of variable inputs, conversion costs, selling expenses and capital related charges. The main technical parameters for fixing the retention prices of urea are the norms of capacity utilization of ammonia plants, the normative plant capacity and the consumption norms for raw material and utilities.

Those whose retention prices were fixed below the farm gate price were required to credit the difference to a Fertilizer Price Fund Account set up under the Fertilizer Industry Coordination Committee which administers the scheme. Units with retention prices higher than the farm gate prices were to receive the difference from the fund account on a monthly basis. The fertilizer producing units are paid apart from the difference between retention price and controlled selling price minus dealer's margin, equated freight to cover the cost of movement from production to consumption points. The system of determination of retention prices under RPS is fairly complex, requires voluminous data and intricate calculations to sustain it.[1]

There were no changes in the scheme till July 1991 when the low analysis nitrogenous fertilizers like Calcium Ammonium Nitrate (CAN), Ammonium Sulphate

(AS) and Ammonium Chloride (AC) were excluded from RPS; these were brought back to the RPS in August 1992 and excluded again from June 1994. Phosphatic fertilizers including DAP were excluded in August 1992 and continue to be decontrolled. At present, RPS covers only urea.

The High Powered Fertilizer Pricing Policy Review Committee set up by the Government of India in 1997, has brought out, in its report (Ministry of Chemicals & Fertilizers, 1998), both positive and negative features of RPS. The positive aspects of RPS include the following:

(a) the secure environment provided by the RPS showed up in the increased capacity installation and production of fertilizers;
(b) the system of equated freight by covering movement costs up to block headquarters led to large scale extension of fertilizer distribution network in difficult and remote areas of the country; and
(c) the centralized allocation system led to building up of a monitoring system regarding fertilizer availability not only at the state but also at the block level and helped in reducing response time for dealing with fertilizer scarcity to a few days with emphasis on farmer satisfaction.

The negative aspects or deficiencies of the RPS arise because the system is non-competitive, is administered in nature and has a combination of norms and actuals. Further, a unit-wise price system tends to place emphasis upon proving and claiming of costs rather than in controlling costs and enhancing efficiency. Several studies have shown that under RPS, there is a tendency to inflate capital costs by fertilizer units. The burden of over capitalization is transferred to the government exchequer through higher capacity utilization and subsidy outgo per unit of output. These apart, the administered pricing system under RPS is characterized by delays, arbitrariness and lack of transparency in fixation of retention prices and related decisions.

DECOMPOSITION OF FERTILIZER SUBSIDIES

With a view to searching for options to reduce fertilizer subsidies, it is relevant to look for components of fertilizer subsidy. In this connection, following observations are relevant.

(i) Of the total increase of Rs. 10,771 crore in budgeted fertilizers subsidies between TE 1979–80 and TE 1999–2000, as much as 96.8 per cent (Rs. 10,432 crore) was accounted for the subsidies on indigenous fertilizers.

(ii) Of the total increase in fertilizer subsidies on indigenous fertilizers between TE 1979–80 and TE 1990–2000, 26.6 per cent increase occurred during the 1980s (between TE 1979–80 and TE 1989–90) and as much as 73.4 per cent occurred during the 1990s (between TE 1989–90 and TE 1990–2000).

(iii) Between TE 1979–80 and TE 1989–90, out of the total incremental budgeted fertilizer subsidy on indigenous fertilizer, 24.4 per cent was due to increased production/ consumption of fertilizers and remaining 75.6 per cent was due to increase in per unit subsidy. However, during 1990s (between TE 1989–90 and TE 1998–9), increase in production/consumption of fertilizers accounted for 33.6 per cent of the incremental subsidy. Thus, during the entire two-decade period, 30 per cent (Rs. 3,250 crore) of the incremental subsidy on indigenous fertilizers was on account of increased consumption/production of fertilizers, remaining 70 per cent (Rs. 7,182 crore) being on account of increase in rate of subsidy per unit.

(iv) Out of Rs. 7,182 crore of incremental subsidy on indigenous fertilizers due to increase in rate of per unit subsidy, Rs. 2,098 crore was during the 1980s and Rs. 5,084 crore was during the 1990s. Thus, nearly half of the increase in budgeted subsidy during the last two decades is on account of rise in the per unit subsidy on indigenous fertilizers during the 1990s. One should look for the reasons for this element of subsidy on fertilizers.

(v) It may be noted that during the 1990s, the farm gate prices of fertilizers in India were raised sharply with the index of prices of fertilizers going up from 99 during 1990–1 (and also the preceding two years) to 228 during 1998–9, registering a compound growth rate of 11 per cent per annum (Table 3.4). In the absence of hike in farm gate prices of fertilizers, the subsidy would have been even higher.

Table 3.4: Index of Prices Received and Paid by the Farmers (1981–2 to 1999–2000)

(1981–2 = 100)

Year	Prices Paid by Farmers			Prices Received by Farmers	Ratio of Fertilizers Prices to Prices Recieved	Terms of Trade for Agricultural Sector
	Fertilizer	All Inputs	All Commodities Purchased			
1981–2	100	100	100	100	1.00	100
1982–3	105	104	103	106	0.99	102
1983–4	99	110	112	116	0.85	104
1984–5	98	111	119	123	0.80	104
1985–6	101	116	125	124	0.81	99
1986–7	108	124	134	137	0.79	103
1987–8	108	131	148	155	0.70	105
1988–9	99	132	156	162	0.61	104
1989–90	99	135	168	176	0.56	104
1990–1	99	151	187	203	0.49	109
1991–2	124	177	211	236	0.53	112
1992–3	161	205	230	241	0.67	104
1993–4	182	229	245	269	0.68	110
1994–5	196	256	272	301	0.65	111
1995–6	211	267	295	321	0.66	109
1996–7	212	283	308	345	0.61	112
1997–8	224	310	327	350	0.64	107
1998–9	228					
1999–2000	233					

Source: For column 2, Ministry of Finance, *Economic Survey*, Government of India, New Delhi, (various issues). For other columns: Ministry of Agriculture, *Reports of Commission for Agricultural Costs and Prices*, Government of India, New Delhi (various issues) (base converted by us).

(vi) The main reason for the increase in fertilizer subsidy during the 1990s, despite the increase in farm gate prices of fertilizers, is the increase in subsidy on urea and ad-hoc concessions given on decontrolled P and K fertilizers.

(vii) The subsidy on urea which was Rs. 371 crore during 1980–1, increased to Rs. 2,278 crore during 1990–1 and went up sharply to Rs. 7,567 crore during 1997–8. It can be seen that the increase in budgeted subsidy on urea has been very sharp since 1994–5 because of both the increase in subsidy on imported urea and increase in retention prices of this fertilizer. The subsidy on imported urea increased from Rs. 599 crore during 1993–4 to Rs. 1,950 crore during 1997–8. This happened, despite hikes in farm gate prices of urea by 32.6 per cent from Rs. 2,760 per metric ton to Rs. 3,660, due to spurt in the landed price of imported urea. The import price of urea which was Rs. 3,792 per metric ton during 1993–4, went up to Rs. 5,626 during 1994–5, Rs. 7,771 during 1995–6 and Rs. 7,308 per metric ton during 1996–7. During all these three years, the price of imported urea exceeded the retention price paid to domestic producers of urea (Table 3.5). The weighted average of retention prices of urea went up from Rs. 3,749 per metric ton during 1989–90 to Rs. 5,098 per metric ton during 1993–4, and further to Rs. 6,826 during 1997–8.

Table 3.5: Trend in Prices and Subsidy on Urea (1977–8 to 1997–8)

Year	Farmgate Price (Rs. PMT)	Retention Price (Rs. PMT)	Import Price (Rs. PMT)	Total Subsidy (Rs. in Crores)		
				Indigenous	Imported	Total
1977–8	1550	1340	1240	25	–46	–21
1978–9	1450[a]	1496	1419	83	120	203
1979–80	1450	NA	1794	152	282	434
1980–1	2000	1900	1708	36	335	371
1981–2	2350	2100	1424	78	100	178
1982–3	2350	NA	1309	333	55	388
1983–4	2150	NA	1771	588	142	730
1984–5	2150	3804	1620	673	727	1400
1985–6	2350[b]	NA	1318	841	324	1165
1986–7	2350	NA	1490	1088	197	1285
1987–8	2350	3717	2010	1312	114	1426
1988–9	2350	3814	1915	1673	201	1874
1989–90	2350	3749	2614	1881	771	2652
1990–1	2350	4036	3086	1619	659	2278
1991–2	3300[c] 3060[d]	4332	3426	1506	1300	2806
1992–3	2760[e]	4929	3273	2365	1000	3365
1993–4	2760	5098	3792	3030	599	3629
1994–5	3320[f]	5297	5626	3249	1166	4415
1995–6	3320	5369	7771	3503	1935	5438
1996–7	3660[g]	5584	7308	4058	1648	5706
1997–8	3660	6826	5855	5617	1950	7567

Notes: * Weighted average; (a) with effect from 10 March 1979; (b) w.e.f. 31 Jan. 1986; (c) w.e.f. 25 July 1991; (d) w.e.f. 14 Aug. 1991; (e) w.e.f. 25 Aug. 1992; (f) w.e.f. 10 June 1994; (g) w.e.f. 21 Feb. 1997.

Source: Ministry of Chemicals and Fertilizers, *Fertilizer Pricing Policy – Report of High Powered Review Committee*, Government of India, New Delhi, 1998, p. 86.

(viii) It may be recalled that following the decontrol of distribution and pricing of P and K fertilizers in 1992, the prices of these two nutrients had shot up by more than 100 per cent (Gulati and Narayanan, 2000) and the NPK balance tilted against these two nutrients. For correcting the imbalance by containing the rising prices of P and K fertilizers, the Government in September 1992 decanalized the imports of DAP which adversely affected the economic viability of domestic DAP plants. Eight out of eleven DAP plants in the country had almost come to a grinding halt due to unrestricted imports of DAP. This compelled the Government to introduce a flat rate concession on the sale of these two nutrients. Later, in June 1993, imports of Muriat of Potash (MOP) were also decanalized. The ad-hoc (flat-rate) concessions on P and K fertilizers have continued on year-to-year basis with adjustments in rate and modalities. The ad-hoc concessions were introduced not to subsidize the fertilizers for the farmers but mainly to correct the imbalance in the use of NPK and provide some protection to the domestic production base of the country.

FERTILIZERS AND FERTILIZER SUBSIDIES IN THE FARM ECONOMY

Before searching for options to reduce fertilizer subsidies, it is relevant to look at the place of fertilizers in the farm economy.

(i) In the middle of the 1960s when the imports of foodgrains had reached 16 per cent of the availability in the country, a new strategy of agricultural development was launched with the objective of building a sound agricultural economy and maximizing foodgrains production. One of the three foundations chosen for achieving these objectives was assurance of a stable and remunerative price environment (both for inputs and outputs) for the farmers, the other two being provision of a more productive technology package (seed-fertilizer-irrigation package) and an efficient system of delivery of inputs and services including credit (Ministry of Agriculture, 1986). The main objective of agricultural price policy launched as a part of agricultural development strategy was to keep the prices of foodgrains at a level consistent with the income level of the masses for improving economic access to food and yet provide a reasonable return to the farmers for inducing them to invest in new technology and increase the production. The provision of such critical inputs as fertilizers at reasonable prices has been one of the important instruments of agricultural development policy and strategy for increasing food security, both at macro and household level. The provision of inputs subsidies helped the country in several ways:[2]

(a) it helped in achieving self-sufficiency in cereals (there was considerable increase in production of wheat and rice, which consume around 60 per cent of the total fertilizer used in the country);
(b) it helped in increasing the economic access to food (the proportion of average per capita income required to buy a unit quantity of staple cereals went down considerably since the early 1970s);
(c) with the increased availability of staple cereals at reasonable prices, the farmers in several regions diverted the resources to other agricultural commodities which improved the availability of edible oils, sugar, fruits, vegetables and spices in the country and helped in improving Balance of Payment (BOP) position; and
(d) it prevented the possible high disprotection to the agricultural sector.

(ii) The share of fertilizer subsidies in the total value of fertilizers used in the country is quite considerable. As per National Accounts Statistics, the value of subsidized fertilizers used in agriculture in India has gone up from Rs. 2,308 crore during 1980–1 to Rs. 6,398 crore during 1990–1 and further to Rs. 10,247 crore during 1995–6. Had there been no subsidies, the value of fertilizers used in India would have been Rs. 2,813 crore during 1980–1, Rs. 10,787 crore during 1990–1 and Rs. 16,982 crore during 1995–6. Thus the proportion of fertilizer subsidies in the aggregate value of fertilizer used in the country has ranged between 35.5 per cent and 42.7 per cent during the 1990s, which was considerably higher than that during the 1980s (Table 3.6). Interpreted differently, complete withdrawal of fertilizer subsidies would have raised the aggregate outgo of farmers on fertilizers by 64 per cent during the 1990s.[3]

Table 3.6: Fertilizer Subsidy and Total Value of Fertilizers Used in Indian Agriculture (1980–1 to 1995–6)

(Rs. in crore)

Year	Value of Fertilizers as per NAS	Subsidy on Fertilizers	Total Estimated Value of Fertilizers	Fertilizer Subsidy as % of Total Value
1980–1	2308	505	2813	18.0
1981–2	3010	375	3385	11.1
1982–3	3064	605	3669	16.5
1983–4	3351	1042	4393	23.7
1984–5	3930	1927	5857	32.9
1985–6	4261	1924	6185	31.1
1986–7	5051	1897	6948	27.3
1987–8	4232	2164	6396	33.8
1988–9	5752	3201	8953	35.8
1989–90	6062	4542	10604	42.8
1990–1	6398	4389	10787	40.7
1991–2	8740	4800	13540	35.5
1992–3	8229	6136	14365	42.7
1993–4	7969	4927	12896	38.2
1994–5	9557	5769	15326	37.6
1995–6	10247	6735	16982	39.7

Source: For column 2: Central Statistical Organization, *National Accounts Statistics of India*, Government of India, New Delhi (various issues). For Column 3: Ministry of Finance, *Economic Survey*, Government of India, New Delhi (various issues).

(iii) The increase in farmers' outgo on fertilizer of this magnitude would have considerably affected the cost of production and thereby the net incomes of the farmers, which in any case are considerably lower than the income of their urban counterparts. At the aggregate level, fertilizers account for 29.1 per cent of the total value of inputs at current prices (Table 3.7). The increase in outgo on fertilizer by 64 per cent would have raised the aggregate costs in farming by 18.6 per cent.

(iv) Fertilizers account for considerable proportion of cost of cultivation of major crops. The fertilizer cost (subsidized rates) as proportion of total cost (cost A1) is

Table 3.7: Share of Fertilizer in Total Inputs in Indian Agriculture (1980–1 to 1995–6)*

(Rs. in crore)

Year	At Current Prices			At Constant Prices (1980–1)		
	Fertilizers	Total Inputs	Fertilizer as % of Total Inputs	Fertilizers	Total Inputs	Fertilizer as % of Total Inputs
1980–1	2308	7122	32.41	2308	7122	32.41
1990–1	6398	20908	30.60	5291	12470	42.43
1991–2	8740	25562	34.19	5648	13050	43.28
1992–3	8229	27574	29.84	5506	13181	41.77
1993–4	7969	30102	26.47	5153	13262	38.86
1994–5	9557	35052	27.27	5798	14386	40.30
1995–6	10274	38885	26.42	6233	15160	41.11

Note: * Inputs other than feed of livestock.
Source: Central Statistical Organization, *National Accounts Statistics of India*, Government of India, New Delhi (various issues).

26.7 per cent for wheat, 19.0 per cent for paddy, 15.0 per cent for cotton, 20.6 per cent for sugarcane, 24.2 per cent for rapeseed-mustard and 14.6 per cent for maize (Table 3.8). If the fertilizer subsidies are withdrawn by raising the farm gate prices by 64 per cent, the cost of production of major crops would go up substantially bringing down the returns for farmers' own resources further.

(v) It is argued in some circles that due to the availability of subsidized fertilizers, farmers have reduced the use of organic manures which is adversely affecting the organic content of the soil. In this connection, the value of fertilizers and organic manures used in agriculture as estimated in National Accounts Statistics (Table 3.9) reveal that while during the 1980s there has been considerable decrease in the share of organic manures in the total value of manures and fertilizers, during 1990s, the proportion has nearly remained the same at around 12 per cent. Indian farmers know the importance of organic fertilizers. They also know that chemical fertilizers are not substitutes for organic manure.

(vi) The price elasticity for fertilizer use in India has been estimated by several scholars.[4] There is considerable variation in estimates of the elasticities owing to the differences in methodology used by the scholars. Generally, the estimates vary between –0.42 to –1.2. Considering the price elasticity co-efficient on the lower side of this range, other things remaining the same, a raise in farm gate prices of fertilizers to wipe off the subsidies would have a perceptible effect on the use of fertilizers and, in turn, on the levels of production.

(vii) It took considerable time and effort to create awareness of fertilizers and persuade the farmers to 'adopt' fertilizer use. The awareness of fertilizers among 105 million Indian farmers is almost universal. However, adoption is not yet universal. By the late

Table 3.8: Fertilizer Cost as Proportion of Total Cost (A1) (c. 1995–6)

Crop	State	Year	Fertilizer Cost (Rs./ha.)	Total Cost A1 (Rs./ha.)	Fertilizer as % of Cost A1
Wheat	Punjab	1996–7	2307	7992	28.9
	Haryana	1996-7	1840	7072	26.0
	Uttar Pradesh	1995-6	1532	6063	25.3
	Average	-	-	-	26.7
Paddy	Punjab	1996–7	1724	9261	18.6
	Haryana	1996-7	1738	8665	20.1
	Uttar Pradesh	1996-7	939	5050	18.6
	Andhra Pradesh	1996-7	2080	11160	18.6
	Average		-	-	19.0
Cotton	Andhra Pradesh	1996–7	2223	12938	17.2
	Maharashtra	1996-7	1127	7693	14.6
	Rajasthan	1996-7	655	5013	13.1
	Average		-	-	15.0
Sugarcane	Maharashtra	1994–5	4562	21304	21.4
	Uttar Pradesh	1994-5	1393	6996	19.9
	Karnataka	1994-5	3465	15989	21.7
	Haryana	1995-6	903	4690	19.3
	Average		-	-	20.6
Rapeseed & Mustard	Uttar Pradesh	1995–6	1028	3361	30.6
	Rajasthan	1995-6	696	2977	23.4
	Haryana	1996-7	778	3661	21.3
	Madhya Pradesh	1996-7	752	3483	21.6
	Average		-	-	24.2
Maize	Andhra Pradesh	1996–7	593	3528	16.8
	Madhya Pradesh	1996–7	370	2978	12.4
	Average		-	-	14.6

Source: Commission for Agricultural Costs and Prices, *Report for Various Years*, Ministry of Agriculture, Government of India, New Delhi.

Table 3.9: Value of Fertilizers and Organic Manures Used in Agriculture (1980–1 to 1995–6)

(1980–1 = 100)

Year	Fertilizers	Organic Manure	Total	Fertilizer as % of Total
1980–1	2308	679	2987	77.27
1990–1	5291	726	6017	87.93
1991–2	5648	711	6359	88.82
1992–3	5506	720	6226	88.44
1993–4	5153	721	5874	87.73
1994–5	5798	716	6514	89.01
1995–6	6233	743	6976	89.35

Source: Central Statistical Organization, *National Accounts Statistics of India*, Government of India, New Delhi (various issues).

1980s, about 80 per cent farmers had adopted fertilizer use. Nearly two-thirds of the cultivated land is fertilized which means nearly half of the unirrigated area was receiving some fertilizer application (Desai, 1993; Desai and Vaidynathan, 1994). Rice and wheat account for 60 per cent, and these along with sugarcane and cotton account for the bulk of the total use of fertilizers in India. Geographically, the use of fertilizers is concentrated in districts with high levels of irrigation and with better development of physical and institutional infrastructure. Availability of credit for buying fertilizers is an important factor in fertilizer use. Given this situation, any raise in fertilizer prices would adversely affect the potential benefits of fertilizer use to relatively resource-poor farmers of dry land areas because fertilizer use on nearly half of the unirrigated lands is yet to be extended. In these areas, there is a considerable periodic discontinuance of use by a significant proportion of farmers. Further, there is sub-optimal use by several farmers of these regions. No-doubt, there is a perceived over use of fertilizers by some farmers. But the solution to this lies in wide spread expansion in the soil testing facilities and making the results available to the farmers. A raise in fertilizer prices for reducing the budgeted subsidy on fertilizers is not the panacea to solve the problem of over use in some cases. The larger issue of need for extension of fertilizer use on rainfed crops, in far-flung difficult areas and on such crops as pulses and coarse cereals has also to be kept in view while considering the withdrawal of fertilizer subsidies.

(viii) It is argued by some that a raise in prices of fertilizers would not significantly reduce the net returns from the cultivation of crops. The linkage of profitability with the fertilizer subsidies can be seen by the share of subsidies in the net income. Withdrawal of fertilizer subsidies will immediately result in reduction in net income of farmers. For example, under the assumption that fertilizer use after the withdrawal of subsidy remains the same, family labour income (return over cost B2) from wheat would decline by 14 to 21 per cent in Punjab and Haryana, from paddy by 29 per cent in Punjab and 10 per cent in West Bengal, and from sugarcane by 14 per cent in Maharashtra (Acharya, 1994 and 1997). The squeeze in the net income of farmers would not only adversely affect the standard of living of farmers but also affect private investment in agriculture through decreased savings. If this decrease in private investment is not fully compensated by the increase in the public investment, the net investment may decline. As regards the option of squeezing farm incomes, it ought to be recognized that there already exists a limit on farmers' incomes owing to the provision of ceiling on land holdings. The argument that, in case the fertilizer subsidies are withdrawn, the income of the farmers would get further squeezed is not without a basis.

(ix) Sometimes, it is suggested that the adverse effect of rise in fertilizer prices on production and farm incomes can be mitigated by compensating the farmers through increase in output prices. In this connection, it needs to be pointed out that nearly 59.4 per cent (63.4 million out of 106.6 million) of the farmers in the country operate less than one hectare of land. These farmers are net buyers of grains. In addition, more than 18.8 per cent farmers (20.1 million) are small farmers operating between one to two hectares of land, with almost no or very little marketed surplus. For them the level of output prices is of little consequence. These farmers account for nearly 39 per cent of the net irrigated area in the country. What matters for them is the fertilizer prices and not the prices of output. As fertilizers constitute a considerable part of their paid out costs, with a rise in fertilizer prices, the cost of living of these farmers would go up and levels of living go down as the mechanism of raising output prices would not be able to compensate their increased cost of production. However, as a majority of them are net buyers of foodgrains, they would be paying higher prices for the grains purchased by them.

(x) It also needs to be noted that the prices released by the farmers for all the products and in all the areas are not administered prices. Compensation for increase in input prices can at best be provided for only such crops as wheat and paddy and in areas where the need for price support operations often arises. However, there is no mechanism to compensate the growers of other crops or of those areas where price support operations are not required.

OTHER OPTIONS TO REDUCE FERTILIZER SUBSIDIES

Given the likely adverse impact of raise in farm gate prices of fertilizers on national food security, economic access to food, agricultural growth and in turn economic growth, disparities between irrigated and dry land farmers and, overall, rate of inflation in the economy, the option of raising farm gate prices to wipe off the fertilizer subsidies from the Union budget is very limited. The option of transferring the burden to feed stock supplying agencies is also now ruled out with the dismantling of the administered price mechanism in the petroleum sector (Gulati and Narayanan, 2000). In this context, Gulati (1999) has suggested decanalization of imports of urea, abolition of RPS and giving away a flat rate subsidy of Rs. 1,500 per metric ton on imported urea and Rs. 2,000 per metric ton on domestically produced urea. This would help in putting the fertilizer industry on the road to efficiency by signaling low cost plants to expand their production and high cost plants to contract, which would help reduce the overall cost of the industry. It will also free the system from normative cost plus pricing route for individual plant to market oriented system based on demand-supply. It is estimated that it would save about Rs. 1,500 crore on fertilizer subsidy. While creating an environment which induces the plants to improve the efficiency, the guiding principle of the policy ought to be availability of fertilizers to farmers at reasonable prices and some degree of self-reliance in fertilizer production.

In this connection, it is important to recall the recommendations of the High Powered Committee on Fertilizer Pricing Policy (Ministry of Chemicals and Fertilizers, 1998) which has suggested that the mechanism for pricing fertilizers in India should: (a) ensure availability of fertilizers to farmers at affordable prices and help realize balanced fertilizer application, (b) sustain existing capacities and promote future investment in the fertilizer sector subject to desired level of efficiency in plant operations, and (c) provide incentives for technological development, resource use efficiency and cost reduction through competition within the domestic industry. The specific recommendations of the committee *inter alia* include:

(a) Ceiling Farm Gate Prices (FGP) be notified annually to make fertilizers available to farmers at affordable prices;
(b) the fertilizer industry be deregulated and the units be allowed to fix their retail prices subject to ceiling FGP;
(c) a Normative Referral Price (NRP) be determined based on Long Run Marginal Cost method (LRMC) for existing units for the purpose of arriving at subsidy to be paid;
(d) subsidy be disbursed through the manufacturers uniformly per metric ton of fertilizer sold to the extent of gap between (NRP + dealer's margin + average freight) and FGP;
(e) fertilizer units using naphtha/coal and Fuel Oil/Low Sulphur High Suphur be reimbursed feed stock differential cost for a period of five years;
(f) the imports of urea be canalized for a period of five years;
(g) additional freight and inventory cost be reimbursed to units in respect of fertilizers distributed in remote and inaccessible areas;
(h) output from new urea plants set up on strategic considerations be given an additional subsidy to cover their higher cost of production based on LRMC; and
(i) gap between FGP of MOP and its border price be bridged in stages.

These recommendations, if accepted, would go a long way in ensuring availability of fertilizers to farmers at reasonable prices, providing incentives to the industry to improve efficiency, sustaining existing capacities and promoting future investment in the industry.

CONCLUDING OBSERVATIONS

There is certainly a need for creating a new fertilizer policy environment but the task is much larger than reduction in fertilizer subsidies. Fertilizer will continue to be an important input for sustained growth of the agricultural sector and in turn of the whole economy and for the food security of the country. Further, as the small country assumption does not hold good for India, domestic fertilizer production capacity will need to be sustained. However, 15 to 20 per cent of our requirements should be met through trading which will also force the domestic industry to face competition and improve efficiency. There is no justification for plant specific retention prices. Further, the defects in the existing RPS should be corrected. In this context, the recommendations of the High Powered Committee on Fertilizer Pricing Policy need to be expeditiously implemented. Further, with a view to improving the fertilizer use efficiency, the facilities of soil testing need to be expanded in all the areas of the country.

It also needs to be recognized that given the organizational structure of the Indian farm sector and the current macro policy environment, the majority of the farmers will not be able to bear the burden of withdrawal of fertilizer subsidies. Investment in fertilizer subsidy should be treated as an investment for food security (macro as well as household level) and for accelerated growth of the agricultural sector and in turn of the entire economy. As regards the burden of fertilizer subsidies on the Union exchequer, it must be recognized that there are several non-merit subsidies, many times more in volume than the fertilizer subsidies, which should be withdrawn first where the pains of withdrawal are not borne by those who are least able to bear them.

NOTES

1. Several committees were appointed by the Government of India from time to time to suggest changes in RPS. For details see Ministry of Chemicals and Fertilizers (1998).
2. For more details see Acharya (1997 and 2000).
3. This is based on the assumption that fertilizers used by the farmers would have remained the same even after withdrawal of subsidies and consequent rise in fertilizers prices. This is possible if other factors which affect fertilizer use decisions of the farmers move appropriately.
4. For a review of these studies, see Srivastava (1993) and Vidya (1993). Also see Sidhu and Sidhu (1993).

REFERENCES

ACHARYA, S.S., 'Economic Reforms Programme. A Case of Agricultural Policies in India', in Surjit S. BHALLA (ed.), *New Economic Policies for a New India*, Indian Council of Social Science Research, Har-Anand Publications, New Delhi, 2000, pp. 154–73.

——, 'Input Subsidies in Indian Agriculture. Some Issues', in V.S. VYAS and Pradeep BHARGAVA, *Policies for Agricultural Development*, Institute of Development Studies, Rawat Publications, New Delhi, 1994, pp. 87–119.

——, 'Agricultural Price Policy and Development. Some Facts and Emerging Issues', *Indian Journal of Agricultural Economics*, Vol. 52, No. 1, Jan.–March 1997, pp. 1–52.

CENTRAL STATISTICAL ORGANIZATION, *National Accounts Statitics of India*, Government of India, New Delhi (various issues)

DESAI, G.M., 'Fertilizer Policies. A Perspective on Price and Subsidy Issues', in Sagar VIDYA, *Fertilizer Pricing. Issues Related to Subsidies*, Institute of Development Studies, Classic Publishing House, Jaipur, 1993, pp. 29–42.

DESAI, G.M. and A. VAIDYANATHAN, *Strategic Issues in Future Growth of Fertilizer Use in India*, ICAR and IFPRI, New Delhi, 1994, pp. 1–21.

FERTILIZER ASSOCIATION OF INDIA, *Fertilizer Statistics. 1997–1998*, New Delhi, 1998.

GULATI, A., 'Towards Rationalization of Fertilizer Subsidy. Case of Urea under an Open Environment', Paper presented at the NCAER–IEG–WB Workshop on *Reforms in Indian Agriculture for Growth, Efficiency, Equity and Sustainability*, India Habitat Centre, New Delhi, 15–16 April 1999.

GULATI, A. and S. NARAYANAN, 'Demystifying Fertilizer and Power Subsidies in India', in Surjit S. BHALLA (ed.), *New Economic Policies*, 2000, pp. 118–53.

MINISTRY OF AGRICULTURE, *Agricultural Price Policy. A Long Term Perspective*, Government of India, New Delhi, 1986.

MINISTRY OF CHEMICALS AND FERTILIZERS, *Fertilizer Pricing Policy. Report of the High Powered Review Committee*, Government of India, Department of Fertilizers, 1998.

MINISTRY OF FINANCE, *Economic Survey 1999–2000*, Government of India, 2000.

SIDHU, D.S. and J.S. SIDHU, 'Demand for Fertilizer and Foodgrain Production in India', in Sagar VIDYA (ed.), *Fertilizer Pricing*, 1993, pp. 115–30.

SRIVASTAVA, U.K., 'Price Elasticity of Fertilizer Demand in India. A Review', in Sagar VIDYA (ed.), *Fertilizer Pricing*, 1993, pp. 131–46.

VIDYA, Sagar, *Fertilizer Pricing. Issues Related to Subsidies*, Institute of Development Studies, Classic Publishing House, Jaipur, 1993, pp. 1–12 and pp. 147–60.

4

Political Economy of Water Pricing

SOME LESSONS FROM INDIA

A. VAIDYANATHAN

INTRODUCTION

Water pricing has been under discussion in India for nearly four decades. The National Irrigation Commission, successive Finance Commissions and practically every five-year Plan document has drawn attention to the need for improving cost recovery from the beneficiaries of public irrigation works. They have all endorsed the suggestion that water rates should be adequate to recover the costs incurred on operation and maintenance of the systems as well as a nominal one per cent of the amount invested in them. But little has been done to implement these suggestions. On the contrary, governments have been most reluctant to raise water rates even to keep pace with rising costs. Only a few states have raised these rates, that too, infrequently and inadequately. Most have not. There is reason to believe that laxity in billing and collections of dues has increased. As a result, the magnitude of unrecovered costs has risen both in absolute and relative terms.

In 1977–8, gross revenue from irrigation accounted for about a fifth of the estimated total cost (including 'full' capital charges) of providing it. In 1993–4, the recovery rate is estimated to have fallen below 10 per cent. In 1977–8, public sector plan outlay on irrigation and flood control was nearly 2.7 times the amount of unrecovered costs. By 1993–4, this ratio had fallen to 0.85. Losses on account of irrigation are also an important factor contributing to the growing fiscal deficits of the states. In 1993–4, they amounted to one fifth of the gross fiscal deficit of Indian states (Table 4.1).

Faced with the deteriorating fiscal situation and the shortage of resources for financing new public investment, the Planning Commission of the Government of India appointed, in 1991, a seventeen member committee to examine afresh the question of water rate revision. The committee included official and non-official engineers, administrators, economists and agriculturists from different parts of the country. They visited several states and project sites and discussed with officials and farmers the need for and implications of raising irrigation rates. Officials and state level politicians, while recognizing the necessity for higher cost recovery, were extremely sceptical about its political feasibility. Farmers on the other hand expressed a willingness to pay more for water, provided the quality of service (in terms of adequacy, timeliness and

Table 4.1: Estimate of Unrecovered[a] Costs on Account of Major and Medium Government Irrigation Works (1977–8, 1987–8 and 1993–4)

(Rs. million)

	1977–8	1987–8	1993–4
1. Recurring expenses[b]	1270	4930	9500
2. Depreciation[c]	560	1580	4210
3. Interest charges[c]	2790	11500	41680
4. Total costs (1+2+3)	4520	18010	55390
5. Gross revenue[d]	1050	1670	5490
6. Unrecovered costs (4–5)	3470	16340	49900
7. Public sector plan outlay on irrigation	9720	33470	58540
8. Gross fiscal deficit of states	63700	246200	290840

Notes: (a) Unrecovered costs = Recurring expenses plus depreciation at 1 per cent of cumulative capital stock plus interest on capital stock at the average rate paid on state Governments public debt minus gross revenue.

(b) Figures for 1977–8 and 1987–8 from Government of India, *Report on Irrigation Pricing Committee*, New Delhi, 1992. Figures for 1993–4 estimated by inflating 1987–8 figure with index of emoluments per public sector employee.

(c) Estimated from cumulative value of public sector plan outlays major and medium irrigation and Command Area Development and average interest rate on state public debt computed from data published in Reserve Bank of India, *Reports on Currency and Finance*, New Delhi, various issues:

(Rs. Million)		1977–8	1987–8	1993–4
a.	Cumulative investments	557608	1575580	420990
b.	Outstanding public debt of states	165180	671620	1600750
c.	Interest payments	8220	48790	1600750
d.	((c/b) x 100)	5	7.3	9.9

(d) Figures for 1977–8 and 1987–8 from Central Water Commission, *Water Related Statistics*, Government of India, New Delhi, 1998, and figures for 1993–4 from D.K. Srivatsava and Tepas K. Sen, *Government Subsidies in India*, National Institute of Public Finance and Policy, New Delhi, 1997 (basis not specified).

predictability of water supplies) was satisfactory. They also raised questions about the fairness of raising irrigation charges, and thereby increasing the economic burden on the 'poor' peasantry, without trimming the huge subsidies given to the urban population through under-pricing of water for domestic use, sanitation, urban transport, electricity and other public services.

The committee could not go into broader issues of inter-sectoral equity. But it was in no doubt that the magnitude of unrecovered costs in public irrigation systems is unconscionable, fiscally unsustainable and inimical to the use of water efficiently and in a sustainable manner. The need for effective action to remedy the situation was recognized and accepted. The question they had to address was to determine the basis for fixing the rates and suggest how to go about implementing a more rational pricing policy. In addressing these questions, the committee rejected the mechanistic approach of focusing exclusively on the magnitude of revisions in rates required to cover reported costs. Their report[1] took a much broader view of the problem, and, in doing so, departed radically from their predecessors in several respects.

COMMITTEE ON IRRIGATION PRICING

APPROACH

In the first place, while endorsing the view that rate revisions should cover Operation and Maintenance (O&M) costs and a modest capital charge, this was considered as a short-term objective. The committee asserted that, eventually, the aim should be full recovery of both recurring and capital costs, the latter including depreciation and the costs of capital invested valued at the average interest paid on public debt of the respective state Governments.

Second, they were against fixing rates on the basis of costs incurred by the Government. They recognized that costs actually incurred may be much higher than necessary if the works were designed, constructed and managed with reasonable efficiency. Users, they argued, should not be made to pay for higher-than-necessary investment due to poor design, unrealistically large command areas, 'leakages' and other factors; or for high operational costs and poor quality service due to inefficient management. The establishment of an independent authority to ensure this was among their important recommendations.

Third, they envisaged a drastic revision in the basis of levying rates. In the short run, they suggested replacing the present system of crop-specific area rates with season-specific area rates as a first step towards full volumetric pricing. In the long run, they proposed a two-part tariff comprising a fixed flat rate per unit area as payment for entitlement to get water and a variable rate related to the volume of water used.

Fourth, the Committee drew pointed attention to laxity in billing and collection of water rates; and to the fact that at present, the system management is under no obligation to specify and observe the quality of service which users can expect.

Therefore, and finally, the Committee came to the conclusion that questions of pricing cannot be considered independently of the way public irrigation systems are organized and managed. The systems as presently managed cannot provide better irrigation services (in terms of quantum and timing of water supply according to specified schedules) at reasonable cost. A number of far-reaching measures were considered necessary to make them better equipped for this purpose.

RECOMMENDATIONS

The package suggested by the Committee included the following:

(1) Physical improvements in the systems to permit more effective regulation of the timing and quantum of water supplies to different segments of the command.
(2) Replacing crop pattern regulations with specifications of volume entitlements by season.
(3) Replacing the existing system whereby Government assumes responsibility for

maintenance, monitoring deliveries and use of water, as well as assessment and collection of water charges at the level of individual farmers, with one where its responsibility for these functions is limited to the outlet level.

(4) Creation of water user associations to manage all these functions below the outlet and increasing their stake (through contributions to the cost of improvement) in the systems.

(5) Introducing a system of contracts, backed by legal sanctions for their enforcement, between the system management and the outlet level associations regarding the timing and quantum of water deliveries.

(6) Vesting the powers and the responsibility for making, monitoring and enforcing managerial decisions (including policy decisions) at all levels of each public irrigation system in autonomous bodies with elected users representatives participating in all tiers, and requiring them to be financially self-reliant.

Conscious that changes of such far-reaching scope and depth in the existing arrangements cannot be accomplished easily or soon, the committee recommended a phased approach to the reform. They further suggested that Government should offer incentives and also exert pressure to generate impetus for such reform. Proposed incentives included smaller increases in water rates and the offer of preferential and liberal financial assistance for improvement of distribution networks and facilities to outlet groups which are willing to undertake group management. The 'pressure' was to be generated by declaring the Governments' intention to reduce its sphere of responsibility, create autonomous organizations for managing the systems, turn over management below the outlet level to user groups and insist on their attaining financial self-reliance within a designated period.

That these recommendations were made unanimously by such a large committee of diverse membership is indicative of an emerging consensus, at the professional level, on the direction of future reform. Consensus among experts, however, does not guarantee its acceptance by governments. The Planning Commission has dutifully circulated the report to the states (who ultimately have to implement the reforms) and discussed its recommendations with their officials. Public discussion has been studiously avoided. In the event there is hardly any sign of a serious effort to act on the suggestions.

NATURE OF OPPOSITION TO REFORM

At one level, the reasons for the reluctance on the part of all concerned—namely the political leadership, the bureaucracy and the farmers—are obvious. For nearly three decades, the politics of populism has held sway. In the name of helping the poor, governments—irrespective of their political hue—have found it expedient not to pass on increasing costs of water, electricity and several other services to their users. Few states have revised rates even to keep pace with the rising costs, though several have levied surcharges on irrigated crops. (These latter are however used for purposes such as education other than maintenance and operation of the irrigation works.) There is no evidence of any improvement in the efficiency with which dues are collected.

The reported collections of irrigation charges show rather large fluctuations from year to year. Estimated collections in 1993–4 are much higher than the average during previous 10–15 years, possibly because special surcharges on irrigated crops are counted as irrigation revenues. Even so, the increase in collections has been slower than that of costs. In 1993–4, estimated revenue was only 30 per cent of the recovery norms suggested by earlier committees and about one-tenth of the amount needed for full cost recovery. Collection will have to be increased 3.2 times to meet the former target and 10 times to meet the latter. Increases in collections (be it from higher rates or stricter enforcement) of this magnitude would be difficult even for autocratic regimes. Governments in a democratic system based on adult franchise and competitive multiparty policies find it far more daunting.

Farmers, like any other set of tax payers, will not readily pay more for their water especially when they have got used to a lax regime of low rates and when the collections are sought to be increased manifold. The argument that low cost recovery cuts seriously into resources available for fresh investment in irrigation and other development benefiting new areas and classes of people does not cut much political ice. Those who stand to lose because of a slower pace of development are too diffused and dispersed to exert pressure strong enough to counter the interests of those already benefiting from public irrigation.

The latter may constitute a small proportion of the total electorate. But their absolute number (20 million households) is large and dispersed. Both in terms of numbers and local power, they can make a significant difference to electoral measures. Their resistance to raising water (or other input) prices can be and has been violent and cause major law and order problems. Consequently political parties are reluctant to risk alienating them. The reluctance is the greater when the partners are numerous and fragmented, with no single party (or a stable coalition of them) having a decisive majority in the state legislatures. Under these circumstances, calculations of electoral advantage and acquisition of power through coalitions of convenience dominate the attitudes of politicians.

The bureaucracy, the other interest group involved, is not uniformly opposed to raising of water rates. In fact, the Finance Departments of most states have always argued strongly in favour of improving cost recovery of irrigation and other public services. But the irrigation/revenue bureaucracies, which are responsible for collecting rates, are not enthusiastic. They have to deal directly with farmers' opposition and its consequences. They are also opposed to restructuring of irrigation organization which gives users an important role and increases pressures for greater accountability in performance.

Organizational changes involving major redeployment of personnel, redefinition of tasks to facilitate flexibility in management, and integration of personnel from different parts of a segmented bureaucracy is a difficult and complex undertaking. Any such proposal evokes strong resistance from middle and lower level functionaries which weak governments with uncertain life spans are unwilling to confront. Furthermore, both politicians and bureaucrats managing irrigation, having got used to the vast opportunities for power and patronage which the present arrangements offer, are loathe to surrender them.

FACTORS FAVOURABLE FOR REFORM

All this would suggest that the atmosphere is hardly propitious for reforms which require determined strong and sustained measures. However, the situation is not entirely hopeless in as much as the deepening fiscal crisis of states compels them to confront the problem of low cost recovery from irrigation, electricity and other services. The parlous condition of states' finances makes it impossible for continued unbridled populism. Fiscal deficits of most states are already large and rising at an alarming rate. The prospects of increased devolution of tax revenue and development assistance from the Centre being extremely limited, their dependence on borrowing (at 10 per cent or more interest rate) has increased. The resulting explosive growth of interest payments aggravates the resource crunch even further.

Under the circumstances, state governments, willy-nilly and sooner rather than later, will be forced to take a hard look at the problem of pricing and efficiency of public services. That these problems and the need for reforms are widely discussed in the media, that at least some state chief ministers are willing to address these issues publicly in their encounters with the people, and that a few have actually managed to raise rates substantially, indicate a welcome change.

At the users end also, there are indications of growing awareness and interest in getting better quality services. The idea that this requires greater involvement and participation of users in the management of government canal systems has now come to be widely accepted. Several states have enacted the necessary legislation and in some (e.g. Tamil Nadu, Karnataka, Andhra, Gujarat and Maharashtra) the number of such associations reported have made farmers more aware of the need for improvements in irrigation systems. They are also becoming vehicles for collectively articulating (with the system management and Government) issues relating to water supply and management effecting their members.

The creation of users associations by itself is not sufficient. Sustained and intense efforts to mobilize farmers are essential if these associations are to be effective in improving the conditions of water supply and its management. A number of NGO-led experiments have demonstrated that, where such an effort is made, significant gains in quality of water management are possible below the outlet levels. Interest in extending such initiatives seems to be spreading in the works of the formation of user associations (for example in Maharashtra and parts of Tamil Nadu). At the same time, their experience is also making it increasingly clear that major restructuring of the management of the system as a whole, with users playing an active role, is imperative if the full potential for improvement is to be exploited. As awareness and understanding of these aspects spreads, farmers' interest in, and pressures for, improvements in the facilities and in system management will also increase. This will create an atmosphere more and more conducive to reforms.

The other favourable circumstance is the low level of efficiency of the existing systems. There is much scope for making water supplies more reliable, reducing waste and allowing greater flexibility in the use of water below the outlet. The potential gains in terms of improvement in the technical aspect of water use and of productivity impact are therefore large. The proposed reforms meant to exploit this potential is thus not a zero sum game. The larger the gains, the greater is likely to be the interest among farmers under the command of public systems to participate in the effort necessary for better management of water. The restructuring of the irrigation bureaucracy and redefining its role *vis-à-vis* users and ensuring autonomy of system management will then become the main challenge to be overcome.

In sum, the fiscal condition of the states will compel them to reduce the magnitude of losses on account of subsidized services of which irrigation is one of the largest; users are showing signs of growing interest in improving irrigation; and the potential gains to irrigators (by way of reduced operational costs and higher productivity of water) are large. This conjecture offers a space potentially favourable for reforms. The challenge is to effectively exploit the space created by their circumstances.

STRATEGY FOR REFORM

None of this is likely to happen without purposeful effort. For reasons mentioned earlier, Governments and their bureaucrats are likely to be slow to take the initiative. Nor will users readily and willingly agree to acquire larger stakes and responsibilities on their own. A combination of public education, incentives and pressure is essential to get both groups out of their present lethargy. A number of steps will be needed.

First, it is necessary to prepare the ground through vigorous and sustained campaign in the public media to highlight the deficiencies in the way irrigation is currently managed, the case for raising water rates (both on grounds of equity and of fiscal necessity), and the rationale for proposed reforms (autonomous organizations managed with full user participation) and ways in which they would improve water supply and increase productivity of irrigation. The experience of successful past and ongoing experiments need to be widely publicized and discussed.

Second, the latent (and growing) interest and concern among farmers in improved irrigation should be consciously mobilized as widely as possible to exert informed pressure for reform on the Government. Spreading greater understanding of the way water is managed now and how it could be better managed so as to make existing irrigators better off (and possibly extend the benefits to cover more areas), as well as the rationale and implications of active user involvement in management, are essential for this purpose. NGOs, concerned citizen groups and the media will have a key role in this process and deserve active encouragement and support of the states and funding agencies. Incentives such as smaller increases in water rates, preferential treatment in financing of improvements taken up by user groups (especially for those willing to share in costs of improvement) can then be expected to stimulate more active involvement of user groups in the reform process.

An essential condition for this process to move forward is a clear and firm commitment from governments to a time-based programme for reform. Here again, the compulsions created by the fiscal crises need to be used by NGOs, activists and funding agencies to persuade governments to accept such a commitment. A major problem is the lack of clarity as to the appropriate form and structure of the organizations; the problems which might be created and how they could be tackled; the relation between systems management users and Government. It is important to recognize that there is no single 'right' solution, that we do not know enough about these matters, and that flexibility and learning from experience are essential. Again, all stake holders must be encouraged and helped to discuss and resolve these through open and frank discussions based on the experience of efficiently run systems elsewhere in the country or even in other countries.

ROLE OF FUNDING AGENCIES

Apart from NGOs, concerned academics and other constituents of 'civil society', funding agencies could play an important role in nudging governments to move faster on reforms and helping them better understand alternative models and their implications. At present, irrigation works are financed by the state governments' internal borrowings, and partly by borrowing from the World Bank, the Asian Development Bank (ADB) and other foreign funding agencies, specifically for extending and improving irrigation. Between 1970 and 1990, the World Bank alone has lent over 4 billion US$ for irrigation, flood control and drainage projects in India. This constitutes nearly one eighth of the Bank's total lending to the country.

Loan agreements invariably include specific covenants regarding design standards, project appraisal and monitoring, and cost recovery. In addition, the Bank has sent a number of missions to review the experience and problems of developing and managing water resources generally and irrigation in particular. Based on these reports, it has attempted to respond by adjusting the choice and structuring of projects and through its technical assistance and policy dialogue.

However accomplishments have been lower than the expectations. Specifically on cost recovery, a review[2] notes: 'Cost recovery covenants in individual project agreements have been 'seldom enforced, often relaxed, and sometimes ignored' and the Bank's standard response has been to provide for further studies; which has enabled on 'easy-way out': despite its substantial lending for irrigation [...], the Bank has been unable or unwilling or both, to pursue effectively the issue of large and mounting subsidies.'

Insistence on observance of cost recovery covenants, and willingness to suspend fresh lending for such projects in case of non-compliance, might well have made some difference. The Bank chose not to do this perhaps because maintaining a high level of lending to India—one of its largest clients—was considered more important.

Nor were efforts by the Bank to improve the efficiency of irrigation planning and management particularly successful. Guhan[3] notes that

> there is much evidence in supervision reports, project completion reports, Post Project Appraisal Reports (PPARs) and sector reviews to indicate that project design has often been based on incomplete data and field investigations. In a number of cases, appraisal preceded the completion of detailed design. In some, more cost effective alternatives were not explored in project choice or design. In others, inappropriate sequencing has been the problem. Funding for time—slice and sectoral investments—has paid insufficient attention to technical soundness and economic viability. Managerial problems have included shortages and high turnover of key staff, inadequate co-ordination and poor supervision of construction. Problems of reliability and environmental issues have been inadequately appreciated nor given insufficient attention.

The Bank's approach has no doubt been evolving: starting with conventional, highly engineering-centred projects, it has moved to addressing problems of efficient utilization (Command Area Development), to proper management of water (National Water Management Project), improving the basic hydrological data needed for planning (National Hydrology Project) and, more recently, encouraging a more comprehensive approach to improving the quality of water resource planning and management. But these approaches have been techno-centric, focusing on particular aspects rather than on the whole system. It has done relatively little to systematically assess the manner in which the projects have in fact worked, or their impact. Changes in orientation and emphasis tend to be rather ad hoc, reflecting more the ideas currently fashionable among 'experts' rather than solid learning from actual experience.

Though increasing attention is being given to institutional issues—as evident in recent Water Resources Consolidation Projects, their emphasis on integrated river basin planning and user participation in management—there are still serious weaknesses. The technical and organizational problems of improving main system management still remain neglected. The specifics of institutional reform are, namely, the form and structure of organization involving user participation, the principles and mechanisms governing water allocation, the relation among user and user groups, between users (user groups) and the system management, and between the system managements and the Government. There is moreover a palpable tension within the Bank (and related organizations) between those advocating institutional reforms of the kind discussed above with those who see tradeable property rights in water and water markets as the appropriate solutions.

The Bank, like everyone else, cannot claim to know the 'right' solution to the problems of water pricing and management. The solutions have to be adapted to the needs of specific situations which vary a great deal. The current status of the knowledge being inadequate, learning from experience and adaptation to changing needs is essential. Unfortunately, this approach is not appreciated much less followed by the Bank and other lending agencies or for that matter any of the governments. The Bank has not shown much interest in assessing the actual performance of its projects and the reasons behind successes and failures. Reports of the Operations Evaluation Department, being more in the nature of project completion reports, do not provide an adequate basis to compare the actual with expected impact in terms of water use efficiency or agricultural productivity or how well the performance of the system is sustained after the project is commissioned.

Learning from experience in the case of irrigation and water control works generally calls for systematic and long-term post-evaluation studies. But surprisingly, the Bank has invested little in such studies on its own. Nor has it encouraged and supported such studies by institutions in the countries concerned. Learning from experience takes resources and effort. Without it, there is a danger of suggesting innovations which are half baked, ill thought out and based on inadequate understanding of their ramifications and problems. The resources needed for such studies is very small relative to the scale of investments involved and the potential gains which applications of their lessons will yield.

A more open minded approach, with a conscious and purposeful effort at learning from experience would therefore enhance the effectiveness of the Banks' efforts to promote a more efficient, equitable and sustainable use of water. Being spread over a large number of countries, the experience of its projects would be valuable for all countries. One would hope that the Bank will commit sufficient resources for systematic and objective post-evaluations of projects, work in close collaboration with professionals in the borrowing countries, promote closer professional interchange and interaction among them to benefit from comparative studies of diverse experiences and, not the least, improve its own institutional memory.

NOTES

1. Planning Commission, *Report of the Committee on Pricing of Irrigation Water*, Government of India, New Delhi, 1992.
2. S. Guhan, *The World Banks' Lending in South Asia*, Brooking Occasional Papers, The Brookings Institute, Washington, DC, 1990, pp. 44–5.
3. Ibid.

PART II

PRODUCTIVITY AND COMPETITIVENESS REVISITED

5

Productivity Gains in Indian Agriculture

FORMATION AND DISTRIBUTION (1980–1996)

BRUNO DORIN, NATHANAËL PINGAULT,
JEAN-MARC BOUSSARD

INTRODUCTION

Since the early 1980s, India has been self-sufficient in cereals thanks to the Green Revolution initiated fifteen years earlier and to the public support it received. This public support played a major role in steering productivity gains in agriculture, as well as in influencing the way they were formed and distributed within the sector. This paper aims not only at presenting the characteristics of the prevailing situation, but also at examining how it has evolved, particularly since the 1991 Structural Adjustment Plan.

In comparison with the work already undertaken in India on the subject, the originality of this research lay in the fact that it uses the Surplus Accounting Method (SAM). Stemming from Solow and Kendrick's research, as well as from the work on technical progress undertaken by Hicks from as far back as in 1932, this method goes much further than the Total Factor Productivity (TFP) calculation method, since it makes it possible to link the formation of productivity gains to their distribution, in the form of price and compensation variations.

In France, SAM was developed in the early 1970s under the aegis of the *Centre d'Étude des Revenus et des Coûts* (CERC, Planning Commission). It represented the outcome and synthesis of the work on total factor productivity, constant price accounts and distribution of productivity gains carried out by L.A. Vincent, R. Courbis and P. Massé in the 1950s and 1960s. Since then, SAM has been applied to French agriculture (Boussard et al., 1981) as well as to agriculture in different European countries (Bureau et al., 1991), but not yet to Indian agriculture unlike TFP which has to solve an important paradox (Murgai, 2000) presented and discussed by the authors elsewhere (Dorin et al., 2001).

This paper therefore has a threefold objective: (1) to present the principles, the relevance and the limitations of SAM; (2) to show how SAM can be applied to Indian agriculture; and (3) to present and comment on the initial results obtained on the basis of national-level data collected between 1980–1 and 1995–6.

SURPLUS ACCOUNTING METHOD

This introduction to the Surplus Accounting Method or SAM, which is based on Saint-Guilhem's (1980) and Courbis et al's (1975) works, first defines the theoretical concept of 'Total Productivity Surplus' and its links with other total factor productivity indicators. It then describes the method's relevance and practical limitations. X and w designate the volumes and the prices for 'primary' (land, capital, work, etc.) or 'secondary' (inputs: raw materials, energy, etc.) factors (charges); Q and p designate outputs (products) volumes and prices. Indexes i and j refer respectively to charges and products, t and 0 to the current period and the reference period. Matrixes and vectors are given in bold characters. Prices are given in row vectors and volumes in column vectors. Finally, h represents a particular sub-set of charges or products, for instance those that are specific to a social category such as 'farmers', 'consumers' or 'land owners'.

THEORETICAL PRINCIPLES

Total Productivity Surplus

The surplus analysis is based on a very simple principle. Let us start with the profit and loss account for a unit of production at current prices, and present this account in the form of a balanced equation so that the value of inputs equals that of outputs:

$$\boldsymbol{pQ} = \boldsymbol{wX} \tag{1}$$

This relationship can be differentiated in two ways: in continuous time and in discrete time. In continuous time, we get:

$$\boldsymbol{P}(\mathrm{d}\boldsymbol{Q} / \mathrm{d}t) - \boldsymbol{w}(\mathrm{d}\boldsymbol{X} / \mathrm{d}t) = -(\mathrm{d}\boldsymbol{p} / \mathrm{d}t)\boldsymbol{Q} + (\mathrm{d}\boldsymbol{w} / \mathrm{d}t)\boldsymbol{X} \tag{2}$$

The left-hand side of this relationship, designated as s, is called the 'Total Productivity Surplus'. It indicates how the productivity gains were produced. Two productivity growth modes can be distinguished (Bureau et al., 1991): the progressive mode based on an increase in production ($\mathrm{d}Q_j / \mathrm{d}t > 0$ and $\mathrm{d}X_i / \mathrm{d}t = 0$), and the recessive mode based on limitation of costs ($\mathrm{d}Q_j / \mathrm{d}t = 0$ and $\mathrm{d}X_i / \mathrm{d}t < 0$).

The right-hand side of the equation (2) indicates the distribution of the surplus between the unit of production's various partners through a variation in prices. In this break-up, the share of the h-class factor supplier or product user is written as a_h and is called 'advantage' (Saint-Guilhem, 1980) or 'legacy' (Courbis et al., 1975). Thus,

$-\sum_{i \in h} (\mathrm{d}p_j / \mathrm{d}t)Q_j$ represents the advantage or gain that clients earn from a fall in the product prices $\boldsymbol{p}_h$

$+\sum_{f \in h} (\mathrm{d}w_i / \mathrm{d}t)X_i$ represents the advantage or gain that suppliers earn from an increase in factor prices $\boldsymbol{w}_h$

An advantage is, of course, algebric. If it is negative, we can say that the concerned group of partners 'brings in a surplus'; otherwise, it 'receives, inherits or consumes a surplus'. The absolute value of a negative advantage is called a 'disadvantage'. Finally, as in:

$$\alpha_h = -\sum_{j \in h} (\mathrm{d}p_j / \mathrm{d}t) Q_j + \sum_{i \in h} (\mathrm{d}w_i / \mathrm{d}t) X_i$$

the equation (2) can be re-written as:

$$s = \alpha = \sum_h \alpha_h \qquad (3)$$

In practice, however, the above-mentioned relationships need to be adapted since statistics are used with discrete time units, such as months, quarters or years. In the latter case, annual variations (which we hope are low) can be equated with differentials. The balance of accounts over two years thus gives us $\boldsymbol{pQ} = \boldsymbol{wX}$ for the first year, and $(\boldsymbol{p}+\Delta\boldsymbol{p})(\boldsymbol{Q}+\Delta\boldsymbol{Q}) = (\boldsymbol{w}+\Delta\boldsymbol{w})(\boldsymbol{X}+\Delta\boldsymbol{X})$ for the second year. By combining these two relations, we get:

$$\boldsymbol{p}\Delta\boldsymbol{Q} - \boldsymbol{w}\Delta\boldsymbol{X} = -\Delta\boldsymbol{p}\,(\boldsymbol{Q} + \Delta\boldsymbol{Q}) + \Delta\boldsymbol{w}\,(\boldsymbol{X} + \Delta\boldsymbol{X}) \qquad (4)$$

We therefore get back to relation (3), provided that:

$$s = \boldsymbol{p}\Delta\boldsymbol{Q} - \boldsymbol{w}\Delta\boldsymbol{X} \qquad (5)$$

$$\alpha = -\Delta\boldsymbol{p}\,(\boldsymbol{Q} + \Delta\boldsymbol{Q}) + \Delta\boldsymbol{w}\,(\boldsymbol{X} + \Delta\boldsymbol{X}) \qquad (6)$$

Total Factor Surplus and Productivity

So as to be called a 'Total Productivity Surplus', s must by one way or another be linked with a measurement of the total factor productivity. It is this relationship that we will now try to clarify, by first deducing it in continuous time.

Let $\boldsymbol{F}$ be the production relationship, $\boldsymbol{\Phi}$ the matrix for marginal physical productivity (which gives $\Phi_{ji} = \partial Q_j / \partial X_i$) and $\boldsymbol{G}$ its derivative in relation to time ($G_{ji} = \mathrm{d}\Phi_{ji}/\mathrm{d}t$). Let us suppose that the function of production comes with constant output. In that case, $\boldsymbol{F}$ is homogeneous to 1 degree and, according to Euler's theorem, $\boldsymbol{Q} = \boldsymbol{F}(\boldsymbol{X}) = \boldsymbol{\Phi X}$. Moreover, in the context of a balanced account (equation 1), we know that the profits defined by $\pi = \boldsymbol{pQ} - \boldsymbol{wX} = (\boldsymbol{p\Phi} - \boldsymbol{w})\boldsymbol{X}$ are nil to the optimum level. We obtain $\boldsymbol{p\Phi} = \boldsymbol{w}$ and, consequently:

$$s = \boldsymbol{pGX} \qquad (7)$$

The surplus is, therefore, closely linked to the primary derivative of marginal physical productivity with respect to time. It is a variable that can be equated with a velocity. Indeed, the more productive the unit of production is, the greater this variable is, with X being a good indicator of size since with equal productivity (same Φ), two companies producing the same outputs (Q) use the same quantity of factors (X). In order to compare two companies of different sizes, or to follow the growth of a company over time, we can use the 'surplus rate' τ that corrects the size effect, as shown after:

$$\tau = s \big/ pQ = s \big/ wX \tag{8}$$

In discrete time, the literature available offers several total factor productivity indicators. For instance, L.A. Vincent defines a level of total productivity n_t, as follows:

$$n_t = p_0Q_t \big/ w_0X_t \tag{9}$$

On the basis of equations (1) and (5), with $n_0 = 1$, we can write:

$$s = p_0(Q_0 + \Delta Q) - w_0(X_0 + \Delta X) = p_0Q_t - w_0X_t = (n_t - n_0)\, w_0X_t \tag{10}$$

With this equation (10) in discrete time, we find a very close relationship between the variation over time of the total factor productivity (n) and the surplus (s). The company 'size' effect (calculated here by Σw_0X_t) can also be corrected by measuring a surplus rate defined as given below:

$$\tau = s \big/ p_0Q_0 = s \big/ w_0X_0 \tag{11}$$

RELEVANCE AND PRACTICAL LIMITATIONS

A relationship (equation 1) and a surplus (the Total Productivity Surplus s) have been associated with the company's profit and loss account. Similarly, it is possible to link an equation and a surplus notion to any structured accounts, as in the case in the chart below.

Account	
Employment (X)	Resource (Q)
Balance (B)	

Hence, SAM offers indisputable advantages; it provides an indicator that can be calculated readily on the basis of a company's or a nation's accounts, and the indicator in question can be interpreted without too many theoretical considerations and without much theoretical knowledge. It is also a powerful synthesis tool since it considers all the accounting items of a production unit, hence reflecting the various dimensions of the productive process, without neglecting their interactions. Finally, since a surplus is a primary derivative in relation to time, it reflects changes in productivity or distribution of wealth fairly effectively, rather than representing their levels.

However, SAM also has its own limitations, which are closely linked to the hypotheses on which the method is based and which are widespread hypotheses in production economics: neoclassical equilibrium,[1] constant returns to scale[2] and neutrality of technical progress as described by Hicks.[3] We shall refer to more thorough explanations (especially Saint-Guilhem, 1980 and Courbis et al., 1975) than the following one in order to properly assess the nature and extent of SAM's limitations.

(a) *Surplus and Advantages.* Before going any further, one must clearly understand that surplus and advantages are calculated simultaneously in equation (4). In other words, surplus calculation depends on the aggregates selected to show how they are distributed, and vice versa (the calculation of advantages depends on the aggregates selected to show how they are formed). It is therefore impossible to establish a causality link between a surplus and its corresponding advantages.

(b) *Surplus and Technical Progress.* The hypothesis of the neutrality of technical progress as described by Hicks assures us that surplus does represent the total factor productivity, but it tends to make us ignore the different effects of technical progress on each factor. However, SAM does calculate the contribution of each of the factors in the formation of the total surplus.

(c) *Surplus and Profit.* The balance in equation (1) presupposes that each element of the sale price corresponds to compensation for a factor of production. In other words, either profit is null, or it provides compensation for a more-or-less fictitious 'hidden' factor (own capital, entrepreneur's competence, risk taken by shareholders, etc.). In both cases, we reach a so-called exhaustive definition of the surplus, i.e. a definition that includes a possible profit among the factors. But this definition raises the problem of the dissociation of profit in volume and in prices. Moreover, profit does not constitute a price-determining cost, but, rather, an income determined by prices and measured *a posteriori*. That is why CERC excludes profits from the list of factors and defines a surplus with a balance B (see chart above). With this balance, relationships (1) and (3) successively become:

$$pQ = wX + B \tag{12}$$

$$s = \alpha + \Delta B \text{ (in discrete time)} \tag{13}$$

It can thus be observed that out of the two components of Total Productivity Surplus, only one corresponds to productivity gains, while the other is explained by the profit evolution. In some extreme cases, the surplus may no longer be related to total factor productivity.

(d) *Surplus, Taxes and Duties.* According to our hypotheses, the value of outputs equals the value of factors, and the market cost (price of factors or products) is equal to the real cost (marginal factor productivity or marginal cost of outputs). However, in the real world, some 'distortions', such as taxes and duties, lead to dissociate the market cost (p) from the real cost (c). In the case of a tax such as VAT (t), the two variables remain proportional: $p = c(1+t)$. But, in general, there is no such direct relationship between p and c, particularly with other kinds of taxes, which are nonetheless included in the definition and calculation of the surplus. We can thus consider that taxes (price p) compensate the supply of public goods (cost c) that are, in fact, factors of production. However, they also cover social transfers that cannot be compared with factors of production, even though they are useful. The link between p and c thus proves to be very ambiguous, and since a part of the products does not correspond to the compensation for a factor of production, it leads us back to our previous discussion on profits. In fact, some authors have already suggested that taxes and duties should also be dealt with in the same manner (see above).

(e) *Surplus and Real Cost*. More generally speaking, according to Saint-Guilhem, two approaches may be adopted concerning the difference between the market cost and the real cost. The first one consists of calculating a so-called 'apparent' surplus on the basis of the market costs that firms are faced with. The second calculates a 'fictitious reference surplus', by using real costs (equal to competitive prices) as a system of prices. This method can be useful for a community that wishes to study the actual productivity gains in its economy and to compare its real situation with the kind of situation it is likely to face under perfect competition. However, the fictitious surplus is useless to us as far as the analysis of the actual distribution of productivity gains is concerned.

(f) *Surplus and Externalities*. Externalities, irrespective of whether they are positive (research, education, training, health, etc.) or negative (depletion of natural resources, pollution, congestion, noise, etc.), constitute another reality that fails to address the notion of surplus which is formed on the basis of an accounting of commercial transactions. And yet, these externalities have a noticeable effect on productivity, both in the short and in the long term. We can, therefore, make all efforts to internalize these externalities. However, we will face major problems in doing so, both in determining these externalities (since it cannot be done exhaustively) and in selecting price and volume indexes (which are always more or less arbitrary). Hence, once again, we often limit ourselves to calculating an 'apparent' surplus, which does not take externalities into account.

(g) *Surplus and Time*. As highlighted by Boussard et al. (1981), a surplus can be compared to a velocity. Hence, in discrete time, it is necessary to calculate the surplus between two close and, if possible, consecutive periods (in practice two years), in order to obtain relevant results. However, this calculation method leads to results that may be highly variable and very disconcerting from one year to the next, especially in agriculture. In fact, weather variations make the production relationship much more uncertain in agriculture than in any other sector. Moreover, agricultural production decisions are made before the season starts. Consequently, there is no instantaneous adjustment of supply to demand. This results in a high degree of instability in agricultural prices (Cobweb, Ezekiel model) on the one hand, and in variations in apparent productivity on the other. Actually, the latter starts decreasing as the price changes, and then rises when the process of production gradually adjusts to the new situation. In this case, there is no technical progress. Hence, such productivity shifts are not very significant. However, in the long term, the unpredictable nature of such variability becomes less evident. In order to circumvent this problem, we can calculate instantaneous surpluses (between two consecutive years) over a long series of periods, and then study the cumulative evolution of such instantaneous surpluses.

In all cases, SAM only determines the formation of productivity gains, without explaining why they appear. Likewise, it can describe the possible distribution of such a surplus and assess the effects of price and income policies, but it does not give us any information with regard to optimal distribution or the means that can be used to achieve such an optimal situation. Thus, SAM is a purely descriptive method. Within the limitations mentioned above, it simply provides a method for interpreting the data in order to assess an economic situation or policy. Under no circumstances can it replace analyses or models aimed at explaining a given economic situation.

Within this clearly defined framework, SAM has two major applications. The remainder of this paper is devoted to the first of these applications:

— Retrospectively, to examine developments in a firm or a branch of activity, the formation of productivity gains, transfers of purchasing power, consequences of public policies, etc.
— Prospectively, to cancel unpredictable variations in instantaneous surpluses with the help of long-term projections, to examine the congruence of the formation and distribution model used for productivity gains, to anticipate variations in prices, etc.

APPLICATION OF SAM TO INDIAN AGRICULTURE

In India, the National Accounts Statistics (NAS) offer the appropriate accounting environment within which the application of SAM in the agricultural sector can be envisaged. The NAS are prepared and published every year by the Central Statistical Organisation (CSO, Department of Statistics, Ministry of Planning and Programme Implementation, Government of India). These national accounts primarily indicate the value of production of different economic sectors, the income of various factors and their distribution amongst economic players, final consumption expenses as well as the accounting relationships that link all these measurements.

These accounts are now available on CD-ROM (EPWRF, 1998) and floppy disks (CSO, 1998), although with all their various time gaps and lacunae.[4] They have been prepared using data and a method (CSO, 1998) that may be severely criticized (Kulshreshtha et al., 1997; Lalvani, 1997; Pradhan et al., 1998) and which, like any other accounting environment, impose or lead us to certain options in our application of SAM to Indian agriculture.

NATIONAL-LEVEL ACCOUNTS

With the national accounts (NAS) available for the Indian Union taken as a whole, a factor/products table can be prepared for the agricultural sector (Table 5.1). The following relationship may be associated with this table, taking equation (1) as the basis:

$$\sum_{j=1}^{16} OUT_j + VAI = \sum_{i=1}^{10} INP_i + WEM + REN + INT + WIN + CFC \quad (14)$$

Table 5.1: Surplus Account at the Union Level

Factors	Outputs
Value of agricultural input (INP_i = *Value of input from...*)	Value of agricultural outputs
Compensation of agricultural employees (*WEM*)	(OUT_j = *Value of output from...*)
Rent incomes from land (*REN*)	
Interest incomes (agriculture) (*INT*)	Gross value added from operation of irrigation system (*VAI*)
Compensation of self-employed in agriculture (*WIN*)	
Consumption of fixed capital (*CFC*)	

Table 5.1 has been drawn up on the basis of the following relations (GDP = Gross Domestic Product from agriculture; NDP = Net Domestic Product from agriculture):

$$\Sigma \text{INP} = \text{INP}_{\text{seeds}} + \text{INP}_{\text{chemical fertilizers}} + \text{INP}_{\text{organic manure}} + \text{INP}_{\text{pesticides and insecticides}} + \text{INP}_{\text{feed of livestock}} + \text{INP}_{\text{current repairs and maintenance}} + \text{INP}_{\text{irrigation charges}} + \text{INP}_{\text{market charges}} + \text{INP}_{\text{electricity}} + \text{INP}_{\text{diesel oil}}$$

$$\text{WIN} = \text{NDP} - \text{WEM} - \text{REN} - \text{INT}$$

$$\text{CFC} = \text{GDP} - \text{NDP}$$

$$\Sigma OUT = OUT_{paddy} + OUT_{wheat} + OUT_{jowar} + OUT_{other\ cereals} + OUT_{gram} + OUT_{other\ pulses} + OUT_{groundnut} + OUT_{rapeseed/mustard} + OUT_{other\ oilseeds} + OUT_{coconut} + OUT_{cotton\ kapas} + OUT_{other\ fibres} + OUT_{sugar\ cane} + OUT_{fruits\ and\ vegetables} + OUT_{other} + OUT_{livestock\ sector}$$

Several remarks can be made concerning this first stage.

(a) The factorization of inputs in 10 categories and of outputs in 16 categories is the result of a compromise between the desired level of accuracy and the prevailing level of accuracy in the NAS.

(b) WIN or 'Compensation of own-account workers' designates all primary factors other than WEM, REN and INT. This heading groups has in fact three factors that are not separated by the CSO, i.e. the wage income of own-account workers, rent of own land, profits and dividends of organized and unorganized enterprises.

(c) Profits and dividends, which are not entered separately in the NAS, are considered as a factor of production by the CSO. We have adopted *a fortiori* the so-called 'exhaustive' definition of the surplus.

(d) We will also measure a so-called 'apparent' surplus, in the sense that neither externalities nor taxes and duties (basically imposed on products) are taken into account since no direct monetary equivalent can be found in the NAS.

(e) Finally, our surplus calculations will be carried out 'from the producer's viewpoint'. In fact, the prices the CSO used to estimate the value of various agricultural products are the average wholesale prices prevailing in primary markets. Moreover, the primary factors (WEM, REN, INT, WIN) and their resulting variables (GDP and NDP) have not been calculated at market prices, but 'at factor cost'—the producer's real costs—before taking into account any downstream indirect taxes or duties on the one hand, and subsidies on the other.

Once the table is drawn up, volume indexes (***X*** for factors, ***Q*** for products) and price indexes (***w*** for factors, ***p*** for products) become necessary for each item. The CSO provides values at current prices and constant prices (base year 1980–1) for all of them, except for primary factors (WEM, REN, INT, WIN: see infra). The values at constant price may serve as volume indexes (*X* and *Q*), so they are expressed in a single unit: the 1980–1 Rupee. As for price indexes (*w* or *p*), they may be deduced from the values at current and constant prices. Conventionally, their value is put to 1 for the base-year 1980–1. However, since the value of a currency unit changes over time, it becomes necessary to correct this bias from the very beginning, rather than after calculating the surplus, since the data and its increase in discrete time are not true derivatives with respect to time. The general index we used for this exercise is the inflation index of India's GDP at factor cost. In the following, surplus and advantages will therefore always be stated in 1980–1 rupees.

As far as primary factors are concerned, their value is unfortunately available only at current prices, since the CSO has not made any estimates at constant prices so far. Hence, we decided to estimate the price indexes of these factors by using three costs of production per hectare, provided annually by the Ministry of Agriculture (1991–6; 1991–9) for various crops in the states in which they are produced the most:

— the 'Total human labour' for our 'Compensation of employees' factor (WEM);
— the 'Rental value of owned land' for our 'Rent paid for lease in land' factor (REN);[5]
— the sum of 'Interest on working capital' and of 'Interest on fixed capital' for our 'Interests' factor (INT).

In order to reach an estimate of these costs at the scale of the Indian Union, we first selected the 11 crops[6] about which information is provided every year (or almost every year[7]) for the period under study (1980–1 to 1995–6). We then calculated, for each of these crops, an average annual cost per hectare for India as a whole, this average being weighted by the area under cultivation (Ministry of Agriculture, 1990–5) in the various states for which information was given. Finally, the average annual cost for all crops taken together is no more than the average of the cost per crop, weighted by the share of these crops in the Indian gross cropped area. From that point on, it is fairly easy to deduce:

— a price index for WEM, REN and INT from their respective average cost per hectare in 1980–1;
— a price index (w_{WIN}) for the 'Compensation of self-employed' factor (WIN), on the basis of $w_{NDP}X_{NDP} = w_{WEM}X_{WEM} + w_{REN}X_{REN} + w_{INT}X_{INT} + w_{WIN}X_{WIN}$, where w_{WIN} is the only unknown factor.

Once all the volume and price indexes are known, calculating the total productivity surplus *s* (equation 5) for two consecutive years is simple, as is calculating the total advantages α (equation 6). All that remains, then, is to check the equation (3).

STATE-LEVEL ACCOUNTING

The Indian Union is a federation of 25 states and 7 Union Territories with their own governments, and therefore a certain degree of autonomy, particularly as far as agricultural policies are concerned. The application of SAM to aggregate accounts at Union level cannot, therefore, reflect any regional specificities, which can also be seen in their natural (relief, soil, climate, etc.) and socio-economic features (techniques, institutions, regional planning, entrepreneurial network, etc.). The formation and sharing of productivity gains may, therefore, differ between these vast regions, but also within them. But in the absence of statistics at the district level for example,[8] there does not exist any other computational possibility but to apply SAM to the state-wise NAS.

Unfortunately, the state-wise NAS series, for which information has been available fairly systematically[9] since 1980–1, do have some major limitations:

— information concerning the value of various agricultural productions is only available up to 1990–1,
— after this date, only some overall variables are available (NDP, GDP), and they do not suffice for the calculation of surpluses.

For the period 1980–1 to 1990–1, a table can be drawn up in the case of most of the states (Table 5.2), with the following relations (GSDP = Gross State Domestic Product from agriculture):

$$CFC = GSDP - NSDP$$

$$\Sigma OUT = OUT_{paddy} + OUT_{wheat} + OUT_{jowar} + OUT_{other\ cereals} + OUT_{pulses} + OUT_{oilseeds} + OUT_{fibres} + OUT_{sugar\ cane} + OUT_{fruits\ and\ vegetables} + OUT_{other} + OUT_{livestock\ sector}$$

Hence, Table 5.2 is unfortunately much less detailed than the one drawn up for the Union as a whole. For this reason, and also because of the absence of data for the period following the major economic reforms that took place in 1991, it was decided not to proceed with the calculation and analysis of surplus accounts at state level, despite the interest of such a study.

Table 5.2: Surplus Account at State Level

Factors	Outputs
Value of agricultural inputs (*INP* = Value of inputs)	Value of agricultural outputs (OUT_j = *Value of output from...*)
Compensation of primary agricultural factors (*NSDP = National State Domestic Product*) Consumption of fixed capital (*CFC*)	Gross state value added from operation of irrigation system (*VAI*)

RESULTS AND COMMENTS

FORMATION OF PRODUCTIVITY GAINS

In 1995–6, the Total Productivity Surplus cumulated since 1980–1 was Rs. 19.66 billion at 1980–1 prices (Table 5.3), which accounts for less than 3 per cent of the GDP in agriculture for the year 1995–6 (668.63 billion at 1980–1 prices). As far as the annual growth of productivity is concerned (instantaneous surplus), it does not go, on an average, beyond 0.25 per cent of the annual agricultural GDP. The Total Productivity Gains for the period under consideration are therefore quite modest. In fact, since the Green Revolution (initiated in the middle of the 1960s) and the realization of its objectives (self-sufficiency in cereals at the end of the 1970s), the strong policy incentives earlier given to scientific research and the pursuit of technical innovations in agriculture have dried up. On their part, Abler et al. (1994) confirm that total factor productivity developed more slowly after an initial phase that saw the rapid spread of techniques (1965–75). However, a moderate increase in total factor productivity does not mean an equally moderate growth in the volume of products (Figure 5.1). In fact, the latter amassed a productivity gain of 337 billion during the period under study. This substantial growth in production followed neither the 'progressive' nor the 'recessive' mode observed in European Union countries between 1968 and 1986 (Bureau et al., 1991), but, rather, an 'intensive' mode (increase in production accompanied by an increase in charges) to which we will return later in this presentation, after examining the contribution of outputs to the surplus.

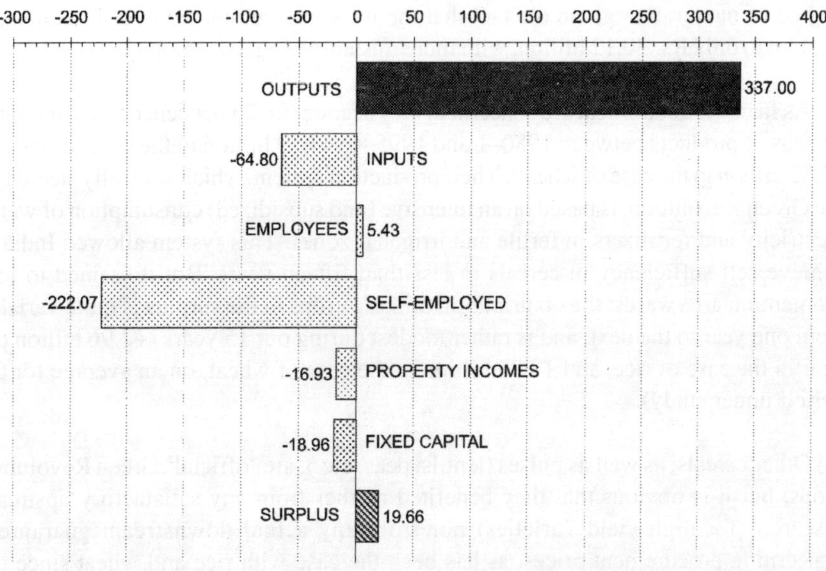

Figure 5.1: Cumulated Surpluses over Fifteen Years (1980–1 to 1995–6) (Rs. billion at 1980–1 prices).

Contribution of Outputs to the Surplus

The whole contribution of outputs to the total cumulated surplus is positive. In the 15-year period (1980–1 to 1995–6), this contribution rose to Rs. 337.00 billion at 1980–1 prices,[10] i.e. 50 per cent of the agricultural GDP in the year 1995–6 (668.63 billion at 1980–1 prices). However, this growth was not steady during the period under consideration (Table 5.3): it first soared, just after the 1982 drought (which affected 32 per cent of the territory), then stabilized (drought covering 29 per cent of the territory in 1985 and 18 per cent in 1986), before accelerating rapidly and regularly (except for the year following the July 1991 financial crisis), soon after the last major drought in 1987 which severely hit 47 per cent of the territory. Thus, Indian agricultural production still remains very sensitive to the uncertainties of the Monsoon. Varying levels of rainfall, however, do not constitute a sufficient explanation for the changing situation between 1980–1 and 1995–6. In fact, there is a fair amount of divergence in the contribution of various outputs to the total surplus (Table 5.3).

(a) The livestock sector remained the most productive. Indeed, over the 15-year period, the cumulated increases in volume, which were very regular from one year to the next (+7.50 billion per year on an average) and much less sensitive to drought than other products, amounted to Rs. 112.47 billion in 1995–6, i.e. 33 per cent of the total gains. Disregarded by those who directed the Green Revolution, this sector, which has been rather 'free' of public intervention, seems to have responded very well to the liberalization of the Indian economy, which was initiated in the beginning of the 1980s and confirmed in 1991, i.e. to markets opening up both on the domestic front (eggs, milk, poultry, etc.) as well as on the export front (aquaculture). A similar observation can be made with respect to the fruit and vegetables sector, with perhaps a greater degree of moderation given the fact that the increases cumulated over fifteen years amount to only Rs. 30.12 billion, with more unstable annual development.

(b) As far as rice and wheat are concerned, they account for 23 per cent of the cumulated surplus of products between 1980–1 and 1995–6 (44.39 billion in the case of rice and 33.82 billion in the case of wheat). Their production system, which was fully steered by the Green Revolution, is based on an intensive (and subsidized) consumption of water, electricity and fertilizers in fertile and irrigated zones. This system allowed India to achieve self-sufficiency in cereals in less than fifteen years. But it seemed to lose momentum afterwards: the contribution of these crops to the surplus is quite variable from one year to the next, and is rather modest during our 15 years (+2.96 billion per year in the case of rice, and +2.25 billion in the case of wheat, on an average for the period under study).

(c) Other cereals, as well as pulses (lentils, peas, etc.), are 'official' Green Revolution crops, but it is obvious that they benefited neither from any satisfactory upstream research (for high-yield varieties) nor from any actual downstream guarantees concerning procurement prices, as has been the case with rice and wheat since the beginning of the 1970s. Indeed, the way the cards have been dealt out does not seem to have changed much since the beginning of the 1980s. On the one hand, rice and wheat continue to receive special attention, and their consumption continues to be

Table 5.3: Cumulated Annual Surpluses (1981–2 to 1995–6)

(Rs. billion at 1980–1 prices)

OUTPUT	1981	1982	1983	1984	1985	1986	1987	1988	1989	1990	1991	1992	1993	1994	1995
Cereals	3.9	–3.8	32.0	22.6	28.4	20.4	17.0	55.8	57.7	63.6	56.6	69.7	78.1	87.7	77.7
– Paddy*	–0.4	–11.9	12.7	9.3	18.0	12.4	6.8	31.2	34.9	36.3	36.9	34.1	46.6	49.5	44.4
– Wheat*	1.8	9.4	13.4	11.0	14.4	11.3	13.2	23.0	17.7	24.5	24.5	26.5	29.9	37.8	33.8
– Jowar*	1.9	–0.1	1.4	0.9	–0.5	–1.6	1.7	–0.4	3.3	2.0	–1.7	5.6	4.1	1.6	2.5
– Other*	0.5	–1.5	4.8	2.0	–2.0	0.1	–3.8	4.6	5.1	4.3	1.2	8.7	4.1	5.7	4.2
Pulses*	2.3	2.7	5.2	2.7	5.8	2.4	0.3	8.9	5.7	10.2	3.0	6.1	7.2	9.6	4.9
– Gram	0.9	2.3	1.5	0.8	4.1	1.2	–0.6	3.2	0.5	4.1	0.6	1.3	3.1	7.6	4.0
– Other	1.5	0.2	4.2	2.3	2.2	1.7	1.8	6.4	5.9	6.8	3.3	5.7	5.1	3.2	2.8
Oil-crops*	7.8	2.6	11.4	13.0	5.8	6.9	13.9	34.7	32.1	37.8	38.9	43.3	47.3	48.5	47.8
– Groundnut	6.0	0.8	7.1	4.7	0.6	2.8	3.0	16.9	12.6	10.9	8.7	13.7	11.6	12.0	11.0
– Rape / Mustard	1.0	–0.4	0.9	2.6	1.2	1.3	4.3	8.1	7.5	11.6	14.0	10.2	11.7	13.4	14.2
– Other	0.7	1.4	3.6	4.3	2.9	2.3	4.9	6.4	8.3	11.0	10.4	14.1	17.6	15.6	15.5
– Coconut	0.1	0.4	–0.3	1.6	1.5	1.3	2.2	4.1	4.9	5.2	6.1	7.2	7.9	8.7	8.3
Fruits & vegetable*	0.7	2.3	6.0	10.8	7.4	13.0	7.3	14.5	13.4	17.5	17.1	24.2	26.8	35.6	30.1
Sugarcane*	7.1	7.9	5.8	4.8	4.4	6.5	8.3	9.8	13.4	16.1	17.7	14.5	15.0	24.0	22.4
Fibres*	2.7	2.6	–0.5	4.4	6.5	2.2	0.8	6.8	12.7	9.9	9.9	13.5	12.5	15.7	17.5
– Cotton	2.6	2.8	–0.3	4.3	4.8	1.5	0.6	6.4	12.0	9.1	8.8	12.9	12.0	14.9	16.9
– Other	0.1	–0.3	–0.1	–0.1	3.9	2.8	2.3	2.5	2.7	3.1	3.5	2.9	2.9	3.1	3.0
Other*	1.6	2.8	5.0	4.2	2.4	1.9	–0.3	10.1	8.0	10.2	6.5	11.6	13.1	14.2	12.0
Livestock*	6.9	12.0	20.8	29.5	37.1	43.8	47.4	54.4	61.3	68.9	75.3	85.2	96.2	105.0	112.5
VA irrigation*	0.3	0.7	1.0	1.2	1.4	1.6	3.2	6.3	5.5	3.3	3.6	3.9	4.2	4.6	4.9
Total (12 items*)	33.2	29.5	87.1	93.8	100.8	100.2	98.8	203.9	213.0	240.7	233.0	277.1	306.9	351.9	337.0

(contd.)

Table 5.3 (contd.)

FACTORS	1981	1982	1983	1984	1985	1986	1987	1988	1989	1990	1991	1992	1993	1994	1995
Inputs	−6.4	−8.7	−13.1	−18.7	−21.9	−25.5	−22.3	−39.3	−42.1	−45.7	−49.0	−51.2	−54.6	−62.8	−64.8
– Seeds*	−0.9	−1.4	−1.5	−1.4	−1.3	−1.1	−1.2	−2.8	−3.4	−3.8	−3.8	−3.9	−4.4	−4.9	−5.2
– Fertilizers*	−3.3	−3.9	−5.5	−10.6	−13.5	−16.8	−12.8	−20.4	−22.0	−23.6	−25.5	−24.7	−22.8	−26.0	−28.2
– Manure*	−0.1	−0.1	−0.3	−0.5	−0.5	−0.3	−0.3	−0.5	−0.4	−0.5	−0.3	−0.4	−0.4	−0.3	−0.7
– Pest. & insect.*	−0.3	−0.4	−0.7	−0.8	−0.8	−0.6	−0.7	−1.4	−0.8	−0.4	−0.4	−0.5	−0.7	−0.8	−1.1
– Electricity*	−0.1	−0.6	−0.7	−0.9	−1.2	−1.8	−2.7	−3.5	−4.0	−4.2	−4.6	−4.7	−5.8	−6.3	−7.0
– Diesel oil*	−0.5	−0.9	−1.3	−1.7	−2.6	−2.9	−3.3	−3.7	−4.4	−5.3	−5.8	−6.1	−6.3	−6.7	−7.4
– Irrigation*	−0.1	−0.2	−0.2	−0.2	−0.3	−0.3	−0.4	−0.6	−0.6	−0.6	−0.6	−0.6	−0.7	−0.7	−0.7
– Market*	−0.3	−0.2	−0.8	−0.8	−0.7	−0.6	−0.5	−1.7	−1.8	−2.1	−1.9	−2.3	−2.5	−3.0	−2.9
– Repairs & M.*	−0.3	−0.3	−1.1	−1.5	−1.6	−1.6	−1.7	−3.2	−3.7	−4.4	−5.1	−6.3	−5.9	−8.1	−8.2
– Feed livestock*	−0.4	−0.7	−1.0	−0.3	0.6	0.3	1.4	−1.5	−1.1	−1.1	−1.1	−1.7	−5.0	−6.0	−3.6
Employees*	−10.2	−14.0	−12.8	−13.1	−24.8	−20.9	−14.4	−15.9	−18.3	−14.8	−8.4	−10.6	10.1	0.4	5.4
Rent incomes*	−0.2	−0.9	−0.8	−0.5	−1.6	−1.5	−0.4	−1.2	−2.3	−1.6	−0.6	−3.1	0.2	−3.8	−3.4
Self-employed*	−14.4	−2.1	−51.8	−50.5	−35.0	−27.1	−43.3	−124.9	−126.0	−159.2	−145.8	−179.3	−230.1	−246.3	−222.1
Interest*	−1.2	−1.9	−4.0	−4.0	−7.8	−11.7	−7.2	−8.1	−8.7	0.2	−5.3	−5.8	2.0	−9.9	−13.5
CF Capital*	−0.9	−2.0	−2.9	−3.8	−4.6	−4.7	−5.2	−6.2	−7.3	−8.4	−10.1	−12.0	−13.4	−15.9	−19.0
Total (15 items*)	−33.2	−29.6	−85.4	−90.5	−95.6	−91.3	−92.8	−195.5	−204.7	−229.6	−219.2	−262.1	−285.7	−338.2	−317.3

TOTAL	1981	1982	1983	1984	1985	1986	1987	1988	1989	1990	1991	1992	1993	1994	1995
Instantaneous	0.00	−0.09	1.85	1.56	1.87	3.77	−2.92	2.36	−0.08	2.85	2.64	1.23	6.14	−7.55	6.03
Cumulated	0.00	−0.09	1.76	3.32	5.19	8.96	6.04	8.40	8.32	11.17	13.81	15.04	21.18	13.63	19.66

promoted through the Public Distribution System (PDS). On the other hand, crops with neither production nor consumption support have only provided marginal cumulated contributions to the Total Productivity Surplus: 7 billion at the end of 15 years for cereals other than rice and wheat, and less than 5 in the case of pulses. The wide variations in their surplus from one year to the next shows that these crops have turned into crops that are grown in dry zones or in dry seasons which are far more sensitive to climatic changes. The same argument was also valid for oilseeds up to 1987.

(d) The increase in the contribution of oilseeds to the surplus after the 1987 drought was due to the inception of the Technology Mission on Oilseeds (TMO), which guaranteed prices to rape/mustard and groundnut producers through the MIO (Market Intervention Operation) that came into effect in April 1989, and lasted up to the early 1990s. Subsequently, the 'Yellow Revolution' dried up too. However, since 1990, annual average gains have stood at +2.62 billion, while they have not risen above +1.49 billion for sugarcane and 0.81 billion for fibres—two sectors that still are protected by the Union and state governments.

Contribution of Charges to the Surplus

The contribution of intermediary consumption (inputs) to the surplus is negative (Figure 5.1). In other words, their use increased, intensified or spread between 1980–1 and 1995–6, to the tune of Rs. 64.80 billion at 1980–1 prices. This was particularly true for the three inputs that formed the basis of the Green Revolution: fertilizers account for 43 per cent of the increase observed in charges over the 15-year period (28.16 billion out of 64.80), water about 35 per cent,[11] and certified seeds 8 per cent. The remaining 14 per cent were essentially divided between cattle-feed (a market that has been consolidating itself since the end of the 1980s) and market charges whose increase translates a growth in the marketing of agricultural products in the country along with lower trade barriers between states. On the other hand, there was no notable hike in the use of pesticides and insecticides (concentrated in certain regions and for some crops, such as cotton), and even less in organic manure, which was gradually replaced by subsidized chemical fertilizers.

The contribution of the four primary factors (WEM, WIN, RENT and INT) to the surplus was also negative (Figure 5.1), and even to a much greater extent since it is almost four times higher than for intermediary consumptions:–233.57 billion cumulated in 15 years. This substantial increase in compensation of primary factors (NDP), which followed the output growth curve quite faithfully (and, all in all, led to a low Total Productivity Surplus), was, in reality, increasingly dedicated to what we call 'compensation of own-account workers' (95 per cent in 1995–6, i.e.–222.07 billion cumulated in 15 years). This factor groups together three components that cannot be dissociated from each other (wage income of own-account workers, rent of own land, profits and dividends of organized and unorganized enterprises), which makes analysis more difficult. However, two things can be deduced by using other statistics. Between 1980–1 and 1995–6, in India, the gross cultivated area increased by 8 per cent (from 173 million hectares to 187) (Ministry of Agriculture, 1990–5): the cultivation of new areas or the extension of cropping after the Monsoon, in the *Rabi* season (thanks mostly to progress in irrigation techniques), played undoubtedly a major role in the increase in Indian agricultural production. But, according to the 1991 census and to some projections for the years after, the growth rate of the farming population (save that of agricultural workers) was +29 per cent (from 91 million active farmers to more than 117). In fact, the increase in the farming population in India was mostly based on an increasing 'family workforce' whose high growth in volume led to a matching reduction in partial productivity. In other words, an increase in family manpower leads to an increase in production which, in fact, is mostly needed to feed the former. This is one of the characteristics of a subsistence economy, where incomes rely largely on good harvests, and where the absence of a real substitution of labour by capital leads to low Total Productivity Gains. Under these conditions, we can not only justify the policies followed by Indian authorities to greatly subsidize 'modern' inputs, but also measure the limitations of this policy thirty years after the Green Revolution.

The contribution of agricultural labour to the Total Productivity Surplus is quite different from those of the factors specific to own-account workers, and even from those of all other factors. Over the 15-year period examined, it was positive, rising to +5.43 billion, which corresponded to a decrease in the volume of personnel compensation. Increasing pressure on land, as described above, definitely leads to the reduction or ousting of salaried incomes in order to secure the incomes of an ever-increasing number of small landowners. This phenomenon can be observed in particular after the 1991 crisis and reforms, especially in 1993. Inflation rates at that time probably pushed seasonal agricultural workers to clamour for increases in wages, which could not be implemented. They then seized any better-paying opportunities that presented themselves, to work in sectors that were more dynamic and less trying than agriculture. According to Bureau et al. (1991), such a quantitative adjustment would be unimaginable in European agriculture, which is characterized by far more rigid agricultural manpower and labour legislation.

The slightly negative contribution of leases in land (–3.42 billion in 15 years) may also be explained, to a certain extent, by its increasing scarcity: the rental market is almost non-existent, above all because available land has to provide for the survival of an increasing number of owners. The almost total stability of this factor, in volume, along with the marked increase in products that has been observed over the same period, therefore suggests a partial increase in land productivity. This observation is in line with the general calculable equilibrium model developed by Abler et al. (1994). According to these authors, productivity increased perhaps even more rapidly between 1974 and 1987 than during the first phase of the Green Revolution (1965–73). As far as the contribution of interests to the surplus is concerned (–13.51 billion in 15 years), it translates a rather moderate rise in the volume of agricultural credit and other investments or bank deposits in the sector, and therefore a weak expansion in the activities of banking institutions during the period under study. At the same time, it may be noted that over the 15-year period, the consumption of fixed capital[12] increased in volume by +18.96 billion at 1980–1 prices. Indeed, in the private sector, labour was not really being replaced by capital, while in the public sector, the state buried an increasing share of its funds in input subsidies, much to the detriment of more long-term investments. Abler et al. (1994) conclude more radically that between 1960 and 1987, in the agricultural sector, there were no incentives to invest, since it was protected and subsidized. Until

1995, the 1991 reforms did not appear to have made any significant changes in these trends.

DISTRIBUTION OF PRODUCTIVITY GAINS

A modest overall surplus may conceal major transfers between economic agents through prices. As compared with the general rate of inflation (measured in this study with the GDP deflator at factor cost), the prices of various products, supplies or compensations may, in fact, follow totally different developments, granting advantages to some agents and disadvantages to others. Indeed, this seems to have been the case with Indian agriculture during the period under study.

In the first place, it was observed that during this 15-year period, the relative price of agricultural products had fallen[13] (Figure 5.2). This constituted an advantage for customers (food processing, consumers, exporters, etc.) but a disadvantage for producers, rising to Rs. 26.85 billion in 1995–6.[14] However, the relative fall in prices was not evenly distributed among the outputs (Table 5.4): the main beneficiaries were cereals (+31.72 billion), oilseed (+14.77 billion) and sugarcane (+12.84 billion) buyers. The advantage customers gained was basically formed in the 1980s. After the 1991 crisis, it stagnated and may even have regressed. Hence, at 1980–1 constant prices, the cumulated advantage until 1995–6 corresponds to the 1986–7 cumulated advantage in the case of cereals, and to the 1981–2 cumulated advantage for sugar. We can see in the cumulated advantage for oilseeds the effect of the fall in production prices observed in 1988–9, which led to the establishment of the MIO (then came inflation, and then a return to this level as of 1992–3). At the same time, in sectors with very little public

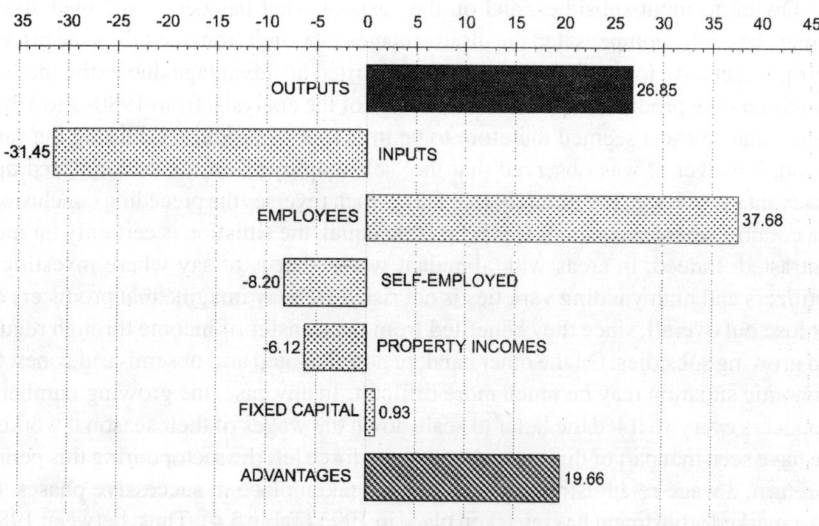

Figure 5.2: Cumulated Advantages over Fifteen Years (1980–1 to 1995–6) (Rs. billion at 1980–1 prices).

regulation such as fruits and vegetables and livestock, producers inherited an advantage since their relative production prices showed a rising trend in the 1980s. This is particularly noticeable in the case of livestock: the cumulated value of disadvantages for clients in this sector rose to 12 billion in 1990–1 (–15.21 billion in 1995–6). However, this must be relativized: the Indian Wholesale Price Index (WPI, 1981–2 base) shows, for the period under study, marked inflation in the case of outputs such as fish, mutton and pork, and even milk, but it must be further qualified in the case of eggs and chicken whose market exploded in India with the emergence of a well-off middle class.

The apparent disadvantage that struck most of cereal, oilseed and sugarcane producers also needs to be taken with care. In fact, these crops are the biggest consumers of inputs, whose relative prices fell in a much higher proportion than those of outputs during the period under study. Indeed, the advantage producers gained rose to Rs. 31.45 billion in 1995–6. If we deduct the Rs. 26.85 billion disadvantage associated with outputs, we get a balance of Rs. 4.60 billion. Accelerating constantly during the period under study (Table 5.4), the advantage gained by farmers basically constitutes a transfer of income by the state, through three inputs (fertilizers, irrigation, seeds) whose use has been heavily subsidized since the Green Revolution in order to contain foodgrain sale prices. The cost of this policy is a heavy burden for the state, all the more so since it is added to upstream subsidies to domestic suppliers of these inputs. However, in spite of attempts to impose financial restrictions in 1991, especially on fertilizers, the latter represent 73 per cent of the cumulated advantage associated with the inputs (–23.03 billion out of –31.45), while water (electricity, petrol and repairs/maintenance for lifting water) represents almost 16 per cent, and seeds 6 per cent. On the other hand, there is an increasing scarcity of traditional inputs such as organic manure: the relative increase in its price resulted in a total cumulated advantage of +5.28 billion in 1995-6.

Owing to input subsidies, and on the basis of what has been explained above, farmers basically compensated the disadvantages inherited from the fall in the price of their products. As for cattle breeders, they inherited an advantage due to the increase in their relative production prices. At this stage of the analysis, from 1980–1 to 1995–6, the balance sheet seemed therefore to be in favour of producers. During the same period, however, it was observed that the 'self-employed' in the sector stored up a disadvantage of Rs. 8.20 billion (Figure 5.2), which reverses the preceding conclusions. In a country where access to water is far from equal, the situation is certainly far more contrasted. Indeed, in areas with abundant water, that is to say where investing in fertilizers and high yielding varieties is not risky, we may imagine that producers did not lose out overall, since they benefited from the transfer of income through regular and growing subsidies. On the other hand, in non-irrigated arid or semi-arid zones, the economic situation may be much more difficult. In any case, the growing number of producers every year led the latter to scale down the wages of their seasonal workers. We have seen that part of this agricultural work-force left the sector during this period. In return, a wage re-adjustment seems to have taken place in successive phases, the most marked adjustment having taken place in 1993 (Table 5.4). Thus, between 1980–1 and 1995–6, agricultural employees amassed an advantage of Rs. 37.68 billion at 1980–1 prices, through the increase in their relative wages.

Table 5.4: Cumulated Annual Advantages (1981–2 to 1995–6)

(Rs. billion at 1980–1 prices)

OUTPUT	1981	1982	1983	1984	1985	1986	1987	1988	1989	1990	1991	1992	1993	1994	1995
Cereals	–0.8	–1.8	8.3	27.2	25.8	30.8	25.9	29.4	37.6	37.2	11.6	22.6	25.6	24.3	24.5
– Paddy*	–3.3	–5.2	–4.7	7.9	10.1	10.9	6.2	10.0	14.4	16.7	6.6	5.3	6.1	5.9	8.8
– Wheat*	0.7	0.1	8.0	10.4	10.6	12.5	11.2	12.6	13.1	11.0	5.1	9.1	9.5	14.6	10.9
– Jowar*	1.5	2.6	2.6	4.0	3.7	4.3	5.4	2.8	5.7	5.8	1.5	5.4	8.4	5.3	6.2
– Other*	0.3	0.5	2.7	5.6	3.1	4.8	3.9	6.7	7.7	7.2	2.8	8.0	8.1	5.6	5.8
Pulses*	5.6	7.6	3.3	1.2	4.3	5.2	1.6	–1.0	–1.5	0.4	2.8	3.3	–1.0	0.2	–3.6
– Gram	3.4	5.2	2.6	0.3	2.4	4.5	2.7	0.2	–0.1	2.2	3.9	2.4	–0.6	4.8	5.7
– Other	2.2	2.1	1.2	1.4	2.4	1.3	–0.2	–0.4	–0.7	–1.0	–0.3	1.8	0.5	–3.3	–7.4
Oil-crops*	3.7	4.3	–0.5	4.8	9.9	–0.8	–1.5	13.8	10.0	–0.3	2.4	10.8	13.0	13.3	14.8
– Groundnut	–0.4	0.6	0.2	0.6	1.8	–1.3	–2.0	6.0	2.7	–2.1	–1.0	4.3	3.8	2.5	5.0
– Rape / Mustard	2.2	2.0	1.2	4.3	3.9	1.2	1.5	6.0	3.0	1.8	4.4	6.2	5.0	4.9	6.4
– Other	0.8	0.7	0.2	2.1	2.5	0.4	0.4	1.8	2.0	–0.8	–1.7	2.7	1.7	1.4	–0.3
– Coconut	1.1	0.6	–2.2	–1.9	2.3	–0.2	–0.9	0.8	3.5	1.6	0.9	–0.5	4.0	5.7	4.9
Fruits & vegetable*	0.9	–0.3	3.0	3.4	–4.7	–5.0	–3.3	–0.7	1.3	–0.6	–1.5	1.4	–0.1	3.9	–3.1
Sugarcane*	13.0	17.5	16.5	14.3	13.0	15.3	16.3	15.8	11.5	14.5	18.6	13.0	7.5	11.0	12.8
Fibres*	1.8	3.8	1.6	–1.0	7.5	6.9	3.1	5.5	7.0	4.6	2.4	9.7	2.3	–3.9	0.5
– Cotton	1.7	3.6	2.3	3.3	7.1	6.1	2.2	5.1	7.1	4.8	2.1	8.6	1.9	–4.4	1.6
– Other	0.1	0.1	–0.7	–4.5	2.7	2.9	3.0	2.6	2.0	2.1	2.7	3.4	2.8	2.9	1.3
Other*	3.3	7.1	4.0	3.2	3.1	5.2	0.5	2.2	5.3	0.7	–4.0	–2.8	1.1	–0.7	–4.5
Livestock *	–3.2	–2.8	–2.4	–5.8	–3.7	–4.4	–8.2	–11.5	–18.5	–11.9	–15.9	–17.3	–17.3	–15.1	–15.2
VA irrigation*	0.2	–0.1	–0.2	–1.2	–2.3	–3.5	–2.1	–1.6	–2.6	–3.6	–4.3	–4.9	–5.0	–5.2	–6.5
Total (12 items*)	24.4	35.0	33.9	46.7	54.5	51.6	33.2	54.7	53.4	44.3	16.3	40.9	32.7	34.9	26.9

(contd.)

Table 5.4 (contd.)

FACTORS	1981	1982	1983	1984	1985	1986	1987	1988	1989	1990	1991	1992	1993	1994	1995
Inputs	-5.7	-14.1	-10.9	-17.1	-19.4	-19.3	-14.5	-21.5	-26.9	-25.1	-24.4	-25.7	-28.9	-31.2	-31.5
– Seeds*	-1.0	-1.4	-1.8	-2.4	-2.2	-2.2	-1.7	-2.6	-2.4	-1.6	-1.2	-2.5	-1.9	-2.1	-1.9
– Fertilizers*	1.0	-1.2	-2.6	-5.3	-8.1	-8.2	-10.3	-12.8	-15.2	-18.2	-14.7	-18.3	-19.8	-20.7	-23.0
– Manure*	0.0	-0.2	-0.8	-0.8	0.0	0.8	1.5	2.1	2.3	2.3	2.6	3.5	4.3	3.2	5.3
– Pest. & insect.*	-0.1	0.0	0.0	-0.2	-0.3	-0.3	-0.4	-0.5	-0.6	-0.7	-0.5	-0.6	-0.7	0.3	-0.3
– Electricity*	0.1	-0.3	-0.5	-0.6	-0.9	-1.2	-1.4	-2.8	-3.0	-3.9	-4.5	-3.9	-3.6	-3.4	-3.9
– Diesel oil*	0.9	0.5	0.4	0.0	-0.4	-0.7	-1.7	-2.1	-2.6	-1.3	-1.7	-0.5	-0.9	-0.8	-1.8
– Irrigation*	0.0	-0.2	-0.1	-0.2	-0.4	-0.3	-0.2	-0.4	-0.6	-0.4	-0.7	-0.3	-0.6	-0.6	-0.5
– Market*	-0.4	-0.5	-0.4	-0.6	-0.7	-0.7	-0.5	-0.7	-0.9	-0.7	-0.4	-0.7	-0.6	-0.6	-0.6
– Repairs & M.*	0.0	0.2	0.2	0.4	1.0	1.1	1.2	1.5	1.7	1.5	0.9	1.1	1.1	0.8	1.1
– Feed livestock*	-6.1	-11.0	-5.1	-7.4	-7.5	-7.7	-1.0	-3.2	-5.7	-2.1	-4.2	-3.5	-6.3	-7.3	-5.7
Employees*	-10.9	-16.6	-8.6	-6.7	-19.4	-15.2	-9.8	-1.1	0.3	7.4	10.5	7.7	36.8	33.6	37.7
Rent incomes*	-0.2	-0.6	0.0	0.3	-0.7	-0.8	-0.1	0.2	-0.4	0.9	1.7	-0.4	3.3	-0.3	0.0
Self-employed*	-7.8	-4.2	-11.3	-20.9	-9.4	-5.1	-5.3	-28.1	-23.4	-25.5	5.8	-11.5	-34.2	-21.1	-8.2
Interest*	0.1	0.0	-1.7	0.7	-2.4	-4.5	0.3	1.9	2.0	6.6	1.4	1.1	9.8	-3.1	-6.1
CF Capital*	0.1	0.4	0.2	0.5	2.0	2.3	2.2	2.5	3.4	2.6	2.5	3.0	1.7	0.9	0.9
Total (15 items*)	-24.4	-35.1	-32.1	-43.3	-49.2	-42.6	-27.1	-46.2	-45.1	-33.1	-2.5	-25.9	-11.5	-21.2	-7.2

TOTAL	1981	1982	1983	1984	1985	1986	1987	1988	1989	1990	1991	1992	1993	1994	1995
Instantaneous	0.00	-0.09	1.85	1.56	1.87	3.77	-2.92	2.36	-0.08	2.85	2.64	1.23	6.14	-7.55	6.03
Cumulated	0.00	-0.09	1.76	3.32	5.19	8.96	6.04	8.40	8.32	11.17	13.81	15.04	21.18	13.63	19.66

CONCLUSION

In spite of the limitations we underlined at the beginning of this paper, the Surplus Accounting Method has proved to be not only an easy-to-use tool, but also a useful way for synthesizing or going into the specifics of existing analyses on the formation and distribution of productivity gains in Indian agriculture. When this method is applied to the national accounts at the Union level, between 1980–1 (fifteen years after the Green Revolution was launched) and 1995–6, it leads to the following general conclusions.

(a) During the period under study, only moderate global productivity gains were observed, in spite of a marked increase in production. The positive contribution of outputs to the surplus has, in fact, been offset first by an increase in the volume of intermediary consumption, which has been increasingly subsidized by the state since the Green Revolution (basically fertilizers and energy for irrigation), and second, and above all, by the compensation of cultivators whose number increased. In other words, the increase in the family labour force essentially contributed to an increase in production that was mainly used for its subsistence. In this subsistence economy in which substituting capital for labour is still extremely difficult, productivity gains are based on two factors: land (with the danger of over-exploitation) and paid labour (which showed a downward trend during the period under study) Thirty years after the Green Revolution, or four years after the 1991 reforms, the formation of productivity gains in Indian agriculture therefore does not seem to have taken any new directions, in particular towards areas where marginal productivity gains could be higher (arid or semi-arid zones, remote regions or with low educational levels, etc.). Indeed, such an attempt would call for a complete reform in the Indian system of agricultural subsidies.

(b) The study of surplus distribution amongst the partners involved in the Indian agricultural sector reveals a major divergence in development, between animal and plant products, between subsidized inputs and other inputs, and between own-account workers and agricultural employees. In real terms, surplus distribution seems to have favoured consumers of plant products (cereals, oilseeds and sugarcane, in particular) and agricultural employees (wage adjustment accompanied by a relative reduction in numbers). In areas with abundant water supply (i.e. where no risks are involved in investments in modern inputs such as fertilizers or high yielding varieties), the disadvantage farmers suffered has been compensated by a relative fall in the price of subsidized inputs (fertilizers, water, electricity, etc.). As for livestock breeders, they have inherited an advantage because of the relative rise in price of their production during the 1980s. On the whole, the equation between producers and consumers seems, therefore, fairly equitable. This is the great success of the Green Revolution. However, this has been achieved to the detriment of public finances and of the producers who cannot enjoy any input subsidies, as also of natural resources (depletion of groundwater tables, soil erosion, pollution by chemical fertilizers, etc.): some negative externalities which we unfortunately could not be assessed with this application of SAM to Indian agriculture.

NOTES

1. The market functions in a perfect state of competition: supply balances demand, the value of production thus corresponds to the value of sales; there are no profits, all sales taken together serving to compensate factors; the marginal productivity of inputs is equal to their price.
2. Production variations are totally proportionate with input variations.
3. The effects of technical progress have the same effect on all factors of production, such as capital and labour.
4. The first official CSO estimations used 1948–9 as the reference year for series with constant prices. They were published in 1956 in a report entitled 'Estimates of National Income' that was published subsequently every year till 1966. Afterwards, the CSO's methodology and fields of investigation were regularly updated and broadened. In order to reflect these changes, the annual CSO report was first renamed 'Estimates of National Product' towards the end of the 1960s, which was changed to 'National Accounts Statistics' (NAS) in January 1975. The NAS series at constant price that we have used take 1980–1 as the base year. However, given the time taken in collecting and processing the data, and for want of adequate computerization, in the mid-1999, no series more recent than 1995–6 were available with the CSO. Furthermore, some of the data was collected directly from CSO since it had not yet been published.
5. Logically, we should retain the 'Rent paid for lease in land', which would represent the farmer's real expenses. Unfortunately, the Ministry of Agriculture has not provided enough information under this heading, which is therefore not representative enough to be able to extrapolate the data at a pan-Indian scale. The 'Rental value of owned lands' series are much more complete as far as data is concerned. Besides, the rental value is calculated from existing rents in the concerned region.
6. Paddy, wheat, jowar, bajra, red gram (tur), groundnut, soyabean, rapeseed/mustard, sugarcane, cotton, jute.
7. There is no such information for red gram in 1980–1, nor for wheat, rapeseed/mustard and jowar for the year 1993–4.
8. District-wise statistics would enable us to work on much less geographically disparate units than the Union states and territories.
9. Information on some states or territories is fairly inadequate or non-existent, because of prevailing political problems (Jammu & Kashmir, Manipur, Nagaland, etc.), or because of their relatively small size (Daman & Diu, Chandigarh, etc.).
10. Including 4.93 billion from gross value added from operation of irrigation system.
11. Water is almost free, but its pumping represents most of the electricity, petrol and repairs/ maintenance charges.
12. Current replacement cost of the consumption of reproducible fixed assets.
13. Reminder: According to equations (2) and (6), in advantage calculations, price variations are designated by a negative sign for outputs.
14. In 1980–1 rupees, and including –6.50 billion for added value on irrigation works.

REFERENCES

ABLER, D.G., G.S. TOLLEY and G.K. KRIPALANI, *Technical Change and Income Distribution in Indian Agriculture*, Westview Press, 1994.

BOUSSARD, J.M., M. NASSEF and I. FOULHOUZE, *Les comptes de surplus de la branche agricole depuis 1960*, Paper presented at the Société Française d'Economie Rurale, Paris, 11 March 1981, 29 p.

BUREAU, J.C., J.P. BUTAULT, D. HASSAN, P. LEROUVILLOIS and J.M. ROUSSELLE, *Formation et répartition des gains de productivité dans les agricultures européennes entre 1967 et 1987*, Cahiers d'Economie et Sociologie Rurales, No. 20, Paris, 1991, pp. 64–90.

COURBIS, R. and P. TEMPLÉ, *La méthode des comptes de surplus et ses applications macroéconomiques*, Les Collections de l'INSEE, No. 35C, Paris, 1975, 100 p.

CSO, *National Accounts Statistics 1998*, Central Statistical Organisation, Government of India, New Delhi, 1998, floppies.

———, *National Accounts Statistics. Sources and Methods*, Central Statistical Organisation, Government of India, New Delhi, 1989.

———, *Statewise and Cropwise Estimates of Value of Output from Agriculture, from 1980–81 to 1990–91*, Central Statistical Organisation, Government of India, New Delhi, 1996, 276 p.

Dorin, B., 'Food Policy and Nutritional Security. Unequal Access to Lipids in India', *Economic and Political Weekly*, Vol. 34, No. 26, 26 June 1998, pp. 1709–17.

Dorin, B. and T. Jullien, *Input Subsidies in India and Agricultural Competitiveness Within and Outside the Country*, Paper presented at the seminar 'Agricultural Incentives and Sustainable Development', India International Centre, New Delhi, 3–4 April 2000, 22 p.

Dorin, B., N. Pingault and J.-M. Boussard, 'Formation et répartition des gains de productivité dans l'agriculture indienne (1980–1996)', *Economie Rurale*, No. 263, Paris, March–June 2001, pp. 78–91.

EPWRF, *National Accounts Statistics of India 1950–51 to 1996–97*, Economic and Political Weekly Research Foundation, Mumbai, 1998, CD-ROM.

Kulshreshtha, A.C., R. Kolli and G. Singh, 'Agriculture in National Accounts. Gaps in Information System', *Economic and Political Weekly*, Vol. 32, No. 27, 1997.

Lalvani, M., 'Some Anomalies in Indian Data Sets. Tackling Inconsistencies', *Economic and Political Weekly*, Vol. 32, No. 23, 1997, pp. 1345–50.

Ministry of Agriculture, *Area and Production of Principal Crops in India*, Ministry of Agriculture, Government of India, New Delhi, 1990–5 (various issues).

———, *Cost of Cultivation of Principal Crops in India*, Ministry of Agriculture, Government of India, New Delhi, 1991–6 (various issues).

———, *Report of the Commission for Agricultural Costs and Prices on Price Policy*, Ministry of Agriculture, Government of India, New Delhi, 1991–9 (various issues).

Murgai, Rinku, *The Green Revolution and the Productivity Paradox. Evidence from the Indian Punjab*, Development Research Group, The World Bank, Washington, D.C., October, 2000, p. 22.

Pingault, N., *Application de la méthode des comptes de surplus à l'agriculture indienne. Disparité inter-régionale des revenus agricoles en Inde*, Mémoire de DEA, Ecole Nationale du Génie Rural des Eaux et des Forêts, Université Paris I, Paris, 1999, 97 p. (typo).

Pradhan, B.K. and M.R. Saluja, 'Industrial Statistics in India. Sources, Limitations and Data Gaps', *Economic and Political Weekly*, Vol. 33, No. 21, 1998, pp. 1263–84.

Saint-Guilhem, C., *Productivité globale et comptes de surplus*, Centre d'Etude des Revenus et des Coûts, No. 55/56, Paris, 1980.

Vincent, L.A., *La mesure de la productivité*, Dunod, Paris, 1968, 305 p.

6

Pseudo Productivity, Pseudo Surpluses

THE SOCIAL AND ECOLOGICAL IMPACT OF GLOBALIZATION OF INDUSTRIAL AGRICULTURE

VANDANA SHIVA

INTRODUCTION

Seventy per cent of the world's food is produced by local farming communities, especially women. Farming systems of the world have been diverse, reflecting diverse agro-climatic features on the one hand and diverse food cultures on the other. In fact, the diversity of agricultural systems and cultural diversity of food systems have co-evolved by a mutual interplay of nature and culture, of biodiversity and cultural diversity. But globalization of ãgriculture had led to the rapid destruction of diverse farming systems and displacement of small farmers worldwide. This is usually justified as necessary in order to produce more food. This destruction of the environment and rural livelihoods is an inevitable consequence of the globalization of industrial agriculture. Industrial agriculture includes the chemicalization and mechanization of agriculture as in the case of the Green Revolution. It also includes genetic engineering.

GLOBALIZATION OF NON-SUSTAINABILITY

HIGH LABOUR PRODUCTIVITY, LOW RESOURCE PRODUCTIVITY

According to the dominant paradigm of industrial agriculture, intensifying chemical inputs and energy use is necessary for feeding a growing population because intensification of inputs and large farms lead to higher productivity and hence more food.

However, productivity is the measure of output per unit input. In terms of resources and energy, the higher the input, the lower the productivity. Since industrial agriculture is resource and energy intensive, it has led to a productivity decline in terms of ecological efficiency and resource use, and to concentrate ownership of land in large farms. Intensification of chemicals and fossil fuels is primarily aimed at substituting labour. When labour is artificially selected as one input among many, and productivity with respect to labour as the only input is treated as productivity in general, an illusion is created of higher overall productivity, and hence greater availability of food, when in reality more resources are being wasted, more livelihoods are being destroyed, and more hunger is being created by destruction of people's entitlements both through resource destruction and livelihood destruction.

Fewer people on farms does not mean more food per unit resources. In fact, by being resource destructive, industrial agriculture destroys food security. Yet labour productivity continues to be used to create an illusion of more food.

This illusion is maintained in the State of Food and Agriculture Report (FAO, 2000) even though the environment costs of industrial agriculture and destruction of small farms is a growing concern across the world. Indeed, according to this FAO report, in the mid-twentieth century, the peoples of the world found themselves in widely differing agricultural situations and engaged in production systems that can be ranked in order of increasing net productivity as follows:

— manual farming, with maximum net productivity of around 1000 kg. of cereal equivalent per worker;
— irrigated rice cultivation, using animal traction, with one harvest per year;
— light cultivation, using animal traction (swing plough, packsaddle, etc.), with fallow, with a maximum net productivity of 2000 kg. per worker;
— heavy cultivation, using animal traction (plough, cart, etc.) with fallow, with a maximum net productivity of 3500 kg. per worker;
— irrigated rice cultivation, using animal traction, with two harvests per year, with a similar net productivity to that of the previous system;
— heavy cultivation, using animal traction, without fallow, with a net productivity of 5000 kg. per worker;
— mechanized cultivation, using animal traction, without fallow, with a net productivity of 10,000 kg. per worker;
— first systems of cultivation using motorized mechanization ('Motorized Mechanization I'), with a maximum net productivity already exceeding 30,000 kg. per worker.

Thus, according to the FAO report, the ratio between the least efficient system (manual farming) and the most productivity of manual farming, which is the least efficient but most widespread type of farming worldwide, is still about 1000 kg. of cereal equivalent per worker, while the net productivity of the most motorized and input-intensive farming system now exceeds 500,000 kg. ('Motorized Mechanization V'). The ratio between these two systems is therefore about 1:500, almost a twenty-fold increase in the ratio over the last fifty years.

However, if instead of labour, energy, natural resources and external inputs are taken into account, industrial agriculture does not have higher productivity compared to ecological alternatives. In fact, the shift from internal input to high external input agriculture reduces productivity from 20 to 0.33, a sixty-six fold decrease in the productivity ratio over the last fifty years. This productivity analysis is based on a study of F. Bray (1994) comparing traditional polycultures with industrial monocultures. It shows that a polyculture system can produce 100 units of food from 5 units of inputs whereas an industrial system requires 300 units of input to produce the same 100 units. The 295 units of wasted inputs could have provided 5,900 units of food. This is a recipe for starving people, not for feeding them.

Besides, the work of the Research Foundation for Science, Technology and Ecology (RFSTE, New Delhi) has shown that farm incomes can increase three-fold by giving up chemicals and using internal inputs produced by on-farm biodiversity, including straw, animal manure and other by-products. Figure 6.1 illustrates the regression in terms of resource productivity of industrial agriculture, and the need to make an ecological transition to produce more food using less resources.

A usual argument used in promoting industrial agriculture like the Green Revolution earlier and genetic engineering in agriculture now is that only industrial agriculture and industrial breeding can keep up with increased food productivity for feeding a growing population. However, increased mouths to feed implies more efficient resource use so that the same resources can feed more people. Since resources, not labour, are the limiting factor in food production, it is resource productivity, not labour productivity, which is the relevant measure. A sixty-fold decrease of food producing capacity in the context of resource use is not an efficient strategy for using limited land, water and biodiversity to feed the world. Further, since food security is based on food entitlements, and entitlements in peasant societies are based on livelihoods and work, increase in labour productivity based on displacement farmers cannot reduce hunger. It decreases food entitlements by destruction of livelihoods.

Contrary to the myth of linear progress as perpetrated by the FAO report, comparative studies of 22 rice-growing systems have shown that indigenous systems are more efficient in terms of yields, and in terms of labour use and energy use (Shiva, 1991). Both from the point of view of food productivity and food entitlements, industrial agriculture is deficient as compared to diversity based internal input sustainable farming systems.

Further, while the partial productivity of industrial agriculture based on resource and energy intensification is non-sustainable both because of environmental externalities and because of livelihood destruction, biodiversity intensification allows increase in food production by increasing resource productivity and increase in employment while improving the environment.

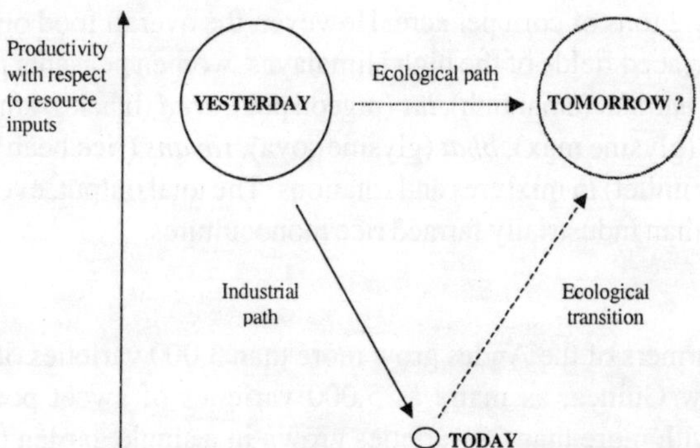

Figure 6.1: Need to Make an Ecological Transition.

HIGH YIELD, LOW OUTPUT

The measure of productivity of industrial agriculture is partial not only because all inputs (including resource and energy inputs) are not taken into account, but also because not all outputs are taken into account. Indeed, ecological agriculture is based on mixed and rotational cropping, and the production of a diversity of crops; traditional farming systems are based on mixed and rotational cropping systems of cereals, pulses, and oilseeds with different varieties of each crop, while the Green Revolution package is based on genetically uniform monocultures.

The polycultures of traditional agricultural systems have evolved because more yield can be harvested from a given area planted with diverse crops than from an equivalent area consisting of separate patches of monocultures. For example, in plantings of sorghum and pigeon pea mixtures, one hectare will produce the same yields as 0.94 ha. of sorghum monocultures and 0.68 ha. of pigeon pea monoculture. Thus one hectare of polyculture produces what 1.62 ha. of monoculture can produce. This is called the Land Equivalent Ratio (LER).

Increased land-use efficiency and higher LER has been reported by Rosset and Altieri (1999) for polycultures of millet-groundnut (1.26), maize-bean (1.38), millet-sorghum (1.53), maize-pigeon pea (1.85), maize-cocoyan-sweet potato (2.08), cassava-maize-groundnut (2.51). The monocultures of the Green Revolution thus actually reduced food yields per acre previously achieved through mixtures of diverse crops. This falsifies the argument often made that chemically intensive agriculture and genetic engineering will save biodiversity by releasing land from food production. In fact, since monocultures require more land, biodiversity is destroyed twice over: once on the farm, and then on the additional acreage required to produce the outputs a monoculture has displaced.

A myth promoted by the one-dimensional monoculture paradigm is that biodiversity reduces yields and productivity, and monocultures increase yields and productivity. However, since yields and productivity are theoretically constructed terms, they change according to the context. Yields usually refers to production per unit area of a single crop. Planting only one crop in the entire field as a monoculture will of course increase its yield. Planting multiple crops in a mixture will have low yields of individual crops, but will have high total output of food (Shiva, 1999).

The Mayan peasants in the Chiapas are characterized as unproductive because they produce only 2 tons of corn per acre. However, the overall food output is 20 tons per acre. In the terraced fields of the high Himalayas, women peasants grow *jhangora* (barnyard millet), *marsha* (amaranth), *tur* (pigeon pea), *urad* (black gram), *gahat* (horse gram), soya bean (glysine max), *bhat* (glysine soya), *rayans* (rice bean), *swanta* (cow pea), *koda* (finger millet) in mixtures and rotations. The total output, even in bad years, is six times more than industrially farmed rice monocultures.

Besides:

— indigenous farmers of the Andes grow more than 3,000 varieties of potato;
— in Papua New Guinea, as many as 5,000 varieties of sweet potato are under cultivation, with more than 20 varieties grown in a single garden (Heywood and Watson, 1999);

— in Java, small farmers cultivate 607 species in their home gardens, with an overall species diversity comparable to a deciduous tropical forest;
— in sub-Saharan Africa, women cultivate as many as 120 different plants in the spaces left alongside the cash crops;
— a single home garden in Thailand has more than 230 species;
— African home gardens have more than 60 species of trees;
— rural families in the Congo eat leaves from more than 50 different species of trees;
— a study in eastern Nigeria found that home gardens occupying only 2 per cent of a household's farmland accounted for half of the farm's total output;
— home gardens in Indonesia are estimated to provide more than 20 per cent of household income and 40 per cent of domestic food supplies (FAO, 1999).

Moreover, industrial agriculture productivity is high only in the restricted context of a 'part of a part' whether it be the forest or of the farm. For example, 'high-yield' plantations pick one tree species among thousands, for yields of one part of the tree (e.g. woodpulp), whereas traditional forestry practices use many parts of many forest species.

'High-yield' Green Revolution cropping patterns select one crop among hundreds, such as wheat, for the use of just one part, the grain. These high partial yields do not translate into high total yields (Figure 6.2), because everything else in the farm system goes to waste. Usually the yield of a single-crop like wheat or maize is singled out and compared to yields of new varieties. This calculation is biased to make the new varieties appear 'high-yielding' even when, at the systems level, they may not be.

Comparative yields of native and Green Revolution varieties in farmers' fields have been assessed by Navdanya, a seed conservation movement. Green Revolution varieties are not higher yielding under the conditions of low capital availability and fragile ecosystems. Farmers' varieties are not intrinsically low yielding and Green Revolution varieties or industrial varieties are not intrinsically high yielding (Shiva et al., 1994) (Tables 6.1 and 6.2).

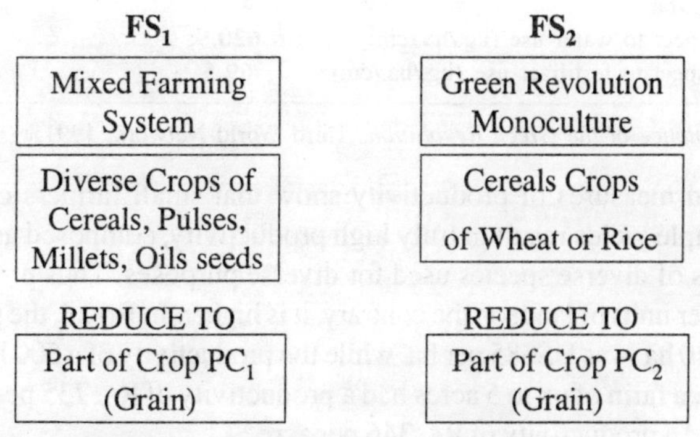

=> The real scientific comparison should be between two farming systems (FS_1 and FS_2) with the full range of inputs and outputs included.
=> This would be the comparison if FS was not given immunity from an ecological evaluation.
=> In the Green Revolution strategy a false comparison is made between PC_1 and PC_2
=> So while $PC_2 > PC_1$ generally $FS_1 > FS_2$.

Figure 6.2: How Green Revolution Makes Unfair Comparison.

Table 6.1: Some High Yielding Indigenous Rice Varieties of the Himalaya

Rice Variety	Yield kg./acre		
	Grain	Straw	Total Biomass
Anjana	1595	6075	7670
Banta puri	1860	5222	7080
Bauk naj	1475	6275	7850
Chama	1820	7695	9515
Dhansari	1295	6885	8180
Chanya naj	1860	7085	8745
Jhayoin	1720	5750	7470
Jhumkya	1720	11945	13665
Laldhan	1295	9030	10325
Lal nalandi	1700	7205	8905
Lal sat	1215	7005	8220
Mansuli	1295	7610	8905
Nagur sat	1780	7570	9350
Padratnachoodi	1375	8015	9390
Ramjawan	1070	7570	9350
Rathwa	1295	7005	8300
Rikhwa	1455	7005	8460
Safed sat	1295	7005	8300
Sita	1400	5585	9685
Jyakanya naj	1820	6235	8055

Source: Navdanya Conservation Centre, Doon Valley.

Table 6.2: Comparison of Productivity of Native Varieties and Borlaug (Green Revolution) Varieties of Wheat

	Native Variety	Borlaug Variety
Yield (kg./ha.)	3291	4690
Water demand (cm)	5.3	16.0
Fertilizer demand (kg./ha.)	47.3	88.5
Productivity with respect to water use (kg./ha./cm)	620.9	293.1
Productivity with respect to fertilizer use (kg./ha./cm)	69.5	52.9

Source: V. Shiva, *Violence of the Green Revolution,* Third World Network, 1991.

Biodiversity-based measures of productivity show that small farmers can feed the world. Their multiple yields result in truly high productivity, composed as they are of the multiple yields of diverse species used for diverse purposes. Thus productivity is not lower on smaller units of land: on the contrary, it is higher. In Brazil, the productivity of a farm of up to 10 ha. was US$85 per ha. while the productivity of a 500 ha. farm was $2 per ha. In India, a farm of up to 5 acres had a productivity of Rs. 735 per acre, while a 35 acres farm had a productivity of Rs. 346 per acre.[1]

Diversity produces more than monocultures. But monocultures are profitable to industry both for markets and political control. The shift from high productivity diversity to low productivity monocultures is possible because the resources destroyed are taken from the poor, while the higher commodity production brings benefits to those with economic power. The polluter does not pay in industrial agriculture both of the chemical era or the biotechnology era. Ironically, while the poor go hungry, it is the hunger of the poor which is used to justify the agricultural strategies which deepen their hunger.

As Yegna Narayan Aiyer reports (Yegna Iyengar, 1944):

> The possibility of obtaining phenomenal and almost unbelievably high yields of paddy in India has been established as the result of the crop competitions organized by the Central Government and conducted in all states. Thus even the lowest yield in these competitions has been about 5,300 pounds per acre (lbs./acre), 6,200 lbs./acre in West Bengal, 6,100 and 7,950 and 8,258 lbs./acre in Thirunelveli, 6,368 and 7,666 lbs./acre in South Arcot, 11,000 lbs./acre in Coorg and 12,000 lbs./acre in Salem.

The measurement of yields and productivity in the Green Revolution as well as in the genetic engineering paradigm is divorced from seeing how the processes of increasing single species, single function, output affect the processes that sustain the condition for agricultural production, both by reducing species and functional diversity of farming systems as well as by replacing internal inputs provided by biodiversity with hazardous agrichemicals. While these reductionist categories of yield and productivity allow a higher measurement of harvestable yields of single commodities, they exclude the measurement of the ecological destruction that affects future yields and the destruction of diverse outputs from biodiversity rich systems.

Productivity in traditional farming practices has always been high if it is remembered that very little external inputs are required. While the Green Revolution has been projected as having increased productivity in the absolute sense, when resource utilization is taken into account, it has been found to be counter productive and resource inefficient.

Perhaps one of the most fallacious myths propagated by Green Revolution protagonists is the assertion that HYVs have reduced the acreage, therefore preserving millions of hectares of biodiversity. India's experience tells us that instead of more land being released for conservation, by destroying diversity and multiple uses of land, the industrial breeding system actually increases pressure on the land since each acre of a monoculture provides a single output and the displaced outputs have to be grown on additional acres.[2]

If we focus on land use in the Green Revolution the industrial breeding strategies increase grain production by 20 per cent under good conditions and lead to a decline of 100 per cent of straw. If traditional varieties produce 1000 kg. per acre grain and 1000 kg. per acre fodder, then traditional varieties produce 2000 kg. grain plus 2000 kg. fodder on 2 acres whereas the industrial strategy only provides 2400 kg. grain on the 2 acres, leading to an increase of 400 kg. grain but a decline of 2000 kg. fodder.

Further, the same reductionist logic of industrial breeding also increases the resource used by cattle. Industrial livestock farming consumes three times more biomass than ecological livestock maintenance. Thus industrial livestock breeding would require three times more acres of land for feed. In fact Europe uses seven times the area of Europe in the Third World countries for cattle feed production. For fodder alone (including that used to produce food products for export) the Netherlands appropriates 100,000 to 140,000 square kilometers of arable land, much from the Third World. This is five to seven times the area of agricultural land in the entire country (Rees and Wackernagel, 1998).

The combination of industrial plant breeding and industrial animal breeding, therefore increases the pressure on land use by a factor of 400 per cent while separately increasing output of grain and milk by only a factor of 20 per cent (Shiva, op. cit.). The extra resources used by industrial systems—either by the Green Revolution or the new biotechnologies—could have gone to feed people. Resources wasted amount to the creation of hunger. By being resource wasteful through intensive external inputs, the new biotechnologies create food insecurity and starvation.

What does all this evidence mean in terms of 'feeding the world'? It becomes clear that industrial breeding has actually reduced food security by destroying small farms and the small farmers' capacity to produce these diverse outputs of nutritious crops. Both from the point of view of food productivity and food entitlements, industrial agriculture is deficient as compared to diversity-based internal input systems. Protecting small farms which conserve biodiversity is thus a food security imperative.

There are a number of strategies which allow this inversion to take place and an illusion of growth to be created. Firstly, a monoculture paradigm looks only at one element of a system and treats an increase in the part as an increase in the whole system. Thus, by focusing only on yield increases of grain of individual cereals like rice or wheat, the reduction in straw availability for fodder is externalized and not accounted for. This is the 'monoculture of the mind' (Shiva, 1993).

A second strategy is to exclude the higher inputs from the resource equation and only focus on the single commodity output. Thus the resource waste is not taken into account and low resource use productivity is converted into high commodity productivity.

To assess the real productivity of a farming system from the perspective of the farmer and the perspective of the soil we need to measure the Biodiversity Based Productivity,[5] not just the price or yield of single commodity or single output. We also need to calculate:

(a) the value of diverse outputs from diverse species and their diverse functions,
(b) the value of internal inputs provided by diverse farm outputs (e.g. straw for organic manure),
(c) the costs of purchased inputs such as fertilizers, pesticides, herbicides, and,
(d) the ecological costs of external chemical inputs.

In a polyculture diversity based internal input farming system the returns the farmers get are the sum of diverse outputs, and internal ihputs:

$$\Sigma O_i + I_j$$

where O_i are the diverse outputs of the farming system and I_j are the internal inputs provided by on farm biodiversity, and where i and j are variables that cover the diversity.

In an industrial monoculture, the returns to the farmer are:

$$O_1 - (\Sigma O_i - O_1) - \Sigma(E_j) - \int M_x(E_j)$$

where O_1 is the single commodity output, $(\Sigma O_i - O_1)$ is the sum of outputs sacrificed for the monoculture O_1. $\Sigma(E_j)$ are the sum of external inputs used, and M_x are the multiple ecological impacts of external inputs.

The higher productivity of diversity based systems indicates that there is an alternative to genetic engineering and industrial agriculture—an alternative that is more ecological and more equitable. This alternative is based on biodiversity intensification in place of chemical intensification. This alternative is also based on small farms, which are far more productive, both in terms of more efficient use of resources and higher output.

LARGE IS WASTEFUL, SMALL IS EFFICIENT

Research done by FAO has shown that small biodiverse farms can produce thousands of times more food than large, industrial monocultures (Figure 6.3). As in the rest of the Third World, agriculture in India has in general been dominated by small farms that produce diverse output using farm resources to the maximum potential (Table 6.3). As the size of the operated holding increases the proportion of the area devoted to a single crop activity (referred to as monoculture) increases proportionately. The biodiversity based productivity of small farms makes the small farm an important component

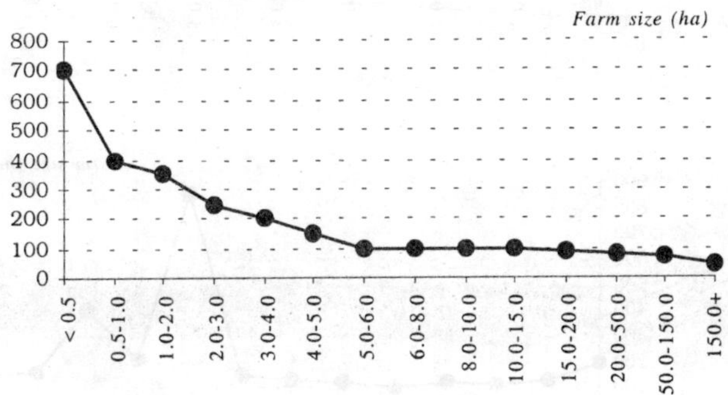

Tanzania

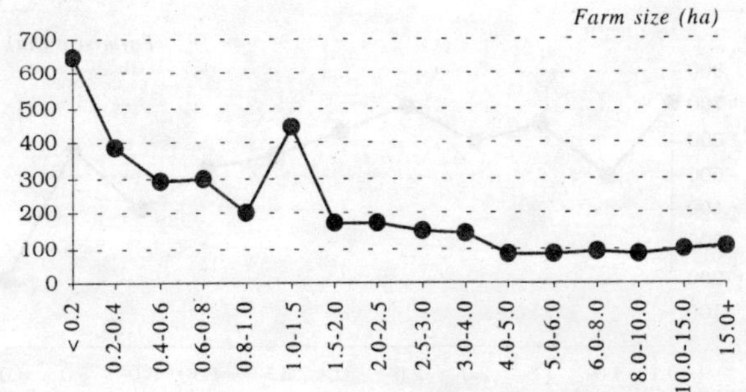

Mexico

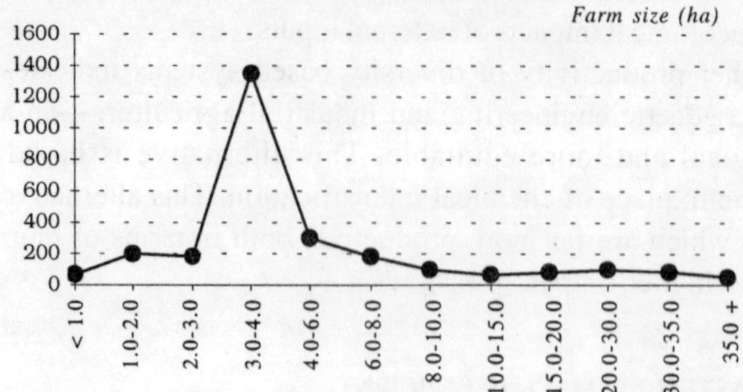

Uganda

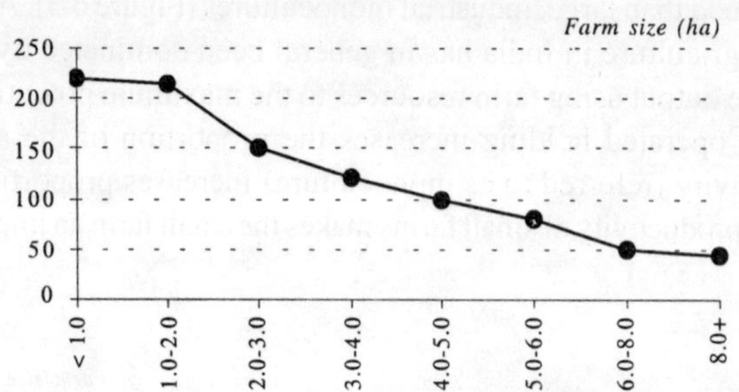

Peru

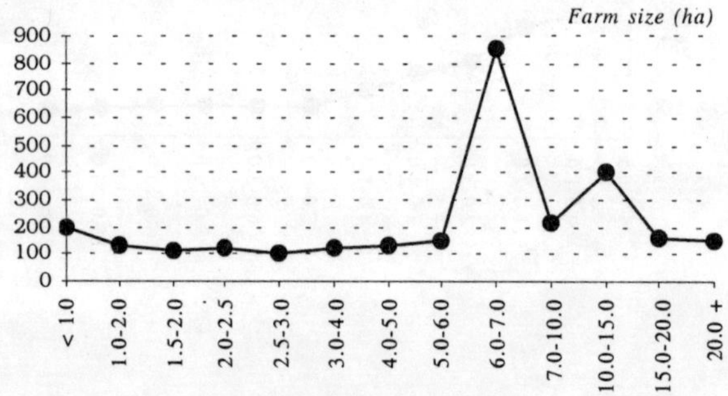

India

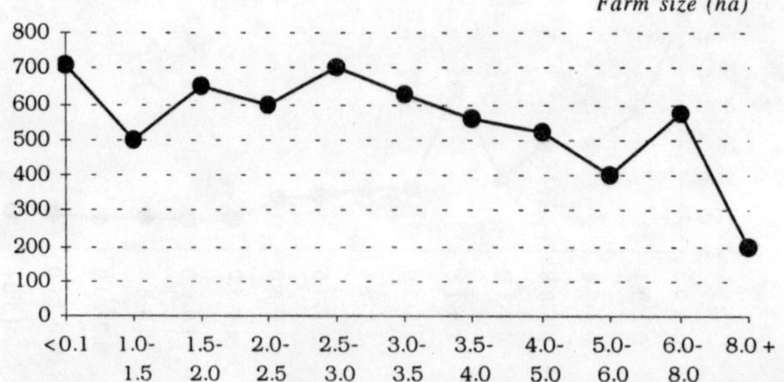

Sudan

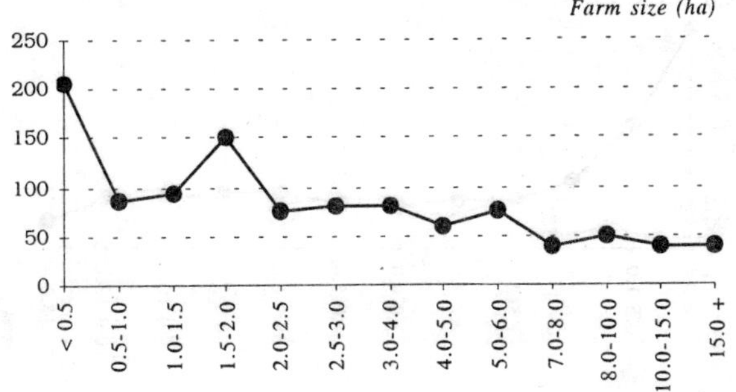

Nepal

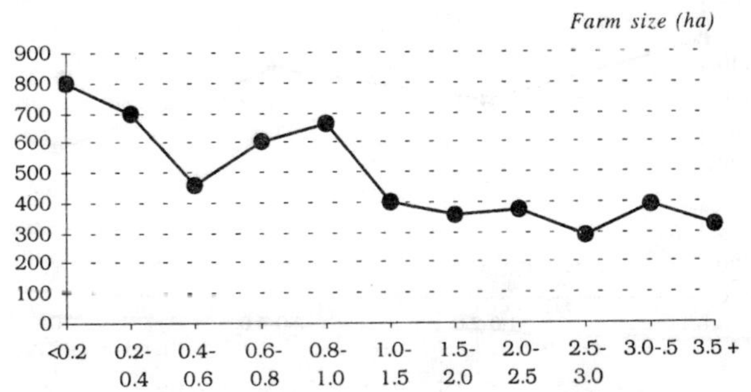

Nigeria

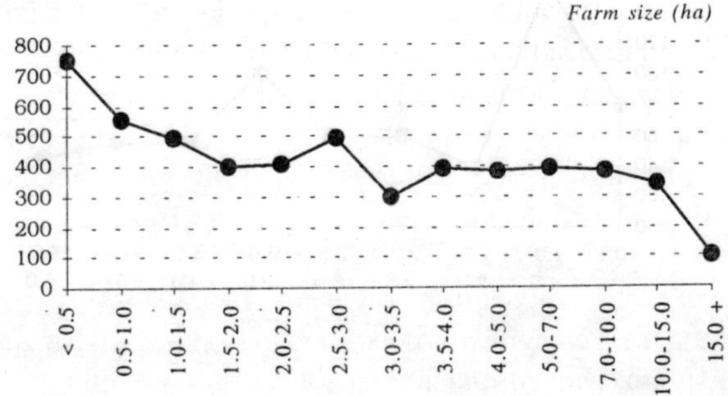

South Korea

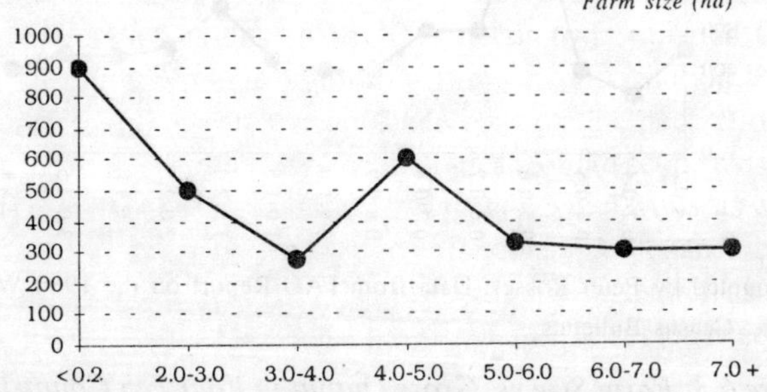

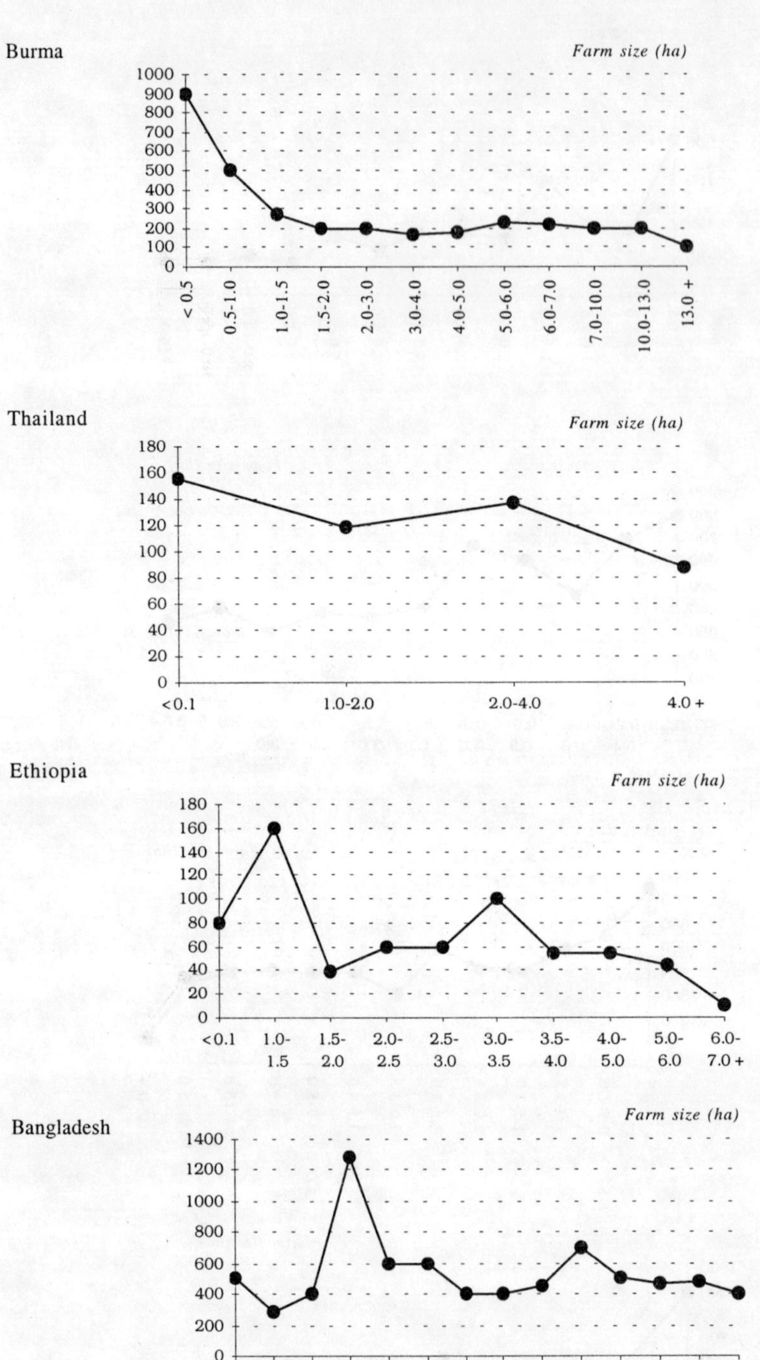

Source: FAO, as quoted by Peter Rosset; Data from FAO Report on the 1980 World Census of Agriculture, Census Bulletins

Figure 6.3: Farm Size vs. Gross Output in Fourteen Countries.

in the larger context of ecological sustainability. Besides, it has been shown that small farms are more efficient in the use of resources towards achieving the potential level of output.

The greater diversification in the cropping pattern of small farms compared to that of large farms is clear from Table 6.4 which shows the percentage distribution of operational holdings by main use of the holding for broad size class of operational holdings. The higher productivity on small farms is explained by A.K. Sen (1962) through the marginal productivity theory. He has shown that: (i) as farm-size (acreage) expands, gross output per acre declines, and (ii) as farm-size (acreage) expands, farm business income per acre declines, though net profits per acre may increase.

Table 6.3: Percentage Distribution of Operational Holdings in India

Category of Holdings	Percentage Distribution of Operational Holdings			
	1960–1 (17th)	1970–1 (26th)	1981–2 (37th)	1991–2 (48th)
Marginal	39.1	45.8	56.0	62.6
Small	22.6	22.4	19.3	17.8
Semi-medium	19.8	17.7	14.2	12.0
Medium	14.0	11.1	8.6	6.1
Large	4.5	3.1	1.9	1.3
All sizes	100.0	100.0	100.0	100.0

Source: National Sample Survey, 48th Round.

Table 6.4: Percentage Distribution of Operational Holdings by Main Use in India (1991–2)

Size of Operational Holdings (ha.)	Crop Production		Hortic Ul. per Orchards	Plantation		Live-stock	Poultry Raising	Pisci-culture	Other	B.P.	All
	Veg.	Others		Forestry	Others						
< 0.002	3.18	7.97	0.37	-	0.26	33.80	6.62	-	8.78	39.03	100
0.002–02	6.51	39.49	2.14	0.16	5.21	25.30	4.94	0.41	5.46	10.38	100
0.21–0.5	2.27	87.50	1.17	0.07	3.83	1.89	0.27	0.32	0.25	2.43	100
0.51–1.00	1.22	92.27	0.85	0.12	2.02	1.30	0.09	0.13	0.33	1.66	100
1.01–2.00	0.86	93.67	0.44	0.09	1.53	1.24	0.12	0.09	0.13	1.84	100
2.01–4.00	0.45	94.74	0.36	0.04	1.00	0.69	0.01	0.06	0.03	2.62	100
4.01–10.00	0.30	94.92	0.39	0.02	0.82	1.29	-	0.07	0.00	2.18	100
>10.01	0.07	95.70	0.45	0.05	0.91	1.32	-	0.29	-	1.20	100
All sizes	2.56	76.68	1.05	0.10	2.79	8.23	1.51	0.21	1.74	2.14	100

Source: National Sample Survey, 48th Round.

These findings are substantiated by the findings of similar studies on the issue of farm size and efficiency (Khusro, 1962). In the Figure 6.4, curve MP represents the schedule of net marginal product of labour applied to a given acre of land, i.e. it shows the amount by which output would increase with the employment of an additional labourer. If cultivation is family-based, as cultivation in the majority of marginal and small farms are, and if the per acre supply of labour in the family equals or exceeds OP, we can expect that OP amount of labour will be applied, given the absence of outside opportunity of employment and provided there is no significant disutility of work in the relevant range of effort. Assuming that the ruling wage rate in the area of OW, the imputation of OW wage to each unit of OP labour, would imply that the total cost of labour is OWAP. A farm cultivating for profits would, thus, hire labour up to the point C where the marginal productivity of labour is equal to the ruling wage rate, i.e. at the point where return from labour are equal to the cost of labour.

Now, in the light of the fact that the proportion of family labour to hired labour falls with the size of farms, it follows that the net profit per acre increases with the size of the holding and that productivity per acre decreases with the size of the holding. However, a closer look at what constitutes net profit suggests that size of the holding after all may have little to do with higher profits and what shows up as greater net profits with the increase in the size of the operated holding is based rather on the nature of definitions. Khusro put forward the observation that net profits per acre is directly proportionate to the size of the operated holding, net profits has been defined as 'net farm business income minus retained cost' (Figure 6.5). Net farm business income, here, refers to gross output minus paid-out cost, where paid-out cost represents the expenditure incurred by farmers in purchasing raw materials, employing non-family labour, borrowing capital (inclusive of depreciation), hiring bullocks and renting in land. Retained costs is defined as the costs of farm-produced inputs—family labour, owned land and farm-produced capital. In other words, retained costs have been defined as costs which farmers aren't interested in minimizing. Given this definition and disregarding the concept of retained costs, small farms turn out as profitable as large farms, if not more were it not for defining net profits on the basis of incorrectly calculated opportunity costs of farm-produced inputs.

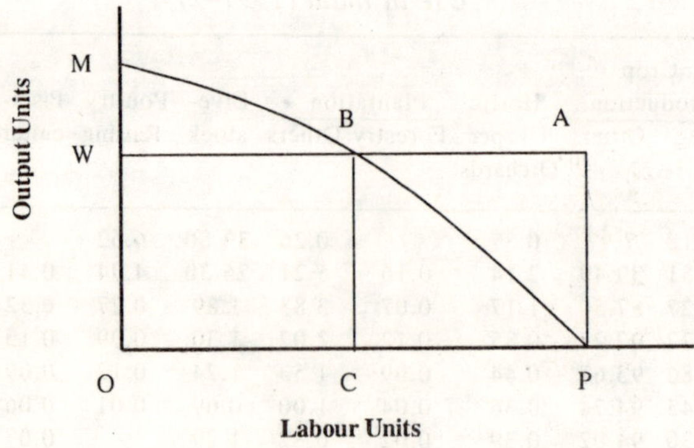

Source: Khusro, 1962.

Figure 6.4: Net Marginal Product of Labour Applied to a Given Acre of Land.

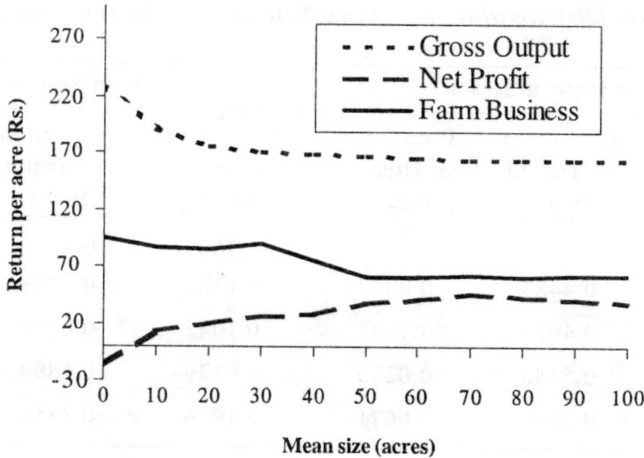

Source: Khusro, 1962.

Figure 6.5: LL-India: Inter-Pollated Returns per Acre by Size of Farm

Besides greater productivity, small farms are characterized by sustainable methods of cropping. Antony (1998) has pointed out that the use of organic manure and family labour is inversely related to the size of the holding, while the proportion of tractor hours increases as the holding size increases (Table 6.5). It is also known that the replacement of indigenous crop varieties with hybrid and high-yielding varieties has been at the cost of genetic diversity.[3] Taking the example of the state of West Bengal alone, the recorded number of folk rice landraces which was about 5,600 until 1970 has declined to no more than 250 today. Further, there has been an acceleration in soil erosion and the declining of water tables in the process of transformation of traditional agriculture. The consequences of vanishing diversity are an enhanced risk of extinction. and increasing vulnerability of farmers.

In the light of the above it would be no exaggeration to say that small farms are the answer to declining agricultural productivity and vanishing bio diversity as is evidenced in the current agricultural scenario of the country. The initial gains of the Green Revolution have been off-set by declining agricultural productivity thereafter. This trend has been observed with respect to almost all major crops (Table 6.6) with rate of growth of foodgrains output being halved from the decade of the 1980s. This trend carries with it adverse implications for food security in the country in a situation where net area under cultivation has more or less stagnated.

Data shows that, everywhere in the world, biodiverse small farms produce more agricultural output per unit area than large farms. Even in the USA, small farms of 27 acres or less have two times greater dollar output per acre than larger farms (Rosset and Altieri, 1999). It is therefore time to switch from measuring monoculture yields to assessing biodiversity outputs in farming systems.

Thus both at the level of individual peasant farms and at the national level, the Green Revolution has led to a decline in food security. What the Green Revolution achieved was an increase in industrial inputs, which, of course, created growth for the agrochemical and fossil-fuel industry. But this increased consumption of toxins and energy by the agricultural sector did not translate into more food.

Table 6.5: Mean Distribution of Factors Across Size-classes of Rice Holdings

	Monsoon Season			Lean Season		
	Proportion of Family Hours	Proportion of Tractor Hours	Proportion of Organic Manure	Proportion of Family Hours	Proportion of Tractor Hours	Proportion of Organic Hours
< 1 ha.	0.2309	0.3344	0.0594	0.2603	0.3550	0.0995
1–2 ha.	0.1345	0.4483	0.0907	0.1386	0.4731	0.0991
2–3 ha.	0.0905	0.4927	0.1045	0.1042	0.8719	0.0381
> 3 ha.	0.0583	0.5185	0.0233	0.0729	0.6899	0.0957
Total	0.1858	0.3856	0.0671	0.1976	0.4551	0.0952

Source: Antony, 1998.

Table 6.6: Compound Annual Growth Rates of Agricultural Output (1980s and 1990s)

	Rice	Wheat	Coarse Cereals	Pulses	Food-grain	Oilseeds	Cotton	Sugar cane	Total Agricul-tural
1980–1 to 1990–1	3.3	4.2	1.2	3.0	3.1	7.0	3.4	4.6	3.8
1990–1 to 1997–8	1.7	2.7	–0.7	–1.3	1.4	3.5	2.2	1.1	1.9

Source: Alternative Economic Survey, 1998–9.

Today, most of the 830 millions people who lack adequate access to food are rural communities whose entitlements have collapsed either due to environmental degradation or due to livelihood destruction and negative terms of trade. Food security is therefore intimately connected to the livelihood security of small rural producers. There are proven alternatives to industrial agriculture and genetic engineering, and these are based on small farms and ecological methods. Sound resource-use combined with social justice is the path of sustainability in agriculture that we should be taking.

Across the world, initiatives that free agriculture of industrial inputs are showing that both food production and farmers' incomes can increase by shifting to ecological agriculture:

— In Indonesia, restrictions were introduced on the use of 57 pesticides in rice-growing, and subsidies for pesticides were eliminated. From 1987 to 1990, the volume of pesticides used on rice fell by over 50 per cent, while yields increased by about 15 per cent. Farmers' net incomes increased by $18 per farmer per season. The government save $120 million per year by ending pesticide subsidies (Thrupp, 1997).

— In Bangladesh the 'No Pest' programme led to pesticide reduction of 76 per cent and yield increases of 11 per cent. Returns increased by an average of 106 per cent in the dry season and 26 per cent in the wet season (Thrupp, 1997).

— In Peru, ecological agriculture has contributed to a net benefit of $162 per ha., $70 from yield increase and $92 from reduced costs of inputs. With wider diffusion, net benefit could reach $288: $183 from yield increase and $105 from cost savings (UNDP, 1995).

— The UNDP project on sustainable agriculture reports tripling and quadrupling of yields from 400 kg. per ha. to 1,200 to 1,600 kg. per ha. in Honduras, and potato yield going up to 8–14 kg. per ha. per year. The project was based on the indigenous Waru-Waru system of raised fields evolved in the high plains of the Peruvian Andes (Thrupp, 1997).

A review of sustainable agriculture projects in 20 countries covering 1.93 million households shows that yields of wheat, maize and sorghum doubled when farmers shifted from high external-input agriculture to biodiversity-based agriculture using low external inputs (Thrupp, 1997).

The higher productivity of diversity-based systems indicates that there is an alternative to genetic engineering and industrial agriculture—an alternative that is more ecological and more equitable. This alternative is based on the intensification of biodiversity—intensifying through integrating diverse species—in place of chemical intensification, which promotes monocultures and, unlike its ecological alternative, fails to take all outputs of all species into account.

In spite of the efficiency of the small farmer, the new Agriculture Policy of India is based on the corporatization of agriculture. Dismantling of the small farmer-centred agricultural civilization and economy is being justified by the myth of the small farmer being unproductive.

The Government policy towards responding to the threats posed to Indian agriculture by removal of Quantitative Restrictions is the uprooting of the small farmer who is the backbone of Indian agriculture and the backbone of Indian society. The recipe for the genocide of the small farmer is based on the false assumption that the small farmer is unproductive.

The resilience to global markets presents itself in the diversified and decentralized small farms that provide food security to millions of households and communities and also to the country. However, the agriculture Ministry sees in small farms an obstruction:

> Limited land holding per capita in India is a great obstacle which inhibits large scale mechanization and adopting other measures for increasing the productivity and bringing down the unit cost of production. It is a well known fact that in the major exporting countries like USA, Canada and Australia, the cost of production is low since the farms are fully mechanized, little human labour is used and natural resources such as soil fertility and rainfall distribution are better.

Similarly, an analysis of Removal of Import Restrictions states:

> If we want our agriculture to become competitive with rest of the world we must go for modernization of this sector with improved technology. This would require huge private investments; but, a large proportion of our farmers in bottom category are resource poor and cannot afford this. The very size of such farms discourages mechanization and use of modern equipments which are essential to increase efficiency. The time has come when something has to be done to put a floor on the bottom size of holding. (Gulati et al., 1994)

Land reforms in independent India put a ceiling on land ownership. Economic reforms of globalization are removing the ceiling on land holdings thus undoing the land reforms based on equity. The new Agriculture Policy also promotes corporate take over of land. However, while inequality creation is justified on grounds of productivity, it is productivity that justifies equity and the need to maintain and protect our small farmers and peasants.

When one recognizes that the small farms of the Third World produce diverse outputs of nutritious crops, it becomes clear that industrial breeding and globalization of agriculture has actually reduced food security by destroying small farms and small farmers. Protecting small farms is a food security imperative. It is also a strategic value to build immunity to the turbulences of globalization.

The bias against the small and the bias towards the large is shifting agriculture out of the hands and control of small farmers and peasants, most of whom are women, and into the control of global corporations.

GLOBALIZATION OF FOOD INSECURITY

PSEUDO SURPLUSES AND THE GROWTH OF HUNGER

The images of butter mountains and oceans of milk are used to display how industrial agriculture has increased productivity and led to build up of surpluses. However, these surpluses more accurately reflect the inability of farmers to sell their produce and the inability of the poor to buy food. Surpluses in a world with 830 million people hungry are not surpluses, they are pseudo surpluses. When every farmer produces the same crop, farmers are unable to sell at a remunerative price. Farmers dumping tomatoes and potatoes in protest against low prices is a growing sight even in India.

The so-called oceans of milk and mountains of butter in Europe are not absolute surpluses but surpluses resulting from imports. Globalization promotes monocultures and monopolies. As Tracy Worcester points out in *Resurgence* (March–April 2000):

> In 1996, Britain exported 111 million litres of milk and imported 173 million litres. It imported 49 million kilograms of butter, but it exported 47 million. Why didn't it just consume its own 47 million kilos and import the shortfall of 2 million, thus saving all the transportation costs? Why? Because not importing and exporting on a grand scale produces no profits for the transnational and their transport fleets. The food giants will fly apples to Britain from 14,000 miles away in New Zealand and bring green beans 4,000 miles from Kenya, although British farmers can easily grow both.

Pseudo productivity based on a faulty monocultural assessment, combined with pseudo surpluses based on flawed rules of economic globalization does create growth for agribusiness, but is pushing farmers everywhere into debt, displacement and destitution. The so-called surpluses are in reality pseudo surpluses because they are based on taking nutrition out of the soil without returning food for the soil-food web. They are also pseudo surpluses because they are based on taking food away from food-producers who starve even as markets are saturated. They are pseudo surpluses because they deny producers a fair share of the price. Finally, they are not real surpluses because imports and exports camouflage real production at national level, giving the illusion of surpluses.

How fictitious 'surpluses' created by free trade can be is well illustrated in the case of wheat in India. Trade liberalization is supposed to bring benefits to national agricultural economies. However, the beneficiaries are neither the farmers nor the governments of the Third World. A recent editorial of a business daily in India ran the headline 'Freeing Wheat'. It is significant to ask from what wheat is being freed and for whom it is being freed. Wheat needs to be freed from chemicals, and wheat farmers need to be freed from their bondage to chemical and seed corporations. However, the reference is to the freedom of corporations and grain-traders.

The freedom to export foodgrain under liberalized trade has already benefited the giant grain traders, Cargill and Continental. They bought wheat at $60 to $100 per ton from India and sold it at $230–40 per ton in the international market in 1996, making a profit of $130–70 per ton, while India lost $100 million in exports because of the concentration of power in the hands of five grain merchants.

The USA grain giants turned to the Indian market because the large-scale wheat monocultures in the USA, the bread-basket of the world, had been affected in nearly 50 per cent of the farmland by a combination of drought and a fungal disease. As a result of the USA crop failure, India's wheat exports increased dramatically.

Exports of wheat, however, led to domestic shortages and a rise in prices. The two million tons that were exported in mid-1996, were in effect re-imported in late 1996 to overcome domestic shortages. Economic growth did take place, but not in terms of food and not for India. The quantity of wheat in the world and in India remained the same, India's foreign exchange expenditure increased, leaving India financially poorer. Because of a domestic price increase there was a decrease in the food entitlements of the poor. Only the 'Merchants of Grain' gained from the liberalization of wheat exports and imports.

The foodgrain stocks held by public agencies in India in 2000 are more than 42 million tons. These stocks are not the result of real surpluses but of declining purchasing power of the people. The large stocks are now being used to allow entry of private traders and Multinational Companies (MNCs) like Cargill in grain procurement and justify exports. The international agencies which are forcing trade liberalization on India, and the Government, have increased domestic prices of food by withdrawing food subsidies. In March 2000, foodgrain prices for the poorest doubled from Rs. 2.5/kg. to Rs. 4.2/kg. This led to decline in offtake and hence a build up of stocks.

Now the government is exporting wheat at Rs. 4,150 or $90 a ton. The cost to the centralized Government controlled food distribution system run by the Food Corporation of India (FCI) is around Rs. 8,000 a ton and the domestic open market price is Rs. 6,000 a ton. Domestic consumers are not allowed to have access to wheat at Rs. 4,150 a ton and are forced to pay much higher prices. People cannot be subsidized but global traders are allowed to be subsidized.

The stocks do not reflect real surpluses but pseudo surpluses. They are indicators of two distortions in the centralized food system built on World Bank and Rockfeller advice in the 1960s and referred to as the Green Revolution.

At the level of production, the rice and wheat stocks are pseudo surpluses because they reflect increases in production of wheat and rice, but do not reflect decrease in production of pulses, oilseeds, millets and maize, all of which are necessary for food security. A shift from diversity to monocultures will of course register an increase in monoculture output but will eclipse the decline in diverse outputs. In Punjab, the area under rice has increased from 227,000 ha. in 1960–1 to 2,250,000 ha. in 1999–2000, a ten-fold increase. The area under wheat increased from 1,400,000 ha. in 1960–1 to

3,300,000 ha. in 1999–2000, a three-fold increase. The area under pulses in the same period has decreased from 2,605,000 ha. to 69,000 ha., a ten-fold decrease. Gram went down from 634,000 ha. in 1966–7 to 108,000 ha. in 1985–6, nearly six-fold decline. The area under maize went down from 327,000 ha. to 185,000 ha. Area under oilseeds also decreased (from 68,000 ha. to 10,000 ha. in the case of groundnut, from 4,000 to 1,000 ha. in the case of linseed), as well as for millets and coarse grains.

Based on rice and wheat monocultures, Punjab's contribution to the central pool has increased from 16 per cent during 1970–1 to 43 per cent in 1998–9 in the case of rice. In the case wheat, it went up from 74 per cent in 1970–1 to 80 per cent in 1997–8, and has dropped to 49 per cent in 1999–2000. This recent decline in wheat contributions is a consequence of FCI procuring less wheat, and it is procuring less because of growing stocks which are caused by the poor buying less food. Hence the growing food stocks are also a pseudo surplus in the sense that they reflect the decreasing purchasing power and entitlements of the poor.

This high cost system was kept afloat with World Bank loans for government subsidies. Now that World Bank is dismantling the public support and subsidy system, the Green Revolution model of short-term food security is coming apart, and is being replaced by the corporate model of agriculture.

As a matter of fact, the Public Distribution System (PDS) was the subsidized food system that allowed food produced at high costs through Green Revolution technologies to reach consumers at low prices. But part of the World Bank reform package was dismantling the PDS to reduce Government expenditure. However, Government expenditure on food subsidies has gone up to Rs. 9,300 crore in 1999–2000 from Rs. 5,166 crore in the mid-1990s. Meantime, offtake has declined dramatically—from 8.52 million tons wheat in 1996–7 to 4.99 million tons in 1999–2000. People are buying less because the food is costlier. In the 2000–1 budget, prices of staples were increased.

The high support price given to farmers is often cited as the primary reason for the rising food prices and hence decreasing offtake from the PDS. The central issue price for wheat, which had increased from Rs. 450 per quintal in January 1995 to Rs. 682 per quintal on first April 1999, shot up again to Rs. 900 per quintal, a 100 per cent of increase over a one year period. However, the support prices are high because costs of production are high in chemical intensive agriculture. In many crops, the costs have become higher than the support price and farmers have been pushed into a negative economy reflected in debts, suicides and kidney sales. The distortions in our food system cannot be removed without addressing the economic implications of capital and chemical intensive agriculture for farmers and consumers.

In today's context, our biggest imports are pulses and edible oils (Table 6.7). The mechanical calculus of competition would suggest this is because we do not have a 'comparative advantage' in pulses and oilseeds.

However, in the pre-Green Revolution period, pulses and oilseeds were exported from India and India was the biggest producer of pulses and oilseeds (Table 6.8). The decline in pulses and oilseeds was a direct result of the Green Revolution, which reduced their cultivation both through monocultures and multiple cropping of rice and wheat. In Punjab alone, the area under pulses went down from 13.38 per cent to 3.48 per cent and the area under oilseeds went down from 6.24 per cent to 2.93 per cent during 1966–7 to 1985–6.

Table 6.7: Commodities Composition of India's Agricultural Imports (1990–1 to 1994–5)

($ million)

Commodity	1990–1	1991–2	1992–3	1993–4	1994–5	5 Years Average	Per cent Share
Pulses	263.8	104.3	115.5	180.8	182.7	169.42	19.60
Sugar	5.3	0.3	0.1	0.1	715.5	144.26	16.69
Cashewnuts	73.8	109.0	129.9	153.9	215.0	136.32	15.77
Vegetable oils	179.6	101.3	35.6	53.1	194.2	117.16	13.55
Fruits and nuts	60.0	40.9	65.2	69.4	99.2	66.94	7.74
Wheat	13.5	0.0	245.2	40.1	1.5	60.06	6.95
Cereal preparation	48.5	66.3	62.9	35.0	30.2	48.58	5.62
Cotton raw & waste	0.0	0.0	74.8	5.9	161.8	48.50	5.61
Raw hides and skins	0.0	18.8	21.7	30.7	40.4	22.32	2.58
Natural rubber	42.5	12.1	16.1	17.5	7.9	19.22	2.22
Rice	21.8	4.5	25.3	17.6	2.7	14.38	1.66
Spices	0.0	0.0	0.0	24.1	17.6	8.34	0.96
Milk and cream	1.9	3.3	15.5	5.3	1.9	5.58	0.65
Oilseed	3.6	3.9	3.7	2.2	1.8	3.04	0.35
Vegetable and animals fats	0.3	0.4	0.5	0.5	0.7	0.48	0.06
Total	714.6	465.1	834.0	636.2	1673.1	846.6	100.0

Source: Economic Survey, Ministry of Finance.

Today's so called 'comparative advantage' is thus based on yesterday's mistakes arising from a focus on a chemically intensive, a monoculture-based agricultural policy which focused only on wheat and rice. Tomorrow's scarcities will similarly be linked to producing for global markets and forgetting that one billion Indians cannot be provided for by any country if we stop producing our own food.

PRICE POLARIZATION: LOW FARM PRICES, HIGH FOOD COSTS

The consumers do not gain from low farm prices, contrary to the claims made by promoters of globalization. The Government's booklet on Quantitative Restrictions (QRs) has stated that 'consumers benefit as they get a wider choice of goods and services at a lower cost'. But the costs of staples have gone up, not down, since policies of liberalization were introduced. As food subsidies are removed, and as price control measures are dismantled, the prices consumers are paying for food increase. In fact, food prices have systematically increased since the economic reforms were started.

The centralized Government controlled food distribution system run by the FCI. This system, dependant on long distance transport, had its flaws, but these flaws can hardly be corrected by replacing it with an 'even more centralized corporate controlled food system which promotes the export of Indian foodgrains and then imports food at

Table 6.8: Value per Ton Realized by Exports (1954–5 to 1963-4)

Crops	1954–5	1955–6	1956–7	1957–8	1958–9	1959–60	1960–1	1961–2	1962–3	1963–4
Pulses	379	407	584	477	790	498	726	784	841	806
Cashew kernels	766	488	834	695	600	745	812	658	587	-
Walnuts	2,930	2,193	2,603	2,452	2,452	2,965	3,160	2,759	2,495	2,648
Onions	257	264	301	344	261	257	213	256	248	283
Garlic	671	784	977	821	1,117	1,113	652	935	1,125	744
Potatoes	N.E.	N.E.	N.E.	522	397	439	383	426	418	407
Sugar*	-	-	-	809	496	489	454	496	397	675
Pepper	3,622	4,992	2,263	2,043	2,058	3,886	4,939	3,737	3,152	3,112
Chillies	1,806	2,157	1,831	1,740	1,225	2,200	2,402	1,875	1,940	2,317
Ginger	4,542	2,311	1,683	929	600	2,100	1,566	1,434	1.975	2.904
Tobacco	2,662	3,017	2,883	3.574	3,180	3,422	3,191	3,159	2,987	3,315
Raw cotton	2,349	2,962	2,821	2,231	1,866	2,209	2,653	2,181	2,085	2,170
Raw wool	6,486	6,150	6,734	6,599	5,649	5,761	5,618	5,785	5,522	5,992
Groundnuts	977	1,084	439	266	848	857	1,019	1,042	998	1,044
Rape & Mustard seed	668	800	N.E.	N.E.	962	913	680	1,143	1,222	1,096
Sesamum seed	939	N.E.	N.E.	N.E.	870	960	1,505	1,102	1,218	1,143
Linseed	1,000	N.E.	1,000	N.E.	1,000	800	500	1,000	1,000	1,000
Groundnut oil	1,237	1,257	1,501	1,750	1,393	1,366	2,129	1,516	1,356	1,384
Rape & Mustard oil	1,556	1,861	2,098	2,370	2,117	2,223	2,313	2,608	2,661	2,600
Sesamum oil	1,358	N.E.	1,655	1,000	1,522	1,677	2,714	3,167	2,667	1,600

(contd.)

Table 6.8 (contd.)

Crops	1954–5	1955–6	1956–7	1957–8	1958–9	1959–60	1960–1	1961–2	1962–3	1963–4
Linseed oil	1,147	1,125	1,657	1,000	1,394	1,392	1,291	2,033	1,721	1,406
Castor oil	1,038	1,080	1,552	1,768	1,463	1,406	1,436	1,606	1,424	1,358
Groundnut cake	368	432	380	279	351	386	357	361	390	377
Sesamum cake	N.E.	N.E.	330	291	331	305	284	314	328	342
Linseed cake	322	309	359	295	375	401	342	383	364	358
Defatted groundnut meal	N.E.	N.E.	302	1,176	335	366	333	350	360	380
Lac	4,038	3,907	3,262	2,541	2,190	2,331	2,341	2,198	2,284	2,238
Lemongrass oil	16,349	17,502	14,535	10,741	7,807	11,869	17,455	25,937	21,197	15,931

Notes: *The figures for 1957–8 relate to the calendar year 1957, and so on; N.E. = nil or negligible exports.
Source: Report of the Study Group (Export Sector) on Agricultural Commodities and Agriculture Industries, Ministry and Commerce, February 1965.

high social economic and political cost to people and the country'. If the Punjab model based on the Green Revolution was flawed as a food security model because it necessitated waste of energy for transportation and public expenditure for a centralized procurement and distribution system, then the Pepsico, Monsanto and Cargill model is even more flawed because it is adding to the 'food miles' and leading to even higher levels of concentration and centralization. The 'competitiveness' of this globalized agriculture is not related to more efficient use of scare resources but to subsidies and externalization of environmental and social costs.

Thus, while Pepsico pays only Rs. 0.75 per kg. to Punjab farmers for growing tomatoes, exporters like Pepsico receive Rs. 10 as subsides for transport. Without these subsidies, non-local supply of food controlled by Transnational Corporations (TNCs) and produced with capital-intensive methods would not be able to displace local food production produced sustainably with low external inputs.

Food security is neither dollar in the pocket, nor food in the elevators of Cargill. Food security lies in ecologically resilient and economically efficient farming systems which provide livelihoods to farmers and self-sufficiency in food at the household, community, regional and national levels, while providing safe and nutritious food to consumers.

Trade liberalization is generating food insecurity at four levels. It is leading to:

— a transfer of resources from peasants to industry;
— a shift of land use from the production of staple foods to luxury and non-food crops (cash crops) such as shrimp and flowers for exports;
— diversion of cereals from domestic markets to exports creating domestic scarcity and price rise;
— a removal of food subsidies, thus lowering domestic consumption, and increasing food exports.

As a result of trade liberalization policies cereal exports have increased from 12 per cent of total agricultural exports in 1991–2 to 27 per cent in 1995–6.[4] The per capita cereal consumption has declined from 17 kg. per capita per month in the 1950s to 13.5 kg. per capita per month in the 1990s. If the right to food has be fully met the Public Distribution System needs to increase by three to four times its present allocation. Instead, food subsidies are being reduced in the name of targeting the poor. While the Targeted Public Distribution System (TPDS) system is justified on the basis of reducing costs, the costs will actually go up from Rs. 51.66 billion under the present PDS to Rs. 76 billion in the TPDS.

Besides increasing food costs for the poor by removing food subsidies, globalization is also polarizing the prices farmers receive and the prices consumers pay. While farm prices are collapsing under the joint impact of removal of domestic support and dumping of artificially cheap agricultural produce, retail prices are increasing due to vertical integration of trading systems and removal of price control.

Under the 'globalization, privatization and liberalization' agenda of the World Bank, World Trade Organization (WTO) and International Monetary Fund (IMF), trade is more unfair than free. Ultimately, it is the poorer sections in every society that are the end-losers. Over the years, there has been an increasing asymmetry between world prices and domestic prices. Index numbers of wholesale prices and consumer prices in India for almost all commodities have increased by over two times between 1990–1 and 1997–8. Over the same period of time, world commodity prices have been on the decline. With the removal of import restrictions, low commodity prices supported by high subsidies are destroying domestic production and livelihoods of the poor. While the incomes of the poor go down, their expenditure is rising due to rising consumer prices.

This growing gap between domestic consumer prices and international commodity prices confutes the claims of the free trade propagandists that the gains from unrestricted trade shall percolate down to consumers in the form of lower prices and high quality. Corporations gain by low commodity prices, people lose with rising costs of essential commodities. That the asymmetric response of domestic commodity prices to world prices is caused by the behavior of international trading companies and is not the result of trade restrictions or bidding processing costs is shown by Morisset (Appendix 1). This has cost the commodity exporting countries over $100 billion a year by limiting the expansion of the final demand for these products in the major consumer markets.[5]

Thus, while consumers in India are paying more for their sustenance, at the macro level, the country is getting less for what it exports, particularly in the face of a sharply declining currency value (Figure 6.6). The World Bank commodity price index (Figure 6.7) shows that the non-fuel commodity price index declined by 45 per cent in constant prices between 1980 and 1998. At the same time, the Manufactures Unit Value (MUV) index rose by 44 per cent over the same period of time. The adverse terms of trade are increasingly reflected in India's worsening trade deficit, which rose from Rs. 10,635 crore in 1990–1 to Rs. 34,495 crore in 1998–9 (Figure 6.9).

The polarization in domestic consumer prices and world prices and the surge in imports with the removal of quantitative restrictions imply that consumers in India end up paying more than they would, had there been no trade. The choices of both farmers and consumers decrease as global corporations increase their control over food and agriculture.

The lack of choice of farmers under conditions of corporate control is well illustrated by the story of 'farmer Jones' dealing with fifteen different companies in fifteen different industries, all of which were owned by one company, Cargill. The story of Jones is the story of any family farmer in USA (Duer, 1997):

> Farmer Jones', a family farmer trying to make ends meet by growing wheat and raising a few head of cattle and some chickens. This is how his year went. Jones bought his wheat seed and his cattle and chicken feed from Nutrena Corporation. His old tractor bit the dust last year, so he had to go to the Bank of Ellsworth and get a loan for a new tractor that he bought from Waycrosse Inc. Jones had another bit of bad luck: his irrigation system fell apart. He called his insurance company, the Horizon Agency, but the agent said it wasn't covered. He has no choice but to buy a new system from Venture Sprinkler Inc. Jones then bought his fertilizer from Cargill, and his wheat was under way. At harvest time, the prices were low, so Jones decided to hold his crop in storage. He shipped his wheat by rail, to the Heinhold Elevator Company. Eventually, Jones sold half of his crop to a Panamanian company called Tradax and the other half to a domestic milling operation owned by the Burrus Company. The Tradax half was shipped on a barge owned by Cargo Carriers Inc., to a giant port elevator owned by Producer Marketing. A few days later, it was loaded on a ship owned by Rogers Shipping and transported to Europe.

Meanwhile, Jones' animals were ready for market. He sold the cattle to a freedlot owned by Caprock Industries, which later sold them to Cargill's meatpacking plant. He sold the chickens to Dean Farms. Unfortunately for farmer Jones, the prices he received were disappointing. By the time he subtracted out the cost of the seed, feed, chemical fertilizer, tractor, sprinkler system, elevator storage, rail transportation, and insurance, he'd lost money and couldn't pay the bank, which, therefore, repossessed his land.

Jones was depressed. He couldn't figure out why the companies wouldn't pay him enough for his products to keep him in business. He thought about suicide, but then he heard about the one-world conspiracy and joined the antigovernment movement instead.

If we looked at this situation as outsiders, we would probably reach the conclusion that farmer Jones had some bad luck and probably was a poor businessman. We would think Jones latched onto a conspiracy theory in his effort to scapegoat someone to alleviate his guilt. What we probably wouldn't consider is that farmers Jones did some investigating on his own and made a startling discovery. The fifteen different companies in the fifteen different industries to which Jones had paid out money during the year were all owned by one company: Cargill. And that is the same company that paid him such a low price for his products and eventually repossessed his land.

The very real companies named in the farmer Jones illustration are just a few of those in Cargill's vertical integration. The company owns many more subsidiaries in still more industries, and Cargill is not unique. The other grain giant are equally integrated.

The myth of consumer choice is equally dramatic. In regions of high cultural diversity of food, free trade is wiping out people's preferred foods as illustrated by the case of edible oil imports to India. That is why women in Delhi dumped soya oil to save their mustard and women in Kerala dumped palm oil to protect their coconut.

Free trade in agriculture in effect establishes a system of total control in which a handful of corporations can fix prices, control production patterns and consumption patterns, and destroy people's basic freedoms as producers and consumers.

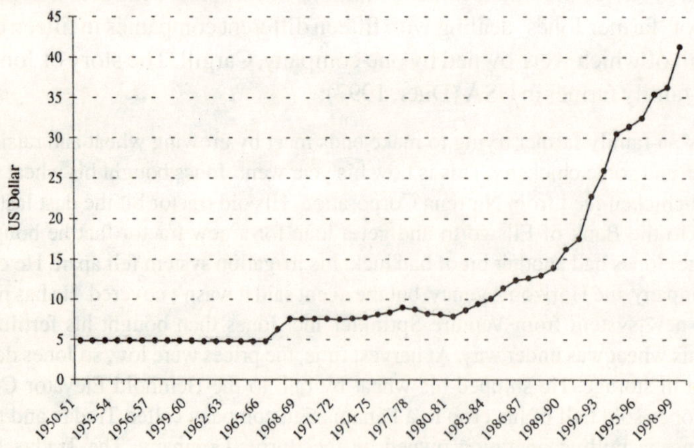

Source: Economic Survey 1998–9, Ministry of Finance.

Figure 6.6: Trends in Exchange Rate of the Rupee against the US Dollar (1950–1 to 1998–9)

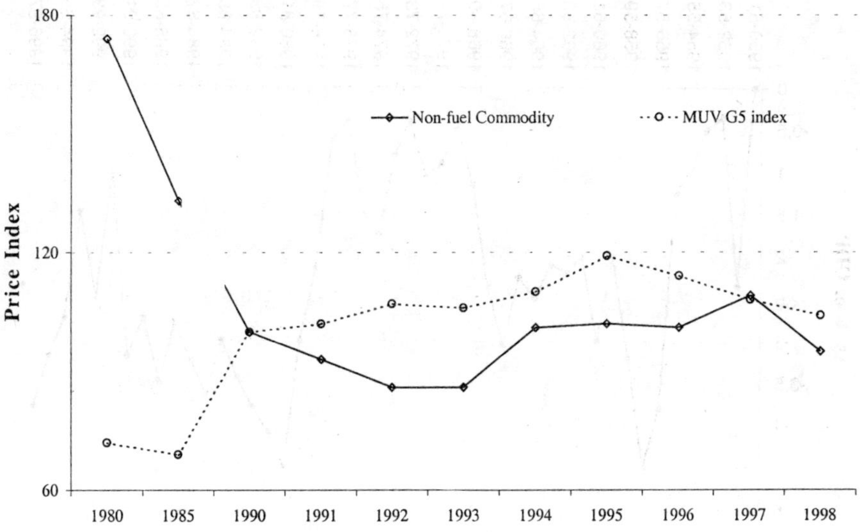

Source: World Bank website.

Figure 6.7: Trends in World Commodity Prices (1980–98).

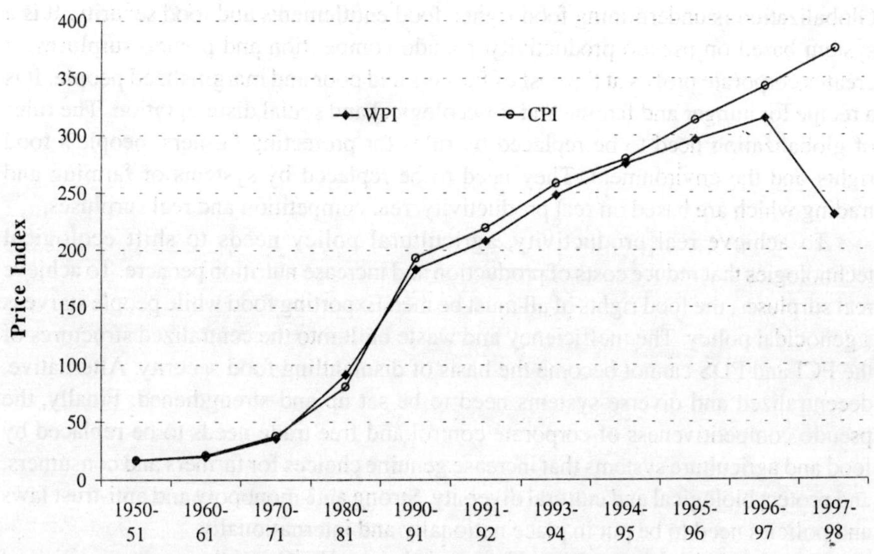

Note: WPI = Wholesale Price Index; CPI = Consumer Price Index.

Source: Government of India, *Economic Survey, 1998–99*, Ministry of Finance, Delhi.

Figure 6.8: Trends in India's WPI and CPI (1950–1 to 1997–8).

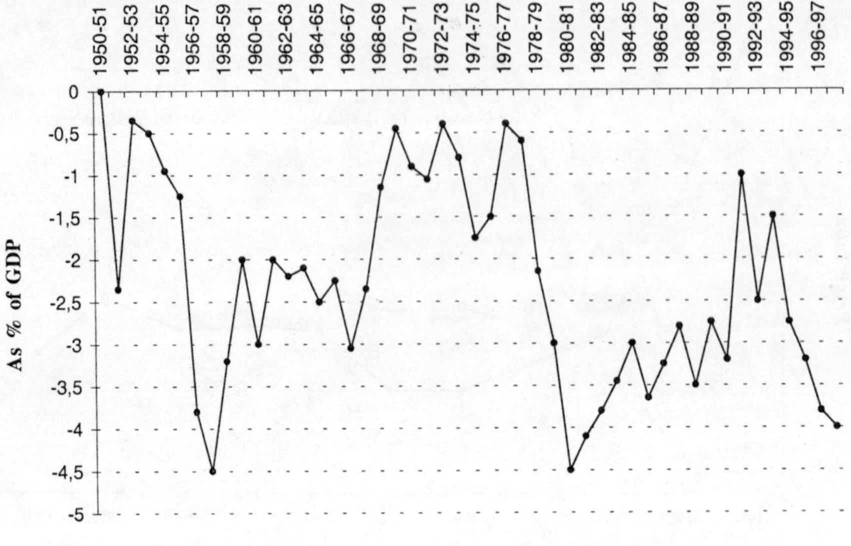

Source: CMIE, July 1999.

Figure 6.9: Trend in India's Trade Deficit (1950–1 to 1996–7) (as percentage of GDP).

CONCLUSION

Globalization is undermining food rights, food entitlements and food security. It is a system based on pseudo productivity, pseudo competition and pseudo surpluses. It creates corporate profits at the cost of farmers and poor and marginalized people. It is a recipe for hunger and famine and for ecological and social disintegration. The rules of globalization need to be replaced by rules for protecting farmers, people's food rights and the environment. They need to be replaced by systems of farming and trading which are based on real productivity, real competition and real surpluses.

To achieve real productivity, agricultural policy needs to shift ecological technologies that reduce costs of production and increase nutrition per acre. To achieve real surpluses, the food rights of all must be met. Exporting food while people starve is a genocidal policy. The inefficiency and waste built into the centralized structures of the FCI and PDS cannot become the basis of dismantling food security. Alternative, decentralized and diverse systems need to be set up and strengthened. Finally, the pseudo competitiveness of corporate control and free trade needs to be replaced by food and agriculture systems that increase genuine choices for farmers and consumers, and protect biological and cultural diversity. Strong anti-monopoly and anti-trust laws and policies need to be put in place nationally and internationally.

A major shift in perception is also needed about WTO and the implementation of its rules and disciplines. Was the removal of QRs inevitable on WTO disciplines, as the Government has argued? No, WTO cannot force a country to remove restrictions on imports. After all, the Europeans have refused to import hormone beef from the

USA in spite of a WTO dispute ruling and in spite of US trade sanctions. The worst that would happen if we maintain QRs is that USA would try and introduce trade sanctions. They have tried that before and it has not hurt us seriously. The best candidate for trade sanctions would be Information Technology (IT) exports, but the USA needs India's IT even more than we do. In any case, no trade sanction could do more economic damage than the destruction being caused to India's economy by the removal of QRs. Implementation of WTO rules has become more costly than non-implementation.

It is therefore imperative that we bring back QRs. If Europe and the USA can impose QRs according to their domestic interests, why shouldn't India? General exceptions to WTO disciplines are in any case allowed in GATT. Article XX of the GATT on general exceptions states that nothing in the WTO Agreement shall be construed to prevent the adoption or enforcement of measures by countries necessary to protect human, animal or plant life or health. The farmers suicides, and the growing hunger among the poor, show that removal of import restrictions and economic reforms have started to threaten the very survival of Indian farmers and poor consumers. The Government has an obligation under the Indian Constitution and a right under WTO to act in defense of the protection of the right to life and livelihood of Indian peasants and the Indian people.

Article XX also allows countries to take action, in exception to WTO rules, relating to the conservation of exhaustible natural resources if such measures are made effective in conjunction with restrictions on domestic production or consumption. Dumping of artificially cheap agricultural products is wiping out our agricultural biodiversity—our mustard and sesame, our coconut and linseed. Conservation of biodiversity is a national and international obligation under the Convention on Biodiversity. The General Exceptions Clause should be invoked to prevent the destruction and erosion and extinction of biodiversity, the most significant natural resource for farming.

Article XI of GATT on Security Exceptions allows us to take steps for the defense of peace and security. Food and livelihood security is at the core of our national security. Defending is the Government's obligation and a national right. The real security threat to the country is not a war—it is the dismantling of our food security and the potential displacement of three quarters of India. A farm leader recently said, '*Jab tak kisan ke hath me kudali hai, who shanth hai. Agar kudali cheeni, who bantook tham lega*' (As long as farmers have their (sickle) livelihoods, they will be peaceful. But if their sickle is snatched from their hands, it will be replaced by a gun).

The Government has no excuse to unleash a war against its own farmers. This is the time to put food security and survival of farmers as the most urgent concern. The price we will pay for not removing QRs pales into insignificance in the face of the price we are paying by removing QRs. Alternatives are available that the government must use. In any case, India is fully justified to take measures to protect its national security, its environment and the lives of its people under Article XX and XI of GATT. These are also duties of the state under our Constitution. The largest democracy must exercise its democratic rights and duties in the defense of its food security and the survival options of its people.

Preventing the destruction of our agriculture and our farmers will be the most severe test for our functioning as a democracy and as a free economy. If we fail this test, we will be wiped out as a society, as a civilization. Far more is at stake in agriculture than trade issues. Agriculture is the expression of our civilization, the basis of our economy, the foundation of our cultural diversity and ecological survival. The defense of our rural communities, our soil and water, our biological and cultural diversity, is the real defense of the country since on it rests the real security of our people. This security agenda must become the national security agenda if India is to survive.

APPENDIX

THE HAND BEHIND DISTORTED PRICES

(Morisset, World Bank, April 1997)

Behind the polarization of domestic consumer prices and world prices is the presence of large trading companies in international commodity markets. The strategic position of these companies between buyers and sellers allows them to influence the transmission of world prices. These companies generally provide information, define the terms of transactions, manage the payments and record keeping for transactions, and also figure out ways for clearing the market. However, without competition they may follow a pricing strategy that will maximize their profits and not those of producers and consumers.

In addition to their trading activities, many companies are vertically integrated and thus close to production. For example, Cargill—the world's largest trading company of cereals—owns plantations, storage facilities, and vessels in many countries around the world. Similarly, Exxon carries out not only mining and refining but also a complex set of activities involving distribution, transportation, inventories, and pricing. Where most of these companies are involved in wholesales—transactions between businesses—there are also many examples in which they also act in the retail sector either directly or indirectly through strategic alliances or intermediary arrangements.

Preliminary evidence indicates that large trading companies have been capable of influencing the transmission of world commodity prices to domestic prices. This is first suggested by the concentration of trading activities in few companies worldwide. UNCTAD has reported that six or fewer trading companies control about 70 per cent of the total international trade, thus obviously limiting the choice of producers and consumers in these markets. As an example, the banana export market is dominated by Del Monte, United Brands, and Standard Fruits, and the wheat export market by Cargill, Continental, Andre, Dreyfuss, and Bunge Born. The suspicion that these companies use their dominant position to control prices is strengthened by the chronic absence of information on their activities. While many people can name retailers, few know wholesalers. These companies are often larger than the economies of many developing countries. For instance, the sale volume of the world's largest trading company, C. Itoh, was as big as Argentina's GDP in 1998. The same company also traded over US$ 20 billion of agricultural products—as much as all the sugar, coffee, beef, rice, and wheat exported by all developing countries at that time.

The trading companies' position of influence on the world market is further implied by the correlation between the variations in the spreads and the variations in the profits of the trading companies. Another indication of the correlation is that the markup in the wheat market grew by 50 per cent over the past two decades, while the sales of Cargill, the world's largest trader of wheat, saw a five-fold increase during this period. In a historical perspective, it is suggestive that this firm has recorded an annual loss in only 3 of its 130 years of existence: 1921, 1936 and 1938.

NOTES

1. Debu Bandopadhyay, 'Food Security and Liberalization', in V. Shiva (ed.), *Globalization and Food Security*, Sage, New Delhi (forthcoming).
2. Brochure 'Feeding the 8 Billion and Preserving the Planet', International Association of Plant Breeders (ASSINSEL), Chemin du Reposoir 7, 1260 Nyon, Switzerland.
3. Seminar 'Sustaining Diversity and Cultivating Diversity', Navdanya, New Delhi, 1993.
4. Speech by Satyamurthy at the workshop on '*Food Security*', Ahmedabad, 6 April 1998.
5. Morisset, World Bank, April 1997.

REFERENCES

Antony, Margaret, M.Phil. Thesis, Centre for Development Studies, Trivandrum, 1998.

Bandopadhyay, Debu, 'Food Security and Liberalization', in V. Shiva (ed.), *Globalization and Food Security*, Sage, New Delhi (forthcoming).

Bray, Francesca, *Agriculture for Developing Nations*, Scientific American, July 1994.

Dyer, Joel, *Harvest of Rage. Why Oklahoma City is only the Beginning*, Westview Press, Boulder, 1997.

FAO, *The State of Food and Agriculture. Lessons from the past 50 years*, Food and Agricultural Organization of the United Nations, Rome, 2000.

———, *Women Feed the World,* Food and Agricultural Organization of the United Nations, Rome, 1999.

Gulati, Ashok, Anil Sharma, Kailash Sharma, Shipra Das, and Vandana Chaabra, *Export Competitiveness of Selected Agricultural Commodities*, National Council of Applied Economic Research, New Delhi, September 1994.

Heywood, V.H., R.T. Watson (eds.), *Global Biodiversity Assessment*, Cambridge University Press, 1995.

Khusro, A.M., *An Analysis of Agricultural Land in India by Size of Holding and Tenure*, Institute of Economic Growth, Delhi, 1962.

Rees, William E. and Mathis Wackernagel, *Our Ecological Footprint. Reducing Human Impact on the Earth,* New Society Publishers, Cabriola Island, 1998.

Rosset, Peter and Miguel Altieri, 'The Productivity of Small-Scale Agriculture', Mimeo, Berkeley, 1999.

Sen, A.K., *Size of Holding and Productivity, The Economic Weekly*, Annual Number, February, 1962.

Shiva, Vandana, *The Violence of Green Revolution. Third World Agriculture, Ecology and Politics*, Malaysia and the Other India Book Store, Goa, 1991.

———, *Monoculture of the Mind. Biodiversity, Biotechnology and the Third World*, Third World Network, Penang, Malaysia, 1993.

——— (ed.), *Sustaining Diversity. Renewing Diversity and Balance through Conservation*, Research Foundation for Science Technology and Ecology, New Delhi, 1994.

———, 'Biodiversity-based Productivity', in Peter Rosset and Miguel Altieri, *The Productivity of Small-Scale Agriculture*.

Thrupp, Lori Ann (ed.), *New Partnerships for Sustainable Agriculture*, World Resources Institute, Washington, DC, 1996.

UNDP, *Agroecology: Creating the Synergisms for a Sustainable Agriculture*, United Nations Development Programme. New York, 1995.

WTO, *Trade Policy Review India*, No. WT/TPR/S/33WTO, World Trade Organization, Geneva, 1999.

Yegna Iyengar, A.K., *Field Crops of India*, BAPPCO, Bangalore, 1944 (reprinted 1980).

7

The Product-Specificity of Indian Input Subsidies

SCOPE AND EFFECTS ON EQUITY AND COMPETITIVENESS

BRUNO DORIN, THOMAS JULLIEN

INTRODUCTION

Since the 1960s and the Green Revolution, India clearly promotes the growth of its agricultural production with input subsidies, mainly to irrigation, to electricity (for pumping groundwater) and to fertilizers. This support to agriculture is not the only one, but many agree that it represents a major and growing part of governmental expenditures. However, since these subsidies are both explicit and implicit (unrecovered costs of public provisions), it is rather difficult to evaluate them. The first attempt of this paper is to bridge this information gap pertaining to a rather long period of the post-Green Revolution era, i.e. from 1980–1 to 1995–6, with the help of previous Indian works on the matter (mainly Gulati et al., 1989, 1995, 2000), and keeping in mind that this kind of exercise always relies on some debatable hypotheses.

The second attempt of this paper is to quantify the actual allocation of these three input subsidies among Indian crops and states (8 crops and 14 states in practice because of statistical gaps) in order to challenge their 'non-specificity'. Indeed, these input subsidies are known to be—in India as well as in the GATT's Agreement on Agriculture (1994)—a 'non-product specific' support to agriculture (unlike 'Minimum Support Prices' for example). In other words, they are not presumed to favour any one crop or region more than another. Consequently, they appear not only to secure a rather equitable income-transfer to regions and farmers—which is a crucial point in India to bypass various market failures—but also to have much less trade-distorting effects than 'product specific' measures, within the borders as well as outside.

Before presenting one's own conclusions on the matter, this study will first detail one's assumptions and calculations by type of input subsidies.

SUBSIDIES TO IRRIGATION

In India, central and/or state governments support agricultural irrigation through three major items (five-year plans or non-plan outlays):

(1) financial expenditures for providing new facilities, within the framework of 'major and medium projects' (dams, canals) or of 'minor projects' (wells, tanks),
(2) financial expenditures to run the Command Area Development (CAD) introduced in 1974–5 for improving the use and the benefit of the already existing major or medium facilities (land leveling, field channel, *warabandi*, etc.),
(3) working expenses for major and medium projects (maintenance and repairs) as well as for minor ones (maintenance of tanks, deepening of wells, etc.).

The cost of these items which are not recovered by central and state 'gross receipts' coming from irrigation can be regarded as a subsidy. However, as far as the first item is concerned, many problems arise in estimating the per hectare and the per year cost of this invested capital (what is the gestation lag between expenditure incurred and potential created? the redemption yield? etc.). Moreover, if one applies Gulati's hypotheses (1989) to the database, it leads to very high estimates for major and medium projects: Rs. 11,677 per hectare of potential created in 1995–6 at all India level (Rs. 1,576 for minor projects). In order to avoid any controversies on the issue, it was then decided to drop this capital cost from all the calculations, for major and medium projects as well as for minor ones. But to compensate a little bit this large underestimation of irrigation subsidies, particularly in the states which enjoy many canals, CAD financial expenditures were not depreciated, and fully charged there annual amount as if they were annual working expenses. This is of course a debatable choice, especially since in Punjab there is no outlay for CAD, but still a large invested public capital: Rs. 208.9 crore during the Sixth Plan (1980 to 1985), Rs. 220.2 during the Seventh Plan (1985 to 1990) and Rs. 257.7 crore during the Eighth Plan (1992 to 1997).

In this study, subsidies to irrigation have therefore even computed as the sum of annual[1] expenditure on CAD programs[2] and of annual working expenses on irrigation,[3] minus annual gross receipts. The annual estimate obtained in billion rupees for each state from 1980–1 to 1995–6 can be itself divided by the prevailing state gross irrigated area in order to get a state average subsidy per ha. It is then easy to multiply the latter by the prevailing gross irrigated area of each crop in order to get an amount of subsidy per crop and state. Similar and other calculations can be done at all-India level, as detailed in Table 7.2a along with an example of calculation at the state and crop levels (rice in Andhra Pradesh). Crop-wise and state-wise results are displayed in Table 7.3 (billion rupees) at current prices, and then in Table 7.4 (billion rupees), Table 7.5 (rupees per ha. of gross cropped area), Table 7.6 (rupees per cultivator) and Table 7.7 (rupees per ton) at 1995–6 constant prices.[4]

SUBSIDIES TO ELECTRICITY

In India, the State Electricity Boards (SEBs) supply electricity to farmers at tariffs which are much lower than the ones which prevail for other consumers (it can even be free). This incentive to irrigation (mainly to groundwater irrigation) varies from state to state according to local politicians as also to the magnitude of cross-subsidization with other sectors (electricity is usually charged above its average cost to industry, commercial users and railway traction). Globally, it is an important grant to agriculture: between 1992–3 to 1997–8, 'while the agricultural sector accounts for nearly one-third of the sales of the SEB and Electricity Departments, the sales revenue realized from this sector has been only 4 to 5 per cent of the total revenue realization' (Planning Commission, 1997: 29).

However, a more precise balance sheet can be drawn up if one multiplies the discount given to farmers (price paid by them minus the cost for supplying power to all categories of consumers) by the agricultural consumption of electricity during the year and in the state concerned. Unfortunately, this option raises a problem: yearly tariffs and costs published by the SEBs can hardly be challenged, which is not so evident for the 'Consumer category-wise sale of power'. Indeed, and to be brief, since there is no meter in the fields to measure the agricultural consumption of power (farmers are billed according to the power of their pump), the latter is estimated by the SEBs as the electricity not consumed by other sectors, minus what it is called 'technical losses' (transmission and distribution failures). The problem is that 'non-technical losses' should also be taken into account since discount tariffs to agriculture lead to many unofficial connections from other sectors. As a matter of fact, in states like Orissa, Haryana and Andhra Pradesh where structural reforms of the SEBs are under way, it is now recognized that real losses range from 40 per cent to 50 per cent for all sectors put together (Salgo, 1998), while official estimates put them respectively at 24, 25 and 19 per cent of the available electricity (generated and purchased) in these states. As far as the present study is concerned, we considered that 45 per cent was the average percentage of total losses in India, and that all these losses pertain to the agricultural sector, which leads to the conclusion that the agricultural consumption of electricity mentioned by the SEBs has to be cut, on an average, by 30 per cent.

In this study, the annual electricity subsidy per unit sold (Rs./kWh.) is therefore the difference between the average unit cost[5] of power supplied and the consumer category-wise average tariff as given (state-wise as well as for all-India) in the various issues of the 'Annual Reports on the Working of State Electricity Boards and Electricity Departments' (Planning Commission). To work out the agricultural subsidies to electricity in billion rupees, this annual per unit subsidy is multiplied by 70 per cent of the annual agricultural consumption estimated by the SEBs in million kWh. per hour. The share of each crop in the gross irrigated area in the respective states and years has then been used to share out the subsidies per crop. Data used and results obtained for all-India as well as for rice in Andhra Pradesh are detailed in Table 7.2b. Crop-wise and state-wise results are displayed in Table 7.3 (billion rupees) at current prices, and then in Table 7.4 (billion rupees), Table 7.5 (rupees per ha. of gross cropped area), Table 7.6 (rupees per cultivator) and Table 7.7 (rupees per ton) at 1995–6 constant prices.

SUBSIDIES TO FERTILIZERS

Subsidies to the domestic production and consumption of nitrogenous (N), phosphatic (P_2O_5) and potassic (K_2O) fertilizers are an important and growing expenditure in the Union budget. Just in 1998–9, they reached Rs. 3.33 billions for imported fertilizers (1.54 million tons of potassic fertilizers, 0.97 and 0.64 respectively for phosphatic and nitrogenous fertilizers), while those to domestic fertilizers and to decontrolled fertilizers (phosphatic and potassic fertilizers since 1992) reached respectively Rs. 74.73 billion and Rs. 37.90 billion, which brings the total up to Rs. 115.96 billion (Ministry of Finance,

2000: 140). These subsidies to fertilizer 'mainly represent the difference between farm sgate price (excluding dealer's margin and sales tax) and Retention Price (inclusive of equated freight) for indigenous fertilizers, and between farm gate price and c.i.f. price plus pool handling charges for imported fertilizers' (Acharya, 2000).

If the above view is rather clear and well shared among economists, a difficult question remains to be answered: how are these subsidies distributed among domestic farmers and domestic industrialists? Without taking into account the so-called 'ad-hoc subsidy' on decontrolled fertilizers which is enjoyed by industrialists since 1992–3, Gulati and colleagues worked out very different figures from 1980–1 onwards (Gulati, 1989; Gulati et al., 1995, 2000). In the latest work, they assumed that farmers' share in fertilizer subsidy was to the tune of 25 per cent in the triennium ending 1983–4, and then increased up to 132 per cent in the triennium ending 1995–6. The authors' annual estimates, which depend on many variables and hypotheses[6] like those of Gulati and colleagues, start at nearly 90 per cent in 1980-1, end at nearly 180 per cent in 1995–6, with a very clear collapse (sometimes negative percentages) between 1982–3 and 1987–8, and a clear increase after 1991. Between these two studies, the present study chose the following middle path: the share of subsidies given to farmers was to the tune of 33 per cent up to the drought of 1987, then increased yearly by 25 per cent up to the reform of 1991, to finally represent 100 per cent of the budget expenditure on fertilizers (without 'ad-hoc subsidy') from 1992–3 onwards. These may appear rather low estimates for the end of our period, but they tally with what India notified to the WTO for the year 1995 (G/AG/N/IND/1, 17 June 1998). Moreover, after a share as high as 132 per cent in the triennium ending 1995–6, Gulati and colleagues assume that there was then a declining trend. The fact is that nobody really knows what is the exact picture in a field which absorbs billions of rupees every year.

Another opaque field in India is the per crop consumption of fertilizers. To ones knowledge, no detailed databases exist or are yearly centralized and systematized. The FAO itself, in its international publication *Fertilizer Use by Crop*, is not able to give better figures for India than its own all-India very rough estimates for the years 1989, 1990 and 1995. It is therefore assumed that the 'fertilizers quantity (kg. nutrients) per ha.' mentioned in the *Cost of Cultivation of Principal Crops in India* or in the *Reports of the Commission for Agricultural Costs and Prices* published by the Indian Ministry of Agriculture is the best source of information, even if this database is very incomplete from crop to crop, state to state, or year to year. In order to work out crop-wise and state-wise estimates of the fertilizer consumption from 1980–1 to 1995–6, we also assumed that the state consumption of fertilizer is better correlated with the annual crop outputs than with the annual gross cropped area of each crop. This assumption has of course its limits, especially as the production of each crop has to be pondered before by a fixed coefficient since, for example, one kilo of wheat does not equal to one kilo of sugarcane in term of fertilizer consumption. However, after many trials, estimates could be worked out to reasonably match the 'fertilizers quantity (kg. nutrients) per hectare' published by the Ministry of Agriculture for some crops, states and years. The best results are obtained if one considers that 80 per cent of the states consumption of fertilizers is used to grow either barley, cotton, groundnut, maize, paddy, rapeseed/mustard, sugarcane and/or wheat. Table 7.1 displays all-India averages of our yearly results for 8 crops (lines 3, 4 and 5) as well as other averages (lines 1 and 2: Ministry of Agriculture) or estimates (line 6: FAO) at the all-India level.

Consequently, in this study, the yearly Union budget expenditure on imported and endogenous fertilizers (Ministry of Finance, various issue) is firstly multiplied by the estimated share going to farmers,[7] and then divided by the corresponding all-India combined consumption of N, P_2O_5 and K_2O (Fertilizer Association of India, various issues) in order to get an idea of the national amount of subsidies per kg. of fertilizer from 1980–1 to 1995–6. This amount is secondly multiplied by our yearly estimates of fertilizers consumption per crop and state, and then by the gross cropped area of the corresponding crop, year and state, in order to get crop-wise and state-wise subsidies to fertilizers from 1980–1 to 1995–6. Data used and results obtained at the all-India level as well as for rice in Andhra Pradesh are detailed in Table 7.2c. Crop-wise and state-wise results are displayed in Table 7.3 (billion rupees) at current prices, and then in Table 7.4 (billion rupees), Table 7.5 (rupees per ha. of gross cropped area), Table 7.6 (rupees per cultivator) and Table 7.7 (rupees per ton) at 1995–6 constant prices.

GROWING BIASED SUBSIDIES

Six main conclusions can be drawn from the above calculations and results.

(1) In 1995–6, the three main subsidies to agricultural inputs (irrigation, electricity and fertilizers) accounted for nearly US$ 6.6 billion (Rs. 21,968 crore). This estimate, which is 16 per cent higher to what India declared to the WTO for the same year as 'Non-Product-Specific Aggregate Measurements of Support' to electricity, fertilizers and irrigation (US$ 5.6 billion or Rs. 18,885 crore) (Table 7.8), represented 2.2 per cent of the total GDP and 8.6 per cent of the agricultural GDP, while it was only 0.9 per cent and 2.7 per cent respectively in 1980–1.

(2) At constant prices, these inputs subsidies have in fact increased more than fivefold since 1980–1. After 1983–4, the trend was rather linear: +12 per cent per year, with an acceleration in 1992–3 but a break the year after (Figure 7.1). The pressure of these subsidies on the Union and states exchequers is thus increasingly heavier, leading to less and less public investments in basic infrastructures while the latter can correct regional imbalances and promote greater equity at farm levels.

(3) The subsidies to irrigation accounted for 66 per cent of the total in 1980–1 (electricity 20 per cent, fertilizers 14 per cent), but less than 30 per cent in 1995–6 (electricity 43 per cent, fertilizers 28 per cent) since after the drought of 1987, they stabilized at around Rs. 6,000 crore at 1995–6 prices. But as far as the subsidies to electricity are concerned, their total value at constant prices was multiplied by more than 10 between 1980–1 and 1995–6 (Rs. 854 to 9,461 crore), just like for fertilizers (Rs. 608 to 6,235 crore). This has undoubtedly encouraged a large-scale degradation of water resources (depletion of groundwater tables, river pollution by fertilizers, etc.) as well as of soils (water-logging, erosion, salinization and so on).

(4) In 1995–6, the all-India average of these inputs subsidies reached US$ 56 per cultivator (Rs. 1,874 as compared to Rs. 465 in 1980–1 at 1995–6 prices) and US$ 35 per hectare of gross cropped area (Rs. 1,178, nearly five times more than in 1980–1). In the

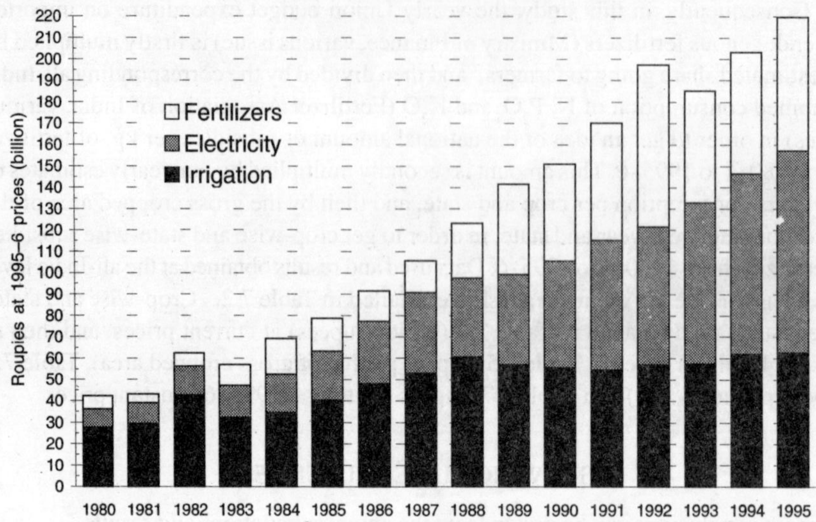

Figure 7.1: Estimated Subsidies to Irrigation, Electricity and Fertilizers (1980–1 to 1995–6).

Indian context, it undoubtedly represents an important income transfer to farmers and rural areas, especially in regions where irrigation permits the cultivation of the same plot twice or thrice a year. But if one compares these figures to those prevailing in developed countries like the USA or the EU, this domestic support remains very low even if one could have brought out highest estimates for India by following other ways to compute them.

(5) The worrying aspect of these input subsidies does not really lie in their absolute amount, but in their unfair distribution: in 1995–6, the per-cultivator subsidies ranged from 6,272 rupees in Punjab (5,009 in Haryana, 4,647 in Gujarat) where there is the highest GDP per capita (Rs. 18,527) in India, to 470 rupees in Orissa (585 in Bihar, 1,104 in Madhya Pradesh) where the GDP per capita is the lowest (Rs. 7,114) along with Bihar.

(6) In 1995–6, 61 per cent of the inputs subsidy benefited 3 crops only: rice (Rs. 6,510 crore), wheat (Rs. 5,527 crore) and sugarcane (Rs. 1,449 crore). Per ha. of gross cropped area, it leads to 3,493 rupees for sugarcane (104 US$), 2,210 for wheat (66 US$) and 1,520 for rice (followed by 1,464 for rapeseed/mustard and 1,370 for cotton). Obviously, the so-called 'non-product specific' subsidies to Indian agriculture are in fact rather 'product specific', which partially explains that India has now more and more often to deal with huge stocks of wheat and rice, while the country is highly deficient in many other basic food items (pulses, oilseeds, fruits and vegetables, etc.).

BIASED EXPORT COMPETITIVENESS

One could try now to compare what can be called 'apparent' and 'real' world competitiveness of Indian agricultural commodities. 'Apparent' competitiveness simply measures the difference (in percentage) between world and domestic prices of a commodity, while 'real' competitiveness calculates the same after having added to domestic prices the total per kilo inputs subsidies at current prices. Such an exercise has of course its own limitations (using, for example, world prices as a reference is a very debatable point), and given all the more reasons for calculating on one's own world prices (f.o.b. values of exports in total world divided by the corresponding quantities as yearly given by the FAO for different commodities) as well as domestic prices (average value of 1981 prices multiplied by the Wholesale Price Index of the corresponding year and product). In Table 7.9a details of these calculations are shown along with the data used, while in Table 7.9b are displayed the results for the 8 crops which could be integrated in this exercise carried out from 1981 to 1995.

These results show first that nearly all the crops appear to be export competitive from 1989 onwards (on the whole, they were 13 per cent below world prices on weighted average from 1990 to 1995), groundnut (36 per cent), rice (19 per cent), barley (16 per cent) and wheat (12 per cent) being in the lead, while sugarcane (–7 per cent) and rapeseed/mustard (–27 per cent) remained above world prices.

But this 'apparent' competitiveness of Indian agricultural products does not integrate input subsidies which lower their costs of production, and therefore their domestic prices. And, as a matter of fact, after having corrected this bias through the calculation of the 'real' competitiveness, Indian agricultural products no more appear to be significantly competitive after 1989 (on the whole, they became 3 per cent below world prices on weighted average from 1990 to 1995), wheat (–4 per cent) and cotton (–2 per cent) being now above world prices while their apparent competitiveness was very positive (12 and 8 per cent respectively). This bias has undoubtedly sent wrong signals to policy makers, especially after 1989, since this bias seems to have increased from that year on (Figure 7.2).

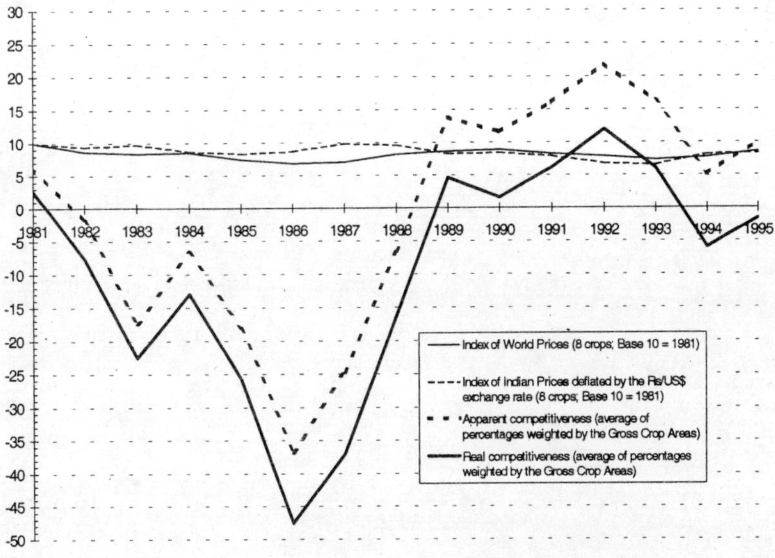

Figure 7.2: Apparent and Real World Competitiveness of Indian Agricultural Products (1981 to 1995).

CONCLUSION

As seen above, sugarcane, wheat and rice are the main beneficiaries of the Indian low-price policy for modern agricultural inputs. This product-specificity of input subsidies has naturally encouraged farmers to go for these crops during the past thirty years, along with specific price schemes which have secured at the farm level the output price of the same crops (Procurement Price for wheat and rice, SMP and SAP for sugarcane). As a result, most of the Indian households are now well secured in rice, wheat and sugar since these foodstuffs are provided to them in sufficient quantity at a rather low price. It is undoubtedly a great achievement.

However, this selective low input/output price policy has also perverse effects whose present extent can no longer be ignored. It has amplified year after year, and has now a powerful capacity to explain problems that the country must today urgently deals with, whether they are financial (ever-increasing input subsidies to the detriment of public investments in basic agricultural infrastructures), or environmental (water-logging and groundwater depletion, soil and biodiversity erosion, pollution by fertilizers), social (concentration of poverty in dryland areas or in states such as Bihar and Orissa), commercial (huge rotting stocks of wheat and rice while some other products cannot express their real competitiveness in domestic and world markets), not to say nutritional (worrying deficiencies in lipids, proteins, vitamins and other micro-nutrients[8] due to ever-increasing prices of vegetable oils, pulses, fruits and vegetables which hardly profit from the present agricultural incentives).

This paper does not suggest any alternatives, but may convince on the need to urgently search for alternatives, all the more given that other Indian agricultural subsidies (towards research, seeds, diesel, credit, prices, buffer-stocks, PDS, etc.) would most probably amplify the statement of this study if one was to try to allot them amongst crops and regions.

Table 7.1: Crop-Wise Consumption of Chemical Fertilizers (1980–1 to 1995–6) (all-India averages)

(kg. of nutrients per ha.)

		1980	1981	1982	1983	1984	1985	1986	1987	1988	1989	1990	1991	1992	1993	1994	1995	1980-95	n =
Barley	(1)	.	12	18	11	25	25	25	26	55	40	56	.	31	.	28	59	32	18
	(2)	.	12	18	11	31	27	27	26	55	53	56	.	31	.	28	66	36	18
	(3)	.	15	17	21	27	34	31	29	45	34	48	.	34	.	15	52	32	18
	(4)	.	15	17	21	29	35	32	29	45	37	48	.	34	.	15	52	33	18
	(5)	22	20	24	27	26	32	31	31	37	36	40	44	40	42	37	49	32	152
Sugarcane	(1)	275	197	174	227	220	191	181	244	326	372	274	79	247	142	307	167	230	65
	(2)	141	135	134	264	134	113	139	188	198	221	200	115	174	142	255	177	171	65
	(3)	166	133	180	211	203	219	211	278	310	358	280	113	252	121	360	239	232	65
	(4)	107	104	136	232	125	147	159	184	177	216	208	118	182	121	259	262	169	65
	(5)	125	124	131	141	157	177	171	185	188	219	236	246	203	211	240	234	191	240
	(6)											167					300		
Cotton	(1)	44	69	79	63	45	51	60	63	85	93	105	100	112	92	105	82	77	58
	(2)	44	69	83	61	45	52	57	62	88	95	108	100	118	90	94	83	76	58
	(3)	75	45	43	42	64	70	73	78	77	120	100	144	122	122	121	100	80	58
	(4)	75	44	43	40	61	73	56	73	83	125	102	144	126	131	120	104	76	58
	(5)	42	46	47	48	73	76	62	72	79	109	105	102	107	115	116	115	82	192
	(6)											103					121		
Groundnut	(1)	58	35	28	28	35	24	31	39	50	46	64	58	48	59	44	43	40	47
	(2)	58	37	33	33	36	29	40	38	49	49	62	56	43	59	47	48	43	47
	(3)	23	29	27	38	25	30	30	50	54	55	52	35	48	19	46	57	40	47
	(4)	23	28	25	36	32	34	31	48	53	51	64	46	55	19	49	52	40	47
	(5)	21	26	23	33	34	28	34	38	49	45	47	42	49	43	47	52	39	227
	(6)											75					60		
Maize	(1)	.	29	23	32	22	27	27	28	33	30	45	38	47	.	18	30	31	35
	(2)	.	29	23	30	22	27	27	30	33	31	46	38	48	.	18	30	32	35
	(3)	.	16	16	24	25	16	21	23	37	46	47	37	45	.	32	43	29	35
	(4)	.	16	16	24	25	16	20	22	37	45	47	37	45	.	32	43	31	35
	(5)	29	30	32	40	47	39	46	42	55	67	66	65	76	69	65	77	53	224
	(6)											70					46		

(contd.)

Table 7.1 (contd.)

		1980	1981	1982	1983	1984	1985	1986	1987	1988	1989	1990	1991	1992	1993	1994	1995	1980-95	n =
Rapeseed	(1)	6	4	11	14	22	25	25	38	64	52	93	101	75	.	54	76	46	38
Mustard	(2)	5	4	16	17	19	18	20	22	60	33	78	90	45	.	48	64	35	38
	(3)	21	18	22	26	36	31	34	45	60	42	61	55	43	.	62	61	43	38
	(4)	21	18	23	27	35	30	32	36	61	35	67	56	38	.	56	55	39	38
	(5)	22	22	25	31	37	34	38	46	49	47	53	56	46	54	59	58	45	209
Rice	(1)	58	70	62	75	85	91	86	105	121	106	113	113	109	183	113	98	93	101
	(2)	47	51	47	57	60	66	68	79	103	78	90	86	82	183	97	84	71	101
	(3)	55	65	63	75	89	103	87	110	125	111	97	108	77	129	114	114	91	101
	(4)	45	47	46	56	67	73	67	79	110	85	87	81	64	129	103	103	72	101
	(5)	49	50	54	66	71	77	78	84	98	103	109	109	105	104	110	111	87	240
	(6)											104					168		
Wheat	(1)	122	73	71	77	89	99	94	96	111	113	135	143	123	.	128	130	102	74
	(2)	99	82	81	86	89	99	104	111	113	116	125	136	120	.	126	129	106	74
	(3)	102	60	64	71	91	103	87	91	120	112	135	150	130	.	147	146	102	74
	(4)	84	69	75	83	92	105	96	99	118	112	125	143	127	.	141	139	105	74
	(5)	64	70	76	85	89	101	95	99	117	111	122	131	122	123	137	137	106	209
	(6)											128					155		
8 crops	(1)	102	78	75	84	85	78	81	97	109	109	129	99	112	114	126	92	95	436
	(2)	64	62	61	65	65	74	76	86	98	92	105	95	92	120	104	95	81	436
	(3)	78	60	70	78	87	89	88	106	109	113	121	96	105	103	142	106	95	436
	(4)	57	52	54	61	70	82	73	83	102	95	103	93	87	98	114	107	81	436
	(5)	49	52	57	66	72	78	76	82	94	99	105	106	102	103	111	113	86	1693

Sources: (1) Ministry of Agriculture states' average.
(2) Ministry of Agriculture states' average weighted by states' gross cropped area.
(3) Authors states' average with states, crops and years of (1).
(4) Authors states' average weighted by states' gross cropped area with states, crops and years of (1).
(5) Authors states' average (14 states) weighted by states' gross cropped area.
(6) Food and Agriculture Organization.

Table 7.2a: Extracts from the Calculation of Irrigation Subsidies (1980–1 to 1995–6)

ALL INDIA - ALL CROPS	Unit	1980	1981	1982	1983	1984	1985	1986	1987	1988	1989	1990	1991	1992	1993	1994	1995
Gross Irrigated Area	1000 ha.	49775	51606	52121	53937	54083	53978	55690	56229	59048	60537	62472	65097	66761	68367	70646	71510
Net Irrigated Area	1000 ha.	38805	39729	39969	41955	41779	41760	43049	43048	45186	45144	47434	48800	50296	51452	52999	53510
Net Irrigated Area - Canal - Public	1000 ha.	14456	14702	14875	15745	15366	15715	15841	14990	15872	15816	16393	16798	16503	16628	16755	16582
Net Irrigated Area - Canal - Private	1000 ha.	836	827	495	495	495	488	479	484	485	501	507	492	483	483	478	560
Net Irrigated Area - Tank	1000 ha.	3198	3511	3112	3783	3330	3070	2933	2806	3302	3200	3245	3348	3179	3152	3249	3111
Net Irrigated Area - Well	1000 ha.	17734	18122	19112	19521	19988	20165	21046	21817	22562	22756	24210	25012	26920	27762	28904	29797
Financial Expend. - CAD Central + States	Rs. crore	100.6	100.6	100.6	100.6	100.6	289.5	289.5	289.5	289.5	289.5	285.6	333.8	322.8	374.9	422.1	437.3
Working Expenses - Major & Medium	Rs. crore	527.2	680.9	1110.4	836.7	969.7	1168.6	1356.9	1400.3	2128.0	2223.8	2476.3	2775.1	3113.2	3582.7	4301.7	4734.5
Working Expenses - Minor - All India	Rs. crore	258.8	267.1	332.9	403.5	406.7	471.8	666.1	1013.6	919.3	915.2	955.3	1050.7	1347.8	1399.4	1509.9	1652.4
Gross Receipt - Major & Medium	Rs. crore	103.4	120.2	117.1	165.1	129.7	223.8	166.7	138.7	166.4	207.6	228.9	232.2	316.1	459.2	437.8	490.9
Gross Receipt - Minor	Rs. crore	24.0	37.0	27.6	33.2	34.7	47.7	42.1	48.3	47.8	38.0	39.0	57.1	94.4	67.9	57.8	61.2
Subsidy - Major & Medium	Rs. billion	5.24	6.61	10.94	7.72	9.41	12.34	14.80	15.51	22.51	23.06	25.33	28.77	31.20	34.98	42.86	46.81

(contd.)

Table 7.2a (contd.)

ALL INDIA - ALL CROPS	Unit	1980	1981	1982	1983	1984	1985	1986	1987	1988	1989	1990	1991	1992	1993	1994	1995
Subsidy - Minor	Rs. billion	2.35	2.30	3.05	3.70	3.72	4.24	6.24	9.65	8.72	8.77	9.16	9.94	12.53	13.32	14.52	15.91
Subsidy - All crops	Rs. billion	7.59	8.91	13.99	11.43	13.13	16.58	21.04	25.16	31.23	31.83	34.49	38.70	43.73	48.30	57.38	62.72
Subsidy - All crops	Rs./ha. GIA	246	282	456	324	417	557	657	407	981	598	680	791	930	1029	1143	1279
Subsidy - Canal	Rs./ha. NIA	343	426	712	475	593	762	907	1002	1376	1413	1499	1664	1837	2045	2487	2731
Subsidy - Well + Tank	Rs./ha. NIA	112	106	137	159	160	183	260	392	337	338	334	350	416	431	452	484
ANDHRA PRADESH - RICE	**Unit**	**1980**	**1981**	**1982**	**1983**	**1984**	**1985**	**1986**	**1987**	**1988**	**1989**	**1990**	**1991**	**1992**	**1993**	**1994**	**1995**
Gross Irrigated Area - All crops	1000 ha.	4342	4678	4518	5058	4470	4337	4360	4298	5440	5454	5369	5378	5085	5020	5185	5304
Gross Irrigated Area - Rice	1000 ha.	3378	3610	3426	3924	3306	3249	3259	3068	4027	4003	3830	3736	3405	3369	3445	3500
Financial Expenditure - CAD	Rs. crore	8.7	8.7	8.7	8.7	8.7	10.0	10.0	10.0	10.0	10.0	5.8	8.0	19.0	26.1	29.3	9.9
Working Expenses - Major & Medium	Rs. crore	86.8	112.1	182.7	137.7	159.6	192.3	223.3	134.6	491.0	315.5	360.3	392.1	430.4	504.4	613.8	711.8
Working Expenses - Minor	Rs. crore	12.6	13.0	16.2	19.6	19.8	42.2	55.4	40.6	39.0	36.5	48.8	40.3	89.3	62.9	54.5	53.6
Gross Receipt - Major & Medium	Rs. crore	1.3	1.5	1.5	2.1	1.6	2.8	2.1	10.4	5.4	35.2	48.8	13.8	65.7	76.3	103.8	94.6
Gross Receipt - Minor	Rs. crore	0.0	0.0	0.0	0.0	0.0	0.3	0.1	0.0	0.8	0.8	0.8	1.0	0.3	0.5	1.2	2.1
Subsidy - All crops	Rs. billion	1.07	1.32	2.06	1.64	1.86	2.41	2.86	1.75	5.34	3.26	3.65	4.26	4.73	5.17	5.93	6.79
Subsidy - All crops	Rs. / ha. GIA	246	283	456	324	417	557	657	406	981	598	680	791	930	1029	1143	1279
Subsidy - Rice	Rs. billion	0.83	1.02	1.56	1.27	1.38	1.81	2.14	1.25	3.95	2.39	2.61	2.96	3.17	3.47	3.94	4.48

Statistical sources: (i) Water and Related Statistics, various issues.
(ii) Land Use Statistics, various issues.
(iii) Indian Agriculture Statistics, various issues.
(v) Centre for Monitoring Indian Economy, various issues.

Table 7.2b: Extracts from the Calculation of Electricity Subsidies (1980–1 to 1995–6)

ALL INDIA ALL CROPS	Unit	1980	1981	1982	1983	1984	1985	1986	1987	1988	1989	1990	1991	1992	1993	1994	1995
Cost of Supply - All India (weight. average)	Paise / kWh	41.90	47.59	54.78	61.77	65.07	74.59	80.37	88.96	91.20	101.50	108.59	116.80	128.15	149.12	163.40	179.60
Tariff - Agriculture (weight. average)	Paise / kWh	18.84	20.19	19.41	20.48	18.64	18.15	17.70	14.89	20.54	21.17	21.39	11.55	16.10	17.90	18.80	19.00
Subsidy - All crops	Rs. / kWh	0.23	0.27	0.35	0.41	0.46	0.56	0.63	0.74	0.71	0.80	0.87	1.05	1.12	1.31	1.45	1.61
Sales - Agriculture (official)	kWh million	14489	15201	17817	18234	20960	23708	28971	34652	38366	41778	50321	58557	63415	68994	77140	84157
Consumption-Agriculture (estimated)	kWh million	10142	10641	12472	12764	14672	16596	20280	24256	26856	29245	35225	40990	44391	48296	53998	58910
Subsidy - All crops	Rs billion	2.34	2.92	4.41	5.27	6.81	9.37	12.71	17.97	18.98	23.49	30.72	43.14	49.74	63.37	78.08	94.61

ANDHRA PRADESH-RICE	Unit	1980	1981	1982	1983	1984	1985	1986	1987	1988	1989	1990	1991	1992	1993	1994	1995
Cost of Supply	Paise / kWh	37.8	42.0	44.2	48.3	47.4	53.3	55.6	62.0	64.5	70.2	78.7	89.2	100.1	109.0	128.9	156.1
Tariff - Agriculture	Paise / kWh	19.8	20.2	15.1	9.6	6.1	5.7	4.9	4.5	4.5	4.5	4.5	3.3	8.6	6.4	5.3	2.8
Subsidy - All crops	Rs. / kWh	0.18	0.22	0.29	0.39	0.41	0.48	0.51	0.58	0.60	0.66	0.74	0.86	0.91	1.03	1.24	1.53
Sales - Agriculture (official)	million kWh	977	1005	1465	1627	2439	2607	3347	3980	4497	4836	6460	7219	7936	9022	10922	11399
Consumption-Agriculture (estimated)	kWh million	684	704	1026	1139	1707	1825	2343	2786	3148	3385	4522	5053	5555	6315	7645	7979
Subsidy - All crops	Rs. billion	0.12	0.15	0.30	0.44	0.70	0.87	1.19	1.60	1.89	2.22	3.36	4.34	5.08	6.48	9.45	12.23

(contd.)

Table 7.2b (contd.)

ANDHRA PRADESH-RICE	Unit	1980	1981	1982	1983	1984	1985	1986	1987	1988	1989	1990	1991	1992	1993	1994	1995
Gross Irrigated Area - All crops	1000 ha.	4342	4678	4518	5058	4470	4337	4360	4298	5440	5454	5369	5378	5085	5020	5185	5304
Subsidy - All crops	Rs. / ha. GIA	28.4	32.8	65.9	87.2	157.7	200.2	272.4	372.8	347.3	407.8	625.1	807.6	999.3	1290.8	1823.0	2306.5
Gross Irrigated Area-Rice	1000 ha.	3378	3610	3426	3924	3306	3249	3259	3068	4027	4003	3830	3736	3405	3369	3445	3500
Subsidy - Rice	Rs. billion	0.10	0.12	0.23	0.34	0.52	0.65	0.89	1.14	1.40	1.63	2.39	3.02	3.40	4.35	6.28	8.07

Statistical sources: (i) Annual Report on the Working of State Electricity Board & Electricity Departments, various issues.
(ii) Financial Performance Review of State Electricity Boards, various issues.
(iii) Land Use Statistics, various issues.
(iv) Indian Agriculture Statistics, various issues.
(v) Centre for Monitoring Indian Economy, various issues.

Table 7.2c: Extracts from the Calculation of Fertilizer Subsidies (1980–1 to 1995–6)

ALL INDIA ALL CROPS	Unit	1980	1981	1982	1983	1984	1985	1986	1987	1988	1989	1990	1991	1992	1993	1994	1995
Subsidy - Indigenous & imported	Rs. billion	5.05	3.75	6.05	10.42	19.27	19.24	18.97	21.64	32.01	45.42	43.89	48.00	57.96	44.00	52.41	62.35
Farmers' Share (estimated)	%	33%	33%	33%	33%	33%	33%	33%	33%	41%	52%	64%	81%	100%	100%	100%	100%
Subsidy - All crops	Rs. billion	1.67	1.24	2.00	3.44	6.36	6.35	6.26	7.14	13.12	23.62	28.09	38.88	57.96	44.00	52.41	62.35
Consumption of NPK- All India	Ton million	5.5	6.1	6.4	7.7	8.2	8.7	8.6	8.8	11.0	11.6	12.5	12.7	12.2	12.4	13.6	13.9
Subsidy - All crops	Rs. / kg. NPK	0.302	0.204	0.312	0.446	0.774	0.727	0.724	0.813	1.189	2.042	2.239	3.055	4.769	3.558	3.864	4.493

ANDHRA PRADESH-RICE	Unit	1980	1981	1982	1983	1984	1985	1986	1987	1988	1989	1990	1991	1992	1993	1994	1995
Subsidy - All crops	Rs. / kg. NPK	0.302	0.204	0.312	0.446	0.774	0.727	0.724	0.813	1.189	2.042	2.239	3.055	4.769	3.558	3.864	4.493
Consumption of NPK- AP - Rice (estimat.)	Kg. / ha. GCA	96	102	122	137	164	158	157	243	205	221	235	229	240	245	256	253
Gross Cropped Area- Rice	1000 ha.	3600	3824	3638	4164	3498	3452	3459	2306	4218	4207	4036	3936	3604	3547	3638	3692
Subsidy - Rice	Rs. billion	0.10	0.08	0.14	0.25	0.44	0.40	0.39	0.47	1.03	1.90	2.13	2.75	4.12	3.09	3.60	4.19

Statistical sources:
(i) Economic Survey, various issues.
(ii) Fertiliser Statistics, various issues.
(iii) Land Use Statistics, various issues.
(iv) Cost of Cultivation of Principal Crops in India, various issues.
(v) Reports of the Commission for Agricultural Costs and Prices, various issues.

Table 7.3a: Crop-Wise Subsidies in Billion Rupees (1980–1 to 1995–6)

(Current Prices)

IRRIGATION	1980	1981	1982	1983	1984	1985	1986	1987	1988	1989	1990	1991	1992	1993	1994	1995
All Crops	7.59	8.91	13.99	11.43	13.13	16.58	21.04	25.68	31.23	31.83	34.49	38.70	43.73	48.30	57.38	62.72
Barley	0.14	0.14	0.19	0.14	0.15	0.21	0.24	0.28	0.31	0.28	0.29	0.35	0.37	0.33	0.45	0.44
Cotton	0.33	0.38	0.62	0.48	0.50	0.70	0.81	0.95	1.30	1.40	1.37	1.54	1.75	1.81	2.22	2.75
Groundnut	0.14	0.17	0.28	0.24	0.29	0.33	0.40	0.59	0.73	0.69	0.86	0.99	1.07	1.15	1.29	1.23
Maize	0.18	0.21	0.33	0.21	0.25	0.33	0.48	0.54	0.65	0.64	0.64	0.79	0.85	0.97	1.02	1.21
Rape / M.	0.15	0.20	0.28	0.27	0.46	0.52	0.59	0.93	1.34	1.40	1.70	2.24	2.19	2.41	2.92	3.63
Rice	2.49	2.96	4.32	3.75	4.37	5.43	6.86	7.65	10.15	10.29	10.75	12.00	13.16	14.65	17.38	18.82
Sugarcane	0.35	0.49	0.62	0.57	0.62	0.79	1.06	1.32	1.58	1.60	1.88	2.15	2.30	2.42	2.96	3.40
Wheat	2.37	2.70	4.58	3.82	4.28	5.31	6.69	7.98	10.02	9.85	10.77	11.63	13.57	15.10	17.90	19.04
Other Crops	1.45	1.67	2.77	1.94	2.21	2.97	3.92	5.46	5.16	5.67	6.23	7.03	8.47	9.46	11.25	12.21
ELECTRICITY	1980	1981	1982	1983	1984	1985	1986	1987	1988	1989	1990	1991	1992	1993	1994	1995
All Crops*	2.34	2.92	4.41	5.27	6.81	9.37	12.71	17.97	18.98	23.49	30.72	43.14	49.74	63.37	78.08	94.61
Barley	0.05	0.05	0.06	0.06	0.07	0.09	0.10	0.14	0.15	0.16	0.18	0.25	0.30	0.31	0.40	0.43
Cotton	0.12	0.13	0.20	0.22	0.27	0.42	0.54	1.06	1.04	1.16	1.47	2.05	2.65	3.04	3.63	4.96
Groundnut	0.07	0.08	0.12	0.18	0.23	0.26	0.37	0.70	0.72	0.77	1.40	1.73	2.04	2.42	3.13	3.78
Maize	0.05	0.06	0.09	0.08	0.10	0.19	0.30	0.38	0.42	0.47	0.60	0.79	0.95	1.15	1.34	1.81
Rape / M.	0.05	0.06	0.07	0.10	0.17	0.18	0.25	0.60	0.70	0.79	1.24	2.02	2.18	2.74	3.17	4.35
Rice	0.71	0.96	1.40	1.67	2.11	2.94	3.91	4.81	5.77	6.88	9.27	11.77	14.16	17.52	21.86	26.78
Sugarcane	0.15	0.23	0.30	0.30	0.40	0.58	0.76	1.11	1.35	1.51	2.24	3.07	3.55	4.09	5.22	6.78
Wheat	0.84	0.94	1.41	1.63	2.03	2.46	3.29	4.66	5.34	6.15	8.06	9.95	12.64	15.61	18.13	21.09
Other Crops	0.32	0.43	0.75	1.02	1.45	2.25	3.19	4.51	3.48	5.62	6.25	11.53	11.28	16.48	21.21	24.64

(contd.)

Table 7.3a (contd.)

FERTILIZERS

	1980	1981	1982	1983	1984	1985	1986	1987	1988	1989	1990	1991	1992	1993	1994	1995
All Crops	1.67	1.24	2.00	3.44	6.36	6.35	6.26	7.14	13.12	23.62	28.09	38.88	57.96	44.00	52.41	62.35
Barley	0.01	0.01	0.01	0.02	0.03	0.03	0.03	0.03	0.05	0.07	0.09	0.13	0.17	0.12	0.12	0.18
Cotton	0.10	0.08	0.12	0.16	0.41	0.42	0.31	0.38	0.69	1.71	1.74	2.38	3.84	3.00	3.52	4.66
Groundnut	0.04	0.04	0.05	0.11	0.19	0.14	0.17	0.21	0.49	0.79	0.87	1.12	1.89	1.29	1.41	1.76
Maize	0.05	0.03	0.05	0.10	0.19	0.15	0.18	0.17	0.35	0.72	0.77	1.04	1.95	1.32	1.41	1.88
Rape / M.	0.03	0.02	0.03	0.05	0.10	0.09	0.09	0.16	0.26	0.44	0.63	1.04	1.26	1.13	1.29	1.61
Rice	0.54	0.38	0.59	1.12	2.07	2.10	2.12	2.34	4.45	8.10	9.55	12.88	19.02	14.39	16.60	19.49
Sugarcane	0.10	0.08	0.14	0.19	0.35	0.36	0.37	0.48	0.73	1.51	1.92	2.85	3.42	2.54	3.55	4.32
Wheat	0.42	0.31	0.55	0.92	1.58	1.65	1.56	1.82	3.28	5.20	6.47	9.08	14.04	10.80	13.36	15.15
Other Crops	0.38	0.30	0.47	0.78	1.43	1.41	1.44	1.55	2.83	5.07	6.05	8.37	12.39	9.43	11.14	13.32

TOTAL

	1980	1981	1982	1983	1984	1985	1986	1987	1988	1989	1990	1991	1992	1993	1994	1995
All Crops	11.60	13.07	20.40	20.13	26.30	32.30	40.01	50.27	53.33	78.94	93.30	120.73	151.43	155.67	187.87	219.68
Barley	0.20	0.20	0.26	0.22	0.24	0.33	0.37	0.45	0.51	0.51	0.56	0.72	0.84	0.76	0.98	1.05
Cotton	0.55	0.59	0.94	0.86	1.18	1.54	1.67	2.38	3.04	4.27	4.58	5.97	8.23	7.85	9.36	12.37
Groundnut	0.25	0.28	0.45	0.53	0.71	0.73	0.94	1.49	1.94	2.25	3.13	3.84	5.01	4.86	5.83	6.76
Maize	0.28	0.30	0.47	0.39	0.55	0.67	0.95	1.08	1.42	1.84	2.02	2.62	3.75	3.45	3.77	4.89
Rape / M.	0.23	0.27	0.38	0.43	0.73	0.79	0.94	1.69	2.30	2.63	3.57	5.29	5.62	6.27	7.39	9.58
Rice	3.74	4.29	6.30	6.54	8.55	10.47	12.89	14.80	20.37	25.26	29.57	36.64	46.34	46.57	55.84	65.10
Sugarcane	0.59	0.80	1.06	1.06	1.37	1.72	2.19	2.91	3.66	4.62	6.04	8.06	9.26	9.05	11.72	14.49
Wheat	3.62	3.95	6.54	6.38	7.89	9.42	11.54	14.46	18.64	21.20	25.30	30.66	40.25	41.51	49.39	55.27
Other Crops	2.15	2.39	4.00	3.73	5.09	6.64	8.54	11.01	11.46	16.36	18.53	26.93	32.13	35.37	43.60	50.17

Note: * With all-India weighted average of the unit cost of power supply computed by the Planning Commission.

Table 7.3b: State-Wise Subsidies in Billion Rupees (1980–1 to 1995–6)

(Current Prices)

IRRIGATION	1980	1981	1982	1983	1984	1985	1986	1987	1988	1989	1990	1991	1992	1993	1994	1995
All India	7.59	8.91	13.99	11.43	13.13	16.58	21.04	25.68	31.23	31.83	34.49	38.70	43.73	48.30	57.38	62.72
Andhra	1.07	1.32	2.06	1.64	1.86	2.41	2.86	1.75	5.34	3.26	3.65	4.26	4.73	5.17	5.93	6.79
Bihar	0.23	0.26	0.43	0.32	0.40	0.63	0.86	1.43	1.56	1.82	1.57	1.51	1.91	2.01	1.81	1.66
Gujarat	0.53	0.64	0.99	0.82	0.93	1.17	2.24	4.78	4.37	4.01	3.61	4.05	4.45	5.88	6.69	8.32
Haryana	0.36	0.47	0.82	0.55	0.70	0.94	1.15	0.99	1.09	1.42	1.30	1.65	1.87	2.29	4.91	2.40
Karnataka	0.71	0.86	1.33	1.09	1.24	1.45	1.68	1.62	1.80	1.91	2.14	2.32	2.82	3.09	3.58	3.87
Kerala	0.13	0.14	0.21	0.19	0.21	0.25	0.32	0.37	0.41	0.45	0.56	0.52	0.60	0.71	0.81	1.03
Maharashtra	0.93	1.14	1.81	1.40	1.63	2.58	2.96	4.21	4.72	5.48	6.40	7.01	8.19	8.98	10.95	13.23
Madhya P.	0.49	0.52	0.69	0.70	0.74	0.71	0.77	0.95	0.92	1.00	1.29	1.28	1.35	1.66	1.46	1.79
Orissa	0.18	0.20	0.29	0.28	0.30	0.24	0.31	0.32	0.34	0.42	0.46	0.63	0.70	0.88	0.92	1.01
Punjab	0.16	0.20	0.39	0.25	0.33	0.44	0.72	0.88	0.90	0.86	0.88	1.01	1.13	1.20	1.15	1.27
Rajastan	0.58	0.71	1.15	0.85	1.01	1.44	1.70	2.08	1.91	1.83	2.34	2.53	3.28	3.87	3.95	4.62
Tamil Nadu	0.34	0.41	0.64	0.52	0.59	0.71	0.82	1.02	0.98	1.10	1.21	1.39	1.61	1.72	1.93	2.13
Uttar P.	1.31	1.46	2.28	1.92	2.21	2.49	3.18	4.41	4.92	6.09	6.13	7.51	8.70	7.81	10.55	11.47
West Bengal	0.46	0.51	0.72	0.72	0.76	0.66	0.73	0.79	0.89	1.01	1.34	1.34	1.38	1.67	1.71	1.94
Other States	0.11	0.09	0.18	0.18	0.19	0.45	0.74	0.07	1.08	1.18	1.62	1.71	1.02	1.37	1.05	1.21

ELECTRICITY	1980	1981	1982	1983	1984	1985	1986	1987	1988	1989	1990	1991	1992	1993	1994	1995
All India*	2.34	2.92	4.41	5.27	6.81	9.37	12.71	17.97	18.98	23.49	30.72	43.14	49.74	63.37	78.08	94.61
Andhra	0.12	0.15	0.30	0.44	0.71	0.87	1.19	1.60	1.89	2.22	3.36	4.34	5.08	6.48	9.45	12.23
Bihar	0.13	0.19	0.30	0.36	0.40	0.65	0.84	1.02	1.12	1.29	1.63	1.93	1.87	1.76	2.08	2.10
Gujarat	0.17	0.11	0.18	0.22	0.35	0.47	0.66	2.08	2.49	2.55	3.52	5.90	7.39	8.44	8.86	11.53
Haryana	0.16	0.26	0.34	0.33	0.47	0.49	0.71	1.40	0.88	1.17	1.40	2.46	3.10	3.78	3.43	4.29
Karnataka	0.01	0.04	0.06	0.10	0.16	0.73	1.01	1.33	1.40	1.44	2.25	2.79	3.46	4.65	6.10	7.76
Kerala	0.00	0.00	0.01	0.02	0.01	0.00	0.03	0.04	0.05	0.06	0.07	0.09	0.10	0.11	0.16	0.25
Maharashtra	0.25	0.37	0.60	0.66	1.21	1.48	1.90	2.24	2.97	3.53	4.55	6.49	7.22	7.89	11.53	15.75

(contd.)

Table 7.3b (contd.)

ELECTRICITY

	1980	1981	1982	1983	1984	1985	1986	1987	1988	1989	1990	1991	1992	1993	1994	1995
Madhya P.	0.04	0.08	0.16	0.16	0.22	0.24	0.36	0.45	0.57	0.67	1.60	2.42	2.95	5.29	7.73	9.91
Orissa	0.01	0.01	0.01	0.02	0.02	0.02	0.05	0.06	0.04	0.06	0.05	0.11	0.15	0.27	0.14	0.21
Punjab	0.28	0.30	0.42	0.47	0.73	0.95	1.44	2.23	2.64	3.29	3.54	3.46	4.81	5.58	5.47	5.80
Rajastan	0.14	0.18	0.28	0.39	0.38	0.53	0.55	0.94	0.94	1.31	1.41	1.80	2.43	3.28	4.36	5.65
Tamil Nadu	0.46	0.61	0.87	0.91	0.88	1.28	1.42	1.59	1.83	2.39	3.00	3.26	4.50	5.32	6.63	7.93
Uttar P.	0.74	0.84	1.07	1.16	1.31	1.51	2.00	2.62	3.41	3.89	4.82	5.97	7.25	8.59	8.93	9.82
West Bengal	0.01	0.01	0.02	0.04	0.04	0.06	0.07	0.14	0.23	0.33	0.42	0.67	0.73	0.81	1.18	1.45

FERTILIZERS

	1980	1981	1982	1983	1984	1985	1986	1987	1988	1989	1990	1991	1992	1993	1994	1995
All India	1.67	1.24	2.00	3.44	6.36	6.35	6.26	7.14	13.12	23.62	28.09	38.88	57.96	44.00	52.41	62.35
Andhra	0.17	0.13	0.23	0.42	0.76	0.65	0.65	0.79	1.61	3.13	3.63	4.83	7.22	5.49	6.35	7.87
Bihar	0.06	0.04	0.06	0.13	0.30	0.36	0.38	0.40	0.72	1.15	1.34	1.84	2.84	2.08	2.52	3.24
Gujarat	0.11	0.08	0.11	0.22	0.39	0.31	0.29	0.36	0.77	1.42	1.58	2.24	3.42	2.38	3.16	3.39
Haryana	0.07	0.05	0.09	0.15	0.26	0.27	0.30	0.32	0.61	1.09	1.31	1.95	2.90	2.39	2.75	3.25
Karnataka	0.10	0.08	0.11	0.22	0.46	0.40	0.31	0.45	0.95	1.59	1.87	2.77	3.72	2.86	3.18	4.21
Kerala	0.03	0.02	0.03	0.06	0.10	0.10	0.11	0.15	0.25	0.43	0.55	0.69	0.97	0.63	0.77	0.91
Maharashtra	0.13	0.11	0.16	0.25	0.45	0.49	0.47	0.60	1.06	2.41	2.95	3.86	5.40	4.25	5.37	6.17
Madhya P.	0.06	0.05	0.08	0.14	0.29	0.31	0.36	0.41	0.81	1.38	1.82	2.47	3.78	2.75	3.46	3.71
Orissa	0.02	0.02	0.03	0.05	0.09	0.10	0.11	0.12	0.19	0.42	0.43	0.60	0.97	0.74	0.85	1.07
Punjab	0.23	0.17	0.28	0.44	0.81	0.80	0.81	0.90	1.33	2.34	2.68	3.80	5.72	4.27	4.97	5.68
Rajastan	0.04	0.03	0.05	0.09	0.16	0.16	0.18	0.18	0.36	0.58	0.83	1.35	2.34	1.79	2.33	2.89
Tamil Nadu	0.15	0.09	0.15	0.26	0.54	0.49	0.49	0.55	0.91	1.60	1.86	2.56	3.81	2.78	3.38	3.39
Uttar P.	0.35	0.26	0.45	0.73	1.25	1.43	1.22	1.30	2.54	4.27	5.02	6.87	10.39	8.16	9.58	11.71
West Bengal	0.09	0.05	0.08	0.17	0.31	0.30	0.36	0.46	0.76	1.37	1.69	2.31	3.49	2.65	2.91	3.81
Other States	0.06	0.06	0.09	0.11	0.20	0.18	0.23	0.15	0.26	0.43	0.54	0.74	0.99	0.78	0.82	1.06

(contd.)

Table 7.3b (contd.)

TOTAL	1980	1981	1982	1983	1984	1985	1986	1987	1988	1989	1990	1991	1992	1993	1994	1995
All India	11.60	13.07	20.40	20.13	26.30	32.30	40.01	50.27	63.33	78.94	93.30	120.73	151.43	155.67	187.87	219.68
Andhra	1.36	1.61	2.59	2.50	3.33	3.93	4.71	4.14	8.84	8.61	10.64	13.43	17.03	17.14	21.73	26.89
Bihar	0.42	0.49	0.79	0.81	1.10	1.64	2.08	2.85	3.40	4.26	4.54	5.29	6.62	5.85	6.41	6.99
Gujarat	0.82	0.83	1.28	1.26	1.67	1.94	3.19	7.22	7.63	7.98	8.71	12.19	15.25	16.71	18.72	23.24
Haryana	0.58	0.78	1.25	1.03	1.43	1.70	2.16	2.71	2.57	3.68	4.01	6.06	7.87	8.46	11.09	9.94
Karnataka	0.83	0.98	1.49	1.41	1.86	2.58	3.00	3.41	4.15	4.94	6.26	7.87	10.00	10.61	12.86	15.84
Kerala	0.16	0.17	0.25	0.27	0.33	0.36	0.46	0.56	0.71	0.94	1.17	1.29	1.67	1.45	1.74	2.19
Maharashtra	1.31	1.61	2.57	2.31	3.29	4.54	5.33	7.06	8.75	11.42	13.90	17.36	20.80	21.12	27.85	35.15
Madhya P.	0.60	0.64	0.92	1.00	1.25	1.26	1.49	1.81	2.31	3.04	4.71	6.17	8.08	9.71	12.65	15.42
Orissa	0.22	0.22	0.33	0.34	0.41	0.36	0.47	0.50	0.58	0.90	0.94	1.34	1.81	1.89	1.91	2.29
Punjab	0.67	0.67	1.09	1.16	1.88	2.19	2.96	4.01	4.86	6.49	7.10	8.27	11.66	11.05	11.58	12.74
Rajastan	0.76	0.91	1.49	1.33	1.55	2.13	2.43	3.19	3.22	3.72	4.58	5.67	8.06	8.94	10.63	13.16
Tamil Nadu	0.95	1.11	1.66	1.70	2.01	2.48	2.73	3.17	3.71	5.09	6.07	7.22	9.92	9.81	11.93	13.46
Uttar P.	2.40	2.56	3.79	3.80	4.78	5.43	6.40	8.32	10.87	14.25	15.97	20.35	26.34	24.56	29.06	32.99
West Bengal	0.56	0.56	0.82	0.92	1.12	1.02	1.15	1.39	1.87	2.71	3.45	4.32	5.60	5.13	5.80	7.20

Note: * With all-India weighted average of the unit cost of power supply computed by the Planning Commission.

Table 7.4a: Crop-Wise Subsidies in Billion Rupees (1980–1 to 1995–6)

(1995–6 Prices)

IRRIGATION	1980	1981	1982	1983	1984	1985	1986	1987	1988	1989	1990	1991	1992	1993	1994	1995
All Crops	27.71	29.42	42.79	32.28	34.54	40.45	48.14	52.98	60.75	57.14	55.81	54.59	56.94	57.43	62.10	62.72
Barley	0.51	0.47	0.58	0.39	0.40	0.52	0.55	0.58	0.61	0.51	0.47	0.49	0.48	0.40	0.49	0.44
Cotton	1.19	1.27	1.90	1.35	1.32	1.71	1.86	1.99	2.53	2.52	2.22	2.17	2.27	2.16	2.40	2.75
Groundnut	0.51	0.55	0.85	0.67	0.77	0.80	0.91	1.23	1.41	1.23	1.40	1.39	1.40	1.36	1.39	1.23
Maize	0.67	0.69	1.00	0.61	0.66	0.81	1.09	1.13	1.26	1.16	1.04	1.11	1.11	1.15	1.11	1.21
Rape / M.	0.55	0.66	0.87	0.77	1.20	1.26	1.36	1.96	2.61	2.52	2.75	3.16	2.85	2.86	3.16	3.63
Rice	9.10	9.75	13.20	10.60	11.49	13.24	15.69	16.10	19.74	18.47	17.39	16.92	17.14	17.42	18.81	18.82
Sugarcane	1.27	1.62	1.90	1.60	1.63	1.92	2.42	2.79	3.08	2.87	3.04	3.03	3.00	2.88	3.20	3.40
Wheat	8.64	8.91	14.01	10.80	11.27	12.96	15.31	16.80	19.49	17.69	17.43	16.40	17.68	17.95	19.37	19.04
Other Crops	5.28	5.50	8.47	5.48	5.81	7.25	8.96	10.41	10.03	10.18	10.07	9.92	11.02	11.25	12.17	12.21
ELECTRICITY	1980	1981	1982	1983	1984	1985	1986	1987	1983	1989	1990	1991	1992	1993	1994	1995
All Crops	8.54	9.62	13.49	14.89	17.93	22.85	29.08	37.83	36.92	42.18	49.70	60.85	64.77	75.36	84.50	94.61
Barley	0.16	0.16	0.17	0.17	0.17	0.21	0.22	0.30	0.29	0.28	0.30	0.35	0.39	0.37	0.44	0.43
Cotton	0.45	0.44	0.62	0.62	0.70	1.03	1.25	2.22	2.03	2.08	2.38	2.89	3.45	3.61	3.93	4.96
Groundnut	0.24	0.25	0.38	0.52	0.60	0.64	0.85	1.47	1.40	1.39	2.27	2.44	2.66	2.88	3.39	3.78
Maize	0.18	0.19	0.28	0.24	0.28	0.47	0.68	0.80	0.82	0.84	0.98	1.12	1.24	1.37	1.45	1.81
Rape / M.	0.18	0.19	0.22	0.30	0.44	0.44	0.58	1.27	1.37	1.41	2.01	2.84	2.84	3.26	3.43	4.35
Rice	2.58	3.15	4.28	4.72	5.55	7.16	8.96	10.13	11.23	12.35	15.01	16.59	18.44	20.84	23.65	26.78
Sugarcane	0.54	0.75	0.92	0.85	1.06	1.40	1.73	2.33	2.62	2.71	3.62	4.33	4.62	4.86	5.65	6.78
Wheat	3.05	3.09	4.33	4.62	5.33	6.00	7.52	9.81	10.39	11.03	13.03	14.03	16.46	18.56	19.62	21.09
Other Crops	1.16	1.40	2.30	2.87	3.81	5.49	7.30	9.50	6.77	10.09	10.11	16.27	14.68	19.60	22.96	24.64

(contd.)

Table 7.4a (contd.)

FERTILIZERS	1980	1981	1982	1983	1984	1985	1986	1987	1988	1989	1990	1991	1992	1993	1994	1995
All Crops	6.08	4.08	6.11	9.71	16.73	15.49	14.33	15.03	25.53	42.40	45.45	54.84	75.47	52.32	56.72	62.35
Barley	0.04	0.02	0.03	0.05	0.07	0.08	0.06	0.06	0.09	0.13	0.14	0.18	0.23	0.14	0.14	0.18
Cotton	0.36	0.25	0.35	0.46	1.09	1.02	0.71	0.80	1.34	3.07	2.82	3.35	5.00	3.57	3.81	4.66
Groundnut	0.16	0.13	0.16	0.31	0.50	0.35	0.39	0.44	0.96	1.43	1.41	1.58	2.46	1.53	1.53	1.76
Maize	0.17	0.11	0.16	0.27	0.50	0.36	0.41	0.36	0.68	1.30	1.25	1.47	2.54	1.57	1.52	1.88
Rape / M.	0.09	0.06	0.08	0.14	0.27	0.22	0.21	0.33	0.50	0.79	1.02	1.47	1.64	1.34	1.40	1.61
Rice	1.97	1.26	1.79	3.15	5.45	5.13	4.84	4.92	8.66	14.54	15.45	18.16	24.76	17.11	17.97	19.49
Sugarcane	0.36	0.26	0.41	0.54	0.93	0.88	0.86	1.02	1.42	2.72	3.11	4.01	4.45	3.02	3.84	4.32
Wheat	1.54	1.02	1.67	2.60	4.16	4.02	3.57	3.84	6.38	9.34	10.47	12.81	18.28	12.84	14.46	15.15
Other Crops	1.39	0.98	1.45	2.19	3.77	3.45	3.29	3.27	5.51	9.11	9.79	11.80	16.13	11.21	12.05	13.32

TOTAL	1980	1981	1982	1983	1984	1985	1986	1987	1988	1989	1990	1991	1992	1993	1994	1995
All Crops	42.33	43.12	62.38	56.88	69.20	78.78	91.55	105.84	123.20	141.72	150.97	170.28	197.18	185.10	203.33	219.68
Barley	0.71	0.66	0.78	0.61	0.64	0.80	0.83	0.94	0.99	0.92	0.91	1.02	1.09	0.90	1.06	1.05
Cotton	1.99	1.96	2.87	2.43	3.10	3.76	3.82	5.01	5.91	7.67	7.41	8.42	10.72	9.33	10.13	12.37
Groundnut	0.91	0.93	1.39	1.50	1.87	1.78	2.15	3.14	3.77	4.05	5.07	5.41	6.52	5.77	6.31	6.76
Maize	1.02	0.99	1.43	1.11	1.43	1.64	2.17	2.28	2.76	3.30	3.26	3.69	4.89	4.10	4.08	4.89
Rape / M.	0.82	0.90	1.17	1.21	1.91	1.92	2.15	3.56	4.47	4.72	5.78	7.47	7.32	7.46	8.00	9.58
Rice	13.65	14.17	19.27	18.47	22.50	25.53	29.48	31.15	39.64	45.35	47.85	51.68	60.34	55.37	60.43	65.10
Sugarcane	2.17	2.63	3.24	2.99	3.61	4.20	5.01	6.13	7.12	8.30	9.77	11.37	12.06	10.76	12.68	14.49
Wheat	13.23	13.02	20.01	18.02	20.76	22.97	26.40	30.45	36.26	38.05	40.93	43.24	52.41	49.35	53.45	55.27
Other Crops	7.83	7.88	12.22	10.54	13.38	16.19	19.54	23.18	22.31	29.38	29.98	37.99	41.83	42.06	47.19	50.17

Table 7.4b: State-Wise Subsidies in Billion Rupees (1980–1 to 1995–6)

(1995–6 Prices)

IRRIGATION	1980	1981	1982	1983	1984	1985	1986	1987	1988	1989	1990	1991	1992	1993	1994	1995
All India	27.71	29.42	42.79	32.28	34.54	40.45	48.14	52.98	60.75	57.14	55.81	54.59	56.94	57.43	62.10	62.72
Andhra	3.89	4.36	6.30	4.63	4.90	5.89	6.56	3.68	10.39	5.85	5.91	6.00	6.16	6.14	6.41	6.79
Bihar	0.83	0.86	1.31	0.91	1.06	1.54	1.97	3.02	3.04	3.28	2.54	2.13	2.48	2.39	1.96	1.66
Gujarat	1.95	2.11	3.03	2.31	2.46	2.85	5.12	10.07	8.51	7.20	5.84	5.71	5.79	7.00	7.24	8.32
Haryana	1.30	1.54	2.52	1.57	1.84	2.29	2.63	2.09	2.12	2.54	2.10	2.33	2.44	2.72	5.31	2.40
Karnataka	2.60	2.85	4.06	3.08	3.25	3.54	3.84	3.42	3.50	3.43	3.46	3.27	3.67	3.68	3.88	3.87
Kerala	0.46	0.47	0.64	0.55	0.56	0.61	0.74	0.78	0.80	0.80	0.90	0.73	0.78	0.85	0.88	1.03
Maharashtra	3.40	3.75	5.55	3.95	4.29	6.28	6.77	8.87	9.18	9.84	10.35	9.88	10.66	10.67	11.85	13.23
Madhya P.	1.80	1.70	2.11	1.98	1.95	1.74	1.75	2.00	1.79	1.79	2.09	1.80	1.76	1.98	1.58	1.79
Orissa	0.67	0.65	0.89	0.78	0.80	0.58	0.71	0.67	0.66	0.75	0.74	0.89	0.91	1.04	0.99	1.01
Punjab	0.57	0.67	1.21	0.69	0.88	1.08	1.65	1.85	1.75	1.55	1.42	1.42	1.48	1.43	1.25	1.27
Rajastan	2.12	2.33	3.52	2.40	2.67	3.51	3.89	4.38	3.72	3.28	3.79	3.56	4.27	4.60	4.28	4.62
Tamil Nadu	1.24	1.35	1.97	1.48	1.56	1.74	1.88	2.15	1.90	1.98	1.97	1.96	2.10	2.04	2.08	2.13
Uttar P.	4.78	4.83	6.97	5.41	5.82	6.08	7.28	9.28	9.57	10.94	9.92	10.59	11.33	9.29	11.41	11.47
West Bengal	1.69	1.67	2.19	2.03	2.01	1.61	1.67	1.67	1.72	1.81	2.17	1.89	1.80	1.98	1.85	1.94
ELECTRICITY	1980	1981	1982	1983	1984	1985	1986	1987	1988	1989	1990	1991	1992	1993	1994	1995
All India	8.54	9.62	13.49	14.89	17.93	22.85	29.08	37.83	36.92	42.18	49.70	60.85	64.77	75.36	84.50	94.61
Andhra	0.45	0.51	0.91	1.25	1.86	2.12	2.72	3.37	3.68	3.99	5.43	6.13	6.62	7.71	10.23	12.23
Bihar	0.47	0.61	0.92	1.00	1.05	1.57	1.92	2.14	2.18	2.31	2.64	2.73	2.44	2.10	2.25	2.10
Gujarat	0.63	0.37	0.55	0.63	0.92	1.14	1.52	4.38	4.85	4.58	5.70	8.33	9.62	10.04	9.59	11.53
Haryana	0.57	0.85	1.04	0.93	1.24	1.20	1.62	2.94	1.71	2.10	2.26	3.47	4.03	4.50	3.71	4.29
Karnataka	0.04	0.14	0.17	0.28	0.43	1.77	2.32	2.80	2.72	2.59	3.65	3.93	4.51	5.53	6.60	7.76
Kerala	0.02	0.01	0.04	0.06	0.04	0.01	0.07	0.08	0.09	0.11	0.11	0.13	0.13	0.14	0.17	0.25

(contd.)

Table 7.4b (contd.)

ELECTRICITY	1980	1981	1982	1983	1984	1985	1986	1987	1988	1989	1990	1991	1992	1993	1994	1995
Maharashtra	0.90	1.21	1.84	1.87	3.18	3.60	4.35	4.73	5.78	6.33	7.36	9.16	9.40	9.38	12.48	15.75
Madhya P.	0.16	0.26	0.47	0.44	0.57	0.58	0.83	0.95	1.12	1.20	2.59	3.42	3.84	6.29	8.36	9.91
Orissa	0.03	0.03	0.04	0.04	0.04	0.04	0.11	0.13	0.09	0.11	0.08	0.15	0.19	0.33	0.15	0.21
Punjab	1.03	0.98	1.29	1.33	1.92	2.32	3.29	4.68	5.13	5.90	5.73	4.88	6.26	6.64	5.92	5.80
Rajastan	0.51	0.58	0.86	1.09	0.99	1.28	1.25	1.97	1.83	2.35	2.28	2.54	3.17	3.90	4.71	5.65
Tamil Nadu	1.66	2.02	2.65	2.58	2.32	3.12	3.25	3.35	3.55	4.29	4.85	4.60	5.86	6.32	7.17	7.93
Uttar P.	2.71	2.77	3.26	3.26	3.46	3.67	4.58	5.51	6.64	6.99	7.80	8.42	9.44	10.22	9.66	9.82
West Bengal	0.03	0.02	0.07	0.11	0.11	0.14	0.15	0.30	0.45	0.59	0.68	0.94	0.95	0.96	1.28	1.45

FERTILIZERS	1980	1981	1982	1983	1984	1985	1986	1987	1988	1989	1990	1991	1992	1993	1994	1995
All India	6.08	4.08	6.11	9.71	16.73	15.49	14.33	15.03	25.53	42.40	45.45	54.84	75.47	52.32	56.72	62.35
Andhra	0.64	0.44	0.70	1.19	2.00	1.57	1.49	1.65	3.14	5.62	5.87	6.82	9.40	6.53	6.88	7.87
Bihar	0.23	0.14	0.20	0.37	0.78	0.89	0.87	0.83	1.40	2.07	2.17	2.60	3.70	2.48	2.73	3.24
Gujarat	0.39	0.27	0.35	0.63	1.03	0.75	0.67	0.76	1.49	2.55	2.56	3.16	4.45	2.83	3.42	3.39
Haryana	0.26	0.17	0.26	0.41	0.69	0.66	0.69	0.67	1.18	1.96	2.12	2.75	3.78	2.84	2.98	3.25
Karnataka	0.38	0.26	0.34	0.61	1.20	0.99	0.70	0.96	1.85	2.86	3.02	3.90	4.84	3.41	3.44	4.21
Kerala	0.11	0.06	0.10	0.16	0.26	0.25	0.25	0.31	0.50	0.78	0.89	0.97	1.26	0.75	0.83	0.91
Maharashtra	0.47	0.36	0.48	0.71	1.18	1.19	1.07	1.27	2.06	4.33	4.77	5.45	7.03	5.05	5.81	6.17
Madhya P.	0.22	0.16	0.23	0.40	0.76	0.77	0.82	0.87	1.58	2.47	2.94	3.49	4.92	3.28	3.75	3.71
Orissa	0.08	0.06	0.09	0.13	0.23	0.25	0.25	0.26	0.37	0.75	0.70	0.84	1.26	0.88	0.92	1.07
Punjab	0.83	0.55	0.85	1.25	2.14	1.95	1.85	1.90	2.58	4.20	4.34	5.36	7.45	5.08	5.37	5.68
Rajastan	0.15	0.09	0.16	0.26	0.42	0.39	0.41	0.37	0.71	1.05	1.34	1.90	3.05	2.13	2.52	2.89
Tamil Nadu	0.54	0.30	0.45	0.74	1.41	1.18	1.12	1.16	1.78	2.87	3.01	3.62	4.96	3.30	3.66	3.39
Uttar P.	1.27	0.86	1.37	2.07	3.29	3.50	2.78	2.73	4.94	7.67	8.12	9.69	13.53	9.70	10.37	11.71
West Bengal	0.31	0.17	0.26	0.47	0.83	0.73	0.83	0.96	1.47	2.46	2.73	3.26	4.54	3.15	3.15	3.81

(contd.)

Table 7.4b (contd.)

TOTAL

	1980	1981	1982	1983	1984	1985	1986	1987	1988	1989	1990	1991	1992	1993	1994	1995
All India	42.33	43.12	62.38	56.88	69.20	78.78	91.55	105.84	123.20	141.72	150.97	170.28	197.18	185.10	203.33	219.68
Andhra	4.98	5.31	7.91	7.07	8.76	9.58	10.77	8.71	17.20	15.46	17.21	18.95	22.17	20.38	23.52	26.89
Bihar	1.53	1.61	2.42	2.28	2.88	4.00	4.77	5.99	6.62	7.65	7.35	7.46	8.62	6.96	6.94	6.99
Gujarat	2.98	2.74	3.92	3.57	4.40	4.74	7.31	15.21	14.85	14.32	14.10	17.19	19.86	19.87	20.26	23.24
Haryana	2.12	2.56	3.81	2.91	3.76	4.15	4.94	5.70	5.00	6.61	6.49	8.54	10.25	10.06	12.00	9.94
Karnataka	3.02	3.24	4.57	3.97	4.88	6.29	6.86	7.17	8.07	8.87	10.12	11.10	13.02	12.61	13.91	15.84
Kerala	0.58	0.55	0.78	0.77	0.86	0.88	1.06	1.17	1.38	1.69	1.90	1.82	2.18	1.73	1.89	2.19
Maharashtra	4.77	5.32	7.87	6.54	8.66	11.07	12.19	14.86	17.02	20.50	22.49	24.49	27.09	25.11	30.14	35.15
Madhya P.	2.18	2.12	2.81	2.82	3.29	3.08	3.40	3.82	4.49	5.46	7.62	8.71	10.52	11.54	13.69	15.42
Orissa	0.78	0.73	1.02	0.95	1.07	0.87	1.07	1.06	1.12	1.62	1.52	1.89	2.35	2.25	2.06	2.29
Punjab	2.43	2.21	3.34	3.27	4.93	5.35	6.78	8.43	9.46	11.64	11.49	11.66	15.19	13.14	12.54	12.74
Rajastan	2.78	3.00	4.54	3.76	4.07	5.18	5.55	6.73	6.26	6.68	7.42	8.00	10.49	10.63	11.51	13.16
Tamil Nadu	3.45	3.67	5.07	4.79	5.29	6.05	6.25	6.67	7.23	9.14	9.82	10.18	12.92	11.67	12.91	13.46
Uttar P.	8.76	8.46	11.59	10.74	12.57	13.25	14.65	17.52	21.16	25.59	25.83	28.70	34.30	29.20	31.45	32.99
West Bengal	2.03	1.86	2.51	2.61	2.94	2.48	2.64	2.92	3.64	4.86	5.58	6.09	7.29	6.09	6.28	7.20

Table 7.5a: Crop-Wise Subsidies in Rupees per Hectare of Gross Cropper Area (1980–1 to 1995–6)

(1995–6 Prices)

IRRIGATION	1980	1981	1982	1983	1984	1985	1986	1987	1988	1989	1990	1991	1992	1993	1994	1995
All Crops	161	166	247	179	196	228	273	308	336	314	300	300	307	308	330	336
Barley	280	273	392	283	321	377	449	506	566	513	492	514	524	498	552	539
Cotton	151	157	242	175	178	227	268	308	345	327	299	284	302	295	305	305
Groundnut	75	74	118	89	107	112	131	180	165	141	168	161	171	164	177	163
Maize	111	116	175	103	114	140	184	203	214	195	176	189	187	193	180	202
Rape / M.	134	149	227	200	301	315	364	425	539	507	475	481	460	455	522	554
Rice	227	240	345	257	279	322	381	415	473	438	407	397	410	410	439	439
Sugarcane	477	507	567	515	551	673	786	850	924	836	823	788	839	841	828	819
Wheat	388	402	594	438	478	564	662	728	808	752	721	705	719	714	754	761
ELECTRICITY	1980	1981	1982	1983	1984	1985	1986	1987	1988	1989	1990	1991	1992	1993	1994	1995
All Crops	49	54	78	83	102	129	165	220	204	231	267	334	349	404	449	507
Barley	91	92	113	122	138	153	182	265	264	282	308	365	423	462	489	519
Cotton	57	55	79	80	94	137	179	344	277	270	319	377	457	493	499	549
Groundnut	35	34	53	69	83	89	121	214	165	159	273	281	326	346	432	502
Maize	30	32	49	40	47	81	114	143	139	143	165	191	208	229	236	302
Rape / M.	44	42	57	76	110	111	156	274	283	285	348	434	458	518	567	664
Rice	64	77	112	114	135	174	218	261	269	293	352	389	441	490	552	625
Sugarcane	201	234	275	272	359	492	563	709	786	787	982	1126	1293	1421	1460	1634
Wheat	137	140	184	187	226	261	325	426	431	470	539	603	669	738	763	843

(contd.)

Table 7.5a (contd.)

FERTILIZERS

	1980	1981	1982	1983	1984	1985	1986	1987	1988	1989	1990	1991	1992	1993	1994	1995
All Crops	35	23	35	54	95	87	81	88	141	233	244	301	407	281	302	334
Barley	23	13	22	34	53	56	50	53	85	127	143	185	246	176	151	215
Cotton	46	31	45	60	148	135	102	124	183	399	378	438	663	487	484	516
Groundnut	23	18	22	41	70	49	56	64	112	164	169	182	301	184	195	234
Maize	29	18	27	46	86	61	68	64	115	219	212	250	425	263	248	314
Rape / M.	23	14	22	36	68	55	56	72	103	158	177	224	264	213	231	246
Rice	49	31	47	76	132	125	118	127	208	345	362	426	593	402	420	455
Sugarcane	134	82	123	175	314	307	278	311	427	791	845	1044	1245	881	993	1041
Wheat	69	46	71	105	177	175	154	166	265	397	433	551	743	511	563	606

TOTAL

	1980	1981	1982	1983	1984	1985	1986	1987	1988	1989	1990	1991	1992	1993	1994	1995
All Crops	245	243	360	316	392	444	518	616	680	778	812	934	1063	993	1081	1178
Barley	395	379	527	439	512	586	681	824	914	923	943	1064	1192	1135	1192	1272
Cotton	255	243	365	315	420	499	549	776	804	996	996	1098	1421	1275	1287	1370
Groundnut	134	125	192	199	261	250	308	458	442	464	610	624	799	694	804	899
Maize	170	166	250	189	247	283	367	410	468	557	553	630	820	684	664	818
Rape / M.	200	205	306	311	479	482	577	771	925	950	1000	1139	1182	1186	1320	1464
Rice	340	348	504	448	547	621	716	803	950	1076	1121	1212	1444	1302	1411	1520
Sugarcane	813	823	965	962	1223	1472	1627	1871	2137	2413	2651	2958	3377	3143	3280	3493
Wheat	594	588	849	730	881	999	1141	1320	1504	1619	1694	1859	2131	1963	2080	2210

Table 7.5b: State-Wise Subsidies in Rupees per Hectare of Gross Cropped Area (1980–1 to 1995–6)

(1995–6 Prices)

IRRIGATION	1980	1981	1982	1983	1984	1985	1986	1987	1988	1989	1990	1991	1992	1993	1994	1995
All India	161	166	247	179	196	228	273	308	336	314	300	300	307	308	330	336
Andhra	317	334	494	346	402	487	561	303	790	441	448	455	483	484	502	520
Bihar	74	81	136	89	103	146	189	292	289	314	242	210	265	245	199	166
Gujarat	182	193	277	220	237	294	493	970	889	668	552	543	526	656	647	825
Haryana	237	264	474	275	333	409	465	445	352	450	355	419	417	468	891	401
Karnataka	244	254	364	268	279	317	325	279	296	283	294	264	296	296	323	323
Kerala	160	161	224	191	196	214	257	269	268	266	299	241	257	278	288	335
Maharashtra	173	191	278	189	210	306	336	451	429	457	470	491	507	500	555	620
Madhya P.	84	78	95	87	87	76	79	88	79	80	87	78	74	80	64	72
Orissa	77	74	107	81	91	63	76	74	72	80	77	91	96	107	102	105
Punjab	84	97	174	100	125	151	228	252	237	210	189	189	195	187	162	163
Rajastan	122	125	192	127	154	194	220	329	197	183	196	197	212	239	210	235
Tamil Nadu	192	196	326	212	220	255	289	320	294	290	296	281	297	285	297	340
Uttar P.	194	195	282	216	232	243	289	380	379	432	389	419	441	364	443	445
West Bengal	221	226	313	259	256	202	203	199	207	215	250	218	211	229	212	271
ELECTRICITY	**1980**	**1981**	**1982**	**1983**	**1984**	**1985**	**1986**	**1987**	**1988**	**1989**	**1990**	**1991**	**1992**	**1993**	**1994**	**1995**
All India	49	54	78	83	102	129	165	220	204	231	267	334	349	404	449	507
Andhra	37	39	71	93	152	175	232	278	280	301	412	464	519	607	800	938
Bihar	43	57	95	98	102	150	184	207	207	222	252	269	261	215	228	209
Gujarat	59	34	50	60	88	118	146	422	507	424	539	793	874	941	857	1143
Haryana	104	146	196	164	225	215	287	627	284	372	382	622	689	773	622	717
Karnataka	4	12	15	25	37	159	196	228	230	214	310	317	363	445	549	649

(contd.)

Table 7.5b (contd.)

ELECTRICITY

	1980	1981	1982	1983	1984	1985	1986	1987	1988	1989	1990	1991	1992	1993	1994	1995
Kerala	5	5	12	20	13	4	25	27	32	35	36	42	44	45	57	81
Maharashtra	46	61	92	90	155	175	216	241	270	294	334	455	447	439	584	738
Madhya P.	7	12	21	19	26	25	37	42	49	53	108	148	161	254	339	396
Orissa	3	3	5	4	5	5	12	14	9	12	8	15	20	33	16	22
Punjab	153	142	186	191	274	324	456	639	693	798	764	648	829	871	769	748
Rajastan	30	31	47	58	57	71	71	148	97	131	118	140	157	202	231	287
Tamil Nadu	257	292	440	371	328	458	500	498	551	629	731	659	829	883	1021	1266
Uttar P.	110	112	132	130	138	147	182	225	263	276	306	333	367	400	375	381
West Bengal	4	2	10	14	14	18	18	35	55	70	79	109	111	110	147	202

FERTILIZERS

	1980	1981	1982	1983	1984	1985	1986	1987	1988	1989	1990	1991	1992	1993	1994	1995
All India	35	23	35	54	95	87	81	88	141	233	244	301	407	281	302	334
Andhra	52	34	55	89	164	130	128	136	238	424	445	517	737	515	538	603
Bihar	20	13	20	36	75	85	84	81	133	198	207	256	395	254	277	323
Gujarat	37	25	32	60	99	77	64	73	155	236	242	301	404	265	306	336
Haryana	47	29	49	72	124	118	121	144	196	347	359	493	646	489	500	544
Karnataka	36	23	30	53	103	88	59	78	157	236	257	315	390	274	286	352
Kerala	38	22	35	57	91	88	87	108	167	258	293	320	414	246	273	298
Maharashtra	24	18	24	34	58	58	53	65	96	201	217	270	334	236	272	289
Madhya P.	10	7	10	18	34	33	37	38	70	110	123	151	207	132	152	148
Orissa	10	6	11	14	26	27	27	29	41	79	73	86	134	90	95	110
Punjab	123	80	123	179	304	272	256	260	349	568	578	713	986	666	699	732
Rajastan	9	5	9	14	24	22	23	28	37	58	69	105	151	110	124	147
Tamil Nadu	84	43	75	106	199	174	172	173	276	420	454	518	702	462	521	542
Uttar P.	52	34	55	83	131	140	110	112	196	302	319	383	527	380	403	454
West Bengal	41	24	37	59	105	91	101	115	177	292	315	376	532	363	361	531

(contd.)

Table 7.5b (contd.)

TOTAL	1980	1981	1982	1983	1984	1985	1986	1987	1988	1989	1990	1991	1992	1993	1994	1995
All India	245	243	360	316	392	444	518	616	680	778	812	934	1063	993	1081	1178
Andhra	405	407	620	528	717	792	921	716	1308	1166	1305	1436	1739	1606	1840	2062
Bihar	137	151	251	223	280	380	457	580	629	734	701	735	921	714	703	698
Gujarat	278	252	358	340	425	489	704	1465	1551	1328	1333	1637	1805	1861	1811	2305
Haryana	388	439	719	511	683	741	873	1217	832	1169	1096	1534	1752	1730	2013	1663
Karnataka	283	289	410	346	419	564	580	585	683	732	861	896	1049	1014	1158	1325
Kerala	203	188	271	268	299	306	370	404	467	559	628	603	715	568	619	715
Maharashtra	243	271	394	313	423	539	606	757	796	952	1021	1216	1288	1175	1411	1648
Madhya P.	102	97	126	124	147	134	153	168	198	243	319	377	442	465	554	616
Orissa	90	83	123	99	122	94	115	116	122	171	158	192	250	231	212	237
Punjab	360	318	483	469	703	747	940	1151	1279	1575	1532	1551	2011	1724	1630	1644
Rajastan	160	161	247	199	236	286	315	505	332	373	383	442	520	552	565	669
Tamil Nadu	533	531	841	690	747	887	960	991	1120	1339	1481	1459	1828	1630	1838	2147
Uttar P.	356	341	469	429	500	530	581	717	838	1010	1014	1135	1336	1143	1222	1279
West Bengal	266	252	359	332	375	310	322	349	439	576	644	703	854	702	720	1003

Table 7.6: *State-Wise Subsidies in Rupees per Cultivator (1980–1 to 1995–6)*

(1995–6 Prices)

	1980	1981	1982	1983	1984	1985	1986	1987	1988	1989	1990	1991	1992	1993	1994	1995
IRRIGATION																
All India	304	318	454	337	354	407	476	514	579	535	513	493	507	504	537	535
Andhra	528	589	845	617	649	775	857	478	1340	750	753	761	766	751	771	801
Bihar	93	95	142	97	111	158	200	300	297	314	239	191	218	206	167	139
Gujarat	480	512	726	544	572	653	1157	2240	1866	1555	1243	1213	1212	1443	1471	1664
Haryana	799	939	1520	935	1084	1336	1518	1191	1194	1418	1158	1275	1306	1430	2733	1209
Karnataka	504	546	767	574	599	643	689	605	612	592	590	553	611	603	626	615
Kerala	522	527	713	601	609	654	778	812	816	812	903	717	764	816	839	973
Maharashtra	405	440	638	446	476	684	724	931	946	995	1028	972	1029	1011	1102	1208
Madhya P.	176	163	198	182	177	154	152	170	149	146	167	140	133	147	115	128
Orissa	167	159	217	187	189	136	163	153	149	167	163	194	194	220	206	208
Punjab	323	379	676	386	485	592	893	994	934	819	745	740	759	723	622	624
Rajastan	336	362	536	357	388	501	543	599	497	429	486	435	512	539	491	519
Tamil Nadu	234	253	366	273	286	317	340	387	338	351	346	346	366	353	357	360
Uttar P.	255	255	362	277	293	301	356	446	453	510	455	481	504	405	488	480
West Bengal	375	364	465	420	406	318	320	313	315	324	378	324	303	328	301	310
ELECTRICITY																
	1980	1981	1982	1983	1984	1985	1986	1987	1988	1989	1990	1991	1992	1993	1994	1995
All India	94	104	143	155	184	230	287	367	352	395	457	550	577	662	731	807
Andhra	61	68	122	166	246	279	355	438	474	512	692	776	824	942	1229	1444
Bihar	53	67	100	107	110	162	195	213	213	222	249	244	215	181	191	175
Gujarat	156	89	131	149	213	261	342	975	1064	989	1214	1770	2014	2070	1948	2305
Haryana	349	520	627	556	732	702	937	1677	963	1171	1247	1894	2160	2360	1909	2161
Karnataka	8	26	32	53	79	322	416	496	476	447	622	664	750	906	1065	1235
Kerala	17	15	39	63	40	11	76	82	96	108	109	124	130	131	167	237

(contd.)

Table 7.6 (contd.)

ELEXTRICITY

	1980	1981	1982	1983	1984	1985	1986	1987	1988	1989	1990	1991	1992	1993	1994	1995
Maharashtra	108	141	212	212	353	392	465	496	595	641	731	900	907	889	1161	1438
Madhya P.	16	25	45	41	52	51	72	80	93	98	207	265	291	469	611	710
Orissa	7	7	10	10	10	10	26	29	19	25	18	33	40	69	32	44
Punjab	588	557	722	740	1060	1269	1784	2519	2735	3119	3006	2543	3220	3363	2954	2855
Rajastan	81	90	130	163	144	183	175	270	245	308	292	311	379	457	541	635
Tamil Nadu	313	378	493	476	426	569	588	602	634	760	852	812	1023	1093	1227	1343
Uttar P.	145	146	169	167	174	182	224	265	314	326	358	382	420	446	413	411
West Bengal	6	3	14	23	22	28	29	56	83	105	119	161	160	158	208	231

FERTILIZERS

	1980	1981	1982	1983	1984	1985	1986	1987	1988	1989	1990	1991	1992	1993	1994	1995
All India	67	44	65	101	171	156	142	146	243	397	418	495	672	459	491	532
Andhra	86	60	94	159	265	207	195	215	404	720	747	864	1170	799	826	929
Bihar	25	15	21	39	82	92	88	83	137	198	204	233	326	214	232	271
Gujarat	97	66	83	149	239	171	151	168	326	551	545	672	932	584	695	677
Haryana	157	103	157	245	405	385	397	385	664	1094	1171	1500	2024	1492	1532	1640
Karnataka	73	49	64	115	222	179	125	169	324	493	514	660	806	558	556	669
Kerala	123	72	111	179	282	270	264	325	508	789	885	952	1229	722	797	865
Maharashtra	55	42	55	80	131	129	115	133	212	438	474	535	678	479	541	564
Madhya P.	21	15	22	37	69	68	71	74	132	202	235	270	374	244	274	266
Orissa	21	14	22	31	55	58	58	59	84	167	153	184	270	186	192	219
Punjab	473	313	477	695	1178	1065	1003	1023	1378	2219	2276	2797	3828	2571	2684	2794
Rajastan	24	14	25	39	61	56	57	50	94	137	172	232	365	249	289	324
Tamil Nadu	102	56	84	137	258	216	202	209	317	508	530	638	867	571	626	575
Uttar P.	68	45	71	106	166	173	136	131	234	357	373	440	602	423	443	490
West Bengal	69	38	54	96	167	143	159	180	269	440	476	557	764	521	512	609

(contd.)

Table 7.6 (contd.)

TOTAL

	1980	1981	1982	1983	1984	1985	1986	1987	1988	1989	1990	1991	1992	1993	1994	1995
All India	465	466	662	593	709	793	905	1027	1174	1327	1388	1538	1756	1625	1760	1874
Andhra	675	717	1061	942	1159	1260	1407	1131	2218	1982	2192	2401	2760	2492	2826	3174
Bihar	172	178	263	244	302	412	483	596	647	735	693	668	759	602	591	585
Gujarat	733	666	939	842	1024	1086	1650	3384	3256	3094	3002	3654	4157	4096	4113	4647
Haryana	1305	1562	2303	1736	2221	2423	2852	3252	2821	3683	3575	4669	5491	5281	6174	5009
Karnataka	585	621	863	742	899	1144	1231	1270	1412	1532	1726	1877	2168	2068	2247	2519
Kerala	662	614	862	843	931	936	1119	1219	1420	1709	1896	1792	2123	1670	1803	2075
Maharashtra	568	623	905	738	960	1205	1303	1560	1754	2074	2234	2407	2614	2379	2803	3210
Madhya P.	213	203	264	260	297	273	295	325	374	445	609	675	799	860	999	1104
Orissa	195	180	249	228	254	205	247	241	253	359	334	411	505	475	430	470
Punjab	1385	1248	1876	1822	2722	2926	3679	4537	5047	6158	6026	6080	7807	6657	6261	6272
Rajastan	441	466	691	559	594	739	775	919	837	874	950	978	1255	1245	1321	1479
Tamil Nadu	649	687	943	885	971	1101	1130	1198	1289	1618	1727	1797	2257	2017	2210	2278
Uttar P.	468	446	602	549	633	657	715	842	1002	1193	1186	1303	1526	1274	1344	1382
West Bengal	451	405	534	540	595	489	508	549	668	869	973	1042	1226	1008	1021	1150

Table 7.7: Crop-Wise Subsidies in Rupees per Ton (1980–1 to 1995–6)

(1995–6 Prices)

IRRIGATION	1980	1981	1982	1983	1984	1985	1986	1987	1988	1989	1990	1991	1992	1993	1994	1995
Barley	221	237	311	214	259	263	329	367	355	342	290	289	317	301	383	294
Cotton	995	945	1486	1245	910	1150	1587	1834	1703	1296	1328	1316	1173	1181	1186	1260
Groundnut	103	76	161	94	120	155	155	210	146	152	186	196	163	174	173	162
Maize	96	100	152	76	78	122	144	197	153	120	116	137	111	120	124	127
Rape / M.	239	276	394	297	391	468	520	568	595	611	526	538	593	537	549	605
Rice	170	183	280	176	197	207	259	283	280	251	234	227	235	217	230	244
Sugarcane	8	9	10	9	10	11	13	14	15	13	13	12	13	13	12	12
Wheat	238	238	327	238	256	275	345	364	360	355	316	295	309	300	295	307
ELECTRICITY	1980	1981	1982	1983	1984	1985	1986	1987	1988	1989	1990	1991	1992	1993	1994	1995
Barley	72	80	90	92	111	107	134	192	166	188	182	205	256	279	340	283
Cotton	376	330	484	570	482	697	1061	2049	1367	1070	1419	1750	1779	1978	1942	2270
Groundnut	48	35	72	73	93	124	144	250	145	171	302	344	311	368	420	498
Maize	26	28	43	30	33	71	89	139	100	87	109	139	124	143	163	190
Rape / M.	78	78	98	113	142	165	223	367	312	343	385	485	590	612	596	725
Rice	48	59	91	78	95	112	148	178	159	168	202	222	253	260	289	348
Sugarcane	3	4	5	5	6	8	9	12	13	12	15	17	20	21	20	24
Wheat	84	83	101	102	121	127	170	213	192	221	236	252	288	310	298	340

(contd.)

Table 7.7 (contd.)

FERTILIZERS

	1980	1981	1982	1983	1984	1985	1986	1987	1988	1989	1990	1991	1992	1993	1994	1995
Barley	19	12	18	25	43	39	37	38	53	85	84	104	149	106	105	117
Cotton	302	184	274	426	754	685	603	736	904	1583	1682	2030	2578	1952	1885	2130
Groundnut	31	18	30	44	78	69	66	75	99	176	187	223	287	195	190	232
Maize	25	16	24	34	59	53	53	62	82	134	140	182	254	164	171	197
Rape / M.	40	25	38	53	89	82	80	96	114	191	195	251	341	251	243	268
Rice	37	24	38	52	93	80	80	87	123	198	208	243	340	213	220	253
Sugarcane	2	1	2	3	5	5	5	5	7	12	13	16	20	13	14	15
Wheat	42	27	39	57	94	85	80	83	118	187	190	230	319	215	220	244

TOTAL

	1980	1981	1982	1983	1984	1985	1986	1987	1988	1989	1990	1991	1992	1993	1994	1995
Barley	311	328	419	332	412	409	500	597	574	616	556	598	722	687	828	694
Cotton	1673	1459	2244	2240	2146	2532	3250	4619	3974	3948	4429	5096	5530	5112	5014	5660
Groundnut	182	129	263	211	290	348	366	536	391	499	675	763	761	737	783	892
Maize	147	143	219	140	170	247	286	399	335	342	364	458	489	427	459	513
Rape / M.	357	380	530	463	622	715	823	1031	1021	1144	1106	1274	1524	1400	1389	1598
Rice	254	266	409	307	386	400	487	548	562	616	644	692	828	690	739	846
Sugarcane	14	14	17	17	21	25	27	31	35	37	41	45	53	47	46	52
Wheat	364	348	468	396	471	488	596	659	670	763	742	776	916	825	813	890

Table 7.8: Official Domestic Supports (1995–6)
Notification of the Delegation of India to the WTO (G/AG/N/IND/1, June 1998)

			Rs. (million)	US$ (million)
DS:1	Measures exempt from the reduction commitment – 'Green Box'		73,437	2,196
	Public stockholding for food security purposes	- Buffer stock operations	52500	1570
	General services	- Research	5451	163
		- Pest and disease control	276	8
		- Training services	158	5
		- Extension and advisory services	-	-
		- Command area development programme	404	12
		- Marketing and promotion services	3506	105
		- Inspection services	n.a.	n.a.
		- Provision for livestock health facilities	3505	105
		- Infrastructural services	1	0
	Relief from natural disasters	- Scarcity relief and natural calamities	4182	125
	Structural adjustment assistance provided through investment aids	- Dry land farming	1880	56
		- Scheme for reclamation of alkaline soils	100	3
		- Drought-prone area programme	n.a.	n.a.
	Payment under environmental programmes	- Soil conservation in catchment areas of river valley projects	651	19
		- Integrated watershed management in the catchment areas of flood prone rivers in the Indo-Gangetic basin	300	9
		- Fodder grassland and pasture development	n.a.	n.a.
		- Desert development programmes	n.a.	n.a.
		- Control of shifting cultivation	160	5
	Income insurance	- Crop insurance scheme	364	11
	Domestic food aid		n.a.	n.a.
DS:2	Measures exempt from the reduction commitment– 'Development Programmes'		8,506	254
	Agricultural input subsidies to low income or resource poor producers	- Small farmer development assistance	5000	149
		- Assistance to small-holders for easy access to inputs	-	-
	Investment subsidies generally available to agriculture	- On farm development work (field channels, land leveling, shallow wells, etc.)	3506	105
		- Subsidies for asset formation (provision of milch animals, farm implements to resource poor farmers, etc. under certain schemes)	-	-
DS:3	Measures exempt from the reduction commitment – 'Exempt Direct Payments'		0	0

Table 7.8 (contd.)

DS:5	Product-Specific Aggregate Measurements of Support: Market Price Support		-990,667	-29,619
	Sugar cane	[(12.7 US$/ton - 11.64 US$/ton) x 174 million tons] - 0	6154	184
	Tobacco	[(605.43 US$/ton - 912.82 US$/ton) x 0.59 million tons] - 0	-6054	-181
	Soya bean	[(191.35 US$/ton - 229.77 US$/ton) x 4.99 million tons] - 0	-6422	-192
	Jute	[(146.5 US$/ton - 417.55 US$/ton) x 1.43 million tons] - 0	-12977	-388
	Rapeseed and mustard toria	[(257.12 US$/ton - 513.37 US$/ton) x 6.59 million tons] - 0	-56492	-1689
	Pulses (gram, urad, moong and tur)	[(231.71 US$/ton - 345.66 US$/ton) x 14.97 million tons] - 0	-57061	-1706
	Groundnut	[(269.08 US$/ton - 514.58 US$/ton) x 7.37 million tons] - 0	-60506	-1809
	Cotton	[(343.83 US$/ton - 1292.64 US$/ton) x 2.22 million tons] - 0	-70439	-2106
	Coarse cereals (bajra, jawar, maize and barley)	[(89.69 US$/ton - 238.57 US$/ton) x 30.43 million tons)] - 0	-151515	-4530
	Rice	[(168.923 US$/ton - 262.51 US$/ton) x 80.96 million tons] - 0	-253428	-7577
	Wheat	[(113.61 US$/ton - 264 US$/ton) x. 64 million tons] - 0	-321927	-9625
DS:6	Product-Specific Aggregate Measurements of Support: Non-Exempt Direct Payments		0	0
DS:9	Non-Product-Specific Aggregate Measurements of Support		193,058	5,772
	Subsidy on electricity	[2436.64 US$ million + 0 - 0]	81498	2437
	Fertilizer subsidy	[1864.14 US$ million + 0 - 0]	62350	1864
	Irrigation subsidy	[1345.41 US$ million + 0 - 0]	45000	1345
	Credit subsidy (1994–5)	[101.95 US$ million + 0 - 0]	3410	102
	Subsidy on average supply of seeds	[23.92 US$ million + 0 - 0]	800	24

Notes: DS5 - Total Market Price Support relevant 'product-specific' de minimis level Article 6.4b = [(Applied Administered Price at the average exchange rate prevalent in the base period – External Reference Price) x Eligible Production] - Associated fees/levies- Applied Administered Price is procurement price for common paddy. (For converting to the equivalent price of rice a coefficient of 1.5 has been used.)

- Exchange rate US$ to Rs.: 1995-6 = 33.447, 1986-89 Average = 13.409.

DS9 - Total Non-Product-Specific Support = [Non-Product-Specific Budgetary Outlays + Other Non-Product-Specific Support (include calculation details) - Associated Fees/ Levies].

- Total Non-Product-Specific Support in 1995-6 represents 7.52 per cent of the corresponding total value of Indian agricultural production (US$ million 76,736) and is accordingly below the relevant 'non-product-specific' de minimis level (Article 6.4(b) refers).

Table 7.9a: Basic Statistics Used for Assessing Export Competitiveness (1980-1 to 1995-6)

GROSS CROPPED AREA (1000 ha.)	1981	1982	1983	1984	1985	1986	1987	1988	1989	1990	1991	1992	1993	1994	1995
Barley	1728	1483	1386	1253	1369	1225	1143	1081	991	962	954	916	794	891	824
Cotton	8057	7871	7721	7382	7533	6948	6459	7343	7695	7440	7661	7542	7321	7871	9035
Groundnut	7429	7215	7539	7168	7125	6982	6844	8529	8710	8309	8668	8166	8322	7849	7524
Maize	5936	5720	5859	5800	5797	5923	5560	5897	5915	5904	5859	5963	5995	6136	5979
Rape / Mustard	4399	3827	3874	3987	3980	3719	4619	4832	4967	5782	6553	6193	6289	6060	6546
Rice	40708	38262	41244	41159	41137	41167	38806	41736	42167	42687	42649	41775	42539	42814	42837
Sugarcane	3193	3358	3110	2953	2849	3079	3279	3329	3438	3686	3844	3572	3422	3867	4147
Wheat	22144	23567	24672	23565	22997	23131	23063	24109	23502	24167	23262	24589	25147	25700	25011

PRODUCTION (1000 t.)	1981	1982	1983	1984	1985	1986	1987	1988	1989	1990	1991	1992	1993	1994	1995
Barley	1993	1867	1834	1556	1962	1669	1577	1722	1486	1632	1699	1512	1313	1283	1510
Cotton	1340	1281	1086	1446	1484	1174	1085	1486	1942	1673	1651	1938	1826	2021	2186
Groundnut	7223	5282	7086	6436	5121	5875	5854	9659	8101	7515	7095	8565	7829	8062	7579
Maize	6897	6549	7922	8442	6644	7593	5721	8229	9651	8962	8064	9992	9601	8884	9534
Rape / Mustard	2382	2207	2608	3073	2681	2605	3455	4377	4125	5229	5863	4803	5328	5758	6000
Rice	53248	47116	60097	58337	63825	60557	56863	70489	73573	74291	74678	72868	80298	81814	76975
Sugarcane	186358	189506	174076	170319	170648	186090	196737	203037	225569	241046	253995	228033	229659	275540	281100
Wheat	37452	42794	45476	44069	47052	44323	46169	54110	49850	55135	55690	57210	59840	65767	62097

(contd.)

Table 7.9a (contd.)

WHOLESALE PRICE INDEX	1981	1982	1983	1984	1985	1986	1987	1988	1989	1990	1991	1992	1993	1994	1995
Barley	100.0	104.1	119.6	119.2	134.9	129.9	137.2	169.1	168.0	184.7	239.6	248.2	231.2	326.7	299.1
Cotton (raw)	100.0	87.5	94.9	116.7	99.5	81.9	122.7	143.0	148.6	139.1	217.1	229.7	215.1	360.0	403.2
Groundnut	100.0	96.8	107.7	117.9	111.7	132.1	159.3	165.1	146.1	192.5	240.3	245.6	226.4	263.7	316.2
Maize	100.0	106.3	117.5	97.7	112.5	134.1	129.0	153.4	150.9	143.3	206.6	237.5	184.7	258.6	295.8
Rape / Mustard	100.0	89.3	109.3	117.9	96.7	116.6	171.9	154.4	131.3	179.9	223.3	211.4	214.6	238.3	270.4
Rice	100.0	113.5	128.7	122.3	125.3	132.1	141.2	159.1	167.7	172.6	203.2	248.2	259.7	286.9	311.6
Sugarcane	100.0	100.3	104.2	109.0	115.8	133.0	138.6	148.3	159.2	*178.4*	181.9	205.1	255.2	281.8	310.3
Wheat	100.0	106.4	115.9	111.3	115.7	126.0	130.3	148.5	154.9	156.9	193.7	230.0	242.5	272.0	271.0

INDIAN PRICES (Rs./t.)	1981	1982	1983	1984	1985	1986	1987	1988	1989	1990	1991	1992	1993	1994	1995
Barley (1981: n=5 std=10)	1214	1264	1452	1447	1637	1577	1665	2053	2039	2243	2909	3014	2806	3967	3631
Cotton (1981: n=16 std=448)	15088	13202	14313	17606	15015	12352	18518	21580	22413	20980	32756	34651	32461	54309	60841
Groundnut (1981: n=6 std=100)	4946	4788	5326	5832	5526	6533	7877	8167	7228	9520	11885	12147	11198	13044	15637
Maize (1981: n=12 std=14)	1509	1604	1773	1474	1697	2024	1947	2315	2277	2162	3118	3584	2786	3902	4463
Rape/Must (1981: n=6 std=52)	4694	4192	5132	5535	4540	5474	8071	7249	6162	8443	10482	9925	10074	11185	12694
Rice (1981: n=55 std=68)	2500	2838	3217	3056	3132	3302	3529	3977	4193	4315	5080	6206	6491	7174	7789

(contd.)

Table 7.9a (contd.)

INDIAN PRICES (Rs./t.)	1981	1982	1983	1984	1985	1986	1987	1988	1989	1990	1991	1992	1993	1994	1995
Sugarcane (1981: n=7 std=1)	146	146	152	159	169	194	202	216	232	260	266	299	373	411	453
Wheat (1981: n=28 std=26)	1597	1699	1851	1777	1848	2012	2081	2371	2474	2505	3093	3673	3872	4344	4328

INPUTS SUBSIDIES (Rs./t.)	1981	1982	1983	1984	1985	1986	1987	1988	1989	1990	1991	1992	1993	1994	1995
Barley	100	137	117	157	168	218	284	295	343	344	424	554	577	765	694
Cotton (lint)	442	734	793	816	1038	1420	2194	2042	2199	2737	3613	4247	4299	4633	5660
Groundnut	39	86	75	110	143	160	254	201	278	417	541	585	620	723	892
Maize	43	72	50	65	101	125	189	172	190	225	325	376	359	424	513
Rape / Mustard	115	173	164	236	293	360	490	525	637	683	903	1170	1178	1283	1598
Rice	81	134	109	147	164	213	260	289	343	398	491	636	580	682	846
Sugarcane	4	6	6	8	10	12	15	18	20	25	32	41	39	43	52
Wheat	105	153	140	179	200	260	313	344	425	459	551	704	694	751	890

Table 7.9b: Apparent and Real Export Competitiveness (1980–1 to 1995–6)

WORLD PRICES ($/t.)	1981	1982	1983	1984	1985	1986	1987	1988	1989	1990	1991	1992	1993	1994	1995
Barley	181	161	143	146	120	107	101	129	145	153	131	139	129	113	141
Cotton (lint)	1719	1443	1521	1693	1457	1134	1236	1571	1457	1647	1615	1325	1160	1431	1808
Groundnut (shelled)	964	697	655	733	587	601	600	572	616	666	773	602	654	731	724
Maize	153	128	142	149	126	117	105	128	132	136	132	134	128	132	139
Rape + Mustard seed	331	329	340	350	313	299	293	351	344	360	322	293	241	280	298
Rice, Paddy	308	291	238	207	184	156	256	287	310	273	293	282	263	268	254
Sugar (total), Raw Equiv. / 30	16.8	12.3	12.2	11.6	10.6	11.6	11.0	12.1	13.2	14.7	11.4	10.2	9.9	10.9	12.2
Wheat	188	172	162	157	144	135	114	140	170	162	130	153	141	135	168

EXCHANGE RATE (Rs.)	1981	1982	1983	1984	1985	1986	1987	1988	1989	1990	1991	1992	1993	1994	1995
US dollar	8.968	9.666	10.34	11.889	12.235	12.778	12.966	14.482	16.649	17.943	24.474	30.649	31.366	31.399	33.450

APPARENT COMPETITIVENESS	1981	1982	1983	1984	1985	1986	1987	1988	1989	1990	1991	1992	1993	1994	1995
Barley	25%	19%	2%	17%	–11%	–15%	–27%	–10%	16%	18%	9%	29%	31%	–12%	23%
Cotton	2%	5%	9%	13%	16%	15%	–16%	5%	8%	29%	17%	15%	11%	–21%	–1%
Groundnut	43%	29%	21%	33%	23%	15%	–1%	1%	30%	20%	37%	34%	45%	43%	35%
Maize	–10%	–30%	–21%	17%	–10%	–35%	–44%	–25%	–4%	11%	4%	13%	31%	6%	4%
Rape / Mustard	–58%	–32%	–46%	–33%	–19%	–43%	–113%	–43%	–8%	–31%	–33%	–11%	–33%	–27%	–27%
Rice	10%	–1%	–31%	–24%	–39%	–66%	–6%	4%	19%	12%	29%	28%	21%	15%	8%
Sugarcane	3%	–23%	–21%	–15%	–30%	–31%	–41%	–23%	–6%	1%	5%	4%	–19%	–20%	–11%
Wheat	5%	–2%	–11%	5%	–5%	–17%	–41%	–17%	13%	14%	3%	22%	12%	–3%	23%
Average	3%	–4%	–12%	1%	–9%	–22%	–36%	–13%	8%	9%	9%	17%	12%	–2%	7%
Weighted average by the GCA	6%	–2%	–17%	–7%	–18%	–37%	–25%	–7%	14%	12%	16%	22%	16%	5%	10%

(contd.)

Table 7.9b (contd.)

REAL COMPETITIVENESS	1981	1982	1983	1984	1985	1986	1987	1988	1989	1990	1991	1992	1993	1994	1995
Barley	19%	10%	–6%	8%	–22%	–31%	–48%	–26%	2%	6%	–4%	16%	17%	–33%	9%
Cotton	–1%	0%	4%	8%	10%	5%	–29%	–4%	–1%	20%	8%	4%	–1%	–31%	–10%
Groundnut	42%	28%	20%	32%	21%	13%	–5%	–1%	27%	17%	34%	31%	42%	40%	32%
Maize	–13%	–36%	–24%	13%	–17%	–44%	–58%	–35%	–13%	2%	–6%	4%	22%	–4%	–7%
Rape / Mustard	–62%	–37%	–51%	–39%	–26%	–53%	–125%	–53%	–19%	–41%	–45%	–24%	–49%	–42%	–43%
Rice	7%	–6%	–35%	–30%	–47%	–77%	–14%	–3%	12%	4%	22%	21%	14%	6%	–1%
Sugarcane	0%	–28%	–25%	–21%	–38%	–39%	–52%	–34%	–15%	–9%	–6%	–9%	–32%	–33%	–23%
Wheat	–1%	–11%	–19%	–5%	–16%	–32%	–62%	–34%	–2%	–2%	–14%	7%	–3%	–20%	7%
Average	–1%	–10%	–17%	–4%	–17%	–32%	–49%	–24%	–1%	0%	–1%	6%	1%	–15%	–5%
Weighted average by the GCA	3%	–8%	–23%	–13%	–26%	–48%	–37%	–16%	5%	2%	6%	12%	6%	–6%	–2%

Sources / Methodology:

Gross Cropped Area (GCA):	Ministry of Agriculture, Government of India.
Production:	Ministry of Agriculture, Government of India.
Wholesale Price Index (WPI):	Ministry of Industry, Government of India.
Indian Prices:	1981: Ministry of Industry, Government of India, 1982–95: 1981 price x (WPI / 100).
Inputs subsidy:	Authors (irrigation + electricity + fertilizers subsidies).
World prices:	FAO, United Nations: Value (f.o.b.) / Quantity Exported in total world (former 1991).
Exchange rate:	Ministry of Finance, Government of India.
Apparent competitiveness:	[World price – (Indian price / Exchange rate)] / World price.
Real competitiveness:	[World price – (Indian price + Inputs subsidy / Exchange rate)] / World price.

NOTES

1. Up to the Seventh Plan, financial expenditures on CAD programmes are not given by year, but for the whole concerned period. To get annual estimates, these plan outlays have been simply divided by the concerned number of years.
2. State-by-state expenditures on CAD+state-by-state releases of the central assistance to CAD.
3. Inclusive interest on capital, but keeping aside soil and water conservation schemes, *Ayacut* development programme, development on hills area and dryland development programmes.
4. The deflator being the national deflator of the GDP at factor cost.
5. Fuel, power purchase, operational and maintenance cost, establishment and administration, other miscellaneous expenditure, interest and depreciation on fixed costs.
6. Unit prices of imported N, P_2O_5 and K_20, exchange rate of the rupee against the US dollar, unit costs of delivering imported N, P_2O_5 and K_20, unit charges of pool and handling, Indian inflation rate, unit price of N, P_2O_5 and K_20 paid by farmers, etc.
7. 33 per cent up to 1987–8, 41 per cent in 1988–9, 52 per cent in 1989–90, 64 per cent in 1990–1, 81 per cent in 1991–2, 100 per cent in 1992–3 onwards.
8. See for more details Dorin, 1999 and Dorin and Landy, 2002.

REFERENCES

ACHARYA, S.S., 'Fertilizer Subsidies in Indian Agriculture. Some Issues', Paper presented at the seminar 'Agricultural Incentives and Sustainable Development. Past Trends and Futures Scenarios', CSH – India International Centre, New Delhi, 3-4 April 2000.

BANSIL, P.C., *Agricultural Statistical Compendium. Foodgrains* (Vol. I), Techno Economic Research Institute, New Delhi, 1990.

DORIN, Bruno, 'Food Policy and Nutritional Security. Unequal Access to Lipids in India', *Economic and Political Weekly*, Vol. 34, No. 26, Mumbai, 26 June 1998, pp. 1709–17.

DORIN, Bruno and Frédéric LANDY, *Agriculture et alimentation de l'Inde*, INRA Editions, Paris, 2002.

CENTER FOR MONITORING INDIAN ECONOMY, *Agriculture*, Mumbay, various issues (1995, 1998, 1999)

CENTRAL WATER COMMISSION, *Water and Related Statistics*, New Delhi, various issues (1992, 1993, 1994, 1995, 1996, 1998)

FERTILISER ASSOCIATION OF INDIA, *Fertiliser Statistics*, New Delhi, various issues

———, *International Prices and Related Statistics*, New Delhi, various issues

GULATI, Ashok, 'Input Subsidies in Indian Agriculture. A Statewise Analysis', *Economic and Political Weekly*, Vol. 24, No. 25, Bombay, June 1989, pp. A57–65.

GULATI, Ashok and M. SVENDSEN (eds.), *Strategic Change in Indian Irrigation*, International Food Policy Research Institute, New Delhi, 1995.

GULATI, Ashok and Sudha NARAYANAN, 'Demystifying Fertiliser and Power Subsidies in India', *Economic and Political Weekly*, Vol. 35, No. 10, Mumbai, March, 2000, pp. 784–94

GULATI, Ashok and Anil SHARMA, 'Subsidy Syndrome in Indian Agriculture', *Economic and Political Weekly*, Mumbai, 30 September 1995, pp. A93–102.

MINISTRY OF AGRICULTURE, *Indian Agriculture Statistics*, Government of India, New Delhi, various issues

———, *Land Use Statistics*, Government of India, New Delhi, various issues

MINISTRY OF FINANCE, *Economic Survey*, Government of India, New Delhi, various issues

PLANNING COMMISSION, *Annual Report on the Working of State Electricity Boards & Electricity Department* (February), Government of India, New Delhi, 1997.

———, *Annual Report on the Working of State Electricity Boards & Electricity Department* (November), Government of India, New Delhi, 1997.

SALGO, H. (ed.), *Power Sector Reform and the Privatization of Distribution in India*, Internal document, World Bank, New Delhi, March 1998.

8

India in World Agricultural Trade

BEFORE AND AFTER MARRAKECH

BRUNO DORIN[1]

INTRODUCTION

India—a founding member of the GATT (1947) and of the World Trade Organization (WTO) that succeeded it on 1 January 1995—signed the Agreement on Agriculture concluding the Uruguay Round (AAUR, 1994) in Marrakech. Like many other countries, India believes that it has a comparative advantage in agriculture, and that it has not been able to assert this comparative advantage fully in view of the multiple import barriers erected elsewhere, or due to production and export subsidies prevailing in other countries. However, India itself subsidizes its agriculture, strongly protects its food market and tries to boost its exports in various ways.

This paper, in its attempt to access AAUR's first effects on India, limits its scope to the sections of the multilateral agreement dealing with imports and exports.[2] In both these fields, it recapitulates the policy followed by the country since its independence in 1947, recalls the AAUR's obligations and exceptions, examines how India put its commitments into practice and how its agricultural foreign trade has evolved from 1990–1 to 1997–8, especially with the European Union.

The UNCTAD's (United Nations Conference on Trade and Development) Trade Analysis and Information System (TRAINS)[3] has been used in this study or, failing that, the CMIE (Centre for Monitoring Indian Economy) databases which have a poorer export format for *sui generis* statistical analyses. The TRAINS data on India offer the great advantage of providing not only the import break-up up to 6 digits of the Harmonized System of commodity classification (HS) for the years 1990–6 (except 1992), but also import origins, as well as the main tariff and non-tariff measures for each product for the years 1990, 1992 and 1997. As indicated in Table 8.1, the sum of the value of these import lines matches official Indian statistics (DGCI&S)[4] rather well, even more so since, unlike the former, the latter include defence expenses.

Thus, the present study is based on the 6-digit level of the HS, and except where otherwise indicated, it classifies and considers as 'agricultural products' those that are strictly listed by the AAUR.[5] On the other hand, it was not possible to establish any connection between the values and quantities provided in TRAINS for about 5,000 products imported each year by India (including about 700 lines of agricultural

Table 8.1: Value of Total Imports According to Different Sources (1990–7)

Year	Exchange Rate	Value of Total Imports					
		Currency	DGCI&S	TRAINS	Diff. (%)	CMIE	Diff. (%)
1990	17.943	US$ (million)	24,075	9,135	–62%		
		INR (10 million)	*43,198*	*16,391*			
1991	24.474	US$ (million)	19,411	17,585	–9%		
		INR (10 million)	*47,851*	*43, 037*			
1992	30.649	US$ (million)	21,882				
		INR (10 million)	*63,375*				
1993	31.366	US$ (million)	23,306	19,258	–17%		
		INR (10 million)	*73,101*	*60,406*			
1994	31.399	US$ (million)	28,654	26,735	–7%	25,148	
		INR (10 million)	*89,971*	*83,945*		*78,963*	*–12%*
1995	33,450	US$ (million)	36,678	29,959	–18%	28,923	
		INR (10 million)	*122,678*	*100,212*		*96,747*	*–21%*
1996	35.500	US$ (million)	39,133	36,863	–6%	28,831	
		INR (10 million)	*138,920*	*130,864*		*102,349*	*–26%*
1997	37.165	US$ (million)	40,779			36,791	
		INR (10 million)	*151,553*			*136,733*	*–10%*

products) and the domestic prices of these products. The Indian monthly Wholesale Price Index database, for instance, from which it is possible to deduce the annual average of domestic prices, lists only 565 product lines (including some lines of aggregation) that are rather difficult to link in any way to the 5,000 product lines mentioned above. In fact, this prevented this study from making various interesting calculations, such as assessing price or tax elasticities of imports, converting non-tariff barriers into tariffs on the basis of an official method, estimating some welfare growth indicators, etc. Finally, if it needs to be specified, the present study reviews the conditions of implementation of India's multilateral commitments within the very strict framework of the AAUR—these exclude issues relating to the Trade Related Intellectual Property Rights Agreement (TRIPs), the Trade Related Investment Measures Agreement (TRIMS) and the General Agreement on Trade in Service (GATS), although these issues too are the subject of intense debate in the country.

MARKET ACCESS

EFFECTIVE IMPORT RESTRICTIONS

The first Five-Year Plan, initiated in 1950–1, did not merely present agricultural development as the maximization of a production function with two variables: irrigation and fertilizers. A second major principle of economic policy, rooted in the 1930s' *swadeshi* movement under British colonization, asserted itself at the same time: to protect the domestic market from foreign competition and imperialism in order to master one's own development fully. The 1947 Imports and Exports Control Act, followed by the 1955 Import Trade Control Order, for instance, banned the import of consumer goods, a part of intermediate goods and capital equipment, as well as all products considered 'non-essential' for a mostly poor population. As regards foreign capital, Indira Gandhi brought in the famous Foreign Exchange and Regulation Act (FERA) in 1973, limiting the share of multinationals in Indian subsidiaries to 40 per cent. However, two decades later, the financial collapse in July 1991, and the Structural Adjustment Programme that succeeded it, dealt a fatal blow to this closed economy model and to its aspirations for endogenous development. The assertion of liberal orthodoxy was no less slow and intensely debated, especially in the field of the Exim policy. Tariff and non-tariff barriers remained numerous and at a high level, although the system was greatly simplified as compared to the preceding decade.

A Battery of Tariff and Non-Tariff Barriers

As far as non-tariff barriers (NTBs) are concerned, i.e. products that India has put on the 'Negative List', even in the year 2000, three regimes can be distinguished (Table 8.2) in addition to the Open General Licence (OGL) provided for free import.

Regarding tariff barriers, many duties were introduced or withdrawn as different budgets came into being (see Table 8.3 for 1999–2000). Their levels remain one of the most disadvantageous in the world. Countervailing and antidumping duties are naturally applied, along with significant national tariffs. In fact, each state in the Indian federation can add their own to these, since they are authorized to fix taxes for a certain number of products (alcohol in particular). However, various customs duties can also

Table 8.2: Non-Tariff Import Barrier Regimes (1999–2000)

PROHIBITED	Import of products under this regime is totally prohibited in order to protect, for instance, certain wild species, or for security reasons linked to religion, such as beef tallow
CANALIZED	Products under this regime can only be imported through State Trading Enterprises (STE) whose role consists of monitoring, supervising and intervening in markets for the generally proffered motive of food security • FCI – Food Corporation of India (cereals and, from time to time, other PDS items) • STC – State Trading Corporation (cereals, pulses, edible oils, fertilizers...) • NAFED – National Agriculture Co-operative Federation of India (PDS products) • STCL – Spices Trading Corporation (spices) • NSC – National Seeds Corporation (seeds) • NDDB – National Dairy Development Board (milk and milk products and, from time to time, edible oils, fruits and vegetables...) • HVOC – Hindustan Vegetables Oils Corporation (edible oils)
RESTRICTED	For a variety of official reasons, the import of these products is only possible after obtaining a licence or a notification, which cannot be transferred to a third-party and is only valid for one transaction (precious or semi-precious stones, consumer goods, products whose manufacture has been reserved for the Small Scale Industry, etc.).
SIL	However, the Special Import Licences (SIL) are also included in this 'restricted' regime. They are issued for specific products to exporters by way of a bonus, worth 2 to 5 per cent of the value of exports realized in foreign exchange,[6] and can be transferred to third parties against payment of a premium that ranged between 6 and 12 per cent in 1999.

Table 8.3: Import Duty Regimes (1999–2000)

BCD	The Basic Custom Duty can take the form of high tariff peaks for certain products
SCD	The Special Custom Duty was announced in the 1999–2000 budget in the form of a 10 per cent surcharge on the basic duty
ACD	The Additional Custom Duty has been in force for the past several years
SAD	The Special Additional Duty amounts to 4 per cent since the 1998–9 budget (2 per cent and later 3 per cent in preceding years); it is aimed at compensating taxes imposed on the sale of domestic products

be exempted or reduced according to the country of origin, within the framework of more or less old agreements, such as the Bangkok Agreement (Bangladesh, Korea, India, Laos, New Guinea, Papua and Sri Lanka) and the SAPTA (South Asia Preferential Trade Agreement) ratified in 1995 by the SAARC (South Asian Association for Regional Co-operation: Bangladesh, Bhutan, India, Maldives, Nepal, Pakistan and Sri Lanka). Other exemptions can also be granted, depending on the imported product, its use, or its value.

Targeted Products / Suppliers

The protection of the Indian domestic market after its independence has turned it into a largely secondary partner in international trade. In 1997–8 for example, the value of Indian imports continued to account for only 0.7 per cent of world imports, that is to say US$ 40.1 billion out of a total of 5,554.5 (Ministry of Finance, 1999: 80, S85). More than half of these imports concern 43 product lines (6 digits) worth more than US$ 100 million (Dorin, 1999c: Appendix 7), including 4 agricultural products (as understood by the GATT): palm oil (1 million tons), wheat (1.5 million tons), cashew nuts intended to be re-exported after processing (227 thousand tons) and chickpeas (381 thousand tons). These last figures are a good reflection of one of the consequences of the Indian agricultural policy, highly focused on wheat and rice. After the Green Revolution, the country imported cereals only occasionally, but is now faced with costly imports of vegetable oils and pulses, which, furthermore, should be higher since they do not meet the domestic needs of the population.[7] More generally, out of the total value of Indian imports in 1997–8, the share of agricultural products only amounts to US$ 2.6 billion, that is to say 6.3 per cent, a percentage that has remained the average since 1994–5 as far as agricultural products are concerned. Thus, since the Marrakech agreement, agricultural imports have not changed in any significant manner (Figure 8.1), even if there seem to be some slight signs of diversification (Dorin, 1999c: Appendices 8, 9, 10).

With regard to the source of imports (Figure 8.2 and Figure 8.3), the European Union (EU), with its fifteen member countries, indisputably appears to be India's primary trade partner, meeting approximately a quarter of its import requirements (1.5 per cent of European exports). France's share, however, only accounted for US$ 597 million in 1996–7 out of the 8.9 billion worth of products that India imported from the EU that year (according to the TRAINS database), that is to say 6.7 per cent. As far as agricultural products are concerned, the French share is slightly higher (10.3 per cent in 1996–7), with agricultural imports from the EU only accounting for 5 per cent of

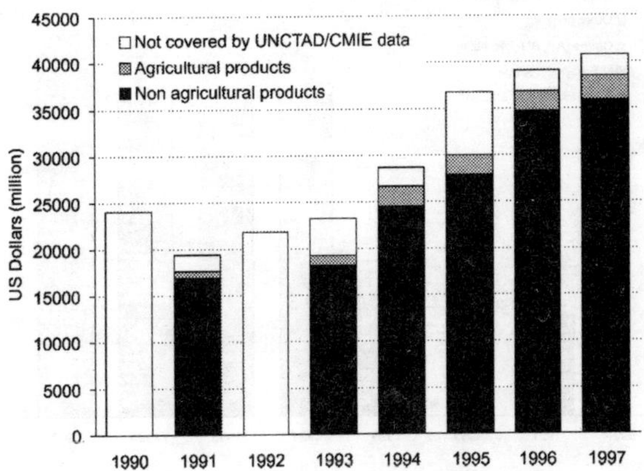

Sources of data: UNCTAD (1996, 1999), CMIE (1999), Ministry of Finance (1999).

Figure 8.1: Importance of Agricultural Imports in Total Indian Imports (1990–1 to 1997–8).

total Indian agricultural imports. Indeed, in this sector, it is North America (United States and Canada) and, above all, four important members of the Cairns group (Argentina, Australia, Brazil and New Zealand), that come out on top. In this connection, the commercial performance of the Cairns group in a sector that is so protected in India appears much better than that of France. In 1996–7 for instance, both these entities exported an appreciably equal number of products (76 and 68 respectively with 6 digits), but the former achieved a figure that is worth more than 40 times that of France (Dorin, 1999c: Appendices 11, 12).

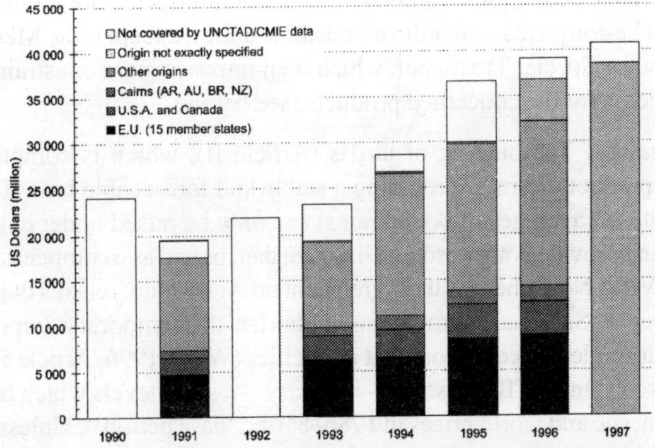

Sources of data: UNCTAD (1996, 1999), CMIE (1999), Ministry of Finance (1999).

Figure 8.2: Total Indian Imports by Origin (1990–1 to 1997–8).

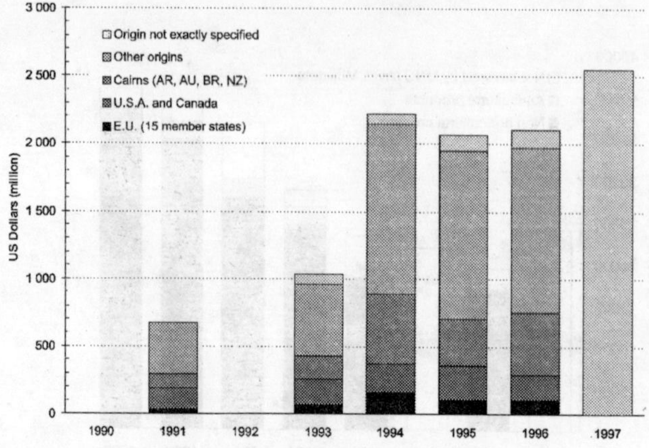

Sources of data: UNCTAD (1996, 1999), CMIE (1999).

Figure 8.3: Agricultural Indian Imports by Origin (1990-1 to 1997-8).

FROM MARRAKECH TO SEATTLE

An Agreement and Some Exceptions

The section on 'Market Access', which is within the direct scope of GATT 1994, lays down or asserts a certain number of rules, including non-discriminatory treatment at borders thanks to the Most Favoured Nation principle (MFN) (Article I), and the application of national treatment to imported products as soon as they are inside the country (Article III). Four other significant elements also appear in the agreement.

Prohibition of quantitative restrictions and their conversion into tariffs. Non-tariff barriers (NTBs), also called QRs (quantitative restrictions, minimum prices, discretionary licensing, variable deductions, voluntary export restraints, etc.) are prohibited (Article XI) and must be converted into ordinary customs duties, except if the Member can and wishes to opt for Special Treatment, which then imposes more constraints in term of minimum access for the concerned products (see below).

Binding of tariffs. The binding of tariffs (Article II), which is compulsory for all agricultural products, aims at providing a stable and foreseeable base for trade since the bound rate (or averages of bound rates) can only be raised under extremely strict conditions, and provided other product lines are then bound as a compensatory measure (Article XXVIII). Nevertheless, the Agreement on Agriculture relaxes these provisions since its Special Safeguard Provisions authorize the temporary imposition of an additional duty under the conditions that it specifies (WTO, 1996: Article 5). Otherwise, bound rates replacing NTBs must not—in theory—exceed levels which bridge the gap between domestic and world prices in 1986–8[8] (the 'base period'). Unless, of course, a country opts for the 'ceiling binding' option, as did India and many developing countries that had not bound their tariffs in the past, or only to a minor extent. In this second formula, a ceiling duty is fixed for each product line without any necessary relationship with past protections, the only obligations being, firstly, to preserve the general market access level of the base period as far as tariffs are concerned (i.e. the arithmetic mean of bound rates must not exceed the arithmetic mean of basic customs duties applied during the base period), and secondly, to not exceed the ceiling bound rate announced for each product in any way in the future (the first formula allowed it as long as the duty average[s] applied did not exceed the average[s] or 'binding levels' of bound rates). This last point is, of course quite restrictive, but it generally results in exemption from having to impose tariff reductions and minimum access, which are otherwise in force for agricultural products.

Reduction of tariffs. The tariffs that stemmed from the conversion of NTBs, along with those that were already applied on agricultural products, have to be reduced by an arithmetic mean of 36 per cent (minimum of 15 per cent per product) over a 6–year period. However, developing countries can apply a reduction of only 24 per cent (minimum of 10 per cent per product), and spread out the reduction over a period of 10 years. As for the Least Developed Countries (LDC) and the developing countries like India that had chosen the option of 'ceiling binding', they are exempted from tariff reductions for agricultural products, their only obligation being not to increase their duties beyond the bound levels.

Minimum access commitment. The minimum access commitment is designed to offset those cases where the NTBs led to zero imports. It must be 3 per cent of domestic consumption at the beginning of the period and move up to 5 per cent at the end. It takes the form of tariff quotas with reduced rates of duty granted on a Most Favoured Nation basis. In the case of the Special Treatment (referred to as 'ST – Annex 5') reserved for countries that fulfil strict conditions (imports less than 3 per cent of consumption in the 1986–8 base period, absence of export subsidy, effective production-restricting), the minimum access must be 4 per cent at the beginning and 8 per cent in the end. However, in the case of developing countries that had not opted for 'ceiling binding' (unlike India which, for this reason, has no minimum access commitment till 2005), the conditions in this respect are more conciliating: in the case of 'a primary agricultural product that is the predominant staple in the traditional diet of a developing country' (WTO, 1996: Annex 5:B), the levels of minimum access can be reduced to 1 per cent at the beginning, reaching 2 per cent in the fifth year, and 4 per cent in the tenth.

Finally, under certain conditions, exceptions to the various obligations mentioned under this section on 'Market Access' are tolerated for reasons of balance of payments (BoP) (Article XII of the 1947 GATT), economic development (Article XVIII, which includes more special provisions with respect to the BoP of developing countries in its section B), serious injury to domestic producers (Article XIX), public morals, human/animal/plant life or health (Article XX), essential security interests (Article XXI), customs unions or free-trade areas (Article XXIV).

A Disputed Dismantling of Quantitative Restrictions

As of April 1994, India accepted the principle of dismantling its NTBs. Since it adopted the Structural Adjustment Programme in 1991, it has already achieved a great deal: nearly 11,000 product lines were still under the 'negative list' at that time, and ever since, this number has been reduced to approximately a quarter.

However, this is not a full commitment, in its strictest sense: India did hope to retain its remaining NTBs on account of Article XVIII:B, '*in order to safeguard its external financial position and to ensure a level of reserves adequate for the implementation of its programme of economic development*', an article to which it has been regularly appealing since 1947. But, the situation became a bit embarrassing when the Reserve Bank of India itself noted that the amount of monetary reserves (including gold and SDRs) had attained the level of US$ 29.9 billion on 14 August, 1997, that is to say the equivalent of seven months of imports and much more than what is regarded as a more-or-less sufficient level (three months of imports) (WTO, 1999). India then proposed a 9-year dismantling plan, which was well accepted by the majority of the developing countries, but rejected by the other Members. In November 1997, based on lists of specific interests presented by Members, India submitted a new programme, spread out over 6 years in three phases (3, 2 and 1 years), from 1 April, 1997 to 31 March, 2003, for 2,714 product lines. The European Union, Switzerland, Canada, Japan and Australia accepted this proposal, although the lifting of NTBs concerning sensitive products would obviously be dealt with in the last phase.

Be that as it may, the United States still objected, and in the same month, left it to the WTO Dispute Settlement Body (DSB). According to them, the 2,714 products had to be removed from the negative list much faster (according to India, there were only 2,296 at that time under GATT Article XVIII:B), including 710 agricultural products (26 per cent). In April 1999, the DSB passed its sentence, and condemned India. In the bulky report in which it gives an account of its judgment (WTO, 1999), the Panel appreciated, nonetheless, 'the courage and sagacity with which India has carried out its economic reform programme', particularly the extensive liberalization of foreign trade and financial transactions, the unification of the exchange rate and the introduction of current account transactions convertibility. India, on its part, remained convinced that there were some significant contradictions between the financial stringency enforced by the IMF in the framework of the Structural Adjustment Programme and the immediate elimination of all quantitative restrictions advocated by the WTO. It then tried to extend the 15-month compliance deadline imposed by the USA, while continuing to place new product lines under OGL, in an effort to prove its goodwill. It also tried to negotiate an increase of the customs duties bound somewhat too quickly to 0 per cent in 1994 (see below). However, since 28 December 1999, in the framework of the mutual agreement that it concluded and signed with the USA, the modalities for dismantling the remaining QRs on 8-digit 1,429 products are clear: 715 must be withdrawn before 31 March 2000 and the other 714 before 31 March 2001.

Thus, by April 2001, India had completed the dismantling process that has been underway since 1994—a process that it has tried to delay as much as possible on account of Article XVIII:B, in particular for agricultural products (Table 8.4).[9] On the other hand, in accordance with Articles XX and XXI for example, it does hope to maintain—as it had announced in Marrakech—quantitative restrictions on 632 customs headings (Gulati, 1999: 2932), including nearly 200 agricultural products (dairy products, meats and animal products, seedlings and bulbs, seeds, oils, etc.). However, discussions and consultations continue with India proposing the elimination of some of these NTBs in exchange for an upward adjustment of tariffs bound at very low levels in 1994 (see below). Likewise, to obtain some concessions reinforcing its tariff

Table 8.4: Product Lines under Negative List (1994, 1997 and 1999)

		1994	1997	1999–2000		
		Boncorps (1999)	UNCTAD (1999)	Gulati et al. (1999)		Mehta (2000)
		Total	Total	Total	under XVIII:B	under XVIII:B
Forbidden	Agricultural products	18	33	51	0	
	Other products		12			
	Total		44			0
Canalized	Agricultural products	66	62	37	8	
	Other products		22			
	Total		84			44
Restricted	Agricultural products	629	474	518	336	
	Other products		1,778			
	Total		2,086			1,385
TOTAL	Agricultural products	713	569 (81%)	606 (43%)	344	
	Other products	1,997	1,645 (37%)	1,508 (17%)	1,138	
	Total	2,710	2,214 (43%)	2,114 (26%)	1,482	1,429

Notes: Percentages expressed in relation to the total number of lines considered (agricultural products and others).
1994: HS digits and positions are not specified for agricultural products.
1997: 6 digits, agricultural products as per GATT definition; the total for each NTB category only includes a single motive per product for a restriction that could be imposed under several headings (Dorin, 1999c: Appendices 13 to 15).
1999–2000: 6 digits, agricultural products = positions 1 to 24 in the HS (Gulati et al., 1999); 8 digits (Mehta, 2000).

weapons, or to justify certain market access restrictions, it plans to condemn, in the forthcoming WTO negotiations, several provisions that it regards as disguised and insuperable NTBs for developing countries (phytosanitary norms, packaging and labeling standards, etc.). In the same vein, but in a much more severe manner, India will reject any attempt to link trade issues with social or environmental concerns.

A Comfortable Binding of Customs Duties

In 1994, India bound 3,374 product lines (6 digits), that is to say 67 per cent of its tariff headings, whereas only 6 per cent had been bound before the Agreement. As regards 2,701 non-agricultural lines, with a few exceptions, the binding is done at 40 per cent *ad valorem* concerning finished products, and at 25 per cent as far as intermediate goods, machines and equipment are concerned (Ministry of Commerce, 1999b). The tariff average of 108 per cent (2,695 lines) over the base period (1986–90 for India) has been rolled back to 37 per cent (2,701 lines) in the process, with a commitment to reduce this later level by 40 per cent over a period of 10 years, i.e. between March 1995 and 2005. However, as far as textiles are concerned, India reserves the right to return to the customs duties that were in force in 1990 if the process of integration planned within the framework of the Multifibre Arrangement on textiles is not completely implemented, or is delayed.

As regards agricultural products, India having chosen the option of 'ceiling binding', no commitment has been made until 2005 as far as minimum market access and tariff reduction are concerned. On the other hand, a ceiling has truly been established for each of the 673 agricultural lines (6 digits) existing in 1994, with *ad valorem*[10] rates whose general average exactly matched that of basic customs duties applied during the base period, i.e. 114 per cent. With the exception of 109 lines that were bound at less than 100 per cent, or even zero (11 lines), the others could thus be generally fixed at 100 per cent for basic commodities, 150 per cent for processed products, and 300 per cent for oils and fats (Table 8.5).

Table 8.5: Summary of Binding Rates for Agricultural Products

Rate	Products
0%	Rice, common wheat, sorghum, maize, millet, milk and cream in solid form...
10%	Vegetable seeds, bulbs, tubers, roses, trees and plants with roots...
25%	Skins and leathers...
35%	Animal feed products...
40%	Milk products, including butter, casein, cheeses...
45%	Olive, soya and rape/mustard oils...
55%	Soups and vegetable preparations...
60%	Concentrates for non-alcoholic drinks, dessert powders...
85%	Fruit juices...
100%	Living animals, fresh, frozen, deep-frozen or dried meat, certain fruits and vegetables, durum wheat, barley, rye, oats, edible nuts, sugar, chocolate and chocolate-based preparations, plant seeds...
150%	Coffee, tea, spices, bakery products, many processed food products, drinks, tobacco...
300%	Edible oils and animal fats with a few exceptions (olive, soya, rape/mustard)...

Sources: Boncorps (1999), Dorin (1999c: Appendices 16, 17).

This binding of agricultural rates is thus extremely comfortable and for many of the products, it could be used, if need be, to substitute the NTBs that must be dismantled. In the meantime, India applies lower basic customs duties than those it could legally enforce. Indeed, their average for agricultural products was 'only' 34 per cent in 1997 (30 per cent for other products). In order to include in this figure the effects of the 2,214 NTBs listed at that time out of 2,191 lines (of agricultural and other products at 6 digits), and in the absence of data that would allow their conversion into tariffs according to official methodologies, one can consider very arbitrarily that a customs duty of 200 per cent is applied when the import is prohibited and that this duty is 100 per cent when it is canalized or restricted.[11] Thus, the tariff average goes up from 34 per cent to 79 per cent in the case of agricultural products (from 30 per cent to 50 per cent for other products), which is considerable but still well below 114 per cent (Table 8.6).

India's comfortable situation with regard to agricultural products can also be measured by successively applying the MFN duties prevailing in 1986–90 (Ministry of Commerce, 1999b), 1992 (UNCTAD, 1996), 1997 (UNCTAD, 1999), as well as the rates as they were bound in 1994 (Ministry of Commerce, 1999b), to the annual import value of the products imported from 1991–2 to 1997–8. These tax earnings (deducted every year for each imported product, and then added) are fictitious since they suppose a tax elasticity of import equal to zero, but they nonetheless speak for themselves (Figure 8.4 and Figure 8.5).

Table 8.6: Basic Tariffs and Tariff Eqivalents (1997)

		Basic customs duties (MFN)	Tariff equivalents for NTBs	Total
Agricultural products	Average rate (%)	33.7	89.4	79.2
	No. of 6-digit lines	371	515	644
Other products	Average rate (%)	30.0	84.0	50.2
	No. of 6-digit lines	3,770	1,676	4,178
Total	Average rate (%)	30.4	85.3	54.1
	No. of 6-digit lines	4,141	2,191	4,822

Source of data: UNCTAD (1999)

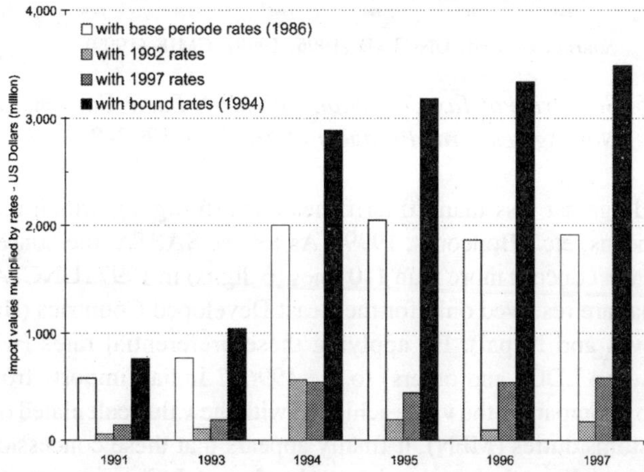

Sources of data: UNCTAD (1996, 1999), CMIE (1999).

Figure 8.4: Scenarios of Tax Collection with MFN Basic Customs Duties Agricultural Products (1991–2 to 1997–8).

Regarding customs duties on imports, it has also to be said that they represent a sizeable income for the coffers of the state: US$16 billion in 1997–8, inclusive of all products and taxes combined, i.e. 27 per cent of the total receipts of the Indian Union government (Table 8.7). Of this amount, basic customs duties bring in at least 7 billion according to available data,[12] with 4 billion from less than 100 lines (6 digits), including 12 agricultural products with more than 10 million dollars. Palm oil alone accounts for 187 million (Dorin, 1999c: Appendix 18). The case of palm oil is particularly illustrative of a dimension of Indian foreign trade that should not be neglected, and which is controlled by powerful lobbies: certain imports are not only an essential and expected source of income for the central government, but also for State Trading Enterprises, as well as for corrupt industrialists and/or civil servants who, in collusion or not, draw significant profits from internal subsidy systems based on these imports (Dorin, 1999a).

As for tax collection losses resulting from the application of regional trade agreements, which were not taken into account in all preceding estimates, they are, in reality—and for the moment—almost negligible. Within the framework of the Bangkok Agreement, these concessions are generally only 5 per cent below ordinary duties,

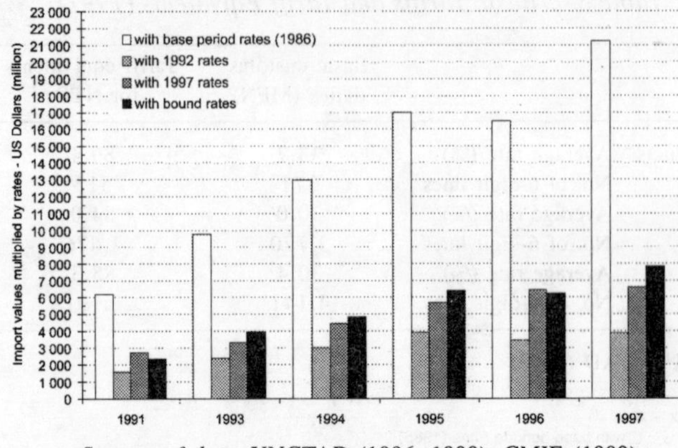

Sources of data: UNCTAD (1996, 1999), CMIE (1999).

Figure 8.5: Scenarios of Tax Collection with MFN Basic Customs Duties Non-Agricultural Products (1991–2 to 1997–8).

and are granted against less than 70 tariff headings (6 digits), which include fish, cloves, cocoa beans, etc. (Boncorps, 1999). As for the SAPTA, the concessions are greater, but did not concern more than 110 lines (6 digits) in 1997 (UNCTAD, 1999), including 66 that are reserved only for the Least Developed Countries (Bangladesh, Bhutan, Maldives and Nepal). By applying these preferential rates (without any distinction between LDCs and others) to the 1996–7 Indian imports from SAPTA countries, and by comparing the value achieved with the value calculated on the basis of ordinary customs duties (MFN), it finally appears that these concessions did not exceed 5 million dollars that year—the total value of Indian imports from the 6 neighbouring countries being only US$ 69 million in 1996–7, of which U$ 50 million concerned the above-mentioned preferential treatment (all agricultural products if one includes fish) (Dorin, 1999c: Appendix 19).

In conclusion, as far as the next WTO round is concerned, India appears to be in a rather strong bargaining position concerning import tariff measures, since we can imagine the following 5-point demonstration.

(a) In 1997, all Indian product lines were placed under the Generalized System of Preferences whereas only about half were under it in 1990, that is to say just before the 1991 Structural Adjustment Programme. During the same period, India has divided by three the average of its basic customs duties for agricultural products, since it was 100 per cent in 1990 (68 per cent for MFN duties) and amounted to only 30 per cent in 1997 (Table 8.8). Only few countries in the world have implemented such big reforms in a so short time.

(b) Moreover, in accordance with the 1994 GATT, all Indian agricultural lines have been bound, at a level that is undoubtedly high (the average is 114 per cent), but with an effective ceiling for each customs position (the 'binding ceiling' option), unlike many other countries which can regularly revise their bound rates as long as group-wise averages remain constant. And yet, this last option, and its widespread practice

Table 8.7: Central Government Receipts (1991–2 to 1998–9)

	(April to March)	1991	1992	1993	1994	1995	1996	1997	1998 (Bud. Est.)
RUPEES (RBI, 1999)									
Total receipts (net)	(billion)	691.00	775.73	796.50	955.49	1,149.14	1,316.86	1,457.73	1,703.00
- of which customs duties on gross imports	(billion) (%)	219.27 32%	241.36 31%	226.55 28%	271.45 28%	361.33 31%	435.13 33%	398.50 27%	441.37 26%
EXCHANGE RATE (Min. Fin., 1999)	(Rs. for 1 US$)	24.474	30.649	31.366	31.399	33.450	35.500	37.165	41.859
US DOLLARS									
Total receipts (net)	(million)	28,234	31,696	32,545	39,041	46,954	53,806	59,562	69,584
- of which customs duties on gross imports	(million) (%)	8,959 32%	9,862 31%	9,257 28%	11,091 28%	14,764 31%	17,779 33%	16,283 27%	18,034 26%
- of which basic customs duties MFN (UNCTAD, 1999)	(million) (%)			2,493 27%				7,176 44%	

Source of data: RBI (1999), Ministry of Finance (1999), UNCTAD (1996, 1999).

Table 8.8: Changes in Import Duties (1986, 1990, 1992, 1997)

BASIC DUTY (general)		1986	1990	1992	1997
Agricultural products	Average (%)		100	65	30
	Min – Max		0 – 355	0 – 355	0 – 260
	Coef. of Var.		0.60	0.58	1.01
	No. of lines		696	692	704
Other products	Average (%)		96	63	30
	Min – Max		0 – 300	0 – 110	0 – 190
	Coef. of Var.		0.53	0.18	0.32
	No. of lines		4,355	4,355	4,439
Total	Average (%)		96	63	30
	Min – Max		0 – 355	0 – 355	0 – 260
	Coef. of Var.		0.54	0.27	0.47
	No. of lines		5,051	5,047	5,143

BASIC DUTY (MFN)		1986	1990	1992	1997	Binding
Agricultural products	Average (%)	114	68	25	30	114
	Min – Max	0 – 240	0 – 200	0 – 100	0 – 260	0 – 300
	Coef. of Var.	0.38	0.48	0.95	1.03	0.47
	No. of lines	659	353	343	704	671
Other products	Average (%)	108	60	26	30	37
	Min – Max	0 – 345	0 – 155	0 – 65	0 – 50	0 – 55
	Coef. of Var.	0.40	0.51	1.03	0.33	0.18
	No. of lines	2,695	2,147	1,990	4,439	2,701
Total	Average (%)	109	61	26	30	52
	Min – Max	0 – 345	0 – 200	0 – 100	0 – 260	0 – 300
	Coef. of Var.	0.39	0.50	1.02	0.48	0.76
	No. of lines	3,354	2,500	2,333	5,143	3,372

PREFERENTIAL DUTY-SAARC		1986	1990	1992	1997	LDC
Agricultural products	Average (%)				22	0
	Min – Max				3 – 36	0 – 0
	Coef. of Var.				0.49	
	No. of lines				18	1
Other products	Average (%)				19	10
	Min – Max				2 – 23	0 – 20
	Coef. of Var.				0.19	0.60
	No. of lines				26	68
Total	Average (%)				21	10
	Min – Max				2 – 36	0 – 20
	Coef. of Var.				0.37	0.61
	No. of lines				44	69

Table 8.8 (contd.)

ADDITIONAL DUTY		1986	1990	1992	1997
Agricultural products	Average (%)			42	5
	Min – Max			0 – 45	0 – 5
	Coef. of Var.			0.29	0.31
	No. of lines			679	704
Other products	Average (%)			43	5
	Min – Max			0 – 45	0 – 5
	Coef. of Var.			0.19	0.15
	No. of lines			4,355	4,439
Total	Average (%)			43	5
	Min – Max			0 – 45	0 – 5
	Coef. of Var.			0.20	0.18
	No. of lines			5,034	5,143

Notes: (1) The statistics provided above concern all product lines for which a tariff had been entered in the concerned years, and not just for lines that had been imported in those years. Thus, these estimates may be slightly different from others presented in this study.
(2) The 'auxiliary' (1992) or 'additional' duty (1997) does not, in any case, cover all import taxes (ACD, SCD, SAD) applied after the basic customs duty (BCD)

Sources of data: UNCTAD (1996, 1999, ad hoc), Ministry of Commerce (1999b).

(as well as the use of tariff quotas), does not provide a stable and foreseeable basis for trade, which is the principle on which binding has been based.

(c) In 1994, certain Indian products were even bound at 0 per cent or at relatively low levels. At the time, India did not suspect that it would have to dismantle so quickly the NTBs that it had to maintain for BoP reasons (Article XVIII:B). In view of the fact that a fast pace of dismantling has been imposed at the end of the 1990s, which could make its liberalization policy—initiated only in 1991—go off the rails after forty years of a planned economy, it now wishes that it will at least be able to rationalize these low tariff bindings (milk, cereal, soya and rapeseed oils, etc.). In return, India may agree to lower the ceiling fixed for some products (yoghurt, avocados, grapes, peaches, strawberries, apricots, cherries, fats, etc.).

(d) While India is a priori against a minimum access commitment for agricultural products since agriculture is still a basic activity in this developing country, it can, however, commit itself to guaranteeing a food market access of 2 per cent, as it has already filled that quota since the Marrakech agreement (Table 8.9).

(e) The reduction of Indian customs duties is another issue altogether. The AAUR has allowed developed countries, thanks to loose tariff bindings and reductions, to circumvent the pursued objectives. As long as the 1994 commitments will not have been suitably implemented, India will therefore keep opposing any new reform in the matter as far as developing countries are concerned.

Table 8.9: Access to the Food and Agricultural Market (1991–2 to 1996–7)

		1991	1993	1994	1995	1996
Imports of agricultural products	US$ (million)	669	1,032	2,219	2,064	2,106
PFCE Food, drink, tobacco	US$ (million)	86,236	86,550	98,633	101, 887	110,544
PFCE Food only	US$ (million)	78,471	78,595	89,369	90,709	98,064
Access / Food, drink, tobacco	%	0.8%	1.2%	2.2%	2.0%	1.9%
Access / Food only	%	0.9%	1.3%	2.5%	2.3%	2.1%

Notes: (1) PFCE = Private Final Consumption Expenditure
(2) The Rupee/Dollar exchange rates used are those provided by Ministry of Finance (1999)

Sources of data: EPWRF (1999), UNCTAD (1999).

EXPORT COMPETITIVENESS

A PROGRESSIVE OPENING OF INDIAN HORIZONS

A Controlled Takeoff of Agri-Exports

After achieving independence, India withdrew into its own market, or was drawn into special agreements with the Soviet Union, which altogether greatly marginalized the country from international trade. But now that it has plunged into the era of liberalization and globalization that is boosting its imports, it has a long way to go: in 1997–8, in spite of being the second most populated country, it accounted for only 0.6 per cent of world exports (US$ 34 billion out of 5,451), with not more than 4 products exceeding 10 per cent of world trade: rice (12 per cent), spices (11.2 per cent), tea and mate (11.1 per cent), pearls and precious or semi-precious stones (10.8 per cent) (Ministry of Finance, 1999). However, since the 'Rajiv Gandhi years' and the deep economic reforms that followed the financial bankruptcy in 1991, Indian business milieus have become more active, although they realize that international ambition does not automatically provide international clientele.

Agriculture too has a say in this new deal. The share of the 'agriculture and allied products' sector in total exports has certainly decreased by more than half since the Green Revolution (44 per cent in 1960–1, 19 per cent in 1997–8), but since the middle of the 1980s, this sector's trade balance had achieved a comfortable surplus, even if structural imports of edible oils had replaced those of cereals from the end of the 1970s onwards. The key to this success is diversification. Traditional exports of coffee, tea and spices, which had accounted for practically 90 per cent of agricultural exports (excluding cashew nuts[13]) at the beginning of the 1960s, today account for less than 30 per cent (Figure 8.6 and Figure 8.7). On the other hand, the Green Revolution in cereals now permits the production of basmati rice for export.[14] Similarly, the Blue Revolution, which hoisted India to the rank of the 7th world producer of fish and seafood, now enables the country to send approximately 6 per cent of its production abroad. As for the more recent Yellow Revolution of oilseeds, it essentially boosted the export of oil cakes, in particular that of soya which is not intensively used in India for domestic animal feeding despite a marked development of Indian animal productions over the past fifteen years.

However, the White Revolution of milk has not been translated into a surge in exports, although India is today amongst the first world producers of milk, as also of many other products. The reason behind is the huge requirement of the domestic

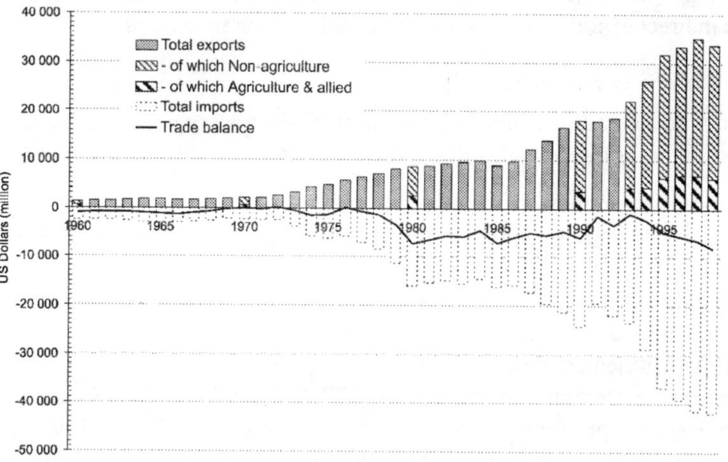

Source of data: Ministry of Finance (2000).

Figure 8.6: Exports and Trade Balances (1960–1 to 1998–9).

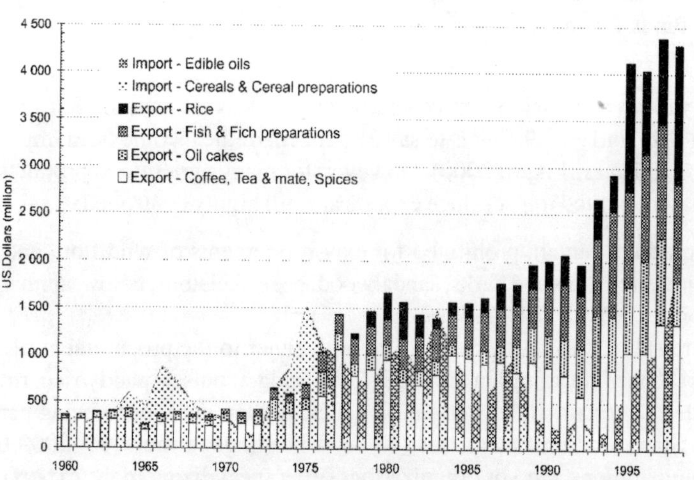

Source of data: Chandhok et al. (1990), Ministry of Finance (2000).

Figure 8.7: Foreign Agricultural Trade above 500 Million US$ (1960–1 to 1998–9).

market which still needs to be met to fight against widespread under- and malnutrition (Dorin, 1999b). The Indian government is well aware of this, and it is with great prudence that it gradually lifts or reduces various restrictions imposed for the past several decades on the movement and export of foodstuffs (cereals especially), despite the need for foreign exchange or the pressure exerted by some producers well aware of the fact that their profit margins could now be much better outside. But then, does the country really have the requisite facilities (roads, railways, harbours, airports, cold chains, quality controls, etc.) to become a significant provider of perishable goods on the international markets? Obviously not, and until it seriously attacks this bottleneck, various indirect export subsidies will have to do for the time being.

A Rash of Indirect Export Subsidies

India does not offer direct subsidies to exporters. On the other hand, it is increasingly taking recourse to indirect subsidies to help them, such as (Boncorps, 1999):

— income tax exemption on export revenues since 1961;
— permission to import—within the limits of SIL transferable quotas—goods that are otherwise subject to restrictions;
— import customs duty reductions or exemptions as long as some export obligations are fulfilled (see EPCG or 'Export Promotion Capital Goods' for capital goods, DEPB or 'Duty Entitlement Pass Book' for other goods);
— export-oriented free and industrial zones (FTZ, EPZ, and now SEZ, 'Special Economic Zone'), whose privileges were consolidated or extended by the 1999–2000 and 2000–1 budgets, but which still remain limited in size (800 to 1000 ha.) and in number (6 in the year 2000), unlike China for example;
— credit at subsidized rates which, fixed by the Reserve Bank of India, is distributed by commercial banks required to direct at least 12 per cent of their net credit to export activities;
— special insurance through the 'Export Credit Guarantee Corporation' (ECGC);
— support funds for Indian brands;
— etc.

These various concessions represented a revenue loss estimated at Rs. 155 billion (3.7 billion US$) in 1998–9, that is to say 38 per cent of the income from import taxes (*The Economic Times*, 1 April 2000). However, for many agricultural products, these measures have a limited impact since exports are still highly controlled:

— certain products remain prohibited for export (elements of wild flora and fauna, exotic birds, tree roots and barks, sandalwood, beef, skeletons, tallow, animal grease and/or oils except for fish oils, etc.);
— others are subject, as in the case of imports, either to the procurement of special licences (90 tariff headings in the 1999–2000 budget: pulses, paddy/rice, rice bran, vegetable oil, certain algae, certain oil cakes, fodder, etc.), or to being canalized through State Trading Enterprises (23 tariff headings in the 1999–2000 budget: karoya gum, onions, etc.) or organizations often specializing in the export of food products collected or produced by tribal populations living in forests;
— some products are also subject to export quotas (wheat and wheat products, cereals and secondary cereal flour, milk, including powder and baby milk, sugar, brown

algae, etc.) since most of them receive significant domestic support for production;

— according to international prices, certain food items must sometimes also be exported at a minimum price—that is still the case for wheat and its products, as for secondary cereals, but no longer for rice (basmati or not), cotton and its soft or hard waste.

FROM MARRAKECH TO SEATTLE

An Agreement for Leveling the Playing Field

Export subsidies, in theory prohibited by the GATT, have been subject to strict discipline since the 1994 Agreement (if one disregards the direct or indirect export tax rebates that are especially generously granted in the USA). Indeed, over a period of 6 years (10 years for developing countries), except for the Least Developed Countries, the subsidized volumes must be reduced by 21 per cent (14 per cent for developing countries) and related budgetary expenditure (listed in Article 9) must be decreased by 36 per cent (24 per cent for developing countries). To prevent the circumvention of these commitments, Article 10 specifies in particular that international food aid: (a) shall not be tied directly or indirectly to commercial exports of agricultural products to recipient countries; (b) shall be carried out in accordance with the FAO 'Principles of Surplus Disposal and Consultative Obligations'; and (c) shall be provided to the extent possible in fully grant form. Lastly, within the framework of Article 12, it is also stipulated that a Member instituting a new export prohibition or restriction on foodstuffs shall give due consideration to the effects of such a prohibition or restriction on the importing Members' food security, and that the provisions of this Article shall not apply to any developing country Member unless it is a net-food exporter of the specific foodstuff concerned.

In 1994, it is believed that India did not set up a sufficiently substantial export subsidy mechanism to enforce any reduction in this matter. And even if it has not notified all the new measures it has instituted since then, this export subsidy mechanism remains incomparable with that of developed countries such as Europe, considered as an artificial exporter: that is what India has hammered and will continue to hammer out, zeroing-in severely on the production and export subsidies of the old continent, which, according to India, limit its exporting capacities, stifle its real competitiveness, lower world prices of raw materials and, consequently, the principal export income of poor countries like itself. While this position is more or less honest as regards the export-related vocation and proficiency of Indian agriculture, it nevertheless remains very popular, especially in a context where the general trade balance seems to be suffering a chronic deterioration since the middle of the 1990s. It is thus along these lines that India should present its arguments, during the next round of negotiations, in order to eliminate all forms of subsidy or protection likely to limit its access to coveted markets like the European market.

An Attack on 'Fortress Europe'?

The European Union is India's primary supplier, but also its primary customer: in 1997–8, 25.2 per cent of Indian exports were directed towards the old continent (2.2 per cent towards France), that is to say nearly US$ 9 billion[15] of export earnings for India (Ministry of Finance, 1999), or nearly US$ 11 billion of import expenditure for Europe

(UNCTAD, 1999). Of these 11 billion, practically 15 per cent (US$ 1.6 billion) were imports of agricultural products (within the meaning of GATT), including 9 products above US$ 50 million each, with coffee (not roasted), cashew nuts, rice (brown or basmati) and black tea heading the list (Dorin, 1999c: Appendix 20).

After a period of stagnation between 1990 and 1993, these European imports of Indian agricultural products saw significant progress since 1994, at an average rate of 16 per cent until 1997, that is to say, faster than European imports of other Indian products (+12 per cent between 1994 and 1997). The Indian market share in total European imports of agricultural produce (US$ 69 billion in 1997) thus went up from 1.6 per cent at the beginning of decade to 2.3 per cent in 1997 (from 1.0 to 1.4 per cent with all products combined together). The trade balance between the two Unions is thus moving in favour of India: on the basis of available data, the deficit for Europe went up from 811 million dollars in 1991 to 1.177 billion in 1996 (Table 8.10) with regard to trade in agricultural products. If one wants to consider that this development is the direct effect of the AAUR, then one can conclude that the latter is favourable to India. It remains to be seen whether these recent trends will be confirmed, in other words, whether India's policies, production and processing apparatus as well as infrastructure will make it possible to increase a market share in Europe that is by no means negligible, but still fairly modest with regard to agricultural products.

CONCLUDING SUMMARY

The Indian Union may well have become the second-largest world food producer, but its share in international trade remains lower than 1 per cent, an isolation that is reinforced by its exclusion from significant regional free trade zones. This GATT Founder Member may thus find it advantageous, in principle, to strengthen the powers of what has since become the WTO. This led India to sign the Agreement on Agriculture in Marrakech in 1994, in spite of the fact that the subject was a highly sensitive one within the country. But it also had other reasons to do so: after having concentrated on its Green Revolution for thirty years and launched wide-ranging economic reforms in 1991, the increase in exports of certain agricultural products is a viable option for a country whose economic liberalization now weighs heavily on imports. And then, with regard to the three sections of the Agreement, in spite of its significant direct and indirect domestic support to agriculture, India is committed to reform only some of its market access restrictions. However, the latter has constrained the country—after a 'ceiling-binding' of its customs duties ranging from 0 to 300 per cent according to the sensitivity of the products concerned—to dismantle its numerous non-tariff agricultural barriers. Calling upon Article XVIII:B of the GATT relating to the balance of payments of developing countries, India then tried to delay this commitment. It managed to do so until 1997, but could no longer avoid it from that point on: the settlement of the dispute lodged by the United States sentenced India to complete the operation by April 2001. However, except for some products, the dismantling of these QRs is not very disturbing since tariff protection could replace—if need be—the old barriers. In 1997, for instance, the average basic customs duties already stood at 30 per cent for agricultural products (80 per cent after a conversion of prevailing QRs into their tariff equivalents), and could, if necessary, be raised up to the binding level of 114 per cent. The Indian market

Table 8.10: European Imports, Indian Market Share and Trade Balances (1990 to 1997)

AGRICULTURAL PRODUCTS		1990	1991	1992	1993	1994	1995	1996	1997
European Imports									
Of all origins	US$ (million)	53,038	53,365	57,100	50,752	58,711	66,807	65,312	68,993
Annual growth	%		0.6%	7.0%	–11.1%	15.7%	13.8%	–2,2%	5.6%
Number of lines	6 digits	693	694	694	694	693	691	702	703
Of Indian origin	US$ (million)	827	842	853	882	1,088	1,167	1,283	1,601
Annual growth	%		1.9%	1.3%	3.4%	23.4%	7.2%	10.0%	24.8%
Number of lines	6 digits	208	204	218	230	245	250	257	267
India's market share	%	1.6%	1.6%	1.5%	1.7%	1.9%	1.7%	2.0%	2.3%
Indian Imports									
Of all origins	US$ (million)		669		1,032	2,219	2,064	2,106	2,505
Number of lines	6 digits		232		304	334	352	361	359
Of European origin	US$ (million)		32		69	160	106	106	
Number of lines	6 digits		150		178	221	247	253	
Europe's market share	%		4.7%		6.7%	7.2%	5.1%	5.0%	
EU-IN Trade Balance	US$ (million)		–811		–813	–928	–1,061	–1,177	

(contd.)

Table 8.10 (contd.)

ALL PRODUCTS		1990	1991	1992	1993	1994	1995	1996	1997
European Imports									
Of all origins	US$ (million)	573,752	597,566	602,342	555,034	600,579	638,415	576,144	773,081
Annual growth	%		4.2%	0.8%	–7.9%	8.2%	6.3%	–9.8%	34.2%
Number of lines	6 digits	5,008	5,010	5,012	5,012	5,010	5,007	5,105	5,107
Of Indian origin	US$ (million)	5,662	5,798	6,224	6,775	8,122	9,663	9,731	10,861
Annual growth	%		2.4%	7.3%	8.9%	19.9%	19.0%	0.7%	11.6%
Number of lines	6 digits	1,992	2,092	2,267	2,389	2,495	2,623	2,711	2,871
India's market share	%	1.0%	1.0%	1.0%	1.2%	1.4%	1.5%	1.7%	1.4%
Indian Imports									
Of all origins	US$ (million)		17,585		19,258	26,735	29,959	36,863	38,516
Number of lines	6 digits		3,228		3,708	3,860	3,985	4,037	4,151
Of European origin	US$ (million)		4,680		6,284	6,494	8,400	8,887	
Number of lines	6 digits		3,319		3,690	4,068	4,365	4,326	
Europe's market share	%		26.6%		32.6%	24.3%	28.0%	24.1%	
EU-IN Trade Balance	US$ (million)		–1,119		–491	–,628	–1,263	–844	

Source of data: UNCTAD (1996, 1999), CMIE (1999).

access for agri-food products should, therefore, barely exceed the current 2 per cent in the future, all the more so since import duty earnings still account for more than a quarter of the central government's revenues. On the other hand, India does hope to use the WTO platform to condemn, a few years after Marrakech, the various forms of domestic support, export subsidies and import barriers that are still being practised, in its view, by the rich countries in order to prevent poor countries from using their comparative advantages. Hence, for the moment, the question of expanding an agenda (environment, labour, etc.) that has not yet been respected does not arise. The European Union is a special target for such charges of unequal competition. However, between 1994 and 1997, the Indian market share in European agricultural imports rose from 1.7 per cent to 2.3 per cent, increasing the European agricultural trade deficit between the two Unions from US$ 813 million to nearly 1.2 billion.

NOTES

1. The author is solely responsible for this paper, which does not reflect the opinion of the organization or the country to which he belongs.
2. As for domestic support, according to the total Aggregate Measurement of Support (AMS), it is in India much below 10 per cent of the value of agricultural production, which exempts the country from any commitment to reduce it (Dorin, 1999c: 9–18). The fact remains that agricultural input subsidies amounted alone to at least 6.57 billion dollars in 1995–6 (irrigation, electricity and fertilisers), and that these so-called 'non-product-specific' subsidies, which are in fact highly concentrated on wheat, rice and sugarcane, greatly influence the competitiveness of Indian farmers, both within as well as outside the country (Dorin et Julien, 2000).
3. I would like to express my gratitude to Hiroaki Kuwahara and his team at Geneva for not only responding very diligently to requests for information and data, but also for providing some statistics relating to India, which were not recorded on the TRAINS CD-ROM Versions 4 (1996) and 6 (1999).
4. Except for 1990, which had been excluded from certain calculations for this reason.
5. That is to say, HS Chapters 1 to 24, less fish and fish-based products (HS positions 0301–0305 and 1604), plus HS codes or headings 290543 (mannitol), 290544 (sorbitol), 3301 (essential oils), 3501 to 3505 (albuminoidal substances, modified starches, glues), 380910 (finishing agents), 382360 (sorbitol n.e.p.), 4101 to 4103 (hides and skins), 4301 (raw furskins), 5001 to 5003 (raw silk and silk waste), 5101 to 5103 (wool and animal hair), 5201 to 5203 (raw cotton, waste and cotton carded or combed), 5301 (raw flax), 5302 (raw hemp).
6. While India does not officially apply quantitative restrictions in the form of fixed quotas, these limits nevertheless constitute an implicit form.
7. In the middle of the 1990s, two-thirds of the Indian population was estimated to be suffering from lipid deficiency (including approximately 20 per cent from a deficiency in essential fatty acids), and half from proteins (Dorin, 1999a).
8. This principle leaves a considerable margin for estimation that developed countries have used to a large extent, such as European Union countries which applied F.O.B. and intervention prices (Matthews et al., 1995: 8).
9. It is difficult to specify accurately how the dismantling of Indian NTBs progressed. There are a number of technical reasons that explain this: (a) the total number of import lines goes up every year, which makes it difficult to use percentages, (b) accessible electronic databases on tariff and non-tariff barriers are rare and provide information for few years only (UNCTAD, 1999), (c) analysts carrying out manual calculations on the basis of budgets generally do not use a rigorous and constant digit level of the HS from one year to the next (4, 6, 8 and even 10 digits) (Boncorps, 1999) (Gulati et al., 1999) (Mehta, 2000), (d) the same analysts, when they do specify the digit level, consider as agricultural products those that range from 1 to 24 in the HS (Gulati, 1999), which is not exactly the definition given by the 1994 Agreement on Agriculture, (e) etc.

10. Except for 2 products (080211 and 080212), for which binding has been expressed in Rs./kg.
11. These two arbitrary rates are either halved when it is indicated in the TRAINS database (UNCTAD, 1999) that the non-tariff measure is only applicable to a part of the 6-digit line, or divided by the number of NTBs on the line. In that later case, the rates calculated for each NTB are then added up. Finally, in case the lines under consideration are also subject to tariff measures, the highest rate between the basic customs duty (MFN) and duties resulting from the conversion of non-tariff measures is retained.
12. UNCTAD (1999) for the basic tariffs in 1997, and CMIE (1999) for the value of imports in 1997–8; the concatenation of these two files leads to leave in the blank the tariff of some 6 digit product lines.
13. Cashew nuts occupy a significant position both in Indian agricultural exports as well as in imports. They are, in fact, mostly imported raw to be re-exported after processing.
14. Less than 1 per cent of rice production was exported in 1995–6, but of this, 90 per cent of basmati production is sent abroad.
15. According to Indian statistics (Ministry of Finance, 1999), the value of exports to the European Union was Rs. 317.9 billion in 1997–8, i.e. US$ 8.6 billion if we use a conversion rate of 37.165 rupees for 1 US$.

REFERENCES

Boncorps, Bertrand, *Point détaillé de la position de l'Inde dans les futures négociations commerciales multilatérales sur l'agriculture*, Note de synthèse, April, Ambassade de France en Inde, New Delhi, 1999.

Bureau, Dominique and Jean-Christophe Bureau, *Agriculture et négociations commerciales*, Rapport Conseil d'Analyse Economique, La Documentation Française, Paris, 1999.

Bureau, Jean-Christophe, Luca Salvatici and Linda Fulponi, *Measuring Trade Liberalisation. European Commitments under the Uruguay Agreement on Agriculture*, Paper presented at the IX European Congress of Agricultural Economics, 24–8 August, Warsaw,1999.

CENCUS, *Customs Tariff of India. Budget Edition*, Census Publications, New Delhi, 1999.

———, *Central Excise Tariff 1999–2000. Budget Edition*, Cen-sus Publications, New Delhi, 1999.

Chandhok, H.L., The Policy Group, *India Database. The Economy. Annual Time Series Data in Two Volumes*, Living Media, New Delhi, 1990.

Chandiramani, Nilima M., *World Trade Organisation and Globalisation. An Indian Overview*, Shroff Publishers & Distributors Pvt. Ltd., Mumbai, 1999.

CMIE, *TRADES*, Centre for Monitoring Indian Economy, New Delhi, Electronic database 1999.

Dorin, Bruno, 'Food Policy and Nutritional Security. Unequal Access to Lipids in India', *Economic and Political Weekly*, Vol. 34, No. 26, 26 June–2 July 1999a, pp. 1709–17.

———, 'Inde. Qui répondra aux besoins du marché alimentaire ?', *Courrier de la Planète*, Solagral, No. 50, Montpellier, April, 1999b, pp. 46–8.

———, *Les conditions de mise en œuvre des accords multilatéraux de Marrakech en Inde*, Rapport pour le Ministère de l'Agriculture, Paris, 1999c.

———, 'L'Inde dans le commerce agricole international. Conditions et bilan de mise en œuvre des accords de Marrakech', *Notes et Etudes Economiques*, No. 13, Paris, March 2001, pp. 49–84.

Dorin, Bruno and Thomas Jullien, *Input Subsidies in India and Agricultural Competitiveness Within and Outside the Country*, Paper presented at the seminar 'Agricultural Incentives and Sustainable Development', India International Centre, New Delhi, 3–4 April 2000.

EC, *Report on Market Access Barriers in India*, Report DGI/D/4 (113 Committee), European Commission, Brussels, 1997.

EPWRF, *National Accounts. India*, Economic and Political Research Foundation, Mumbai, 1999, CD-ROM.

Gulati, Ashok, Rajesh Mehta and Sudha Narayanan, 'From Marrakesh to Seattle. Indian Agriculture in a Globalising World', *Economic and Political Weekly*, Vol. 34, No. 41, October 1999, pp. 2931–42.

MATTHEWS, Alan and P. MAHE, *L'accord agricole de Marrakech et les politiques agricoles nationales des PED : poursuite de l'ajustement structurel ?*, Paper presented at the seminar 'Le GATT et les échanges agricoles méditérranéens', CIHEAM-CHANIA, Greece, 3–5 December 1995.

MEHTA, Rajesh, 'Removal of QRs and Impact on India's Imports', *Economic and Political Weekly*, Vol. 35, No. 19, 6–12 May, Mumbai, 2000, pp. 1667–71.

MINISTRY OF COMMERCE, *WTO Agreement on Agriculture and its Implications*, India & the WTO, Vol. 1, No. 5, May, Ministry of Commerce, New Delhi, 1999a (http://www.nic.in/commin/).

———, *Schedule XII. India. Part I. Most Favoured Nation Tariff*, Ministry of Commerce, New Delhi, 1999b (http://www.nic.in/commin).

MINISTRY OF FINANCE, *Economic Survey*, Ministry of Finance, New Delhi, 1999 and 2000 (http://www.nic.in/finmin/).

OECD, *Preliminary Report on Market Access Aspects of Uruguay Round Implementation*, Report COM/AGR/APM/TD/WP(99)50, Organization for Economic Cooperation and Development, Paris, 1999.

———, *Non-OECD Countries and Multilateral Trade Liberalisation. A Background Note on Some Key Issues*, Paper TD/TC(99)18/FINAL, Organization for Economic Cooperation and Development, Paris, 1999.

RBI, *Report on Currency and Finance 1997–98*, Reserve Bank of India, Vol. 2, Mumbai, 1999.

UNCTAD, *Trade Analysis and Information System*, United Nations Conference on Trade and Development, Geneva, 1996, CD-ROM Version 4.0 (Autumn 1996).

———, *Trade Analysis and Information System*, United Nations Conference on Trade and Development, Geneva, 1999, CD-ROM Version 6.0 (Winter 1998/99) (http://cs.usm.my/untrains/trains.html).

WB, *Implications of the Uruguay Round Agreement for South Asia. The Case of Agriculture*, South Asia Rural Development Series, World Bank–Allied Publishers, New Delhi, 1999.

WTO, *Agreement on Agriculture*, 1996, pp. 43–71 (http://www.wto.org/english/docs_e/legal_e/14-ag.pdf).

———, *Inde. Restrictions quantitatives à l'importation de produits agricoles, textiles et industriels*, Rapport du Groupe spécial (WT/DS90/R), World Trade Organization, Geneva, 6 April 1999 (http://www.wto.org/wto/ddf/ep/public.html).

PART III.

FUTURE SCENARIOS AND PROSPECTIVE MODELLING

9

Food Production and Demand Projections for South Asian Countries

POLICY IMPLICATIONS FOR INDIAN AGRICULTURE

RAJENDRA SINGH PARODA, PRADUMAN KUMAR

INTRODUCTION

South Asia, comprising Bangladesh, Bhutan, India, Nepal, Pakistan and Sri Lanka, has high population pressure on land and other resources to meet its food and development needs. These countries, with a total population of about 1,291.2 million people in 1998, accounted for 21.8 per cent of the world population and shared only 3 per cent of the world land area. Half of the land area of South Asia is under arable and permanent crops, while in the world as a whole only 11 per cent of the total land area is under arable land and permanent crops.

During the last three decades, the South Asia region has made remarkable progress in its food and agricultural production. The near famine conditions prevailing in the early 1970s in several pockets of this region have largely been eliminated, and it has not only generally overcome the food crisis but also has been transformed from a food deficit to a self-sufficiency status. For instance, foodgrains production in India increased from about 72 million tons in 1965 to 195 million tons in 1997–8. Bangladesh, a country which imported annually about 3.5 million tons foodgrains in the mid-1960s, became self-sufficient in foodgrains in 1993 despite a doubling of her population during this period.

These changes have been triggered by the Green Revolution in South Asia involving the development and diffusion of high yielding varieties (HYVs), especially of rice and wheat, from the mid-1960s onwards. This has been accompanied by increased levels of use of inputs, particularly irrigation, fertilizers and tractors, and policy support. Government investment in infrastructure, research and extension, price and other policies along with crop, livestock and fisheries production strategies have significantly helped to increase food production and its availability.

However, the emerging production scenario, higher economic growth, population explosion and shifts in dietary patterns, will change the supply and demand prospects for food in the next century. Producing additional food with limited and degraded natural resources and providing economic access to food at the household level for ensuring food security for all will be the major challenge for South Asian nations. This calls for a review of the current food situation and future needs in the next century in South Asia.

This paper is an attempt to (i) examine the current situation and trends in food production, (ii) project the food consumption pattern and nutritional status of the households, (iii) project the demand for foodgrains, livestock, fisheries and horticultural products from 2000 to 2030, and (iv) estimate yield and production targets to maintain self-reliance among South Asian countries and suggest the strategies to enhance food production.

FOOD PRODUCTION

AGRICULTURAL PRODUCTION AND PRODUCTIVITY

Crop production strategies followed during the post-green revolution period have significantly helped to increase food production. The current situation and trend in food production based on the data provided in various publication of the Food and Agriculture Organization of the United Nations (FAO, Rome)/Regional Office for Asia and the Pacific (RAP, Bangkok) (FAO/RAP, 1994) is examined. The average production and yield figures are given for the year 1994–6 in Tables 9.1 and 9.2. The growth analysis for area, production and yield pertaining to the period 1986–96 is presented in Table 9.3. The average annual growth rates in area, production and yield of major cereals for South Asian countries for the periods 1969–78, 1978–87 and 1987–96 are also given in Table 9.4.

Cereals

With an annual production of about 106 million tons of rice, South Asia accounts for about 29 per cent of the world's rice production, although its average yield is only 1.87 tons per hectare against about 2.5 to 4.7 tons per ha. in several of the developing countries of the Asia-Pacific region. The gains in rice output have come essentially from the steady increase in yield per unit area in India and Bangladesh. This has come essentially from the better utilization of the existing infrastructure, stepping-up the use of modern inputs, and extending the techniques of Green Revolution to new areas. Yield growth rates of 2.5 and 2.2 per cent in India and Bangladesh were quite satisfactory, but there is scope to accelerate the pace in other countries. India and Bangladesh are the major rice producing countries of the South Asia region, contributing 77 per cent and 16 per cent respectively.

With an annual production of about 80 million tons, the region accounted for about 14.5 per cent of the world's wheat production. India and Pakistan are the major wheat producing countries in South Asia, contributing 77 per cent and 20 per cent respectively. The production in these two countries during the last ten years has followed more than 3 per cent annual growth, resulting from both yield and area growth. Bangladesh started wheat production about 25 years ago and now produces about 1.2 million tons. Nepal is historically a wheat growing country, sharing about 1.2 per cent in the production of the region, and its production is also increasing steadily at an annual growth rate of 3 per cent. Bhutan is a marginal producer and the production has declined in the last ten years. Sri Lanka does not produce any wheat.

Table 9.1: Share of Country in the Production of South Asia (1994–6)

(%)

Items	Bangladesh	Bhutan	India	Nepal	Pakistan	Sri Lanka	South Asia	
							Prod. (Mts)	Share in World (%)
Staple food crops								
Rice	16.37	0.03	76.57	1.98	3.45	1.60	105.7	28.85
Wheat	1.54	0.01	76.85	1.18	20.43	0.00	80.1	14.54
Maize	0.02	0.33	77.70	10.86	10.82	0.26	12.0	2.16
Coarse grains	0.21	0.30	89.11	4.65	5 50	0.11	34.4	3.97
Cereals	8.45	0.09	78.67	2.11	9.95	0.78	220.3	12.35
Pulses	3.27	0.01	90.68	1.14	4.66	0.24	16.4	28.81
Roots & Tuber	6.41	0.19	84.60	3.35	4.96	0.49	29.4	4.72
Horticultural crops								
Vegetables	2.04	0.02	90.05	1.73	5.32	0.85	71.3	12.75
Fruits	3.02	0.14	83.69	1.22	11.44	1.70	45.6	11.32
Other crops								
Edible Oil crops	1.62	0.00	88.16	0.44	7.17	2.60	9.6	10.65
Sugar cane	2.39	0.01	81.60	0.49	15.06	0.45	302.9	26.34
Livestock products								
Milk	2.25	0.04	74.76	1.16	21.44	0.34	84.0	15.69
Meat	5.48	0.12	63.94	2.66	26.51	1.29	6.5	3.15
Eggs	4.17	0.01	78.00	0.98	14.21	2.62	1.9	4.37
Fisheries products								
Fish	16.58	0.00	71.10	0.28	8.59	3.44	6.7	6.11

Source: Computed by the authors.

Table 9.2: Average Yield for Staple Food and Other Crops in South Asia (1994–6)

(t./ha.)

Items	Bangladesh	Bhutan	India	Nepal	Pakistan	Sri Lanka	South Asia	World Average
Rice	1.74	1.11	1.92	1.46	1.70	1.96	1.87	2.47
Wheat	1.88	0.80	2.45	1.49	2.00	-	2.31	2.48
Maize	0.89	0.99	1.52	1.67	1.46	1.04	1.53	4.01
Coarse grains	0.73	1.22	0.95	1.51	1.00	0.94	0.97	2.63
All cereals	1.74	1.08	1.74	1.49	1.79	1.91	1.74	2.55
Pulses	0.75	1.00	0.60	0.61	0.50	0.77	0.60	0.81
Roots & Tubers	10.64	10.75	16.95	7.70	14.23	8.90	14.60	12.75
Sugarcane	40.58	31.22	67.36	35.53	46.62	58.59	61.3	62.12

Source: Computed by the authors.

Table 9.3: Average Annual Compound Growth Rate in Area, Production and Yield for Various Food Commodities in South Asia (1986–96)

(%)

Items		Bangladesh	Bhutan	India	Nepal	Pakistan	Sri Lanka	South Asia	World
Staple food crops									
Rice	Area	–0.6	–1.3	0.5	0.0	0.9	1.3	0.3	0.4
	Prod	1.6	–1.9	3.0	1.0	1.5	–0.5	2.7	1.8
	Yield	2.2	–0.6	2.5	1.0	0.6	–1.8	2.4	1.4
Wheat	Area	1.8	–3.5	1.1	1.2	1.2	-	1.0	–0.1
	Prod	2.4	–4.2	3.7	3.0	3.0	-	3.4	0.9
	Yield	0.6	–0.6	2.6	1.8	1.8	-	2.4	1.0
Maize	Area	–0.5	–1.0	0.3	0.5	0.6	–0.5	0.2	0.7
	Prod	–1.9	–2.7	2.3	2.6	1.5	–2.4	1.9	2.4
	Yield	–1.4	–1.7	2.0	2.1	0.9	–1.9	1.7	1.7
Cereals	Area	–0.5	–1.0	–0.4	0.6	0.9	1.1	–0.3	–0.1
	Prod	1.6	–2.0	2.9	1.8	2.5	–0.6	2.8	1.1
	Yield	2.2	–1.0	3.2	1.2	1.6	–1.7	3.1	1.3
Puises	Area	–0.3	–7.2	0.8	2.2	0.9	–2.6	0.7	0.3
	Prod	0.2	–4.3	2.2	2.7	0.5	–3.2	1.6	0.7
	Yield	0.5	1.6	1.3	0.5	–0.5	–0.4	0.9	0.3
Roots and Tubers	Area	1.0	–3.2	1.6	2.7	3.1	–4.2	1.8	1.1
	Prod	1.4	0.8	3.5	5.9	6.5	–4.8	4.0	1.3
	Yield	0.4	3.3	1.9	3.1	3.2	–0.5	2.2	0.2
Other crops									
Sugarcane	Area	1.0	0.3	2.4	6.5	2.4	1.3	2.5	1.7
	Prod	0.8	1.0	3.8	10.4	4.9	6.7	4.0	2.2
	Yield	–0.2	0.7	1.3	3.9	2.4	5.3	1.5	0.4

(contd.)

Table 9.3 (contd.)

Items		Bangladesh	Bhutan	India	Nepal	Pakistan	Sri Lanka	South Asia	World
Edible oil crops									
Edible oil equivalent	Prod	0.9	-	6.2	2.0	2.0	0.3	5.0	3.7
Horticultural crops									
Vegetables	Prod	2.5	0.7	2.2	5.2	5.1	–0.1	2.8	2.9
Fruits	Prod	0.3	3.6	4.7	4.8	4.6	–0.4	4.5	2.2
Livestock products									
Milk	Prod	4.1	0.4	3.3	–1.8	2.4	5.4	3.4	0.2
Meat (slaughtered)	Prod	3.6	1.7	3.1	1.8	6.3	7.1	3.3	2.9
Hen eggs	Prod	6.9	4.0	5.2	4.5	6.0	1.7	5.4	3.2
Fish	Prod	7.0	0.0	4.9	7.9	1.2	4.3	5.2	3.7

Source: Computed by the authors.

South Asia's share in the world maize production is only 2.2 per cent with about 12 million tons of annual production. India accounts for about 78 per cent of the region's maize production, and Nepal and Pakistan for 11 per cent each. Maize yields in this region are generally low, ranging between 0.89 tons to 1.67 tons per ha. against world average of 4.0 tons per ha. India and Nepal recorded yield growth rate of about 2 per cent per year during 1986–96. In Bangladesh, Sri Lanka and Bhutan, maize production declined due to fall in area and yield during the last ten years. In Nepal, India and Pakistan, production increased annually by 2.6 per cent, 2.3 per cent and 1.5 per cent respectively due to positive growth in area and per ha. yield.

Declining trends in area growth were observed for rice, wheat and other cereals. However, the increasing trend in production was because of yield growth (Table 9.4). Future growths in cereal production have to be essentially achieved as a result of yield growth. South Asian cereals account for about 12.4 per cent of the total world cereal production. India accounts for about 79 per cent of the region's cereal production, which is followed by Pakistan (10 per cent), Bangladesh (8 per cent) and other South Asian countries (3 per cent). Average cereal yield in the region was 1.7 tons per ha. against 2.6 tons in the world as a whole. Sri Lanka recorded the highest average yield of about 1.9 tons per ha. followed by 1.8 tons in Pakistan, 1.7 tons in Bangladesh and India, and less than 1.5 tons in the remaining countries. The yield growth rate in the region, at 3.1 per cent, was more than double that of the world's growth rate (1.3 per cent). Despite the large area, India scored the highest growth rate in cereal production of 2.9 per cent, followed by 2.5 per cent in Pakistan, 1.8 per cent in Nepal and 1.6 per cent in Bangladesh, whereas Bhutan and Sri Lanka registered negative growth rates. Share of rice in total cereal production was 48 per cent, followed by wheat (36 per cent) and coarse grains (15 per cent). Maize accounted for one-third of the total coarse grains production, and is gaining importance as feed.

Pulses

South Asia, with an annual production of 16.4 million tons, accounted for about 29 per cent of the world's pulse production while its share in world area was 40 per cent in the recent past. The average yield was only 600 kg. per ha. against 847 kg. per ha. at the world level. India accounted for 91 per cent of the region's pulse production. During the decade ending 1996, India registered a yield growth of 1.2 per cent per annum while for the other major countries yield gains were generally low; in Pakistan it was even negative.

Roots and Tubers

South Asia, which produces over 29 million tons of roots and tubers, mainly potato, accounted for about 4.7 per cent of the world's production, and registered an average yield of about 14.6 tons per ha. against about 12.8 tons for the world as a whole. India's share in the region's production was 82 per cent and its yield of 17 tons per ha. was one of the highest in the whole of Asia-Pacific. The increase in production accrued through an increase in area as well as yield in all the major countries.

Sugarcane

With an annual production of 303 million tons, South Asia comprises 26 per cent of the

Table 9.4: Average Annual Growth Rates in Area, Production, Yield of Major Cereals in South Asia (1969–96)

(%)

	Rice			Wheat			Coarse Grains		
	1969–78	1978–87	1987–96	1969–78	1978–87	1987–96	1969–78	1978–87	1987–96
Bangladesh									
Area	0.18	0.31	–0.63	5.13	13.83	1.11	–2.46	8.33	–4.14
Production	1.37	2.17	1.49	12.10	16.28	0.94	–4.27	9.87	–3.96
Yield	1.19	1.86	2.13	6.96	2.45	–0.17	–1.81	1.55	0.18
Bhutan									
Area	2.02	2.67	–2.66	3.14	4.30	–5.08	2.12	1.52	–1.01
Production	2.07	2.22	–3.50	3.14	4.10	–4.86	2.32	0.55	–2.84
Yield	0.05	–0.45	–0.83	0.00	–0.20	0.22	0.20	–0.97	–1.83
India									
Area	0.74	0.29	0.44	3.33	1.30	0.90	–1.07	–0.97	–2.22
Production	2.16	2.78	3.16	5.77	5.03	3.56	1.17	–0.57	1.28
Yield	1.41	2.48	2.72	2.44	3.73	2.66	2.24	0.40	3.50
Nepal									
Area	1.03	1.63	–0.21	6.85	5.59	1.03	0.77	6.20	1.03
Production	1.16	2.61	0.51	7.34	7.07	2.61	–0.04	4.44	2.94
Yield	0.13	0.98	0.72	0.49	1.48	1.59	–0.81	–1.76	1.91
Pakistan									
Area	2.20	0.13	0.78	0.34	1.88	1.18	–0.79	0.04	0.11
Production	3.70	0.59	1.61	3.91	3.89	3.19	0.77	1.27	1.36
Yield	1.50	0.46	0.83	3.57	2.01	2.01	1.56	1.23	1.25

(contd.)

Table 9.4 (contd.)

	Rice			Wheat			Coarse Grains		
	1969–78	1978–87	1987–96	1969–78	1978–87	1987–96	1969–78	1978–87	1987–96
Sri Lanka									
Area	2.30	0.63	1.40	-	-	-	9.03	–3.11	–2.06
Production	0.54	4.72	1.06	-	-	-	7.18	1.36	–3.25
Yield	–1.76	4.09	–0.34	-	-	-	–1.85	4.47	–1.19
South JAsia									
Area	0.71	0.33	0.26	2.62	1.64	0.97	–1.03	–0.76	–2.03
Production	2.01	2.60	2.72	5.35	4.96	3.42	1.11	–0.28	1.32
Yield	1.30	2.27	2.46	2.73	3.32	2.45	2.14	0.49	3.35
World									
Area	1.13	0.08	0.28	0.81	–0.12	–0.18	0.48	–0.06	–0.37
Production	2.58	3.04	1.73	3.08	2.69	0.95	2.73	1.58	0.79
Yield	1.45	2.96	1.45	2.27	2.81	1.14	2.25	1.64	1.16

Source: Computed by the authors by using *FAOSTAT* data based on three years moving averages.

world's sugarcane production. India and Pakistan are the major sugarcane producers in the region, with 82 and 15 per cent of the South Asian production respectively. India's average yield of about 67 tons per ha. which is highest in the region, exceeded the world average by about 5 tons per ha. Yield levels in other south Asian countries are rather low. During 1986–96, sugarcane production in the region showed an increase of 4 per cent per annum.

Oil Crops

Measured in terms of oil equivalent, South Asia, with an annual production of 9.6 million tons, constitutes 11 per cent of the world's vegetable oil production. India accounted for 88 per cent of the regional production, and has achieved a very high growth rate, exceeding 6 per cent per annum during 1986–96. The breakthrough in Indian oilseed production has occurred due to a synergistic interplay of appropriate technologies, development and institutional support, and marketing and price support through a mission mode project.

Horticultural Crops

Having achieved significant increase in the production of cereals, oilseeds, and other food crops, the countries of the region, especially India and Pakistan, have in the recent past started paying much greater attention to the horticultural crops. South Asia produced about 46 million tons of fruits (excluding melon), which was about 11 per cent of the world's production. India accounted for about 84 per cent of the region's fruit production, followed by Pakistan (11 per cent), Bangladesh (3 per cent) and a small share by other countries. India and Pakistan, the major producers, achieved impressive growth rates of about 4.5 per cent per annum. Banana, citrus and mango are predominant fruits in the region, with about 18 per cent, 15 per cent and 23 per cent share in the total fruit production respectively.

South Asian countries, with an annual production of about 71 million tons of vegetables, accounted for about 13 per cent of the world production of vegetables (including melon). India accounted for 90 per cent of the total regional production, showing an annual growth of 2.2 per cent at this high base level. All the countries, except Sri Lanka, witnessed considerable increase in vegetable production, and the rate of growth during the decade ending 1996 varied from 0.7 per cent in Bhutan, 2.2 per cent in India, 2.5 per cent in Bangladesh, 5.1 per cent in Pakistan and 5.2 per cent in Nepal.

Livestock and Fisheries Production

With an annual production of only 6.5 million tons of total meat, 1.9 million tons of eggs, 6.7 million tons of total fish, the share of the region in the world production of these foods has been rather small, being 3.2 per cent, 4.4 per cent and 6.1 per cent respectively. Compared to this region's share in the world, milk production is quite high (15.7 per cent), with an annual production over 84 million tons. The average rate of growth of production of various products during the decade 1986–96 in the South Asian countries was 3.4 per cent for milk, 3.3 per cent for meat, 5.4 per cent for eggs (hen) and 5.2 per cent for fish. These rates generally far exceed the corresponding rates of growth at the world level. India, Pakistan and Bangladesh are the major producers of livestock and fisheries products in South Asia. The share of India in South Asian production ranged between 64 to 78 per cent, being minimum for meat and maximum for eggs. Pakistan accounted for 21 per cent of milk, 27 per cent of meat and only 9 per cent of fish, and Bangladesh for 2.2 per cent of milk, 4.2 per cent of eggs, 5.5 per cent of meat and 16.6 per cent of fish in the production of South Asia.

DEMAND FOR FOOD

The conditions relating to food, agriculture and the environment in South Asia appear to be improving. Foodgrains production has been growing faster than the population. It is true for India, Pakistan and Bangladesh. The per capita availability of foodgrains in the region, except pulses, is rising. On account of the increasing availability of a wide variety of food items and increase in per capita income and urbanization, the demand for better quality and nutritional food is now greater than in the past. This has led to changes in the composition of the food basket, with consumers moving from coarse grains to superior grains (rice and wheat), and from grains to livestock and horticultural products (milk, meat, vegetable, fruits, etc.). The structural changes have brought about major shifts in consumption of milk, fruits and vegetables (Table 9.5). This has significant implications for food demand, research priority setting and resource allocation to achieve food security. Diversification into the production of livestock, fisheries and horticultural products needs to be the current research priority with additional resource support.

DEMAND ELASTICITY

Demand elasticities are crucial parameters in food demand projections. The elasticities based on Food Characteristic Demand System (Bouis and Haddad, 1992) are given in Table 9.6. The expenditure elasticities are in accordance with *a priori* expectations. The magnitude of expenditure elasticity for each food item across countries varies in a narrow range. The expenditure elasticity for all food items, except cereals, is found to be positive and significantly high in magnitude. Expenditure elasticities for cereals were negative and highly inelastic. The results follow the historical trend in per capita consumption of cereals as observed in the household consumer survey data available for various South Asian countries. The commodities with the cheapest source of calories have the lowest income elasticities. The magnitude of food demand elasticity is about 0.4 while the calorie-income elasticities were not significantly different from zero because of a structural shift in the dietary pattern from low-cost calorie food to high-cost calorie food items. This observation confirms the empirical evidences presented by Behrman and Deolalikar (1989) for India and by Bouis and Haddad (1992) for Philippines. Thus, the demand elasticities estimated in this study are likely to give the most reliable demand projections for cereals and other food commodities.

FUTURE CONSUMPTION PATTERN

Expenditure elasticities are used to project the annual per capita consumption for each South Asian country under the alternative scenarios of 3.5 per cent and 5.5 per cent

Table 9.5: Annual Per Capita Consumption of Various Food Items in South Asia

(kg.)

Items	Bangladesh		India		Pakistan		Nepal	Sri Lanka
	1981	1988	1977	1993	1979	1993	1993	1993
Rice	98.0	163.5	82.7	79.9	14.9	16.1	62.9	72.5
Wheat	23.9	21.5	52.4	54.5	134.4	124.6	33.3	46.6
Coarse grains & cereal substitutes	3.8	3.4	48.4	19.8	10.5	8.2	61.6	18.2
Pulses	5.3	8.2	9.3	9.5	8.0	8.1	8.3	5.5
Vegetables	38.9	78.6	27.7	55.8	31.8	65.4	79.3	34.7
Fruits	2.9	5.0	3.3	12.5	7.8	15.7	22.6	36.1
Milk	7.1	8.2	26.6	55.8	59.0	87.3	35.9	27.1
Meat, eggs & fish	11.4	12.9	3.1	4.6	8.1	8.1	9.9	18.7
Edible oils	2.3	3.7	3.1	5.0	4.3	10.8	3.9	2.7
Sugar	3.5	3.4	14.2	9.9	14.5	17.8	6.0	19.3

Sources: Computed by the authors.

Data Source: For Bangladesh: *Report on the Household Expenditure Survey, 1988–1989,* Bureau of Statistics, Dhaka; for Pakistan: *Household Income and Expenditure Survey*, Federal Bureau of Statistics, Islamabad; for India: NSSO, *Household Expenditure Consumer Survey, 50th National Sample Survey Round,* New Delhi; for Nepal and Sri Lanka: *FAOSTAT*, Food and Agriculture Organization of the United Nations, Rome (Note: availability of food is taken).

Table 9.6: Expenditure Elasticities based on FCDS

Items	Bangladesh	India	Nepal	Pakistan	Sri Lanka
Rice	–0.078	–0.016	0.016	0.025	–0.071
Wheat	0.004	–0.109	–0.111	–0.121	–0.027
Coarse grains	–0.129	–0.147	–0.152	–0.172	–0.170
Roots & Tubers	0.250	0.336	0.300	0.282	0.371
Pulses	0.227	0.214	0.263	0.189	0.224
Vegetable oils	0.191	0.176	0.199	0.186	0.128
Sweeteners	0.116	0.115	0.107	0.094	0.097
Vegetables	0.500	0.673	0.599	0.565	0.748
Fruits	0.666	0.702	0.698	0.710	0.556
Milk	0.581	0.589	0.634	0.575	0.689
Meat, fish & eggs	0.822	0.892	0.860	0.670	0.866
Other food	1.139	1.200	1.177	1.242	1.196
Non-food	1.883	1.959	1.919	2.052	1.977
Income elasticities of					
- Calories	0.044	0.050	0.048	0.047	0.058
- Food	0.448	0.399	0.424	0.342	0.402

Notes: FCDS = Food Characteristic Demand System (Bouis, 1992).

Source: Computed by the authors.

Data Source: For Bangladesh: *Report on the Household Expenditure Survey. 1988–1989*, Bureau of Statistics, Dhaka; for Pakistan: *Household Income and Expenditure Survey*, Federal Bureau of Statistics, Islamabad; for India: NSSO, *Household Expenditure Consumer Survey. 50th National Sample Survey Round*, New Delhi; for Nepal and Sri Lanka: *FAOSTAT*, Food and Agriculture Organization of the United Nation, Rome.

growth in per capita Gross Domestic Product (GDP), under the alternative assumptions that the magnitude of expenditure elasticities as observed in 1993 will decline to a level of 75 per cent, 50 per cent, and 25 per cent by 2025. Historically, the first scenario is the most plausible rate and the second scenario is considered to be the most likely in the long term. The prediction for per capita consumption of food becomes asymptotic to the year axis in the long term under the assumption of 50 per cent decline in elasticities. In reality, it is expected that the consumption level of food items in the long run will stabilize, income elasticity for food will converge to zero, and income effect on per capita consumption will be negligible. Fifty per cent decline in expenditure elasticities by 2025 is considered the most likely scenario to explain the consumption behaviour in the long term. Average annual per capita availability of different commodities during 1992–4 given in Table 9.7 is used as the base consumption for predicting the dietary pattern in the years 2000 to 2030 (Tables 9.8 and 9.9).

Rice is the staple food of Bangladesh, wheat of Pakistan, rice and wheat of India and Sri Lanka, and rice, coarse grains and wheat of Nepal. The per capita consumption of edible oils is the highest in Pakistan (14 kg. per year) as compared to other countries where it ranges between 2.3 kg. in Sri Lanka to 8 kg. in India. Pakistan also dominates the consumption of sugar, milk and meat. India dominates in the consumption of vegetables and sugar. Bangladesh and Sri Lanka dominate fish consumption. A wide variation in the consumption of various foods is observed across the countries. This is so because the dietary pattern of a country is highly influenced by its domestic production pattern and food availability.

EXPENDITURE GROWTH EFFECTS ON CONSUMPTION

The expenditure growth effects on the dietary pattern at constant prices are shown in Table 9.10. The income effect on consumption is negative and mild for cereals, and is positive and significant for livestock, horticultural and fisheries products. South Asia will continue to exhibit a more diversified food basket with significantly higher levels of per capita consumption of milk, fruits, vegetables, meat, eggs and fish. The increasing trends in consumption towards high value commodities will generate high growth in demand for pulses, livestock, fisheries and horticultural products.

The impact of the structural shift in the dietary pattern on nutrition in terms of energy, as expected, has shown a marginal improvement (Table 9.11). The average daily calorie intake varied from 2,017 calories in Sri Lanka to 2,347 calories in Pakistan in the year 1995 which will improve by the year 2030 to a calorie level of 2,059 in Bangladesh, 2,151 in Sri Lanka, 2,235 in Nepal, 2,389 in India and 2,554 in Pakistan. These calorie levels are higher than recommended for each of the South Asian countries. In spite of a shift in diet from cereals to non-cereals, foodgrains will continue to dominate and maintain their share of 62–65 per cent in total energy. In spite of a steady growth in the consumption of horticultural, livestock and fisheries products, and other food items, their share in total energy will improve only from 5.2 per cent in 1995 to 7–8 per cent in 2030 for horticultural product, from 7.5 per cent in 1995 to 10–12 per cent in 2030 for livestock and fisheries products, while there will be marginal improvement for other food items. This pattern is true for each South Asian country. The importance of cereals and pulses will not decline in the future to meet the nutritional requirement.

Table 9.7: Average Annual Per Capita Availability of Food Items in South Asia (1992–4)

(kg.)

Items	Bangladesh	India	Pakistan	Nepal	Sri Lanka
Cereals	184.9	157.7	155.6	190.0	159.1
Rice	165.0	76.1	18.1	94.4	108.7
Wheat	19.9	55.7	128.9	33.3	46.6
Maize	0.0	8.4	5.5	49.3	3.1
Other coarse grains	0.0	17.5	3.1	13.0	0.8
Pulses	4.7	13.1	4.6	7.7	5.5
Roots & Tubers	13.4	20.7	8.6	33.0	17.7
Potato	9.9	13.6	5.8	27.2	3.3
Other roots & tubers	3.5	7.1	2.8	5.8	14.4
Horticultural products	21.8	99.8	58.1	74.9	67.7
Vegetables	10.1	64.2	20.9	52.2	31.6
Fruits	11.7	35.6	37.2	22.7	36.1
Edible oils	4.2	7.1	12.6	3.9	2.1
Sweeteners	7.3	23.7	25.0	2.8	19.3
Livestock products	18.0	64.9	118.9	46.4	32.8
Milk	14.1	57.7	102.1	36.1	27.1
Meat	3.0	4.4	11.9	8.2	2.8
Eggs	0.7	1.4	1.6	0.9	2.5
Animal fats	0.2	1.4	3.3	1.2	0.4
Fisheries products	8.5	4.1	2.3	0.8	15.8

Source: Computed by the authors.

Table 9.8: Projected Annual Per Capita Food Consumption in South Asia with Low Income Growth (1995, 2000, 2015, 2030)

(kg.)

Items	Year	Bangladesh	India	Nepal	Pakistan	Sri Lanka	South Asia
Rice	1995	164.1	76.0	94.5	18.1	108.2	78.9
	2000	162.3	75.8	94.7	18.2	107.1	78.2
	2015	159.0	75.5	95.1	18.3	105.1	76.6
	2030	157.7	75.4	95.3	18.4	104.3	75.7
Wheat	1995	19.9	55.3	33.1	127.9	46.5	59.4
	2000	19.9	54.4	32.5	125.7	46.3	58.9
	2015	19.9	52.9	31.6	121.6	46.0	58.4
	2030	19.9	52.3	31.2	120.1	45.9	58.8
Maize	1995	0	8.3	48.8	5.4	3.1	7.8
	2000	0	8.1	47.7	5.3	3.0	7.7
	2015	0	7.8	45.8	5.1	2.9	7.4
	2030	0	7.7	45.1	5.0	2.8	7.3
Other coarse grains	1995	0	17.3	12.9	3.1	0.8	13.7
	2000	0	17.0	12.6	3.0	0.8	13.4
	2015	0	16.3	12.1	2.9	0.7	12.7
	2030	0	16.1	11.9	2.8	0.7	12.2
Total cereals	1995	184.0	157.0	189.2	154.5	158.6	159.9
	2000	182.2	155.4	187.6	152.1	157.2	158.2
	2015	178.9	152.5	184.6	147.9	154.7	155.1
	2030	177.6	151.5	183.5	146.3	153.7	154.0
Total pulses	1995	4.8	13.3	7.8	4.7	5.6	10.8
	2000	4.9	13.7	8.1	4.8	5.8	11.1
	2015	5.2	14.5	8.7	5.0	6.1	11.6
	2030	5.4	14.8	9.0	5.1	6.3	11.7

(contd.)

Table 9.8 (contd.)

Items	Year	Bangladesh	India	Nepal	Pakistan	Sri Lanka	South Asia
Foodgrains	1995	188.8	170.2	197.1	159.1	164.1	171.2
	2000	187.2	169.1	195.7	156.9	162.9	169.8
	2015	184.1	167.0	193.4	152.9	160.8	167.2
	2030	183.0	166.3	192.5	151.4	160.0	166.2
Potato	1995	10.1	13.9	27.7	5.9	3.4	12.7
	2000	10.4	14.6	29.0	6.2	3.6	13.3
	2015	11.2	16.0	31.4	6.6	3.9	14.4
	2030	11.4	16.5	32.4	6.8	4.1	14.8
Other roots and tubers	1995	3.5	7.2	5.9	2.8	14.6	6.4
	2000	3.6	7.4	6.0	2.9	15.0	6.6
	2015	3.7	7.7	6.2	3.0	15.7	6.8
	2030	3.8	7.8	6.3	3.0	16.1	6.8
Total roots and tubers	1995	13.6	21.1	33.6	8.7	18.0	19.2
	2000	14.0	21.9	34.9	9.0	18.5	19.9
	2015	14.9	23.7	37.6	9.6	19.7	21.2
	2030	15.2	24.4	38.7	9.9	20.1	21.5
Edible oils	1995	4.3	7.2	4.0	12.8	2.1	7.4
	2000	4.4	7.4	4.1	13.1	2.2	7.6
	2015	4.6	7.7	4.3	13.8	2.2	8.1
	2030	4.7	7.9	4.4	14.0	2.3	8.3
Vegetables	1995	10.4	67.5	54.3	21.7	33.2	56.2
	2000	11.2	75.3	59.1	23.5	36.9	62.2
	2015	12.8	92.2	69.4	27.3	45.1	74.8
	2030	13.5	99.8	73.8	29.0	48.7	79.2
Fruits	1995	12.2	37.3	23.8	39.0	37.5	34.8
	2000	13.4	41.2	26.3	43.1	40.5	38.5
	2015	16.1	49.7	31.6	52.1	47.0	46.4
	2030	17.2	53.4	34.0	56.1	49.8	49.7

(contd.)

Table 9.8 (contd.)

Items	Year	Bangladesh	India	Nepal	Pakistan	Sri Lanka	South Asia
Sweeteners	1995	7.4	23.9	2.8	25.2	19.4	22.0
	2000	7.5	24.3	2.9	25.5	19.7	22.4
	2015	7.7	25.0	2.9	26.1	20.2	23.0
	2030	7.8	25.3	3.0	26.4	20.4	23.2
Milk	1995	14.7	60.0	37.7	106.1	28.4	59.9
	2000	15.9	65.2	41.2	115.1	31.3	65.4
	2015	18.6	76.3	48.8	134.2	37.6	77.5
	2030	19.7	81.1	52.1	142.4	40.4	83.2
Animal fats	1995	0.2	1.5	1.3	3.4	0.4	1.5
	2000	0.2	1.6	1.4	3.7	0.4	1.7
	2015	0.3	1.9	1.6	4.3	0.5	2.0
	2030	0.3	2.0	1.7	4.6	0.6	2.2
Meat	1995	3.2	4.7	8.7	12.4	3.0	5.4
	2000	3.6	5.3	9.8	13.7	3.4	6.2
	2015	4.4	6.7	12.3	16.4	4.2	7.9
Eggs	1995	0.8	1.4	1.0	1.7	2.6	1.4
	2000	0.9	1.6	1.1	1.9	3.0	1.6
	2015	1.1	2.1	1.4	2.2	3.8	2.0
Fish	1995	9.0	4.3	0.8	2.4	16.7	4.7
	2000	10.1	4.9	1.0	2.6	18.9	5.3
	2015	12.5	6.3	1.2	3.1	23.8	6.6
	2030	13.7	6.9	1.3	3.3	26.0	7.1

Source: Computed by the authors.

Table 9.9: Projected Annual Per Capita Food Consumption in South Asia with High Income Growth (1995, 2000, 2015 and 2030)

(kg.)

Items	Year	Bangladesh	India	Nepal	Pakistan	Sri Lanka	South Asia
Rice	1995	163.6	76.0	94.6	18.1	107.9	78.8
	2000	160.8	75.7	94.9	18.3	106.2	77.9
	2015	155.6	75.2	95.5	18.4	103.0	76.1
	2030	153.6	75.0	95.8	18.5	101.9	75.0
Wheat	1995	19.9	55.1	32.9	127.3	46.5	59.2
	2000	19.9	53.7	32.1	123.8	46.2	58.2
	2015	20.0	51.3	30.6	117.7	45.7	56.7
	2030	20.0	50.4	30.1	115.4	45.5	56.7
Maize	1995	0	8.3	48.5	5.4	3.0	7.8
	2000	0	8.0	46.9	5.2	2.9	7.5
	2015	0	7.5	44.0	4.8	2.7	7.1
	2030	0	7.3	42.9	4.7	2.7	7.0
Other coarse grains	1995	0	17.2	12.8	3.0	0.8	13.7
	2000	0	16.7	12.4	2.9	0.8	13.2
	2015	0	15.7	11.6	2.7	0.7	12.1
	2030	0	15.3	11.3	2.6	0.7	11.6
Total cereals	1995	183.6	156.5	188.8	153.9	158.2	159.4
	2000	180.7	154.1	186.2	150.2	156.0	156.8
	2015	175.6	149.7	181.7	143.7	152.1	152.1
	2030	173.6	148.1	180.1	141.3	150.7	150.3
Pulses	1995	4.8	13.4	7.9	4.7	5.6	10.9
	2000	5.1	14.1	8.4	4.9	5.9	11.4
	2015	5.6	15.4	9.4	5.3	6.5	12.3
	2030	5.8	15.9	9.3	5.5	6.7	12.5

(contd.)

Table 9.9 (contd.)

Items	Year	Bangladesh	India	Nepal	Pakistan	Sri Lanka	South Asia
Foodgrains	1995	188.4	169.9	196.7	158.5	163.8	170.8
	2000	185.8	168.1	194.6	155.1	162.0	168.7
	2015	181.1	165.1	191.1	149.0	158.6	164.9
	2030	179.4	164.0	189.9	146.7	157.4	163.4
Potato	1995	10.2	14.1	28.1	6.0	3.4	12.9
	2000	10.7	15.2	30.0	6.4	3.7	13.8
	2015	11.9	17.5	34.0	7.2	4.4	15.7
	2030	12.4	18.5	35.7	7.5	4.6	16.4
Other roots and tubers	1995	3.5	7.2	5.9	2.8	14.7	6.5
	2000	3.6	7.5	6.1	2.9	15.3	6.7
	2015	3.8	8.1	6.5	3.1	16.6	7.1
	2030	3.9	8.3	6.6	3.2	17.1	7.2
Total roots and tubers	1995	13.7	21.3	34.0	8.8	18.1	19.4
	2000	14.4	22.7	36.1	9.3	19.0	20.5
	2015	15.8	25.5	40.5	10.3	20.9	22.8
	2030	16.4	26.7	42.4	10.7	21.7	23.6
Edible oils	1995	4.3	7.2	4.0	12.8	2.1	7.4
	2000	4.5	7.5	4.2	13.4	2.2	7.8
	2015	4.8	8.1	4.5	14.5	2.3	8.5
	2030	5.0	8.3	4.7	14.9	2.4	8.8
Vegetables	1995	10.6	69.5	55.5	22.2	34.1	57.8
	2000	11.9	82.3	63.5	25.1	40.3	67.9
	2015	14.7	113.1	81.5	31.8	55.0	91.3
	2030	15.9	128.0	89.8	34.9	62.2	101.1
Fruits	1995	12.5	38.3	24.4	40.0	38.2	35.7
	2000	14.5	44.7	28.5	46.9	43.3	41.8

(contd.)

Table 9.9 (contd.)

Items	Year	Bangladesh	India	Nepal	Pakistan	Sri Lanka	South Asia
	2015	19.2	60.0	38.1	63.0	54.6	55.9
	2030	21.4	67.2	42.7	70.7	59.8	62.6
Sweeteners	1995	7.4	24.0	2.8	25.2	19.5	22.1
	2000	7.6	24.6	2.9	25.8	19.9	22.7
	2015	8.0	25.8	3.0	26.8	20.8	23.7
	2030	8.1	26.3	3.1	27.2	21.1	24.1
Milk	1995	15.0	61.3	38.6	108.4	29.1	61.2
	2000	17.0	69.9	44.4	123.2	33.9	70.1
	2015	21.7	89.4	57.8	156.6	45.2	90.7
	2030	23.9	98.4	64.1	172.0	50.6	100.8
Animal fats	1995	0.2	1.5	1.3	3.5	0.4	1.6
	2000	0.2	1.7	1.5	4.0	0.5	1.8
	2015	0.3	2.2	1.9	5.1	0.6	2.4
	2030	0.3	2.4	2.1	5.6	0.7	2.6
Meat	1995	3.3	4.8	9.0	12.8	3.1	5.6
	2000	3.9	5.9	10.8	14.8	3.7	6.8
	2015	5.5	8.5	15.5	19.6	5.3	9.8
	2030	6.3	9.8	17.8	21.8	6.1	11.4
Eggs	1995	0.8	1.5	1.0	1.7	2.7	1.5
	2000	1.0	1.8	1.2	2.0	3.3	1.8
	2015	1.3	2.6	1.7	2.7	4.7	2.5
	2030	1.5	3.0	2.0	3.0	5.5	2.9
Fish	1995	9.3	4.5	0.9	2.4	17.3	4.8
	2000	11.1	5.5	1.1	2.8	20.9	5.8
	2015	15.6	7.9	1.5	3.7	30.0	8.3
	2030	17.9	9.2	1.7	4.2	34.5	9.4

Source: Computed by the authors.

Table 9.10: Annual Per Capita Food Consumption in South Asia (2000 and 2030)

(kg.)

Items	Low income growth			High income growth		
	2000	2030	Changes	2000	2030	Changes
Cereals	158.2	154.0	−4.2	156.8	150.3	−6.5
Pulses	11.1	11.7	+0.6	11.4	12.5	+1.1
Roots & tubers	19.9	21.5	+1.6	20.5	23.6	+3.1
Edible oils	7.6	8.3	+0.7	7.8	8.8	+1.0
Sugar	22.4	23.2	+0.8	22.7	24.1	+1.4
Vegetables	62.2	79.2	+17.0	67.9	101.1	+33.2
Fruits	38.5	49.7	+11.2	41.8	62.6	+20.8
Milk	65.4	83.2	+17.8	70.1	100.8	+30.7
Meat	6.2	8.8	+2.2	6.8	11.4	+4.6
Eggs	1.6	2.2	+0.6	1.8	2.9	+1.1
Fish	5.3	7.1	+1.8	5.8	9.4	+3.6

Notes: Low income growth = 3.5 per cent GDP growth, High income growth = 5.5 per cent GDP growth.
Source: Computed by the authors.

Table 9.11: Demand for Energy from Major Food Groups as Percentage of Total Dietary Energy Demand (1995 and 2030)

Food group	Year	Per capita GDP	Bangladesh	India	Nepal	Pakistan	Sri Lanka	South Asia
Food grains								
	1995	3.5%	85.9	68.6	81.9	61.0	75.8	69.8
	2030	3.5%	83.2	64.0	77.6	55.5	71.4	65.0
	2030	5.5%	81.2	60.6	74.3	51.7	68.1	61.7
Horticultural product								
	1995	3.5%	2.4	5.8	5.9	3.4	6.3	5.2
	2030	3.5%	3.0	7.6	7.2	4.4	7.9	6.7
	2030	5.5%	3.4	8.9	8.3	5.1	9.1	7.8
Livestock and fisheries products								
	1995	3.5%	3.0	7.3	6.5	13.0	5.7	7.5
	2030	3.5%	4.4	9.8	9.1	16.8	8.2	10.3
	2030	5.5%	5.5	11.7	11.1	19.7	10.2	12.3
Other food items								
	1995	3.5%	8.7	18.2	5.7	22.7	12.2	17.5
	2030	3.5%	9.4	18.6	6.1	23.3	12.5	18.0
	2030	5.5%	9.8	18.8	6.8	23.5	12.5	18.3
Daily per capita energy in kcal								
	1995	3.5%	2,048	2,188	2,107	2,347	2,017	2,188
	2030	3.5%	2,048	2,294	2,172	2,455	2,086	2,289
	2030	5.5%	2,059	2,389	2,235	2,554	2,151	2,375

Source: Computed by the authors.

Thus, there is no scope for complacency for any nation and they must refrain from diluting their efforts in increasing the foodgrains production along with serious efforts to increase the production of non-cereal food items.

POPULATION PROJECTIONS

In the region as a whole, the population growth rate has shown a slowing down trend. Each country is in a different stage of demographic transition. Sri Lanka's birth rate has almost declined to a point where it is just replacing the present population, and India and Bangladesh are at the point where both the fertility and mortality rates are beginning to decline. Nepal and Pakistan still have high population growth rates. According to the United Nations Population Fund (UNFP, 1998), the population growth during 1995–2000 will be 1.6 per cent for India and Bangladesh, 2.8 per cent for Bhutan, 2.5 per cent for Nepal, 2.70 per cent for Pakistan, and 1 per cent for Sri Lanka. The population of South Asia is estimated at 1,291.2 million in 1998 and it is projected to reach 1,847.2 million by 2025. Using the predicted population growth during the years 1995–2000 and 1998–2025, the population for each country is predicted for different years during 1995 to 2030 (Table 9.12). These population figures are used in projecting the demand for food. The population growth, which is estimated at 1.73 per cent for South Asia during 1995–2000, will decline to 1.55 per cent during the period 2000 to 2010, 1.21 per cent during 2010–20 and 0.94 per cent in the subsequent decade. The population of South Asia, which was 1,226.3 million in 1995, will increase to 1,929.9 million by the year 2030. India is the most populous country in South Asia, accounting for 76 per cent of the total population in 1995 which would decline to 71 per cent in the year 2030. Population shares of Pakistan, Bangladesh, Nepal and Bhutan in South Asia show an increasing trend due to higher population growth rate.

HOUSEHOLD DEMAND FOR FOOD

Food demand for human consumption is arrived at by multiplying the projected per capita consumption with projected population. The results of human food demand predictions corresponding to the scenario of low and high per capita GDP growth at constant prices (1981) are given in Tables 9.13 and 9.14. The household demand for foodgrains in South Asia is projected to grow annually by 1.2 per cent and increase from 210 million tons in 1995 to 321 million tons by 2030 with a break-up of about 145–146 million tons for rice, 110–114 million tons for wheat, 13–14 million tons for maize, 22–24 million tons for other coarse grains (sorghum, pearl millet, barley, etc.) and 24–25 million tons for pulses (Table 9.15). The demand for roots and tubers (mainly potato) and edible oils will be 42–46 million tons and 16–17 million tons respectively by 2030. The demand by the year 2030 for horticultural, livestock and fisheries products will grow three to four times from the level in the year 1995, and will attain the maximum level at 161–195 million tons for milk, 153–195 million tons for vegetables, 96–121 million tons for fruits, 45–46 million tons for sweeteners (mainly sugar), 17–22 million tons for meat, 14–18 million tons for fish, and 4.2–5.6 million tons for eggs.

India predominates in the food demand in South Asia and accounts for a higher share than its population share (71 per cent) for maize (75 per cent), pulses (91 per

Table 9.12: Projected Population in South Asia (1995–2030)

	Bangladesh	Bhutan	India	Nepal	Pakistan	Sri Lanka	South Asia
Projected population (millions)							
1995	118.2	1.7	930.4	21.5	136.4	18.0	1,226.3
1998*	124.0	1.9	975.8	23.2	147.8	18.5	1,291.2
2000	128.0	2.0	1,007.3	24.4	155.9	18.9	1,336.4
2005	138.2	2.3	1,084.6	27.4	177.2	19.8	1,449.6
2010	148.6	2.6	1,156.0	30.6	199.4	20.8	1,558.0
2015	159.1	2.9	1,221.0	33.9	222.2	21.8	1,660.9
2020	169.6	3.3	1,279.1	37.2	245.5	22.8	1,757.4
2025*	180.0	3.6	1,330.2	40.6	268.9	23.9	1,847.2
2030	190.4	4.0	1,374.4	43.9	292.3	25.0	1,929.9
Share of population as percentage of total South Asian population							
1995	9.5	0.1	75.9	1.8	11.1	1.5	100
2000	9.5	0.2	75.4	1.8	11.7	1.4	100
2015	9.6	0.2	73.5	2.0	13.4	1.3	100
2030	9.9	0.2	71.2	2.3	15.1	1.3	100
Population growth (per cent per annum)							
1995–2000	1.60	2.80	1.60	2.50	2.70	1.00	1.73
2000–10	1.50	2.66	1.39	2.29	2.49	0.96	1.55
2010–20	1.33	2.41	1.02	1.97	2.10	0.92	1.21
2020–30	1.16	1.94	0.72	1.67	1.76	0.92	0.94
1995–30	1.37	2.47	1.12	2.06	2.20	0.94	1.30

Note: *The State of World Population*, UNFPA, 1998.

Source: Computed by the authors.

Table 9.13: Projected Human Demand for Food with Low Income Growth Scenario (1995, 2000, 2015, 2030)

(Thousand tons)

Items	Year	Bangladesh	India	Nepal	Pakistan	Sri Lanka	South Asia
Rice	1995	19407	70729	2036	2473	1943	96724
	2000	20776	76398	2308	2837	2021	104496
	2015	25286	92207	3226	4070	2291	127305
	2030	30016	103622	4187	5367	2606	146098
Wheat	1995	2353	51446	712	17440	835	72890
	2000	2549	54832	793	19589	874	78756
	2015	3172	64550	1071	27030	1003	96997
	2030	3797	71851	1372	35112	1146	113510
Maize	1995	0	7739	1051	742	55	9600
	2000	0	8203	1164	827	56	10265
	2015	0	9559	1555	1125	62	12323
	2030	0	10599	1983	1454	70	14135
Other coarse grain	1995	0	16122	277	418	14	16856
	2000	0	17089	307	466	15	17904
	2015	0	19914	410	634	16	21012
	2030	0	22080	523	820	18	23489
Total cereals	1995	21760	146036	4076	21072	2847	196071
	2000	23325	156522	4571	23719	2966	211421
	2015	28458	186230	6261	32860	3373	257636
	2030	33813	208152	8065	42753	3840	297233
Total Pulses	1995	564	12364	169	635	100	13852
	2000	631	13802	198	746	109	15509
	2015	833	17713	296	1119	133	20130
	2030	1020	20384	394	1500	156	23503

(contd.)

Table 9.13 (contd.)

Items	Year	Bangladesh	India	Nepal	Pakistan	Sri Lanka	South Asia
Foodgrains	1995	22324	158400	4245	21708	2947	209923
	2000	23956	170323	4770	24465	3075	226930
	2015	29291	203943	6557	33979	3507	277766
	2030	34834	228536	8459	44254	3997	320736
Potato	1995	1190	12941	598	806	61	15618
	2000	1335	14700	706	959	67	17794
	2015	1774	19492	1064	1475	86	23933
	2030	2179	22716	1422	1997	102	28474
Other roots and tubers	1995	417	6681	126	386	262	7883
	2000	460	7409	146	450	283	8760
	2015	591	9394	211	666	343	11225
	2030	716	10759	278	888	401	13069
Total roots and tubers	1995	1607	19621	724	1192	323	23501
	2000	1795	22109	852	1409	350	26555
	2015	2365	28886	1275	2140	429	35157
	2030	2895	33475	1700	2885	503	41543
Edible oils	1995	503	6684	85	1740	38	9063
	2000	560	7421	99	2043	41	10178
	2015	732	9428	145	3060	49	13438
	2030	893	10807	192	4102	57	16084
Vegetables	1995	1235	62827	1170	2960	596	68886
	2000	1435	75800	1441	3666	697	83165
	2015	2038	112556	2353	6074	983	124222
	2030	2568	137098	3244	8469	1216	152909
Fruits	1995	1446	34700	512	5318	673	42710
	2000	1721	41507	640	6724	765	51434
	2015	2553	60648	1072	11577	1025	77011
	2030	3273	73406	1493	16386	1244	95999

(contd.)

Table 9.13 (contd.)

Items	Year	Bangladesh	India	Nepal	Pakistan	Sri Lanka	South Asia
Sweeteners	1995	870	22221	61	3432	349	26971
	2000	957	24457	70	3975	372	29876
	2015	1227	30570	100	5810	441	38216
	2030	1486	34821	131	7716	510	44757
Milk	1995	1733	55826	811	14468	509	73452
	2000	2037	65723	1004	17947	590	87434
	2015	2956	93197	1654	29815	820	128668
	2030	3756	111488	2289	41614	1008	160483
Animal fats	1995	25	1355	27	468	7	1883
	2000	29	1595	33	580	8	2248
	2015	42	2261	55	964	11	3339
	2030	53	2705	76	1345	14	4202
Meat	1995	375	4342	187	1697	53	6663
	2000	456	5335	239	2133	63	8238
	2015	705	8196	418	3634	92	13068
	2030	918	10118	592	5123	115	16900
Eggs	1995	91	1332	21	232	48	1726
	2000	111	1637	26	292	56	2126
	2015	171	2515	46	498	82	3318
	2030	223	3104	65	702	103	4206
Fish	1995	1061	4046	18	324	300	5758
	2000	1291	4971	23	407	357	7059
	2015	1996	7637	41	693	519	10906
	2030	2601	9428	58	977	651	13742

Source: Computed by the authors.

Table 9.14: Projected Human Demand for Food with High Income Growth Scenario (1995, 2000, 2015 and 2030)

(Thousand tons)

Items	Year	Bangladesh	India	Nepal	Pakistan	Sri Lanka	South Asia
Rice	1995	19348	70686	2037	2475	1937	96621
	2000	20581	76251	2313	2845	2003	104150
	2015	24752	91804	3240	4098	2247	126364
	2030	29248	103073	4210	5412	2545	144784
Wheat	1995	2354	51231	709	17359	834	72590
	2000	2551	54114	732	19305	872	77740
	2015	3175	62650	1039	26148	996	94174
	2030	3802	69291	1322	33726	1136	109502
Maize	1995	0	7695	1045	737	55	9545
	2000	0	8058	1142	810	55	10081
	2015	0	9181	1491	1074	59	11826
	2030	0	10092	1885	1373	66	13445
Other coarse grain	1995	0	16031	276	415	14	16760
	2000	0	16788	301	456	14	17587
	2015	0	19127	393	605	15	20176
	2030	0	21026	497	774	17	22360
Total cereals	1995	21702	145643	4066	20986	2840	195516
	2000	23132	155211	4538	23416	2945	209557
	2015	27927	182763	6163	31925	3318	252541
	2030	33050	203482	7914	41286	3764	290090
Total Pulses	1995	569	12465	170	640	101	13965
	2000	648	14159	205	763	112	15911
	2015	886	18773	318	1178	142	21334
	2030	1100	21878	430	1597	168	25225

(contd.)

Table 9.14 (contd.)

Items	Year	Bangladesh	India	Nepal	Pakistan	Sri Lanka	South Asia
Foodgrains	1995	22271	158108	4237	21626	2941	209481
	2000	23780	169370	4743	24179	3056	225468
	2015	28813	201536	6481	33103	3460	273875
	2030	34150	225360	8344	42882	3933	315315
Potato	1995	1202	13106	605	815	62	15811
	2000	1376	15299	731	992	70	18497
	2015	1899	21351	1154	1592	95	26136
	2030	2366	25382	1570	2192	116	31691
Other roots and tubers	1995	419	6724	127	388	264	7932
	2000	467	7560	148	457	289	8935
	2015	611	9833	220	692	361	11738
	2030	747	11375	292	931	427	13799
Total roots and tubers	1995	1621	19830	731	1202	325	23744
	2000	1843	22859	880	1450	359	27431
	2015	2510	31184	1374	2284	456	37875
	2030	3113	36757	1862	3123	542	45490
Edible oils	1995	507	6729	86	1752	38	9125
	2000	573	7579	101	2088	41	10398
	2015	771	9890	154	3218	50	14108
	2030	951	11455	205	4362	59	17068
Vegetables	1995	1258	64630	1196	3023	613	70821
	2000	1522	82869	1547	3918	760	90753
	2015	2332	138042	2764	7071	1200	151676
	2030	3027	175948	3948	10195	1553	195071
Fruits	1995	1482,0	35618	526	5461	687	43835
	2000	1860,0	45065	694	7307	817	55827
	2015	3053,0	73199	1293	14002	1191	92901
	2030	4071,0	92372	1877	20673	1493	120733

(contd.)

Table 9.14 (contd.)

Items	Year	Bangladesh	India	Nepal	Pakistan	Sri Lanka	South Asia
Sweeteners	1995	874,0	22319	61	3444	350	27086
	2000	971,0	24796	71	4020	376	30280
	2015	1267	31543	103	5961	453	39395
	2030	1545	36174	136	7960	527	46436
Milk	1995	1771	57068	830	14783	523	75081
	2000	2182	70438	1082	19204	640	93686
	2015	3456	109183	1961	34800	986	150652
	2030	4545	135257	2818	50257	1263	194539
Animal fats	1995	25	1385	28	478	7	1925
	2000	31	1709	36	621	9	2409
	2015	49	2649	65	1125	13	3909
	2030	64	3282	94	1624	17	5092
Meat	1995	386	4488	193	1740	55	6871
	2000	501	5918	264	2308	70	9074
	2015	878	10396	526	4350	116	16293
	2030	1201	13534	784	6380	153	22097
Eggs	1995	94	1377	21	238	49	1782
	2000	122	1816	29	316	62	2349
	2015	214	3190	58	596	104	4168
	2030	292	4153	86	874	137	5553
Fish	1995	1094	4182	19	332	310	5945
	2000	1421	5514	26	440	395	7807
	2015	2486	9687	51	830	654	13733
	2030	3402	12611	76	1217	863	18207

Source: Computed by the authors.

Table 9.15: Share of Food Demand by Country in Total Human Demand for South Asia (2030)

(%)

Items	Bangladesh	India	Nepal	Pakistan	Sri Lanka	South Asia Demand (Mts)
Low growth in Income (3.5% per capita GDP growth)						
Rice	20.5	70.3	2.8	3.6	1.7	146.1
Wheat	3.3	63.2	1.2	30.2	1.0	113.5
Maize	0	74.9	14.0	10.2	0.4	14.1
Other coarse grains	0	94	2.2	3.4	0.1	23.5
Total cereals	11.3	70.0	2.7	14.3	1.2	297.2
Total Pulses	4.3	86.7	1.6	6.3	0.6	23.5
Total foodgrains	10.8	71.2	2.6	13.7	1.2	320.7
Potato	7.6	72.7	4.2	7.0	0.3	28.5
Other R&T	5.4	82.3	2.1	6.7	3.0	13.1
Total R&T	6.9	80.5	4.0	6.9	1.2	41.5
Edible oils	5.5	67.0	1.1	25.5	0.3	16.1
Vegetables	1.6	89.6	2.1	5.5	0.7	152.9
Fruits	3.4	76.5	1.5	17.0	1.2	96.0
Sweeteners	3.3	77.8	0.2	17.2	1.1	44.8
Milk	2.3	69.4	1.4	25.9	0.6	160.5
Animal fats	1.2	64.3	1.8	32.0	0.3	4.2
Meat	5.4	59.8	3.5	30.3	0.6	16.9
Eggs	5.3	73.7	1.5	16.6	2.4	4.2
Fish	18.9	68.6	0.4	7.1	4.7	13.7

(contd.)

Table 9.15 (contd.)

Items	Bangladesh	India	Nepal	Pakistan	Sri Lanka	South Asia Demand (Mts)
High growth in Income (5.5% per capita GDP growth)						
Rice	20.0	71.1	2.9	3.7	1.8	144.8
Wheat	3.5	63.0	1.2	30.7	1.0	109.5
Maize	0	75.0	14.0	10.0	0.4	13.4
Other coarse grains	0	94	2.2	3.4	0.1	22.3
Total cereals	11.3	70.0	2.7	14.3	1.3	290.1
Total Pulses	4.3	86.7	1.7	6.3	0.6	25.2
Total foodgrains	10.8	71.8	2.6	13.5	1.2	315.3
Potato	7.4	80.0	4.9	6.9	0.3	31.7
Other R&T	5.4	82.4	2.1	6.7	3.0	13.8
Total R&T	6.8	80.8	4.0	6.8	1.1	45.5
Edible oils	5.5	67.0	1.2	25.5	0.3	17.1
Vegetables	1.5	90.2	2.0	5.2	0.7	195.1
Fruits	3.3	76.5	1.5	17.0	1.2	120.7
Sweeteners	3.3	77.9	0.2	17.1	1.1	46.4
Milk	2.3	69.5	1.4	25.8	0.6	194.5
Animal fats	1.2	64.4	1.8	31.8	0.3	5.1
Meat	5.4	61.2	3.5	28.8	0.6	22.1
Eggs	5.2	74.7	1.5	15.7	2.4	5.6
Fish	18.6	69.2	0.4	6.6	4.7	18.2

Notes: R&T = Roots and tubers, Mts = Million tons.
Source: Computed by the authors.

cent), potato (80 per cent), vegetables (90 per cent), fruits (77 per cent), sugar (78 per cent) and eggs (75 per cent). India's share of demand is lower than its population share for wheat (63 per cent), edible oils (67 per cent), milk (69 per cent), animal fats (64 per cent), meat (61 per cent), and fish (69 per cent). Pakistan accounted for a very high share in demand as compared to its population share (15 per cent) for wheat (31 per cent), edible oils (25 per cent), milk (26 per cent), meat (29 per cent), fruits (17 per cent), sugar (17 per cent), and eggs (16 per cent). Except for rice and fish, the share of Bangladesh's demand in South Asia is lower than its population share (10 per cent). More than 90 per cent food demand in South Asia will be governed by Bangladesh and India for rice and fish, and India and Pakistan for wheat, edible oil, fruits, sugar, milk, meat, and eggs. India alone will account for a little more than 90 per cent of South Asia's demand for pulses and fruits. As India accounts for the largest share in total food production and consumption in the South Asian region, any change in the surplus/deficit status of India for food production will have significant implications on South Asia's food security and trade.

DOMESTIC DEMAND PROJECTIONS

Apart from the demand for direct human consumption, an increasingly important component is the indirect demand for seed, feed, industrial uses and wastage. FAO estimates (FAOSTAT reports, various issues) of other uses (1992–4 averages given in Table 9.16) and human demand are used to derive the domestic food demand for South Asian countries. Domestic demand projections for food corresponding to low (3.5 per cent) and high (5.5 per cent) per capita GDP growth are presented in Tables 9.17 and 9.18, respectively. The annual growth rates in food demand under low and high income scenarios for various time intervals are given in Table 9.19. The results are discussed by country/region in the subsequent sections. This projected demand may be somewhat higher (5 to 10 per cent) if one accounts for export, buffer stock needs and risk factors.

Bangladesh

In Bangladesh, demand for foodgrains of 24 million tons in 1995 is projected to grow to about 37 million tons in 2030 with a break-up of 32 million tons for rice, 4 million tons for wheat and 1.1 million tons for pulses. These projections are close to the earlier estimates given in the IRRI-IFPRI country study (Shahabuddin and Zohir, 1995). For the year 2000, the demand for roots and tubers works out to about 2.2 million tons, edible oil 0.6 million tons, vegetables 1.6 million tons, fruits 2.0 million tons, sugar 1.2 million tons, milk 2.6–2.7 million tons, meat 0.5 million tons, eggs 0.14–0.15 million tons, and fish 1.3–1.4 million tons. In the year 2030, this demand will grow to a level of 3.6–3.8 million tons of roots and tubers, 1 million tons of edible oils, 2.8–3.3 million tons of vegetables, 3.6–4.5 million tons of fruits, 1.9 million tons of sweeteners, 4.7–5.7 million tons of milk, 0.9–1.2 million tons of meat, 0.28–0.37 million tons of eggs, and 2.6–3.4 million tons of fish.

India

In India, the domestic demand for foodgrains is estimated at 197 million tons in the year 2000, 232–235 million tons in the year 2015 and 260–264 million tons in the year 2030. In

Table 9.16: Share of Seed, Feed, Wastage and Other than Food Uses as Percentage of Total Supply (1992–4 averages)

Items	Bangladesh	India	Nepal	Pakistan	Sri Lanka
Rice	7.58	9.50	15.27	11.17	8.81
Wheat	7.21	13.48	18.82	8.65	2.87
Coarse grains	57.14	21.62	17.80	41.86	17.91
Roots & Tubers	17.41	19.99	24.62	13.28	30.24
Pulses	7.71	16.85	10.80	25.84	3.90
Vegetable oils	11.44	8.95	16.32	6.19	29.7
Sweeteners	20.13	2.08	0.00	2.63	0.77
Vegetables	8.97	9.10	9.92	9.81	11.69
Fruits	9.77	12.96	9.85	12.96	10.94
Milk	20.54	14.57	22.31	20.19	3.13
Animal fats	59.90	5.46	2.60	16.40	34.48
Meat	0.00	0.00	0.97	0.00	0.00
Eggs	20.20	12.95	10.00	15.24	7.04
Fish	0.44	9.73	0.00	9.73	6.91

Source: Computed by the authors.

Data source: *FAOSTAT*, Food and Agriculture Organization of the United Nations, Rome.

Table 9.17: Projected Demand for Food in South Asia with Low Growth in Per Capita Income (1995, 2000, 2015 and 2030)

(Thousand tons)

Items	Year	Bangladesh	India	Nepal	Pakistan	Sri Lanka	South Asia
Rice	1995	20998	78154	2402	2784	2130	106620
	2000	22480	84417	2724	3193	2216	115204
	2015	27360	101886	3807	4582	2513	140395
	2030	32478	114499	4942	6042	2858	161149
Wheat	1995	2536	59462	877	19091	860	82944
	2000	2747	63375	977	21444	900	89578
	2015	3418	74607	1319	29590	1033	110161
	2030	4092	83045	1690	38437	1180	128708
Maize	1995	0	9873	1279	1275	67	12512
	2000	0	10466	1415	1422	69	13392
	2015	0	12196	1891	1936	76	16127
	2030	0	13522	2413	2501	85	18560
Other coarse grains	1995	0	20569	337	719	17	21674
	2000	0	21803	373	802	18	23030
	2015	0	25407	499	1091	20	27064
	2030	0	28171	636	1410	22	30301
Total cereals	1995	23535	168058	4895	23869	3074	223750
	2000	25227	180061	5490	26861	3203	241204
	2015	30778	214096	7516	37199	3641	293747
	2030	36570	239238	9681	48390	4145	338718
Pulses	1995	611	14869	189	857	104	16655
	2000	684	16599	222	1006	113	18652
	2015	903	21303	332	1508	139	24228
	2030	1106	24515	442	2023	163	28306

(contd.)

Table 9.17 (contd.)

Items	Year	Bangladesh	India	Nepal	Pakistan	Sri Lanka	South Asia
Foodgrains	1995	24146	182928	5085	24726	3179	240405
	2000	25911	196659	5712	27867	3316	259856
	2015	31681	235399	7848	38707	3780	317975
	2030	37676	263752	10123	50413	4308	367024
Potato	1995	1477	17523	812	989	87	20918
	2000	1657	19905	959	1177	96	23831
	2015	2202	26394	1446	1809	123	32031
	2030	2704	30760	1932	2450	147	38070
Other roots and tubers	1995	505	8350	167	445	375	9856
	2000	557	9260	194	519	405	10951
	2015	716	11740	280	767	492	14021
	2030	867	13447	369	1024	575	16316
Total roots and tubers	1995	1982	25873	980	1434	462	30775
	2000	2214	29165	1153	1696	501	34781
	2015	2918	38135	1726	2577	615	46051
	2030	3571	44207	2301	3474	721	54386
Edible oils	1995	568	7341	102	1855	54	9934
	2000	632	8151	118	2177	58	11153
	2015	826	10355	174	3262	69	14712
	2030	1009	11870	230	4373	80	17597
Vegetables	1995	1356	69117	1299	3282	675	75836
	2000	1577	83388	1600	4065	789	91557
	2015	2239	123824	2612	6735	1113	136763
	2030	2821	150823	3601	9391	1377	168358
Fruits	1995	1602	39867	568	6106	755	48968
	2000	1907	47688	710	7720	859	58972
	2015	2830	69678	1189	13292	1151	88295
	2030	3627	84336	1656	18813	1397	110055

(contd.)

Table 9.17 (contd.)

Items	Year	Bangladesh	India	Nepal	Pakistan	Sri Lanka	South Asia
Sweeteners	1995	1089	22693	61	3524	352	27758
	2000	1199	24977	70	4083	375	30749
	2015	1537	31220	100	5967	444	39337
	2030	1861	35561	131	7925	514	46086
Milk	1995	2181	65347	1044	18129	526	87350
	2000	2564	76932	1292	22487	609	104042
	2015	3720	109092	2129	37357	846	153415
	2030	4727	130502	2946	52141	1041	191749
Animal fats	1995	61	1433	28	559	11	2095
	2000	72	1687	34	694	12	2503
	2015	105	2392	56	1153	17	3729
	2030	133	2861	78	1609	21	4712
Meat	1995	375	4342	189	1697	53	6665
	2000	456	5335	241	2133	63	8240
	2015	705	8196	422	3634	92	13072
	2030	918	10118	598	5123	115	16906
Eggs	1995	114	1530	23	274	51	1996
	2000	139	1880	29	345	61	2457
	2015	215	2889	51	587	88	3837
	2030	280	3566	72	828	111	4867
Fish	1995	1066	4482	18	353	323	6242
	2000	1297	5507	23	444	383	7653
	2015	2005	8460	41	756	558	11820
	2030	2612	10444	58	1065	699	14878

Source: Computed by the authors.

Table 9.18: Projected Demand for Food in South Asia with High Growth in Per Capita Income (1995, 2000, 2015 and 2030)

(Thousand tons)

Items	Year	Bangladesh	India	Nepal	Pakistan	Sri Lanka	South Asia
Rice	1995	20935	78106	2404	2787	2124	106508
	2000	22269	84255	2730	3203	2197	114826
	2015	26782	101441	3824	4613	2464	139370
	2030	31646	113893	4968	6092	2791	159718
Wheat	1995	2537	59213	873	19002	859	82602
	2000	2749	62545	964	21133	897	88421
	2015	3422	72411	1280	28624	1025	106951
	2030	4097	80087	1629	36920	1169	124157
Maize	1995	0	9818	1271	1267	67	12440
	2000	0	10281	1390	1393	67	13151
	2015	0	11714	1814	1847	72	15474
	2030	0	12876	2294	2362	81	17649
Other coarse grains	1995	0	20453	335	714	17	21550
	2000	0	21419	366	785	17	22622
	2015	0	24403	478	1041	19	25987
	2030	0	26825	605	1331	21	28841
Total cereals	1995	23472	167589	4884	23770	3067	223100
	2000	25018	178500	5449	26514	3179	239019
	2015	30204	209969	7396	36125	3581	287781
	2030	35744	233681	9495	46706	4062	330364
Pulses	1995	617	14991	191	863	105	16791
	2000	702	17028	229	1029	116	19134
	2015	960	22578	356	1588	148	25675
	2030	1192	26312	482	2153	175	30377

(contd.)

Table 9.18 (contd.)

Items	Year	Bangladesh	India	Nepal	Pakistan	Sri Lanka	South Asia
Foodgrains	1995	24089	182580	5075	24633	3172	239891
	2000	25720	195528	5679	27543	3295	258153
	2015	31164	232547	7752	37713	3728	313456
	2030	36936	259993	9977	48859	4237	360741
Potato	1995	1491	17747	821	1000	88	21178
	2000	1708	20716	994	1217	101	24773
	2015	2357	28911	1568	1953	136	34987
	2030	2937	34370	2133	2689	166	42382
Other roots and tubers	1995	508	8403	168	447	378	9919
	2000	565	9448	197	527	414	11169
	2015	740	12290	292	798	518	14664
	2030	904	14217	388	1073	612	17228
Total roots and tubers	1995	1999	26150	990	1447	466	31096
	2000	2273	30165	1191	1745	515	35942
	2015	3097	41201	1860	2751	654	49650
	2030	3841	48587	2521	3763	777	59610
Edible oils	1995	572	7390	103	1868	54	10002
	2000	646	8324	121	2226	59	11394
	2015	870	10863	183	3431	72	15446
	2030	1074	12581	246	4650	84	18673
Vegetables	1995	1382	71100	1328	3352	694	77967
	2000	1672	91165	1717	4345	861	99910
	2015	2562	151861	3068	7840	1359	166985
	2030	3325	193562	4383	11304	1759	214773
Fruits	1995	1643	40921	583	6269	771	50258
	2000	2062	51774	770	8389	917	64009
	2015	3383	84099	1434	16076	1337	106516
	2030	4512	106126	2082	23735	1677	138415

(contd.)

Table 9.18 (contd.)

Items	Year	Bangladesh	India	Nepal	Pakistan	Sri Lanka	South Asia
Sweeteners	1995	1094	22793	61	3537	353	27877
	2000	1216	25323	71	4129	379	31164
	2015	1586	32213	103	6122	456	40551
	2030	1934	36942	136	8175	531	47816
Milk	1995	2228	66801	1069	18522	539	89287
	2000	2746	82451	1392	24062	661	111479
	2015	4349	127805	2525	43604	1018	179616
	2030	5720	158325	3627	62971	1304	232423
Animal fats	1995	63	1465	28	572	11	2141
	2000	77	1808	37	742	13	2682
	2015	122	2802	67	1345	21	4365
	2030	161	3471	96	1943	26	5709
Meat	1995	386	4488	195	1740	55	6873
	2000	501	5918	267	2308	70	9077
	2015	878	10396	531	4350	116	16299
	2030	1201	13534	791	6380	153	22104
Eggs	1995	118	1582	24	281	53	2060
	2000	153	2086	32	373	67	2715
	2015	268	3664	64	703	111	4819
	2030	366	4770	96	1031	147	6423
Fish	1995	1099	4632	19	362	333	6445
	2000	1427	6108	26	480	424	8465
	2015	2497	10731	51	904	703	14887
	2030	3417	13971	76	1326	927	19717

Source: Computed by the authors.

Table 9.19: Annual Growth in Domestic Demand for Food in South Asia (2000–30)

(%)

Items	Low income growth			High income growth		
	2000–15	2015–30	1995–2030	2000–15	2015–30	1995–2030
Rice	1.3	0.9	1.2	1.3	0.9	1.2
Wheat	1.4	1.0	1.3	1.3	1.0	1.2
Maize	1.2	0.9	1.1	1.1	0.9	1.0
Other coarse grains	1.1	0.8	0.9	0.9	0.7	0.8
Total cereals	1.3	0.9	1.2	1.3	0.9	1.1
Pulses	1.8	1.0	1.5	2.0	1.1	1.7
Food grains	1.4	1.0	1.2	1.3	0.9	1.2
Roots & Tubers	1.9	1.1	1.6	2.2	1.2	1.9
Edible oil	1.9	1.2	1.6	2.1	1.3	1.8
Vegetables	2.7	1.4	2.3	3.5	1.7	2.9
Fruits	2.7	1.5	2.3	3.4	1.8	2.9
Sweeteners	1.7	1.1	1.5	1.8	1.1	1.6
Milk	2.6	1.5	2.3	3.2	1.7	2.8
Animal fat	2.7	1.6	2.3	3.3	1.8	2.8
Meat	3.1	1.7	2.7	4.0	2.1	3.4
Eggs	3.0	1.6	2.6	3.9	1.9	3.3
Fish	2.9	1.6	2.5	3.8	1.9	3.2

Notes: Low income growth = 3.5 per cent annual per capita GDP growth,
High income growth = 5.5 per cent annual per capita GDP growth.

Source: Computed by the authors.

the year 2030, the domestic demand in India will be about 114 million tons for rice, 80–83 million tons for wheat, 13 million tons for maize, and 24–26 million tons for pulses. These projections are lower than the earlier estimates (Kumar, 1998 and Bansil, 1998) on account of lower population projections as given by a recent study of UNFP and steady decline in per capita consumption of cereals which has not yet stabilized. In the year 2000, the demand for milk is estimated to be 77–82 million tons, vegetables 83–91 million tons, fruits 48–52 million tons, meat 5.3–5.9 million tons, eggs 1.9–2.1 million tons and fish 5.5–6.1 million tons. In the year 2030, total demand is likely to be 130–158 million tons for milk, 151–193 million tons for vegetables, 84–106 million tons for fruits, 10–14 million tons for meat, 3.6–4.8 million tons for eggs, and 10–14 million tons for fish. Income effect on the demand for these non-grain food items is high. The demand for edible oils of 7.3 million tons in 1995 will grow to about 12 million tons in 2030. Similarly the demand for sweeteners will increase from a level of 22.7 million tons in 1995 to 35.6–36.9 million tons in the year 2030.

Nepal

The domestic demand for foodgrains in Nepal is estimated at 5.7 million tons in 2000, 7.8 million tons in 2015, and 10 million tons in the year 2030. Demand for horticultural products comprising fruits, vegetables and roots and tubers is estimated at 3.5 million tons in 2000, and will increase to 7.6 million tons in 2030 with low income growth, and to 9 million tons with high income growth. In the year 2030, livestock products' demand is likely to be 2.9–3.6 million tons for milk, and 0.73–0.96 million tons for other livestock and fisheries products.

Pakistan

Foodgrains demand in Pakistan, which was 25 million tons in 1995, will grow to a level of 38 million tons in 2015 and 49–50 million tons in 2030 with a break-up of 37–38 million tons for wheat, only 6 million tons for rice, 3.7–3.9 million tons for coarse grains and 2.0–2.1 million tons for pulses. These projections are close to the earlier estimates given in the IRRI-IFPRI country study (Naqvi et al., 1995). For the year 2030, the demand for roots and tubers was worked out to about 3.5–3.8 million tons, edible oil 4.4–4.7 million tons, vegetables 9–11 million tons, fruits 19–24 million tons, sweeteners 7.9–8.2 million tons, milk 52–63 million tons, meat 5.1–6.4 million tons, eggs 0.83–1.0 million tons, and fish 1.1–1.3 million tons.

Sri Lanka

Demand for foodgrains in Sri-Lanka was estimated at 3.3 million tons in 2000, 3.8 million tons in 2015 and 4.3 million tons in the year 2030. Demand for other food items in 2030 will be 1.0–1.3 million tons for milk, 1.4–1.7 million tons for vegetables, 1.4–1.7 million tons for fruits, 0.9–1.2 million tons for meat, eggs, and fish, 80–84 thousand tons for edible oils, and 514–531 thousand tons for sugar.

South Asia

The domestic demand for food in South Asia is derived by adding the demand for individual countries, namely, Bangladesh, India, Nepal, Pakistan and Sri Lanka. The demand is adjusted upward in proportion to Bhutan's population share in the South Asia.

For the year 2000, the demand for rice worked out to about 115 million tons, wheat 88–89 million tons, maize 13 million tons, pulses 19 million tons, roots and tubers 35–36 million tons, edible oils 11 million tons, vegetables 92–100 million tons, fruits 59–64 million tons, sweeteners 31 million tons, milk 104–111 million tons, animal fats 2.5–2.7 million tons, meat 8.2–9.1 million tons, eggs 2.5–2.7 million tons, and fish 7.7–8.5 million tons.

For the year 2030, the total foodgrains demand will grow to a level of 361–367 million tons, comprising 160–161 million tons of rice, 124–129 million tons of wheat, 17–19 million tons of maize, 29–30 million tons of other coarse grains and 28–30 million tons of pulses. IFPRI projections for 2020 (Rosegrant et al., 1995) for cereals in South Asia appear to be very high (417 million tons) as compared to present projections and do not seem to be justified by the cereal consumption pattern prevailing in South Asia. Also the projected population is on the higher side in the IFPRI study as compared to recent population projections given by UNFP. By the year 2030, South Asia will need about 54–60 million tons of roots and tubers, 18–19 million tons of edible oils, 168–215 million tons of vegetables, 110–138 million tons of fruits, 46–48 million tons of sweeteners, 192–232 million tons of milk, 4.7–5.7 million tons of animal fat, 17–22 million tons of meat, 4.9–6.4 million tons of eggs, and 15–20 million tons of fish to meet its domestic requirements.

A deceleration in the growth rate of total domestic demand for food was observed (Table 9.19). The demand for cereals during 1995–2030 will grow at a slightly lower rate (1.1–1.2 per cent) than the population growth. Demand for roots and tubers, edible oils, and sugar will grow at a moderately higher rate (1.5–1.8 per cent) than the population growth. Annual growth in demand for horticultural, livestock and fisheries products will grow much faster than the growth in population and foodgrains demand. High growth in livestock product's demand will put a pressure on foodgrains and oilcakes to meet the feed demand for livestock. Fast growth of income will diversify the dietary pattern in favour of non-foodgrains crops, livestock and fisheries products. The per capita availability of arable land in South Asia is quite low and declining over time. Diversification towards these high value commodities which are labour intensive also can provide adequate income and employment to the agricultural labourers and farmers who depend on small size farms, and will thus help the weaker sections to attain household food security.

SCENARIO FOR FOOD SECURITY

Future increase in the production of food and non-food agricultural commodities have to be essentially achieved through increases in productivity as possibilities of area and livestock population expansion are minimal. To meet the future demand, the required level of yield and production targets are presented in Tables 9.20 and 9.21 respectively. To meet the projected demand for food and to attain self-reliance, the South Asia region must attain a per hectare average yield of 2.82 tons for rice, 3.72 tons for wheat, 2.36 tons for maize and 0.99 tons for pulses by the year 2030. This means an improvement

Table 9.20: Target Yield Levels to Meet the Domestic Demand for Food Grains (1994–2030)

(Tons per ha.)

Crops	Year	Bangladesh	India	Nepal	Pakistan	Sri Lanka	South Asia
Rice	1994–96	1.74	1.92	1.46	1.70	1.96	1.87
	1999–00	2.26	2.00	1.91	1.48	2.56	2.03
	2014–15	2.75	2.41	2.66	2.13	2.90	2.48
	2029–30	3.18	2.70	3 48	2.82	3.23	2.82
Wheat	1994–96	1.88	2.45	1.49	2.00	-	2.31
	1999–00	4.19	2.52	1.54	2.62	-	2.59
	2014–15	5.22	2.97	2.08	3.61	-	3.18
	2029–30	6.25	3.30	2.67	4.69	-	3.72
Maize	1994–96	-	1.52	1.67	1.46	1.04	1.53
	1999–00	-	1.71	1.82	1.60	2.29	1.70
	2014–15	-	2.00	2.43	2.18	2.53	2.05
	2029–30	-	2.21	3.10	2.81	2.85	2.36
Cereals	1994–96	1.74	1.74	1.49	1.79	1.91	1.74
	1999–00	2.35	1.80	1.76	2.19	3.54	1.90
	2014–15	2.87	2.15	2.41	3.04	4.03	2.32
	2029–30	3.41	2.40	3.10	3.95	4.59	2.67
Pulses	1994–96	0.75	0.60	0.61	0.50	0.77	0.60
	1999–00	0.96	0.67	0.73	0.66	2.18	0.65
	2014–15	1.26	0.86	1.08	0.99	2.67	0.85
	2029–30	1.55	0.99	1.44	1.32	3.13	0.99

Source: Computed by the authors.

Table 9.21: Required Increase in Production by the Year 2030 over the Existing 1994–6 Average Production to Meet the Domestic Demand in South Asia

(%)

Items	Bangladesh	India	Nepal	Pakistan	Sri Lanka	South Asia
Low income growth (3.5 per cent per capita GDP growth)						
Rice	88	41	136	65	69	52
Wheat	232	35	79	135	-	61
Maize	-	45	85	93	167	55
Other coarse grains	-	32	109	101	341	35
Total cereals	96	38	109	121	140	54
Pulses	107	64	137	165	307	65
Total food grains	97	40	110	122	144	55
Roots & Tubers	92	67	134	132	64	75
Edible oils	545	40	447	534	–68	83
Sweeteners	145	37	–11	68	236	48
Vegetables	94	135	192	148	127	136
Fruits	164	121	198	261	80	142
Milk	150	108	201	189	265	128
Meat	156	142	244	196	37	159
Eggs	261	146	297	214	127	162
Fish	137	121	215	86	206	124

(contd.)

Table 9.21 (contd.)

Items	Bangladesh	India	Nepal	Pakistan	Sri Lanka	South Asia
High income growth (5.5 per cent per capita GDP growth)						
Rice	83	41	137	67	65	51
Wheat	232	30	73	125	-	55
Maize	-	38	76	82	152	47
Other coarse grains	-	25	99	90	317	29
Total cereals	92	35	105	113	135	50
Pulses	123	76	158	182	338	77
Total food grains	93	38	107	115	140	52
Roots & Tubers	106	84	156	151	77	92
Edible oils	587	48	485	574	–66	94
Sweeteners	154	42	–8	73	247	54
Vegetables	129	202	255	198	190	201
Fruits	228	178	275	355	116	204
Milk	202	152	271	249	358	177
Meat	235	224	355	268	82	238
Eggs	372	229	425	291	201	246
Fish	210	195	317	132	305	197

Source: Computed by the authors.

in average yield, at regional level, by 52 per cent in rice, 61 per cent in wheat, 55 per cent in maize, and 77 per cent in pulses. Foodgrains, which contribute more than two-thirds of the calorie intake in the South Asian diet, require substantial increases of 144 per cent in Sri Lanka, 122 per cent in Pakistan, 110 per cent in Nepal, 97 per cent in Bangladesh and 40 per cent in India, by 2030, over the production level in 1995. The required rice yield levels for 2030 in Bangladesh (3.2 t./ha.), Nepal (3.48 t./ha.), and Sri Lanka (3.2 t./ ha.) seem to be an unattainable task. Even for India and Pakistan, to attain the rice yield levels of 2.7 t./ha. and 2.8 t./ha., respectively, compared to the existing level of 1.7–1.9 t./ha., will not be easy under the existing environment of decelerating growth in factor productivity. For wheat and pulses also, the required yield target seems to be difficult to attain by most of the South Asian countries. For India, average yield of wheat must reach to 3.3 t./ha. by 2030 from the existing level of 2.45 t./ha. Looking at the past performance of wheat productivity, the target may be possible to attain. To meet the non-foodgrains demand for South Asia, production of livestock, fisheries and horticultural products must be improved by 177–246 per cent by 2030 over the base year 1995. The required improvement in productivity levels needs serious efforts by the National Agricultural Research System, policy makers and farmers. Average yields of most of the commodities are low. The vast agricultural potential still remains highly under-realized. The emphasis for yield improvement must be placed on regions where the current yield levels are low. This requires micro level planning with location specific emphasis.

Average annual increment in the production of major cereals and other food items during 1965–95 and annual incremental demand during 2000–30 (Table 9.22) reveals that the additional annual requirement during 2000–15 and 2015–30 for foodgrains, livestock, horticultural and fisheries products will be much more than that achieved in the past by the South Asian countries, especially in the next one decade, except for rice, wheat, maize, pulses and milk in India. In the past, major sources of growth in production were area and yield. However, future growth in food production has to essentially come through increase in yield. In the past, rapid growth in public investment in irrigation and other infrastructure, research and extension along with crop production strategy and policy support, helped to expand agricultural production. The slackness in investment on agricultural research and technology development as seen during the early 1990s is a matter of concern in the context of continued increase in population, diminishing land and fresh water resources, expanding biotic and abiotic stresses, increasing soil salinity, water logging problems and decelerating productivity growth. Such Asian nations are likely to experience a deficit in production to match their domestic need for most of the food commodities. India is the major producer and consumer of food in the South Asia region and has huge potential which remains highly under-realized. Therefore, India has to play a major role not only to maintain its own self-sufficiency in food production but also to meet the additional requirements of its neighbouring countries. The right research priorities and production strategies will promote future growth in agriculture and ensure sustainable food and nutrition security.

Table 9.22: Average Annual Increment in Production during 1965–95 and Domestic Demand during 2000–30 for Food in South Asia

(Thousand tons)

Items	Production		Demand (3.5% PCGDP)		Demand (5.5% PCGDP)	
	1965–80	1980–95	2000–15	2015–30	2000–15	2015–30
Rice						
Bangladesh	226.5	171.4	325.3	341.2	300.9	324.3
India	1537.8	1556.7	1164.6	840.9	1145.7	830.1
Nepal	11.4	19.7	72.2	75.7	72.9	76.3
Pakistan	121.0	46.0	92.6	97.3	94.0	98.6
Sri Lanka	61.2	24.6	19.8	23.0	17.8	21.8
South Asia	1957.9	1818.4	1679.4	1383.6	1636.3	1356.5
Wheat						
Bangladesh	52.5	25.1	44.7	44.9	44.9	45.0
India	1727.0	1713.0	748.8	562.5	657.7	511.7
Nepal	20.1	33.5	22.8	24.7	21.1	23.3
Pakistan	417.8	409.7	543.1	589.8	499.4	553.1
Sri Lanka	-	-	8.9	9.8	8.5	9.6
South Asia	2217.4	2181.3	1372.2	1236.5	1235.3	1147.1
Maîze						
India	142.3	189.5	115.3	88.4	95.5	77.5
Nepal	–7.5	37.3	31.7	34.8	28.3	32.0
Pakistan	28.7	20.3	34.3	37.7	30.3	34.3
Sri Lanka	0.7	0.7	0.5	0.6	0.3	0.6
South Asia	164.2	247.8	182.0	162.5	154.9	145.0

(contd.)

Table 9.22 (contd.)

Items	Production		Demand (3.5% PCGDP)		Demand (5.5% PCGDP)	
	1965–80	1980–95	2000–15	2015–30	2000–15	2015–30
Pulses						
Bangladesh	23.9	–5.8	14.6	13.4	17.2	15.5
India	46.0	112.0	313.6	214.1	370.0	248.9
Nepal	2.4	3.2	7.3	7.3	8.5	8.4
Pakistan	–8.2	–8.3	33.5	34.3	37.3	37.7
Sri Lanka	2.4	–0.1	1.7	1.6	2.1	1.8
South Asia	66.5	101.0	371.7	271.9	436.1	313.5
Roots & Tubers						
Bangladesh	57.8	11.1	46.9	39.9	54.9	49.6
India	620.3	464.0	598.0	404.8	735.7	492.4
Nepal	7.8	38.5	38.2	38.3	44.6	44.1
Pakistan	26.7	62.3	58.7	59.8	67.1	67.5
Sri Lanka	19.8	–17.0	7.6	7.1	9.3	8.2
South Asia	732.4	558.9	751.3	555.7	913.9	664.0
Edible oils						
Bangladesh	5.0	1.2	12.9	12.2	14.9	13.6
India	71.3	212.5	146.9	101.0	169.3	114.5
Nepal	1.0	0.9	3.7	3.7	4.1	4.2
Pakistan	8.7	21.3	72.3	74.1	80.3	81.3
Sri Lanka	–1.8	1.5	0.7	0.7	0.9	0.8
South Asia	84.2	237.4	237.3	192.3	270.1	215.1
Sweetener						
Bangladesh	8.1	–7.0	22.5	21.6	24.7	23.2

(contd.)

Table 9.22 (contd.)

Items	Production		Demand (3.5% PCGDP)		Demand (5.5% PCGDP)	
	1965–80	1980–95	2000–15	2015–30	2000–15	2015–30
India	245.4	449.1	416.2	289.4	459.3	315.3
Nepal	1.4	0.9	2.0	2.1	2.1	2.2
Pakistan	60.6	60.7	125.6	130.5	132.9	136.9
Sri Lanka	1.5	0.8	4.6	4.7	5.1	5.0
South Asia	317.0	504.5	572.5	449.9	625.8	484.3
Vegetables						
Bangladesh	5.1	25.3	44.1	38.8	59.3	50.9
India	1273.5	1364.9	2695.7	1799.9	4046.4	2780.1
Nepal	24.7	50.5	67.5	65.9	90.1	87.7
Pakistan	63.3	140.8	178.0	177.1	233.0	230.9
Sri Lanka	17.1	12.7	21.6	17.6	33.2	26.7
South Asia	1383.7	1453.4	3013.7	2106.3	4471.7	3185.9
Fruits						
Bangladesh	–7.2	5.1	61.9	53.1	88.1	75.3
India	361.0	1237.3	1466.0	977.2	2155.0	1468.5
Nepal	2.3	28.7	31.9	31.1	44.3	43.2
Pakistan	72.9	131.1	371.5	368.1	512.5	510.6
Sri Lanka	59.7	–45.9	19.5	16.4	28.0	22.7
South Asia	488.8	1356.3	1954.9	1450.7	2833.8	2126.6
Milk						
Bangladesh	14.1	52.5	77.1	67.1	106.9	91.4
India	820.9	2196.0	2144.0	1427.3	3023.6	2034.7
Nepal	11.5	17.4	55.8	54.5	75.5	73.5

(contd.)

Table 9.22 *(contd.)*

Items	Production		Demand (3.5% PCGDP)		Demand (5.5% PCGDP)	
	1965–80	1980–95	2000–15	2015–30	2000–15	2015–30
Pakistan	157.1	661.5	991.3	985.6	1302.8	1291.1
Sri Lanka	6.8	2.9	15.8	13.0	23.8	19.1
South Asia	1010.4	3940.7	3291.5	2555.6	4542.5	3520.5
Meat						
Bangladesh	6.5	10.6	16.6	14.2	25.1	21.5
India	56.7	107.3	190.7	128.1	298.5	209.2
Nepal	3.3	5.2	12.1	11.7	17.6	17.3
Pakistan	20.5	72.6	100.1	99.3	136.1	135.3
Sri Lanka	–0.13	0.50	1.9	1.5	3.1	2.5
South Asia	86.9	196.1	322.1	255.6	481.5	387.0
Eggs						
Bangladesh	1.6	3.4	5.1	4.3	7.7	6.5
India	25.1	68.4	67.3	45.1	105.2	73.7
Nepal	0.3	0.4	1.5	1.4	2.1	2.1
Pakistan	5.9	12.7	16.1	16.1	22.0	21.9
Sri Lanka	1.0	1.3	1.8	1.5	2.9	2.4
South Asia	33.9	86.2	92.0	68.7	140.3	106.9
Fish						
Bangladesh	–11.0	28.6	47.2	40.5	71.3	61.3
India	74.1	134.4	196.9	132.3	308.2	216.0
Nepal	0.2	0.9	1,2	1.1	1.7	1.7
Pakistan	10.6	19.6	20.8	20.6	28.3	28.1
Sri Lanka	5.7	1.9	11.7	9.4	18.6	14.9
South Asia	79.5	185.4	277.8	203.9	428.1	322.0

Note: PCGDP = Per capita GDP.

Source: Computed by the authors.

STRATEGIES FOR ENHANCING FOOD PRODUCTION

SUSTAINING GREEN AREAS AS GREEN

Managing Rice-Wheat Production System

Most of the countries had concentrated on enhanced production of few commodities like rice and wheat which could quickly contribute to their total food and agricultural production. This resulted in considerable depletion of natural resources and the rain-fed dry areas having maximum concentration of resource poor farmers remained ignored, aggravating problems of inequity and regional imbalances. This also led to a high concentration of malnourished people in these rain-fed low productive areas. This era also witnessed rapid loss of soil nutrients, agro-biodiversity including indigenous land races and breeds.

The rice-wheat based cropping system, spread in the most fertile parts, is the backbone of food security in South Asia. This system is stagnating and hard-pressed to maintain even the modest growth in demand for these two commodities. Smaller growth in yield and decelerating growth in Total Factor Productivity (TFP) in some high input use areas is a matter of concern. An integrated approach of developing crop varieties with greater efficiency in utilization of nutrients and other natural resources, ameliorating soil related problems, incorporation of legumes in the cropping systems, and enhancing water use efficiency will be required to develop location specific management practices to improve the factor productivity growth in the rice-wheat system. Countries must strengthen multidisciplinary researches and refine their technologies towards the goal of sustaining the productivity of rice-wheat system (Singh and Paroda, 1994). Legumes play an important role in improving the sustainability of the system. Ironically, rice and wheat have replaced the principal legumes over a period of time. With the availability of high-yielding and short duration varieties of improved legumes, there is a need to incorporate them in the rice-wheat cropping system to improve the sustainability of the system so as to meet the future foodgrains demand without degradation of the natural resource base (Kumar et al., 1998).

Enhancing Yield of Major Commodities

Yield of major crops and livestock in the region is much lower than in the rest of the world. Considering that the frontiers of expansion of cultivated areas are almost closed in the region, the future increase in food production to meet the continuing high demand must come from increase in yield. There is a need to strengthen adaptive research and technology assessment, refinement and transfer capabilities of the countries so that the existing wide technology transfer gaps are bridged. For this, an appropriate network of extension services needs to be created to stimulate and encourage both top-down and bottom-up flows of information between farmers, extension workers, and research scientists to promote the generation, adoption, and evaluation of location specific farm technologies. Ample scope exists for increasing genetic yield potential of a large number of vegetables and fruits, as well as other food crops and livestock and fisheries products. Besides maintenance breeding, greater effort should be made towards developing hybrid varieties as well as varieties suitable for export purposes. Agronomic and soil researches in the region need to be intensified to address location specific problems as factor productivity growth is decelerating in major production regimes. Research on coarse grains, pulses and oilseeds must achieve a production breakthrough. Hybrid rice, single cross hybrids of maize and pigeonpea hybrids offer new opportunities. Soybean, sunflower and oil palm will help in meeting future oil demands successfully. Forest cover must be preserved to keep off climatic disturbances and to provide enough of fuel and fodder. Milk, meat and draught capacity of our animals needs to be improved quickly through better management practices.

Arresting Deceleration in Total Factor Productivity

More and more cases of deceleration in total factor productivity growth are being reported (Kumar and Mruthyunjaya, 1992; Rosegrant and Evenson, 1992; Kumar and Rosegrant, 1994; Kumar et al., 1998). All the efforts in future have to be concentrated on breaking of yield plateau, by conserving natural resources and promoting ecological integrity of the agricultural system. Producing more with less of inputs will be the major challenge in the next two decades. Most often suggested measures to accelerate and sustain growth in TFP are jacking up investment in research and infrastructural facilities, and increasing input use efficiency (Kumar and Rosegrant, 1994). Biotechnology research to address biotic and abiotic stresses should pay more attention. Given the declining trend in public investment in agriculture which can hardly be reversed, the only option to accelerate growth in TFP is increasing yield potential by developing appropriate technology, both for irrigated and rain-fed areas. Research problems in rain-fed unfavourable ecosystems and breaking of the current irrigated yield ceilings are more complex and challenging. To make headway in them will require mobilizing the best of science and the best of scientists in National Agricultural Systems in partnership mode. This needs higher agricultural research investment which is convincingly justified by several studies (Kumar and Rosegrant, 1994; Jha et al., 1995; Pal and Singh, 1998).

Integrated Nutrient Management

Attention should be given to balanced use of nutrients. Phosphorus deficiency is now the most widespread soil fertility problem in both irrigated and unirrigated areas. Correcting the distortion in relative prices of primary fertilizers could help correct the imbalances in the use of primary plant nutrients (nitrogen, phosphorus and potash) and use of biofertilizers. To improve efficiency of fertilizer use, what is really needed is enhanced location-specific research on efficient fertilizer practices (such as balanced use of nutrients, correct timing and placement of fertilizers, and, wherever necessary, use of micro-nutrient and soil amendments), improvement in soil testing services, development of improved fertilizer supply and distribution systems, and development of physical and institutional infrastructure (Kumar and Desai, 1995).

MAKING GREY AREAS AS GREEN

Resource-poor farmers in the rain-fed ecosystems practice less-intensive agriculture, and since their incomes depend on local agriculture, they benefit little from increased food production in irrigated areas. To help them, efforts must be increased to disseminate available dry land technologies and to generate new ones. For technology-based agricultural growth, it will be necessary to enlarge the efforts for promoting available dry land technologies, increasing the stock of this knowledge, and removing pro-irrigation biases in public investment and expenditure, as well as credit flows. Attention should also be paid on watershed development for raising yields of rain-fed crops, as on the widening of the seed revolution to cover oilseeds, pulses, fruits and vegetables. Farming system research to develop location specific technologies must be intensified in the rain-fed areas. Strategy to make grey areas green will lead to second Green Revolution which would demand a three pronged strategy: watershed management, hybrid technology and small farm mechanization.

DIVERSIFICATION OF AGRICULTURE AND VALUE ADDITION

In the face of shrinking natural resources and ever increasing demand for larger food and agricultural production arising due to high population and income growths, agricultural intensification is the main course of future growth of agriculture in South Asia. Research for product diversification should be yet another important area. Besides developing technologies for promoting intensification, South Asians' countries must give greater attention to the development of technologies which will facilitate agricultural diversification, particularly towards intensive production of fruits, vegetables, flowers and other high value crops which are expected to increase income growth and generate effective demand for food. The per capita availability of arable land in South Asia is quite low and declining over time. Diversification towards these high value and labour intensive commodities can provide adequate income and employment to the farmers dependent on small size of farms. Due importance should be given to quality and nutritional aspects. High attention should be given to develop post-harvest handling and agro-processing and value addition technologies, not only to reduce the heavy post-harvest losses but also to improve quality through proper storage, packaging, handling and transport. The role of biotechnology in post-harvest management and value addition deserves to be enhanced.

SUMMARY AND CONCLUSION

South Asia has the highest population pressure on land and other resources to meet its food and development needs. These countries, with a total population of about 1,291.2 million people in 1998, account for 21.8 per cent of the world's population and share only 3 per cent of the world's land area. During the last three decades, the South Asia region has made remarkable progress in its food production. Public investment in infrastructure, research and extension along with production strategy and policy support have significantly helped to expand food production. This had witnessed Green, White and Blue (fish) Revolutions.

A structural shift in dietary patterns towards livestock, fisheries, horticultural products is already underway and is being predicted to intensify further. Decline in per capita consumption of cereals and rapid increase in consumption of fruits, vegetables, milk, meat, eggs and fish is observed. The level of consumption of high value foods is much lower than the recommended levels. Cereals contribute two-thirds of the total calorie intake and hold the key to nutritional food security in South Asia.

The population of South Asia will be 1,930 million in the year 2030. At 3.5 per cent growth in per capita annual real income, demand for cereals will be 339 million tons, pulses 28 million tons, roots and tubers 54 million tons, edible oils 18 million tons, vegetables 168 million tons, fruits 110 million tons, sugars 46 million tons, milk 191 million tons, animal fats 4.7 million tons, meat 17 million tons, eggs 4.9 million tons and fish 15 million tons. To meet the projected demand for food and to attain self-reliance, the South Asian region must attain per ha. yield of 2.82 tons for rice, 3.72 tons for wheat, 2.36 for maize and 0.99 tons for pulses by the year 2030. This means an improvement in average yield, at regional level, by 55 per cent in rice, 61 per cent in wheat, 55 per cent in maize, and 77 per cent in pulses. The production of livestock, fisheries and horticultural products must be improved by 177–246 per cent by 2030 over the year 1995. Additional annual requirement of food in the next century is predicted to be much more than achieved in the past by South Asian countries. Future increases in the production of cereals and non-cereal agricultural commodities will have to be essentially achieved through increases in productivity as the possibilities of expansion of area and livestock population are minimal. Vast uncommon opportunities to harness agricultural potential still remain and could be tapped to achieve future targets. There are serious gaps both in yield potential and technology transfer as the national average yields of most of the commodities are low: they could be harnessed if addressed properly.

Food demand challenges ahead are formidable considering the non-availability of favourable environment of past growth, declining factor productivity in major cropping systems and shrinking resource base. Also the capital investment in agriculture is invariably declining in most of South Asian countries. Managing the rice-wheat production system, enhancing the yields of major commodities, arresting deceleration in factor productivity, improving productivity in rain-fed agriculture, integrated nutrient management and post harvest management and value addition will be the major strategies for these nations. These strategies need to be blended with the farmers' wisdom to ensure the adoption of improved technologies. Thus, to ensure food security, more capital investments in agriculture, including greater support to research and development, will be critical in future. However, the serious efforts by the National Agricultural Research System, policy makers and the hard working innovative farmers of South Asia would, no doubt, be able to convert these formidable challenges into great successes in the future as well.

REFERENCES

BANSIL, P.C., 'Demand for Foodgrains by 2020', *Economic and Political Weekly*, Vol. 31, No. 10, 1998, pp. 546–8.

BEHRMAN, Jere and Anil DEOLALIKAR, 'Is Variety the Spice of Life? Implications for Calorie Intake', *Review of Economics and Statistics*, 1989, pp. 331–98.

BOUIS, Howarth and Lawrence HADDAD, 'Are a Estimates of Calorie-Income Elasticities too High: A Recalibration of the Plausible Range', *Journal of Development Economics*, Vol. 39, No. 2, 1992, pp. 333–64

FAO/RAP, *Selected Indicators of Food and Agriculture Development in Asia-Pacific Region. 1983–1993*, Food and Agriculture Organization of the United Nations, Regional Office for Asia and the Pacific, Bangkok, 1994.

JHA, D., P. KUMAR, S. PAL, S. SELVARAJAN and Alka SINGH, *Research Priorities in Indian Agriculture*,

National Centre of Agricultural Economics and Policy Research, Policy Paper 2, New Delhi, 1995, pp. 1–120.

KUMAR, Praduman and MRUTHYUNJAYA, 'Measurement and Analysis of Total Factor Productivity Growth in Wheat', *Indian Journal of Agricultural Economics*, Vol. 17, No. 3, 1992, pp. 451–8.

KUMAR, Praduman and Mark W. ROSEGRANT, 'Productivity and Sources of Growth for Rice in India', *Economic and Political Weekly*, Vol. 29, No. 53, 1994, pp. 183–8.

KUMAR, Praduman and Gunvant M. DESAI, 'Fertiliser Use Pattern in India During the Mid-1980s. Micro-Level Evidence on Marginal and Small Farms', in Gunvant M. DESAI and A. VAIDYANATHAN (eds.), *Strategic Issues in Future Growth of Fertiliser Use in India*, Macmillan India Ltd., New Delhi, 1995.

KUMAR, Praduman, *Food Demand and Supply Projections for India*, Agricultural Economics Policy Paper, Indian Agricultural Research Institute, New Delhi, January 1998.

KUMAR, Praduman, P.K. JOSHI, C. JOHANSEN and M. ASOKAN, 'Sustainability of Rice-Wheat Based Cropping Systems in India. Socio-Economic and Policy Issues', *Economic and Political Weekly*, Vol. 33, No. 39, 1998, pp. 152–7.

NAQVI, S.N.H., A.M. AHMED and R. SIDDIQUI, *Projections and Policy Implications of Medium and Long-Term Supply and Demand of Major Food Crops in Pakistan*, Collaborative Research on 'Projections and Policy Implications of Medium and Long Term Rice Supply and Demand', International Rice Research Institute (Los Banos) and International Food Policy Research Institute (Washington, DC), 1995.

PAL, Suresh and Alka SINGH, 'Strengthening the Indian Agricultural Research and Extension Systems: Funding and Institutional Imperatives', *Indian Journal of Agricultural Economics*, Vol. 53, No. 3, 1998, pp. 488–503.

ROSEGRANT, Mark W. and Robert E. EVENSON, 'Agricultural Productivity and Sources of Growth in South Asia', *American Journal of Agricultural Economics*, No. 74, August 1992, pp. 757–61.

ROSEGRANT, Mark W., Mercedita AGCAOILI-SOMBILLA and Nicostrato D. PEREZ, 'Global Food Projections to 2020. Implications for Investment', in *Food, Agriculture and the Environment Discussion Paper Number 5*, International Food Policy Research Institute, Washington, DC, 1995.

SHAHABUDDIN, Quazi and Sajjad ZOHIR, *Projections and Policy Implications of Rice Supply and Demand in Bangladesh*, Collaborative Research on 'Projections and Policy Implications of Medium and Long Term Rice Supply and Demand', International Rice Research Institute (Los Banos), and International Food Policy Research Institute (Washington, DC), 1995.

SINGH, R.B. and R.S. PARODA, 'Sustainability and Productivity of Rice-Wheat Systems in the Asia-Pacific Region. Research and Technology Development Needs', in R.S. PARODA and R.B. SINGH (eds.), *Sustainability of Rice-Wheat Production Systems in Asia*, FAO/RAP Publication Number 11, Oxford & IBH Publishing Co. Pvt. Ltd, New Delhi, 1994.

SINGH, R.B., *Agricultural Research Priorities in South Asia,* The Asia-Pacific Association of Agricultural Research Institutions, Bangkok, 1996.

UNPF, *The State of World Population,* United Nations Population Fund, New York, 1998.

10

Agricultural Production System and Environmental Sustainability

LESSON FROM LONG TERM MANURIAL EXPERIMENTAL TRIALS

BHUBAN C. BARAH[1]

INTRODUCTION

The total agricultural production in India at the time of independence was about 50.82 million tons. But on account of several policy measures including the expansion of cropped area from 99 million ha. in 1950–1 to 118 million ha. in 1964–5, the agricultural production increased to the tune of 89.64 million tons. Subsequently, the area under foodgrains touched about 124 million ha. in 1997. More importantly, the launching of the Green Revolution along with policy support measures such as irrigation, fertilizers, plant protection measures, intensive research and development initiatives, gave a boost to production, increasing to 199.32 million tons in 1997. During the period, crop productivity improved from 553 kg./ha. in 1951 to 1,601 kg. in 1997. The per capita availability of foodgrains also improved substantially (Table 10.1).

The rapid increase in production, however, also brought about a change in the composition of the food basket. In the per capita food configuration, wheat gained substantially, while gram, pulses and other cereals lost their previous position. Coarse

Table 10.1: Per Capita Availability of Foodgrains in India (1951, 1964, 1980, 1990)

(kg. per year)

Year	Rice	Wheat	Other cereals	Cereals	Gram	Pulses	Foodgrains
1951	158.9	65.7	109.6	334.2	22.5	60.7	394.9
1964	201.4	90.1	109.5	401.0	20.3	51.0	452.0
1980	166.1	126.8	86.6	379.5	10.7	30.9	410.4
1990	212.2	132.6	86.8	451.1	13.4	41.9	493.4
1999	205.4	163.9	64.0	433.5	14.7	36.9	470.4

Source: Government of India, *Agricultural Statistics at a Glance*, Ministry of Agriculture, New Delhi, 2001, pp. 88–9.

cereals lost rapidly in favor of the rice-wheat system and other short-run remunerative systems. The Green Revolution ushered in the mid-1960s made a tremendous impact on the availability of superior cereals such as rice and wheat, but ignored several other commodities. The cereal-cereal rotations are getting more prominence and the ecologically beneficial cereal-pulse rotation has declined drastically. This development raises doubts not only on the adequacy from the nutritional point of view, but also several other ecological questions. While the goal of feeding the hungry millions has been achieved to an extent by increasing production, the policy initiative to tackle the problem of feeding hungry plants has not received the desired attention. The growth has become a single tracked production oriented approach, which has marginalized the crucial question on sustainability.

What kind of sustainable agricultural production system should be promoted (Table 10.2): the modern input-intensive, traditional stable low input but low unit productivity system, or a mix of both? The traditional agricultural methods are non-degrading, low external input, more local input-using and stable. But their productivity is low. The modern system is high and fast yielding, promising return to investment and uses intensive external inputs, but it is highly resource degrading, affecting the long-term stability. Therefore, a desirable agricultural system is one which yields more, sustained net returns and conserves natural resources such as land, water and biodiversity.

Table 10.2: Evolution of Alternative Technology

Mode of production		Special characteristics
Inherited: Traditional Agriculture	=>	Non-degrading, stability at low level, utilizes low external inputs and more local inputs. But amenable to acute poverty.
Developed: Modern agriculture	=>	High and fast yielding, more promising returns to investment but intensive external input using, resulted in resource depletion. It affects long run sustainability.
Required: Sustainable agricultural	=>	Yields more and sustained net return, low production cost, conserve natural resources and preserve bio-diversity.

It is in this context that the declining responses to the factors of production in recent times as compared to the earlier period has been threatening long term food security. It implies that more additions of external inputs are necessary to obtain constant levels of yield, resulting in adverse impact on net farm incomes, which is a symptom of unsustainability. In fact, ensuring sustainability requires in-depth study on cause and effect relationship, as well as an objective evaluation of alternative options. It is also necessary to choose the method of more efficient economic analysis of scientifically designed micro-level production systems for enhanced value addition and agricultural policy (Abraham and Rao, 1966; Heady and Dillon, 1961; Anderson, 1968). There is also need to integrate socioeconomic dimensions to modern agricultural technology by using on-station data on experimental trials.

In this context, studies in fertilizer use efficiency assume crucial importance in view of the growing concern for intensive use of chemical fertilizers in crop production in the recent period. Past studies looked at issues such as fertilizer use, geographical spread of fertilizer consumption, fertilizer response at on-station as well on farmers' field (Kumar and Deasi, 1995; Gandhi et al., 1995; Desai and Vaidyanathan, 1995). The level of adoption, nature of fertilization, extent of area and rate of fertilization, etc., are some of the questions examined in these studies. But the long term impact of continuous and intensive modern technology on sustainability as well as on the health of the natural resource, have not been not examined adequately.

This paper focuses on the long-term fertilizer experimental trials on rice-wheat system in order to examine the change in fertilizer responses and address the complex problems of measurement of unsustainability. It attempts to study in-depth the trends in response to on-station fertilizer treatments and to test the hypothesis that indiscriminate use of external inputs endangers environmental sustainability. The trend in crop productivity is analysed to develop efficient choice criteria of alternative technologies and indicators of long-term sustainability of the cropping system. The following questions are particulary addressed: what is the long-term impact of use of external input on sustainable agriculture as well as on natural resources? Or, alternatively, could the productivity be sustained through intensive modern methods in agriculture?

The productivity trend in long-term fertilizer experiments is examined using appropriate trend equations. The incremental gain in net return is calculated to measure the partial factor productivity of fertilizer. The gain in net return over the control level is considered advantageous as it combines the biological yield and market prices. The net return to that of physical yield, of socioeconomic and agronomic decision criteria is an efficient indicator. The alternative technological options are ranked on the basis of stability and adaptability parameters and compared with the ranking based on biological yield characteristics. The physical and biological characteristics are used to build the indicator of sustainability. The panel regression technique is particularly useful to sharpen the choice of the options.

Finally, the rice-wheat system is purposively chosen in this study as its contribution is significantly and is in the forefront in Indian agriculture. Rice and wheat together cover about 55 per cent of total cropped area and contribute about 75 per cent of total foodgrains production. Moreover, the area under the sequential rotation is expanding rapidly in recent times due to its popularity among the farmers. However, despite its importance in increased food economy, the system is highly vulnerable to environmental degradation.[2] The externality of continuos cropping resulted in resource degradation and endangered food security (Kumar et al., 1999). The rapid productivity decline is signalling a disturbing trend in the long-term food production. The rate of productivity growth has been relatively modest and had stagnated in the recent past. It is therefore pertinent and timely to address the sustainability question. A unified socioeconomic approach to biological experimentation would provide a realistic approach to understanding the actual behavioural pattern of a sustained agricultural production system.

LONG-TERM EXPERIMENTAL TREATMENTS

Long-term experiments are those which are continued on the same set of plots over a series of years, with a pre-planned sequence of treatments, or crop, or both (Panse and Sukhatme, 1967). It aims to study the long-term effects of treatments and crops on soil fertility and on economic returns.

The on-station long-term fertilizer experimentations are designed to accomplish the following objectives: (a) to develop suitable nutrient supply and management system and to study the long-term effect of fertilizer in combination with organic manure on the productivity of cereal based crop sequence and on soil health; (b) to improve fertilizer use efficiency in the cropping system.

The micro level long-term experimental trials are conducted on permanent plots at multi-locations and over multi-period under the All Indian Co-ordinated Agronomic Research Project of the Indian Council of Agricultural Research (ICAR). The Project Directorate of Cropping System Research (PDCSR) at Modipuram monitors the agronomic trials at a large number of experimental stations.

The multi-location experiments on integrated nutrient supply in cereal based crop sequences and on long range effects of continuous cropping and manuring look at the soil fertility and yield stability on permanent plot experiments. The permanent plots long-term fertilizer trials data are collected from 11 centres from 4 agro-ecological regions, viz., semi arid, sub humid, humid and coastal (Table 10.3). It covers physical data of the experiments for over 10 years, from the year 1985–6 to 1995–6.

Table 10.4 lists twelve treatments of various mix of organic and inorganic fertilizers (numbered serially from 1 to 12, the first one being the control treatment and the last the farmers' practices). The well laid experiments using randomized block design are conducted with these treatments in four replicated plots in each of the stations. Since the analysis deals with rice-wheat system, the treatments represent sequential application of fertilizer doses in the order of kharif and rabi season.

Table 10.3: A Multi-Location Experiments

No.	Ecosystem	Stations
1.	Semi-arid	Kanpur, Ludhiana
2.	Sub-humid	Palampur, R.S. Pura, Pantnagar, Sabour, Vranasi, Faizabad, Jabalpur
3.	Humid	Kalyani
4.	Coastal	Navasari

Table 10.4: Treatment Combination for On-Station Long Term Fertilizer Trials

S.No.	Treatments		Kharif (Rice)	Rabi (Wheat)
Tr1	NPK0	Control (zero fertilizer)	Control	Control
Tr2	NPK1	Chemicals	50% NPK	50% NPK
Tr3	NPK2	Chemicals	50% NPK	100% NPK
Tr4	NPK3	Chemicals	75% NPK	75% NPK
Tr5	NPK4	Chemicals	100% NPK	100% NPK
Tr6	NPK1FYM1	Organo-chemical	50% NPK + 50% FYM	100% NPK
Tr7	NPK2FYM2	Organo-chemical	75% NPK + 25% FYM	75% NPK

(contd.)

Table 10.4 (contd.)

Tr8	NPK1CR1	Organo-chemical	50% NPK + 50% CR	100% NPK
Tr9	NPK2CR2	Organo-chemical	75% NPK + 25% CR	75% NPK
Tr10	NPK1GM1	Organo-chemical	50% NPK + 50% GM	100% NPK
Tr11	NPK2GM2	Organo-chemical	75% NPK + 25% GM	75% NPK
Tr12	Farmer practices	Farmers' practices		

Notes: NPK denotes Recommended Fertilizer Dose (RFD) of nitrogen (N), phosphorous (P) and potash (K); FYM = farm-yard manure; CR = crop residues (straw); GM = green manure.

LONG TERM FERTILIZER RESPONSE

The average yield of paddy as well as wheat are the lowest in Navasari as compared to the same in other locations (Table 10.5). It varies from 14.7 Q./ha. for control treatments to 33.7 Q./ha. for treatment NPK4 for paddy. Barring the yield of control treatment, the yields for other treatments are found to be highest in Ludhiana. In the case of wheat also, the yield for the treatments NPK1, NPK3 and NPK1FYM1 was highest in Ludhiana, for treatments NPK4, NPK2FYM2, NPK1CR1 and NPK2CR2 in Kanpur, and for control treatment and NPK2 in Pantnagar. The control yield of 7.1 Q./ha. is the lowest at Palampur, followed by Kalyani (7.6 Q./ha.), the highest yield being 15.5 Q./ha. for control treatment in Pantnagar. The response to the treatment NPK1FYM1 is the highest at 47.2 Q./ha. in Ludhiana, and for the treatment NPK1GM1 in Kanpur (47.5 Q./ha).

The comparison of treatment performances across locations shows that the gain in yield due to fertilization is over 100 per cent of the control yield for all the treatments in paddy as well as wheat. However, the gain varies from treatment to treatment. The variation of productivity gain arises primarily due to local factors other than fertilization; the differential responses to treatments depend on how the physical and environmental condition of the locality reacts to external inputs. As a matter of fact, as many as 21 cases show significant improvement in yield of paddy in the range of 50–100 per cent, none in the 'below 50 per cent' class and the rest in the 'above 100 per cent' class. Barring Navasari, the improvement in yield in most of the locations exceeds by over 100 per cent of control level for wheat and there are only 7 cases in the '50–100 per cent' category. Being situated in the coastal ecosystem, the crop yield is severely constrained in Navasari. The yield for other treatments in Navasari is also lowest and in Ludhiana it is the highest.

Wheat follows paddy in the experimental plot and, thus, the yield of paddy as well as wheat is influenced by their joint effect. The functional relationship of yield response can be written as:

$$Y_{paddy} = f(NPK, Y_{wheat}) \text{ and } Y_{wheat} = f(NPK, Y_{paddy})$$

The treatment NPK1 doubles the yield of paddy in 44 per cent of the stations and the Farmers' Practices in 6 out of 12 experimental stations. The yield of paddy is more evenly distributed in all the yield improvement class intervals across the locations, the highest performing treatment being for the treatment NPK1GM1. The response to organo-chemical, i.e. treatments 6 to 11, is higher than that of the pure chemical fertilizer. These treatments help upkeep of the soil health; they also sustain the soil quality, particularly under anaerobic condition (Singh et al, 1991).

Table 10.5a: Mean Experimental Yield of Paddy for Different Fertilizer Treatment

(Q./ha.)

S.No.	Treatments	Navasari	Faizabad	Jabalpur	Kalyani	Kanpur	Pantnagar	Sabour	Ludhiana
Tr1	Control	14.7	19.1	24.1	14.6	15.7	30.3	16.2	20.3
Tr2	NPK1	24.4	32.8	33.0	25.8	29.9	38.0	28.8	42.4
Tr3	NPK2	25.8	34.3	36.7	26.1	31.6	37.3	29.4	45.9
Tr4	NPK3	28.7	40.5	46.7	30.7	35.2	40.9	34.4	54.9
Tr5	NPK4	33.7	43.8	46.8	33.1	40.2	46.6	41.7	64.9
Tr6	NPK1FYM1	29.5	42.0	43.7	33.9	34.9	40.6	39.0	59.5
Tr7	NPK2FYM2	31.0	46.2	48.2	36.5	37.3	41.5	42.2	61.6
Tr8	NPK1CR1	32.5	40.8	38.7	33.2	33.3	40.5	37.4	53.7
Tr9	NPK2CR2	31.8	40.6	40.5	34.2	35.5	42.4	39.1	60.4
Tr10	NPK1GM1	32.7	41.9	48.9	36.6	35.9	44.4	37.3	65.2
Tr11	NPK2GM2	33.2	42.6	49.1	37.4	39.1	46.0	39.0	66.7
Tr12	Farmers' Practices	20.5	31.1	29.0	24.4	32.6	44.4	29.4	24.7

DECLINING FERTILIZER RESPONSE

Chemical fertilizer increases the productivity of paddy and wheat significantly. But indiscriminate use of agro-chemicals results in accumulation of nutrients in the soil, which affect crop productivity adversely. The excessive build up of the micro-nutrients distorts the proper balance and the soil becomes sensitive to crop yield. Thus, appropriate policy intervention is necessary to ensure minimum balance of required nutrients in the soil to sustain yield.

The depletion of soil nutrients due to the application of agro-chemicals perpetuates the application of more external artificial inputs to maintain minimum quantity of soil nutrients. Even if the external inputs such as chemical fertilizers stabilize the crop yield whereby the adopter farmers reap substantial short-term gains, the practice is associated with a heavy cost of deleterious soil health in the long run. The soil fertility experiment shows that, at the beginning of the crop cycle, nutrients improve the yield and maintain soil balance within the range of carrying capacity. But on continuous cropping, these nutrients gradually get depleted, and hence result in loss of crop yield.

Yadav (1998) has shown that continuous cropping of rice-wheat system depleted soil micro-nutrients and reduces the partial factor productivity of fertilizer treatment to the extent of –32 per cent for control and –40 per cent for full dose of recommended fertilizer in Pantnagar. Similarly, the results of the on-station experiment on fertilizer treatments show that the response to fertilizer improves in the initial period, but stagnates or declines in the long run (Figure 10.1). It is also observed that the decline of yield is more widespread than the gains, and that the frequency distribution of decline in yield is more skewed in case of pure chemical fertilizer treatment (Table 10.6), which confirm the hypothesis of productivity decline.

The loss of yield due to combination treatments is milder and also occurs in fewer stations (lowest being in 2 stations for NPK1FYM1). The decline in yield response to pure chemical fertilizers as a group (Treatments 2–5) is more than that of the organo-

Table 10.5b: Mean Experimental Yield of Wheat for Different Fertilizer Treatments

(Q /ha.)

S.No.	Treatments	Faizbad	Jabalpur	Kalyani	Kanpur	Navasari	Pannagar	Sabour	Ludhina	Palmpur	Raipur	Varanasi
Tr1	Control	8.2	10.2	7.6	12.0	10.5	15.5	9.0	12.7	7.1	12.5	9.0
Tr2	NPK1	24.7	18.1	15.7	29.4	16.8	28.4	18.1	31.7	15.9	20.3	18.1
Tr3	NPK2	33.7	26.3	26.2	37.7	19.1	36.1	29.1	43.2	21.2	26.0	29.1
Tr4	NPK3	30.3	25.3	24.7	35.4	18.9	32.1	26.7	38.5	20.3	25.6	26.7
Tr5	NPK4	34.6	28.6	25.5	45.4	21.5	38.8	30.7	44.4	24.6	27.7	30.7
Tr6	NPK1FYM1	36.2	26.8	29.3	46.0	20.3	37.4	33.0	47.2	27.5	29.0	33.0
Tr7	NPK2FYM2	30.7	24.6	26.3	43.2	18.9	35.1	29.2	43.0	21.8	28.3	29.2
Tr8	NPK1CR1	32.9	27.1	30.7	45.4	20.1	34.9	30.4	43.6	22.0	29.0	30.4
Tr9	NPK2CR2	31.0	22.8	25.9	42.5	17.4	34.3	28.5	38.8	20.9	28.2	28.5
Tr10	NPK1GM1	33.3	27.0	29.7	47.5	19.9	38.4	29.6	43.0	20.3	30.5	29.6
Tr11	NPK2GM2	30.7	24.5	24.8	45.4	17.6	35.4	29.2	39.1	20.9	29.9	29.2
Tr12	Farmers' Practices	22.1	18.2	21.5	32.7	17.8	37.2	18.9	24.3	16.2	25.3	18.9

chemical treatment group (Treatments 6–11). More than 50 per cent of the stations showed yield decline in paddy as well as wheat for pure chemical fertilizer treatments. On the contrary, 8 out of 10 stations show yield improvement for the organo-chemical treatments. It is thus inferred that the organo-chemical treatments are more yield-sustaining as against the pure chemicals. The analysis of micro level experimental data confirms that the indiscriminate application of pure chemical fertilizer without the use of organic fertilizers potentially endangers food production, besides degrading the land resources. Thus, the judicious use of external inputs and choice of optimal input mix is the kingpin to negate the environmental externality and improve the overall efficiency of the system. An effective synergy among biological scientists and social scientists is thus necessitated in problem solving on the sphere of technology generation and its dissemination.

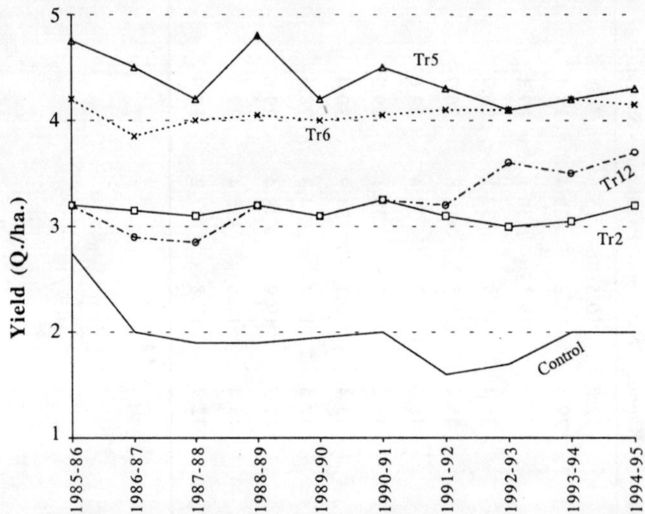

Figure 10.1: Trend of Average Experimental Yields of Paddy for Selected Treatments.

Table 10.6: Frequency of Distribution of Experimental Station Showing Decline in Crop Yield for Each of the Treatments

S.No.	Treatment	Yield of Paddy		Yield of Wheat	
		Decline	No decline	Declined	No decline
Tr1	Control	5 stations	5 stations	7 stations	3 stations
Tr2	NPK1	5	5	4	6
Tr3	NPK2	5	4	5	5
Tr4	NPK3	3	7	5	5
Tr5	NPK4	4	6	4	6
Tr6	NPK1FYM1	2	8	5	5
Tr7	NPK2FYM2	3	7	3	7
Tr8	NPK1CR1	3	7	6	4
Tr9	NPK2CR2	3	7	3	7
Tr10	NPK1GM1	3	7	4	6
Tr11	NPK2GM2	3	7	4	6
Tr12	Farmers' practices	2	8	2	8

IMPACT OF SEQUENTIAL CROPPING ON LAND QUALITY

Table 10.7 compares the initial values to the nutrient status after the application of reatments at different locations. It appears that pure fertilizers are more nutrient depleting than combination fertilizers. The organo-chemical treatments potentially replenish the depleted nutrients (figures in italic in Table 10.7) and help maintain a desired level of soil balance. The loss of nutrient (as indicated by downward arrow in Table 10.7) occur more frequently for Control, NPK1 and NPK4 than for organo chemical treatments. In fact, Paroda et al. (1994) already observed that the application of green manure (*sesbania*) increases the organic carbon in soil after rice cultivation, from 0.29 per cent (initial) to 0.45 per cent at the experimental stations. With incorporation of crop residues (wheat straw in rice field), the organic carbon increases from 0.38 per cent to 0.47 per cent. The multi-location analysis of the change of soil status over the years reconfirms these long run deleterious effects on productivity.

Table 10.7 shows that the depletion of Organic carbon content (OC-content), Nitrogen content (N-content), Phosphorous content (P-content) and Potash content (K-content) is occurring more frequently than the accumulation across the stations and over time. The organic carbon depleted by as much as 50 per cent to 75 per cent of the initial values in majority of the station. The OC-content of soil has improved the initial value in a couple of stations, which indicates the need for regionally differentiated fertilizer policy. The behaviour of N-content in soil is similar to that of OC-content. The FYM+NPK and GM+NPK contribute more towards the nitrogen balance of soil as compared to the pure chemicals. The accumulation of the nutrient ranges from 52 per cent to 56 per cent above the level of control treatment. Similarly, mixed fertilizers (organic manure based) improves the P-content of soil in almost all the locations, the magnitude of this improvement varying widely from location to location. The phosphorous enrichment in the soil is extremely essential to maintain a nutrient balance, but the farmers often neglect it and substitute it by overuse of nitrogenous fertilizers. Finally, the combination treatments are, by and large, more K-enriching in the soil, particularly after rice-wheat rotation. Perhaps the artificial application of fertilizer treatment releases the nutrients rapidly in the field in most of the stations. The crop meet the potash demand from the soil rather than from the externally applied potassic fertilizer in these locations, which result in the to reduction of the stock of soil nutrient.

This analysis of rice-wheat sequence brings out important issues:

(a) The cultivation of continuous cereal crop rotations depletes soil nutrients such as organic carbon, native nitrogen, phosphorous and potash content in majority of cases. The magnitude of the changes depends on the basic properties of the soil. It has thus emphasized the need for a region specific policy of induced fertilization of soil through artificial chemicals to maintain the nutrient balance.

(b) As compared to pure chemical fertilizer treatments, the combination is more efficient in upkeep of the nutrient balance in soil, quality and health of soil.

(c) The relationship between degradation of soil quality in the experimental stations (where all factors of production are controlled) and the declining trends in yield of cereal crops is an important indicator of unsustainability, particularly in the case of pure chemical fertilizers.

Table 10.7: Depletion of Soil Nutrients for Selected Treatments due to Continuous Cropping of Rice-Wheat Sequence

Station		Initial Values	Control No fertilizer	NPK1 NPK 50%	NPK4 NPK 100%	NPK1FYM1 NPK+FYM	NPK1CR1 NPK+CR	NPK1GM1 NPK+GM
Organic Carbon (%)								
Navasari	1992–3	0.62	0.43↓	0.44↓	0.44↓	0.45↓	0.45↓	0.44↓
Sabour	1993–4	0.44	0.30↓	0.30↓	0.40↓	*0.50*	*0.50*	*0.50*
Palampur	1993–4	0.60	0.60	0.60	*0.70*	*0.70*	*0.60*	*0.70*
Kanpur	1993–4	0.40	0.10↓	0.10↓	0.30↓	0.20↓	0.20↓	0.30↓
Faizabad	1993–4	0.40	0.20↓	0.40	0.40	*0.50*	*0.50*	*0.50*
Nitrogen (kg./ha.)								
Navasari	1992–3	286	256↓	255↓	270↓	248↓	239↓	253↓
Palampur	1993–4	675	477↓	*702*	*752*	*671*	*665*	*727*
Faizabad	1993–4	102	78↓	90↓	114	127	117	122
Phosphorous (kg./ha.)								
Navasari	1992–3	42	22↓	21.5↓	22.5↓	23↓	23.5↓	22.5↓
Sabour	1993–4	21.5	5↓	6↓	10↓	11↓	10↓	10↓
Palampur	1993–4	21.9	27	41	48	43	42	41
Kanpur	1993–4	16	8↓	10↓	*22*	*18*	*17*	*21*
Faizabad	1993–4	14	8↓	13↓	*20*	*21*	*17*	22
Potash (kg./ha.)								
Navasari	1992–3	336	204↓	190.5↓	167↓	209↓	209↓	192↓
Sabour	1993–4	165	108↓	117↓	138↓	152↓	145↓	136↓
Palampur	1993–4	221	247	258	219	*241*	*219*	*247*
Kanpur	1993–4	182	115↓	122↓	155↓	169↓	191	179↓
Faizabad	1993–4	355	277↓	292↓	290↓	305↓	324↓	298↓

COST ADVANTAGE OF ORGANIC MANURE BASED TREATMENTS

The organic-manure based treatments not only exhibit gain in yield but also have definite cost advantages. The estimated marginal value addition to fertilizer treatments of various fertilizer treatments is presented in Figure 10.2. The profit per rupee investment on pure fertilizer (Rs. 1.11 for NPK4) is lesser than that on combination treatments (Rs. 1.38 for treatment NPK1FYM1, Rs. 1.62 for treatment NPK1CR1 and Rs. 1.91 for treatment NPK1GM1).

Besides the higher economic gains, the combination fertilizers also potentially mitigate the environmental externality of continuous cereal-cereal sequential cropping. The economic return as well as the intangible ecological benefits (as explained earlier) clearly indicate that combination fertilizer is more income enhancing as well as ecologically beneficial in the long-run.

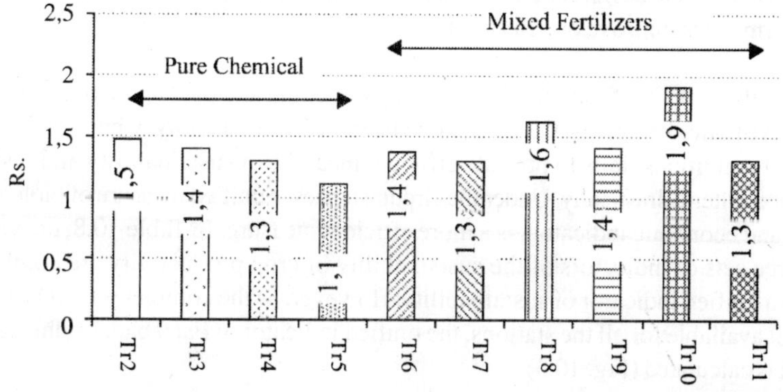

Figure 10.2: Marginal Value Product of Fertilizer Treatments.

INDICATORS OF SUSTAINABLE PRODUCTION SYSTEM

The indiscriminate use of chemical fertilizers gives rise to problems with long run implications on environmental sustainability. The irreversible effects of the externality on the environmental resources such as hidden secondary salinity, ground water contamination, etc., are more serious than the economic ones. The scientific investigation has shown the traces of increasing levels of nitrate (N) in ground water in north India due to intensive fertilizer use in the farm fields (Singh and Singh, 1997; Kanwar and Katyal, 1997).[3] But the problem is complex, which requires carefully developed measures such as the sustainability indicator of technology options and monitoring mechanisms.

Experimental data on three treatments (representing major groups of treatments) have been used to develop the following three sets of sustainability indicators:

(a) The economic indicator, which combines biological factors, market factors, and productivity growth, is derived from the First Difference of the net Return (FDR) of the t and t–1 year. It is a measure of sustainability in the short run. A positive FDR shows an improved performance in the current year as compared to the previous one, and a negative denotes deterioration. Since the short run weather risk influences the FDR, an average of FDR over the entire period, along with its standard deviation, could be a measurement of long run sustainability. Ideally, a positive average value and lower standard deviation denote a state of sustainability, and a negative the unsustainability of the production process.

(b) The physical indicator is derived from the status of soil health after continuous crop rotation. Indeed, changes in soil status (depletion/accumulation of soil nutrients such as OC-content, N-content, P-content and K-content) influence the land productivity. Some nutrients deplete rapidly upon continuous crop rotation, and others get built-up at the same time depending upon the soil condition.

(c) The biological indicator is derived from measurements of plant characteristics. A positive First Difference of physical Yields (FDY) shows whether the yield performances improve or deteriorate over time.

But as the yield is a function of factors of production, and the factors of production are sensitive to market prices, the physical indicator of sustainability, when confronting the market conditions, fails to provide efficient measure of sustainability and distorts the decision criteria drastically. Under the circumstances, a unified measure of biological, physical and economic indicators is a more efficient measure. In Table 10.8, the values of the three sets of indicators of the sustainability of crop productivity are combined to derive a unified indicator of sustainability. However, as the complete set of required data is not available for all the stations, the unified indicator of Faizabad, Ludhiana and Navasari is calculated (Fig. 10.3).

Table 10.8: Integrated Sustainability Indicators

Location	Indicator	NPK 100%	NPK1FYM1	NPK1CR1
Faizabad	FDR	+ (R3)	+ (R1)	+ (R2)
	Std. (FDR)	(R1)	(R3)	(R2)
	CARG	(R3)	(R1)	(R2)
	OC-content	Unchanged	Improvement	Improvement
	N	Improvement (+12%)	Improvement (+25%)	Improvement (+15%)
	P	Improvement (+42%)	Improvement (+50%)	Improvement (+21%)
	K	Depletion (–18%)	Depletion (–9%)	Depletion (–8%)
	FDY	+ (R2)	+ (R1)	+ (R3)
	Std. (FDY)	(R1)	(R3)	(R2)
Faizabad	Sust. Index	10.36	11.85	13.56
Ludhiana	Sust. Index	3.0	5.0	4.0
Navasari	Sust. Index	0.65	0.40	3.13

Notes: Bracketed figures with R prefix denoted rank of the treatments. % denote per cent of improvement or depletion of the soil nutrients. Alternatively appropriate binary scales can also used to denote the soil status.

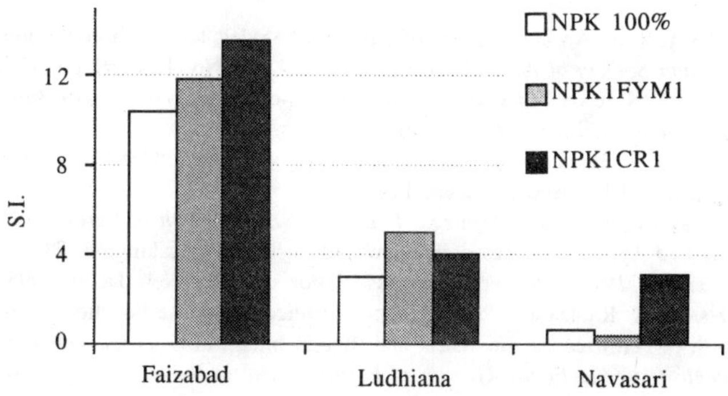

Figure 10.3: Indicator of Sustainability.

The measure indicates that the treatments NPK+FYM and NPK with Crop Residues are more sustainable than the pure chemical fertilizers. In general, the organo-chemical treatments as a group is more yield sustaining and cost effective than the pure chemical fertilizers. The soil enriching property of the treatments and improved productivity property make the mixture treatments more efficient and eco-friendly. Thus, the balance mixture of organic and inorganic fertilizers must be promoted effectively in order to ensure sustainability. In this context, it is important to spread the message of comparative and sustainable benefits of technological options and convince the users to pursue dynamic cropping pattern with greener crop system as well as re-using the residues. Maximizing both social and ecological benefits in the long run through sustainable cropping practice is preferred to short-term private gain.

NOTES

1. The author is grateful to Dr. Dayanatha Jha, Director, NCAP, New Delhi, and Prof. A. Vaidyanathan for their encouragement to me to work on the subject. The study has been supported and funded by the NCAP. I am thankful to Dr. R.L. Yadav, Director, Project Directorate of Cropping System Research (Modipuram) for his comments and several suggestions me in the formulation and conduct of the study.
2. See, on declining trends in productivity of Rice-Wheat System, Chaudhury and Harrington (1993) in Haryana, Kumar and Joshi (1999) in Indo-Gangatic region, Hobbs and Morris (1996) in South Asia, Paroda et al. (1994) in Asia, and Sinha (1998) in Punjab and Haryana.
3. The gaseous oxides of N derived from N fertilization are highly reactive and pose threats to stability of ozone layer. The consumption of N-fertilizer in India was 9.5 million tons (FAI, 1995) with urea as dominant group. Urea, when applied to soil, transforms into the NH_4^+ and NO_3^- form, which are plant usable form. Water plays an important role in NH_4^+ and NO_3^- use and loss by affecting their mobility towards or away from plant roots. It also influences the evolution of gaseous N product N_2O, which lends to readily escape the soil and soil-plant system (Katyal, J.C., Reddy, K.C.K., 'Plant Nutrients Needs. Rain-fed Food Cros', in J.S. Kanwar and J.C. Katyal, 1997, pp. 1–13).

REFERENCES

ABRAHAM, T.P. and V.Y. RAO, 'An Investigation of Functional Models for Fertilizer Response Surface', *Journal of Indian Society of Agriculture Statistics*, Vol. 18, No. 1, 1966, pp. 45–61.

ANDERSON, Jock R., 'A Note on Some Difficulties in Response Analysis', *Australian Journal of Agriculture Economics*, June 1968, pp. 46–53.

CHAUDHURY, M.K. and L.W. HARRINGTON, *Rice-Wheat System in Haryana*, Chaudhury Chandra Sekhar Haryana Agricultural University, Hissar, 1993.

DESAI, G. and A. VAIDYNATHAN (eds.), *Strategic Issues in Future Growth of Fertilizer Use in India*, Indian Council of Agricultural Research (New Delhi), IFPRI (Washington, DC), 1995.

FAI, *Fertilizer Statistics. 1995–1996*, Fertilizer Association of India, New Delhi, 1995.

GANDHI, V., G. DESAI, S.K. RAHEJA and Prem NARAIN, 'Fertilizer Response Function Environment and Future Growth of Fertilizer Use on Wheat and Rice in India', in G. DESAI and A. VAIDYNATHAN (eds.), *Strategic Issues in Future Growth of Fertilizer Use in India*, 1995, pp. 59–100.

HEADY, E.O. and John DILLON, *Agricultural Production Functions*, Iowa State University Press, Ames, 1961, pp. 475–525.

HOBBS, Peter and M. MORRIS, *Meeting South Asia's Future Food Requirement from Rice-Wheat Systems*, NRG Paper, Mexico, 1996.

KANWAR, J.S., KATYAL, J.C., *Plant Nutrient Needs, Supply, Efficiency and Policy Issues*, National Academy of Agricultural Sciences, New Delhi, 1997, 329 p.

KUMAR, P. and G. DEASI, 'Fertilizer Use Pattern in India during the Mid-1980s. Micro Level Evidence on Marginal and Small Farms', in G. DESAI and A. VAIDYANATHAN (eds.), *Strategic Issues in Future Growth of Fertilizer Use in India*, 1995, pp. 101–32.

KUMAR, P., P.K. JOSHI, C. JOHANSEN, and M. ASOKAL, 'Sustainability of Rice-Wheat Based Cropping Systems in India. Socio-Economic and Poilcy Issues', in P. OINGANI (ed.), *Sustaining Rice-Wheat Production Systems: Socio Economics and Policy Issues*, Rice-Wheat Consortium Paper Series 5, Rice-Wheat Consortium, New Delhi, 1999, pp. 61–77.

KUMAR, P. and P.K. JOSHI, *Sustaining Rice-Wheat Based Cropping System in Indo-Gangatic Plains in India*, Rice-Wheat Consortium, New Delhi, 1999, pp. 61–77.

MINISTRY OF AGRICULTURE, *Basic Agricultural Statistics in India*, Bureau of Economics and Statistics, Government of India, New Delhi, 1994.

PANSE, V.G. and P.V. SUKHATME, *Statistical Methods for Agricultural Workers*, Indian Council of Agricultural Research, New Delhi, 1967, pp. 331–57.

PARODA, R.S., T. WOODHEAD and R.B. SINGH, *Sustainability of Rice-Wheat System in Asia*, FAO, Rome, 1994.

ROY, J.K., *ICAR/IRRI Rice Consortium Meeting*, Indian Council of Agricultural Research (New Delhi), IFPRI (Washington, DC), New Delhi, 1998.

SINGH, Yadvinder, C.S. KHIND and Bijay SINGH, *Efficient Management of Leguminous Green Manure in Wetland Rice*, Advances in Agronomy, Vol. 45, 1991, pp. 135–88.

SINGH, Bijay and Yadvinder SINGH, 'Green Manuring and Biological N Fixation. North Indian Perspectives', in J.S. KANWAR and J.C. KATYAL (eds.), *Plant Nutrient Needs, Supply, Efficiency and Policy Issues, 2000–2025*, National Academy of Agricultural Sciences, New Delhi, 1997, pp. 29–44.

SINHA, S.K., *Decline in Crop Productivity in Punjab and Haryana. Myth or Reality*, Indian Council of Agricultural Research, New Delhi, 1998.

WOODHEAD, T., Robert HUKE, Eleanor HUKE, and Laarni BALABAB, *Rice-Wheat Atlas of India*, IRRI (Manila, Philippines) / CYMMYT (Mexico) / ICAR (New Delhi), February 1994, pp. 1–8.

YADAV, R.L., *Analysis of Productivity Trends in Rice-Wheat System*, Experimental Agriculture, Vol. 38, 1998, pp. 1–18.

11

Price Policies and Agricultural Supply

THE RESPONSES OF PRODUCERS IN HARYANA

VÉRONIQUE ALARY, DANIEL DEYBE

INTRODUCTION

Before as also after its independence, India experienced serious food problems: in the 1950s, the daily average cereal consumption per individual was not more than 460 gm. However, in 1966, after three difficult years (deficit of the balance of payment, devaluation of the rupee, moratorium on planning), the Government launched the Green Revolution to achieve food self-sufficiency. High yielding varieties, input subsidies, low banking interest rates, procurement prices and development of the public distribution system led to spectacular results: wheat and rice production more than doubled between 1970 and 1985, and cereal self-sufficiency was achieved at the end of the 1970s.

But the cost of these policies was high and contributed to the financial bankruptcy of the country in 1991: a Structural Adjustment Program (SAP) had to be implemented with the International Monetary Fund (IMF), including reforms and restructuring of the agricultural sector. The initial steps towards the liberalization of this sector stirred-up wide-ranging debates between liberal economists, supporters of a further development of the Green Revolution through direct and indirect subsidies on wheat and rice, and supporters of the diversification of the production system and food intake in India (Alhuwalia, 1995; Gulati, 1990; Bansil, 1997). However, few SAP reforms have actually been implemented in the agricultural sector, as many fear that such reforms would jeopardize the objective of food security on which depend the survival of millions of households. Financial problems are thus still pending, and have even become worse, besides an ever-increasing depletion and over-exploitation of natural resources.

In such a context, there is obviously a need to evaluate the outcome of alternative agricultural policies and to test their degrees of acceptability before implementing them in the field. Simulation models can help in the matter, by assessing the socioeconomic and environmental impacts of alternative scenarios of agricultural policies. They can take into account the agro-social-economic context, the numerous actors and their complex interactions, as also the trade-off between private and public objectives. Among the available sector-based models, the Multilevel Analysis Tool for Agriculture (MATA) (Gérard et al., 1994) can provide useful indicators to decision-makers, like the reactions of producers to liberalization, its impact on the country's self-sufficiency, and the effects of alternative ways to reduce state expenditures without discouraging producers and/or affecting consumers. This paper presents an application of MATA to Haryana.

METHODOLOGY AND CHOICE OF REPRESENTATIVE FARMERS

MATA requires to identify relatively homogenous regions which play an important role in agriculture. Moreover, in order to be able to specify the differentiated effects of a policy on various agents, it is necessary to built a typology of the farming systems in each area according to their endowment and the main characteristics of their production structure.

CHOICE OF THE AREAS

India is a large country, with very various agro-ecological, social and institutional situations. In order to have a more integrated vision of this territory, the Agro-Climatic Regional Planning (ACRP) (ARPU, 1998) developed a classification which combines the natural endowments (soil, climate, topography) as well as the main production systems (crops, cattle-breeding, etc.) and the non-agricultural activities in several regions. This classification is all the more suitable for our analysis as, at the operational or political level, it takes into account the local institutions, and thus the scale of planning in each state. It also incorporates a rather new political issue: the conservation and replenishment of natural resources (land, water, etc.).

Given the importance of food security in Indian policy, the northern part of the country (Trans-Gangetic zone, zone VI) appears to be of particular importance and, within it, the state of Haryana:

— Haryana is one of the most important producers of rice and wheat together with Punjab; for the Government and the food security of the country, it is a key region which has to sustain a high level of agricultural production in the context of liberalization.

— Haryana is sub-divided into 3 large zones: arid, semi-arid and humid. The arid and semi-arid zones are representative of two large systems of production: (1) the wheat/rice system which covers a large part of the alluvial zones in the north of India, and (2) a more diversified system comprising traditional food crops (millet, sorghum) and cash crops (cotton, rapeseed, mustard) characteristic of the arid zones of a large part of India. These two zones also distinguished themselves by their irrigation system (canal, private well and tube-well and, or the combination of canal/tube-well) and their environmental constraints: in the semi-arid zone, excessive uses of fertilizer during the post-Green Revolution period has generated problems of water table pollution, while in the arid zone, the inadequate management of irrigation have led to important problems of soil salinity.

— Finally, Haryana, close to New Delhi, benefits from the regular and reliable data of the numerous surveys carried out by the Indian Agricultural Research Institute (ICAR).

DATA AND FARMERS' CLASSIFICATION

The two agro-climatic zones selected in Haryana (arid zone and semi-arid zone) are sub-divided into districts which are the administrative regions of the state (basic planning unit at the state level). In the semi-arid zone, the district of Karnal has a wheat/rice system of crop rotation which requires higher inputs of fertilizer and capital for irrigation. The density of population is about 355 inhabitants per sq. km., and more than 87 per cent of the land is cultivated, with nearly 70 per cent in foodgrains (cereals and pulses) and 9 per cent in fodder crops. The arid zone is represented by two districts: Hisar and Bhiwani. With a population density of 236 and 180 inhabitants per sq. km. respectively, Hisar and Bhiwani's districts cultivate on an average 67 per cent of their agricultural lands for foodgrains, and 17,3 per cent for fodder crops. Hisar's district was chosen because of the importance of its canal irrigation, wheat and cotton being the dominant crops, and the district of Bhiwani because it is badly affected by the problems of salinity which are tolerated only by wheat, rapeseed and millet/bajra.

Our survey in these three districts covered 123 farmers (50 in Karnal and 73 in Hisar and Bhiwani) spread out in 12 villages. The survey was designed to include all the families who work in the agricultural sector, i.e. owners, tenants and landless farmers. Since it was not possible to select these families through a random sampling using demographic or economic data of the Census of India (1991) and of the Ministry of Agriculture (1994) (these data are compiled at the district level), 10 farmers have been chosen randomly in each village according to their farm size: 2 'large' (above 6 ha.), 2 'medium' (4–6 ha.), 2 'semi-medium' (2–4 ha.), 2 'small' (0–2 ha.) and 2 'landless' (subdivisions provided by the Agricultural Census).

The data collected was then analysed with the Multiple Correspondence Analysis (MCA) in order to built a typology of the farmers. The MCA is a descriptive analysis, designed to analyse multidimensional tables and based on a geometric approach. It explores the structural relations between qualitative variables and identifies the main factors (or axis) of differentiation of the farmers within the surveyed population. Each factor is a combination of modalities which characterize the population. These factors lead to the construction of a synthetic and visual representation of the different types of farmers, and are determined by variables called 'active' variables. The types of individuals can be described with these variables, but also with 'illustrative' variables.[1]

The surveys revealed how important was the access to the factors of production (land, water and mechanical traction) to understand the working modes of the farms. These productive factors have thus been considered as active variables enabling the measuring of the land endowments (cultivated area, mode of access to the land) and the means of production (irrigation, etc.). For animal breeding, the principle variables were the cattle-owned and part of the land allotted to fodder crops, besides, the level of investment in capital required to take into account the investments in intermediary goods (seeds, fertilizers, tools, processed goods including feed for animals). Data relating to crop practices, method of management or even the manner in which the producers understand environmental and institutional constraints have also been collected and used to give a more complete description of the factors of differentiation of the sample studied.

The analysis of the first two axes (Appendix 1), explained by land endowment and by the type of irrigation (13 and 7.2 per cent respectively of the variability), reveals some characteristics of the Indian agricultural economy. On the first axis, a close relationship is noticed between the ownership of a private well/tractor and the total cultivated land area. In fact, this axis opposed two types of farms: (1) small farms of less than two hectares with low investment in agricultural equipment and capital (irrigation, buffalo), and (2) medium and large farms of more than 6 ha. called 'capital intensive' due to their technological level (important set of implements, certified seeds for rice, utilization of zinc in soils). The second axis opposes two crop systems: the wheat/rice system in the semi-arid region, and wheat/cotton in the arid region, largely irrigated by the canal. It therefore opposes two regions possessing two distinct systems of irrigation (private tube-well and canal). In other words, it classifies the farms according to the availability of a public irrigation system, which is itself correlated with the regional agro-climatic conditions.

The projection of the 'illustrative' variables (characteristics of the head of the household, geographical location of the farm, etc.) enables one to observe a very strong relationship between the initial endowments of the farmers (area cultivated, means of irrigation, tractor) and the social characteristic of the household head (caste, educational level). Small and medium farms are held by a majority of low caste people, whose level of education rarely exceeds primary school, whereas large farms are mainly held by high caste people having generally a secondary education. Therefore, the first axis can also be defined as the axis of social differentiation.

The variable 'access to credit' is projected near the center of gravity of the first two axes, suggesting that this variable is not socially discriminatory in Haryana in view of the multiple sources of credit existing in this state. However, there is a regional difference as far as the reasons for asking a credit are concerned: the farmers in the arid zone usually ask credit from the cooperative for financing their agricultural activities, whereas the farmers of the Karnal's region use it for social needs.

The first two factors well represent the population structures in terms of farm endowments and regional characteristic (climate, type of soil, presence of a canal). However, an Ascending Hierarchical Classification (AHC) taking into account some other structural features appearing on other axes helps to get more homogeneous and stable groups of producers than the ones already identified with the first two axes. As in the case of the MCA, the AHC is obtained with 'active' variables (Figure 11.1), and uses 'illustrative' variables to better describe the typology (Appendix 2).

The field survey, just like the MCA and the AHC, confirms the importance of the land inheritance as an explaining factor of the level of intensification (irrigation system, implements, etc.). However, if the level of mechanization and the irrigation system appear to be a function of the land endowment, it does not seem to have a strong impact on the cropping systems. This hypothesis, which suggests that a liberalization of the sector could increase the technological gap between farmers without affecting their cropping patterns, needs to be confirmed and better examined through a model able to predict, in this context, the most probable evolution of food supplies according to different scenarios of agricultural policy.

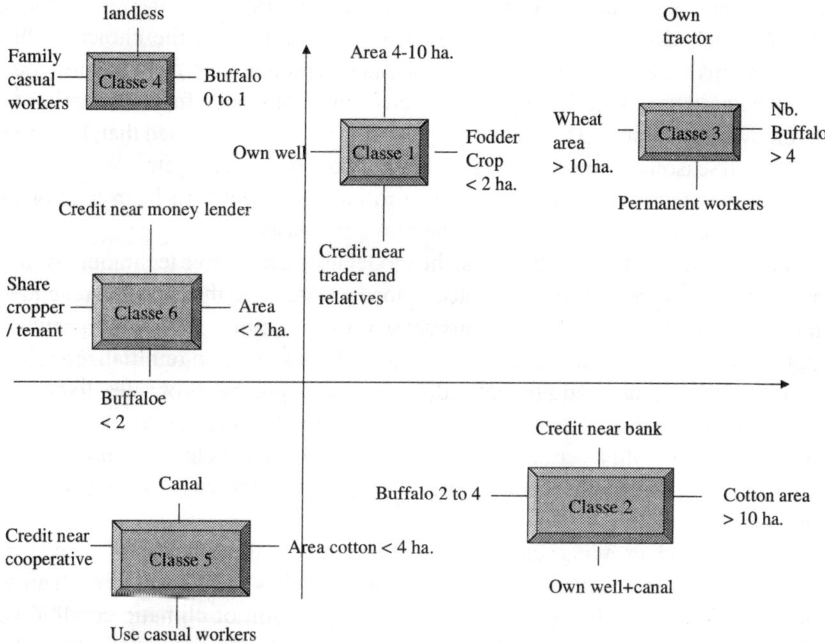

Figure 11.1: Factorial Map of the Multiple Correspondence Analysis.

STRUCTURE OF THE MODEL

The previous analysis helped to identify, in each respective zone (arid and semi-arid), one farm-type in each of the five following classes: landless and marginal ('farm 0' of 0–1 ha.), small ('farm 1' of less than 2 ha.), semi-medium ('farm 2' of 2–4 ha.), medium ('farm 3' of 4–6 ha.) and large ('farm 4' above 6 ha.). These 5 farm-types were used to calibrate the MATA model in each zone, along with the following exogenous parameters/variables whose values are those collected in the field: land and implements (tractor, tools) endowments, livestock characteristics (species, age, sex, etc.) and irrigation (system and potential), as well as initial stocks of dry fodder and cereal, and initial levels of cash flow and savings.

The main objectives of the model are: (1) to reproduce the individual economic decision-making process of the farmers thanks to an optimization procedure subject to individual and regional constraints; (2) to reproduce the exchanges between farmers; and (3) to consider different technologies for each crop in order to test the possibilities of their adoption under different scenarios. Besides these objectives, the principal assumption is that there is no automatic adjustment between agricultural supply and demand; even with policy incentives, the response will not be automatically in equilibrium. The reasoning and the basic hypotheses are as follow.

When farmers have to decide their future productions, they evaluate different alternatives based on yields and price expectations, expectations that might not be fulfilled at the moment of harvesting. There is thus a gap between their expected income and their real income, the difference between the two affecting their investment capacity and their decision for the next season. In other words, the choice of the cropping systems are adjusted according to the (lack of) actual income in the previous seasons (the kharif season from July to October, or the rabi season from November to March) or, eventually, during the rest of the year since it is also assumed that, between these two main seasons, farmers can either let their field rest or cultivate it with fodder (sorghum or cow peas) or sunflower in order to increase the return from land or to compensate for a low wheat outcome in the previous season.

To take into account these time gaps, the model uses a recursive technique. At the beginning of the growing season, the choice is done according to the expected revenues and the cash available from the previous season. After having solved the growing season, the income is estimated and some of the variables are then reinitialized (cash flow, food availability, land, equipment and, in the case of purchases or sales, livestock and fodder stocks) and used as initial conditions for the following season. It is thus possible to describe, with a certain level of precision, the actual changes taking place over time, and to compare production trends in the medium term as a consequence of the simulated policies.

Besides, for each growing season, the model simulates the behavior of each farm type and aggregates their revenues and agricultural supplies at the level of each area. For each farm type, the choice between crops is a function of climatic conditions, irrigation system, monthly water stock possibilities and fodder requirement for animals. The technical coefficients of each crop vary according to the source of traction (tractor, bullock, etc.), the seed variety and the level of fertilization.

A special attention is paid to cattle activities, since they are an important source of income, manure, traction, reserve of cash and consumer goods in Indian farms. The animals are differentiated by categories (specie, purpose, age), and it is considered that the decision to keep the animals on the farm is taken at the beginning of the season, according to the farmers' expectations on fodder availability, the value of the animals at the end of the year, their productivity and cash requirements. For each type of breeding (buffalo, cattle, crossbred), the milk productivity is a function of the animal diet (content of dry matter and composition of the diet in fodder and concentrates). The annual changes in livestock are a function of the natural reproduction rate and of the sales and purchases.

It is also considered that the farms interact among themselves through exchanges of manpower, manure, animal traction or hours of tractor. These exchanges take into consideration not only the 'actual' local availability of resources, but also some social rules. For example, it is supposed that the small or medium farms cannot employ agricultural workers from the large farms, or that few women participate in the agricultural work of the large farms.

As for credit, each type of farm can use different sources to finance its activities or its social requirements. However, the duration of the credit depends on the expected uses: short-term for seasonal inputs or for carrying out activities during a bad season, medium-term credit for livestock investment, and long-term credit for investment (like tube-well or tractor). Each type of credit requires a certain level of guarantees and is limited, at the regional level, by the national governmental credit planning. Besides, if farmers could also require advances from traders or private lenders, informal credit channels have been omitted in the model for three main reasons: (1) there is a large availability of formal credit in Haryana at official interest rates, (2) formalities to get credit from traders are much easier than from banks, but the interest rates are not so different from each other, (3) it is rather complicated to take into consideration the latter aspects in a model.

The model imposes also a minimum level of human consumption, which is calculated according to the recommended minimum intake of calories and protein per capita in India, and which reflects also the preferences of the household that can be overcome either through self production or purchases.

As for farmers' attitude concerning agronomic and economic risk, the Target MOTAD methodology (Tauer, 1983) is applied. It introduces a set of minimum income constraints along with possible deviations from these minimums in order to quantify farmers' attitude towards risk. This representation, based on the minimization of the net incomes gap, has the advantage of considering the risk in the form of a linear constraint (Appendix 3).

Finally, the objective-function links explicitly the short-term liquidity goal with the long-term households' objective of increasing the level of savings and assets. The surplus can be used to increase the productive capacities of the farm (strategy of accumulation), or to improve the short-term return by increasing the use of inputs such as fertilizers or concentrates (strategy of maximization of the short-term cash flow availability).

Optimization is based on expected prices, but farm results are estimated after each growing season with observed prices: 1998 prices are used for the reference year, and these prices are supposed unchanged in the following years since farmers can hardly predict what will be their future evolution. It reflects also their optimistic view: according to them, the Indian Government will not run out on them. However, different price policies will be analysed: maintenance of the present system of support prices besides market prices, or elimination of these support prices (all crops sold at market prices).

ROBUSTNESS OF THE MODEL

After building a model, its robustness must be checked, first by verifying retrospectively that its simulated results tally more or less with the real ones (the greater is the accuracy of the model in this field, greater trust can be laid on its future simulated farmers' responses to alternative scenarios of agricultural policy); otherwise the model has to be improved. In our case, for the two agro-climatic zones, the model reproduces rather accurately farmers' choices, since, for each type of farm, real and simulated values of important parameters (cropping system, use of fertilizer and of mechanical traction, dairy activity) are close to each other (Appendix 4).

Besides accuracy, the stability of the model has also to be checked. Indeed, even with our baseline scenario defined as a repetition of the same external conditions (without major changes in the macro- and micro-economic context), results can vary a lot over time due to ill-defined dynamic transfers: with alternative scenarios (change of context), it will be difficult to distinguish results coming from the new scenario from those coming from the endogenous characteristics of the model itself.

Simulations over a planning horizon of 15 years show a good stability of the

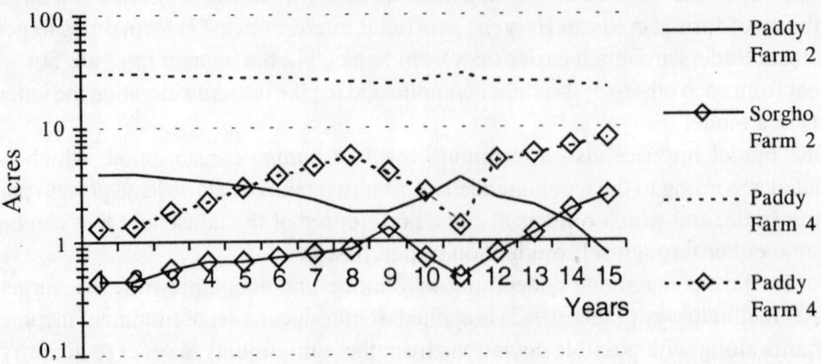

Figure 11.2: Paddy and Sorghum Acreage in Kharif Season – Semi-Arid Area – Small and Large Farms (over 15 years) (logarithmic scale).

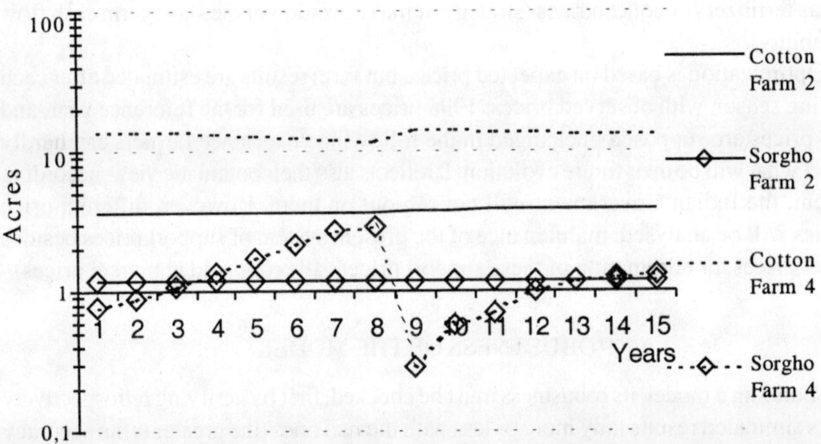

Figure 11.3: Cotton and Sorghum Acreage in Kharif Season – Arid Area – Small and Large Farms (over 15 years) (logarithmic scale).

model (Figures 11.2 and 11.3). However, during the first 8 years, there is a regular increase of the fodder area along with its concomitant increase of livestock, followed by a rapid decrease over the next 3 years, and then a new increase to finally reach the initial situation (1998–9). Obviously, the evolution of the mixed farming system seems to be determined by the natural growth cycle of one of the main farm assets: the livestock. As a matter of fact, our model does not take into account unexpected natural accidents which may explain that the livestock is usually constant or sometimes declining, but it rather well simulates the common farmers' strategy of progressive accumulation whose degree is dictated by the natural reproduction of the livestock as also by the initial level of endowment. In this connection, no change in livestock is observed over the 15 years for the small farms (less than 2 ha.) as well as for the semi-medium farms (Figure 11.3, farm 2 of 2–4 ha.) in the arid area: livestock accumulation seems to be limited to large farms in the arid area whereas there is apparently no such limitation in the semi-arid zone (Figure 11.2). Low opportunity of capitalization in arid area and difficulties in obtaining credit below a certain level of assets can partly explain this restriction for the smaller farms.

The robustness of the model is well-tested, but some deviations from field observations need now to be mentioned. They concern fodder crops, secondary crops like mustard or coarse cereal (*bajra*), and milk production.

In the semi-arid area, the intensity of fodder activities is lower (or higher) than the real ones, which might be explained by the following reasons: (1) in the case of small farms (Appendix 4, Figure 11.13), especially those below 1 ha., the availability of fodder for animals might have been underestimated due to problems faced, during the data collection, to know precisely how the plot and its borders were used; (2) on the contrary, in the case of larger farms (Appendix 4, Figures 11.12 to 11.16), potential losses at harvesting time encourage farmers to grow more fodder, which the model does not take into consideration. However, this difference between real and simulated areas of fodder does not exceed 14 per cent in the case of very small farms (less than 0.12 ha.), and 2 per cent (4.5 per cent in arid area) for farms a little bit bigger (less than 0.3 ha.).

In the arid area, other differences are observed (Appendix 4, Figures 11.17 to 11.20). In the cases of small and large farms, farmers grow more leguminous crops than what is simulated, for preventing or reducing the soil degradation generated by over-irrigation as well as rapid and permanent rotations without fallow. Similarly, in the rainy season (*kharif*), farmers grow more coarse cereals than what is simulated, most probably to cover their potential losses of wheat production during the dry season (*rabi*), while the model, which works season after season, solves the minimum level of food requirements without considering the potential losses of the next season, and so on (semi-medium farms cultivate more wheat to the detriment of mustard, etc.): in fact, different strategies are observed according to the differences of soils and of agro-climatic conditions, which are especially contrasted in this arid area, between Bhiwani and Hisar districts. This observation reinforces the necessity to work with very homogeneous areas, even if, in this study, the differences between real and simulated areas do not exceed 0.7 ha.

Finally, as far as the milk is concerned, green fodder production of small farms is slightly lower (minus 3 to 9 kg. per day) in the model than in real life, which is compensated by higher feed intakes of dry fodder (plus 3 to 9 kg. per day). This difference might be explained as follow: without regular weighing, it is difficult for such farmers to tell what is the exact quantity of feed they give per animal. Moreover, the survey was carried out at the end of the dry season, when the quantity of dry fodder per animal is increased.

However, the simulated milk production for each type of farm is quite in conformity to the reality (Table 11.1). The 'relative deviation'[2] is higher for large farms in the two zones, as well as for the landless but only in the semi-arid area. In the semi-arid area, the landless (farm 0) invest in one milking cow in the model in order to secure a regular revenue and also to increase their assets, which reflects quite accurately the main objective of landless farmers to face their lack of land. For large farms (farm 4), differences are higher in dry and summer seasons, and milk production can be overestimated up to 30 per cent in the semi-arid area. These gaps can be explained by many reasons, among them: (a) in the model, farmers use more concentrates in order to increase their production, but some other inputs may be available on the market, and (b) during the survey, estimate of milk production at the farm level was often provided by the head of the household although only women milk the animals in this part of India.

All in all, simulated results with the baseline scenario (reference situation) provide a relative good picture of the cropping and breeding systems observed during the survey, and well reflect the farm practices in this region of India. The model should therefore be able to offer useful insights into how farmers can react to possible alternative policies.

Table 11.1: Real and Simulated Milk Productions per Type of Farm

(liters per year)

		Arid area			Semi-arid area		
		Kharif	Rabi	Summer	Kharif	Rabi	Summer
Farm 0	Survey	0	0	0	840	1050	630
	Simulation	0	0	0	1265	1581	678
Farm 1	Survey	960	1200	720	1920	2400	1440
	Simulation	875	1094	502	1750	2188	1902
Farm 2	Survey	600	750	450	2280	2850	1710
	Simulation	603	753	434	2353	2942	1439
Farm 3	Survey	1920	2400	1440	2400	3000	1800
	Simulation	1750	2188	1005	2353	2942	1439
Farm 4	Survey	2520	3150	1890	4800	6000	3600
	Simulation	2353	3917	1439	4375	7908	2996

WHICH AGRICULTURAL POLICY FOR WHICH PURPOSE?

SCENARIOS

Several changes might affect the Indian agricultural sector in the near future. As part of the Structural Adjustment Program, the Government of India has already been encouraged to decrease its fertilizer subsidies: the fertilizer price policy has been modified five times since 1991 (Appendix 5) and, nowadays, only nitrogenous fertilizers are still heavily subsidized. But this process led to an unbalanced use of N (Nitrogen), P (Phosphorus) and K (Potassium) nutrients by the farmers, which has harmful effects on soil fertility. According to Desai (1997), a 30 per cent increase in nitrogen prices would be necessary to re-establish a balanced use of these three nutrients. This will be the object of our first scenario, called 'urea' because its price increased by 30 per cent.

Another issue is the farm price of irrigation, which is still just a tax paid on the pump power (without any connection with the real consumption of water and of electricity) for fear of generating social and political unrest in case of alternative policy. This tax is especially low if we take into account that agricultural electricity is heavily subsidized in India, which has now persuaded the Government of India to privatize the State Electricity Board (SEBs) since the situation is no more sustainable. In this context, for minor irrigation systems (tube wells and bore wells), Dhawan suggested to first fix a tariff of Rs. 4 per kWh. consumed, and later to increase it to Rs. 8 per kWh. (Dhawan, 1995, 1997a, 1997b). Per cubic metre of water, it means about Rs. 0.3 and Rs. 0.6 respectively according to the current characteristics of Indian agricultural wells and pumps. This will be our second scenario, with two options respectively called 'water 0.3' and 'water 0.6'.

Besides, in the baseline scenario, farmers sell their produce either to the free market where prices fluctuate according to supply and demand, or in the public market where support prices are fixed seasonally by the Government. In our third scenario, called 'liberal', these support prices are eliminated along with subsidies on urea. Moreover, in this scenario, farmers must also pay their water at Rs. 0.4 per cubic metre.[3] However, as already mentioned, all over our planning horizon, expectations on crops prices are based on 1998–9 market prices, which are in fact slightly higher than support prices and international prices: liberalization might imply a drop in their level, but this scenario is not simulated. Be that it may, our third scenario will undeniably test the sensibility of farmers' reactions to price instability.

All simulations are tested over a planning horizon of 5 years, which gives time to the cropping systems to adjust, as also to the livestock systems since all types of cattle require 3 years to be productive. Annual results will be added up over the five-year period at the regional level, and then compared with those of the reference situation (support prices, subsidy on urea and fixed tax on water pump).

RESULTS

Regional Revenues

The results are strongly linked to the agro-ecological conditions and to the irrigation system in each area. However, all scenarios proved to have a positive impact on regional agricultural revenues, except for the 'water 0.6' scenario in the semi-arid area, and for the 'urea' scenario in the arid area (Figure 11.4).

In the semi-arid area, this increase in revenues cannot be explained by an increase in the total area of paddy and wheat—the main crops—since this area decrease by 1 to 4 per cent according to the scenarios. In the 'liberal' scenario, the mechanical increase of wheat and paddy prices (13 per cent and 8 per cent respectively on average) plays undoubtedly a role, but for the other scenarios it seems more linked with the cattle activities: (a) farmers keep their male calves for draft power but also as capital assets (which increase accordingly) in the 'urea' scenario, and, (b) in the 'water 0.3' scenario, they reduce their purchases of concentrates but grow more fodder to increase their production and sale of milk (Figure 11.9). In this respect, milk production increases in most cases (by 1 to 2 per cent on general average, but up to 8 per cent in case of 'water 0.3'), the exception being for the 'water 0.6' scenario.

In the arid area, farmers grow slightly more cotton and wheat. But these increases do not exceed an average of 2 per cent of the area, since it concerns mainly small farms (38 per cent of the rural households in this zone), which explains the rather low increase of regional revenue for the two 'water' scenarios. In the 'liberal' scenario, the increase is mainly due to the cotton price, which in the free market is 15.7 per cent above the support price.

In short, a change in input or output prices increases the production of cotton and wheat in the arid area (to the detriment of fodder crops), but these changes affect mainly small farms (less than 2 ha.) and reduce their livestock not directly involved in dairy production (bullocks and bulls). In the semi-arid area, responses are more various. For farms 1 and 2 (less than 4 ha.), a change in urea or water prices discourages the production of wheat and rice for the benefit of fodder. Then these farmers increase their animal stock to have a safety net, which generally leads to milk more animals, and therefore to increase the milk production.

Finally, if a change in input or output prices plays an important role, it is not the sole one. Indeed, impacts of alternative policies vary also according to the zone. In the arid area for example, the crop production is more affected by an increase in fertilizer prices, while the semi-arid area is more affected by a change in water policy. In other words, farmers response is not just a mechanical adjustment to an increase or fall in prices.

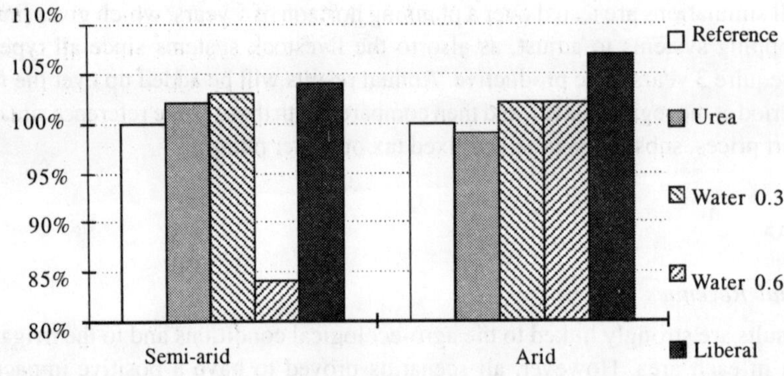

Figure 11.4: Variations in Regional Farming Revenue as per Four Alternative Policies (over 5 cumulative years of simulation).

Regional Supplies of Food Crop

Alternative policies of inputs prices—especially of water pricing—have an effect on wheat and rice supplies (Figure 11.5) since, for example, farmers in the semi-arid zone have to reduce their paddy area in order to limit their irrigation costs. In fact, all sales of irrigated crops (wheat and rice) are negatively affected by such changes, which might have an impact on urban food consumption. But this impact might be more important in rural areas, especially in the semi-arid zone with the 'liberal' scenario, since higher prices of wheat and rice lead farmers to sell their productions instead of

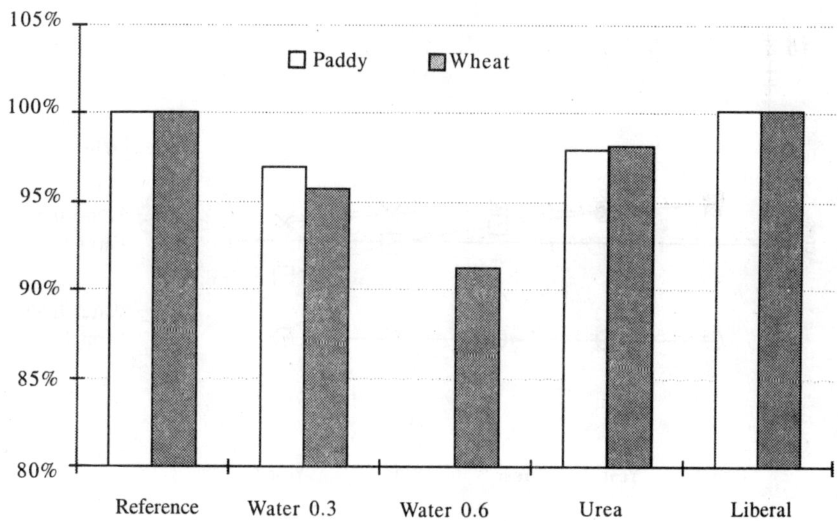

Figure 11.5: Variations in Paddy and Wheat Rural Sales as per Four Alternative Policies – Semi-Arid Area (ending of the 5-year simulation).

consuming them, and, all in all, to not have a balanced consumption of wheat and rice as compared to the baseline scenario (Figure 11.6).

In the semi-arid area, a groundwater tariff of Rs. 0.6 per cubic metre appears to be too high since no more paddy is produced and consumed by farmers whereas rice is of special importance in Indian food habits. In such a context, exploring lower price options for maintaining some paddy production is of special relevance. These options vary a little according to farm size. For small farms (Figure 11.7), as well as for semi-medium farms, paddy acreage decreases even at Rs. 0.3 per cubic metre, but up to

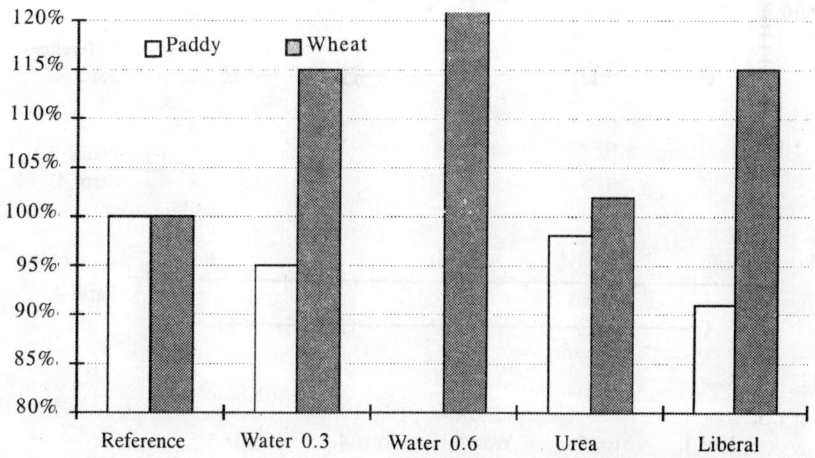

Figure 11.6: Variations in Paddy and Wheat Rural Consumptions as per Four Alternative Policies – Semi-Arid Area (ending of the 5-year simulation).

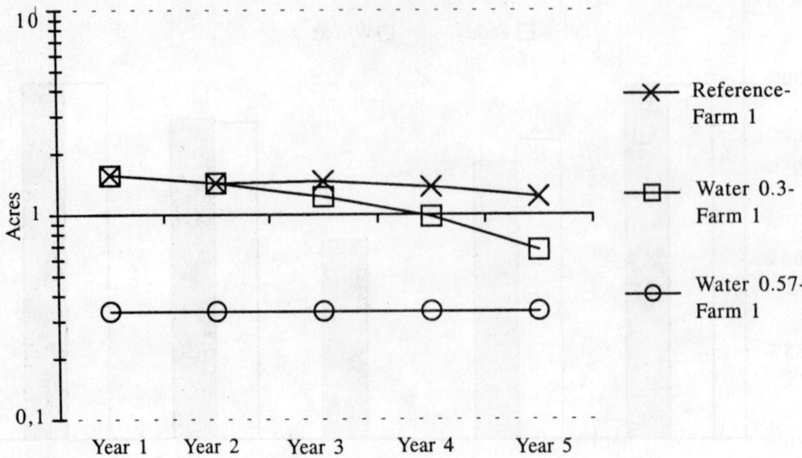

Figure 11.7: Paddy Acreage as per Two Alternative Policies of Water Pricing – Semi-Arid Area – Small Farms (over 5 years of simulation) (logarithmic scale).

Rs. 0.56 per cubic metre, farmers still continue to grow some, while above this level (Rs. 0.57 per cubic metre), they will definitely prefer to cultivate cotton which requires less water. For large farms (Figure 11.8), as well as for medium ones, some paddy production is maintained up to a tariff of Rs. 0.58 per cubic metre. On the other hand, in the arid area where canal irrigation under public control prevails, few changes are observed due to the lack of alternatives.

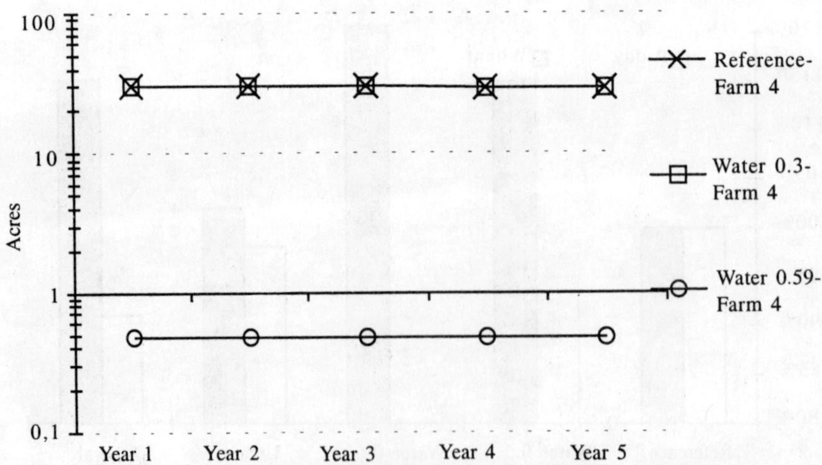

Figure 11.8: Paddy Acreage as per Two Alternative Policies of Water Pricing – Semi-Arid Area – Large Farms (over 5 years of simulation) (logarithmic scale).

Milk Supplies

Apart from 'water 0.6' scenario, no impact on milk production is observed since there is, in animal diets, a substitution of dry fodder for green fodder in the semi-arid area, and a bigger cotton cake availability in the arid area. However, for 'water 0.6' scenario, the importance of green fodder i n animal diets, especially of *berseem* which requires more water than sorghum, explains why an important change in groundwater tariff has an impact on milk production.

Besides, in the semi-arid area, farmers sell more milk to secure their revenue in the context of the 'liberal' scenario, which is more risky for the sale of their crop productions. Similar (but lower) changes in milk sales are observed for other scenarios (Figure 11.9), except for the 'urea' one which affects the cropping system less.

Cash Flow

Alternative scenarios of agricultural policies have various impacts on farmers cash flow, which is in all cases strongly affected by an increase in water prices. In the arid area, semi-medium and medium farmers witness a reduction of their cash flow (by 20 per cent on average for the five-year period of simulation), which affects each season their capacity to invest. Besides, along with smaller farmers (i.e. all farm types except the large one), they face more variations in their cash flow from one year to the next (Figure 11.10). The risk is thus increasing. In the semi-arid area, even if support prices are maintained, all farms have their cash flow strongly affected by insecure or higher input prices, especially semi-medium ones (Figure 11.11). When support prices are eliminated ('liberal' scenario), the situation seems to improve, but its is only for the first three years.

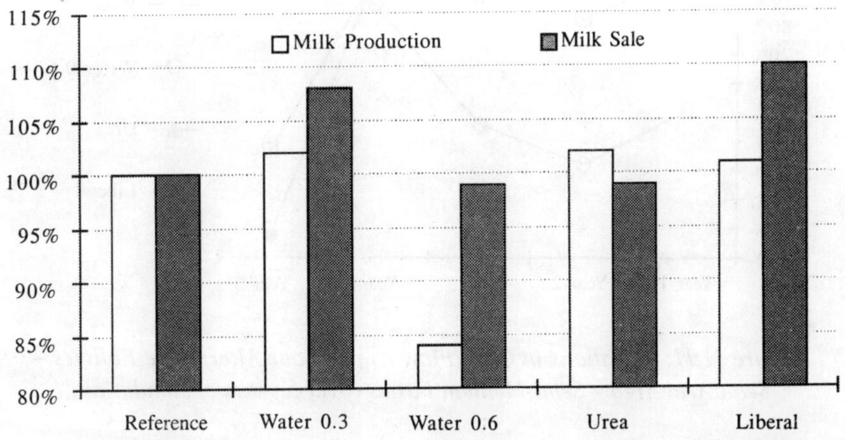

Figure 11.9: Variations in Production and Sale of Milk as per Four Alternative Policies – Semi-Arid Area (ending of the 5-year simulation).

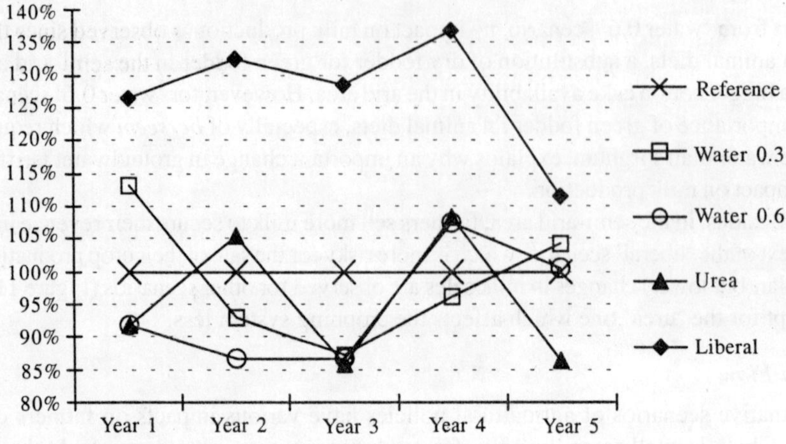

Figure 11.10: Variations in Cash Flow as per Four Alternative Policies – Arid Area – Semi-Medium Farms (over 5 years of simulation).

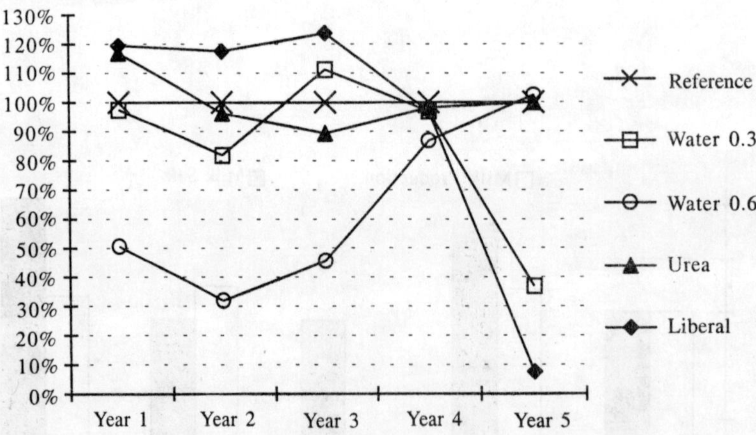

Figure 11.11: Variations in Cash Flow as per Four Alternative Policies – Semi-Arid Area – Semi-Medium Farms (over 5 years of simulation).

CONCLUSION

Our MATA model in Haryana tends to show that the most favoured regions and farms might be able to capture the positive impacts of a liberalization of the agricultural sector in India. But it also shows that such alternative policy might turn food diets in rural areas upside down (much less rice), as also deepen the gap between farms and regions which are able to take advantage of the situation, and those which are not for agronomic or other reasons. This calls for very special attention, along with the fact that farmers, when confronted with higher market risks, tend to develop strategies to limit them, which limits at the same time their positive responses to the three scenarios of alternative policies which have been tested here.

All in all, the results show that modeling the agricultural sector at a province level can provide useful insights on the impact of agricultural and economic policies, even if the model has to be improved to better represent farmers' decision making processes and their interactions.

APPENDIX 1

CHARACTERISTICS OF THE FARMS SURVEYED

Table 11.2: Socioeconomic Characteristics of the Farms Surveyed in Karnal, Hisar and Bhiwani Districts (1998–9)

District	sample	castes	mode of farming	farm size
Karnal	50 farms	Low: 22%	Heir: 72%	0 to 1 ha.: 24.0%
	(40.6%)	High: 70%	Buyer: 10%	< 2 ha.: 8.0
		Unknown: 8%	Tenant farmer: 4%	< 4 ha.: 18.0%
			Tenant: 6%	< 6 ha.: 28.0%
			Landless: 8%	> 6 ha.: 22.0%
Hisar	37 farms	Low: 35.1%	Heir: 81.1%	0 to 1 ha.: 18.9%
	(30.1%)	High: 56.7%	Buyer: 0%	< 2 ha.: 8.1%
		Unknown: 9.2%	Tenant farmer: 5.4%	< 4 ha.: 29.8%
			Tenant: 13.5%	< 6 ha.: 13.5%
				> 6 ha.: 29.75
Bhiwani	36 farms	Low: 19.45	Heir: 72.2%	0 to 1 ha.: 30.6%
	(29.3%)	High: 50.0%	Buyer: 8.3%	< 2 ha.: 13.9%
		Unknown: 30.6%	Tenant farmer: 8.4%	< 4 ha.: 16.6%
			Tenant: 11.1%	< 6 ha.: 16.7%
				> 6 ha.: 22.2%

Table 11.3: Main Features of Each Type of Farm in the Arid and Semi-Arid Areas Surveyed in Haryana (1998–9)

Zones	Classes	Population	Farm size	Main rotation	Fodder area per Season	Wells	Tractors
Semi-arid							
(Karnal)		50 households					
	0. Landless	8.0%	0		0 ha.	No	No
	1. Small	22.0%	0–2 ha	Wheat/Rice	0 ha.	Community	Hired
	2. Semi-medium	16.0%	2–4 ha	Wheat/Rice/Fodder	< 1 ha.	Individual	Hired
	3. Medium	34.0%	4–6 ha	Wheat/Rice/Fodder	1–2 ha.	Individual	Hired
	4. Large	20.0%	> 6	Wheat/Rice/Fodder	> 2 ha.	Individual	Own
Arid							
(Hisar & Bhiwani)		70 households					
	0. Landless & marginal	19.2%	0–1	Wheat/Cotton		Canal	No
	1. Small	13.7%	1–2	Wheat/Cotton		No	No
	2. Semi-medium	26.7%	2–4	Wheat/Cotton		Canal	Hired
	3. Medium	24.7%	4–6	Wheat/Cotton		Canal	Own
	4. Large	17.8	> 6	Wheat/Cotton/Mustard	< 1 ha.	Well 1	Own

APPENDIX 2

FACTORIAL ANALYSIS

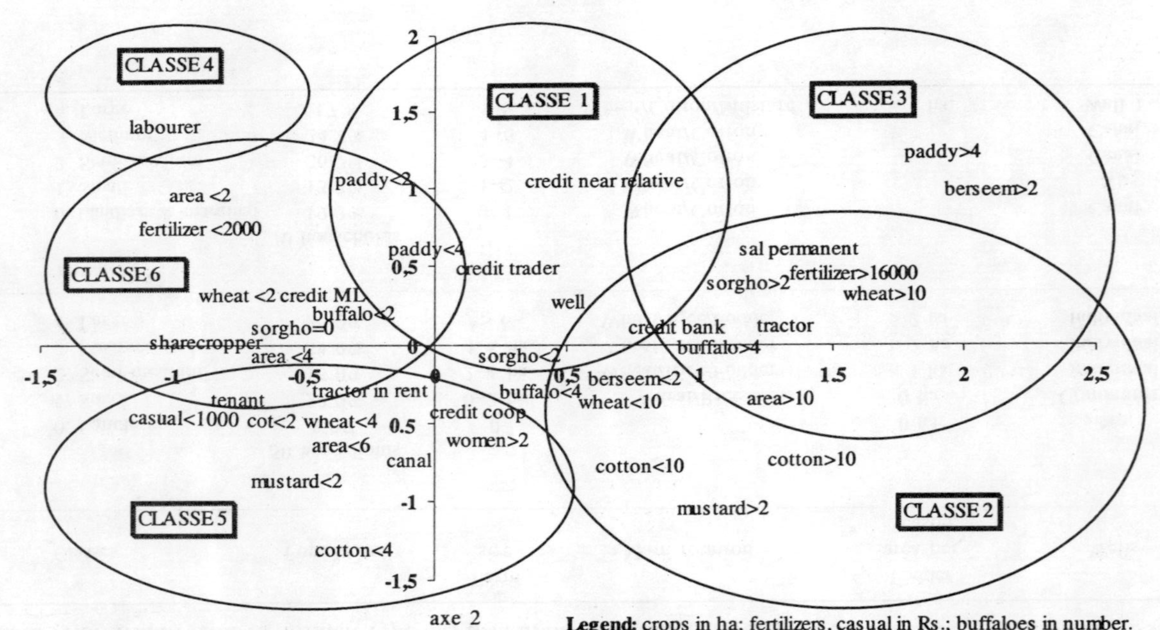

Legend: crops in ha; fertilizers, casual in Rs.; buffaloes in number.

Figure 11.12: Map of the Factorial Analysis – Axes 1 and 2.

APPENDIX 3

STRUCTURE OF THE MODEL

For each farmer, the level of each activity j ($x_{j,i}$) is limited by the amount of resources b_i as follow:

$$\sum_j a_{j,i}.x_{j,i} \leq b_i \qquad (1)$$

where ($a_{j,i}$) are the technical coefficients of each activity j of the farm i.

Resources (*bb*) that can be hired in or out to other farmers, like labour, animal traction, equipment, etc., are represented as follow for each type of farm i:

$$\sum_j a_{j,i}.x_{j,i} \pm \sum_{j,k} aa_{j,i,k}.xx_{j,i,k} \leq bb_{j,i} \qquad (2)$$

where k are the other types of farms, (*aa*) the technical coefficient of the activity, and (*xx*) the level of use of the resource of farm k by farm i or of farm i by farm k.

Then, with n the number of farms within each farm group, the equilibrium between supply and demand of resources is reached through:

$$\sum_{j,i,k} aa_{j,i,k}.xx_{j,i,k}.n_i = \sum_{j,k,i} aa_{j,k,i}.xx_{j,k,i}.n_k \qquad (3)$$

With $Credtot_{cr}$ the total availability of each type of credit (*cr*) in the region, the optimal allocation ($Cred_{j,i,cr}$) is obtained through:

$$\sum_{j,i} x_{j,i}.Cred_{j,i,cr}.n_i \leq Credtot_{cr} \qquad (4)$$

The model takes into account a minimum level of nutrition (*CalMin*), that can be satisfied through self production or purchases (*purch*) of food products containing a certain number of calories and proteins (cal_{nu}):

$$cal_{nu,j}.x_{j,i}.y_{j,i,ye} + purch_{j,i}.cal_{nu,j} \geq CalMin_{nu,i,ye} \qquad (5)$$

Besides, constraints of minimum income (*MinInc*) are introduced along with possible deviation (*Dev*) from these minimum to quantify farmers attitude (*Devtot*) towards risk:

$$\sum_j y_{j,i,ye}.x_{j,i}.p_{j,ye} + Dev_{i,ye} \geq MinInc_{i,ye}$$

$$\text{with } \sum_{ye} Dev_{i,ye} \leq Devtot_i \qquad (6)$$

The objective function aggregates the maximization of the net income (7) (revenue less cost *co*) at the community level, plus the cost-effectiveness of investment in livestock (*livestock*):

$$MaxZ = \sum_i \left\{ \left[\sum [p_j.(y_{j,i}.x_{j,i} - cons_{j,i}) - co_{j,i} \right] + \left[\sum livestock_{l,i} \right] \right\} \qquad (7)$$

APPENDIX 4

COMPARISON BETWEEN REAL AND SIMULATED RESULTS

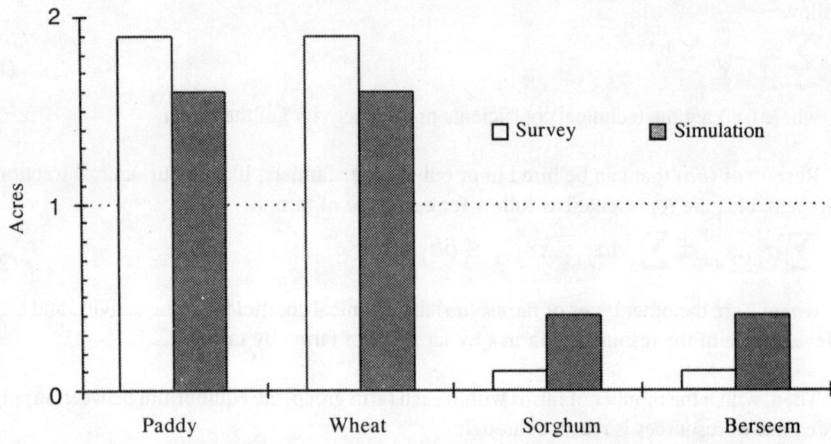

Figure 11.13: Real and Simulated Acreage of Paddy, Wheat, Sorghum and Berseem – Semi-Arid Area – Small Farms.

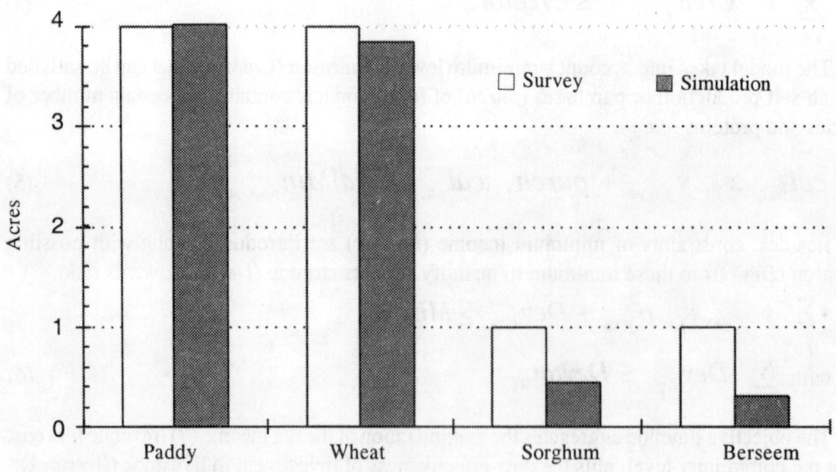

Figure 11.14: Real and Simulated Acreage of Paddy, Wheat, Sorghum and Berseem – Semi-Arid Area – Semi-Medium Farms.

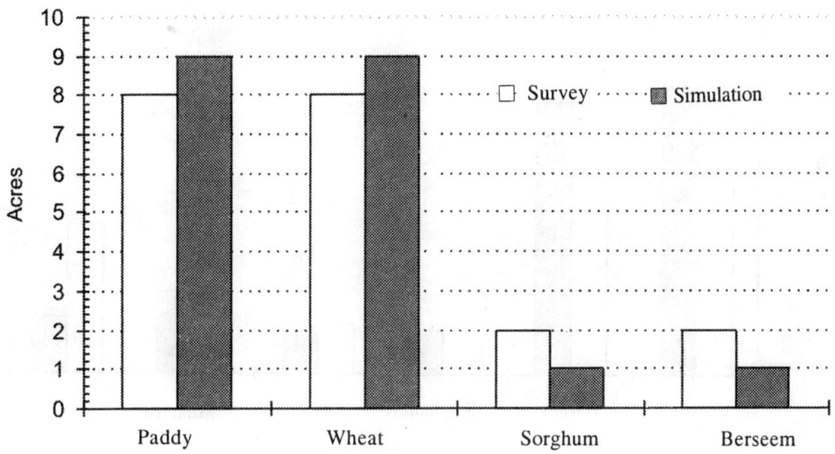

Figure 11.15: Real and Simulated Acreage of Paddy, Wheat, Sorghum and Berseem – Semi-Arid Area – Medium Farms.

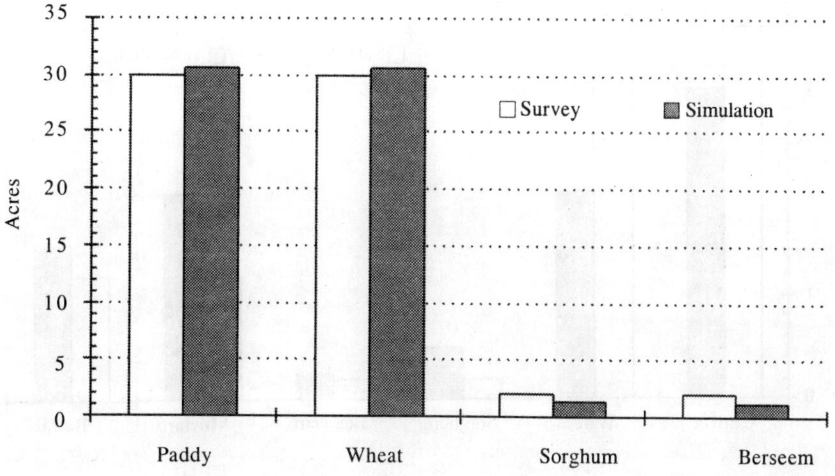

Figure 11.16: Real and Simulated Acreage of Paddy, Wheat, Sorghum and Berseem – Semi-Arid Area – Large Farms.

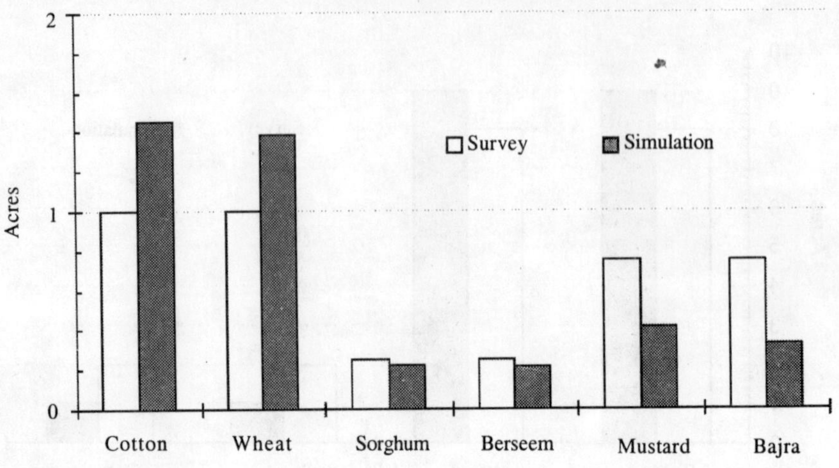

Figure 11.17: Real and Simulated Acreage of Cotton, Wheat, Sorghum, Berseem, Mustard and Bajra – Arid Area – Small Farms.

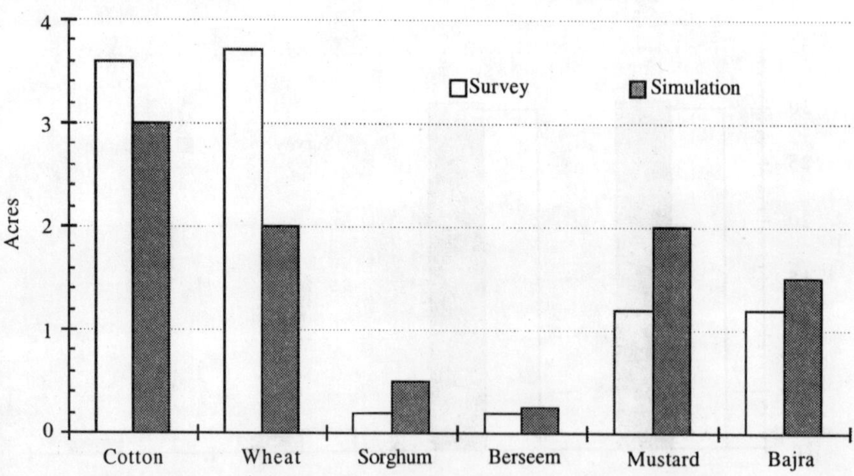

Figure 11.18: Real and Simulated Acreage of Cotton, Wheat, Sorghum, Berseem, Mustard and Bajra – Arid Area – Semi-Medium Farms

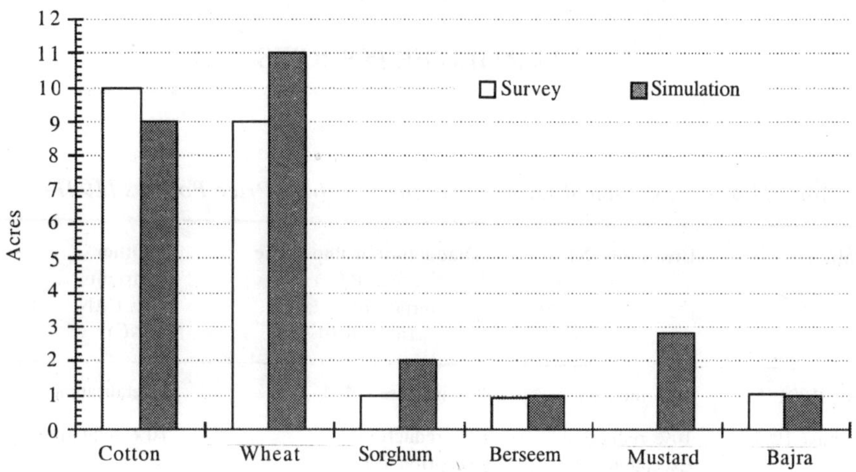

Figure 11.19: Real and Simulated Acreage of Cotton, Wheat, Sorghum, Berseem, Mustard and Bajra – Arid Area – Medium Farms.

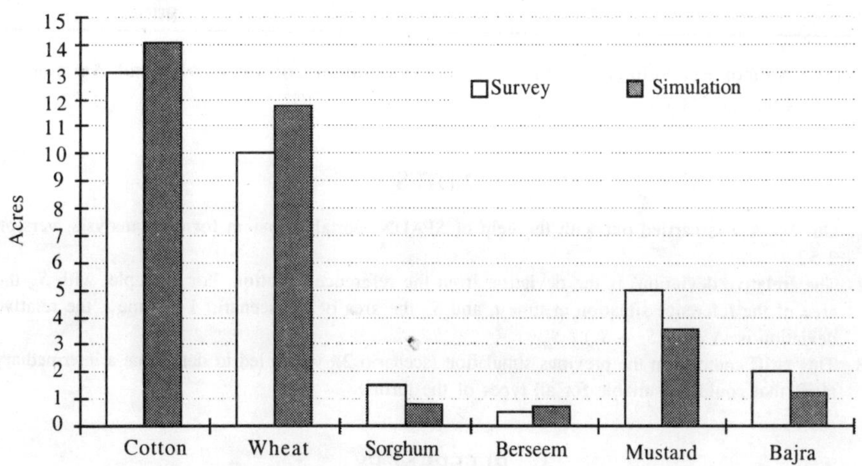

Figure 11.20: Real and Simulated Acreage of Cotton, Wheat, Sorghum, Berseem, Mustard and Bajra – Arid Area – Large Farms.

APPENDIX 5

FERTILIZER POLICIES

Table 11.4: Chronological Overview of the Fertilizer Price Policies (1991–4)

Policies	Urea (46% N)	Diammonium Phosphate ($46\%\ P_2O_5$) Muriate of Potash ($60\%\ K_20$)	Other nutrients (SA, CAN, AC)*
July 1991	40% increase	40% increase	Liberalization
August 1991	10% reduction Exception for small farmers (< 1 ha.)	10% reduction Exception for small farmers (< 1 ha.)	10% reduction
August 1992	10% reduction	Liberalization Import taxes on K_20 Subsidy on transport Less import taxes on raw materials	Controlled prices
September 1992		Fixed subsidy per ton	
June 1994	20% increase		Decontrolled prices

Notes: * Ammonium Sulphate (SA), Calcium Ammonium Nitrate (CAN) and Ammonium Chloride (AC)

NOTES

1. Our MCA was carried out with the help of SPADN, portable system for data analysis, version 2.52.
2. The 'relative deviation' is the deviation from the reference situation. For example, with S_{0t} the area of the reference situation at time *t*, and S_{1t} the area of the scenario 1 at time *t*, the relative deviation is: $\Delta S_t = (S_{0t} - S_{1t}) / S_{0t}$.
3. This tariff come from the previous simulation (scenario 2), which led to determine a intermediary tariff that could be suitable for all types of the farms.

REFERENCES

AHLUWALIA, M.S., 'India's Economic Reforms', in R. CASSEN and Joshi VIJAY (ed.), *India: The Future of Economic Reform*, Oxford University Press, 1995.

ARPU, *Agro-Climatic Regional Planning. Recent Developments*, Working Paper, No. 10, August 1998, pp. 50–9.

BANSIL, P.C., *Concurrent Evaluation of IRDP (5th Round) in Haryana*, Techno Economic Research Institute, New Delhi, 1997.

CANDLER, W., J. FORTUNATY-AMAT and B. MCCARL, 'The Potential Role of Multilevel Programming in Agricultural Economics', *American Journal of Agricultural Economics*, No. 63, 1981, pp. 521–31.

CENSUS OF INDIA, *General Population Tables and Primary Census Abstract*, Serie 8, Haryana, Part II-A & Part II-B, Director of Census Operations, Haryana, 1991, pp. 1–140.

DESAI, G.M., 'Fertilizer policy reforms and agricultural growth', in M. Desai BHUPAT (ed.), *Agricultural Development Paradigm for the Ninth Plan Under next Economic Environment,* Oxford & IBH Publishing Co., 1997, pp. 456–75.

DHAWAN, B.D., *Groundwater Depletion, Land Degradation and Irrigated Agriculture in India*, Institute of Economic Growth, Studies in Economic Development and Planning, No. 60, Commonwealth Publishers, New Delhi, 1995.

———, 'Latent threats to irrigated agriculture', in M. Desai BHUPAT (ed.), *Agricultural Development Paradigm for the Ninth Plan Under next Economic Environment*, Oxford & IBH Publishing Co., 1997a, pp. 492–507.

———, *Studies in Minor Irrigation With Special Reference to Groundwater*, Institute of Economic Growth, Studies in Economic Development and Planning, No. 56, Commonwealth Publishers, New Delhi, 1997b.

FAI, *Fertilizer Statistics 1997–1998*, Fertiliser Association of India, New Delhi, 1998.

GÉRARD, F., J.-M. BOUSSARD and D. DEYBE, *MATA Prototype of Multilevel Analysis Tool. Policies and Markets Program*, CIRAD-URPA, Working Paper, No. 23, 1994.

GULATI, A., 'Fertilizer Subsidies. Is the Cultivator Net Subsidised?', *Indian Journal of Agricultural Economics*, Vol. 45, No. 1, 1990, pp. 106–21.

MINISTRY OF AGRICULTURE, *Indian Agricultural Statistics. 1989–90 to 1990–91*, Vols. 1 & 2, Dept. of Agriculture and Co-operation, Government of India, New Delhi, 1994.

TAUER, L.W., 'Target MOTAD', *American Journal of Agricultural Economics*, No. 65, 1983, pp. 606–10.

12

Exploring Land Use Options for Sustainable Food Production

METHODOLOGICAL FRAMEWORK

P.K. AGGARWAL, N. KALRA, S. KUMAR, A.K. VASISHT[1]

INTRODUCTION

There has been large increase in food production in India during last three decades that made the country self-sufficient and contributed tremendously to its food security. This, however, is now at risk due to the continuously increasing population. By 2050, India's population is expected to grow to 1.6 billion people and it will outgrow China as the most populous country in the world (United Nations, 1997). This rapid and continuing increase in population implies a greater demand for food. Although the world as a whole may still have sufficient food for everyone, it would need to be produced where needed due to socioeconomic and political compulsions (Rabbinge, 1999). The food will have to be produced from the same or even shrinking land resource because in India there is no additional land available for cultivation.

Although there is now large pressure to increase production, there has lately been a significant slow down of the growth rate in area, production as well as yield. The rate of annual growth of food production and yield showed a peak during the early years of the Green Revolution but since then there has been a decline in its growth rate. Increase in area under cultivation does not look feasible. However, due to increasing population, food production per person has become constant during the 1980s. Continuously increasing population pressure, urbanization and income growth could further lower the availability unless corrective measures are taken (Pinstrup-Anderson and Pandya-Lorch, 1995). Further production increase has to come, therefore, from an increase in productivity per unit area.

At the same time, grain yields in experimental farms are stagnating and there are increasing reports of deteriorating soil health of intensive cropping regions. The yield potential of rice in tropics has not increased above 10 tons per ha. since the IR8 variety was released 30 years ago, despite significant achievements in attaining yield stability, increasing per day productivity and improving grain quality (Aggarwal et al., 1996). The yield of major cereals in north-western India has also not shown any significant increase during the last few decades (Sinha, 1999). The gradual increase of environmental degradation is compounding the problem, early signs of which are becoming visible in areas that benefited largely from the Green Revolution technologies.

There is now a great concern about decline in soil fertility, change in the water table, rising salinity, resistance to many pesticides and degradation of irrigation water quality, for example in north-western India (Sinha et al., 1998).

Analysis of options for regional development generally focuses on defining the best strategy. However, when there are potentially conflicting objectives such as maximizing production, sustaining environment and maximizing farmers' income, it is difficult to define *the best* solution. Information, therefore, needs to be generated to determine the consequences and trade-offs of different sets of policy aims on agriculture. The economically viable optimal solutions should be based on the consideration of the biophysical potential of the resources available and the socioeconomic constraints. Thus, a systems approach is needed where it is possible to translate policy goals into objective functions integrated into a biophysical land evaluation model.

We have recently started studies with a goal to develop a decision support framework to explore the future land use options based on our present technical knowledge, and anticipated future objectives and constraints. The major objectives are to determine the magnitude of production possibilities, associated environmental risks and the inputs required to attain the targeted production levels. The study is being illustrated for the state of Haryana in northern India which has contributed tremendously to the success of the Green Revolution in India. Rice and wheat, commonly grown in double cropping rotation, are the major food crops of the region and their current total production is 10 million tons. Regional stakeholders of Haryana are interested to find optimal agricultural land use plans that can meet this goal as well as maximize employment and income from agriculture, while minimizing pesticide residues, nutrient losses and groundwater withdrawal.

METHODOLOGY

Implications of various conflicting scenarios relating to multiple goals of maximizing food production and income and minimizing environmental degradation are evaluated by symphonic use of simulation models, GIS (Geographical Information Systems) and optimization techniques. The overall approach of our study is diagrammatically shown in Figure 12.1. Resource characterization is done using GIS and spatial databases. Crop and livestock modeling are used for estimation of primary productivity and milk. Multiple goal linear programming is used for optimization.

RESOURCE CHARACTERIZATION

Agro-climatic units were developed based on considerations of soil and weather properties. The basic soil map of Haryana was reclassified into 17 homogenous soil units based on particle size, organic carbon, sodicity and salinity, considered as critical characteristics for this study. Annual rainfall in the state varies from 300 mm. in western to 1,200 mm. in the north-eastern regions. A rainfall map was prepared based on data of 58 stations in and around Haryana. Overlaying this over the reclassified soil map yielded 59 homogeneous units. Rain-fed area, which occurs mainly in western parts of Haryana, was mapped indirectly through satellite scan (IRS 1C and 1D images) and other conventional resource inventory. Addition of this layer consisting of irrigated and rain-fed areas resulted in 87 agro-ecological units.

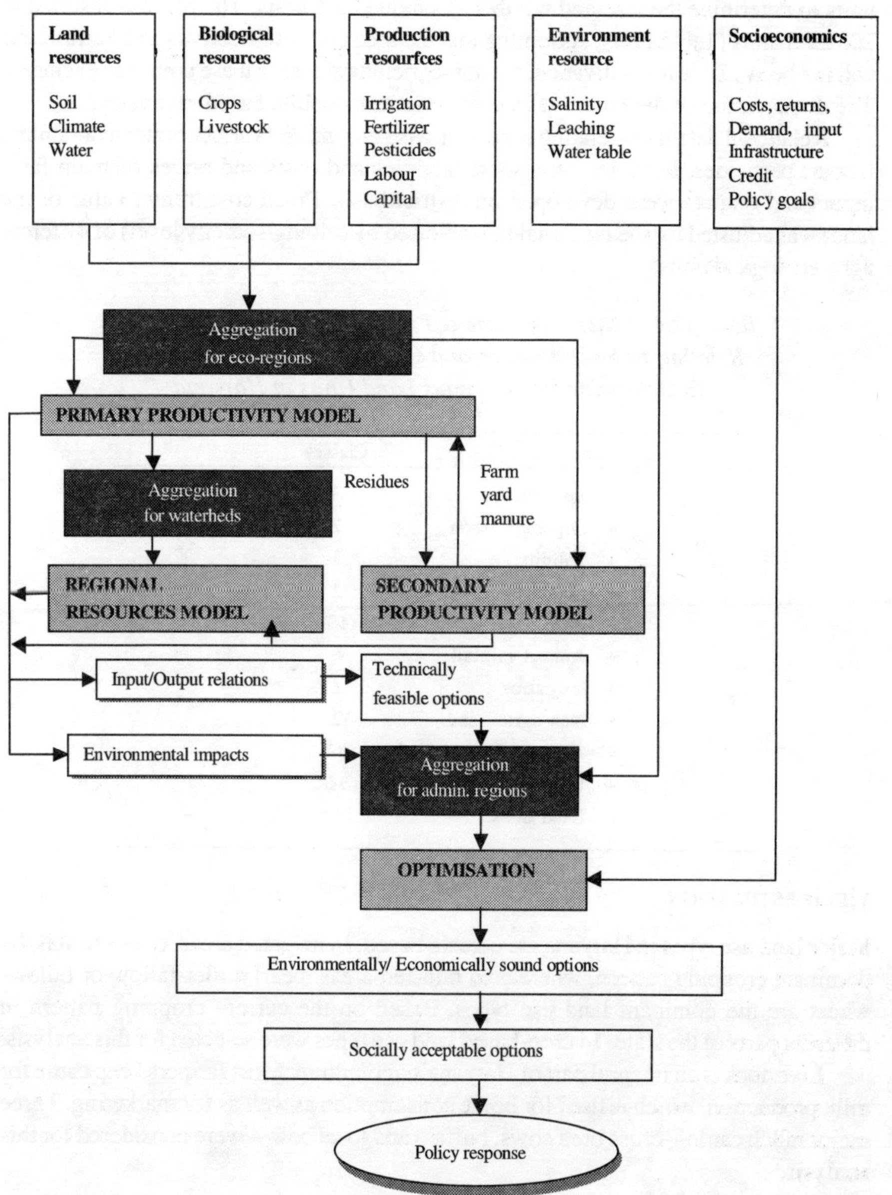

Notes: Dark rectangles indicate the use of GIS, light-shaded rectangles indicate use of models.

Figure 12.1: A Framework for Integrated Land Evaluation.

The boundaries of the 16 districts in the state were overlaid on these agro-ecological units to determine the size and number of unique land units. The process resulted in 208 land units (Table 12.1). Assuming that the area under settlements and barren land will not be available for cultivation, the non-agricultural area in these units was excluded. The demarcation of the latter was based on recent satellite based inventory.

Relational databases relating to season-wise ground and surface water availability, labour, pesticides, fertilizers, transport facilities and costs and prices of main farm inputs and outputs were developed on district basis. Fixed cost (rental value of the land) was adjusted for the land quality (indicated by salinity/sodicity level) of different agro-ecological units.

Table 12.1: Classes of Various Production Systems Attributes Relating to Soil, Weather and Geographical Distributions to Determine Homogenous Land Units in Haryana

Zone	Classes
Soil texture	3
+ Organic carbon	2
+ Salinity	3
+ Sodicity	2
= Soil	17
+ Annual rainfall	6
+ Irrigation	2
+ Non-agric. land	2
= Agro-ecological	87
+ Districts	16
= Land units	208

YIELD ESTIMATION

Major land use types in Haryana are cereals-based. In irrigated areas, rice-wheat is the dominant cropping pattern, whereas in rain-fed areas, pearl millet-fallow or fallow-wheat are the dominant land use types. Based on the current cropping pattern in different parts of the state, 14 crop-based land use types were selected for this analysis.

Livestock is an integral part of Haryana's agriculture. Most farmers keep cattle for milk production, which is used for home consumption as well as for marketing. Three major milch cattle—cross bred cows, buffalo and local cow—were considered for this analysis.

A major goal of our study is to explore options for increasing production. Considering this, we have used five technology levels: current yield level, potential yield and three levels between the two for irrigated areas. The techniques for these levels were assumed to become more and more site-specific and capital-intensive. In the absence of good data, only the current level of technology used is considered for rain-fed areas and for livestock.

Yield levels were determined for different technology levels of each land use types for each of the 208 land unit indicated earlier using crop simulation models, reduction factors for salinity/sodicity and other production losses (Aggarwal et al., 2000). The current yields of different crops were obtained from the state average (Government of Haryana, 1998a). Since it was not possible to get current yield data at land unit level, a weighting criteria dependent upon the area, approximate area of different land use types and level of salinity/sodicity was developed to allocate state average yield to different land units. The difference between the adjusted potential yield and the calculated current yield level in each land unit was considered the attainable yield gap for that unit. Target yield levels were set at bridging 25, 50 and 75 per cent yield gaps at three different technology levels. Since rain-fed areas are resource poor as well, only the current level of technology was considered feasible for such areas.

The average yield per head was considered to be 4.0, 5.5 and 6.4 litres of milk per day for cows, buffalo and hybrid cows, respectively.

INPUT-OUTPUT TABLES

Required inputs and other ancillary resultant outputs of various land use types were calculated for the specified yield levels as described earlier by Aggarwal et al. (2000) and summarized in Figure 12.2. Various approaches, including simulation models, technical coefficient generator, surveys and thumb rules, were used for the estimation. The basic data for these were collected from a literature survey.

A recent survey conducted by Haryana Agricultural University, Hissar provided information related to human labour and machine labour used at current technology level (Government of Haryana, 1998b). It was assumed that labour requirement would increase in proportion to increase in yield up to technology level 3. Technology levels 4 and 5 were assumed to be knowledge based, precise and mechanized. Costs of fertilizers and FYM (Farm Yard Manure), human and animal labour, hiring of tractor, and procurement price of produce as well as residues were derived from the Government statistics (Government of Haryana, 1998a). Fixed cost was considered to be formed by the rental value of land. Total cost of production included costs of seeds, human labour, machine labour, irrigation, fertilizers, farm-yard manure, zinc sulfate, biocides, fixed costs and miscellaneous costs (10 per cent of operational costs). Gross income is the value of main produce (grain/cane) and residue. Net income of the farmers was calculated as the difference between gross return and total costs.

OPTIMIZATION MODEL

Various policy views concerning future land use in the state were distilled from policy documents and in consultation with regional policy makers. These views have been operationalized by means of objective functions. The important future agricultural development goals for Haryana were considered to be maximization of production of cereals (rice+wheat), pulse (gram), fibre (cotton), oilseed (mustard) and milk production accompanied by maximization of employment in and income from agriculture, reclamation of salinity, and minimization of pesticide residues, nitrogen leaching and groundwater use. From these, the following four important goals were selected for this preliminary study, based on their relative importance and considering the limited availability of data:

— Maximize food production
— Maximize income
— Minimize groundwater pumping (maximize water use efficiency)
— Minimize nitrogen losses

Several subsets of scenarios were identified for the important goal of maximizing food production as indicated in Table 12.2. Since doubling of food production by 2010 is an important policy goal for Haryana, in one scenario this has been considered as the major goal.

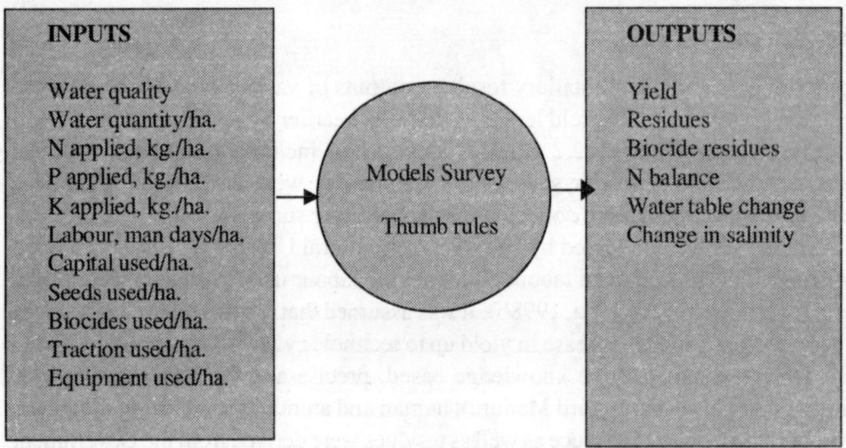

Figure 12.2a: Input-Output Relationships for Various Crop Production Systems of Each Land Evaluation Unit

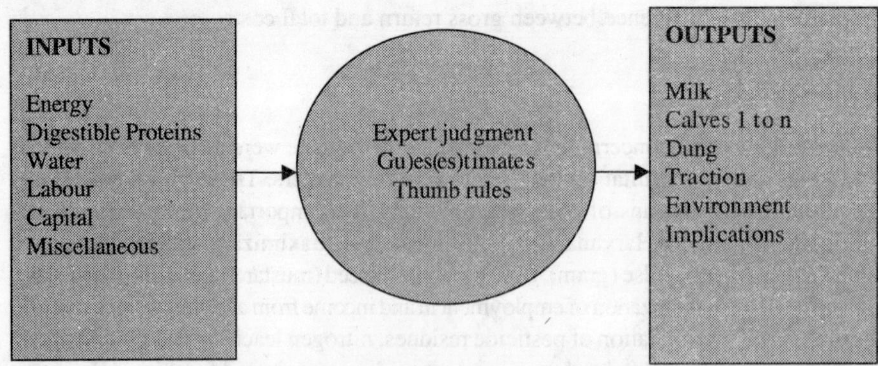

Figure 12.2b: Input-Output Relationships for Various Livestock Production Systems of Each Land Evaluation Unit.

OPTIMIZATION

The constraints set for the optimization exercise with current constraints and technologies included the availability of land, water, capital and labour. Adoption of different technologies is at present generally constrained by the size of land holding which indirectly relates to the farmers' capital base. Small farmers are generally not able to adopt capital intensive technologies. Since this constraint was not directly operational in the model, a land based constraint was built in the model to restrict the use of capital intensive technologies 4 and 5 to 25 per cent of the area (rich farmers with large holdings) in Haryana. All farmers (small, medium, and large) could use technologies 1 and 2 whereas only medium and large farmers (50 per cent of the area) could use technology 3. Another similar sub-constraint of land-water was built in the model to restrict the amount of water used by area under various technologies.

Land and water constraint was specified per district and per land unit. Labour constraint was specified for each district on a monthly basis whereas capital constraint was operational on district basis only.

RESULTS AND DISCUSSION

SUSTAINED GROWTH IN FOOD PRODUCTION

In the first scenario, the objective was to maximize food production with land, technology adoption and no possibility of irrigation water to currently rain-fed areas as the only constraint set to determine the maximum possible opportunity for food production. The results indicated a maximum food production of 44.4 million tons is theoretically possible in Haryana (Table 12.2), including 30.5 millions tons of rice and 13.9 million tons of wheat. Eighty-four per cent of the area was under summer rice-rice and wheat, and the rest was under fallow-wheat. For realizing this, however, 97.6 billion cubic metre of water, Rs. 119.2 billion of capital and 1190 million labour days will be needed in rural areas (Table 12.2). These numbers are several times more than the current availability. Obviously, it is impossible to ever mobilize such huge resources. There was a substantial amount of nitrogen loss as well through leaching.

In the second scenario, water was included as a constraint, together with land (including rain-fed area and technology adoption sub-constraints as well). In this scenario, maximum possible food production was only 16 million tons, including 3.4 million tons of rice and 12.8 million tons of wheat. All land and water was used. Area under rice-based cropping systems was drastically curtailed, because of excessive irrigation requirement by this system. The cropping patterns shifted from summer rice-rice-wheat to irrigated pearl millet-wheat (IPW, 56 per cent), followed by rice-wheat (RW, 20 per cent), fallow-wheat (FW, 16 per cent) and basmati-rice-wheat (BrW, 0.2 per cent). Milk production in this scenario increased to 5.8 billion tons as compared to current production of 4.2 billion tons. Forty-four per cent of the land was cultivated with the current level of technology (technology level 1) and 53 per cent with technology levels 2 and 3. Only 3 per cent area was cultivated with technology 5. This area was largely rice-wheat and was related to increased water use efficiency of this system in high technologies. The farmers' income in this scenario was Rs. 31.0 billion, which was lower than the current because in this scenario there was no emphasis on maximization of income and hence cash crops were not grown.

In the third scenario, constraints on labour and capital were added. This did not result in any change in food production compared to the last scenario where water alone was the constraint. There were only marginal changes in cropping pattern and use of various resources. This indicates that water is the main constraint in Haryana. There was a slight increase in area under technologies 4 and 5. There was a slight reduction in milk production due to reduced fodder availability in higher technologies allocated to wheat.

In the fourth scenario, optimization was further tightened by introducing lower bounds for all crops under all constraints. This resulted in a significant drop in food production to 13.0 million tons (Table 12.2). There was a large increase in the production of basmati rice in this scenario compared to the previous one (1.23 million tons from 0.24 million tons) at the expense of traditional rice (2.9 million tons from 3.6 million tons). There was a substantial decrease in the area under irrigated pearl millet (IPM, from 53 per cent in the previous scenario to only 18 per cent in the current one). Basmati rice-wheat, maize-mustard, cotton-wheat and sugarcane-wheat occupied the space vacated. Because of decreased wheat residue availability, milk production came down significantly. Since cotton and potato were grown in this scenario, biocide residue index increased but it was still within the tolerance limits (Table 12.2).

DOUBLING FOOD PRODUCTION WITH LAND AND TECHNOLOGY ADOPTION AS CONSTRAINTS

In order to double food (rice and wheat) production with all other crop production levels at least at the current level and with land constraints, there will be a need for Rs. 91 billion capital against the currently available Rs. 54 billion. In addition, doubling of food will require 90 per cent more labour and irrigation, and 56 per cent more nitrogenous fertilizer than used today. Only 10 per cent area will be under the current level of technology, whereas 72 per cent area will be under technologies 2 and 3. Since the resource requirement, particularly of water, is far more than the state can ever afford to provide, doubling food production, even if increased capital and labour are made available, appears to be an unfeasible goal.

CONCLUSIONS AND FUTURE WORK

The methodology described here provides a useful tool to explore the window of opportunities for land use planning. The key advantage is that it integrates the knowledge base of several scientists from different disciplinary backgrounds, and attempts to address some real issues put forward by the stakeholders. It also helps us in analysing scientifically whether many of our ambitious goals for development are feasible and, if yes, at what costs.

Our preliminary results point out that availability of water will remain a major constraint to increasing food production in Haryana. Since we considered availability of water resources at today's level, the analysis needs to be redone more critically in view of the total water resources likely to become available in future. The latter is not

Table 12.2: Results of Optimization for Different Scenarios Relating to Maximizing Food Production in Haryana

Constraints				Target	Production		Farmers income	Resources needed					Environmental impact	
Land	Water	Capital	Labour		Food	Milk		Labour	Land Used	Irrig.	N Fertilizers	Capital	P Index	N Loss
					10^6 tons	10^9 tons	10^9 Rs.	10^9 labour days	%	10^9 m^3	10^3 tons	10^9 Rs.		10^3 tons
*				Nil	44.4	4.6	62	1.19	100	97.6	1894	119.2	96	88.7
*	*			Nil	16.0	5.8	31	0.41	94	18.2	644	53.7	35	27.0
*	*	*	*	Nil	16.0	5.1	32	0.41	92	18.2	594	51.8	35	25.6
*	*	*	*	Current other crops	13.0	3.3	37	0.45	94	18.2	398	50.0	80	21.9
*				Double food	21.0	6.8	55	0.65	100	34.8	1014	91.1	370	34.8

easily available at the land unit or even district level. Therefore, the availability of quality data at the desired scale is an important limitation in making progress.

The results of the Interactive Multiple Goal Programming (IMGLP) model are exploratory and only a single (higher) level of stakeholders is considered while the interest of the other stakeholders in the region, e.g. farmers, village level managers, has been ignored. Consequently, although the model illustrates the opportunities at the regional level, these may not be implementable by the primary land use decision-maker—the farmer. There will be a large capital requirement at the state level to finance the equipment needed for implementing capital intensive technologies 4 and 5. These also need to be considered in the future analysis. Finally, although IMGLP is meant to assist in resolving conflicts among competing goals, we could not implement this at present due to paucity of quantitative information on various goals. Nevertheless, the framework developed in this paper should be able to address such issues once such data become available.

NOTE

1. We would like to acknowledge the various officials of the Haryana Government for their generous provision of data and for numerous discussions and support throughout the study.

REFERENCES

AGGARWAL, P.K., M.J. KROPFF, P.S. TENG and G.S. KHUSH, 'The Challenge of Integrating Systems Approach in Plant Breeding. Opportunities, Accomplishments and Limitations', in M.J. KROPFF, P.S. TENG, P.K. AGGARWAL, B. BOUMAN, J. BOUMA, H.H. van LAAR (eds.), *Applications of Systems Approaches at the Field Level*, Kluwer Academic Publishers, Netherlands, 1996, pp.1–24.

AGGARWAL, P.K., N. KALRA, S. KUMAR, S.K. BANDYOPADHYAY, H. PATHAK, A.K. VASISHT, C.T. HOANH and R. ROETTER, 'Exploring Land Use Options for Sustainable Increase in Food Grain Production in Haryana. Methodological Framework', in R. ROETER et al. (eds.), *Land Use Options for Future*, IRRI, Philippines, 2000.

GOVERNMENT OF HARYANA, *Statistical Abstract of Haryana. 1997–1998*, Government of Haryana, Chandigarh, 1998a.

———, *Economics of Important Kharif and Rabi Crops in Haryana during 1997*, Farm Management Extension Service, Department of Agricultural Economics, Government of Haryana, 1998b.

PINSTRUP-ANDERSON, P. and R. PANDYA-LORCH, 'Scenarios for World Food Security and Distribution', in J. BOUMA, A. KUYVENHOVEN, B.A.M. BOUMAN, J.C. LUYTEN, H.G. ZANDSTRA (eds.), *Eco-Regional Approaches for Sustainable Land Use and Food Production,* Kluwer Academic Publishers, The Netherlands, 1995, pp. 89–111.

RABBINGE, R., 'The Role of Asia in World Food Security', in P.S. Bindraban, H. van Keulen, A. KUYENHOVEN, R. RABBINGE, P.W.J. UITHOL (eds.), *Food Security at Different Scales. Demographic, Biophysical and Socioeconomic Considerations*, Research Institute of Agrobiology and Soil Fertility and Graduate School for Production Ecology, Series on Quantitative Approaches in Systems Analysis, No. 21, Wageningen, 1999, pp. 153–7.

SINHA, S.K., G.B. SINGH and M. RAI, *Decline in Crop Productivity in Haryana and Punjab. Myth or Reality*, Indian Council of Agricultural Research, New Delhi, 1998.

SINHA, S.K., *Indian Agriculture in the Next Millennium. Time for a Change*, 29th LB Shastri Lecture, Indian Agricultural Research Institute, New Delhi, 1999.

UNITED NATIONS, *World Population Prospects. 1950–2050. The 1996 Revision*, Population Division, United Nations, New York, 1997.

Contributors

S.S. Acharya, former Chairman of the Commission of Agricultural Cost and Prices (CACP, Ministry of Agriculture, New Delhi), is Director of the Institute of Development Studies (IDS, Jaipur).

Paroda Aggarwal, N. Kalra, S. Kumar and A.K. Vasisht are Scientists from the Indian Agricultural Research Institute (IARI, New Delhi), Centre for Applications of Systems Simulation.

Véronique Alary is Economist at the Centre de Coopération Internationale en Recherche Agronomique pour le Développement (CIRAD, Montpellier).

P.C. Bansil, ex-Senior Staff Member of the Indian Economic Service and of the Food and Agriculture Organization (FAO, Rome), is Founder Member and Head of the Techno-Economic Research Institute (TERI, New Delhi).

Bhuban C. Barah is Principal Scientist (Economics) at the National Centre for Agricultural Economics and Policy Research (NCAP, ICAR, New Delhi). His paper has also been presented at the Third Asian Conference of Agricultural Economists (18–20 October 2000, Jaipur).

Jean-Marc Boussard is Senior Economist at the Institut National de la Recherche Agronomique (INRA, Paris).

Daniel Deybe heads the programme 'Economy, Policies and Markets' at the Centre de Coopération Internationale en Recherche Agronomique pour le Développement (CIRAD, Nogent-sur-Marne).

Bruno Dorin (brunodorin@yahoo.com), Ph.D. in Economics and graduate in Agricultural Engineering, is former Director of the Centre de Sciences Humaines (CSH, New Delhi).

Thomas Jullien (thomas_jullien@yahoo.com), Postgraduate in Economics, was Research Assistant at the Centre de Sciences Humaines (CSH, New Delhi) at the time of the seminar.

Praduman Kumar is Principal Scientist and Head of the Division of Agricultural Economics at the Indian Agricultural Research Institute (IARI, New Delhi). His paper written with R.S. Paroda is a revised version of a paper presented in the Study Week on Food Needs of the Developing World in the Early Twenty-First Century (27–30 January 1999, Vatican City).

R.S. PARODA is Secretary of the Department of Agricultural Research and Education (DARE, Government of India, New Delhi) and Director General of the Indian Council of Agricultural Research (ICAR, New Delhi).

NATHANAËL PINGAULT, postgraduate from the Ecole Nationale du Génie Rural des Eaux et Forêts (ENGREF, Paris), is Staff Member of the Bureau de l'Analyse Economique et de la Prospective (BAEP, Paris) of the French Ministry of Agriculture.

ABUSALEH SHARIFF is Chief Economist and Head of Human Resource Development at the National Council of Applied Economic Research (NCAER, New Delhi).

VANDANA SHIVA is Founder Member and Head of the Research Foundation for Science, Technology and Ecology (RFSTE, New Delhi).

A. VAIDYANATHAN is Emeritus Professor of the Madras Institute of Development Studies (MIDS, Chennai). His paper was presented at a conference on Political Economy of Water Pricing held by the World Bank (Washington DC) in November 1998.

Alfio Quarteroni

Luca Formaggia

Alessandro Veneziani

Complex Systems in Biomedicine

A. Quarteroni (Editor)
L. Formaggia (Editor)
A. Veneziani (Editor)

Complex Systems in Biomedicine

With 88 Figures

Alfio Quarteroni
MOX, Dipartimento di Matematica
Politecnico di Milano
Milan, Italy
and
CMCS-IACS
École Polytechnique Fédérale de Lausanne
Lausanne, Switzerland
alfio.quarteroni@epfl.ch

Luca Formaggia
MOX, Dipartimento di Matematica
Politecnico di Milano
Milan, Italy
luca.formaggia@mate.polimi.it

Alessandro Veneziani
MOX, Dipartimento di Matematica
Politecnico di Milano
Milan, Italy
alessandro.veneziani@mate.polimi.it

The picture on the cover shows an integration of a synapsis (bottom right), the computational domain for a pulmonary artery bifurcation (top right), the human heart (top left), and the wall shear stress in a pulmonary artery (bottom left).

Library of Congress Control Number: 2006923296

ISBN-10 88-470-0394-6 Springer Milan Berlin Heidelberg New York
ISBN-13 978-88-470-0394-1 Springer Milan Berlin Heidelberg New York

This work is subject to copyright. All rights are reserved, whether the whole or part of the material is concerned, specifically the rights of translation, reprinting, reuse of illustrations, recitation, broadcasting, reproduction on microfilm or in other ways, and storage in data banks. Duplication of this publication or parts thereof is permitted only under the provisions of the Italian Copyright Law in its current version, and permission for use must always be obtained from Springer. Violations are liable to prosecution under the Italian Copyright Law.

Springer is a part of Springer Science+Business Media
springer.com
© Springer-Verlag Italia, Milano 2006
Printed in Italy

Cover-Design: Simona Colombo, Milan
Typesetting: PTP-Berlin Protago-TEX-Production GmbH, Germany
Printing and Binding: Signum, Bollate (Mi)

Preface

Mathematical modeling of human physiopathology is a tremendously ambitious task. It encompasses the modeling of most diverse compartments such as the cardiovascular, respiratory, skeletal and nervous systems, as well as the mechanical and biochemical interaction between blood flow and arterial walls, and electrocardiac processes and electric conduction in biological tissues. Mathematical models can be set up to simulate both vasculogenesis (the aggregation and organization of endothelial cells dispersed in a given environment) and angiogenesis (the formation of new vessels sprouting from an existing vessel) that are relevant to the formation of vascular networks, and in particular to the description of tumor growth.

The integration of models aimed at simulating the cooperation and interrelation of different systems is an even more difficult task. It calls for the setting up of, for instance, interaction models for the integrated cardio-vascular system and the interplay between the central circulation and peripheral compartments, models for the mid-to-long range cardiovascular adjustments to pathological conditions (e.g., to account for surgical interventions, congenital malformations, or tumor growth), models for integration among circulation, tissue perfusion, biochemical and thermal regulation, models for parameter identification and sensitivity analysis to parameter changes or data uncertainty – and many others.

The heart is a complex system in itself, where electrical phenomena are functionally related to wall deformation. In its turn, electrical activity is related to heart physiology. It involves nonlinear reaction-diffusion processes and provides the activation stimulus to heart dynamics and eventually the blood ventricular flow that drives the haemodynamics of the whole circulatory system. In fact, the influence is reciprocal, since the circulatory system in turn affects heart dynamics and may induce an overload depending upon the individual physiopathologies (for instance, the presence of a stenotic artery or a vascular prosthesis).

Virtually all the fields of mathematics have a role to play in this context. Geometry and approximation theory provide the tools for handling clinical data acquired by tomography or magnetic resonance, identifying meaningful geometrical patterns and producing three-dimensional geometric models stemming from the original patient's data. Mathematical analysis, fluid and solid dynamics, stochastic analysis are used to set up the differential models and predict uncertainty. Numerical analysis and high performance computing are needed to solve the complex differential models numerically. Finally, methods from stochastic and statistical analysis are exploited for the modeling and interpretation of space-time patterns.

Indeed, the complexity of the problems at hand often stimulates the use of innovative mathematical techniques that are able, for instance, to capture accurately those processes that occur at multiple scales in time and space (such as cellular and systemic effects), and that are governed by heterogeneous physical laws.

In this book we have collected the contributions of several Italian research groups that are successfully working in this fascinating and challenging field. Each chapter deals with a specific subfield, with the aim of providing an overview of the subject and an account of the most recent research results.

Chapter 1 addresses a class of inverse mathematical problems in biomedical imaging. Imaging techniques (such as tomography or magnetic resonance) are a powerful tool for the analysis of human organs and biological systems. They invariably require a mathematical model for the acquisition process and numerical methods for the solution of the corresponding inverse problems which relate the observation to the unknown object.

Chapter 2 addresses those biochemical processes which are composed of two phases, generation (nucleation, branching, etc.) and subsequent growth of spatial structures (cells, vessel networks, etc), which display , in general, a stochastic nature both in time and space. These structures induce a random tessellation as in tumor growth and tumor-induced angiogenesis. Predictive mathematical models which are capable of producing quantitative morphological features of developing tumor and blood vessels demand a quantitative description of the spatial structure of the tessellation that is given in terms of the mean densities of interfaces.

A preliminary stochastic geometric model is proposed to relate the geometric probability distribution to the kinetic parameters of birth and growth. For its numerical assessment, methods of statistical analysis are proposed for the estimation of the geometric densities that characterize the morphology of a real system.

Chapter 3 presents a review of models of tumor growth and tumor treatment. One family of models concerns blood vessels collapsing in vascular tumors, another is devoted to the modeling of tumor cords (growing directly around a blood vessel), highlighting features that are relevant in the evolution of solid tumors in the presence of necrotic regions. Tumor cords are also taken as an example of how to deal with certain aspects of tumor treatment.

The aim of Chapter 4 is the description of models that were recently developed to simulate the formation of vascular networks which occurs mainly through the two different processes of vasculogenesis and angiogenesis. The results obtained by mathematical models are compared with in vitro and in vivo experimental results. The chapter also describes the effects of the environment on network formation and investigates the possibility of governing the network structure through the use of suitably placed chemoattractants and chemorepellents.

Chapter 5 deals with mathematical models of cardiac bioelectric activity at both cellular and tissue levels, their integration and their numerical simulation. The so-called macroscopic bidomain model of the myocardium tissue is derived by a two-scale homogenization method, and is coupled with extracardiac medium and extracardiac potential. These models provide a base for the numerical simulation of anisotropic cardiac excitation and repolarization processes.

In Chapter 6 the authors discuss the role of delay differential equations for describing the time evolution of biological systems whose rate of change depends on their configuration at previous time instances. A noticeable example is the Waltman model which describes the mechanisms by which antibodies are produced by the immune system in response to an antigen challenge.

In the last Chapter the authors illustrate recent advances on the modeling of the human circulatory system. More specifically, they present six examples for which numerical simulation can help to provide a better understanding of physiopathologies and a better design of medical tools such as vascular prostheses and even to suggest possible alternative procedures for surgical implants. Each example provides the conceptual framework for introducing mathematical models and numerical methods whose applicability, however, goes beyond the specific case addressed.

This chapter aims as well to provide an account of successful interdisciplinary research between mathematicians, bioengineers and medical doctors.

We are well aware that this is simply a preliminary contribution to a mathematical research field which is growing impetuously and will attract increasing attention from medical researchers in the years to come.

We kindly acknowledge the Italian Institute of Advanced Mathematics (INDAM) whose scientific and financial support has made this scientific cooperation possible.

Milan,
February 2006

Alfio Quarteroni
Luca Formaggia
Alessandro Veneziani

Contents

List of contributors

- **Sergey Astanin,** Dipartimento di Matematica, Politecnico di Torino, Corso Duca degli Abruzzi 24, 10129 Turin, Italy
- **Alfredo Bellen,** Dipartimento di Matematica e Informatica, Università di Trieste, Via Valerio 12/b, 34127 Trieste, Italy
- **Mario Bertero,** Dipartimento di Informatica e Scienze dell'Informazione, Università di Genova, Via Dodecaneso 35, 16146 Genoa, Italy
- **Alessandro Bertuzzi,** Istituto di Analisi dei Sistemi ed Informatica "A. Ruberti", Viale Manzoni 30, 00185 Rome, Italy
- **Vincenzo Capasso,** Dipartimento di Matematica, Università di Milano, Via Saldini 50, 20133 Milan, Italy
- **Piero Colli Franzone,** Dipartimento di Matematica, Università di Pavia, Via Ferrata 1, 27100 Pavia, Italy
- **Antonio Fasano,** Dipartimento di Matematica "U. Dini", Università di Firenze, Viale Morgagni 67/A, 50134 Florence, Italy
- **Luca Formaggia,** MOX, Dipartimento di Matematica, Politecnico di Milano, Piazza L. da Vinci 32, 20133 Milan, Italy
- **Alberto Gandolfi,** Istituto di Analisi dei Sistemi ed Informatica "A. Ruberti", Viale Manzoni 30, 00185 Rome, Italy
- **Nicola Guglielmi,** Dipartimento di Matematica e Informatica, Università di Trieste, Via Valerio 12/b, 34127 Trieste, Italy
- **Stefano Maset,** Dipartimento di Matematica e Informatica, Università di Trieste, Via Valerio 12/b, 34127 Trieste, Italy
- **Alessandra Micheletti,** Dipartimento di Matematica, Università di Milano, Via Saldini 50, 20133 Milan, Italy
- **Luca F. Pavarino,** Dipartimento di Matematica, Università di Milano, Via Saldini 50, 20133 Milan, Italy
- **Michele Piana,** Dipartimento di Informatica, Università di Verona, Ca` Vignal 2, Strada le Grazie 15, 37134 Verona, Italy
- **Luigi Preziosi,** Dipartimento di Matematica, Politecnico di Torino, Corso Duca degli Abruzzi 24, 10129 Turin, Italy
- **Alfio Quarteroni,** MOX, Dipartimento di Matematica, Politecnico di Milano, Piazza L. da Vinci 32, 20133 Milan, Italy
  and
  CMCS-IACS, École Polytechnique Fédérale de Lausanne, Station 8, 1015 Lausanne, Switzerland

- **Giuseppe Savaré,** Dipartimento di Matematica, Università di Pavia, Via Ferrata 1, 27100 Pavia, Italy
- **Alessandro Veneziani,** MOX, Dipartimento di Matematica, Politecnico di Milano, Piazza L. da Vinci 32, 20133 Milan, Italy

Inverse problems in biomedical imaging: modeling and methods of solution

M. Bertero, M. Piana

Abstract. Imaging techniques are a powerful tool for the analysis of human organs and biological systems and they range from different kinds of tomography to different kinds of microscopy. Their common feature is that they require mathematical modeling of the acquisition process and numerical methods for the solution of the equations relating the data to the unknown object. These problems are usually named inverse problems and their main feature is that they are ill-posed in the sense of Hadamard, so that their solutions require special care. In this chapter we sketch the main issues which must be considered when treating inverse problems of interest in biomedical imaging.

Keywords: inverse problems, tomography, image deconvolution, regularization and statistical methods, iterative reconstruction methods.

1 Introduction

The invention of Computed Tomography (CT) by G. H. Hounsfield at the beginning of the seventies was a breakthrough in medical imaging comparable to the discovery of X-rays by W. C. Roengten in 1895. Even if CT and radiography derive from the same physical phenomenon, the conception of CT was based on ideas which opened new and wide perspectives. Indeed, CT requires mathematical modeling of X-ray absorption, in order to provide equations which relate the observed data to the unknown physical parameters, and methods for the solution of these equations. In such a way it is possible to exploit the tremendous amount of information contained in radiographic data: a 3D image of the human body can be obtained, descerning variations in soft tissue such as the liver and pancreas, and measuring in a quantitative way the density variations of the different tissues. An accuracy of few percent can be obtained with a resolution of the order of 1 mm.

The new ideas introduced in CT were soon transferred to other methods for imaging human tissues. The first was Magnetic Resonance (MR), which is based on the detection of the signals emitted by the magnetic moments of hydrogen nuclei when polarized by means of suitable static magnetic fields and excited by radiofrequency signals under resonance conditions. Moreover, earlier scintigraphic methods evolved into the functional imaging techniques known as Positron Emission Tomography (PET) and Single Photon Emission Computed Tomography (SPECT). In these cases a radio-pharmaceutical is administered to the patient and its distribution, due to metabolic processes, is investigated by detecting the γ-rays emitted by the radionuclides. As we briefly discuss at the end of this chapter, the development of other techniques, based, e.g., on microwaves and on infrared radiation, is in progress.

In general, the new techniques of medical imaging are based on the interrogation of the human body by means of radiation transmitted, reflected or emitted by the body: the effect of the body on the radiation is observed, a mathematical model for the body-radiation interaction is developed and the equations provided by this model are solved in post-processing of the observed data. The same approach applies to cell imaging by means of fluorescence or electron microscopy.

We emphasize a specific requirement of medical imaging, namely, the need for a solution in almost real time. In general a refined model of body-radiation interaction leads to complex non-linear equations, whose solution may require hours of computation time on a powerful computer. Hence the need to develop sufficiently accurate linear models, whenever this is possible, or also to design the observation process in such a way that a linear approximation is feasible. For this reason linearity is the first issue we discuss in this chapter (Sect. 2).

A second specific feature of biomedical imaging is that the problems to be solved are ill-posed in the sense of Hadamard. As we discuss in Sect. 3, being ill-posed implies that it is meaningless to look for exact solutions and that, nevertheless, the set of approximate solutions is too broad to be significant. In other words, although the data at our disposal can contain a tremendous amount of information, the fact that the problem is ill-posed, combined with the presence of noise, implies that the extraction of this information is not trivial.

A very important consequence of being ill-posed is that mathematical modeling of the medical imaging process cannot uniquely consist in establishing the equations relating the data to the solution; it must also include a model of the noise perturbing the data and, as far as possible, a model of known properties of the solution. Indeed the modeling of the noise is needed in order to clarify in what sense one is looking for approximate solutions; on the other hand the modeling of the solution properties must be used for extracting meaningful solutions from the broad set of approximate ones. Therefore noise and "a priori" information on the solution are two other important issues to be considered in biomedical imaging. These are discussed in Sect. 4 and Sect. 5 respectively. In Sect. 6 we outline the main computational issues concerned with the solution procedure and the solution methods which are most frequently used in practice and, lastly, in Sect. 7 we provide a brief description of some of the current medical imaging techniques in progress.

Before concluding this introduction we briefly describe two important examples which can be used as reference cases for the general treatments described in subsequent sections: the first is X-ray tomography and the second fluorescence microscopy.

1.1 X-ray tomography

In the case of X-ray tomography we adopt a tutorial approach which does not correspond exactly to the data acquisition geometry in CT scanners. Therefore we assume that we have a source S emitting a well collimated X-ray beam; the beam crosses the body to be imaged and, at exit, its intensity is measured by a detector D. The attenuation of the X-rays across the body is described by the following simple model:

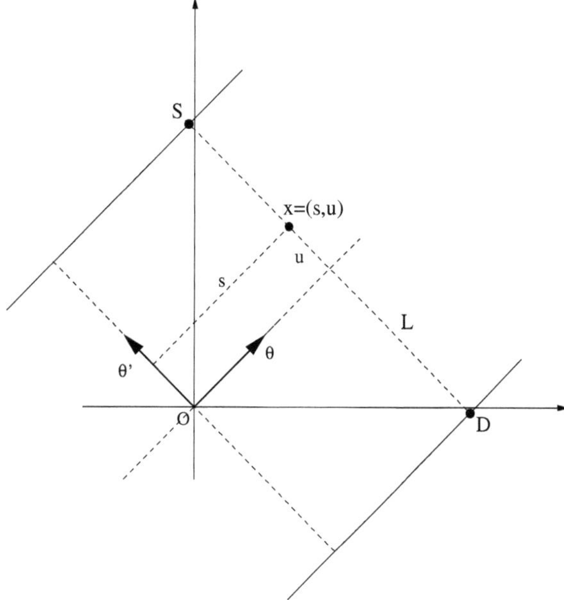

Fig. 1. Geometry of data acquisition in X-ray CT. The source S and the detector D move along two parallel straight lines with direction θ. The line L, joining S and D, is the integration line, with direction θ', orthogonal to θ. A point x of L has coordinates $\{s, u\}$ with respect to the system formed by θ, θ'

let $f(x)$ be the attenuation coefficient at point x (roughly proportional to the density of the tissue at x); then, if u is a coordinate along the straight line L joining S and D (see Fig. 1), the intensity loss at x is given by:

$$\frac{dI}{du}(x) = -f(x)I(x),$$

where I is the intensity measured by D.

It follows that, if I_0 is the intensity emitted by S, then

$$I = I_0 \exp\left\{-\int_L f(x)du\right\},$$

so that the logarithm of the ratio between the intensities of the emitted and detected radiation is just the line integral of the attenuation coefficient. By moving the S-D system along two parallel lines, the plane to be imaged is defined, and, by measuring the intensity for all the positions, one gets what is called a projection of the unknown function $f(x)$. More precisely, if θ is the unit vector in the direction of the movement of the S-D system (linear scanning), s the distance (with sign) of L from the origin of the coordinate system (see Fig. 1), and θ' the unit vector in the orthogonal direction, then the *projection of f in the direction θ* is given by

$$(P_\theta f)(s) = \int f(s\theta + u\theta')du. \tag{1}$$

By rotating the S-D system and repeating the linear scanning for all possible angles (angular scanning) one obtains all possible projections and the result is the (two-dimensional) *Radon transform* of the function f: $(Rf)(s,\theta) = (P_\theta f)(s)$. These are just the data of X-ray CT, obtained by combining the linear and angular scanning as described above. Then, in order to get the function f, one has to solve the linear equation

$$g(s,\theta) = (Rf)(s,\theta),$$

where $g(s,\theta)$ denotes the measured data. This problem was solved by Radon in 1917 [59] and its inversion formula in the 2D case can be written as follows [57]:

$$f(x) = \frac{1}{4\pi^2} \int_{S^1} P \int_{R^1} \frac{1}{x \cdot \theta - s} \frac{\partial g}{\partial s}(s,\theta) ds d\theta, \tag{2}$$

where P denotes the principal value. This formula clearly shows that the inversion of the Radon transform is an ill-posed problem since it requires the computation of the derivative of (noisy) data. Moreover the *filtered backprojection algorithm*, first introduced by Bracewell and Riddle [9] in radio astronomy and now widely used in medical imaging, is just a clever implementation of this formula.

The 3D imaging is obtained by repeating the previous procedure for different planes, namely, by scanning in the z-direction also, orthogonal to the imaging plane. Therefore the data of the problem depend on the variables $\{s, \theta, z\}$, which essentially characterize the position of the S-D system. These data can be called the *image* of f, as provided by the CT scanner. For a given z the representation of g in the plane $\{s, \theta\}$ is the so-called *sinogram*. We give an example in Fig. 2. It is obvious that the interpretation of these data without the help of a reconstruction algorithm is impossible. As text books in tomography we mention the books of Kak and Slaney [42] and Natterer [56].

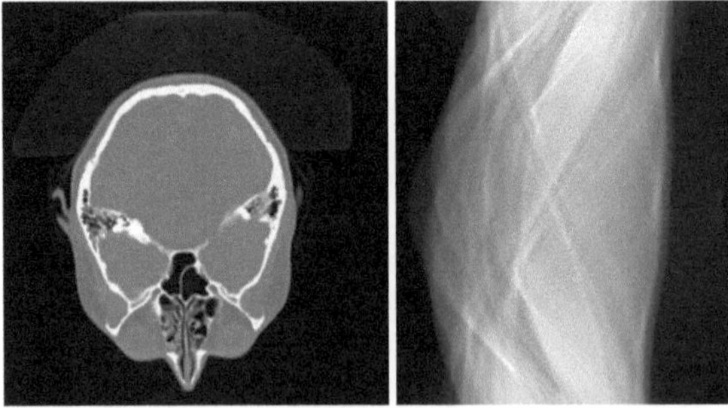

Fig. 2. Left-hand panel: tomographic reconstruction of a section of a human head. Right-hand panel: the corresponding sinogram

1.2 Fluorescence microscopy

Fluorescence microscopy is a technique which is used for the investigation of living cells. The cell is treated with a fluorescent marker and its 3D image is formed by means of a technique known as *optical sectioning* [70], which can be applied to different kinds of optical microscopes (wide-field, confocal, multiphoton, etc. [28]). In all cases a 3D image is formed by acquiring a set of 2D images corresponding to different planes of focus.

Thanks to geometric optics there is a one-to-one correspondence between the points of the image plane where the detector is located and the points of the focal plane where the section of the object to be imaged is located. From this correspondence we can identify a point of the image plane with the corresponding point of the focal plane, hence with a point $x \in \mathbb{R}^3$ of the volume of the sample. In such a way the 3D image can also be considered as a function of x. However the image at one point receives contributions not only from the corresponding point in the focal plane but also from neighboring points both in the plane which is in focus and in the other planes. The result is an integral relationship between the image and the object.

By assuming perfect spatial incoherence of the radiation emitted by the sample it turns out that the relationship between the intensities of the detected and emitted radiation is linear. Moreover, by neglecting the spherical aberration of lenses, this relationship is also translation invariant. In such a case the imaging system is called *isoplanatic* or *space invariant*. As a result, the image $g(x)$ is given by the convolution product of the object $f(x)$ (proportional to the density of fluorescent atoms at x) with a function $h(x)$, which is the image of a point-source and is called the Point Spread Function (PSF) of the imaging system:

$$g(x) = \int_{\mathbb{R}^3} h(x - x')f(x')dx'. \qquad (3)$$

The effect of the PSF is usually denoted as *blurring*; moreover the data are obviously corrupted by noise. It may be interesting to remark that (1) has a similar structure but with a PSF which is a distribution and is not translation invariant.

In conclusion, the problem of reconstructing the object is equivalent to that of solving the convolution equation of (3). Such a problem is called *deconvolution* or *deblurring* and is another classical example of an ill-posed problem. An introduction to deconvolution methods is given in [4].

2 Linearity

As outlined in the introduction, a basic requirement in medical imaging is the availability of a linear relationship between the properties of the tissues to be imaged and the data provided by the medical equipment. All the algorithms implemented in commercial machines are based on such an assumption, because, in spite of the increasing power of computers, only in the case of linear problems it is possible to get a solution in almost real time for large-scale problems such as those arising in medical imaging. Of course the situation can change in the future and, for this reason, the investigation of non-linear models, which provide a more accurate description of radiation-body interaction, is an exciting research topic.

In the mathematics of image processing the object is that property which is distributed in a volume and which has to be estimated by means of indirect measurements; it belongs to a well-characterized functional space (typically, for sake of simplicity, a Hilbert space) and, in what follows, it is denoted by f, a function of the variable $x \in \mathbb{R}^3$. On the other hand, the image, denoted from now on by g, is that measurement (again, in a suitable Hilbert space) which is provided by the imaging device and is regarded as best representing the object given the specific hardware. The image g is a function of the parameters, overall denoted by ξ, which characterize the acquisition process. The two examples outlined in the introduction show that these parameters can have quite different physical meanings: in the case of X-ray CT they characterize the position of the detector and the direction of the incoming radiation, while in the case of microscopy they can be identified with the coordinates of a point in the object volume. In the first case the 3D image is a set of sinograms, one for each section of the volume; in the second case, it is a "blurred" version of the object.

The relationship between object and image can be obtained by a mathematical model of the physical phenomena which provide the basis of the acquisition process. The most general representation of such a process is given by the non-linear integral equation

$$g(\xi) = \int h(\xi, x, f(x))dx, \tag{4}$$

where h is a continuous function of all its variables. This equation provides the solution of the so-called *direct* problem, namely, the problem which must be solved for computing the data g related to a given object f. The continuous dependence on f is just due to the fact that this problem is well-posed, namely, the solution exists, is unique and depends continuously on the data (in this case, the object f). The inverse problem is obtained by exchanging the roles of the data and the solution: in such a case one must find the object f for a given image g.

The solution of (4) is very difficult from both the theoretical and practical points of view. No general theory exists for such a non-linear integral equation: each problem requires specific analysis. Moreover the problem may be ill-posed and no general regularization theory exists for wide classes of non-linear problem. A few general results applying to non-linear problems are described in the book of Engl et al. [29]. Most of these results are inspired by classical Newton-like optimization schemes, according to which stable approximate solutions of (4) can be obtained by stopping an iterative procedure appropriately initialized. The idea is to assume that a non-linear differentiable map $\mathcal{F}$ can be defined between two functional spaces so that

$$g = \mathcal{F}(f), \tag{5}$$

where $\mathcal{F}$ is here the integral operator on the right-hand side of (4). With $\mathcal{F}'$ denoting the Frechet derivative of $\mathcal{F}$, (5) can be replaced by the linearized equation

$$\mathcal{F}(f) + \mathcal{F}'(f)q = g,$$

which has to be solved for the perturbation q to obtain the upgrade $f+q$. The method consists in solving iteratively the equations:

$$\mathcal{F}'(f^{(k)})q^{(k+1)} = g - \mathcal{F}(f^{(k)}),\ f^{(k+1)} = f^{(k)} + q^{(k+1)}, \quad k = 0, 1, 2, \ldots, \tag{6}$$

starting from an initial guess $f^{(0)}$ for the function f. This approach has two fundamental drawbacks. First of all, it requires, at each step of the iteration procedure, the solution of the direct problem in order to compute the map $\mathcal{F}$ (this point is clarified by the example of the inverse scattering problem discussed below) as well as its Frechet derivative; therefore it is extremely demanding from the numerical point of view and an accurate approximation of the solution can be achieved only after such a long computational time that no realistic technological application is possible at the moment. Second, the sequence of iterates defined in (6) may converge only if a sufficiently accurate initialization of f is to hand. This can be a serious bias in applications to medical imaging.

The situation is quite different if the non-linear equation (4) can be approximated by a linear one:

$$g(\xi) = \int h(\xi, x) f(x) dx, \tag{7}$$

where h is the impulse response of the imaging instrument, which can be a distribution, as in tomography, or a function, as in microscopy. Indeed, in the case of linear inverse problems one has at one's disposal very powerful theoretical tools.

However, the key question is to understand what kind of approximation can lead to (7). Indeed, two possibilities may occur. In one case, (7) is a brand new model where linearity is obtained by a precise technological realization or is the consequence of physical approximations. For instance, in Magnetic Resonance (MR), data acquisition is designed in such a way that the data are just the values of the Fourier transform of the object to be imaged. On the other hand, in fluorescence microscopy, linearity is obtained by neglecting a physical phenomenon. Indeed, it is well-known that an optical system, namely, a system of lenses, is linear in the amplitude of the wave field in the case of perfectly coherent radiation and in intensity in the case of perfectly incoherent radiation (see, e.g., Goodman [32]). In the case of fluorescence, the emitted radiation is partially coherent but the degree of coherence is not very high so that the approximation of perfectly incoherent radiation is assumed to be satisfactory and one gets a linear integral equation relating the intensities of the emitted and detected radiation. Even the model used in X-ray CT corresponds to a rather simplified (but sufficiently accurate) description of the absorption of photons due to the interaction with the material and moreover, in this case, linearity is obtained by considering the logarithm of the data and not the data itself.

A completely different situation occurs when (7) is obtained by means of a sort of perturbation theory applied to (4), i.e., through a linearization of Eq. (4) around zero or an approximate object $f^{(0)}$. This approach can be illustrated by the following inverse scattering problem which represents a reasonable model for microwave tomography. For simplicity we consider a two-dimensional approximation and we assume that the body is illuminated by means of a set of plane waves with fixed wave number k, coming from different directions given by unit vectors θ. The body is characterized by a refractive index $n(x) \neq 1$ and is immersed in a medium with $n(x) = 1$; therefore it is described by the function

$$f(x) = n^2(x) - 1, \tag{8}$$

whose support Ω is the domain of the body. The direct problem consists in determining the wave function $u(x) \in C^2(\mathbb{R}^2 \setminus \partial D) \cap C^1(\mathbb{R}^2)$ which solves the problem:

$$\Delta_2 u(x) + k^2 n^2(x) u(x) = 0, \tag{9}$$

$$u(x) = e^{ik\theta\cdot x} + u^s(x), \tag{10}$$

$$\lim_{r\to\infty} \sqrt{r}\left(\frac{\partial u^s}{\partial r} - iku^s\right) = 0. \tag{11}$$

In (10) the first term is the incident plane wave with direction θ and the second term is the scattered wave. Potential theory allows us to prove that the elliptic differential problem (8)–(11) is equivalent to the integral equation [21]:

$$u(x) = e^{ik\theta\cdot x} - k^2 \int_{R^2} \Phi(x, y) f(y) u(y) dy, \quad x \in R^2, \tag{12}$$

known as the Lippmann-Schwinger equation, in which $\Phi(x, y)$ is the fundamental solution of the Helmholtz equation in $\mathbb{R}^2$ given by

$$\Phi(x, y) = \frac{i}{4} H_0^{(1)}(k|x - y|), \tag{13}$$

where $H_0^{(1)}$ is the zero-order Hankel function of the first kind. From this integral formulation, together with considerations based on the unique continuation principle, one can prove that the direct problem is well-posed. Furthermore, the Sommerfeld radiation condition (11) implies that the scattered field, propagating in the direction θ', can be asymptotically factorized in the form

$$u^s(r, \theta', \theta) = \frac{e^{ikr}}{\sqrt{r}} u_\infty(\theta', \theta) + O(r^{-3/2}),$$

where $u_\infty(\theta', \theta)$ is the far-field pattern of the scattered field u^s. By using Green's theorems one can easily show that

$$u_\infty(\theta', \theta) = \frac{e^{i\pi/4}}{\sqrt{8\pi k}} \int_{\mathbb{R}^2} e^{-ik\theta'\cdot y} f(y) u(y) dy, \tag{14}$$

where the dependence on θ on the right-hand side comes from the fact that u is given by (10). Equation (14) is an actual realization of (4) and allows us to define the non-linear inverse problem we are interested in. In such a problem the data function g is represented by the measured values of the far-field pattern u_∞ (with

ξ represented by the unit vectors θ, θ'), the unknown object is the contrast function (8) and the non-linearity is due to the fact that $u(y)$ depends on $f(y)$ itself. In other words the mapping $\mathcal{F}$ of (5) is defined by the right-hand side of (14) and therefore its computation requires the solution of the direct problem (9)–(11).

However, under the low-frequency, weak-scattering assumption,

$$kMa < \pi, \tag{15}$$

where $f(x) \neq 0$ for $|x| \geq a$ and $M = \sup_{|x|\leq a} |f(x)|$, a linearization of (14) can be obtained by applying the successive approximations method to the Lippmann-Schwinger equation. If we take the first-order approximation, i.e. we neglect the scattered wave with respect to the incident plane wave, we obtain the so-called Born approximation and the result is the diffraction tomography equation [27]:

$$u_\infty(\theta', \theta) = \frac{e^{i\pi/4}}{\sqrt{8\pi k}} \int_{\mathbb{R}^2} e^{-ik(\theta'-\theta)\cdot y} f(y) dy.$$

In other words, in the Born approximation regime, the non-linear image reconstruction problem becomes a linear Fourier transform inversion problem with limited data, where the low-frequency cut-off which determines the data limitation is given by (15).

Linearizations like the one induced by Born approximation are unreliable in those cases, as for microwaves, in which the wavelength $\lambda = 2\pi/k$ of the field propagating in the biological tissue is of the same order of magnitude as the tissue dimensions and therefore diffraction effects are predominant. However, even in this case a linear approach is possible if the aim is the reconstruction of the support of the object f and not the object itself. This is the basic idea underlying visualization techniques based on the so-called *linear sampling method* [20, 23].

The starting point of this approach is to consider a one-parameter family of linear integral equations of the first kind which, in a sense, provide exactly the same estimate of the support of the object as the complete solution of the non-linear inverse scattering problem (8)–(11). In order to clarify this statement, we introduce the parameter x, which is just a point in a region of $\mathbb{R}^2$ (or $\mathbb{R}^3$, if the problem is 3D) containing the support Ω of f and, for each x, we write the far-field equation [18–20]:

$$\int_{S^1} u_\infty(\theta', \theta) g_x(\theta) d\theta = \Phi_{\infty,x}(\theta'), \tag{16}$$

where:

- $u_\infty(\theta', \theta)$ is the far-field pattern of (14); in practical applications of the method it is approximated by means of its measured values;
- $\Phi_{\infty,x}(\theta')$ is the far-field pattern of the fundamental solution of the Helmholtz equation, which, in the 2D case discussed above, is given by the far-field pattern of (13), i.e.,

$$\Phi_{\infty,x}(\theta') = \frac{e^{i\pi/4}}{\sqrt{8\pi k}} e^{-ik\theta'\cdot x}.$$

A solution $g_x(\theta)$ of (16) does not exist for any scattering data and, when it does exist, has no physical meaning. However it can be proved [12] that an approximate solution exists which has the property that is becomes unbounded for x on and in the exterior of the boundary of the scatterer, thus acting as an indicator for the boundary itself. Such behavior naturally inspires a visualization algorithm [23] where, for each point of the grid containing the inhomogeneity, an approximate stable solution of the far-field equation is constructed and its norm is plotted: the contour of the scatterer is given by all points x where this norm is greater than a given threshold.

This approach has two main advantages: first, the implementation is computationally simple and a notable computational speed is achieved (2D objects can be visualized in a few minutes, which become a couple of hours in the case of very complicated 3D objects); second, very little a priori information on the inhomogeneity is necessary for the method to work (no knowledge of the number of scatterers, of their physical nature and of possible boundary conditions is required). A vast literature on the linear sampling method is at our disposal. Besides the more theoretical papers already cited, applications to biomedical imaging problems involve microwave tomography [51], impedance tomography [10] and the detection of leukemia in human bone marrow using microwaves [22]. Furthermore, a variety of similar inversion schemes has been formulated, involving the factorization method of Kirsch [44] and the indicator sampling method of You et al. [71]. Of course, such approaches also have significant drawbacks. As already mentioned, the linear sampling method is not a reconstruction method, in the sense that it simply allows a visualization of the object, without providing information on the point values of the refractive index. Moreover the spatial resolution achieved is not yet satisfactory, particularly if medical imaging applications are considered. However these techniques allow us to visualize, although coarsely, even very complex objects in a reasonable time, and therefore can be very helpful to provide inizializations for Newton-like schemes or information on the support of the inhomogeneity in the case of the application of constrained iterative algorithms.

3 Ill-posed problems and uncertainty of solution

The availability of a reliable linearized mathematical model for image formation is not sufficient for a straightforward solution of the reconstruction problem. A crucial difficulty is due to the fact that most image reconstruction problems arising in biomedical imaging are ill-posed. The concept of being ill-posed was introduced by Hadamard as a mathematical anomaly in the solution of particular boundary value problems for partial differential equations. The discussion of the Cauchy problem for the Laplace equation is a classical example [33]. However, an exhaustive definition of ill-posed is given by a negative characterization: ill-posed problems do not satisfy at least one of the three conditions required for being well-posed, i.e., existence, uniqueness and continuous dependence of the solution on the data. As pointed out in [24], the lack of the third requirement has particularly important consequences for the solution of ill-posed problems modeling physical situations: indeed, in the case of practical applications, the presence of measurement noise in the data may imply the presence of strong numerical instabilities in the solution when obtained by means of a straightforward approach.

Inverse problems are very often ill-posed in the sense of Hadamard. In fact, in most cases of interest for applications, the linear operators modeling the problems are compact and being ill-posed is a consequence of this functional property of the model. This is the case, for example, of the two image reconstruction problems described in the introduction. Indeed, if the object has bounded support, for suitable choices of the source and data functional spaces, both the Radon transform describing X-ray tomography and the convolution operator at the basis of fluorescence microscopy become compact linear operators.

In order to formally discuss the issue of being ill-posed and its relationship with compactness we consider linear inverse problems characterized by the following general structure [3]. The first step is to define the corresponding direct problem, whose solution allows us to define a linear operator A from the (Hilbert) space X containing all functions characterizing the unknown objects to the (Hilbert) space Y containing all functions describing the corresponding measurable images. Therefore the measured image $g \in Y$ is related to the true physical object $f^\circ \in X$ by

$$g = Af^\circ + h, \tag{17}$$

where Af^0 is the (exact, or computable) image of the object f° and h is a term containing all possible experimental errors. When the signal-to-noise ratio associated with g is sufficiently high, the norm of Af° is larger than the norm of h, the noise term in (17) can be neglected and the linear inverse problem that we are interested in reduces to that of finding $f \in X$ such that

$$g = Af. \tag{18}$$

Equations (17) and (18) display a clear pathological feature of ill-posed inverse problems. In fact, if we represent the problem by (17), we have only one equation for two unknowns: the true object f° and the function h. On the other hand, if we use (18), since the correct representation of g is given by (17), then a solution may not exist for all noise realizations.

According to a more formal approach, being ill-posed can be viewed as a property of the triple $\{A, X, Y\}$. Indeed, the data space Y must be broad enough to contain both the exact image Af° and the noisy image g, and therefore the range of A, $\mathcal{R}(A)$, is strictly contained in Y (typically the functions in $\mathcal{R}(A)$ are much smoother than the functions describing noisy images). Furthermore, when the kernel of A is not empty, the solution of the problem is not unique; finally, if A^{-1}, when it can be defined, is an unbounded mapping from Y to X, then the dependence of the solution on the data is not continuous. The only way out from this puzzling situation is, first of all, to give up looking for an exact solution of (18) and, for example, to consider the least-squares problem of determining all functions $f \in X$ such that

$$\|Af - g\|_Y = \min . \tag{19}$$

It is easy to prove that the set of least-squares solutions concides with the set of solutions of the Lagrange-Euler equation

$$A^*Af = Ag,$$

where A^* is the adjoint operator of A. If P is the linear projection operator onto the closure $\overline{\mathcal{R}(A)}$ of the range of A, it can be shown that the Euler equation is in its turn equivalent to

$$Af = Pg.$$

It follows that there exists a class of linear operators for which the set of least-squares solutions is not empty, namely, the operators whose range is closed. Furthermore, the set of least-squares solutions is a closed and convex subset of the Hilbert space X and therefore there exists only one least-squares solution of minimal norm, which is called the generalized solution and denoted by $f^\dagger$. It is natural to introduce the generalized inverse operator $A^\dagger : Y \rightarrow X$ mapping g to $f^\dagger$ and, since $A^\dagger$ is continuous if and only if $\mathcal{R}(A)$ is closed, then, in this case, the problem of determining the generalized solution is well-posed (these results apply to the case of discretized problems, as discussed later). However, as already stated, most inverse problems of interest in biomedical imaging are modeled by compact operators and easy considerations essentially based on the open mapping theorem show that, if A is compact, then $\mathcal{R}(A)$ is not closed (unless it is finite-dimensional). In other words, the search for a generalized solution of problem (18) for compact operators is still an ill-posed problem.

According to a different approach, the best one can do is to look for an f reproducing the given g within a tolerable uncertainty [6]. Since the image (17) can contain a term in $\mathcal{R}(A)^\perp = \ker(A^*)$ due to the noise, then the idea is to look for functions in X such that

$$\|Af - Pg\|_Y \leq \epsilon, \tag{20}$$

where ϵ measures the magnitude of the noise. In order to show that, in the case of compact A, even this attempt is unsuccessful, we introduce the singular system of A [43], defined as the set of triples $\{\sigma_k; u_k, v_k\}_{k=1}^\infty$ solving the shifted eigenvalue problem

$$Au_k = \sigma_k v_k, \quad A^* v_k = \sigma_k u_k. \tag{21}$$

Before going on, we recall that a complete characterization of the singular system of the Radon transform in any dimension is known [26,49] (also see [4,56]). These results show that the singular values tend to zero very slowly, so that thousands of singular values are necessary for an accurate reconstruction. Moreover, the singular functions, which are related to orthogonal polynomials, become highly oscillating when associated to small singular values. This is a rather general property of the singular functions of operators involved in biomedical imaging.

Now, since the set of the singular functions $\{u_k\}_{k=1}^{\infty} \subset X$ is an orthonormal basis in $\ker(A)^{\perp}$ while the set of the singular functions $\{v_k\}_{k=1}^{\infty} \subset Y$ is an orthonormal basis in $\overline{\mathcal{R}(A)} = \ker(A^*)^{\perp}$, elementary computations based on (21) lead to the following expression of (20):

$$\sum_{k=1}^{\infty} \frac{\sigma_k^2}{\epsilon^2} |(f, u_k)_X - \frac{(g, v_k)_Y}{\sigma_k}|^2 \leq 1. \tag{22}$$

This equation defines the set of interior points of a sort of 'ellipsoid' in the infinite-dimensional solution space X. As the singular values of a compact operator are real positive numbers accumulating to zero when $k \to \infty$, then such an ellipsoid is unbounded, and, together with the true object, it also contains completely unreliable approximate solutions which, nevertheless, can reproduce the data within the prescribed accuracy. We observe in Sect. 5 that the main idea, common to most available methods for dealing with ill-posed problems, is just to restrict the class of admissible solutions by selecting a subset of this ellipsoid by exploiting a priori information available about the solution.

We conclude this section devoted to ill-posed problems, by discussing the case of finite-dimensional linear problems obtained from a sort of discretization of the original ill-posed linear inverse problem formulated in the infinite-dimensional Hilbert space setting.

Since, in general, we are interested in 3D images, we assume that the volume of the body is partitioned into N voxels, characterized by an index n, and we denote by f_n the average value of the quantity of interest $f(x)$ in the voxel n. Moreover we assume that the radiation transmitted, reflected or emitted by the body is measured by means of M detectors, characterized by an index m; we denote by g_m the output of the detector m. We denote by $\mathbf{f}$ and $\mathbf{g}$ respectively, the vectors of the unknown parameters (the object) and of the outputs (the image). If a linear model has been developed for the imaging process, then the discretization of this model leads to a matrix A, $M \times N$, relating the unknown object $\mathbf{f}$ to the image $\mathbf{g}$. In practical applications, typically we have $M \geq N$, so that the problem can be overdetermined. Then, in the absence of experimental errors, the output of the detector m should be given by

$$(A\mathbf{f})_m = \sum_{n=1}^{N} A_{m,n} f_n.$$

We can now reformulate a discrete least-squares problem, as in (19), with the norm of Y replaced, for instance, by the usual Euclidean norm of an M-dimensional vector space. The problem of determining the generalized solution is then well-posed in the sense of Hadamard. However, being well-posed is only a necessary condition for numerical stability. Indeed, if we introduce the singular system of the matrix A, the generalized solution is given by

$$f^{\dagger} = \sum_{k=1}^{p} \frac{(\mathbf{g}, \mathbf{v}_k)_2}{\sigma_k} \mathbf{u}_k,$$

where p is the rank of the matrix A and the scalar product is the Euclidean one. The numerical stability of this solution is controlled by the condition number $\alpha = \sigma_1/\sigma_p$ and, since the problem derives from the discretization of an ill-posed problem where the singular values accumulate to zero, it is quite natural to expect that this condition number is quite large. Moreover, it is also clear that the value of the condition number increases for increasing accuracy of the discretization. Lastly, since the instability is due to the propagation of the noise corrupting the components associated with small singular values and since these components are associated with singular functions which are, in general, highly oscillating, as remarked above, it should be clear that the generalized solution is characterized by wild oscillations.

If we now consider that, analogous to (22) in the case of infinite-dimensional spaces, in a finite-dimensional framework the set of approximate solution vectors which reproduce the data vector within an uncertainty ϵ is represented by the ellipsoid

$$\sum_{k=1}^{p} \frac{\sigma_k^2}{\epsilon^2} |(f, u_k)_2 - \frac{(g, v_k)_2}{\sigma_k}|^2 \leq 1,$$

we find that the center of this ellipsoid is the generalized solution and its half-axes have lengths ϵ/σ_k. Since for a refined discretization the singular values of the matrix become closer and closer to those of the corresponding (compact) operator, it follows that, from some principal directions on, the ellipsoid also contains, together with the true and generalized solutions, approximate solutions characterized by huge norms and therefore completely unreliable. Lastly, we note that the ratio between the lengths of the longest and shortest half-axes is just the condition number.

4 Noise modeling

In the previous section we emphasized the idea that, in the case of an ill-posed problem or of a discrete ill-conditioned problem, one must look for approximate solutions, namely, for objects which do not reproduce the detected image exactly. Indeed, an exact reproduction of the image should be a reproduction not only of the signal contained in the image but also of the noise affecting the signal. In other words one must look for objects approximating the image "within the noise" and it is obvious that this expression becomes more significant if the structure of the noise is known. Therefore in this section we concentrate on the noise models which are most frequently used in applications, we provide a definition of the set of approximate solutions and we ignore, for a moment, the difficulty due to the fact that this set is too broad. Moreover, in this and subsequent sections we continue to consider discrete versions of the imaging problems.

We already stated that the outputs of the detectors are affected by perturbations which are usually denoted as *noise*; randomness is their main feature. As a consequence, the output of a detector must be viewed as the realization of a random variable and, if a measurement is repeated several times, the results will always be different. As a consequence of the ill-conditioning of the imaging matrix A, the solutions of the linear equation

$$A\mathbf{f} = \mathbf{g},$$

corresponding to different realizations $\mathbf{g}$ of the image are completely different and this is another way of stating that the set of approximate solutions is too broad; indeed the solution associated to one realization is an approximate solution for another realization of the image of the same object.

These remarks indicate that a statistical approach is a quite natural setting for discussing these questions and, in the following, we provide an attempt at quantifying the concept of approximate solution in a general way, for any given noise; it is based on the so-called *likelihood function*, which is related to the randomness of the images. As we show, the least-squares approach, already discussed in the previous section, is obtained as a particular case.

We assume that $\mathbf{g}$ is the realization of a vector-valued Random Variable (RV) $\mathbf{G}$ and that we know the probability distribution of $\mathbf{G}$ for a given object $\mathbf{f}$; for simplicity, we assume that it can be given in terms of a probability density, which is denoted by $P_{\mathbf{G}}(\mathbf{g}|\mathbf{f})$. If a particular realization $\mathbf{g}$ of $\mathbf{G}$ is given and if we insert this value in $P_{\mathbf{G}}(\mathbf{g}|\mathbf{f})$, we obtain a function of $\mathbf{f}$ which is called the *likelihood*, or the *likelihood function*, and is denoted as

$$L_{\mathbf{g}}(\mathbf{f}) = P_{\mathbf{G}}(\mathbf{g}|\mathbf{f}).$$

A careful discussion of the statistical meaning of this function is beyond the scope of this work; we only observe that, if we consider two objects, $\mathbf{f}_1$ and $\mathbf{f}_2$, and if $L_{\mathbf{g}}(\mathbf{f}_1) > L_{\mathbf{g}}(\mathbf{f}_2)$, then $\mathbf{f}_1$ is "more likely" than $\mathbf{f}_2$ to be the object which has generated the image $\mathbf{g}$. Since, in general, the RVs G_m are independent, so that $P_{\mathbf{G}}(\mathbf{g}|\mathbf{f})$ is the product of a large number of density functions, it is convenient to introduce the logarithm of the likelihood function,

$$J_{\mathbf{g}}(\mathbf{f}) = -\ln\ P_{\mathbf{G}}(\mathbf{g}|\mathbf{f}). \tag{23}$$

Then we can define a set of approximate solutions as the set of objects with a likelihood greater than a given value (obviously smaller than the maximum value of the likelihood) or, equivalently, as the set of objects defined by the condition

$$S_{\epsilon,\mathbf{g}} = \left\{\mathbf{f}\,|\,J_{\mathbf{g}}(\mathbf{f}) \leq \epsilon\right\}. \tag{24}$$

In the two examples we discuss in detail, this second definition is preferable, when combined with a suitable rescaling of the function $J_{\mathbf{g}}(\mathbf{f})$, since, in these cases, this function can also be interpreted as the discrepancy between the computed image $A\mathbf{f}$, associated with the object $\mathbf{f}$ and the detected image $\mathbf{g}$.

Before discussing the two particular models, we point out that in both cases the basic assumption is that the expected value of the RV $\mathbf{G}$ is given by the ideal (computed) image

$$E\{\mathbf{G}\} = A\mathbf{f}.$$

The two models correspond, respectively, to so-called *additive Gaussian noise* and *Poisson noise*.

4.1 Additive Gaussian noise

In this model, the RV **G** is given by

$$\mathbf{G} = A\mathbf{f} + \mathbf{W}, \tag{25}$$

where **W** is a Gaussian vector-valued RV. The noise is called additive just because it is a random process which is added to the deterministic signal coming from the object. If all the RVs W_m have zero expected value and if C is their covariance matrix, then the joint probability density of these RVs is given by

$$P_{\mathbf{W}}(\mathbf{w}) = [(2\pi)^M |C|]^{-\frac{1}{2}} \exp\left\{-\frac{1}{2}(C^{-1}\mathbf{w}, \mathbf{w})_2\right\}, \tag{26}$$

where $|C|$ is the determinant of the covariance matrix. If $C = \sigma^2 I$, with I the identity matrix, then we have so-called *white noise*. From (25) and (26) we obtain

$$P_{\mathbf{G}}(\mathbf{g}|\mathbf{f}) = [(2\pi)^M |C|]^{-\frac{1}{2}} \exp\left\{-\frac{1}{2}(C^{-1}(\mathbf{g} - A\mathbf{f}), \mathbf{g} - A\mathbf{f})_2\right\},$$

and therefore the functional (23), after multiplication by a factor of 2 and addition of a suitable constant, becomes

$$J_{\mathbf{g}}(\mathbf{f}) = (C^{-1}(\mathbf{g} - A\mathbf{f}), \mathbf{g} - A\mathbf{f})_2. \tag{27}$$

In the particular case of white noise, we get

$$J_{\mathbf{g}}(\mathbf{f}) = ||A\mathbf{f} - \mathbf{g}||_2^2, \tag{28}$$

and this is just the discrete version of the least-squares approach discussed in the previous section in a continuous setting. On the other hand the functional (27) is that used in the so-called *weighted* least-squares approach. In other words, from a statistical point of view, all the different forms of least-squares approach derive from a specific assumption on the noise perturbing the data. We remark that these approaches are also the starting points of the regularization theory of ill-posed problems [29].

4.2 Poisson noise

The second model we consider applies to the case of so-called *photon noise*, namely, the noise due to fluctuations in the emission and counting of the photons involved in the imaging process. This noise is relevant both for transmission and emission CT as well as for fluorescence microscopy. The treatment of emission CT and fluorescence microscopy is very similar and, for simplicity, we discuss a model which applies to both cases. The model for transmission CT is discussed, e.g., by Lange and Carson [47].

The basic assumption is that each voxel n is a source of photons. This is a statistical process and we denote by F_n the RV describing the statistical distribution of the number of photons emitted at voxel n and collected by the detectors of the CT scanner or of the microscope during a given acquisition time T. Then the *first basic assumption* is the following:

- F_n is a Poisson RV, with expected value f_n, i.e., the probability of the emission of k photons at voxel n is given by

$$P_{F_n}(k) = \frac{e^{-f_n} f_n^k}{k!}, \quad k = 0, 1, 2, \ldots;$$

- the RVs F_n and $F_{n'}$, corresponding to different voxels, are statistically independent.

Next, we denote by $A_{m,n}$ the probability that a photon emitted at voxel n is collected by the detector m. This probability is a crucial quantity in the modeling of the imaging process. Its computation must take into account both the geometry of the acquisition system (for instance, the geometry of the collimating devices) and the physical processes perturbing the photon before arriving at the detector m. In the case of emission tomography, for instance, one must take into account the scattering of the photons by the constituents of the tissues (generating the effects known as attenuation and scatter in PET and SPECT imaging), while, in the case of microscopy, one must consider the diffraction effects. In the latter case the matrix $A_{m,n}$ is given essentially by the PSF of the optical system.

Let $F_{m,n}$ be the RV corresponding to the number of photons emitted at voxel n and collected by the detector m. Then, the *second basic assumption* is the following:

- $F_{m,n}$ is a Poisson RV with expected value given by $A_{m,n} f_n$;
- for any fixed n and $m \neq m'$, the RVs $F_{m,n}$ and $F_{m',n}$ are statistically independent.

If we now denote by G_m the RV corresponding to the number of photons collected by the detector m, and if we assume efficiency 1 (i.e., all the photons arriving at the detector are detected), then it is obvious that G_m is given by

$$G_m = \sum_{n=1^N} F_{m,n}.$$

Thanks to the previous assumptions this RV is also a Poisson process with an expected value given by

$$E\{G_m\} = \sum_{n=1}^{N} A_{m,n} f_n = (\mathbf{Af})_m.$$

Moreover, the RVs associated to different detectors are statistically independent. It follows that the probability distribution of the vector-valued RV $\mathbf{G}$ is given by

$$P_{\mathbf{G}}(\mathbf{g}|\mathbf{f}) = \prod_{m=1}^{M} e^{-(\mathbf{Af})_m} \frac{(\mathbf{Af})_m^{g_m}}{g_m!}, \tag{29}$$

where we denote by $\mathbf{g}$ the set of whole numbers corresponding to the outputs of the detectors.

In the framework of the likelihood approach outlined above, it is easy to see that the functional (23) associated with (29) is equivalent to

$$J_{\mathbf{g}}(\mathbf{f}) = \sum_{m=1}^{M} \left\{ g_m ln \frac{g_m}{(A\mathbf{f})_m} + (A\mathbf{f})_m - g_m \right\}, \tag{30}$$

since their difference does not depend on $\mathbf{f}$. This is the Csiszár I-divergence which has the properties of a discrepancy functional [25]. It is a convex and non-negative functional. Its level sets can be used for defining the sets of approximate solutions. Their investigation is not easy. However experimental results obtained on the minimum points of this functional indicate that these sets of approximate solutions are also presumably very broad.

We conclude this section with a generalization and refinement of the previous model. In the case of fluorescence microscopy where photons are detected by means of a Charge-Coupled Device (CCD), in addition to photon noise, described above, one should also take into account so-called Read-Out Noise (RON) [64]. This is a white additive Gaussian noise and is statistically independent of the photon noise, so that we have a combination of the two types of noise described above. Since each G_m is the sum of two independent RVs, one with a Poisson distribution and the other with a Gaussian one, it follows that the probability density of the detected signals is given by

$$P_G(\mathbf{g}|\mathbf{f}) = \prod_{m=1}^{M} \sum_{k=0}^{+\infty} e^{-(A\mathbf{f})_m} \frac{(A\mathbf{f})_m^k}{k!} P_{\mathrm{RON}}(g_m - k),$$

with

$$P_{\mathrm{RON}}(u) = \frac{1}{\sqrt{2\pi}\sigma} e^{-\frac{(u-r)^2}{2\sigma^2}}$$

if the read-out noise has expected value r and variance σ^2.

We remark that the functional (23) derived from this probability density is bounded from below but is not convex. The utility of such an approach has still to be demonstrated; however, its investigation is an interesting mathematical problem.

5 The use of prior information

In the framework of the likelihood approach outlined in the previous section, it is quite natural to consider the object which most likely reproduces the detected image $\mathbf{g}$ as a possible approximate solution of the inverse problem. This is the Maximum Likelihood (ML) *estimate* and, according to (23), it can be defined by

$$\mathbf{f}_{\mathrm{ML}} = \arg \min_{\mathbf{f}} J_{\mathbf{g}}(\mathbf{f}).$$

In the case of additive white noise, we again obtain the least-squares problem discussed in Sect. 3. As we know, the solution of this problem (which, in general, is unique since the problem is overdetermined) is ill-conditioned and therefore is affected by strong noise propagation from the data to the solution. The situation is not so clear in the case of photon noise. However there are strong experimental indications that the minimization of the Csiszár I-divergence (30) also does not provide sound solutions. Indeed the minimum (or the minima; uniqueness is not proved) lies on the boundary of the closed cone of the non-negative vectors and, as a result, minima must have several zero values. This effect appears in the use of iterative methods converging to these minima and is known as *checkerboarding effect*. From these remarks one can draw the conclusion that, if one defines a set of approximate solutions as (24), then this set is too broad. Indeed, it contains both the minima of the Csiszár I-divergence and the correct solution (if ϵ is correctly chosen) and therefore it contains very different objects.

A very general idea underlying all approaches to the definition of meaningful approximate solutions of inverse problems consists in introducing criteria for extracting these solutions from the broad set of all approximate solutions by means of additional information on the solution itself. This additional information, which is sometimes called *a priori information*, derives from knowledge of expected properties of the solution. For instance, in almost all the problems considered in this chapter, the solutions must be non-negative; we also know that they cannot be too large and so on. This information can be expressed in the form of constraints on the solution of the minimization problems outlined above. For example, constraints on a norm of derivatives of the solution lead, through the method of Lagrange multipliers, to the minimization of functionals which are the sum of the discrepancy and of a regularization functional derived from the smoothness conditions. This is just the basis of Tikhonov regularization theory. In general terms, the problem becomes the minimization of a functional with structure

$$\Phi_{\mathbf{g},\mu}(\mathbf{f}) = J_{\mathbf{g}}(\mathbf{f}) + \mu\Omega(\mathbf{f}), \tag{31}$$

where $\mu > 0$ is the so-called *regularization parameter* controlling the trade-off between data fitting (the first term) and smoothness of solution (second term).

However we prefer to provide here a probabilistic justification of this approach, based on Bayes formula, which is probably more general than the regularization approach even if its formulation is, in general, restricted to the discrete case in order to avoid excessive mathematical technicalities.

The basic point in this approach is that the object $\mathbf{f}$ is also considered as a realization of a vector-valued RF $\mathbf{F}$; moreover, the probability density $P_{\mathbf{G}}(\mathbf{g}|\mathbf{f})$, introduced in the previous section, is viewed as the *conditional probability of* $\mathbf{g}$, *given* $\mathbf{f}$. Therefore, if the marginal probability density of $\mathbf{F}$, $P_{\mathbf{F}}(\mathbf{f})$, is also given, then the conditional probability density of $\mathbf{F}$ for a given $\mathbf{g}$ can be obtained by means of *Bayes formula*:

$$P_{\mathbf{F}}(\mathbf{f}|\mathbf{g}) = \frac{P_{\mathbf{G}}(\mathbf{g}|\mathbf{f})P_{\mathbf{F}}(\mathbf{f})}{P_{\mathbf{G}}(\mathbf{g})}, \tag{32}$$

where $P_{\mathbf{G}}(\mathbf{g})$ is the marginal probability density of $\mathbf{G}$, which can be obtained from the joint probability density $P_{\mathbf{FG}}(\mathbf{g},\mathbf{f}) = P_{\mathbf{G}}(\mathbf{g}|\mathbf{f})P_{\mathbf{F}}(\mathbf{f})$. Equation (32) is the basis of the so-called *Bayesian approach* to inverse problems; the marginal density $P_{\mathbf{F}}(\mathbf{f})$ is usually called the *prior*, while the conditional probability $P_{\mathbf{F}}(\mathbf{f}|\mathbf{g})$ is also called the *a posteriori* conditional probability of the object for a given image. This function provides a complete solution of the inverse problem in the sense of the Bayesian approach. Indeed, from (32) one can compute, in principle, everything about the unknown object corresponding to the detected image: expected value, maximum probability value, probability of subsets of objects, etc.

The difficulty in this approach is that the marginal probability distribution of $\mathbf{F}$, the prior, is not known, even if in some specific medical applications (e.g., the image of a human organ), one could use data bases of previously obtained images for estimating the prior. This example suggests that the prior is just what is needed for expressing our *a priori information* about the object. In other words we must use our knowledge, or ignorance, about the object for selecting this marginal distribution which restates, in a probabilistic setting, the need, mentioned above, of criteria to be used for selecting meaningful objects from the broad set of all objects compatible with the given image.

However, in any application of inverse problems and, in particular, in medical imaging, it is necessary to show at least one reconstructed object and this can be provided by the Maximum A Posteriori (MAP) estimate, which is an object maximizing the a posteriori conditional probability

$$\mathbf{f}_{\text{MAP}} = \arg\max_{\mathbf{f}} P_{\mathbf{F}}(\mathbf{f}|\mathbf{g}).$$

It is possible to replace this problem with a minimization problem by taking the logarithm of the a posteriori density and changing its sign, so that, on recalling the definition (23), we obtain

$$\mathbf{f}_{\text{MAP}} = \arg\max_{\mathbf{f}} \left\{ J_{\mathbf{g}}(\mathbf{f}) - \log P_{\mathbf{F}}(\mathbf{f}) \right\},$$

where we have neglected the contribution of $P_{\mathbf{G}}(\mathbf{g})$ since it is independent of $\mathbf{f}$. Therefore the term $-\log P_{\mathbf{F}}(\mathbf{f})$ plays the role of a regularization functional.

The most frequently used priors are of Gibbs type

$$P_{\mathbf{F}}(\mathbf{f}) = C \exp\{-\mu \Omega(\mathbf{f})\},$$

where $\Omega(\mathbf{f})$ is, in general, a convex and non-negative functional expressing prior information about the object and μ is a positive parameter (which is just the regularization parameter in regularization theory). In such a case we find that the functional to be minimized for determining the MAP estimate is just that given in (31). In particular we obtain precisely the standard functional of Tikhonov regularization theory if we assume that the image is perturbed by additive white noise, so that the discrepancy functional is given by (28), and also that the prior of the object corresponds to a white Gaussian process with zero expected value and variance $1/2\mu$, i.e., $\Omega(\mathbf{f}) = ||\mathbf{f}||_2^2$. As a result the functional (31) becomes

$$\Phi_{\mathbf{g},\mu}(\mathbf{f}) = ||A\mathbf{f} - \mathbf{g}||_2^2 + \mu||\mathbf{f}||_2^2.$$

As is well-known, for each value of μ this functional has a unique minimum which is just the classical Tikhonov regularized solution

$$\mathbf{f}_\mu = \left(A^T A + \mu I\right)^{-1} A^T \mathbf{g}, \tag{33}$$

where I denotes the identity matrix. This regularized solution, however, is not frequently used in medical imaging, firstly, because it is not suitable for the solution of large-scale problems, as we discuss in the next section and, secondly because, in the case of tomography, it is affected by aliasing effects, as discussed, e.g., in [14].

6 Computational issues and reconstruction methods

If a linear model is available, then one has at one's disposal the very powerful theoretical tools outlined in the previous Sections for the analysis and the solution of the problem. However its practical solution can require a considerable computational burden both in the 2D and in the 3D case. If the problem is treated as a sequence of 2D problems as in standard X-ray CT and also, in general, in emission tomography, then, for each section, the number of unknowns is 256×256 or 512×512. If the problem is genuinely 3D, then the number of unknowns becomes $256 \times 256 \times 64$ or $512 \times 512 \times 64$ and therefore it is of the order of millions. In such a situation it is clear that the discretization of (7) leads to a matrix which, in general, cannot be stored, even if it is sparse. Therefore further approximations are, in general, introduced in the model in order to make the problem tractable from the numerical point of view.

For instance, in the case of SPECT imaging, one neglects the so-called "collimation blur". The discrimination of the photons coming from a given direction is obtained by means of a hole in a slab; the detector counts all the photons crossing the hole and therefore integrates over the acceptance cone of this hole. It follows that the problem is moderately 3D. However, if one assumes that it is reasonable to approximate the cone with a circular cylinder, one has a situation close to that of X-ray CT (integration over straight lines or, more precisely, over tubes). In such a case the problem can be approximated by a sequence of 2D problems (as in CT) and one can use the fast algorithm of filtered back-projection. A more refined model, also leading to a sequence of 2D problems, is the so-called 2D+1 model developed in [8]. However, in such a case, the matrix does not come from the discretization of the Radon transform and, since it is too large, it must be computed whenever it is required. It is also worth mentioning the case of Magnetic Resonance (MR), where the acquisition process is designed in such a way that the reconstruction process can be reduced to Fourier transform inversion and therefore is extremely fast (and well-posed as well). For an early but excellent tutorial on MR we suggest the paper of Hinshaw and Lent [35], while, for a good description of the physics of radioisotope imaging such as PET and SPECT, see [69].

Another example is provided by fluorescence microscopy. Given that, in such a case, the PSF in (7) describes the spatial dependence of the point process, this dependence is, in general, space variant, i.e., it is not the same for all locations of the point source in the object space. Such an effect is a consequence, for instance, of the spherical aberration of the lenses of the optical system. However, the manufacturor attempts to correct this effect as far as possible. As a consequence, it is reasonable to assume that the PSF is space-invariant so that (7) can be replaced by (3). Moreover, practitioners approximate the convolution operator by means of a 3D circulant matrix, so that the Fast Fourier Transform (FFT) can be used for the computation of the matrix. It is also obvious that the storage of the matrix can be reduced to the storage of the PSF and therefore is just that of one image. If the condition of space-invariance is not satisfied, then one in general assumes that it is satisfied in subdomains of the image volume so that in each of them the reconstruction techniques developed for the space-invariant case can be used.

In general it is assumed that the problem can be solved in almost real time if one of the following conditions is satisfied:

- the 3D problem can be approximated by a sequence of 2D problems, each implying a Radon transform inversion;
- the matrix A is sparse and is not stored since it can be computed by means of a look-up table of given values or by means of simple rules;
- the matrix A is not sparse but is given by a space-invariant PSF which can be stored.

In the first case the standard algorithm is Filtered Back-Projection (FBP) while, in the other cases, the most frequently used approaches are based on iterative algorithms with regularization properties. For the convenience of the reader we briefly describe FBP, also because it provides terminology which is frequently used in medical imaging.

As discussed in the introduction the 2D Radon transform is given by $(Rf)(s,\theta) = (P_\theta f)(s)$, with $(P_\theta f)(s)$ defined in (1). The inversion formula (2) can be decomposed in two steps: the first is the computation of the Hilbert transform of the derivative, with respect to s, of the Radon transform of f; the second consists in applying to the result the *back-projection operator* defined by

$$(R^{\#}g)(x) = \int_{S^1} g(x\cdot\theta,\theta)d\theta .$$

This operator is, in a sense, the dual of the Radon operator R because, while R corresponds to integrating over the points of a line, $R^{\#}$ corresponds to integrating over the lines through a point. It is also the (formal) adjoint of R and, for this reason, in medical imaging it is usual to denote the imaging matrix A, introduced in the previous sections, as the *projection matrix* and the matrix A^T as the *back-projection matrix*.

Now, the FBP algorithm consists of the following two steps, corresponding to the two steps indicated above.

- Step 1 (filtering). Compute the 1D Fourier transform of each projection $g_\theta = (Pf)_\theta(s)$, namely, $\hat{g}_\theta(\omega)$, multiply the result by the ramp filter $|\omega|$ and take the inverse Fourier transform; the result is the filtered projection in the direction θ:

$$G_\theta(s) = \frac{1}{2\pi}\int_{-\infty}^{+\infty} |\omega|\; \hat{g}_\theta(\omega)e^{is\omega}\,ds .$$

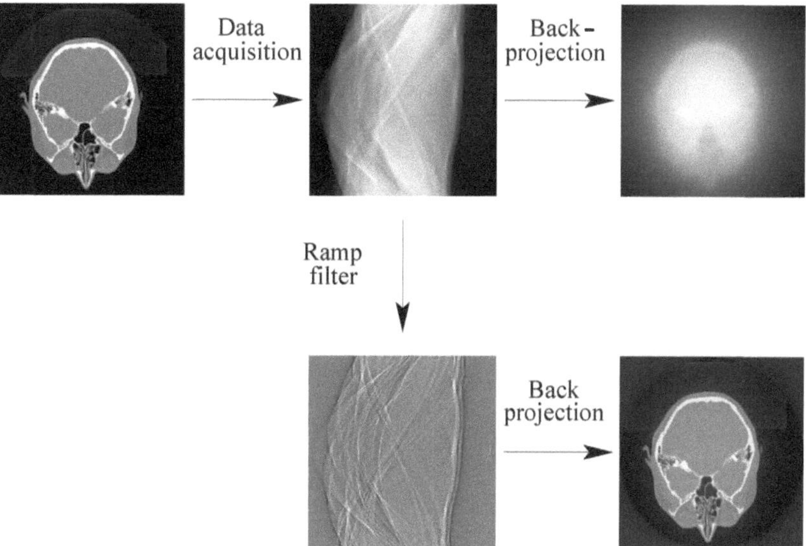

Fig. 3. Pictorial representation of the FBP algorithm

This step is just the computation, except for a constant, of the Hilbert transform of the derivative of g_θ. The set of filtered projections provides the filtered sinogram, which is the representation of the 2D function $G(s, \theta) = G_\theta(s)$.

- Step 2 (back-projection). This step is just the application of the back-projection operator to the filtered projections; it provides the function f:

$$f(x) = \frac{1}{4\pi}(R^{\#}G)(x).$$

It is easy to verify that this is a different way of writing (2).

In Fig. 3 we give a pictorial representation of the FBP algorithm, also showing that, if we apply the back-projection operator directly to the projections (without the ramp filter), then we get a blurred version of the object. In this figure the sharpening of the sinogram provided by the ramp filter is also evident.

It is obvious that the FBP algorithm, here described in a continuous setting, allows for fast implementations. Indeed the computation of the filtered projections can be performed by means of the FFT algorithm and therefore its computational cost is of the order of $M \log_2 M$, if M is the number of data. More expensive is the computation of the back-projection even though fast algorithms have also been designed in this case (see, e.g., [57]). We also note that the ill-posed nature of the problem manifests itself in the multiplication of $\hat{g}_\theta(\omega)$ by the ramp filter. Indeed this filter amplifies the high-frequency noise, i.e., the noise affecting the values of $\hat{g}_\theta(\omega)$ for large values of ω. This effect is corrected, in practice, by attenuating the ramp filter at the higher frequency. An example is provided by the *Shepp-Logan filter* [42] and such a procedure is basically a regularizing one.

The efficiency of the FBP algorithm makes clear why it is the favorite in commercial machines: very often the acquisition processes are designed in such a way that the data approximately provides line integrals of the unknown object. If this approximation is not satisfactory or if one intends to improve the results provided by FBP, then one has to deal with a large-scale projection matrix A and one must solve one of the large-scale minimization problems discussed in the previous sections. It is obvious that direct methods such as those provided by the Tikhonov regularized solution (33) are not feasible in practice. Therefore it is quite natural to look for iterative methods such that at each iteration the main computational burden is a matrix-vector multiplication.

It is also important to observe that the minimization of regularized functionals, such as those introduced in the previous section, faces the problem of the choice of regularization parameter. Several criteria have been investigated in the case of Tikhonov regularization method (see, e.g., Engl et al. [29], Bertero and Boccacci [4]) but not in the other cases. In general one should experiment with the method on sets of simulated images for establishing the best possible values of these parameters. Such an approach is very costly from the computational point of view.

A practical solution to these problems is provided by iterative methods for the minimization of a discrepancy functional, such as that defined in (28) or (30), with a property which is called *semiconvergence*: if we define a restoration error as a distance between the result of iteration k and the true object, then the restoration error first decreases, goes through a minimum and then increases up to very large values (remember that the minima of the discrepancy functionals are strongly affected by noise propagation). It turns out that the solution provided by the iteration corresponding to the minimum of the restoration error is, in general, a reliable solution of the reconstruction problem. In such a case, we do not have the problem of choice of regularization parameter, but rather a problem of optimal stopping. Again criteria can be obtained by experimenting with the algorithm.

In the case of the least-squares functional (28) the prototype of these methods is a gradient method known, in the inverse problem literature, as the *Landweber method*:

$$\mathbf{f}^{(k+1)} = \mathbf{f}^{(k)} + \tau A^T (\mathbf{g} - A\mathbf{f}^{(k)}),$$

where τ is a relaxation parameter (the step in the direction of steepest descent) satisfying the condition

$$0 < \tau < \frac{2}{\sigma_1^2},$$

where σ_1 is the maximum singular value of the matrix A. It is easy to prove (see, e.g., [4]) that this iterative method, initialized with the zero object, converges to the minimum norm least-squares solution and that the effect of the iteration k is essentially a filtering of this solution, keeping the components corresponding to the largest singular values and dropping the others: as the iteration goes on, the components corresponding to smaller singular values also appear, hence introducing the instability due to noise amplification. By such an analysis it is possible to prove the semiconvergent behavior of the restoration error.

The main drawback of this method is that it is too slow: the minimum of the reconstruction error can be reached only after hundreds of iterations. A much more convenient method is provided by the Conjugate Gradient (CG). It is proved that this method also has the semiconvergence property (see, e.g., Engl et al. [29]), but it is much faster than the Landweber method, especially if preconditiong techniques are used. For instance, it was shown, in an application to SPECT imaging [8], that CG, equipped with a very simple preconditioner, provides reliable solutions after 10 to 15 iterations.

However, in the case of emission tomography or fluorescence microscopy, the least-squares discrepancy (28) is not the appropriate one and one must use that derived from the appropriate noise model, namely, (30). An iterative method for the minimization of this discrepancy in the case of emission tomography was proposed by Shepp and Vardi [63]. It is called Expectation Maximization (EM) since it is a particular case of a general method, with the same name, for the solution of maximum likelihood problems. It must be remarked that the same iterative algorithm was previously proposed by Richardson [60] and Lucy [50] for deconvolution problems and, for this reason, the algorithm is known as the Richardson-Lucy Method (RLM) in astronomy and microscopy. The algorithm is as follows:

$$\mathbf{f}^{(k+1)} = \mathbf{f}^{(k)} A^T \frac{\mathbf{g}}{A\mathbf{f}^{(k)}},$$

where product and quotient of vectors are to be interpreted component-wise (Hadamard product and quotient). We note an interesting property of this algorithm: since the elements of the matrix A and the components of the image $\mathbf{g}$ are non-negative, if the initial guess is also non-negative, then all the iterates are automatically non-negative. It must also be pointed out that, in general, a positive initial guess is chosen (as a rule of thumb, a constant vector), because, as a consequence of the multiplicative structure of the algorithm, if a component of the initial guess is zero, then the same component of all the iterates is zero.

It was proved by Shepp and Vardi (a more complete proof is given by Lange and Carson [47]) that, for any positive initial guess, the iterates converge to a maximum of the likelihood function, hence, to a minimum of the Csiszár I-divergence. However, after a number of iterations, the iterates show the checkerboard effect mentioned earlier, indicating that the minima of the Csiszár I-divergence are not reliable solutions. The utility of the algorithm is due to the fact that it has the semiconvergence property (see [4] for a discussion), an experimental result derived from numerical practice. Some theoretical insight is derived from an analysis of the filtering effect of the iterations [58]. As a consequence reliable solutions can be obtained by suitable stopping of the iteration.

EM, as Landweber, is very slow and, in general, requires a large number of iterations. In the case of emission tomgraphy, an accelerated version known as Ordered Subset – Expectation Maximization (OS-EM) was proposed by Hudson and Larkin [36]. This approach improves considerably the efficiency of EM; it was implemented for the reconstruction of SPECT data based on the 2D+1 model [8] and it provides reconstructions in almost real time (a few minutes) so that it is currently used

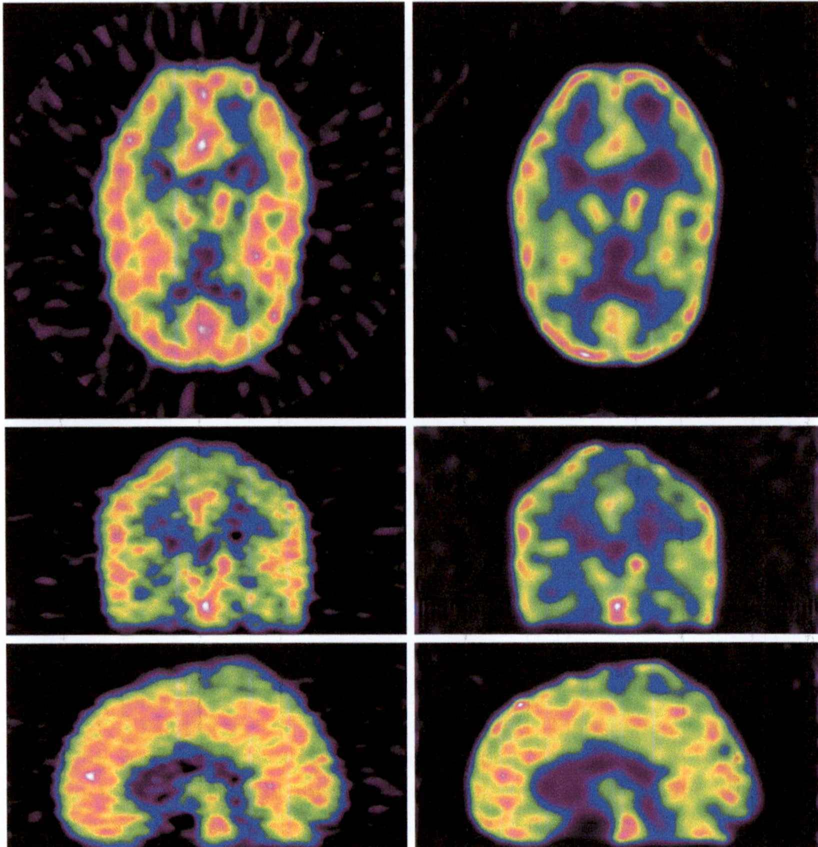

Fig. 4. Left-hand panels: reconstructions provided by the FBP algorithm, using a Gaussian filter in the axial direction. Right-hand panels: reconstructions obtained by means of the 2D+1 model for the collimator blur and the OS-EM algorithm for data inversion. Upper panels: transaxial sections. Middle panels: coronal sections. Lower panel: sagittal sections

by medical doctors of the Universities of Genoa and Florence for the reconstruction and analysis of SPECT data.

In Fig. 4 we give an example of the improvement which can be obtained by means of this method with respect to FBP, by comparing the results obtained with the two methods in the reconstruction of a 3D brain image. The comparison is performed by showing the results for a transaxial section (upper panels), a coronal section (middle panels) and a sagittal section (lower panels). The left-hand panels correspond to FBP. Since the FBP reconstruction is simply a set of 2D reconstructions, a smoothing in the axial direction is needed and this is obtained by means of a suitable Gaussian filter. In the right-hand panels we show the result of a 3D reconstruction, based on the $2D + 1$ model for the collimator blur, obtained by grouping the projections into 12 subsets and using 9 $OS\text{-}EM$ iterations. The improvement is evident not only in the coronal and sagittal sections, as is obvious since we use a 3D method, but also in the transaxial section. This improvement is mainly due to the improved model of the imaging matrix.

7 Perspectives

The different kinds of tomography and microscopy, as well as MR, represent well-established imaging techniques, characterized by a high degree of technological development and validated through a large variety of biomedical applications. Nonetheless, recent decades have seen an impressive growth of the investigation of new modalities able to provide information on both the structural characteristics and the functional status of the tissues under examination. The theoretical modeling of these procedures often utilizes sophisticated mathematical tools, thus providing actual motivations to investigations in important areas of pure mathematics. On the other hand, the necessity of reliable reconstruction methods on a real-time scale, has inspired the formulation of numerical algorithms of a more general flavor and has significantly contributed to the development of a new applied science, characterized by the integration of concepts and techniques originally formulated in different areas of mathematics, physics, biomedicine and computer science. In the following we provide outlines of some of these new imaging techniques, indicating both their roles as diagnostic tools and the main theoretical connections with different aspects of applied mathematics.

7.1 Electrical Impedence Tomography

The inverse problem at the basis of Electrical Impedance Tomography (EIT) is that of imaging the electrical properties in the interior of a body given measurements of electric currents and voltages at the boundary. Medical problems for which knowledge of internal electrical properties are helpful are related to: lung medicine, for the detection of pulmonary emboli or blood clots in the lung; non-invasive monitoring of heart functions; non-invasive detection of breast cancer. Within the highly idealized framework of the continuum model [16], the electric potential u in the domain Ω satisfies the boundary value problem:

$$\nabla \cdot \gamma(x, \omega) \nabla u = 0, \quad x \in \Omega,$$
$$\gamma(x) \frac{\partial u(x)}{\partial \nu} = j(x), \quad x \in \partial\Omega,$$

where

$$\gamma(x, \omega) = \sigma(x, \omega) + i\omega\epsilon(x, \omega)$$

is the admittivity (proportional to the inverse of the impedance of the body), σ the electric conductivity, i the imaginary unit, ω the angular frequency of the applied current, ϵ the electric permittivity, j the surface current density and ν the inward-pointing normal unit vector. This Neumann problem is well-posed under very general conditions on γ, provided that the charge conservation is assured and a choice for $\int_{\partial\Omega} u$ is fixed. The corresponding inverse problem is much more complicated and can be related to that of determining γ when the Dirichlet-to-Neumann map relating the voltage and the current on the surface is known. Many important mathematical issues arise from this problem, essentially concerned with both the injectivity of the forward map and the stability of the inverse operator [46, 55, 67]. In [13] an imaging algorithm is provided for the linearized problem, which was numerically tested in [37] in the case of an elementary geometry. However, in [17] and [65] it is pointed out that the continuum model is very poor for real experiments, where the current is known only in wires attached to discrete electrodes and significant electrochemical effects at the contact between the electrodes and the body must be taken into account. On the basis of a more realistic and complete model, an adaptive current tomography system has been designed and realized by the EIT group at the Rensselaer Polytechnic Institute [16] and reconstruction algorithms have been implemented and tested in the case of both simulated and real experiments.

7.2 Optical Tomography

Optical Tomography (OT) [2] is a functional medical imaging modality whose aim is to reconstruct images of optical coefficients, such as tissue absorption, from boundary measurements of light transmission by using light in the infrared band between 700 and 1000 nm. The medical motivations of this technique [15] are essentially related to brain function monitoring, as in the case of the early detection of after-birth oxygen deficiency or localization of cortical activation areas during stimulated tasks, although structural reconstructions can be obtained in the case of the detection of brain and breast cancer. According to a very schematic description, the mathematical model for OT is based on the integro-differential equation for photon transport theory [39], which, in the frequency domain, reduces to the elliptic equation [48]:

$$\nabla \cdot (A \nabla \hat{\Phi}) - (B + iC)\hat{\Phi} = \hat{q}.$$

Here $\hat{\Phi}$ and $\hat{q}$ are, respectively, the Fourier transform of the isotropic photon density Φ and of a function q of space and time representing the source distribution; A and B are coefficients related to the tissue absorption and scattering properties and C is the reciprocal of the speed of light. The inverse problem of OT is that of determine estimates of A and B, given C and complete data at the boundaries for two different frequencies. A proof of uniqueness of the solution can be found in [38] while a rather complete review of the most frequently used reconstruction methods is in [2]. A prototype scanner for OT has been constructed at the Biomedical Optics Research Laboratory, University College, London (http://www.medphys.ucl.ac.uk/research/borg/index.htm). The primary clinical aim of this device is to detect oxygenation failures in newborn infant brains.

7.3 Microwave Tomography

Microwave Tomography (MT) is a structural imaging technique whose aim is to reconstruct the electrical parameters of the tissues from measurements of the scattered electromagnetic field in the gigahertz frequency range (see [53]). This low-resolution, high-contrast methodology is particularly helpful both when a low invasivity is required, as in mammography [30], and when the cancerous tissues provoke significant variations of the refractive index, as in the diagnosis of leukemia in human bone marrow [1]. From a mathematical point of view, in a two-dimensional setting MT can be described by the scattering problem introduced in Sect. 2 (Eqs. (8)–(11)) and, in three dimensions, by an inverse problems for Maxwell equations [11]. In MT no linearization can be performed, since the microwave wavelength in biological tissues is of some centimeters and therefore genuine resonance conditions hold. This implies that most image reconstruction methods in this framework must address a notable computational expense which is often reduced by adopting oversimplified approximations in the model. Typical approaches are based on generalization of methods originally formulated for diffraction tomography [68], on Newton-Kantorowich algorithms [40] and on gradient methods [45]. Important experimental results have been obtained at the Biophysical Laboratory, Carolinas Medical Center, Charlotte, NC for various biomedical applications [62]. At the Department of Biocybernetics, Niigata University, Niigata, an MT prototype has been realized, with the aim of obtaining a hardware-type linearization of the image restoration problem by means of a mixing/filtering procedure [52]. In this application, named Chirp-Pulse Microwave Computerized Tomography (CP-MCT), the input is a chirp signal characterized by increasing frequency, typically from 1 GHz to 2 GHz, the geometry adopted is that of X-ray tomography in its parallel or fan beam setup and the basic reconstruction algorithm is filtered back-projection. A refined mathematical model for image formation in CP-MCT was proposed in [7] and validated in [54] while a theoretical computation of the PSF of the device by means of scattering theory methods was performed in [5]. Applications of this technique to the reconstruction of real simple objects and to simulated mammographical experiments are currently in progress.

7.4 Magnetoencephalography

Magnetoencephalography (MEG) [34] is a brain-imaging technique measuring the weak magnetic fields related to neural currents, with the aim of inverstigating cerebral functional behavior, particularly in the study of auditory and visual cortex, for both healthy subjects and for patients affected by epilepsy or developmental dyslexia. When compared to other functional modalities such as PET, SPECT or fMRI, MEG is characterized by very low invasivity and by a high temporal resolution, which allows us to infer information about the temporal hierarchy of physiological responses. Furthermore current MEG scanners provide a spatial resolution of some millimeters, particularly for sources in the brain cortex, and reliable coregistration procedures with MR structural data allows us to reproduce even finer anatomical details. The MEG forward problem [61] is that of computing the magnetic field $\mathbf{B}(\mathbf{r})$ of a given continuous current distribution $\mathbf{J}(\mathbf{r})$ when $\mathbf{B}(\mathbf{r})$ and $\mathbf{J}(\mathbf{r})$ are related by the Biot-Savart law

$$\mathbf{B}(\mathbf{r}) = \frac{\mu_0}{4\pi} \int_{\Omega} \mathbf{J}(\mathbf{r}') \times \frac{\mathbf{r} - \mathbf{r}'}{|\mathbf{r} - \mathbf{r}'|^3} d\mathbf{r}',$$

and Ω is a bounded domain corresponding to the human brain. The MEG inverse problem is ill-posed in the sense of Hadamard and non-linear. Being ill-posed is a consequence of the fact that the integral operator relating the current density and magnetic field has a non-empty kernel [31] while non-linearity arises when one models the neuronal source as a current dipole and wants to determine both its position and amplitude. Standard approaches to the solution of this problem need to deal with time series characterized by a sufficiently high signal-to-noise ratio, and this result is typically accomplished by averaging several magnetic responses to repeated realizations of the same external stimulus. However, the main drawback of this procedure is that the averaging process strongly dissipates the temporal information content of the MEG signal. Furthermore, averaging such signals would be reliable if the stimulus responses were the same for each repetition, but this is not necessarily true, since, in order to avoid habituation effects, typical stimulus realizations involve small modifications of the stimulus characteristics at each repetition. More effective approaches utilizing raw non-averaged time series are based on a probabilistic Bayesian approach [66], whereby the source localization at each sampled time is obtained by building up a density probability function conditioned on the experimental measurements, and the time evolution of the process is realized by exploiting the form of the density function constructed at the previous time step.

Acknowledgements

We thank our colleagues Patrizia Boccacci and Piero Calvini for providing some of the images displayed in this chapter and for discussing related topics.

References

[1] Albanese, R.A., Medina, R.L., Penn, J.W.: Mathematics, medicine and microwaves. Inverse Problems **10**, 995–1007 (1994)

[2] Arridge, S.R.: Optical tomography in medical imaging. Inverse Problems **15**, R41–R93 (1999)

[3] Bertero, M.: Linear inverse and ill-posed problems. In: Hawkes, P. W. (ed.): Advances in electronics and electron physics. Vol. 75. New York: Academic 1989, pp. 1–120

[4] Bertero, M., Boccacci, P.: Introduction to inverse problems in imaging. Bristol: Institute of Physics 1998

[5] Bertero, M., Conte, F., Miyakawa, M., Piana, M.: Computation of the response function in chirp-pulse microwave computerized tomography. Inverse Problems **17**, 485–500 (2001)

[6] Bertero, M., De Mol, C., Viano, G.: The stability of inverse problems. In: Baltes, H.P. (ed.): Inverse scattering problems in optics. Berlin: Springer 1980, pp. 161–214

[7] Bertero, M., Miyakawa, M., Boccacci, P., Conte, F., Orikasa, K., Furutani, M.: Image restoration in chirp-pulse microwave CT (CP-MCT). IEEE Trans. Biomed. Eng. **47**, 690–699 (2000)

[8] Boccacci, P., Bonetto, P., Calvini, P., Formiconi, A.R.: A simple model for the efficient correction of collimator blur in 3D SPECT imaging. Inverse Problems **15**, 907–930 (1999)

[9] Bracewell, R.N., Riddle, A.C.: Inversion of fan-beam scans in radio astronomy. Astrophys. J. **150**, 427–434 (1967)

[10] Brühl, M., Hanke, M.: Numerical implementation of two noniterative methods for locating inclusions by impedance tomography. Inverse Problems **16**, 1029–1042 (2000)

[11] Bulyshev, A.E., Souvorov, A.E., Semenov, S.Y., Svenson, R.H., Nazarov, A.G., Sizov, Y.E., Tatsis, G.P.: Three-dimensional microwave tomography. Theory and computer experiments in scalar approximation. Inverse Problems **16**, 863–875 (2000)

[12] Cakoni, F., Colton, D.: On the mathematical basis of the linear sampling method. Georgian Math. J. **10**, 411–425 (2003)

[13] Calderon, A.-P.: On an inverse boundary value problem. In: Seminar on numerical analysis and its applications to continuum physics. Rio de Janeiro: Soc. Brasileira de Matematica 1980, pp. 65–73

[14] Caponnetto, A., Bertero, M.: Tomography with a finite set of projections: singular value decomposition and resolution. Inverse Problems **13**, 1191–1205 (1997)

[15] Chance, B., Delpy, D.T., Cooper, C.E., Reynolds, E.O.R. (eds.): Near-infrared spectroscopy and imaging of living systems. Philos. Trans. Roy. Soc. London Ser. B **352**, 643–761 (1997)

[16] Cheney, M., Isaacson, D., Newell, J.C.: Electrical impedance tomography. SIAM Rev. **41**, 85–101 (1999)

[17] Cheng, K.-S., Isaacson, D., Newell, J.C., Gisser, D.G.: Electrode models for electric current computed tomography. IEEE Trans. Biomed. Engr. **36**, 918–924 (1989)

[18] Colton, D., Coyle, J., Monk, P.: Recent developments in inverse acoustic scattering theory. SIAM Rev. **42**, 369–414 (2000)

[19] Colton, D., Haddar, H., Piana, M.: The linear sampling method in inverse electromagnetic scattering theory. Inverse Problems **19**, S105–S138 (2003)

[20] Colton, D., Kirsch, A.: A simple method for solving inverse scattering problems in the resonance region. Inverse Problems. **12**, 383–393 (1996)

[21] Colton, D., Kress, R.: Inverse acoustic and electromagnetic scattering theory. 2nd ed. Berlin: Springer 1998

[22] Colton, D., Monk, P.: A linear sampling method for the detection of leukemia by using microwaves. SIAM J. Appl. Math. **58**, 926–941 (1998)

[23] Colton, D., Piana, M., Potthast, R.: A simple method using Morozov's discrepancy principle for solving inverse scattering problems. Inverse Problems **13**, 1477–1493 (1997)

[24] Courant, R., Hilbert, D.: Methods of mathematical physics. II. Partical differential equations. New York: Interscience 1962

[25] Csiszár, I.: Why least squares and maximum entropy? An axiomatic approach to inference for linear inverse problems. Ann. Statist. **19**, 2032–2066 (1991)

[26] Davison, M.E.: A singular value decomposition for the Radon transform in n-dimensional Euclidean space. Numer. Funct. Anal. Optim. **3**, 321–340 (1981)

[27] Devaney, A.J.: Diffraction tomography. In: Boerner, W.-M. et al. (eds.): Inverse methods in electromagnetic imaging. Part 2. Boston: Reidel 1985, pp. 1107–1136

[28] Diaspro, A. (ed.): Confocal and two-photon microscopy: foundations, applications, and advances. New York: Wiley 2001

[29] Engl, H.W., Hanke, M., Neubauer, A.: Regularization of inverse problems. Dordrecht: Kluwer 1996

[30] Fear, E.C., Li, X., Hagness, S.C., Stuchly, M.A.: Confocal microwave imaging for breast cancer detection: localization of tumors in three dimensions. IEEE Trans. Biomed. Eng. **49**, 812–823 (2002)

[31] Fokas, A.S., Kurylev, Y., Marinakis, V.: The unique determination of neuronal currents in the brain via magnetoencephalography. Inverse Problems **20**, 1067–1082 (2004)

[32] Goodman, J.W.: Introduction to Fourier optics. San Francisco: McGraw-Hill 1968

[33] Hadamard, J.: Lectures on Cauchy's problem in partial differential equations. New Haven: Yale University Press 1923

[34] Hämäläinen, M., Hari, R., Ilmoniemi, R., Knuutila, J., Lounasmaa, O.V.: Magnetoencephalography – theory, instrumentation and applications to noninvasive studies of the working human brain. Rev. Mod. Phys. **65**, 413–497 (1993)

[35] Hinshaw, W.S., Lent, A.H.: An introduction to NMR imaging: From the Bloch equation to the imaging equation. Proc. IEEE **71**, 338–350 (1983)

[36] Hudson, H M., Larkin, R.S.: Accelerated image reconstuction using ordered subsets of projection data. IEEE Trans. Med. Imag. **13**, 601–609

[37] Isaacson, D., Isaacson, E.: Comments on A.-P. Calderon's paper: 'On an inverse boundary value problem'. Math. Comp. **52**, 553–559 (1989)

[38] Isakov, V.: Inverse problems for partial differential equations. New York: Springer 1998

[39] Ishimaru, A.: Wave propagation and scattering in random media. New York: Academic Press 1978

[40] Joachimowicz, N., Pichot, C., Hugonin, J.P.: Inverse scattering: an iterative numerical method for electromagnetic imaging. IEEE Trans. Antennas Propagat. **39**, 1742–1753 (1991)

[41] Kaipio, J.P., Somersalo, E.: Statistical and computational inverse problems. Berlin: Springer 2004

[42] Kak, A.C., Slaney, M.: Principles of computerized tomographic imaging. Philadelphia: SIAM 2001

[43] Kato, T.: Perturbation theory for linear operators. Berlin: Springer 1980

[44] Kirsch, A.: Characterization of the shape of a scattering obstacle using the spectral data of the far field operator. Inverse Problems **14**, 1489–1512 (1998)

[45] Kleinman, R.E., van den Berg, P.M.: A modified gradient method for two-dimensional problems in tomography. J. Comput. Appl. Math. **42**, 17–35 (1992)

[46] Kohn, R.V., Vogelius, M.: Determining conductivity by boundary measurements. Comm. Pure Appl. Math. **37**, 289–298 (1984)

[47] Lange, K.L., Carson, R.: EM reconstruction algorithms for emission and transmission tomography. J. Comput. Assist. Tomogr. **8**, 306–316 (1984)

[48] Lionheart, W.R.B, Arridge, S.R., Schweiger, M., Vaukhonen, M., Kaipio, J.P.: Electrical impedance and diffuse optical tomography reconstruction software. In: Proceedings of the 1st World Congress on Industrial Process Tomography. CD. Wilmslow, UK: Process Tomography 1999, pp. 474–479

[49] Louis, A.K.: Laguerre and computerized tomography: consistency conditions and stabiloity of the Radon transform. In: Brezinsky, C. et al. (eds): Polynomes orthogonaux et applications. (Lecture Notes in Math **1171**) Berlin: Springer 1985, pp. 524–531

[50] Lucy, L.: An iterative technique for the rectification of observed distributions. Astron. J. **79**, 745–754 (1974)

[51] Mazzone, F., Coyle, J., Massone, A.M., Piana, M.: FIST: A fast visualizer for fixed-frequency acoustic and electromagnetic inverse scattering problems. Simulation Modeling Practice Theory **14**, 177–187 (2006)

[52] Miyakawa, M.: Tomographic measurements of temperature change in the phantoms of the human body by chirp radar-type microwave computed tomography. Med. Biol. Eng. Comput. **31**, S31–S36 (1993)

[53] Miyakawa, M., Bolomey, J.C.: Non-invasive thermometry of human body. Boca Raton: CRC Press 1996

[54] Miyakawa, M., Orikasa, K., Bertero, M., Boccacci, P., Conte, F., Piana, M.: Experimental validation of a linear model for data reduction in chirp-pulse microwave CT. IEEE Trans. Med. Imag. **21**, 385–395 (2002)

[55] Nachman, A.I.: Global uniqueness for a two-dimensional inverse boundary value problem. Ann. of Math. **143**, 71–96 (1996)

[56] Natterer, F.: The mathematics of computerized tomography. Philadelphia: SIAM 2001

[57] Natterer, F., Wübbeling, F.: Mathematical methods in image reconstruction. Philadelphia: SIAM 2001

[58] Prasad, S.: Statistical-information-based performance criteria for Richardson-Lucy image deblurring. J. Opt. Soc. Amer. A**19**, 1286–1296 (2002)

[59] Radon, J.: Über die Bestimmung von Funktionen durch ihre Integralwerte längs gewisser Mannigfaltigkeiten. Ber. Verh. Königl. Sächs. Ges. Wiss., Leipzig, Math.-Phys. Kl. **69**, 262–277 (1917)

[60] Richardson, W.H.: Bayesian-based iterative method of image restoration. J. Opt. Soc. Amer. **62**, 55–59 (1972)

[61] Sarvas, J.: Basic mathematical and electromagnetic concepts of the biomagnetic inverse problem. Phys. Med. Biol. **32**, 11–22 (1987)

[62] Semenov, S.Y., Svenson, R.H., Boulyshev, A.E., Souvorov, A.E., Borisov, V.Y., Sizov, Y., Starostin, A.N., Dezern, K.R., Tatsis, G.P., Baranov, V.Y.: Microwave tomography: two-dimensional system for biological imaging. IEEE Trans. Biomed. Eng. **43**, 869–877 (1996)

[63] Shepp, L.A., Vardi, Y.: Maximum likelihood reconstruction for emission tomography, IEEE Trans. Med. Imag. **1**, 113–122 (1982)

[64] Snyder, D.L., Hammoud, A.M., White, R.L.: Image recovery from data acquired with a charge-coupled-device camera. J. Opt. Soc. Amer. A**10**, 1014–1023 (1993)

[65] Somersalo, E., Cheney, M., Isaacson, D.: Existence and uniqueness for electrode models for electric current computed tomography. SIAM J. Appl. Math. **52**, 1023–1040 (1992)

[66] Somersalo, E., Voutilainen, A., Kaipio, J.P.: Non-stationary magnetoencephalography by Bayesian filtering of dipole models. Inverse Problems **19**, 1047–1063 (2003)

[67] Sylvester, J., Uhlmann, G.: A uniqueness theorem for an inverse boundary value problem in electrical prospection. Comm. Pure Appl. Math. **39**, 91–112 (1986)

[68] Tabbara, W., Duchêne, B., Pichot, C., Lesselier, L., Chommeloux, L., Joachimowicz, N.: Diffraction tomography: contribution to the analysis of some applications in microwaves and ultrasonics. Inverse Problems **4**, 305–331 (1988)

[69] Webb, S.: The physics of medical imaging. Bristol: Institute of Physics Publishing 1988

[70] Weinstein, M., Castleman, K.R.: Reconstructing 3-D specimens from 2-D section images. In: Herron R.E. (ed.): Quantitative imagery in the biomedical sciences I. (Proceedings SPIE **26**) Redondo Beach, CA: Soc. Photo-Optical Instrumentation Engineers 1971, pp. 131–138

[71] You, Y.X., Miao, G.P., Liu, Y.Z.: A fast method for acoustic imaging of multiple three-dimensional objects. J. Acoust. Soc. Amer. **108**, 31–37 (2000)

Stochastic geometry and related statistical problems in biomedicine

V. Capasso, A. Micheletti

The notion of the importance of pattern is as old as civilization.... Mathematics is the most powerful technique for the understanding of pattern, and for the analysis of the relation of patterns.

A.N. Whitehead

Abstract. Many processes of biomedical interest may be modeled as birth-and-growth processes (germ-grain models), which are composed of two processes, birth (nucleation, branching, etc.) and subsequent growth of spatial structures (cells, vessel networks, etc.), both of which, in general, are stochastic in time and space. These structures induce a random division of the relevant spatial region, known as a random tessellation. A quantitative description of the spatial structure of the tessellation can be given, in terms of the mean densities of the interfaces (the so-called n-dimensional facets).

One significant example in this respect is the mathematical modeling of tumor growth and of tumor-induced angiogenesis. The understanding of the principles and dominant mechanisms underlying tumor growth is an essential prerequisite for identifying optimal control strategies, in terms of prevention and treatment. Predictive mathematical models which are capable of producing quantitative morphological features of developing tumor and blood vessels can contribute to this. The strong coupling of the kinetic parameters of the relevant birth-and-growth (or branching-and-growth) process with underlying biochemical fields, induces stochastic time and space heterogenities, thus motivating a more general analysis of the stochastic geometry of the process. But the formulation of an exhaustive evolutionary model which relates all the relevant features of the real phenomenon dealing with different scales, and the stochastic domain decomposition at different Hausdorff dimensions, is a problem of great complexity, both analytical and computational. Methods for reducing complexity include homogenization at larger scales, thus leading to hybrid models (deterministic at the larger scale, and stochastic at lower scales). As examples we present two simplified stochastic geometric models, for which we discuss how to relate the geometric probability distribution to the kinetic parameters of birth and growth.

On the dual side, in order to compare numerical simulations of proposed mathematical models with existing data, we provide methods of statistical analysis for the estimation of geometric densities that characterize the morphology of a real system.

Keywords: birth-and-growth processes, stochastic geometry, geometric densities, tumor growth, angiogenesis, statistics of stochastic geometries.

1 Introduction

As D'Arcy Thompson pointed out in his pioneering book "On Growth and Form" [52] (also see [26,47]):

> *"there is an important relationship between the form or shape of a biological structure and his function".*

The scope of this presentation is to introduce relevant nomenclature and mathematical tools for the analysis of the morphology of spatial patterns characterized by a random decomposition of the relevant spatial region. More than that, we wish to introduce those mathematical tools for the modeling of the time evolution of such patterns, in a causal description with respect to the parameters/variables which are responsible for the phenomenon itself.

Further we present methods for the statistical analysis of random spatial tessellations, e.g., for the estimation of the geometric quantities that characterize the morphology of a real system. This aspect is essential if we wish to compare numerical simulations of proposed mathematical models with existing real data.

We wish to call attention to two of the most important topics of mathematical interest in this context, for which we refer to the relevant literature. While *morphogenesis (pattern formation)* deals with the mathematical modeling of the causal description of a pattern (a direct problem) [42,53] (also see [44]), *stochastic geometry* deals with the analysis of geometric aspects of "patterns" subject to stochastic fluctuations (direct modeling, and inverse statistical problems) [5,51]. A recent source of case studies related to the topics presented here is [48].

In applications to medicine, it is very important, for either prevention or clinical treatment, to describe the relevant phenomenon in terms of a biomathematical model including all significant features of the system, so as to identify possible control parameters (variables). Examples of interest for biomedicine are provided by tumor growth [20,22,25], angiogenesis [18,41,49], etc.

The understanding of the principles and dominant mechanisms underlying tumor growth is an essential prerequisite for identifying optimal control strategies, in terms of prevention and treatment. Predictive mathematical models which are capable of producing quantitative morphological features of developing tumors and blood vessels can contribute to this [19,27]. In [22], and its references, different models of the growth of tumors are discussed.

In this paper we present and analyze a multicellular model of tumor invasion, based on a spheroidal growth model, coupled with a stochastic birth model (as typical of brain tumors for example) [46]. Ignoring death is a nontrivial modeling simplification, that reduces the applicability of the model to early stages of growth and in vitro experiments [25]. On the other hand, for such a model, we may import a good deal of the experience of similar models in the polymer solidification literature, and we are able to provide significant results about the evolution of the morphology in terms of the kinetic parameters of the system [9].

In [18] the authors focus on the mathematical modeling of capillary networks as they arise in tumor-induced angiogenesis. Angiogenesis is one of the most important

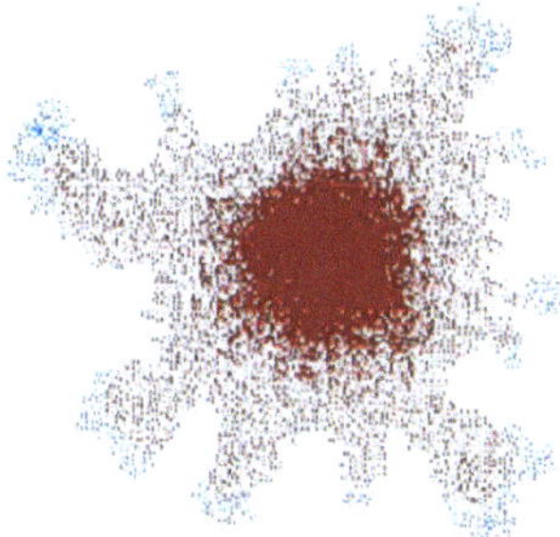

Fig. 1. A simulation of the growth of a tumor mass coupled with a random underlying field (published with permission from [2]). The effective growing zone is the area in light blue at the surface

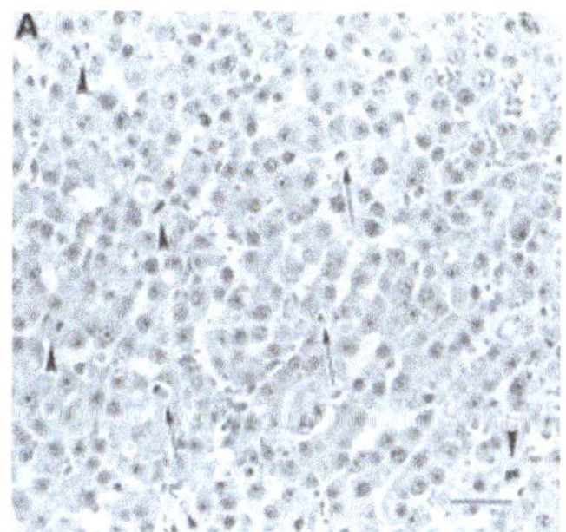

Fig. 2. Section of an hepatocellular adenoma from rat. Numerous mitotic figures (arrowheads) are present throughout the neoplasm (published with permission from [38])

areas of interest in biomedicine, recently renewed by its fundamental link with tumor growth and metastasis, although we already know of the importance of angiogenesis in embryogenesis (development) and wound-healing [45]. The clinical importance of angiogenesis as a prognostic tool is now recognized [17, 28].

The study of small vessel growth – a phenomenon referred to generally as angiogenesis – has such potential for providing new therapies that it has been the subject of countless press stories and has received enthusiastic interest from the pharmaceutical and biotechnology industries. Indeed, dozens of companies are now pursuing angiogenesis-related therapies, and approximately 20 compounds that either induce or block vessel formation are being tested in humans. Although such drugs can potentially treat a broad range of disorders, many of the compounds now under investigation inhibit angiogenesis and target cancer. We therefore focus the bulk of our discussion on those agents. Intriguingly, animal tests show that inhibitors of vessel growth can boost the effectiveness of traditional cancer treatments (chemotherapy and radiation). Preliminary studies also hint that the agents might one day be delivered as a preventive measure to block malignancies from arising in the first place in people at risk of cancer.

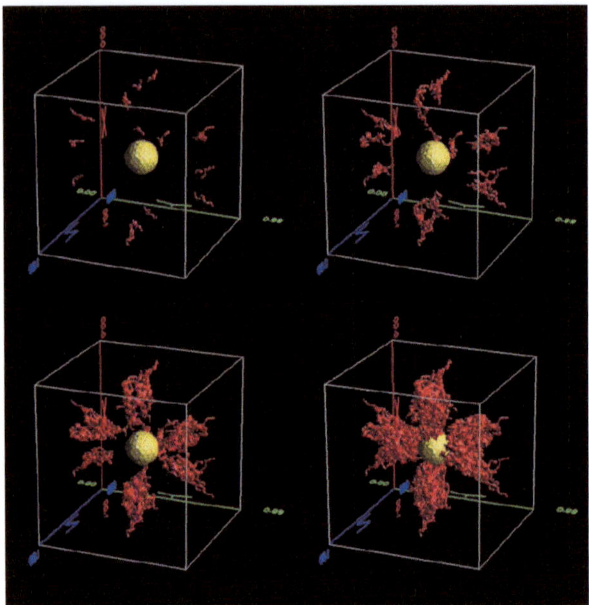

Fig. 3. A 3-d simulation of tumor-induced angiogenesis (published with permission from [18])

In developing mathematical models of angiogenesis, the hope is to be able to provide a deeper insight into the underlying mechanisms which cause the process. It is therefore essential that we develop a predictive mathematical model which is capable of producing the precise quantitative morphological features of developing blood vessels. In order to build a significant mathematical model we need then to take the following facts into account.

Tumor-induced angiogenesis consists of the proliferation of a network of blood vessels that penetrates into cancerous growths, supplying nutrients and oxygen and removing waste products.

A solid tumor secretes a number of chemicals, collectively known as TAF (Tumor Angiogenic Factors) into the surrounding tissue. These factors diffuse through the tissue, thereby creating, in the tissue surrounding the tumor, chemical gradients. When the TAF reach neighboring existing blood vessels, endothelial cells are stimulated to proliferate and migrate, following the chemical gradients. Following this, finger-like capillary sprouts are formed by accumulation of endothelial cells, which are recruited from the parent vessels. This sequence of events, which affects sprout vessels themselves in a branching process, culminates in the tumor being penetrated by a capillary network of blood vessels.

Initially, the sprouts arising from a parent vessel grow essentially parallel to each other. It is observed that once the finger-like capillary sprouts have reached a certain distance from the parent vessel, they tend to incline toward each other, leading to fusions called anastomoses. Such fusions lead to a network of vessels. On

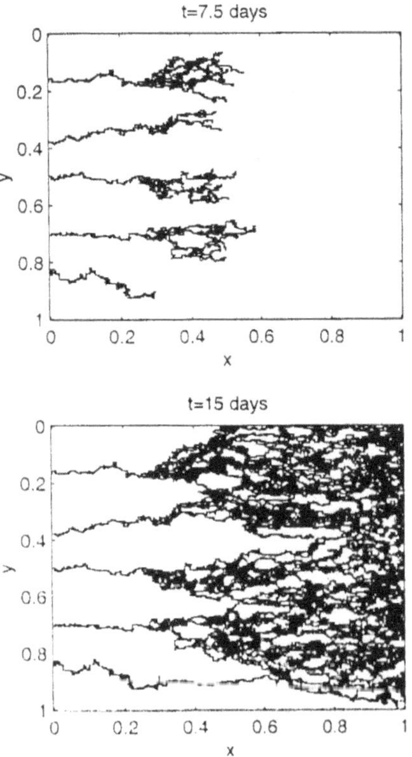

Fig. 4. A simulation of angiogenesis due to a distributed tumor mass at the right-hand edge (published with permission from [18])

the other hand sprout branching dramatically increases as it approaches the tumor mass, eventually resulting in vascularization.

The coupling of the branching-and-growth process to the underlying chemical gradients is limited by the local density of the existing capillary network, thus leading to a mathematical strong coupling between this density and the kinetic parameters of the branching-and-growth process.

Angiogenesis is regulated by both activator and inhibitor molecules. Normally, the inhibitors predominate, blocking growth. Should a need for new blood vessels arise, angiogenesis activators increase in number and inhibitors decrease. On the other hand angiogenesis inhibitors appear to "normalize" tumor blood vessels before they kill the vessels themselves. This normalization can help anticancer agents reach tumors more effectively. The effects of angiogenesis inhibitors on the geometric structure of blood vessels provide a further stimulus towards the statistical shape analysis of capillary networks. In this respect a major problem is to visualize experimental vascularization; a way has been found by working on tumors in the cornea [32, 43].

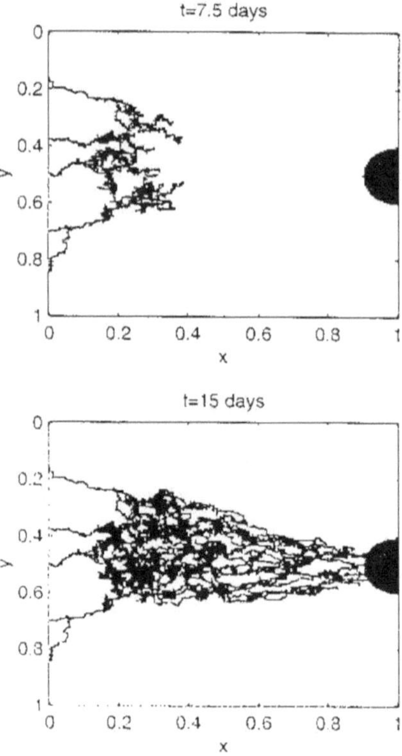

Fig. 5. A simulation of angiogenesis due to a localized tumor mass (black region on the right) (published with permission from [18])

In a detailed description, all these processes can be modelled as birth-and-growth processes strongly coupled with an underlying field. Tumor growth develops thanks to the underlying nutritional field driven by blood circulation (angiogenesis). On the other hand angiogenesis is activated by the presence of chemicals released by the tumor mass. These phenomena are subject to random fluctuations, together with the underlying fields, because of intrinsic reasons or because of the coupling with the stochasticity of the growth process itself.

An important goal is thus the integration of mathematical models for angiogenesis and tumor growth, but the unsolved complexity of the individual models has so far prevented such an integration.

The formulation of an exhaustive evolution model which relates all the relevant features of a real phenomenon dealing with different scales, and a stochastic domain decomposition at different Hausdorff dimensions, is a problem of great complexity, both analytical and computational. Here we propose a method for reducing complexity by means of homogenization at larger scales, thus leading to a hybrid model (deterministic at the larger scale, and stochastic at smaller scales).

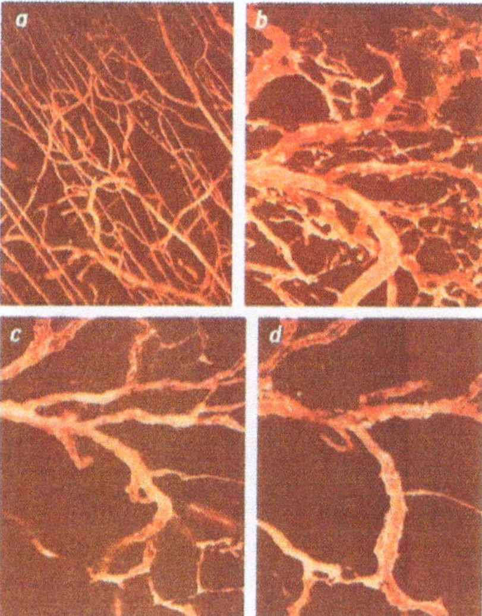

Fig. 6. Response of a vascular network to an antiangiogenic treatment (published with permission from [28])

As an example we present a simplified stochastic geometric model for spatially distributed multicellular tumor growth, strongly coupled with an underlying field, which models nutrients. For this model we discuss how to relate the geometric probability distribution to the kinetic parameters of birth and growth. Such a model might be used for predicting the evolution of tumors (diagnosis) and for identifying optimal control strategies (medical treatment).

On the other hand, modeling angiogenesis is based on the theory of stochastic fibre processes. In this case the theory of birth-and-growth processes, developed for volume growth, cannot be applied to analyze realistic models, due to intrinsic mathematical difficulties.

Notwithstanding, we propose methods of statistical analysis for the estimation of geometric densities of a stochastic fibre process that characterize the morphology of a real system. We apply such methods to real data, taken from the literature, and to simulated data, obtained by existing computational models of tumor-induced angiogenesis. These methods can be used for validating computational models, and for monitoring the efficacy of possible medical treatment.

2 Elements of stochastic geometry

The scope of stochastic geometry is the mathematical and statistical analysis of the spatial structure of patterns which are random in location and shape.

In many biomedical examples a bounded region of interest $E \subset \mathbf{R}^d$ in a space of dimension $d \geq 2$ is decomposed as a random tessellation (see Figs. 8, 9, or random sets, such as tumor masses or random lines, appear in the region E (see Figs. 2, 6).

Given a random object $\Sigma \subset \mathbf{R}^d$, a first question that arises is that of *where* this object is located. As this object is random, the only answer we may provide concerns the probability that a point x belongs to Σ, or else the probability that a compact set K intersects Σ.

In the following we denote by $\mathcal{F}$ the family of closed subsets of $\mathbf{R}^d$, by $\mathcal{K}$ the family of compact sets in $\mathbf{R}^d$, by ν^k the k-dimensional Lebesgue measure (which coincides with the k-dimensional Hausdorff measure when measuring *nice* sets, (see [23, p. 61], [51]), and by $\mathcal{B}_{\mathbf{R}^d}$ the family of Borel sets in $\mathbf{R}^d$. An RACS (Random Closed Set) $\Sigma \in \mathbf{R}^d$ is a random variable

$$\Sigma : (\Omega, \mathcal{A}, P) \longrightarrow (\mathcal{F}, \sigma_{\mathcal{F}}),$$

valued in the measurable space $(\mathcal{F}, \sigma_{\mathcal{F}})$ of the family of closed sets in $\mathbf{R}^d$ endowed with the σ-algebra generated by the hit-or-miss topology [35] with

$$\sigma_{\mathcal{F}} = \sigma\{\mathcal{F}_K, \quad \text{with} \quad K \in \mathcal{K}\},$$

where $\mathcal{F}_K = \{F \in \mathcal{F} | F \cap K \neq \emptyset\}$ is the family of the closed sets which "hit" a given compact set K.

The theory of Choquet-Matheron [35] shows that it is possible to determine a unique probability law associated with an RACS, by assigning its *hitting functional* T_Σ, which is defined as

$$T_\Sigma : K \in \mathcal{K} \longmapsto P(\Sigma \cap K \neq \emptyset).$$

The compact K acts then as an "exploration window" on the RACS Σ. In fact we may consider the restriction of T_Σ to the family of closed balls

$$\{B_\varepsilon(x); x \in \mathbf{R}^d, \varepsilon \in \mathbf{R}_+ - \{0\}\}.$$

Example 1 (Homogeneous Boolean model [5, 51]). The homogeneous Boolean model serves as an initial example that may be used to clarify the above concepts.

Let $N = \{X_1, X_2, \cdots, X_n, \cdots\}$ be a homogeneous spatial Poisson point process in $\mathbf{R}^2$, with intensity $\lambda > 0$ (i.e., λ is the mean number of points per unit area):

$$\forall A \in \mathcal{B}_{\mathbf{R}^2} : P[N(A) = n] = \exp(-\lambda\nu^2(A))\frac{\lambda\nu^2(A)}{n!}, n \in \mathbf{N}.$$

Let $\{\Sigma_1, \Sigma_2, \cdots, \Sigma_n, \cdots\}$ be a sequence of i.i.d. RACS all having the same distribution as a *primary grain* Σ_0, (e.g., a ball with a random radius R_0). We define

$$\Sigma := \bigcup_n \{X_n \oplus \Sigma_n\},$$

where $\oplus$ denotes Minkowski addition between sets ($A \oplus B := \{a + b | a \in A, b \in B\}$, $A, B \subset \mathbf{R}^d$). In this case the hitting functional is given by

$$T_\Sigma(K) = 1 - \exp\{-\lambda \mathbf{E}[\nu^2(\Sigma_0 \oplus \check{K})]\}, K \in \mathcal{K},$$

where

$$\check{K} = \{-x | x \in K\}.$$

Example 2 (A discrete time tumor growth model based on an inhomogeneous Boolean model [20]). A model for tumor growth was proposed in [20] based on a discrete time birth-and-growth model in $\mathbf{R}^d$, $d = 2, 3$. Given an RACS Σ_i representing the region occupied by the tumor mass at time $i \in \mathbf{N}$, a spatial point process $\Phi_{i+1} = \{X_k\}_{k \in \mathbf{N}}$ is generated with an intensity

$$\lambda_{i+1}(x) = \lambda_{i+1} I_{\Sigma_i}(x),$$

where the λ_i's are a family of given positive real numbers, and I_{Σ_i} denotes the indicator function of the RACS Σ_i.

The region occupied by the tumor mass at time $i + 1$ is modelled as a Boolean model

$$\Sigma_{i+1} = \bigcup_k \{Z_{i+1}(X_k) | X_k \in \Phi_{i+1} \cap \Sigma_i\},$$

where the $Z_{i+1}(X_k)$ are i.i.d. balls (disks) with a random radius R_{i+1} centered at X_k (see Fig. 7).

The hitting function of the model is now

$$T_{\Sigma_{i+1}}(K) = 1 - \exp\{-\lambda_{i+1} \mathbf{E}[\nu^d((Z_{i+1} \oplus K) \cap \Sigma_i)]\}, \; K \in \mathcal{K}, \; i \in \mathbf{N}.$$

Based on this model, procedures have been proposed for the estimation of the relevant parameters of birth λ_i and growth R_i [20].

Example 3 (Birth-and-growth model; an inhomogeneous Boolean model [9, 12]). Apart from being discrete in time, the previous model takes into account the dependence of the intensity of the birth process upon space only via the indicator of the region already invaded by the tumor; moreover the cells $Z_{i+1}(s_k)$ are taken as i.i.d. balls independent of location. Hence the above model completely ignores the coupling of the birth-and-growth process with the underlying fields, which is very important for tumor growth (see, e.g., [2]).

In the next example we consider a more significant model that we may assume as an initial model for multicluster tumor growth, i.e., for the activation of tumor clusters, and their growth. Suppose that the process takes place in a subspace $E \subseteq \mathbf{R}^d$ (which may coincide with $\mathbf{R}^d$) and let $\mathcal{E}$ be the Borel σ-algebra restricted to E. In the case in which E is bounded, suitable boundary conditions need to be introduced.

Consider the Marked Point Process (MPP) N defined as a random measure given by

$$N = \sum_{n=1}^{\infty} \epsilon_{X_n, T_n},$$

where:

- T_n is an $\mathbf{R}_+$-valued random variable representing the time of activation of the nth germ (cell or cluster),

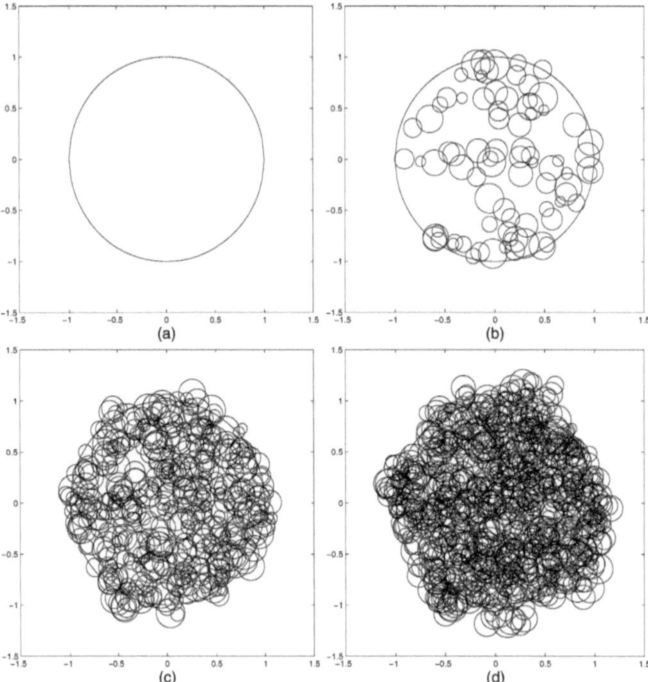

Fig. 7. Simulation of a tumor growth via a Boolean model

- X_n is an E-valued random variable representing the spatial location of the germ born at time T_n,
- $\epsilon_{x,t}$ is the Dirac measure on $\mathcal{E} \times \mathcal{B}_{\mathbf{R}_+}$ such that, for any $t_1 < t_2$ and $B \in \mathcal{E}$,

$$\epsilon_{x,t}(B \times [t_1, t_2]) = \begin{cases} 1 & \text{if} \quad t \in [t_1, t_2], x \in B, \\ 0 & \text{otherwise.} \end{cases}$$

We have

$$N(B \times A) = \sharp\{X_n \in B, T_n \in A\}, B \in \mathcal{E}, A \in \mathcal{B}_{\mathbf{R}_+},$$

is the (random) number of germs born during A in the region B.

We assume that, in the free space (i.e., the spatial region not occupied by growing grains), the birth process (an MPP) is modelled as a space-time inhomogeneous Poisson process; we assume further that the probability of birth of a new cluster (the intensity of the Poisson process) in the free space is given by

$$\begin{aligned} \nu_0(dx \times dt) : &= P(N(dx \times dt) = 1) = E(N(dx \times dt)) + o(dx)o(dt) \\ &= \alpha(x, t) dx dt + o(dx)o(dt). \end{aligned}$$

In this basic model we assume that the rate of birth, which in reality should be somehow coupled with the process itself, is only dependent upon a given underlying field; i.e., $\alpha(x,t)$ is a given deterministic field, known as the *free space intensity*, and is independent of the past history.

Let $\Xi(t; X_n, T_n)$ be the RACS obtained as the evolution up to time $t > T_n$ of a germ born at time T_n in X_n, according to some growth model. The *germ-grain* model is given by

$$\Xi(t) = \bigcup_{T_n < t} [\Xi(t; X_n, T_n)].$$

To complete the model we need to introduce a specific growth model.

Real data [46] and simulations of corresponding models by various authors (see, e.g., Fig. 1) show that the effective space invasion by a tumor cluster may be modelled as a surface growth (which possibly is stopped by impingement of different grains; c.f. Sect. 4).

Here we assume here the *normal growth model* (see, e.g., [4,6,8]), according to which, at almost every time t, at every point of the actual grain surface, i.e., at every $x \in \partial\Xi(t; x_0, t_0)$, growth occurs with a given strictly positive normal velocity

$$v(t,x) = G(t,x)n(t,x),$$

where $G(t,x)$ is a given sufficiently regular deterministic strictly positive growth field, and $n(t,x)$ is the unit outer normal at point $x \in \partial\Xi(t; x_0, t_0)$. In particular, from now on we assume that G is bounded and continuous on $\mathbf{R}_+ \times \mathbf{R}^d$ with $G_0 \leq G(t,x) < \infty$ for a constant $G_0 > 0$.

That the *evolution problem* for the growth front $\partial\Xi(t; x_0, t_0)$ be well-posed requires further regularity. We refer to [6] (also see [7]) to state that, subject to the initial condition that each initial germ is a spherical ball of infinitesimal radius, under suitable regularity assumptions on the growth field $G(t,x)$, each grain $\Xi(t; x_0, t_0)$ is such that

$$\Xi(s; x_0, t_0) \subset \Xi(t; x_0, t_0) \quad \text{for} \quad s < t,$$

and

$$\partial\Xi(s; x_0, t_0) \cap \partial\Xi(t; x_0, t_0) = \emptyset \quad \text{for} \quad s < t.$$

Moreover, for almost every $t \in \mathbf{R}_+$, we have $\dim_H \partial\Xi(t; x_0, t_0) = d-1$, and $\nu^{d-1}(\partial\Xi(t; x_0, t_0)) < \infty$, where we denote the Hausdorff dimension by $\dim_H$.

Under the regularity assumptions made above, and since the whole germ-grain process $\Xi(t)$ is a finite union of individual grains, because of the modeling assumptions, we may assume that, for almost every $t \in \mathbf{R}_+$, there exists a tangent hyperplane to $\Xi(t)$ at ν^{d-1}- a.e. $x \in \partial\Xi(t)$. As a consequence $\Xi(t)$ and $\partial\Xi(t)$ are finite unions of rectifiable sets [24] and satisfy:

$$\lim_{r\to 0} \frac{\nu^d(\Xi(t) \cap B_r(x))}{b_d r^d} = 1 \quad \text{for} \quad \nu^d\text{-a.e. } x \in \Xi(t),$$

$$\lim_{r\to 0} \frac{\nu^{d-1}(\partial\Xi(t) \cap B_r(x))}{b_{d-1} r^{d-1}} = 1 \quad \text{for} \quad \nu^{d-1}\text{-a.e. } x \in \partial\Xi(t),$$

where $B_r(x)$ is the d-dimensional open ball centered on x with radius r, and b_n denotes the volume of the unit ball in $\mathbf{R}^n$.

Further we assume that $G(t, x)$ is sufficiently regular so that, at almost any time $t > 0$, the following holds (also see [15,34,50]):

$$\lim_{r \to 0} \frac{\nu^d(\Xi(t)_{\oplus r}) - \nu^d(\Xi(t))}{r} = \nu^{d-1}(\partial \Xi(t)),$$

where we denote by $\Xi(t)_{\oplus r}$ the parallel set of $\Xi(t)$ at distance $r \geq 0$ (i.e., the set of all points $x \in \mathbf{R}^d$ with distance at most r from $\Xi(t)$).

Definition 1. Let Ξ_n be a subset of $\mathbf{R}^d$ with Hausdorff dimension n. If Ξ is (ν^n, n)-rectifiable and ν^n-measurable (see [24]), we say that Ξ_n is *n-regular.*

In particular, $\nu^n(\Xi_n) < \infty$. Note that, according to the above definition, $\Xi(t)$ and $\partial \Xi(t)$ are d-regular and $(d-1)$-regular closed sets, respectively [15].

Hence, a grain $\Xi(t; x_0, t_0)$ born at time t_0 at location x_0 may be described in terms of the indicator function

$$f(x, t; x_0, t_0) = \begin{cases} 1 & \text{for} \quad x \in \Xi(t; x_0, t_0), \\ 0 & \text{otherwise.} \end{cases}$$

Correspondingly, as far as the boundary $\partial \Xi(t; x_0, t_0)$ is concerned, we use the following generalized function (a geometric Dirac δ-function):

$$u(x, t; x_0, t_0) = \lim_{r \to 0} \frac{\nu^{d-1}(\partial \Xi(t; x_0, t_0) \cap B_r(x))}{\nu^d(B_r(x))}.$$

The growth model for the individual grain may then be expressed as follows (to be understood in a weak sense):

$$\frac{\partial}{\partial t} f(x, t; x_0, t_0) = G(x, t) u(x, t; x_0, t_0), \tag{1}$$

subject to the initial condition that at t_0 the initial germ is described by a spherical ball of infinitesimal radius centered at x_0.

We have assumed here that $\alpha(x, t)$ and $G(x, t)$ are given deterministic underlying fields; a more realistic, but also more sophisticated, model would consider a strong coupling between the birth-and-growth process and an underlying field such as those describing available nutrients (see, e.g., [9] and [41]; see Fig. 8).

As a matter of fact, if the growth model is mathematically well-posed for any initial germ located at x_0 and t_0, and the kinetic parameters $\alpha(x, t)$ and $G(x, t)$ are assumed as given deterministic underlying fields, then it is possible to introduce the causal cone associated with a compact set $K \subset E$ and a time $t > 0$, as defined by Kolmogorov [30]:

$$A(K, t) := \{(y, s) \in E \times [0, t] | K \cap \Xi(t; y, s) \neq \emptyset\}.$$

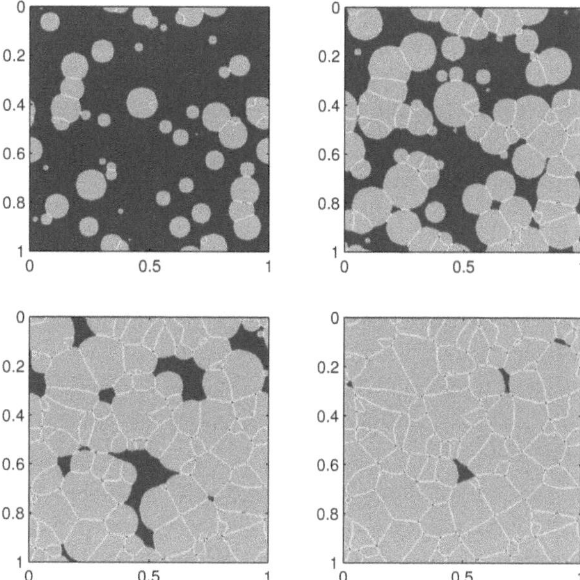

Fig. 8. A simulation of the birth-and-growth model (published with permission from [9])

Under the above modeling assumptions, the hitting functional of $\Xi(t)$ is given by

$$T_{\Xi(t)}(K) = P(K \cap \Xi(t) \neq \emptyset) = 1 - P(N(A(K,t)) = 0) = 1 - e^{-\nu_0(A(K,t))},$$

where $\nu_0(A(x,t))$ is the volume of the causal cone with respect to the intensity measure of the Poisson process

$$\nu_0(A(K,t)) = \int_{A(K,t)} \alpha(y,s) ds\, dy.$$

In particular, the causal cone of a point $x \in E$ can be obtained by reducing the compact set to the singleton $\{x\}$, thus obtaining

$$A(x,t) = \{(y,s) \in E \times [0,t] | x \in \Xi(t; y, s)\},$$

for the causal cone, and

$$P(x \in \Xi(t)) = T_{\Xi(t)}(\{x\}) = 1 - P(N(A(x,t)) = 0) = 1 - e^{-\nu_0(A(x,t))}$$

for the hitting functional of the singleton, which corresponds to the probability that a point has been covered by the tumor mass.

Knowledge of the hitting functional completely characterizes the random region in space that is invaded by the tumor mass, either in the unicellular models, or in the multicellular models.

Under these simplifying assumptions, which are not completely unrealistic for growth in vitro, more can be obtained mathematically. We may relate the hitting functional to the kinetic parameters of birth ($\alpha(x,t)$), and growth ($G(x,t)$).

To this aim we introduce the following mean geometric densities.

2.1 Stochastic geometric measures

Consider the measure spaces $(\mathbf{R}^d, \mathcal{B}_{\mathbf{R}^d}, \nu^d)$ and $(\mathcal{F}, \sigma_{\mathcal{F}}, P_{\Xi})$, where $\mathcal{F}$ is again the family of closed subsets of $\mathbf{R}^d$, $\sigma_{\mathcal{F}}$ is the $\sigma-$algebra generated in $\mathcal{F}$ by the hit-or-miss topology [35], and P_{Ξ} is the probability measure induced by an RACS Ξ on $(\mathcal{F}, \sigma_{\mathcal{F}})$. Correspondingly we denote by E_{Ξ} the expected values computed with respect to this law.

A quantitative description of the RACS Ξ can be obtained in terms of mean densities of volumes, surfaces, edges, vertices, etc., at the various Hausdorff dimensions, in the following way.

Definition 2. Let Ξ be a d-dimensional RACS in $\mathbf{R}^d$, having boundary of Hausdorff dimension $d-1$. The *mean local volume density* and *mean local surface density*, respectively, of the RACS Ξ at the point $x \in \mathbf{R}^d$ are defined as follows:

$$V_V(x) = \lim_{r\to 0} \frac{E_{\Xi}[\nu^d(\Xi \cap B(x,r))]}{\nu^d(B(x,r))},$$

$$S_V(x) = \lim_{r\to 0} \frac{E_{\Xi}[\nu^{d-1}(\partial\Xi \cap B(x,r))]}{\nu^d(B(x,r))},$$

provided that the limits exist and are a.e. finite.

It is easily seen [10, 31] that

$$V_V(x) = T_{\Xi}(\{x\}) = P(\{x \in \Xi\}) = E_{\Xi}(I_{\Xi}(x)), \quad x \in E,$$

where $T_{\Xi}(\{x\})$ is the hitting functional for the singleton (a compact set) $K = \{x\}$.

In the dynamical case, such as a birth-and-growth process, the RACS $\Xi(t)$ depends upon time so that corresponding space and time dependent quantities might be defined, such as $V_V(x,t)$, and $S_V(x,t)$. Indeed this is the case for the above mentioned birth-and-growth model, under sufficient regularity of the parameters.

Moreover, assume that a birth-and-growth process evolves in such a way that germs are born with a birth rate $\alpha(x,t)$ and grains grow according to the normal growth model with a growth rate $G(x,t)$, independently of each other; under the above mentioned regularity assumptions on these parameters, the following quantities are well defined [11].

Definition 3. The *mean extended volume density* at point x and time t is the quantity

$$V_{ex}(x,t) := \lim_{r\to 0} E_{\Xi}\left[\sum_{T_j<t}\left(\frac{\nu^d(\Xi(t;X_j,T_j)\cap B_r(x))}{\nu^d(B_r(x))}\right)\right],$$

which represents the mean of the sum of the volume measures, at time t, of the grains assumed free to be born and grow.

Definition 4. The *mean extended surface density* at point x and time t is the quantity

$$S_{ex}(x,t) := \lim_{r\to 0} E_{\Xi}\left[\sum_{T_j<t}\left(\frac{\nu^{d-1}(\partial\Xi(t;X_j,T_j)\cap B_r(x))}{\nu^d(B_r(x))}\right)\right],$$

which represents the mean of the sum of the surface measures, at time t, of the grains assumed free to be born and grow.

In both definitions, as usual, X_j represents the random location in space of a germ born at the random time T_j.

Recently Capasso and Micheletti [12] showed that

$$\nu_0(A(x,t)) = V_{ex}(x,t).$$

As a consequence,

$$\begin{aligned} V_V(x,t) &= P(x\in\Xi(t)) = 1 - P(x\notin\Xi(t)) \\ &= 1 - P(N(A(x,t))=0) = 1 - e^{-\nu_0(A(x,t))} \\ &= 1 - e^{-V_{ex}(x,t)}, \end{aligned}$$

so that

$$\frac{\partial}{\partial t}V_V(x,t) = (1 - V_V(x,t))\frac{\partial}{\partial t}V_{ex}(x,t).$$

Moreover, under the required conditions for Problem (1) to be well-posed, it was shown [7] that

$$\frac{\partial}{\partial t}\nu_0(A(x,t)) = \frac{\partial}{\partial t}V_{ex}(x,t) = G(x,t)S_{ex}(x,t).$$

The proof includes an explicit evaluation of $S_{ex}(x,t)$, a $(d-1)$-surface in $\mathbf{R}^d$, as:

$$S_{ex}(x,t) = \int_0^t dt_0 \int_{\mathbf{R}^d} dx_0 K(x_0,t_0;x,t)\alpha(x_0,t_0)$$

with

$$K(x_0,t_0;x,t) := \int_{\{z\in\mathbf{R}^d|\tau(x_0,t_0;z)=t\}} da(z)\delta(z-x).$$

Here δ is the Dirac function, $da(z)$ is a $(d-1)$-surface element, and $\tau(x_0,t_0;z)$ is the solution of the eikonal problem:

$$\begin{aligned} \left|\frac{\partial\tau}{\partial x_0}(x_0,t_0,x)\right| &= \frac{1}{G(x_0,t_0)}\frac{\partial\tau}{\partial t_0}(x_0,t_0,x), \\ \left|\frac{\partial\tau}{\partial x}(x_0,t_0,x)\right| &= \frac{1}{G(x,\tau(x_0,t_0,x))}. \end{aligned}$$

We have referred above to a process evolving in $\mathbf{R}^d$. If E is a bounded proper subset of $\mathbf{R}^d$, suitable boundary conditions are needed.

3 Hazard function

In the dynamical case an important question arises, namely, *when* is a point $x \in E$ reached (*captured*) by an invading stochastic region $\Xi(t)$? or vice versa up to when does a point $x \in E$ survive capture?

In this respect the volume density $V_V(x,t) = P(x \in \Xi(t))$ may be seen as the probability of capture of the point x, by time t, and the porosity

$$p_x(t) = 1 - V_V(x,t) = P(x \notin \Xi(t))$$

represents the *survival function* of the point x at time t, i.e., the probability that the point x is not yet covered by the random set $\Xi(t)$.

With reference to a growing region $\Xi(t)$ (such as in the birth-and-growth model), we may introduce the (random) time $\tau(x)$ of survival of a point $x \in E$ with respect to its capture by $\Xi(t)$, so that

$$p_x(t) = P(\tau(x) > t).$$

Correspondingly, a *hazard function* $h(x,t)$ can be defined as the rate of capture by the process $\Xi(t)$, i.e.,

$$h(x,t) = \lim_{\Delta t \to 0} \frac{P(x \in \Xi(t+\Delta t) | x \notin \Xi(t))}{\Delta t}.$$

Under the modeling assumptions made on the birth-and-growth model, if the volume density $V_V(x,t)$ and the survival function $p_x(t)$ are differentiable with respect to time, the random variable $\tau(x)$ admits a probability density function $f_x(t)$ such that [15]

$$f_x(t) = \frac{d}{dt}(1 - p_x(t)) = \frac{\partial V_V(x,t)}{\partial t},$$

so that

$$h(x,t) = \frac{f_x(t)}{p_x(t)}$$

and

$$f_x(t) = p_x(t) h(x,t),$$

from which we immediately obtain

$$\frac{\partial V_V(x,t)}{\partial t} = (1 - V_V(x,t)) h(x,t).$$

In order to provide an expression for $h(x,t)$ in terms of the defining parameters of the process, we need to refer to specific cases. For the birth-and-growth model presented in Example 3 we have

$$h(x,t) = -\frac{\partial}{\partial t} \ln p_x(t) = \frac{\partial}{\partial t} \nu_0(A(x,t)) = \frac{\partial}{\partial t} V_{ex}(x,t) = G(x,t) S_{ex}(x,t).$$

In general, we may study the case of *when a compact set* $K \subset E$ is reached by the invading stochastic region $\Xi(t)$. In this respect we may introduce the (random) hitting time $\tau(K)$ of a compact set $K \subset E$ by $\Xi(t)$. It is such that the corresponding survival function is given by

$$S_K(t) := P(\tau(K) > t) = P(\Xi(t) \cap K = \emptyset) = 1 - T_{\Xi(t)}(K).$$

Correspondingly, a *hazard function* $h(K, t)$ can be defined as the hitting rate of the process $\Xi(t)$, i.e.,

$$h(K,t) = \lim_{\Delta t \to 0} \frac{P(\Xi(t+\Delta t) \cap K \neq \emptyset) | \Xi(t) \cap K = \emptyset))}{\Delta t}.$$

Suppose now that the random variable $\tau(K)$ admits a probability density function $f_K(t)$; then

$$\begin{aligned} h(K,t) &= \frac{1}{P(\tau(K) > t)} \lim_{\Delta t \to 0} \frac{P(t < \tau(K) \leq t + \Delta t)}{\Delta t} = \frac{f_K(t)}{S_K(t)} \\ &= \frac{1}{1 - T_{\Xi(t)}(K)} \lim_{\Delta t \to 0} \frac{T_{\Xi(t+\Delta t)}(K) - T_{\Xi(t)}(K)}{\Delta t} \\ &= -\frac{d}{dt} \ln(1 - T_{\Xi(t)}(K)). \end{aligned}$$

Since $T_{\Xi(0)}(K) = 0$, it follows that

$$T_{\Xi(t)}(K) = 1 - \exp\left\{ -\int_0^t h(K,s)ds \right\}.$$

This expression allows an estimation of the hitting function by means of an estimation of the hazard function. Of great importance is the fact that it holds even though a causal cone cannot be defined, as in the model for angiogenesis in Sect. 6.1.

4 Mean densities of stochastic tessellations

A random subdivision of space needs further information to be characterized. For example, in the birth-and-growth process described above, we may include an additional feature known as impingement, by assuming that, at points of contact of their growth front, grains stop growing. In this case the spatial region in $\mathbf{R}^d$ in which the process occurs is divided into cells (random Johnson-Mehl tessellation [29,39]; also see [40]), and interfaces (n-facets, $n = 0, 1, 2, \cdots, d$) at different Hausdorff dimensions (cells, faces, edges, vertices) appear (for a planar process, see Fig. 9).

As above, we may describe the tessellation quantitatively by means of mean densities of the n-facets with respect to the d-dimensional Lebesgue measure [13].

A *cell* of a random tessellation is any element of a family of RACS's partitioning the region E in such a way that any two distinct elements of the family have empty intersection of their interiors. It is clear that this definition may also be used in the static (time independent) case.

We now introduce a rigorous concept of "interface" at different Hausdorff dimensions.

Definition 5. An *n-facet* at time t ($0 \leq n \leq d$) is a non-empty intersection between $m+1$ cells, with $m = d - n$.

Note that in this definition:

- $d =$ dimension of the space in which the tessellation takes place
- $n =$ Hausdorff dimension of the interface under consideration
- $m + 1 =$ number of cells that form such an interface
- for $n = d$ a d-facet is simply a cell

(also see Fig. 9).

Consider now the union of all n-facets at time t, $\Xi_n(t)$. For any Borel set B in $\mathbf{R}^d$ one can define the *mean n-facet content* of B at time t as the measure

$$\mathcal{M}_{d,n}(t, B) = E_\Xi \left[\lambda_n(B \cap \Xi_n(t))\right],$$

where λ_n is the n-dimensional Hausdorff measure (coinciding with the n-dimensional Lebesgue measure ν^n, $n = d, d-1$). Note that, with the previous definitions, $\Xi_d(t) \equiv \Xi(t)$, so that $\mathcal{M}_{d,d}(t, B)$ is the d-dimensional volume of the portion of the set B occupied by cells at time t.

Suppose that the kinetic parameters of the birth-and-growth process are such that $\mathcal{M}_{d,n}$ admits a density $\mu_{d,n}(t, x)$ with respect to ν^d, the standard d-dimensional Lebesgue measure on $\mathbf{R}^d$, i.e., for any Borel set B,

$$\mathcal{M}_{d,n}(t, B) = \int_B \mu_{d,n}(t, x) dx; \tag{2}$$

then the following definition is meaningful.

Definition 6. The function $\mu_{d,n}(t, x)$ defined by (2) is called the *local mean n-facet density* of the (incomplete) tessellation at time t.

In particular $\mu_{d,d}(t, x)$ is the mean local volume density of the occupied region at time t, and $\mu_{d,d-1}(t, x)$ is the surface density of the cells. It is still an open problem, in general, to obtain evolution equations for these densities.

Under sufficient regularity conditions for the birth-and-growth model analyzed above, the following evolution equations have been obtained [13]:

$$\frac{\partial}{\partial t} \mu_{d,n}(x, t) = c_{d,n} \frac{(h_{m+1}(x, t))}{(m+1)!} (1 - V_V(t, x))(G(x, t))^{-m},$$

where $c_{d,n}$ is a constant depending only on the space dimension, and

$$h_k(x, t) = (h(x, t))^k \quad \text{for} \quad k = 2, 3, \ldots;$$

$h_1(x, t)$ has a different expression (see [13]).

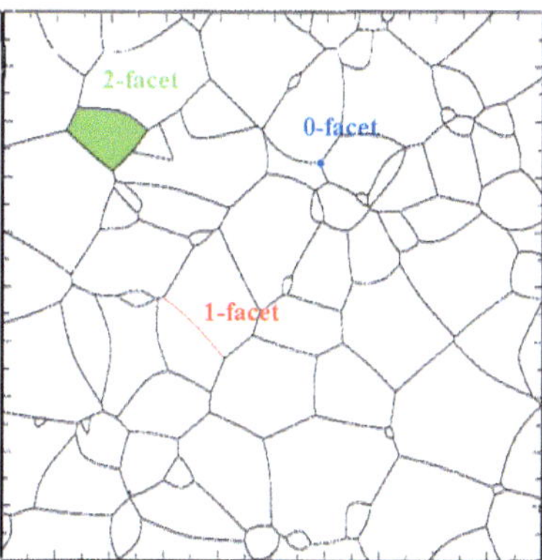

Fig. 9. $n-$facets for a tessellation in $\mathbf{R}^2$

5 Interaction with an underlying field

In most real cases spatial heterogeneities are induced because of the dependence of the kinetic parameters of the birth-and-growth process upon an underlying field $\phi(x, t)$ (chemicals, nutrients, etc.):

$$G(x, t) = \widetilde{G}(\phi(x, t)),$$
$$\alpha(x, t) = \widetilde{\alpha}(\phi(x, t)).$$

Vice versa, the birth-and-growth process may induce a source term in the evolution equation of the underlying field

$$\frac{\partial}{\partial t}\phi = \text{div}\,(\kappa \nabla \phi) + g[\rho, (I_{\Xi(t)})_t], \qquad \text{in} \quad \mathbf{R}^+ \times E,$$

subject to suitable boundary and initial conditions.

Here $(I_{\Xi(t)})_t$ denotes the (distributional) time derivative of the indicator function $I_{\Xi(t)}$ of the growing region $\Xi(t)$ at time t, and ρ denotes a relevant parameter or family of parameters (the above equation has to be understood in a weak sense).

Also the parameters in the evolution equation of the underlying field may depend upon the evolving "phase", i.e., if ρ_1, and κ_1 denote the parameters in the growing mass, and ρ_2 and κ_2 those in "empty" space, one should write:

$$\rho = I_{\Xi(t)}\rho_1 + (1 - I_{\Xi(t)})\rho_2,$$
$$\kappa = I_{\Xi(t)}\kappa_1 + (1 - I_{\Xi(t)})\kappa_2.$$

This equation is now a random differential equation, since all parameters, and the source term, depend upon the stochastic geometric process $\Xi(t)$. A direct consequence is the stochasticity of the underlying field, and vice versa the stochasticity of the kinetic parameters.

This strong coupling between the underlying field and the birth-and-growth process, makes the previous theory for the hazard function, and, consequently, for the evolution of mean geometric densities, not directly applicable, as now the kinetic parameters of the process are themselves stochastic.

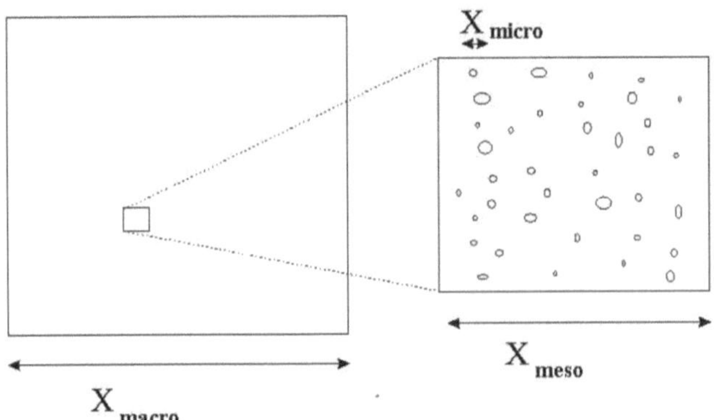

Fig. 10. Typical scales in the birth-and-growth process

Multiple scales and hybrid models. For many practical tasks, the stochastic models presented above, which are able to describe the full process, are too sophisticated. On the other hand, in many practical situations multiple scales can be identified. As a consequence it suffices to use averaged quantities at the larger scale, while using stochastic quantities at the lower scales. The advantage of using averaged quantities at the larger scale is convenient, both from a theoretical point of view and from a computational point of view.

Under typical conditions,we may assume that the typical scale for diffusion of the underlying field (*macroscale*) is much larger than the typical grain size (*microscale*). This allows us to approximate the full stochastic model by a hybrid system. Under these conditions a *mesoscale* may be introduced, which is sufficiently small with respect to the macroscale of the underlying field and sufficiently large with respect to typical grain size. First this means that the substrate may be considered approximately homogeneous at this mesoscale. A typical size x_{meso} on this mesoscale satisfies

$$x_{\mathrm{micro}} \ll x_{\mathrm{meso}} \ll x_{\mathrm{macro}},$$

where x_{micro} and x_{macro} are typical sizes for single grains and for field diffusion. This typical feature of the process is illustrated in Fig. 10.

It makes sense to consider a (numerical) discretization of the whole space in subregions $B_i, i = 1, \ldots, L$, at the level of the mesoscale, i.e., small enough that the space variation of the underlying field ϕ inside B_i may be ignored; essentially this corresponds to approximating the contribution due to the growth process by its local mean value, i.e., by the the mean rate of phase change in the equation for ϕ,

$$I_{\Xi(t)}(x) \simeq E_{\Xi}[I_{\Xi(t)}(x)] = V_V(x,t),$$

(for a more rigorous discussion on this item we refer to [14, 37]).

For the parameters, we take the corresponding averaged quantities:

$$\widetilde{\rho}(x,t) = V_V(x,t)\rho_1 + (1 - V_V(x,t))\rho_2,$$
$$\widetilde{\kappa}(x.t) = V_V(x,t)\kappa_1 + (1 - V_V(x,t))\kappa_2.$$

If we now substitute all stochastic quantities in the equation for the underlying field by their corresponding mean values, we obtain an initial-boundary value problem for a parabolic partial differential equation:

$$\frac{\partial}{\partial t}(\widetilde{\phi}) = \text{div}\,(\widetilde{\kappa}\nabla\widetilde{\phi}) + g[\widetilde{\rho}, \frac{\partial}{\partial t}V_V],$$

in $E \times \mathbf{R}^+$, $d = 1, 2, 3$, supplemented by suitable boundary conditions and initial conditions. We have expressed the essential difference between the (generalized) functions $I_{\Xi(t)}$ and ϕ and their deterministic counterparts in the averaged equations by using V_V and $\widetilde{\phi}$. In order to solve the above equation we need to provide an evolution equation for the mean volume density $V_V(x, t)$.

We may observe that the above system now provides a deterministic field $\widetilde{\phi}(x, t)$ in $E \times \mathbf{R}_+$. Once we approximate ϕ with its deterministic counterpart, we are given deterministic fields for the kinetic parameters

$$\alpha(x, t) = \widetilde{\alpha}(\widetilde{\phi}(x, t)) \quad \text{and} \quad G(x, t) = \widetilde{G}(\widetilde{\phi}(x, t)).$$

With these parameters the birth-and-growth process is now again stochastically simple.

We may then refer to the previous theory to obtain the hazard function $h(x, t)$, and consequently the evolution equations for the mean geometric densities, in terms of the fields $\alpha(x, t)$ and $G(x, t)$. In particular we obtain the required evolution equation for the volume density,

$$\frac{\partial V_V(x, t)}{\partial t} = (1 - V_V(x, t))G(x, t)S_{\text{ex}}(x, t),$$

subject to trivial initial conditions. For $S_{\text{ex}}(x, t)$ we use the expression obtained above for the evolution with deterministic fields G and α.

This approach is called "hybrid", since we have substituted the stochastic underlying field $\phi(x, t)$ given by the full system by its "averaged" counterpart $\widetilde{\phi}(x, t)$. One should check that the hybrid system is fully compatible with the rigorous derivation of the evolution equation for V_V. In fact, once we substitute the deterministic volume density V_V in the equation for ϕ, and the deterministic averaged parameters, we obtain a linear equation for ϕ; we may apply the expectation operator and easily obtain the given equation for $\widetilde{\phi}$.

We refer to [8] for further details on a mathematical theory of the averaged model.

6 Fibre and surface processes

For processes that naturally evolve at Hausdorff dimensions lower than the dimension d of the host space E, it is more convenient to consider the so-called fibre and surface processes.

The theory of fibre and surface processes deals with the study of systems of lines (i.e., sets having Hausdorff dimension 1, called *fibres*) or surfaces (i.e., sets having

Hausdorff dimension $d-1$), or fragments of fibres or surfaces, distributed at random in a subset of $\mathbf{R}^d$. Usually only the cases relevant for applications, $d = 2, 3$, are considered. Fibre and surface processes are particular cases of RACS's, and may often be modeled by suitable Boolean models or point processes. Stochastic models and related statistical techniques have been mainly developed in the case of *stationary* fibre or surface processes, i.e., of RACS's having invariant distribution with respect to rigid motions [51].

As already mentioned, an example of a fibre process is a network of blood vessels formed in an angiogenetic process, while examples of surface processes are the $d-1$ interfaces of a spatial tessellation formed by a birth-and-growth process in $\mathbf{R}^d$ (note that, when $d = 2$, a surface process is also a fibre process, with $d - 1 = 1$). Both these examples stress the fact that statistical techniques for non stationary processes are strongly required by biomedical applications, since the different geometric characteristics of a network of vessels or interfaces of cells are biomedical indicators of a pathology (e.g., presence of tumors); often the fibre processes present in normal tissues are themselves non-stationary, because of the interaction with an underlying field during their formation. Thus what should be identified by the statistical geometric analysis of such processes are both the typical geometric characteristics of the relevant fibre or surface processes in normal tissues, in the absence of pathologies, but possibly in the presence of "usual" local inhomogeneities, and the *deviation from normality* in the presence of pathologies, which would be of great importance for automatic diagnosis.

Since in the majority of applications data are available from two-dimensional images (even obtained by tomographic sections) of biological samples, we restrict the mathematical description to planar fibre processes, that is, fibre processes in $\mathbf{R}^2$, but many results stated here also hold in higher dimensional spaces.

6.1 Planar fibre processes

As mentioned above, planar fibre processes model random collections of curves in the plane $\mathbf{R}^2$ (see [51, Chap. 9] for nomenclature and basic results). A fibre is a sufficiently smooth simple curve in the plane having finite length. That is, a fibre γ is a subset of $\mathbf{R}^2$ which is the image of a parametric curve $\gamma(t) = (\gamma_1(t), \gamma_2(t)),\ t \in [0, 1]$, such that,

1. $\gamma : [0, 1] \to \mathbf{R}^2$ is continuously differentiable;
2. $|\gamma'(t)|^2 = |\gamma_1'(t)|^2 + |\gamma_2'(t)|^2 > 0\ \forall t \in [0, 1]$;
3. the mapping γ is one-to-one, that is, a fibre does not intersect itself.

We associate to a fibre γ its *length measure* $\Gamma(\cdot)$ defined by

$$\Gamma(B) = \nu^1(\gamma \cap B) = \int_0^1 \mathbf{1}_B(\gamma(t))\sqrt{(\gamma_1'(t))^2 + (\gamma_2'(t))^2}dt$$

for any Borel set $B \subset \mathbf{R}^2$. This measure is independent of the particular parametric representation chosen for γ.

Definition 7. A *fibre system* ϕ is a subset of $\mathbf{R}^2$ which may be represented as

$$\phi = \bigcup_{i \in \mathbf{N}} \gamma_i,$$

where $\{\gamma_i\}_{i \in \mathbf{N}}$ is a sequence of fibres such that:

i) $\gamma_i((0, 1)) \cap \gamma_j((0, 1)) = \emptyset$ for any $i \neq j$, that is, fibres are disjoint apart from the extreme points;
ii) any bounded Borel set $B \subset \mathbf{R}^2$ intersects only a finite number of fibres (that is, the process of fibres is locally finite).

The *length measure* Φ associated with the fibre system ϕ is defined on a Borel set B by

$$\Phi(B) = \sum_{i \in \mathbf{N}} \Gamma_i(B),$$

where Γ_i is the length measure associated with the fibre γ_i.

If we denote by $\mathbf{D}$ the family of all planar fibre systems, endowed with the σ-algebra $\mathcal{D}$ generated by sets of the form

$$\{\phi \in \mathbf{D} | \Phi(B) < x\} \quad \text{for } x \in \mathbf{R},\ B \in \mathcal{B}_{\mathbf{R}^2},$$

then we may define a *fibre process* Σ as a random variable

$$\Sigma : (\Omega, \mathcal{A}, P) \rightarrow (\mathbf{D}, \mathcal{D}),$$

i.e., a particular type of RACS. It can be shown that $\mathcal{D}$ is the restriction to $\mathbf{D}$ of the hit-or-miss σ-algebra, so that the hitting functional still characterizes the distribution of a fibre process.

Definition 8. Let Σ be a fibre process. Then the measure

$$\Lambda(B) = E(\Phi(B)) = E\left(\sum_i \nu^1(\gamma_i \cap B)\right),$$

where Φ is the length measure associated with Σ and γ_i are the fibres of the system Σ, is called the *mean length measure* of Σ.

If $\Lambda \ll \nu^2$, then the Radon-Nikodim derivative

$$\lambda(x) = \frac{d\Lambda}{d\nu^2}(x)$$

is defined for almost all $x \in \mathbf{R}^2$ and is called the *local mean density of length* of the fibre process.

If the fibre process is stationary, then $\lambda(x) = \lambda$ is constant and $\Lambda(\cdot) = \lambda\nu^2(\cdot)$.

A mathematical model for describing an angiogenetic process based on the theory of stochastic birth-and-growth processes is very difficult to analyze. The main problem is due to the fact that a causal cone approach for defining the hazard function is not possible, since the birth kinetics needs to be modeled as follows.

Consider the birth-and-growth model introduced in Example 3. Now the marked counting process modeling the birth processes, call it M, refers to the offspring of a new capillary from an already existing vessel, i.e., from points of the stochastic fibre process $\Sigma(t)$, so that the branching rate is given by

$$\mu(dl \times dt) = P(M(dl \times dt) = 1) = \beta(x,t)dldt + o(dl)o(dt) \tag{3}$$

for $x \in \Sigma(t-)$.

This shows the dependence of the branching rate upon the existing stochastic fibre process $\Sigma(t-)$, and the fact that the point of birth belongs to its infinitesimal element dl. Further, the growth process of the fibres needs to be modeled, including a chemotactic dependence upon an underlying field.

Thus, even though we may assume, as above, a multiscale approach leading to a deterministic homogenization of the underlying field, with stochastic branching rate, via its dependence upon the existing capillary network, the causal cone cannot be properly defined, and the theory of Sect. 2 (Example 3) and Sect. 3 cannot be applied.

Nonetheless we may build significant computational models for the simulation of angiogenesis; some results are displayed in Figs. 4, 5, and 19a.

We may then use the statistical theory of stochastic fibre processes to estimate hitting functionals and relevant geometric mean densities, so as to analyze the efficacy of possible treatments of the angiogenesis. Of course statistical methods are also used for analyzing real pictures as in Figs. 15 and 16.

7 Estimate of the mean density of length of planar fibre processes

As was mentioned in the previous section, in many biomedical applications it is relevant to estimate the mean length of the lines forming the fibre process per unit area, that is, its (local) mean density of length, from two-dimensional digitized images, since the number of lines and their "complexity", which can be measured in terms of length, is related to the presence of pathologies. Since images are formed by pixels, that is, two-dimensional sets, while a line is a one-dimensional object, simple "pixel counting" techniques do not lead to accurate estimates of the mean density of length of the fibre process, even with black and white images of high resolution.

Methods based on the Cauchy-Crofton formula have been proposed to estimate the length of a line in a plane [21] or the area of a surface in 3-dimensional space [33], which rely on counting the number of intersections of the fibre (or surface) process with systems of parallel or random lines, spanning the window of observation in many possible directions. Depending also on the resolution of the image, it is sometimes rather difficult to obtain a good estimate of the number of intersections of the fibres with the chosen system of lines. Here we propose an estimation method for the mean density of length of fibre processes in $\mathbf{R}^2$, based on the spherical contact distribution function, and we apply the method to simulated fibre processes, whether stationary-symmetric or not (that is, whether or not they have a constant mean density of length in space), in order to test the properties of the estimator.

7.1 Local mean density of length and the spherical contact distribution function

We next introduce the definition of the local spherical contact distribution of an RACS Θ.

Definition 9. The local spherical contact distribution function H_s of a random set $\Theta \subseteq \mathbf{R}^d$ is defined as

$$H_s(r, x) :\; = P(x \in \Theta \oplus B_r(0) | x \notin \Theta),$$

where $\oplus$ is Minkowski addition (see [10]).

Note that the spherical contact distribution function is a "geometric" analogue of the hazard function, since it refers to the probability that a point x, which is not "captured" by a random set, is captured by its parallel set of radius r, $\Theta_r = \Theta \oplus B_r(0)$ (see Fig. 11). The function H_s can also be defined in a general dynamical setting, for random sets $\Theta(t)$ growing radially with growth rate $G(t)$ depending only on time, by setting

$$H_s(r, x, t) := P(x \in \Theta(t + r) | x \notin \Theta(t)),$$

where $\Theta(t + r) = \Theta(t) \oplus B_r(0)$, and $r = \int_t^{t_1} G(s) ds$ for some $t_1 > t$. In this particular case,

$$\frac{d}{dr} H_s(r, x, t)_{|r=0} = h(x, t).$$

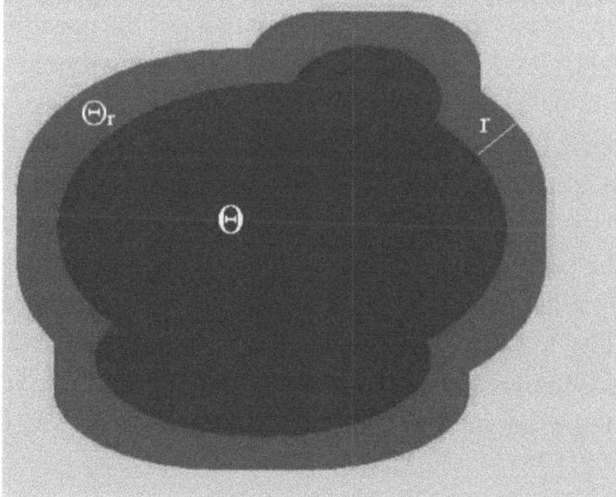

Fig. 11. Θ_r is the parallel set of radius r of the set Θ

Suppose now that Σ is an M-continuous fibre process [16,35], i.e., $P(x \notin \Sigma) = 1$ for all $x \in \mathbf{R}^2$, so that its associated spherical contact distribution function is

$$H_s(r, x) = P(x \in \Sigma \oplus B_r(0)).$$

In addition, the local mean volume density of Σ is zero, having a null 2-dimensional volume. Theorem 3.1 in [10] can then be applied to obtain the following result.

Proposition 1. *Let Σ be a fibre process in $\mathbf{R}^2$. Let Σ_r be its parallel set of radius r,*

$$\Sigma_r = \Sigma \oplus B_r(0).$$

If the condition

$$\lim_{r \to 0} \frac{\nu^2(\Sigma_r \cap B_\varepsilon(x))}{r} = 2\nu^1[\Sigma \cap B_\varepsilon(x)] \tag{4}$$

is satisfied for ν^2-almost all $x \in \mathbf{R}^2$ and for all $\varepsilon > 0$, then the function $H_s(r, x)$ is differentiable at $r = 0$ for ν^2-almost all x, and its derivative satisfies

$$\frac{d}{dr} H_s(r, x)_{|r=0} = 2\lambda(x), \tag{5}$$

where $\lambda(x)$ is the local mean density of length of the fibre process.

Note that Condition (4) is a regularity condition on the boundary of the set Σ, and it is satisfied by a large class of sets of interest for applications. For the proof of Proposition 1 see [36].

7.2 Estimate of the local mean density of length

Equation (5) may be used to estimate $\lambda(x)$ from an estimator of H_s. Note that, by definition, $H_s(r, x)$ equals the local volume density of the parallel set Σ_r of Σ, which is in general nonzero, as Σ_r is a set of Hausdorff dimension 2.

The (local) volume density $V_V(x)$ of a random set Ξ, having Hausdorff dimension 2, can be estimated by considering an observation window $W(x)$, centred at x, sufficiently small so that the volume density may be considered constant inside the window, but not too small with respect to the size of the random set (if possible); this means that the probability that the window is completely occupied by the set or completely empty must be nontrivial, i.e., not 0 or 1 (for example, if the random set Ξ is represented by a 2-dimensional black and white digitized image, the window $W(x)$ must be chosen much larger than the size of a pixel). Then a grid of n points is made to overlay the window $W(x)$ and the local volume density of Ξ is estimated by

$$\hat{V}_V(x) = \frac{1}{n} \sum_{i=1}^{n} \mathbf{1}_\Xi(x_i),$$

where $\mathbf{1}_{\Xi}$ is the indicator function of the set Ξ. This estimator is unbiased with variance

$$\sigma^2 = \frac{1}{n^2}\left(nV_V(x)(1 - V_V(x)) + 2\sum_{i>j} k(r_{ij}) \right),$$

where $r_{ij} = ||x_i - x_j||$ and k is the covariance function of Ξ (see [51, p. 212] for further details). Thus in our case an unbiased estimator of $H_s(x, t)$ in a window $W(x)$, with a grid $x_1, \ldots, x_n$, is

$$\widehat{H}_s(r, x) = \frac{1}{n}\sum_{i=1}^{n} \mathbf{1}_{\Theta_r(t)}(x_i).$$

An estimator of $\frac{d}{dr}H_s(r, x)_{|r=0}$ may be obtained by numerical approximation, by considering that, if the function $H_s(r, x)$ is a.s. right continuous for $r \to 0^+$ for any $x \in \mathbf{R}^2$, then, by a Taylor expansion in a suitable neighborhood of $r = 0$, we obtain

$$H_s(r, x) = H_s(0, x) + r\frac{d}{dr}H_s(r, x)_{|r=0} + \frac{r^2}{2}\frac{d^2}{dr^2}H_s(r, x)_{|r=0} + o(r^2).$$

Then an estimator of $\frac{d}{dr}H_s(r, x)_{|r=0}$ is obtained by estimating $H_s(r, x)$ for different (small) values of r, thus forming a paired sample $(r_i, \hat{H}_s(r_i, x))_{i=1,\ldots,m}$ and fitting a parabola to the data

$$\hat{H}_s(r_i, x) = \beta_0 + \beta_1 r_i + \beta_2 r_i^2. \tag{6}$$

The least squares estimator $\hat{\beta}_1$ of β_1 is an estimator of $\frac{d}{dr}H_s(r, x)_{|r=0}$. An estimator of the local mean density of length of the fibre process is then

$$\hat{\lambda}(x) = \frac{1}{2}\hat{\beta}_1.$$

Note that, by definition,

$$H_s(0, x) = P(x \in \Sigma | x \notin \Sigma) = 0,$$

which means that the coefficient β_0 in (6) should be zero. Unfortunately, since the resolution of the image, and so also the size of a pixel, affects the minimum possible choice of r, usually the least squares estimator $\hat{\beta}_0$ of β_0 is significantly different from zero. This effect is usually known in spatial statistics as the nugget effect [20]. By imposing $\beta_0 = 0$ in the estimate procedure, some bias is induced on the estimate of the other parameters.

7.3 Numerical results

The estimator $\hat{\lambda}(x)$ was first tested on a simulated stationary symmetric 2-dimensional fibre process formed by straight lines, in $[-0.5, 0.5] \times [-0.5, 0.5]$, having constant

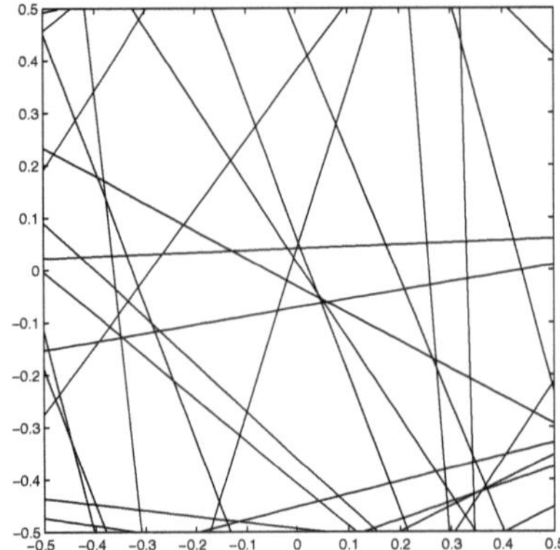

Fig. 12. A simulated stationary symmetric fibre process, with mean density of length $\lambda = 20$

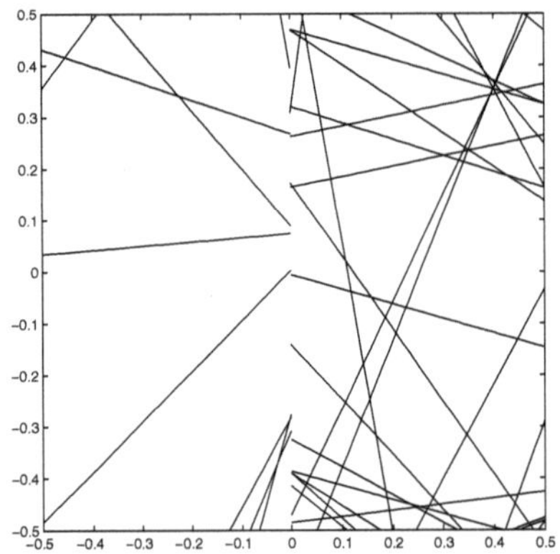

Fig. 13. Simulation of a non-stationary fibre process

mean density of length $\lambda = 20$ (see Fig. 12). The simulation was performed via the chord algorithm described in [33]. The estimator and the nugget effect nug (i.e., the estimated coefficient $\hat{\beta}_0$) were computed on 100 simulations, in the whole window of observation and their empirical mean and standard deviation were computed, with the following results:

$$\bar{\hat{\lambda}} = 21.204 \quad \mathrm{std}(\hat{\lambda}) = 5.5374$$
$$\overline{\mathrm{nug}} = -0.1982 \quad \mathrm{std}(\mathrm{nug}) = 0.0493.$$

A 99% Gaussian confidence interval for λ is $[19.77, 22.73]$, which includes the true value 20, while the nugget effect is significantly different from zero, since a 99% confidence interval for nug is $[-0.21, -0.18]$, which does not include zero.

Next a non-stationary fibre process was simulated, by dividing the window of observation into two halves and by generating independent stationary fibre processes in each half, with mean density of length 20 in the left-hand half, and with mean density of length 40 in the right-hand half (see Fig. 13).

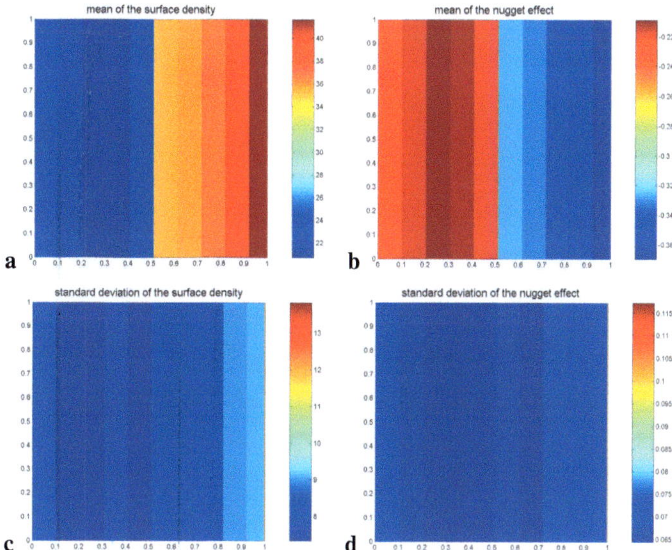

Fig. 14. a–b Mean of the estimated local mean density of length and of the nugget effect, respectively; **c–d** standard deviation of the local mean density of length and of the nugget effect, respectively

The window of observation was then divided into 10 subwindows, formed by equally spaced vertical strips, and the estimator $\hat{\lambda}$ and the nugget effect nug was computed in each subwindow. The results over 100 simulations are reported in Fig. 14. In this case as well the true value of $\lambda(x)$ is included in a 99% Gaussian confidence region, while the nugget effect is significantly different from zero. Note also that the inhomogeneity of the process and the step in its mean density of length is very visible.

From the experimental results, the proposed estimator seems to be rather accurate, and may be unbiased or only slightly biased. In the non-homogeneous case a proper choice of the subwindows for the estimation should "capture" the details in the variation of the mean density of length. The estimator should also be tested in more complicated non-homogeneous cases, but it is usually difficult to retrieve an analytical expression for the true value of the mean density of length in complicated non stationary processes. Also the variation of quality of the estimate with respect to the resolution of the image should be tested, and a comparison with other estimation techniques, such as the Cauchy-Crofton formula, should be performed, both in statistical terms and in computational terms. We leave these analyses to subsequent papers.

7.4 Applications

The estimator described in the previous section has been applied to biomedical images or simulated images (coming from [18]) where fibre processes appear, in order to estimate the local mean density of length of fibres. The results are reported in Figs. 15–18. It is clear that in many cases the estimator is able to capture the details

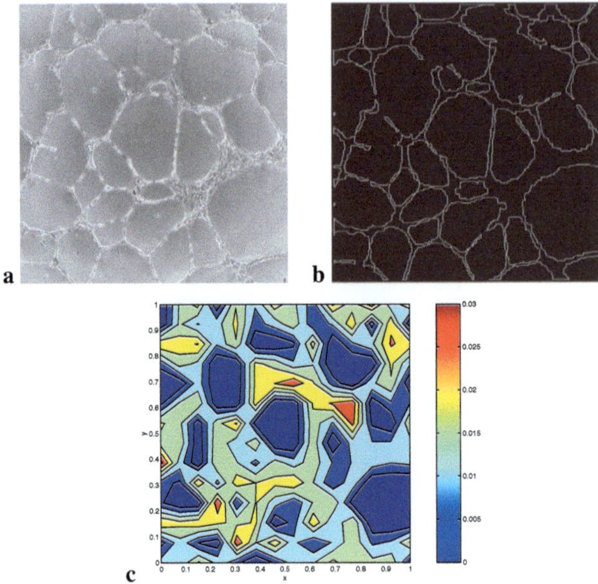

Fig. 15. a A real picture showing a 2D spatial tessellation due to the vascularization of biological tissue: endothelial cells form a network of blood vessels (published with permission from [49]); **b** the result of image analysis to detect the edges of the network in **a**; **c** the estimate of the local mean density of length of capillaries

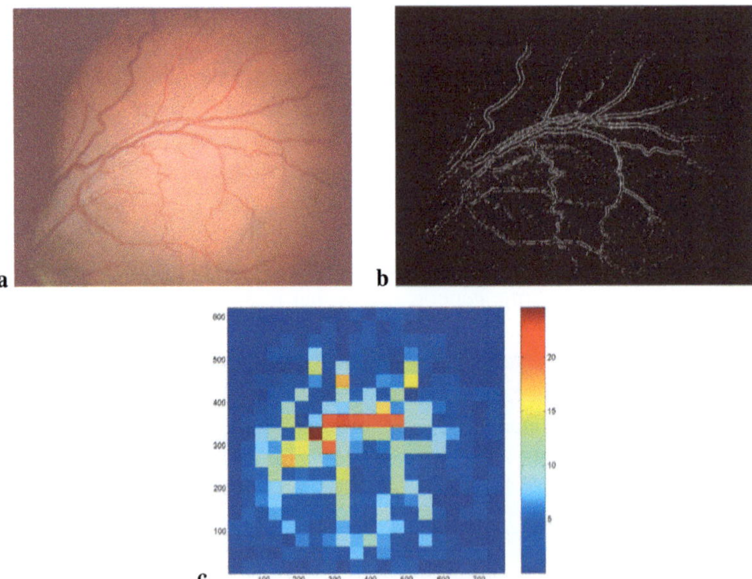

Fig. 16. a An image of capillaries on the retina; **b** the result of image analysis to detect the fibres; **c** the estimate of the local mean density of length of capillaries

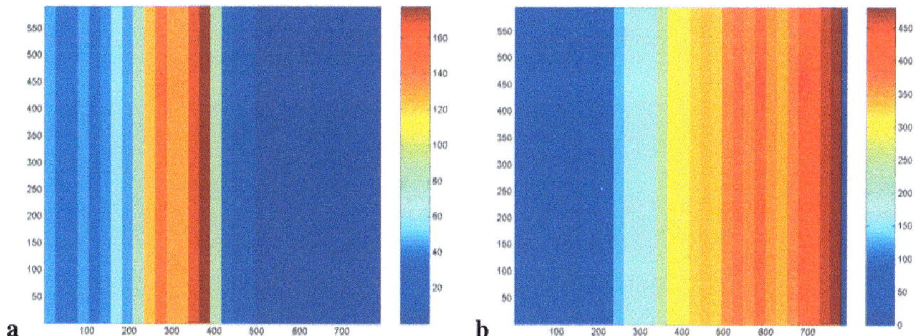

Fig. 17. Estimate of the local mean density of length using vertical windows for the simulation of angiogenesis in Fig. 4; **a** time 7.5 days; **b** time 15 days

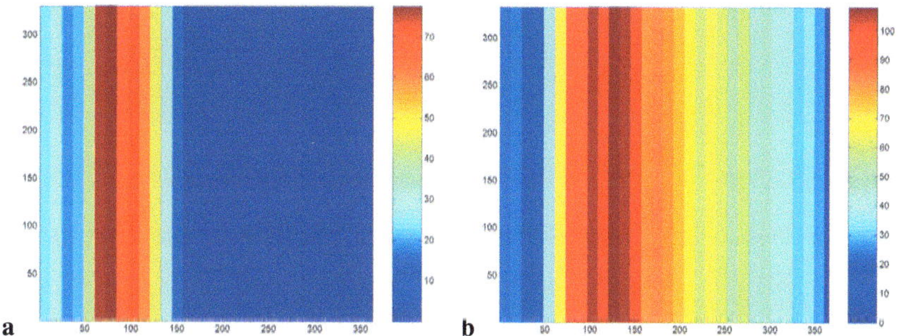

Fig. 18. Estimate of the local mean density of length using vertical windows for the simulation of angiogenesis in Fig. 5; **a** time 7.5 days; **b** time 15 days

of variation of the mean density of length, thus resulting in a good instrument for automatic diagnosis.

A 2-dimensional simulator of an angiogenic process was also implemented. Vessels are born at the left-hand side of the rectangular window according to a 1-dimensional Poisson process, or along existing vessels, according to model (3) and are attracted by the surface of a tumor represented by the right-hand side of the window. The growth of the vessels is driven by a given deterministic growth field; a more realistic simulation for angiogenesis should take into account the coupling of the growth of vessels with the time evolution of the underlying growth field. We took $G(x, y) = 0.1 + 0.1x^2$, constant in time but with a gradient in space in the direction of x, and strictly positive in the whole study region. The vessels have a random inclination, with respect to the horizontal direction. The probability that a vessel produces a new branch at a point $(x, y) \in \mathbf{R}^2$ in a time interval dt is $\beta(x, y)dt$, with $\beta(x, y) = 0.1x^2$, thus with a gradient along the x direction. The time step chosen in the simulation is $dt = 0.1$. The result of a simulation stopped at time $t = 10$ is shown in Fig. 19a.

The local mean density of length of fibres, that is, the mean length per unit area of vessels, was estimated using the estimator described in the previous section, both

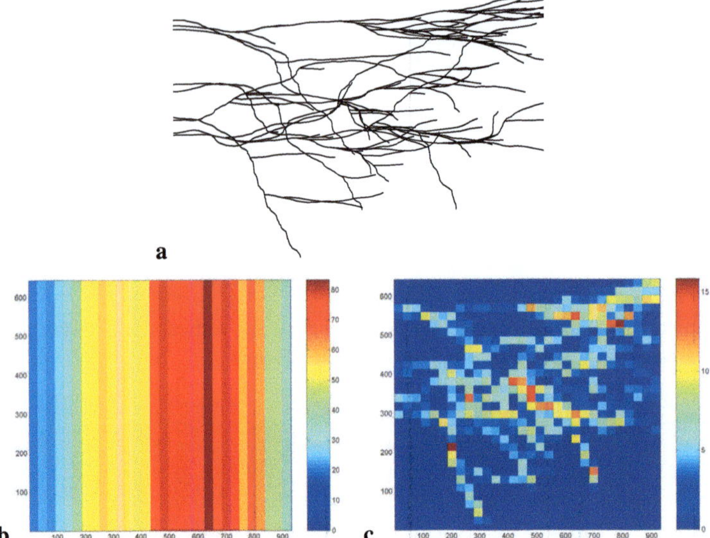

Fig. 19. a A simulation of a branching process of vessels from a tissue (on the right-hand side) to a tumor mass (on the left-hand side); **b** estimate of the local mean density of length using vertical windows, transverse to the main direction of growth; **c** estimate of the local mean density of length using rectangular windows

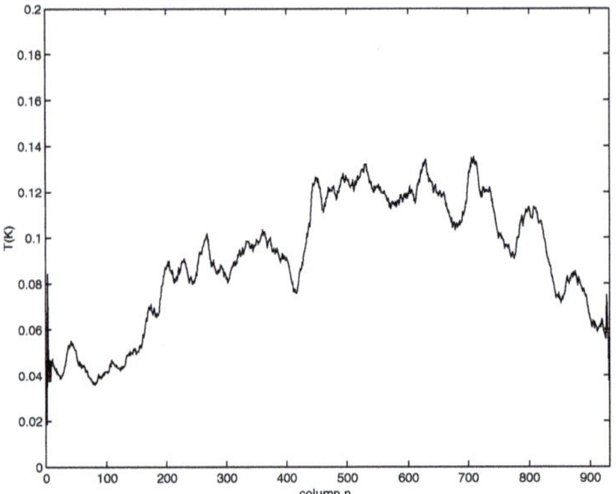

Fig. 20. Estimate of the hitting functional by rectangular windows for the vessel network in Fig. 19a

by dividing the window of observation into vertical strips, transverse to the main direction of growth, and by using a partition of the window into small rectangles. The results are shown in Figs. 19b,c. Also the hitting functional was estimated on the simulated pattern, by using an exploratory compact set formed by a vertical segment composed of 6 pixels. The estimate was performed on all the columns of the pixel matrix of Fig. 19a, and the results are reported in Fig. 20.

Acknowledgements

It is a pleasure to acknowledge the important contribution of M. Burger in Linz and of D. Morale and E. Villa in Milan in the development of joint research projects relevant to this presentation.

References

[1] Ambrosio, L., Fusco, N., Pallara, D.: Functions of bounded variation and free discontinuity problems. Oxford: Clarendon Press 2000

[2] Anderson, A.R.A.: The effects of cell adhesion on solid tumour geometry. In: Sekimura, T. et al. (eds.): Morphogenesis and pattern formation in biological systems. Tokyo: Springer 2003, pp. 315–325

[3] Anderson, A.R.A., Chaplain, M.A.J.: Continuous and discrete mathematical models of tumor-induced angiogenesis. Bull. Math. Biol. **60**, 857–899 (1998)

[4] Barles, G., Soner, H.M., Souganidis, P.E.: Front propagation and phase field theory. SIAM J. Control Optim. **31**, 439–469 (1993)

[5] Barndorff-Nielsen, O.E., Kendall, W.S., van Lieshout, M.N.M. (eds.): Stochastic geometry. Likelihood and computation. Boca Raton, FL: Chapman & Hall/CRC 1999

[6] Burger, M.: Growth fronts of first-order Hamilton-Jacobi equations. SFB Report 02-8. Linz: J. Kepler Universität 2002

[7] Burger, M., Capasso, V., Pizzocchero, L.: Mesoscale averaging of nucleation and growth models. CAM Report 05-19. Los Angeles: Dept. of Math., UCLA 2005

[8] Burger, M., Capasso, V., Salani, C.: Modelling multi-dimensional crystallization of polymers in interaction with heat transfer. Nonlinear Anal. Real World Appl. **3**, 139–160 (2002)

[9] Capasso V. (ed.): Mathematical modelling for polymer processing. Berlin: Springer 2003

[10] Capasso, V., Micheletti, A.: Local spherical contact distribution function and local mean densities for inhomogeneous random sets. Stochastics Stochastics Rep. **71**, 51–67 (2000)

[11] Capasso, V., Micheletti, A.: On the hazard function for inhomogeneous birth-and-growth processes. Quaderno 39/2001. Milano: Dip. di Mat., Uni. degli Studi di Milano 2001

[12] Capasso, V., Micheletti, A.: Stochastic geometry of spatially structured birth and growth processes. Application to crystallization processes. In: Capasso, V. et al. (eds.): Topics in spatial stochastic processes. (Lecture Notes in Math. **1802**) Berlin: Springer 2003, pp. 1–39

[13] Capasso, V., Micheletti, A., Burger, M.: Densities of n-facets of incomplete Johnson-Mehl tessellations generated by inhomogeneous birth-and-growth processes. Quaderno 38/2001. Milano: Dip. di Mat., Uni. degli Studi di Milano 2001

[14] Capasso, V., Morale, D., Salani, C.: Polymer crystallization processes via many particle systems. In: Capasso, V. (ed.): Mathematical modelling for polymer processing. Berlin: Springer 2003, pp. 243–259

[15] Capasso, V., Villa, E.: On the evolution equation of mean geometric densities for a class of space and time inhomogeneous stochastic birth-and-growth processes. In: Weil, W. (ed.): Stochastic geometry. (Lecture Notes in Math.) Berlin: Springer 2006

[16] Capasso, V., Villa, E.: On the continuity and absolute continuity of random closed sets. Preprint. Milano: Dip. di Mat., Uni. degli Studi di Milano 2005

[17] Carmeliet, P., Jain, R.K.: Angiogenesis in cancer and other diseases. Nature **407**, 249–257 (2000)

[18] Chaplain, M.A.J., Anderson, A.R.A.: Modelling the growth and form of capillary networks. In: Chaplain, M.A.J. et al. (eds.): On growth and form: spatio-temporal pattern formation in biology. Chichester: Wiley 1999, pp. 225–250
[19] Chen, C.S., Mrksich, M., Huang, S., Whitesides, G.M., Ingber, D.E.: Geometric control of cell life and death. Science **276**, 1425–1428 (1997)
[20] Cressie, N.A.C.: Statistics for spatial data. (Wiley Series in Probability and Mathematical Statistics) New York: Wiley 1993
[21] do Carmo, M.P.: Differential geometry of curves and surfaces. Englewood Cliffs, NJ: Prentice-Hall 1976
[22] Drasdo, D., Hoehme, S.: Individual-based approaches to birth and death in avascular tumors. Math. Comput. Modelling **37**, 1163–1175 (2003)
[23] Evans, L.C., Gariepy, R.F.: Measure theory and fine properties of functions. Boca Raton, FL: CRC Press 1992
[24] Federer, H.: Geometric measure theory. Berlin: Springer 1969
[25] Galle, J., Loeffler, M., Drasdo, D.: Modeling the effect of deregulated proliferation and apoptosis on the growth dynamics of epithelial cell populations in vitro. Biophys. J. **88**, 62–75 (2005)
[26] Ghyka, M.: The geometry of art and life. New York: Dover 1977
[27] Huang, S., Ingber, D.E.: Shape-dependent control of cell growth, differentiation, and apoptosis: switching between attractors in cell regulatory networks. Exp. Cell. Res. **261**, 91–103 (2000)
[28] Jain, R.K., Carmeliet, P.F.: Vessels of death or life. Sci. Amer. **285**(6), 38–45 (2001)
[29] Johnson, W.A., Mehl, R.F.: Reaction kinetics in processes of nucleation and growth. Trans. AIME **135**, 416–458 (1939)
[30] Kolmogorov, A.N.: On the statistical theory of the crystallization of metals. (Russian) Izv. Akad. Nauk SSSR Ser. Mat. **4**, 355–359 (1937)
[31] Kolmogorov, A.N.: Foundations of the theory of probability. New York: Chelsea 1956
[32] Landini, G., Misson, G.: Simulation of corneal neovascularization by inverted diffusion limited aggregation. Invest. Ophtamol. Vis. Sci. **34**, 1872–1875 (1993)
[33] Li, X., Wang, W., Martin, R.R., Bowyer, A.: Using low-discrepancy sequences and the Crofton formula to compute surface areas of geometric models. Comp. Aided Design **35**, 771–782 (2003)
[34] Lorenz, T.: Set-valued maps for image segmentation. Comput. Vis. Sci. **4**, 41–57 (2001)
[35] Matheron, G.: Random sets and integral geometry. New York: Wiley 1975
[36] Micheletti, A.: The surface density of a random Johnson-Mehl tessellation. Quaderno 17/2001. Milano: Dip. di Mat., Uni. degli Studi di Milano 2001
[37] Micheletti, A., Capasso, V.: The stochastic geometry of polymer crystallization processes. Stochastic Anal. Appl. **15**, 355–373 (1997)
[38] Miller, R.T., Anderson, S.P., Corton, J.C., Cattley, R.C.: Apoptosis, mitosis and cyclophilin-40 expression in regressing peroxisome proliferator-induced adenomas. Carcinogenesis **21**, 647–652 (2000)
[39] Møller, J.: Random Johnson-Mehl tessellations. Adv. Appl. Prob. **24**, 814–844 (1992)
[40] Møller, J.: Lectures on random Voronoï Tessellations. (Lecture Notes in Statist. **87**) New York: Springer 1994
[41] Morale, D.: A stochastic particle model for vasculogenesis: a multiple scale approach. In: Capasso, V. (ed.): Mathematical modelling & computing in biology and medicine. Bologna: Esculapio Pub. Co. 2003, pp. 616–621
[42] Murray, J.D.: Mathematical biology. Berlin: Springer 1989
[43] Muthukkaruppan, V.R., Kubai, L., Auerbach, R.: Tumor-induced neovascularization in the mouse eye. J. Natl. Cancer Inst. **69**, 699–708 (1982)

[44] Nanjundiah, V.: Alan Turing and "The Chemical Basis of Morphogenesis". In: Sekimura, T. et al. (eds.): Morphogenesis and pattern formation in biological systems. Tokyo: Springer 2003, pp. 33–44

[45] Pettet, G., Chaplain, M.A.J, McElwain, D.L.S., Byrne, H.M.: On the role of angiogenesis in wound healing. Proc. Roy. Soc. London Ser. B **263**, 1487–1493 (1996)

[46] Santini, M.T., Rainaldi, G., Indovina, P.L.: Apoptosis, cell adhesion and the extracellular matrix in the three-dimensional growth of multicellular tumor spheroids. Crit. Rev. Oncol. Hematol. **36**, 75–87 (2000)

[47] Schiaparelli, G.V.: Studio comparativo tra le forme organiche naturali e le forme geometriche pure. Milano: Hoepli 1898

[48] Sekimura, T., Noji, S., Ueno, N., Maini, P.K. (eds.): Morphogenesis and pattern formation in biological systems. Tokyo: Springer 2003

[49] Serini, G., Ambrosi, D., Giraudo, E., Gamba, A., Preziosi, L., Bussolino, F.: Modeling the early stages of vascular network assembly. EMBO J. **22**, 1771–1779 (2003)

[50] Sokołowski, J., Zolésio, J.-P.: Introduction to shape optimization. Berlin: Springer 1992

[51] Stoyan, D., Kendall, W.S., Mecke, J.: Stochastic geometry and its application. 2nd ed. New York: Wiley 1995

[52] Thompson, D.W.: On growth and form. Cambridge: Cambridge University Press 1917; 2nd ed. 1968

[53] Turing, A.M.: The chemical basis of morphogenesis. Philos. Trans. Roy. Soc. London Ser. A **237**, 37–72 (1952)

[54] Ubukata, T.: Computer modelling of microscopic features of molluscan shells. In: Sekimura, T. et al. (eds.): Morphogenesis and pattern formation in biological systems. Tokyo: Springer 2003, pp. 355–367

Mathematical modelling of tumour growth and treatment

A. Fasano, A. Bertuzzi, A. Gandolfi

Abstract. We review some of the models that have been proposed to describe tumour growth and treatment. A first class is that of models which include the analysis of stresses. Here the question of blood vessel collapse in vascular tumours is treated briefly. Results on the existence of radially- and of non-radially-symmetric solutions are illustrated together with an investigation of their stability. Two sections are devoted to tumour cords (growing directly around a blood vessel), highlighting basic facts that are indeed important in the evolution of solid tumours in the presence of necrotic regions. Tumour cords are also taken as an example to deal with certain aspects of tumour treatment. The latter subject is too large to be treated exhaustively but a brief account of the mathematical modelling of hyperthermia is given.

Keywords: mathematical modelling of tumour growth, mathematical modelling of tumour treatments, free boundary problems, systems of partial differential equations.

1 Introduction

1.1 Why

A glance at the table of contents of this book reveals what a formidable tool mathematics is for a variety of problems in biology and medicine. Oncology is not an exception, but – to be honest – despite the impressively large literature produced since the pioneering papers [24, 44] (for a recent comprehensive overview, see [7]), the mathematical models so far available have not become part of medical practice. The main reason is that, while the oncologists are fighting every day with systems of impressive complexity, possessing a biological, chemical and mechanical evolving structure able to modify the environment to create favourable conditions for their growth, the mathematicians have usually considered only isolated, largely simplified aspects.

Efforts have been intensified in recent years in various directions to provide synthetic descriptions of tumour growth and treatment, as well as to model delicate phenomena such as the mechanics of cell-cell interaction, mutual interactions between cells and the extracellular matrix, cell mutations, etc. Mathematicians have proposed various solutions based on different approaches and techniques, making the "mathematics of cancer" an incredibly rich and diversified subject, although far from embracing its entire complexity. On the other hand, there are serious justifications for a prudent approach in proposing new models. The motivation lies not only in the increasing mathematical difficulty generated by enlarging the sets of equations, but also in a very practical reason: including more phenomena in the general picture requires the knowledge of more and more physical parameters of difficult (if not impossible) experimental determination. This makes the necessity of cooperation with oncologists dramatically evident.

Let us return to the negative statement we started with about the slight impact of the mathematical work on the oncologists. To correct the impression of an exceedingly large gap between mathematics and oncology, we can say that there have also been remarkable contributions by mathematicians as, for instance, the modelling of vasculogenesis (a subject not exclusive to tumour growth, treated by L. Preziosi and S. Astanin in another chapter of this book), as a result of a close cooperation with experimentalists. And we shall see other cases in which mathematicians have produced practical tools, e.g., in therapeutical procedures. But much more has to be done to predict in a reliable way the tumour growth rate, the chances of metastasis and, above all, the effects of specific drugs or of radiation. Doing as much as possible on the computer can make experiments shorter and more effective, and it allows us to perform virtual treatment cycles, helping in deciding the optimal way of administering them. For instance, what is the best arrangement of duration, intensity and frequency of radiation treatments? Even an approximate model can give interesting indications. In fact, the stimulus of envisaging new perspectives was for us a strong motivation for looking into the present condition of mathematical research on cancer with a critical eye, hoping to open an appealing horizon particularly to young researchers.

1.2 How

We have talked so far of "tumours" in a generic way. It is unfortunately very well-known that there are many types of tumours. Most of the models in the literature refer either to spheroids of tumour cells growing in culture or to idealised systems emphasising only a limited number among the many coupled processes occurring in tumours *in vivo*. A real tumour is far more complex than such partial schemes, but adopting simplified views is not useless (provided simplifications are not too radical) if they allow us to draw at least qualitative or semi-quantitative conclusions, for instance, about the effects of treatments.

Even before considering in detail what is going on in a tumour, one is faced with very basic questions having no easy answers. First of all, tumours are colonies of cells and in this sense they have a discrete nature, although they are permeated by liquids, i.e., by a continuum phase. Thus the first question which naturally arises is the following: can they be treated in the framework of a totally continuum model, in which cells are grouped in species represented through densities, or is it preferable to keep the discrete character of the tumour, following the history of each cell (e.g., using cellular automata)? Moreover, each cell is an extremely complex system in itself, hosting a large number of chemical, electrochemical and mechanical processes which are crucial to its life and its behaviour within the tumour. How much of all this has to enter a mathematical model for the tumour? And in addition consider this: the cell cycle has stochastic components; so would it make sense to formulate a totally deterministic model? Finally, if one has to keep track of the phases through which each cell is cycling, cell age becomes an important parameter.

Continuum or discrete? Deterministic or stochastic? Averaging over age or not? Or a compromise? Even before trying to write the most basic equations, all these decisions must be taken. There is no such thing as the best possible approach for all circumstances. It all depends on two fundamental factors: a) the kind of system we are dealing with, b) the kind of target we have in mind. For instance, if we are studying a tumour with relatively few cells the discrete approach looks more natural, while a continuum model would be hardly justifiable. If we are considering a sufficiently large population it makes sense to consider the average properties locally and use the continuum approach. As to age dependence, it may or may not be essential. In a large, unsynchronised population we need not include age structure in our model if we are only interested in determining the global growth; on the contrary, if we want to predict the effect of a chemical which interacts with the cells only in a specific age compartment, monitoring cell age would obviously be absolutely necessary.

1.3 What

What are we talking about? We have anticipated many concepts, assuming that everyone knows what a tumour is, at least approximately. Mathematicians usually refer to an "ideal" tumour, governed by "simple" rules, that are believed to be common to all solid tumours. This approach fits relatively well the spheroids grown in a gel under controlled conditions (pressure, nutrient supply, etc.), whose behaviour is far simpler than that of tumours. The early papers by Greenspan [44, 45] refer to that case, which is still the subject of intense research. On the contrary a tumour in the body interacts with the surrounding tissues and with blood vessels in a number of ways.

First of all, it can hardly be considered a homogeneous population, since cells may be in different states from highly proliferating to quiescent, and the population is constantly evolving because of mutations. Cells may die, either because of lack of oxygen (if they are too far from the primary sources, which are the blood vessels) or because of treatments. Even death mechanisms can be complicated [58]. Depending on the cause, for instance, death may occur through programmed stages leading to the fragmentation of the cell into smaller bodies (apoptosis). Degrading dead cells and waste materials may form sizeable necrotic regions. Sometimes death can be preceded by defensive strategies. For instance, cells for which oxygen becomes scarce may switch to a different metabolic regime for some time. Tumours are also able to modify their surroundings by means of chemical signals. Starving cells secrete chemicals (tumour angiogenic factors) which diffuse and stimulate the lysis of the wall of nearby blood vessels, whose endothelial cells can detect the concentration gradient of the chemical signals and move towards their source (a displacement mechanism known as chemotaxis). In this way a new network of blood vessels is created, providing the tumour with a potentially unlimited supply of nutrients [36]. The structure of the new network is so irregular, however, that the blood flow through it may be very slow. From the hydraulic point of view we can say that the network has a high resistance. A side effect of great importance in chemotherapy is that such high resistance slows down the delivery of drugs to the tumour. Drugs injected in the blood stream will preferably bypass the tumour, reaching other parts of the body. It has been observed [53] that drug efficiency is improved by suitably pruning the tumour vessel network by vessel-targeted drugs.

The network may be partially neutralised by the tumour itself if it builds up a pressure sufficiently high to close the lumen of the vessels, as recently pointed out in [64]. Another interesting feature is the tendency of some tumours to create a "capsule", i.e., a layer of relatively compact material at the boundary [57], favouring its confinement. Capsules are formed in benign or less aggressive tumours. By contrast, more aggressive tumours are able to invade neighbouring tissues and to spread metastases in other parts of the body. Reduced cell-to-cell adhesion facilitates this process. These tumours try to conquer new territory by producing enzymes which "soften" the environment facilitating invasion of the surrounding tissues. Other cell-synthesized proteins exert several actions inside the tumour, giving rise to a complex set of self-interactions deeply influencing its evolution.

It now begins to be clear why modelling the mechanics and chemistry of a solid tumour is such a frightening idea. How to describe the multiple interactions among cells, extracellular matrix, vascular network, and the surrounding tissue? And we have not mentioned the transport of nutrients, the motion of the fluid surrounding the cells (which has a basic role in the global mass balance), the presence of macrophages feeding over the noxious components, the immune response of the body.

When we come to treatments we are faced with a no less complicated scenario. The discovery of tumour angiogenesis suggests that contrasting the factors that stimulate the sprouting of new vessels might be an effective way of controlling tumour growth. Unfortunately, so far this strategy has not been as successful as expected. Thus treatments are still aimed at killing tumour cells either by radiation or by means of cytotoxic drugs. Both techniques have been greatly improved over the years, but they would still greatly benefit from an accurate mathematical description. Independently of the way cells are killed, their massive destruction loosens the internal bonds of the system, altering transport mechanisms and producing less evident – but not less important – effects in the evolution of the necrotic regions.

It is no surprise that most mathematical models, independent of their mathematical structure, are based on greatly simplified pictures. We do not make an exhaustive review, which is not within the scope of this book, but we want to illustrate general ideas about possible future developments, along with results in particular directions. Due to space limitations, our exposition is concise and necessarily not complete. For instance, we omit the analysis of models of tumour invasion (see, e.g., [5,43,69]), or models including angiogenesis (see, e.g., [6,32,46]). Moreover, we confine ourselves to models based on the continuum scheme.

2 Models including the analysis of stresses

The analysis of stresses is of fundamental importance in modelling the spatial evolution of solid tumours when: a) the absence of geometrical symmetry does not allow us to derive the cell velocity field on the basis of the simple assumption of constant density of the total cell population; or b) the effects of stresses on their own, e.g., on the proliferation rate or on the perfusion of vascular tumours, must be accounted for. In this section we summarise some of the models that have been proposed to account consistently for the presence of stresses, using mainly the basic principles of mixture mechanics [65].

2.1 Applying the mechanics of mixtures to tumour growth: the "two-fluid" approach

This subject has been developed in a large number of papers and it is not possible to make a complete review here (for a recent survey, see [13]). We cite only [3, 22, 28, 29, 57] and we summarise the main concepts, following the exposition in [2]. In the above papers the tumour is schematised as a *mixture* having the properties of a deformable porous medium with two components: cells and extracellular liquid. Let $\Omega(t)$ be the spatial domain occupied by an evolving *avascular* tumour. Let δ_C, ν_C denote the mass density and the local volume fraction of the tumour cells and δ_E, ν_E be the corresponding quantities for the extracellular liquid. The mass densities are taken to be constant. Cells are assumed to lose their volume, by degradation into liquid, *instantly* upon death. The two components have the respective velocities $\mathbf{u}$, $\mathbf{v}$.

The mass balance leads to the following equations in $\Omega(t)$:

$$\frac{\partial \nu_C}{\partial t} + \nabla \cdot (\nu_C \mathbf{u}) = \Gamma_C, \tag{1}$$

$$\frac{\partial \nu_E}{\partial t} + \nabla \cdot (\nu_E \mathbf{v}) = \Gamma_E, \tag{2}$$

where Γ_C, Γ_E are the net volume production rates of the respective species, which incorporate cell proliferation and cell death. It is assumed that no voids are created within the system, so that

$$\nu_C + \nu_E = 1. \tag{3}$$

By summing (1)–(2) and taking (3) into account we obtain

$$\nabla \cdot (\nu_C \mathbf{u} + \nu_E \mathbf{v}) = \Gamma_C + \Gamma_E. \tag{4}$$

In order to express the momentum balance we must introduce the partial stresses $\mathbf{T}_C$, $\mathbf{T}_E$ (to be specified), obtaining

$$\begin{aligned} \delta_C \nu_C \left(\frac{\partial \mathbf{u}}{\partial t} + \mathbf{u} \cdot \nabla \mathbf{u}\right) &= \nabla \cdot \mathbf{T}_C + \mathbf{m}_C, \\ \delta_E \nu_E \left(\frac{\partial \mathbf{v}}{\partial t} + \mathbf{v} \cdot \nabla \mathbf{v}\right) &= \nabla \cdot \mathbf{T}_E + \mathbf{m}_E, \end{aligned} \tag{5}$$

where $\mathbf{m}_C$, $\mathbf{m}_E$ are the momentum supply rates due to the interaction between the two species in relative motion. The body forces, that could represent active behaviour of cells such as chemotaxis, are here considered negligible.

Mass is not generated nor destroyed within the system, but simply transferred from one species to the other. Thus,

$$\delta_C \Gamma_C + \delta_E \Gamma_E = 0, \tag{6}$$

and the same is true for momentum; therefore we have

$$\mathbf{m}_C + \mathbf{m}_E + \delta_C \Gamma_C \mathbf{u} + \delta_E \Gamma_E \mathbf{v} = 0.$$

The cumulative behaviour of the mixture can be described by defining the mixture density δ_m and the mixture velocity $\mathbf{w}_m$:

$$\begin{aligned} \delta_m &= \delta_C \nu_C + \delta_E \nu_E, \\ \delta_m \mathbf{w}_m &= \delta_C \nu_C \mathbf{u} + \delta_E \nu_E \mathbf{v}, \end{aligned}$$

which, together with the preceding equations, give the global mass balance

$$\frac{\partial \delta_m}{\partial t} + \nabla \cdot (\delta_m \mathbf{w}_m) = 0,$$

and the momentum balance of the mixture

$$\delta_m \left(\frac{\partial \mathbf{w}_m}{\partial t} + \mathbf{w}_m \cdot \nabla \mathbf{w}_m\right) = \nabla \cdot \mathbf{T}_m, \tag{7}$$

implicitly relating the stress tensor of the mixture $\mathbf{T}_m$ to the quantities already introduced. Since all inertia terms are indeed negligible, Eq. (7) can be replaced by

$$\nabla \cdot \mathbf{T}_m = 0, \tag{8}$$

and Eq. (5) by

$$-\nabla \cdot \mathbf{T}_E = \mathbf{m}_E. \tag{9}$$

Equations (1), (4), (8), (9) are the basic equations for the evolution of the system.

The core of the model consists in the choice of the stress tensors and of the interaction terms. Since the system is at every instant saturated, i.e., Eqs. (3) and (4) hold, it can be shown that the stress $\mathbf{T}_m$ must contain a term $-P\mathbf{I}$ [65], where P can be called the pressure of the mixture. This pressure contributes to each partial stress proportionally to the volume fraction of the species. In other words we can write

$$\mathbf{T}_C = \tilde{\mathbf{T}}_C - \nu_C P\mathbf{I}, \quad \mathbf{T}_E = \tilde{\mathbf{T}}_E - \nu_E P\mathbf{I}.$$

The strategy adopted is to treat both the components as "fluids". The interstitial fluid is considered ideal, while the cells exhibit elastic interactions and viscous drag. In summary,

$$\tilde{\mathbf{T}}_E = 0,$$

and $\mathbf{T}_m$ is taken as

$$\mathbf{T}_m = -[\Sigma(\nu_C) + P - \lambda_C(\nu_C)\nabla \cdot \mathbf{u}]\mathbf{I} + \mu_C(\nu_C)(\nabla\mathbf{u} + (\nabla\mathbf{u})^T), \tag{10}$$

where $\Sigma(\nu_C)$ models the cell-cell interaction, and $\mu_C(\nu_C)$, $\lambda_C(\nu_C)$ are the viscosity and the bulk viscosity of the "cell fluid". On the basis of (10) Eq. (8) takes the form

$$\nabla P = -\Sigma'(\nu_C)\nabla\nu_C + \nabla[\lambda_C(\nu_C)\nabla \cdot \mathbf{u}] + \nabla \cdot [\mu_C(\nu_C)(\nabla\mathbf{u} + (\nabla\mathbf{u})^T)], \tag{11}$$

where the superscript ($'$) denotes the derivative. The momentum transfer rate $\mathbf{m}_E$ is chosen so that Darcy's law is obtained for the interstitial fluid flow:

$$(1 - \nu_C)(\mathbf{v} - \mathbf{u}) = -\kappa\nabla P, \tag{12}$$

where κ represents the hydraulic conductivity of the system, seen as a deformable porous medium. Thus, the model is described by Eqs. (1), (4), (11), (12). Equation (12) may be used to eliminate $\mathbf{v}$ in (4).

Even in this relatively simple setting there are many parameters to be chosen. Particularly important is the selection of the elastic stress component $\Sigma(\nu_C)$. The choice proposed by Ambrosi and Preziosi [3] is to introduce a threshold ν_0 for ν_C below which cells do not interact, setting $\Sigma(\nu_C) = 0$ for $\nu_C \leq \nu_0$, and describing cell-cell interactions for $\nu_C > \nu_0$ as follows

$$\Sigma(\nu_C) = \alpha\frac{(\nu_C - \nu_0)^2(\nu_C - \nu_2)}{(1 - \nu_C)^\beta}, \tag{13}$$

where α, β and ν_2 are positive constants, $\nu_0 < \nu_2 < 1$. The above function has a negative minimum at $\nu_C = \nu_1$ with $\nu_0 < \nu_1 < \nu_2$, and has $\nu_C = 1$ as a vertical asymptote. The function (13) accounts for cell repulsion for $\nu_C \in (\nu_2, 1)$ and attraction for $\nu_C \in (\nu_0, \nu_2)$. The last piece of information to be supplied is the structure of the cell volume production rate, Γ_C. In [29] it is proposed that

$$\Gamma_C(\nu_C) = \begin{cases} \dfrac{\gamma}{1 + \eta\Sigma(\nu_C)}\dfrac{\sigma - \tilde{\sigma}}{1 + \omega\sigma}\nu_C - \mu\nu_C, & \sigma > \tilde{\sigma}, \\ -\dfrac{1 + \epsilon\tilde{\sigma}}{1 + \epsilon\sigma}\mu\nu_C, & \sigma \leq \tilde{\sigma}, \end{cases}$$

where σ is the nutrient concentration and γ, η, ω, ϵ, μ are positive constants. Proliferation takes place when σ exceeds a threshold $\tilde{\sigma}$, and the proliferation rate decreases as the stress Σ increases. Although both oxygen and glucose have been shown to be critical for cell viability and to affect proliferation [62], only one representative "nutrient" is considered here, as usual in the relevant literature. Of course, at this stage of modelling, nutrient dynamics comes into play. Since we refer to an avascular tumour that receives nutrient from the external environment, this dynamics may be described by the equation:

$$\frac{\partial\sigma}{\partial t} + \nabla \cdot (\sigma\nu_E\mathbf{v}) = \nabla \cdot (D\nabla\sigma) - \phi\sigma\nu_C, \tag{14}$$

exhibiting advection in the interstitial space, consumption by cells with rate $\phi\sigma$, and diffusion (in all the tumour space) with diffusivity D.

In [29] the one-dimensional Cartesian case ($\Omega(t) = \{|x| < L(t)\}$) is analyzed, under the simplifying assumptions of $\delta_C = \delta_E$ (so that the formation of new cells and the degradation of dead cells to liquids are not accompanied by volume change) and of no viscosity of the cell fluid. In this case the velocity fields are the scalar fields $u(x,t)$ and $v(x,t)$, and from Eqs. (4) and (6) we have $\nabla \cdot (\nu_C u + \nu_E v) = 0$. Since we must have $u(0,t) = v(0,t) = 0$ because of symmetry, we obtain

$$\nu_C u = -\nu_E v.$$

Thus the model predicts that cells and liquid move in opposite directions. Using the above equation in Darcy's law (12), we can write

$$u = \kappa \frac{\partial P}{\partial x}, \qquad v = -\kappa \frac{\nu_C}{1-\nu_C} \frac{\partial P}{\partial x},$$

so that from Eq. (11) we have

$$u = -\kappa \Sigma'(\nu_C) \frac{\partial \nu_C}{\partial x}.$$

Thus the following equation for $\nu_C(x,t)$ can be derived:

$$\frac{\partial \nu_C}{\partial t} = \frac{\partial}{\partial x}\left(\kappa \Sigma'(\nu_C) \frac{\partial \nu_C}{\partial x} \nu_C\right) + \Gamma_C, \tag{15}$$

which turns out to be forward parabolic for $\nu_C > \nu_1$ (and then in the repulsive case) and backward parabolic for $\nu_0 < \nu_C < \nu_1$ (becoming more delicate at least from the computational point of view). Equation (15) must be solved together with Eq. (14). Of course, the model also contains initial and boundary conditions, which we do not discuss for the sake of brevity, only recalling that the condition for (15) at the outer boundary accounts for tumour-environment mechanical interaction. We stress the fact that the outer boundary of the tumour is a free boundary, subject to the condition of being a material surface moving with the velocity u. Of course, the important case of the growth of a spherical tumour can be treated similarly. We also note that the formation of a necrotic core, as observed during the growth of multicellular spheroids [62], is here represented by a decrease of the fraction ν_C as the centre of the sphere is approached.

2.2 Vascular tumours: models for vascular collapse

The investigation of stresses in vascular tumours has been carried out in more detail to highlight specific features such as the blood vessel collapse. A three-phase model was studied in [23] to include a volume fraction of blood vessels, besides the volume fractions of cells and of extracellular liquid. All three components were considered as inviscid fluids, with the isotropic stress of the "cell fluid" containing a component

Σ that depends on the local cell fraction, as seen in Sect. 2.1. The cell proliferation rate is taken to be proportional to the fraction of vessels, whereas the growth rate of the vessels is regulated by the tumour cell fraction. Thus the model describes the interaction between tumour growth and the development of the supporting vasculature and, in this respect, it might be compared to the models in [32, 46, 50]. Cell death rate increases as the vessel fraction decreases and vessels decay with an assigned rate when the pressure exerted on them by the cells exceeds a critical pressure, so representing vessel collapse. The model was studied in the one-dimensional Cartesian case, and the simulations showed the possible formation of a core deprived of cells, as a consequence of vessel collapse due to an advancing front of increased cell pressure.

A different approach was proposed by Araujo and McElwain [8], in which tumour growth is viewed as the expansion of a single elastic material (see also [2]). The capillary network is assumed to collapse when the circumferential component of the stress exceeds a given threshold. By taking the cell volume fraction constant within the tumour and assuming that the cell proliferation rate is proportional to nutrient concentration, the mass balance yields the following equation for the (radial) cell velocity u:

$$\frac{1}{r^2}\frac{\partial}{\partial r}(r^2 u) = \alpha\sigma - \mu. \tag{16}$$

In Eq. (16) σ is the nutrient concentration, and α, μ are rate constants associated, respectively, with proliferation and volume loss due to cell death. The hypothesis that the cell volume fraction is constant with r has the physical meaning that the system is arranged according to a uniform "optimal" packing of cells.

The radial stress σ_r and the circumferential stress σ_θ, after defining $\beta = \sigma_r - \sigma_\theta$, obey the equations

$$\frac{\partial \sigma_r}{\partial r} + \frac{2\beta}{r} = 0, \tag{17}$$

$$\frac{\partial \beta}{\partial t} + u\frac{\partial \beta}{\partial r} = E\left(\frac{\partial u}{\partial r} - \eta_r \alpha\sigma + \zeta_r \mu\right), \tag{18}$$

where E is the Young's elasticity modulus, and η_r, ζ_r are the anisotropic growth-strain multipliers. Equation (17) expresses the balance of forces, whereas Eq. (18) was derived in [9] for an anisotropic growth process where the daughter cells are moving preferentially in the direction of least stress. The multipliers η_r, ζ_r were chosen in [8, 9] as an increasing and, respectively, a decreasing function of β, that take the same isotropic value (equal to 1/3) when $\beta = 0$, while $\eta_r \to 1$, $\zeta_r \to 0$ as $\beta \to +\infty$, and $\eta_r \to 0$, $\zeta_r \to 1$ as $\beta \to -\infty$. In this way, a steady-state distribution of stress can be achieved, unlike the isotropic case with $\eta_r = \zeta_r \equiv 1/3$ which would imply the indefinite growth of β at the outer boundary. Indeed, the specific behaviour adopted for η_r, ζ_r prevents β from going to $\pm\infty$ along characteristics in Eq. (18).

Instead of coupling the nutrient dynamics to the above equations, the authors follow a simplified approach, in which $\sigma(r, t)$ has a prescribed quadratic profile with a maximum at the tumour centre, in the absence of vascular collapse. When the latter intervenes, i.e., when σ_θ is less than a critical value (compressive stress is taken to be negative), it creates a moving interface $r = r_b(t)$ enclosing the collapsed zone and cutting off the nutrient supply. For $r < r_b(t)$, σ is just transported by diffusion. In this inner zone, on the assumption that diffusion is quasi-steady and the nutrient absorption rate is proportional to σ, the nutrient concentration can be determined explicitly by matching it with the value taken by σ at the collapse front (in the outer, still perfused, zone the nutrient concentration is assumed to keep the given quadratic profile). Knowing σ, one can find the radial velocity profile (in terms of r_b) and derive a first order ordinary differential equation for the motion of the outer boundary of the sphere.

3 About tumour morphology and asymptotic behaviour

Most of the literature about mathematical modelling of tumours deals with spherical or cylindrical geometry (sometimes also one-dimensional Cartesian geometry is considered, for qualitative conclusions). Going beyond such simple geometries leads to heavy complications and, generally speaking, the mechanics plays a greater role. If one manages to keep the mechanics at a very simple level and suitably reduces the number of variables, it becomes relatively easy to find spherically symmetric equilibrium solutions in $\mathbb{R}^3$. However, even in such extremely simplified situations, the following questions pose formidable difficulties. Are there steady state solutions exhibiting more complex morphologies? Are the radially symmetric solutions asymptotically stable with respect to general perturbations? In this section we report some of the answers that have been given in the last few years. Of course, once again, we make a limited selection, confining ourselves to a group of papers [11, 30, 31, 37, 40–42] which are based on the model in [25].

3.1 Radially symmetric solutions and their stability under radially symmetric perturbations

Friedman and Reitich [40] studied the model proposed by Byrne and Chaplain in [25] for the growth of a non-necrotic *vascular* tumour, in the case of absence of treatment. Vascular supply of nutrients is modelled by a source uniformly distributed within the tumour mass. For a spherically symmetric tumour with radius $R(t)$, the model equations are as follows:

$$\frac{\partial \sigma}{\partial t} = \frac{D}{r^2}\frac{\partial}{\partial r}\left(r^2 \frac{\partial \sigma}{\partial r}\right) + \lambda_b(\sigma_b - \sigma) - \lambda\sigma, \tag{19}$$

$$\frac{\partial \sigma}{\partial r}(0, t) = 0, \quad \sigma(R(t), t) = \bar{\sigma}, \tag{20}$$

$$\frac{dR}{dt} = \frac{1}{R^2(t)} \int_0^{R(t)} s[\sigma(r, t) - \tilde{\sigma}] r^2 \, dr, \tag{21}$$

where σ is the nutrient concentration, D the diffusion coefficient, σ_b the (constant) nutrient concentration in the vasculature, λ_b the transfer rate coefficient of the vascular network, and λ the consumption rate constant. The quantities $s\sigma$ and $s\tilde{\sigma}$ on the right-hand side of (21) are the local volume production rate due to cell proliferation and the volume loss rate due to cell death, respectively, and cells are assumed to lose their volume instantly upon death. The cell velocity field u is obtained by assuming that the volume fraction of cells is constant, so that the mass balance yields $\operatorname{div} u = s(\sigma - \tilde{\sigma})$, from which Eq. (21) for the velocity of the boundary follows easily.

After suitable non-dimensionalization and rescaling of the variables and parameters, the above model may be rewritten as follows (keeping the same symbols for the scaled quantities):

$$\alpha \frac{\partial \sigma}{\partial t} = \frac{1}{r^2} \frac{\partial}{\partial r} \left(r^2 \frac{\partial \sigma}{\partial r} \right) - \lambda \sigma, \tag{22}$$

$$\frac{\partial \sigma}{\partial r}(0, t) = 0, \quad \sigma(R(t), t) = \bar{\sigma}, \tag{23}$$

$$\frac{dR}{dt} = \frac{s}{R^2(t)} \int_0^{R(t)} [\sigma(r, t) - \tilde{\sigma}] r^2 dr \tag{24}$$

with spherically symmetric initial conditions for σ satisfying regularity and compatibility assumptions. The parameter α, defined as $\bar{R}^2/(D\bar{T})$, where $\bar{R}$ is a characteristic tumour radius and $\bar{T}$ is the tumour growth time scale, is the ratio of the nutrient diffusion time scale to the tumour growth time scale. In this model the tumour is simply seen as a "background" for the evolution of the quantity σ. By confining attention to spherically symmetric perturbations, the following conclusions were reached by assuming $\bar{\sigma} > \tilde{\sigma} > \lambda$ and, to slightly simplify the calculations, by taking $s = 3$: (i) there is a unique stationary solution $R(t) = R_0 > 0$; (ii) if α is sufficiently small (which is reasonable in the physical case) then, for any initial value of R, the solutions of (22)–(24) tend asymptotically to the stationary solution, exponentially fast; (iii) the function $R(t)$ is always greater than a computable positive constant (the tumour never shrinks to zero). Moreover, conditions are given which guarantee that $R(t)$ remains bounded or goes to infinity exponentially fast for some initial data. A similar analysis was also performed when a second diffusible chemical that enhances nutrient consumption was added to the model [31].

3.2 Looking for non-radially symmetric stationary solutions

A far more complicated morphology is considered in [41] (also see [42]), where the authors develop a method for solving systems of partial differential equations that depend analytically on a parameter ϵ, finding solutions in the form of convergent series of powers of ϵ. The model, given in non-dimensional form, describes the stationary state of a tumour occupying a two-dimensional domain Ω. For the nutrient concentration, quasi-stationary diffusion is assumed:

$$\Delta \sigma = \sigma. \tag{25}$$

The velocity field $\mathbf{u}$ of the cellular material was assumed to be governed by a Darcy-type law

$$\mathbf{u} = -\nabla p, \tag{26}$$

with p representing the "cell pressure". If the volume fraction of cells is still assumed to be constant, and then $\nabla \cdot \mathbf{u} = s(\sigma - \tilde{\sigma})$, then p satisfies the equation

$$\Delta p = -s(\sigma - \tilde{\sigma}). \tag{27}$$

The above equations are completed with the boundary conditions on $\partial\Omega$ (the free boundary):

$$\sigma = \bar{\sigma}, \tag{28}$$

$$\mathbf{n} \cdot \nabla p = 0, \tag{29}$$

$$p = \gamma k, \tag{30}$$

where $\mathbf{n}$ is the outward normal to the boundary, k the curvature of $\partial\Omega$ and γ the "tumour surface tension" (in the spirit of [45]).

In the spherical case it is shown that, for $\tilde{\sigma}/\bar{\sigma} \in (0, \frac{1}{2})$, there is only one solution $(\sigma_0(r), p_0(r), R_0)$, with the radius R_0 depending only on $\tilde{\sigma}/\bar{\sigma}$ (and not on γ). Next is addressed the question of whether there is a bifurcation branch $(\sigma_\epsilon, p_\epsilon, \Omega_\epsilon, \gamma_\epsilon)$ of solutions of the form:

$$R_\epsilon(\theta) = R_0 + \sum_{n=1}^{\infty} \epsilon^n f_n(\theta), \tag{31}$$

$$\sigma_\epsilon(r, \theta) = \sigma_0(r) + \sum_{n=1}^{\infty} \epsilon^n \sigma_n(r, \theta), \tag{32}$$

$$p_\epsilon(r, \theta) = p_0(r) + \sum_{n=1}^{\infty} \epsilon^n p_n(r, \theta), \tag{33}$$

$$\gamma_\epsilon = \gamma_0 + \sum_{n=1}^{\infty} \epsilon^n \gamma_n, \tag{34}$$

bifurcating from a radial solution for $\gamma = \gamma_0$ (the branching point). The procedure for determining the above series is by no means simple. First the boundary perturbation

$$R_\epsilon(\theta) = R_0 + \epsilon f_1(\theta)$$

is considered, by fixing $f_1 = \cos l\theta$ for an integer $l \geq 2$, and it is shown that the corresponding perturbations of σ, p in the form $\sigma = \sigma_0(r) + \epsilon\sigma_1(r) f_1(\theta)$, $p = \gamma_1/R_0 + p_0(r) + \epsilon p_1(r) f_1(\theta)$ can be calculated, provided γ_0 satisfies an equation (the bifurcation equation, here omitted for the sake of brevity) which is shown to have a unique solution. Then the general system (31)–(34) is studied by seeking the functions σ_n, p_n in (32), (33) in the form of series

$$\sigma_n = \sum_{k=0}^{\infty} \sigma_{nk}(\theta)(r - R_0)^k, \quad p_n = \sum_{k=0}^{\infty} p_{nk}(\theta)(r - R_0)^k.$$

The procedure is successfully performed, obtaining enough estimates to show that all the series converge (in particular it is shown that all the odd terms in (34) vanish).

An alternative approach to the same question of the branching of non-radially symmetric solutions is presented in [37], this time in three dimensions. Instead of using the technique of mapping the free boundary into a fixed circle, as in the previous paper, here the Hanzawa transformation [47] is employed to reformulate the problem in a fixed domain. The new technique proves advantageous, and establishes the existence of a bifurcation branch of the form:

$$R_\epsilon = R_0 + \epsilon Y_{l,m}(\theta, \phi) + O(\epsilon^2), \quad l \geq 2, \quad m = 0, 1 \ldots, l,$$

$$\gamma_\epsilon = \gamma_0 + \epsilon \gamma_1 + O(\epsilon^2),$$

where $Y_{l,m}(\theta, \phi)$ are the spherical harmonics and the bifurcation point $\gamma = \gamma_0$ is uniquely determined for each l.

3.3 The general problem of the stability of radially symmetric solutions

The question of stability is resumed in [11] under a point of view much more general than that in [40]. The authors' goal is now to investigate the stability of radially symmetric solutions with respect to small perturbations which are not necessarily radially symmetric. If the stability of spherically symmetric solutions is to be investigated with respect to non-radially symmetric perturbations, the tumour dynamics must be considered in a multidimensional setting (also see [27]). As in the study reported in the previous section, the cell velocity is described by Eq. (26) through the cell pressure field given by (27) with the boundary condition (30). For the evolution problem in $\Omega(t)$, Eq. (29) must be replaced with

$$V = -\mathbf{n} \cdot \nabla p \quad \text{on} \quad \partial\Omega(t), \tag{35}$$

where V is the normal velocity. The authors also decided to describe the nutrient diffusion not just by the quasi-steady equation (25) but with the parabolic equation

$$\sigma_t - \Delta\sigma + \sigma = 0,$$

with the boundary condition (28). Initial conditions $\Omega(0) = \Omega^0$ and $\sigma|_{t=0} = \sigma^0$ in Ω^0 are prescribed.

The main result of [11] is that, if s is sufficiently small, then, for Ω^0, σ^0 sufficiently near a radially symmetric stationary solution with radius R_0, one can state the following: (i) there exists a unique global solution of the evolution problem (the Hanzawa transformation is again used); (ii) there exists a point $\mathbf{x}^*$ near the origin such that $\partial\Omega(t)$ converges exponentially fast, as $t \to \infty$, to the circle $|\mathbf{x} - \mathbf{x}^*| = R_0$. A remark which is relevant to those interested in the general theory of free boundary problems is that the above results also apply for $s = 0$ (no proliferation), in which case the problem (27), (30), (35) for the pressure becomes decoupled and identical to the problem of the Hele-Shaw cell.

3.4 Asymptotic regimes and vascularisation

In the paper [30] basically the same model for a vascular non-necrotic tumour is considered, as seen in the previous section, with the aim of investigating the influence of the level of vascularization on the asymptotic behaviour in more than one dimension. By suitably redefining the nutrient concentration σ and the pressure p, the authors were able to formulate the model in terms of two decoupled problems:

$$\begin{aligned} \Delta\sigma &= \sigma \quad \text{in} \quad \Omega(t), \\ \sigma &= 1 \quad \text{on} \quad \partial\Omega(t), \end{aligned}$$

and

$$\begin{aligned} \Delta p &= 0 \quad \text{in} \quad \Omega(t), \\ p &= k - AG\frac{|\mathbf{x}|^2}{2d} \quad \text{on} \quad \partial\Omega(t), \end{aligned}$$

where d is the number of space dimensions. The normal velocity at the tumour surface is expressed by

$$V = -\mathbf{n}\cdot\nabla p + G\mathbf{n}\cdot\nabla\sigma - AG\frac{\mathbf{n}\cdot\mathbf{x}}{d} \quad \text{on} \quad \partial\Omega(t).$$

As usual, k is the curvature of $\partial\Omega$ and $\mathbf{x}$ is the position in space, while the constants A, G are combinations of the physical parameters introduced in Eqs. (19)–(21),

$$A = \frac{\tilde{\sigma}/\bar{\sigma} - B}{1 - B}, \qquad G = \frac{s\bar{\sigma}}{\lambda_R}(1 - B),$$

where λ_R is a rate of relaxation (synthesizing the mechanical properties of the system) and B is the main parameter in this study, defined by

$$B = \frac{\lambda_b}{\lambda_b + \lambda}\cdot\frac{\sigma_b}{\bar{\sigma}}.$$

The analysis is first performed in the radially symmetric case and three possible regimes of growth are identified as follows. *Low vascularisation*: $B \le 1$ and $B < \tilde{\sigma}/\bar{\sigma}$. The tumour evolves monotonically to a stationary state. This result parallels that in [40]. Avascular growth ($B = 0$) also belongs to this regime. *Moderate vascularisation*: $\tilde{\sigma}/\bar{\sigma} < B < 1$. For any initial radius the tumour grows indefinitely. *High vascularisation*: if $\tilde{\sigma}/\bar{\sigma} > B > 1$ the tumour evolves to extinction. If $B > 1$ and $B > \tilde{\sigma}/\bar{\sigma}$, when growth occurs (depending on the initial radius) it is unbounded.

Moreover, a particular case of unbounded growth is described, in which the initial shape is non-spherical and the non-spherical shape is preserved (self-similar growth). For $d = 3$ this may only occur in the regime of moderate vascularisation. The authors find a critical value $A_c < 0$ for which the shape is preserved, that discriminates between decaying ($A < A_c$) and growing ($A > A_c$) deviations from the spherical shape. The latter case is of some importance because it may be associated with invasion with fingering.

4 Models with cell age or cell maturity structure for tumour cords

4.1 Tumour cords

If we move from the whole tumour to its microarchitecture, we find a very complex organisation, with necrotic regions possibly mixed with regions of viable cells, in relation to the irregular and poorly effective vascular network of the tumour. In some human and experimental tumours, however, tumour cells arranged in cylindrical structures around central blood vessels were observed. These structures are called *tumour cords* [48,61,70]. The radial growth of a cord is limited by the availability of nutrients. At a distance of the order of 100 μm from the vessel they become insufficient to sustain life. Therefore, cords can be surrounded by a necrotic region, merging with the necrotic regions of the neighbouring cords (see Fig. 1). Cords are not of equal size and are not parallel, but it is difficult to resist the temptation of considering an array of identical and parallel cords in the same spirit as that of Krogh's blood perfusion model [55], in which blood vessels through a portion of tissue are taken to be parallel and of the same size. In this situation each cord inside the array, together with its necrotic region, behaves as a system which does not exchange any material with the surrounding cords. Thus, a model for a single cord and its evolution can be considered. Moreover we can push our idealisation a bit further by supposing that the cord has cylindrical symmetry around the vessel. This is not a minor step. The main concern is not so much about altering the geometry of the external boundary of the necrotic region by setting it at $r = B$, as seen in Fig. 2 (an hexagonal cross-section would be more appropriate), as about the fact that some of the fundamental parameters in the system (blood pressure and oxygen concentration in blood) vary along the central vessel producing a situation potentially incompatible with cylindrical symmetry. Is

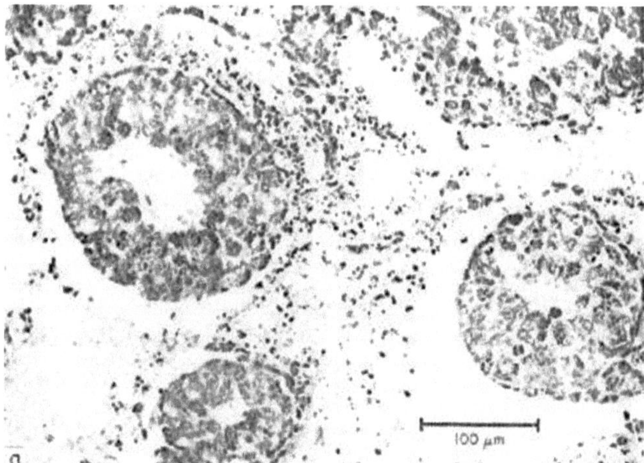

Fig. 1. Histological section showing tumour cords within a region of necrosis (Reproduced, with authorization, from: [48], p. 35

there a size of blood vessel such that the relative changes of the two above mentioned quantities may be negligible? The question is not trivial because a small decrease in oxygen concentration requires a sufficiently fast flow, but the latter is driven by a sufficiently large pressure gradient. Fortunately, there is room for a compromise as shown in [18].

Having simplified the geometry is fundamental to carry out a rigorous analysis of whatever model we want to formulate, but does not eliminate the complexity of the many coupled processes we have to describe. In this respect the study of tumour cords is good training for undertaking more ambitious projects. Although we are treating a single cord, the presence of the whole tumour is felt through the boundary conditions. Note that the physical situation of a tumour cord is somehow dual to that of the avascular spheroid, in which nutrients come from outside and the necrotic region is inside.

Roughly speaking, we can divide the mathematical models so far proposed for tumour cords into two classes: 1) models in which the population of proliferating cells is structured by cell age or other quantities reflecting the position of cells in the cell cycle; 2) models with no age or equivalent structure, in which, as in the mixture approach of Sect. 2.1, cell subpopulations are represented by their local volume fractions. Depending on the specific target of the investigation, cell age may be a relevant factor or excess information, complicating an already largely complex scenario. Therefore, models with or without age structure can be regarded as complementary.

4.2 Age and maturity structured models

In a first attempt to model the cell population within a tumour cord [14], the population of proliferating cells was structured by cell *age* in order to represent the different cell cycle phases. We denote by r_0 the radius of the central blood vessel, by r the radial distance from vessel axis, and by ρ_N the cord radius (see Fig. 2). Cords were considered to be surrounded by necrosis, so that ρ_N identifies the cord/necrosis interface. According to experimental observations in untreated tumours, ρ_N was assumed constant and the model was focussed on the *stationary* state. All the variables describing the cord were assumed to be independent of the axial coordinate.

The population of viable tumour cells was viewed as a continuum composed of proliferating (cycling) cells and quiescent cells. The population of cycling cells was described by the density $n(a, r, t)$, where $n(a, r, t)\, da$ is the number of cycling cells with age between a and $a + da$ in a unit volume, at position r and time t. Further, $n_Q(r, t)$ gives the number of quiescent cells in a unit volume. Cell motion was assumed to be radially directed and was represented by a velocity field u common to all the cells. Thus, rearrangements among cell subpopulations due to cell motions of diffusive type were excluded. The total cell density was assumed constant. Measurements of the number of cells in histological sections of untreated tumour cords [59–61] support this assumption.

The effect of different concentrations of oxygen and/or nutrients was assumed to affect only the transition of cells to quiescence. If the concentration profile of

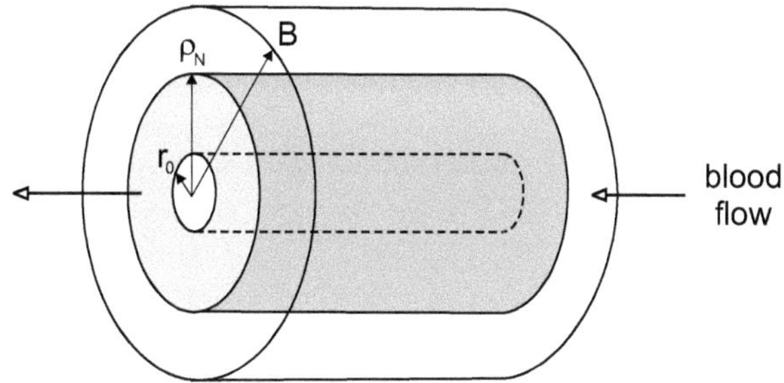

Fig. 2. Geometry of a tumour cord (symbols explained in the text)

chemicals does not change with time, the cell cycle parameters may be regarded as functions of the radial distance. Thus in [14] we assumed that a fraction $\theta(r) \in [0, 1]$ of the cells born at position r enters the cycle, and a fraction $1 - \theta(r)$ becomes quiescent. In view of the observed decrease of proliferation along the cord radius, the fraction θ is a non-increasing function of r. The recruitment of quiescent cells into the cycle was considered to be negligible in the untreated cords. All the proliferating cells were instead assumed to traverse the cycle in the same time T_c, and thus for the cell age we have $0 \le a \le T_c$. All cells die at $r = \rho_N$, and possible random cell death within the cord was neglected.

The conservation equations for the cell densities $n(a, r, t)$ and $n_Q(r, t)$, $r \in [r_0, \rho_N]$, can be written as:

$$\frac{\partial n}{\partial t} + \frac{\partial n}{\partial a} + \frac{1}{r}\frac{\partial}{\partial r}(run) = 0, \tag{36}$$

$$n(0, r, t) = 2\theta(r)n(T_c, r, t), \tag{37}$$

$$\frac{\partial n_Q}{\partial t} + \frac{1}{r}\frac{\partial}{\partial r}(run_Q) = 2(1 - \theta(r))n(T_c, r, t). \tag{38}$$

In Eqs. (37) and (38), $n(T_c, r, t)$ yields the rate of cell division. By integrating (36) with respect to age, and taking (37) and (38) into account, we have that the total cell density $n_C(r, t)$,

$$n_C(r, t) = \int_0^{T_c} n(a, r, t)\, da + n_Q(r, t),$$

satisfies the equation

$$\frac{\partial n_C}{\partial t} + \frac{1}{r}\frac{\partial}{\partial r}(run_C) = n(T_c, r, t). \tag{39}$$

Assuming $n_C(r, t) = n^\star$, since there is no cell flux across the vessel wall ($u(r_0, t) = 0$), from (39) we see that

$$ru(r, t) = \frac{1}{n^\star}\int_{r_0}^{r} r' n(T_c, r', t)\, dr'. \tag{40}$$

We note that u is positive unless $n(a,r,t) \equiv 0$, and thus the cells move towards the periphery and the necrotic region is continuously fed by cells that die when crossing the interface ρ_N.

At the stationary state, the cell densities $n(a,r)$ and $n_Q(r)$ satisfy the equations:

$$\frac{\partial n}{\partial a} + \frac{1}{r}\frac{\partial}{\partial r}(run) = 0, \tag{41}$$

$$n(0,r) = 2\theta(r)n(T_c,r), \tag{42}$$

$$\frac{1}{r}\frac{d}{dr}(run_Q) = 2(1-\theta(r))n(T_c,r), \tag{43}$$

$$ru(r) = \frac{1}{n^\star}\int_{r_0}^{r} r'n(T_c,r')\,dr'. \tag{44}$$

Existence and uniqueness of solutions of Eqs. (41)–(44) was proved by Webb [72] under the hypothesis that $\theta(r)$ is constant in a right neighbourhood of r_0. The condition $\theta(r_0) > 1/2$ is necessary in order to have $n(a,r)$ nonzero and appears to be biologically meaningful, since it states that there is a portion of the cord (at least close to the vessel) in which cell division produces a number of proliferating cells larger than the number of quiescent cells.

The evolutive problem for the model (36)–(38) and (40), slightly modified in the boundary condition (37), from an assigned initial state $n(r,0)$ and in the fixed domain $r \in [r_0, \rho_N]$, was investigated in [35]. It was shown that, if $\theta(r_0) \le 1/2$ (and $\theta(r)$ remains constant in a right neighbourhood of r_0), the proliferating population asymptotically vanishes, whereas, if $\theta(r_0) > 1/2$, then $n(a,r,t)$ converges to an eventually periodic solution with period T_c. Moreover, it was found that the solutions $n(a,r,t)$ converge to the stationary solution if and only if $n(a,r,0)$ is proportional to it. This dependence of the asymptotic behaviour on the initial data is a consequence of the assumption of uniform duration of cell cycle in the cell population.

The assumption that the cell cycle time is not affected by changes in the microenvironment can be relaxed by describing the cell population in terms of cell *maturity* [66] (the maturity gives the position of the cell along the cycle), and by introducing a maturation velocity dependent on the microenvironment. This representation of the cell population was considered by Dyson et al. [34]. The population of proliferating cells is described, at the stationary state, by a density $n(x,r)$, where $x \in [0,1]$ is the cell maturity, and the maturation rate is a function $w(x,r)$ of cell maturity and radial distance. The conservation equations for the cell densities are given by:

$$\frac{\partial}{\partial x}(wn) + \frac{1}{r}\frac{\partial}{\partial r}(run) = 0,$$

$$w(0,r)n(0,r) = 2\theta(r)w(1,r)n(1,r),$$

$$\frac{1}{r}\frac{d}{dr}(run_Q) = 2(1-\theta(r))w(1,r)n(1,r).$$

The maturation rate was taken as the product of a function of x by a function of r alone. Also in this model, the cell velocity field is determined by imposing the condition that the number of cells in a unit volume does not change with r. The existence and uniqueness of the solution was proved. We note that the age-structured model (41)-(43), in which the cell cycle duration is constant, can be rewritten in terms of maturity by defining $x = a/T_c$ and taking the constant maturation rate $1/T_c$.

The cell cycle was described in [15] by a sequence of M discrete compartments of cell maturity, so that the proliferating cell population in the tumour cord was represented by the functions $n_k(r, t)$, $k = 1, \cdots, M$, where $n_k(r, t)$ is the local number of cells in the kth compartment, in a unit volume. Poisson exit from each compartment with rate constant λ_k was assumed, together with the possibility of cell arrest in a quiescent status after mitosis. As the fraction θ, so also the rate constants λ_k's may be taken as non-increasing functions of r, thus making the progression through the cell cycle dependent on the radial position of the cell. It may be noted that the stochastic counterpart of this model assigns an exponentially distributed residence time to each maturity compartment, so that the deterministic model would describe a population with cell-to-cell variability of the cycle transit time even if the exit rates λ_k's were independent of r.

Instead of assuming that the number of cells per unit volume is constant, we may suppose that the volume fraction occupied locally by the cells, as introduced in Sect. 2.1, is constant within the cord. As the cells have different volumes according to their position in the cycle or to their quiescence, the mean cell volume may change with r and a constant cell volume fraction is not necessarily equivalent to a constant cell density n_C. As in Sect. 2.1, let $\nu_C(r, t)$ denote the volume fraction of cells. We can express ν_C in terms of the age structured model (36)-(38), obtaining

$$\nu_C(r, t) = \int_0^{T_c} \upsilon(a) n(a, r, t)\, da + \upsilon(0) n_Q(r, t),$$

where $\upsilon(a)$ is the volume of a proliferating cell of age a, with $\upsilon(T_c) = 2\upsilon(0)$ (note that $\upsilon(0)$ is also the volume of quiescent cells). By integrating Eq. (36) multiplied by $\upsilon(a)$ with respect to age, and by adding (38) multiplied by $\upsilon(0)$, the following equation for $\nu_C(r)$ is obtained:

$$\frac{\partial \nu_C}{\partial t} + \frac{1}{r}\frac{\partial}{\partial r}\left(r u \nu_C\right) = \int_0^{T_c} \frac{d\upsilon}{da} n(a, r, t)\, da.$$

If ν_C is assumed to be constant and equal to $\nu^\star$, the above equation provides an expression for the velocity field u, different from that in (40), to be used in (36)–(38). A similar computation may also be done when the cell population is structured by maturity.

In the simple case of $\upsilon(a) = \upsilon_0 + (\upsilon_0/T_c)a$, the mean volume of proliferating cells is equal to $\upsilon_0(1 + \langle a \rangle / T_c)$, where $\langle a \rangle$ is the mean cell age. At the stationary state, the mean cell volume in the cord is always between υ_0 (all cells quiescent) and $\upsilon_0/\ln 2$ (all cells proliferating). Therefore, under the hypothesis of constant number of cells per unit volume, $\nu_C(r)$ should decrease with r since the proliferation fraction decreases, but the ratio $\nu_C(\rho_N)/\nu_C(r_0)$ cannot be smaller than ln 2. These considerations suggest that the assumptions of constant ν_C or constant n_C could lead to similar quantitative results, despite the different expressions of the velocity field.

The models that incorporate the cell cycle structure are useful for analysing experimental data from labelling experiments designed to investigate how the cell proliferation changes at different distances from the blood vessel [48,60,70]. Examples of these analyses are reported in [14] and [15], where the curves of the labelling index and of the fraction of labelled mitoses versus time at different radial distances were simulated on the basis of the models reported in this section and compared to experimental data, with the aim of estimating the changes in cell kinetic parameters from the inner zone to the cord periphery.

Moreover, the models that distinguish the cells according to their position along the cell cycle may be useful for accurately describing the effects of chemotherapy or radiotherapy, since many drugs as well as radiation are more effective on cells in specific cell cycle phases. However, the models described above are not able to explain why the stationary radius of the cord attains a particular value, or why the kinetic parameters of the cell population exhibit a particular pattern of change with the radial distance. To this aim, it is necessary to consider the transport within the cord of the chemicals (oxygen and other nutrients) that are critical for cell viability and affect cell proliferation.

5 A tumour cord model including interstitial fluid flow

The main aspects have been favoured when disregarding the age or maturity structure of the cell population are the diffusion and consumption of the nutrient, the cord evolution caused by treatments and the flow of interstitial fluid. Nutrient diffusion and consumption were considered in a model of an isolated tumour cord (a micrometastasis) [21], where the longitudinal growth of the cord was also taken into account. In [67] the growth of the cord was studied through a simulation approach that represented the local interactions of cells.

With the aim of describing the response to anticancer agents, in [17] the cell population in the cord was subdivided into viable and dead cells and was described by the volume fractions occupied by these cells. The following assumptions were made: i) the total volume fraction occupied by the cells (either alive or dead) is constant, ii) oxygen is the only nutrient considered and its transport is diffusive and quasi-stationary, iii) all cells die if oxygen concentration is below a *threshold* value, iv) the cytotoxic drug is transported by diffusion. Under these assumptions taking into account the motion of extracellular fluid is not necessary, and the description of fluid flow was in fact avoided. The first assumption requires that cells take an ideal arrangement even when they die and lose volume. Consistent with this hypothesis, dead cells within the cord are assumed to be transported by the living cells, which form a reasonably uniform structure. Thus the model cannot be valid when treatments cause a massive destruction of cells, making the cord incoherent. The problem of describing the transition to such an incoherent regime is still open. The second assumption, common to other models as we have seen, is quite reasonable in view of the high oxygen diffusivity [70]. The third has been adopted in many models of avascular spherical tumours after Greenspan [44]. The last assumption is adequate for drugs whose molecules are not too large. More precisely, it may be estimated [17] that the diffusivity must be greater than $5 \cdot 10^{-8}$ cm^2/s to disregard convective transport in tumour cords. We must also say that diffusivity and drug concentration were taken to be the same in the fluid and in the cells. This is hardly the case, also because of the role that the cell membrane may play in the process of drug uptake and export by the cells.

Notwithstanding all its limitations, the model of [17] was helpful in understanding the mathematical structure of the evolutive problem. In particular, as described later in this section, it was pointed out that during treatments the evolution of the interface between the cord and the necrotic region is necessarily subjected to a pair of unilateral constraints which intervene in selecting the correct boundary conditions for oxygen diffusion-consumption. In other words the boundary conditions on that interface cannot be described a priori but are determined by the process itself. This fact, which is true irrespective of the geometry and of other simplifications possibly introduced, and which then applies also to spheroids containing a necrotic core, was disregarded in the previous literature. Numerical simulations [16] show that the constraints do come into play and their action is of crucial importance in keeping the model adherent to physics.

The question of the interstitial fluid flow was dealt with more recently [18]. Any approach in modelling the fluid dynamics also requires a reasonable schematisation of the necrotic region, since the fluid seeps from the blood vessel into the tumour tissue, traverses the viable region, and eventually leaves the system at the two extreme cross sections both from the viable region and from the necrotic region. We devote this section to a concise illustration of a model including the fluid flow. The basic reference is [18], but, for the necrotic region, we describe the more refined model in [19].

5.1 Cell populations and cord radius

We keep the assumption of equal mass density for all the components, and we denote by ν_P, ν_Q, ν_A, ν_E the volume fractions of proliferating (P) cells, quiescent (Q) cells, dead cells (in the form of apoptotic (A) bodies), and extracellular (E) liquid respectively. We are not interested in the composition of the liquid, as we only use the fact that it supplies the material needed for cell replication. We assume no voids and constant porosity in the viable region, i.e., for $r_0 < r < \rho_N$, even in the presence of (moderate) cell death. That is,

$$\nu_P + \nu_Q + \nu_A = \nu^\star \tag{45}$$

with $\nu^\star$ constant. That said, the mass conservation equations are:

$$\frac{\partial \nu_P}{\partial t} + \nabla \cdot (\nu_P \mathbf{u}) = \chi \nu_P + \gamma(\sigma)\nu_Q - \lambda(\sigma)\nu_P - \mu_P(r,t)\nu_P, \tag{46}$$

$$\frac{\partial \nu_Q}{\partial t} + \nabla \cdot (\nu_Q \mathbf{u}) = -\gamma(\sigma)\nu_Q + \lambda(\sigma)\nu_P - \mu_Q(r,t)\nu_Q, \tag{47}$$

$$\frac{\partial \nu_A}{\partial t} + \nabla \cdot (\nu_A \mathbf{u}) = \mu_P(r,t)\nu_P + \mu_Q(r,t)\nu_Q - \mu_A \nu_A, \tag{48}$$

$$\nu_E \nabla \cdot \mathbf{v} = \mu_A \nu_A - \chi \nu_P. \tag{49}$$

In Eqs. (46)–(49) **u** is the velocity of the cellular components (the same for all species), and **v** is the velocity of the fluid component. The coefficient χ is the proliferation rate. The transitions $Q \to P$ and $P \to Q$ have the respective rates γ and λ which depend on the oxygen concentration σ. We assume that γ, λ are piecewise smooth functions of σ, nondecreasing and nonincreasing respectively. More precisely, we introduce threshold values of oxygen concentration $\sigma_P > \sigma_Q$, and take $\lambda = \lambda_{\max} > 0$, $\gamma = \gamma_{\min} \geq 0$ for $\sigma \leq \sigma_Q$, $\lambda = \lambda_{\min} \geq 0$, $\gamma = \gamma_{\max} > 0$ for $\sigma \geq \sigma_P$. All cells are assumed to die when the oxygen concentration falls below a threshold σ_N. Clearly $\sigma_Q > \sigma_N$.

The coefficients μ_P, μ_Q are the death rates for the respective species and μ_A is the degradation rate of apoptotic bodies into liquid. In chemotherapy the death rates μ_P, μ_Q are in fact dependent on drug concentration (possibly the intracellular concentration). We already mentioned that accurately describing drug transport and action is generally a very difficult task. The corresponding model should be tailored to the specific drug selected. Here we just prescribe μ_P, μ_Q as functions of the radial coordinate and of time (their expression is given later; see Sect. 6.2), thus bypassing the core of the chemotherapy model. In describing the effect of treatments, some details are neglected for the sake of simplicity: i) the proliferation rate χ could also be decreased by drug uptake in a stage preceding apoptosis or even independently of cell death; ii) the initial rapid volume loss accompanying apoptosis is neglected here, although it was taken into account in [18]; iii) the action of drug (or radiation) is the only cause of death in the cord, so disregarding spontaneous cell death.

Combining Eqs. (45)–(48) and recalling that $\nu^\star$ is constant we obtain

$$\nu^\star \nabla \cdot \mathbf{u} = \chi \nu_P - \mu_A(\nu^\star - \nu_P - \nu_Q).$$

Under the assumption that the cell velocity is radially directed and independent of the axial coordinate z, that is, $\mathbf{u} = (u_r, u_z) = (u(r,t), 0)$, the above equation can be rewritten as

$$\nu^\star \frac{1}{r}\frac{\partial}{\partial r}(ru) = \chi \nu_P - \mu_A(\nu^\star - \nu_P - \nu_Q); \tag{50}$$

this yields the cell velocity field when completed by the boundary condition $u(r_0, t) = 0$.

Concerning the equation for σ, we assume that

$$\Delta\sigma = f_P(\sigma)\nu_P + f_Q(\sigma)\nu_Q, \tag{51}$$

to be complemented by the boundary conditions

$$\sigma(r_0, t) = \sigma_b, \quad \left.\frac{\partial \sigma}{\partial r}\right|_{r=\rho_N(t)} = 0,$$

where $f_P(\sigma)$, $f_Q(\sigma)$ denote the ratios between the consumption rate per unit volume of proliferating and, respectively, quiescent cells and the diffusion coefficient. We set $f_P(\sigma) \geq f_Q(\sigma)$ and require $f_Q(\sigma_N) > 0$. At the inner boundary $r = r_0$, i.e., at the vessel wall, for simplicity we prescribe the (constant) oxygen blood concentration $\sigma_b > \sigma_P$.

The interface $r = \rho_N(t)$ is determined by considering that the necrotic material cannot be converted back to living cells and that cells cannot be viable when σ is smaller than σ_N. Thus the following inequalities must be satisfied:

$$u(\rho_N(t), t) - \dot{\rho}_N(t) \geq 0 \tag{52}$$

$$\sigma(\rho_N(t), t) \geq \sigma_N. \tag{53}$$

Therefore, from (52), two cases are possible: $u(\rho_N, t) - \dot{\rho}_N > 0$ or $u(\rho_N, t) - \dot{\rho}_N = 0$. If $u(\rho_N, t) - \dot{\rho}_N > 0$, that is, if the cells cross the interface $\rho_N(t)$, the cord boundary is defined by the condition

$$\sigma(\rho_N(t), t) = \sigma_N, \tag{54}$$

and the interface is a nonmaterial free boundary. This case occurs, for instance, in the stationary state in the absence of treatment. Otherwise, the cord boundary becomes a material free boundary defined by

$$\dot{\rho}_N = u(\rho_N(t), t). \tag{55}$$

The switch to the material interface regime may intervene when a sudden massive destruction of cells rapidly lowers oxygen consumption, and the interface $\rho_N(t)$ defined by (54) tends to acquire a velocity larger than $u(\rho_N(t), t)$. The material boundary (55) is in turn subjected to the constraint (53) so that, if during cord repopulation $\sigma(\rho_N(t), t)$ tends to drop below σ_N, the free boundary must become nonmaterial again. In other words, the doubly constrained regime of interface evolution can be summarized in the complementarity form (52)–(53) and $(u(\rho_N, t) - \dot{\rho}_N)(\sigma(\rho_N, t) - \sigma_N) = 0$.

5.2 Extracellular fluid flow and the necrotic region

Passing now to the extracellular fluid motion, we assume that the fluid flow is governed by Darcy's law,

$$(1 - \nu^\star)(\mathbf{v} - \mathbf{u}) = -\kappa \nabla P, \tag{56}$$

where $P(r, z, t)$ is the fluid pressure and $\kappa > 0$ is the hydraulic conductivity of the tissue within the viable region of the cord. From (46)–(49) and because of the no-void hypothesis, we deduce the overall incompressibility equation

$$\nabla \cdot (\mathbf{v} + \frac{\nu^\star}{1 - \nu^\star}\mathbf{u}) = 0. \tag{57}$$

Instead of solving the full boundary value problem for P, we introduced an approximation that simplifies the problem. Defining the longitudinal average of the radial component of $\mathbf{v}$,

$$v(r,t) = \frac{1}{2H}\int_{-H}^{H} v_r(r,z,t)\,dz,$$

where $2H$ is the cord length, from the longitudinal average of Eq. (57) we obtain

$$\frac{1}{r}\frac{\partial}{\partial r}(rv) + \frac{1}{2H}\left[v_z(r,H,t) - v_z(r,-H,t)\right] = -\frac{\nu^\star}{1-\nu^\star}\frac{1}{r}\frac{\partial}{\partial r}(ru). \tag{58}$$

Next, the following approximation for the outgoing volumetric current is made:

$$(1-\nu^\star)\left[v_z(r,H,t) - v_z(r,-H,t)\right] = 2\zeta_{\text{out}}(p(r,t) - p_\infty), \tag{59}$$

where ζ_{out} is a positive constant representing the mean conductance of the tissues traversed by the outgoing flux, p_∞ is a "far field" pressure identifiable with the pressure in the lymphatic vessels, and $p(r,t)$ is the longitudinal average of $P(r,z,t)$. On replacing the pressure with its longitudinal average, from (58), (59) and (50) the following equation for $v(r,t)$ can be obtained:

$$\frac{1}{r}\frac{\partial}{\partial r}(rv) = -\frac{1}{1-\nu^\star}\left[\chi\theta\nu_P - \mu_A(\nu^\star - \nu_P - \nu_Q) + \frac{\zeta_{\text{out}}}{H}(p - p_\infty)\right]. \tag{60}$$

At this point, the longitudinal average of the radial component of Darcy's equation (56),

$$(1-\nu^\star)(v-u) = -\kappa\frac{\partial p}{\partial r},$$

yields the following equation for p:

$$p(r,t) = p_0(t) - \frac{1-\nu^\star}{\kappa}\int_{r_0}^{r}[v(r',t) - u(r',t)]\,dr', \tag{61}$$

where $p_0(t) = p(r_0^+,t)$ is the (unknown) pressure immediately outside the vessel wall. The equation for p requires a condition at $r = \rho_N(t)$, that can be established after the necrotic region has been described. Equation (60) for v is complemented by the boundary condition at the vessel wall,

$$(1-\nu^\star)v(r_0,t) = \zeta_{\text{in}}(p_b - p_0(t)),$$

where ζ_{in} is a permeability constant, and $p_b > p_\infty$ represents the longitudinal average of the hydraulic pressure in the blood (corrected according to the jump of the osmotic pressure across the wall).

To represent the necrotic region, in [18] we took the very simplified view of considering this region as completely filled with an incompressible fluid having uniform pressure (opposite to the approach of [16, 17], where degrading cells were assumed to remain densely packed). In [19] we still assumed that the necrotic region is a compartment with uniform properties (and thus uniform pressure) but, more realistically, we distinguished the solid (cellular) from the liquid component, allowing the overall volume fraction of the cellular component to change. Since the necrotic cells retain some structural integrity before degradation [58], we assumed that this fraction cannot exceed a maximal value smaller than one, which we set to $\nu^\star$ for simplicity, an assumption that appears reasonable although rather crude. Necrotic cells are degraded to liquid with rate constant μ_N.

In our geometry the necrotic region has the shape of a hollow cylinder (see Fig. 2) with fixed bases $z = \pm H$ and moving lateral boundaries $r = \rho_N(t)$ and $r = B(t)$, both unknown. Viewing the cord inside an array of parallel and identical cords, no flux takes place through the latter boundary because of symmetry. We denote by V_N^c the volume of the cellular component and by V_N^l the volume of the liquid component. Disregarding the loss of necrotic cells through the ends at $z = \pm H$, we write the mass balance as:

$$\dot{V}_N^c = 4H\pi\rho_N\nu^\star[u(\rho_N, t) - \dot{\rho}_N] - \mu_N V_N^c, \tag{62}$$

$$\dot{V}_N^l = 4H\pi\rho_N(1 - \nu^\star)[v(\rho_N, t) - \dot{\rho}_N] + \mu_N V_N^c - q_{\text{out}}(t), \tag{63}$$

where the volume efflux of liquid at $z = \pm H$, q_{out}, is expressed as follows:

$$q_{\text{out}} = 2\zeta_{\text{out}}^N \frac{V_N^l}{V_N^c + V_N^l}\pi(B^2 - \rho_N^2)(p_N - p_\infty), \tag{64}$$

where ζ_{out}^N is a positive constant and p_N the pressure in the necrotic region. Since we exclude the formation of voids, the volume of the necrotic region, given as $V_N = 2H\pi(B^2 - \rho_N^2)$, is equal to $V_N^c + V_N^l$, and the total volume balance is expressed by

$$\dot{V}_N = 4H\pi\rho_N[(1-\nu^\star)v(\rho_N, t)+\nu^\star u(\rho_N, t)-\dot{\rho}_N]-\frac{\zeta_{\text{out}}^N}{H}(V_N - V_N^c)(p_N - p_\infty). \tag{65}$$

The limitation which we have assumed on the overall cellular fraction,

$$\frac{V_N^c(t)}{V_N^c(t) + V_N^l(t)} \leq \nu^\star, \tag{66}$$

must be imposed as a constraint. In fact, while (62) guarantees that $V_N^c \geq 0$ because of (52), one cannot exclude the possibility that inequality (66) could be violated if too much liquid is removed or too much solid material is supplied. When (66) holds in the strict sense, we assume that the pressure p_N is determined by the reactions to the displacement of the tissue that surrounds the whole tumour, whose size is likely to increase as B increases. Thus we write

$$p_N(t) = \Psi(B(t)), \tag{67}$$

where $\Psi(B)$ is an increasing function with $\Psi(B) \geq p_\infty$. Equation (65) can be rewritten as a differential equation for B^2,

$$\frac{dB^2}{dt} = 2\rho_N[(1-\nu^\star)v(\rho_N,t)+\nu^\star u(\rho_N,t)] - \frac{\zeta_{\text{out}}^N}{H}\Big(B^2-\rho_N^2-\frac{V_N^c}{2H\pi}\Big)(p_N-p_\infty), \tag{68}$$

and Eqs. (62), (68), together with (67), describe the evolution of B. On the contrary, when the equality sign in (66) comes into play, B and V_N^l remain defined as

$$B(t) = \left(\rho_N^2 + \frac{V_N^c}{2H\pi\nu^\star}\right)^{1/2}, \tag{69}$$

$$V_N^l(t) = \frac{1-\nu^\star}{\nu^\star}V_N^c \tag{70}$$

with V_N^c determined via Eq. (62), and the role of Eq. (63) is to provide $q_{\text{out}}(t)$. Hence $p_N(t)$ becomes a function of $B(t)$ no longer through (67) but through (70), (63) and (64):

$$p_N(t) = p_\infty + \frac{H}{\zeta_{\text{out}}^N}\left(\frac{2\rho_N[v(\rho_N,t)-u(\rho_N,t)]}{B^2-\rho_N^2} + \frac{\mu_N}{1-\nu^\star}\right). \tag{71}$$

When the cellular fraction takes the limiting value $\nu^\star$, the action of surrounding tissues becomes supported by the cellular component, while the liquid pressure adjusts itself to preserve the volume balance, necessarily dropping below $\Psi(B)$ and reducing q_{out}. In this situation, the reaction of the tissues surrounding the tumour creates a stress on the cellular component of the whole region $r_0 < r < B$ by the contact with the cellular component of the neighbouring cords at the boundary $r = B$. As a result, a stress is exerted on the central vessel, in addition to the stress generated by cell proliferation, which may have further consequences on the evolution of the tumour. In the tumour cord model, however, this phenomenon is ignored and computation of the stresses is avoided.

The compliance of the surrounding tissues imposes the further constraint

$$p_N(t) \leq \Psi(B(t)). \tag{72}$$

During the evolution, e.g., if too much liquid is supplied, Eqs. (71), (69) may lead to a value of p_N violating the above constraint. Then the system has to switch to the previous regime governed by (67) and (68), with V_N^l evolving according to (63) and constrained by (66). The switch has a clear physical explanation: the liquid resumes the role of the reaction supporting component in the necrotic region. To summarise, the evolution of the necrotic region takes place under a pair of unilateral constraints, i.e., the inequalities (66) and (72) that decide which the correct governing equations are, to which we associate the equation $(p_N - \Psi(B))(V_N^c/V_N - \nu^\star) = 0$.

Assuming that the longitudinal average of the pressure is continuous across $r = \rho_N$, we impose the condition

$$p(\rho_N,t) = p_N(t),$$

which gives the only information missing in order to determine the average pressure field in the cord through Eq. (61).

In summary, the model is given by the basic equations (46), (47), (50), (51), (60), (61), (62), together with the boundary conditions and the equations for p_N, B, defined according to the two pairs of constraints (52), (53) and (66), (72) as previously described. In [18], existence and uniqueness were proved for the steady-state problem in the absence of treatment (and for a "fluid" necrotic region), by means of a suitable combination of a shooting technique (applied to oxygen-diffusion consumption) and of a fixed point argument (applied to the problem for the pressure field). A similar result for the evolution problem has been obtained using time discretisation and with the introduction of "tolerances" for the constraints, tending to zero as the time step tends to zero.

6 Modelling tumour treatment

Mathematical models of spherical tumours have been used mainly to investigate the effect of chemotherapy on tumour growth, following the approach of distributed drug sources outlined in [10] and pioneered in [54]. However, a deeper insight into the effect of treatments could be achieved by taking into account the discrete nature of vasculature, an approach that makes it possible to represent both the drug gradients and the oxygen distribution on the cell scale. Models of tumour cords, although referring to only a very particular structure, may thus be useful. Another direction is that of considering the actual shape of the (clinical) tumour. This aspect has been addressed in the mathematical optimisation of hyperthermia treatment.

6.1 Spherical tumours

Byrne and Chaplain [26] studied the evolution of multicellular spheroids growing *in vitro* under the action of a cytotoxic drug diffusing (as the nutrient) from the external medium. A necrotic core is assumed to be present, where the interface between viable rim and necrotic core is defined as the radius at which the nutrient concentration falls below a critical value. Cells in the necrotic region lose their volume according to a uniform rate constant. Concerning the response of multicellular spheroids to a drug, Ward and King [71] proposed a model in which cell death occurs with a sudden partial loss of volume and the residual volume is maintained over time. Spontaneous cell death is accounted for by a death rate which increases with a sigmoidal pattern as the nutrient concentration decreases, so that in the central region of the spheroid the density of viable cells may be virtually zero. The rate of cell death induced by the drug is proportional to drug consumption and is modulated by nutrient concentration. The model was used to compare the multiple fractionated exposition to a drug with the single exposition. In addition, model equations were adapted to represent the response of a monolayer cell culture.

The paradigm of the multicellular spheroid was also used to describe the response to irradiation [73]. The authors used the so-called linear-quadratic model for the dose-response relationship [39]. Since the radiosensitivity of cells is known to be related to oxygen concentration and to decrease as the oxygen concentration decreases, the parameters of the linear-quadratic model were assumed to depend on the radial distance. Using several approximations to reduce the mathematical description to an ordinary differential equation model, the authors investigated fractionated irradiation and possible optimisation of the scheduling.

In attempting to describe the response to a drug of a vascularised spherical tumour, a distributed source of drug was introduced [25] in the diffusion-consumption equation that describes the transport of the drug. A general model was studied, in which the net volume production due to the balance between cell proliferation and degradation of dead cells is assumed to depend on both the nutrient and the drug concentration according to suitably chosen functional forms. The concept of a distributed source of drug was also adopted in [52], where the cell population was divided into two subpopulations having different proliferation rates and different sensitivities to the drug. The cell velocity was derived by imposing the condition that the total number of cells per unit volume is also constant after the treatment, and by assuming an instantaneous disappearance of dead cells. Constant continuous infusion was compared to bolus administration in different scenarios characterised by a different initial proportion of the resistant subpopulation. This model was further developed in [50] to include a volume fraction of blood vessels, that evolves stimulated by tumour cells and in turn modulates tumour cell growth and the extent of the distributed source of drug. In order to incorporate the cellular pharmacokinetics of the drug Doxorubicin, Jackson [51] reduced the previous model to a single cell population and introduced equations for the drug concentrations in the extracellular space, the cell cytoplasm, and an intracellular non-exchangeable compartment in which the drug is sequestered. The drug extracellular concentration obeys a diffusion-advection equation with a distributed source. The model was used to fit data of the growth of an experimental subcutaneous tumour subjected to bolus-based treatment.

6.2 Tumour cords

The tumour cord model illustrated in Sect. 5 has been used to investigate the effects of an anticancer agent that induces cell death. Here we illustrate the main features of the response to a single dose of the agent, by simply assigning the death rates as functions of time (see Eqs. (46)–(48)). The evolutive problem, that arises when the cord is perturbed by the treatment, was studied by assuming as initial condition the stationary state of the untreated cord. The following simulations refer to a nondimensional setting (see [18, 19]), with equal oxygen consumption of proliferating and quiescent cells, and taking $f_P(\sigma) = f_Q(\sigma) = F\sigma/(K+\sigma)$ in Eq. (51). Moreover, the function $\Psi(B) = e(B-1)^2$ was chosen. Figure 3 shows an example of the stationary state of the cord. Panel A reports the profiles of $\nu_P(r)$, $u(r)$ and $\sigma(r)$. Since $\lambda_{\min} = 0$, $\nu_P(r) = 1$ until r is less than the radius where $\sigma = \sigma_P$, then ν_P decreases as the fraction of quiescent cells increases due to the decay of the oxygen concentration, in agreement with observation [48, 60]. With this choice of model parameters the cell velocity u is increasing (with a maximal dimensional value of 0.87 µm/h, if $r_0 = 20\,\mu\text{m}$ and $\chi = \ln 2/24\,\text{h}^{-1}$), although it can also decrease after a maximum

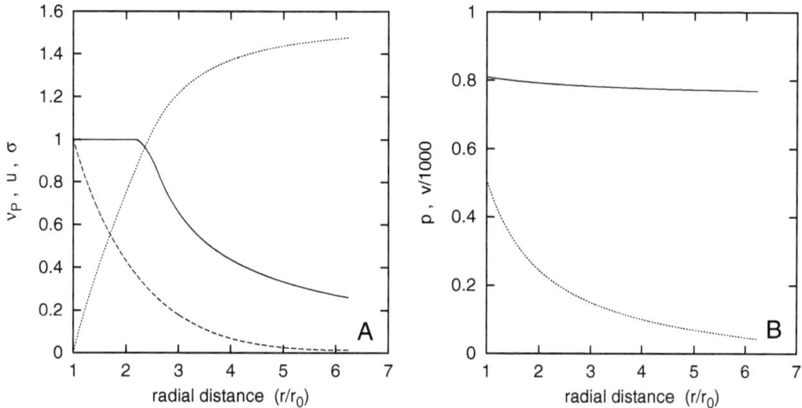

Fig. 3. a Profiles of $\nu_P(r)$ (*solid line*), $u(r)$ (*dotted*) and $\sigma(r)$ (*dashed*). **b** Profiles of $p(r)$ (*solid*) and $v(r)$ (*dotted*). Nondimensional parameters: $\sigma_b = 1$, $\sigma_P = 0.375$, $\sigma_Q = 0.25$, $\sigma_N = 0.0125$, $F = 0.128$, $K = 0.108$, $\chi = 1$, $\lambda_{\max} = 1$, $\gamma_{\max} = 4$, $\lambda_{\min} = \gamma_{\min} = 0$, $\nu^\star = 0.85$, $p_b = 1$, $p_\infty = 0$, $\zeta_{\text{in}} = 400$, $\zeta_{\text{out}} = 2$, $\kappa = 3000$, $e = 12 \cdot 10^{-3}$, $\zeta_{\text{out}}^N = \zeta_{\text{out}}/(1 - \nu^\star)$, $\mu_N = 0.3$

when the fraction of proliferating cells in the outer zone of the cord is smaller. The corresponding profiles of $p(r)$ and $v(r)$ are shown in Fig. 3b. The pressure p_0 is a substantial fraction of p_b and $p(r)$ exhibits a slight decrease with r; thus the model agrees with the experimental observation of a large interstitial pressure in tumours [63]. The slope of p depends on the Darcy coefficient: the assumed value of κ is in the typical range of tumour tissues [63]. The fluid velocity v is very high with respect to the cell velocity.

To simulate the single-dose treatment, we have chosen for $\mu_P(r, t)$ and $\mu_Q(r, t)$ the following space-independent expressions:

$$\mu_P(r, t) = \frac{m_P}{\tau_1 - \tau_2}(\mathrm{e}^{-t/\tau_1} - \mathrm{e}^{-t/\tau_2}),$$

$$\mu_Q(r, t) = \frac{m_Q}{\tau_1 - \tau_2}(\mathrm{e}^{-t/\tau_1} - \mathrm{e}^{-t/\tau_2}),$$

where m_P, m_Q, τ_1 and τ_2 are parameters suitably selected to mimic the effect of a drug delivered as a single bolus. We note that the analysis of the time evolution of ν_P, ν_Q, u, σ, ρ_N and V_N^c can be carried out independently of the study of the time evolution of p, v and B. The numerical solution of the first problem is based on a procedure that suitably extends the one described in [16]. Its basic feature is the computation of the viable cell fractions along a fixed set of characteristic lines of Eqs. (46)–(47). The numerical computation of p, v and B was performed in parallel, guaranteeing that the constraints (66), (72) are satisfied [19].

Figure 4 shows an example of the time evolution of the cord in the case of a cycle-specific drug, that is, a drug affecting mainly the proliferating cells. Panel A

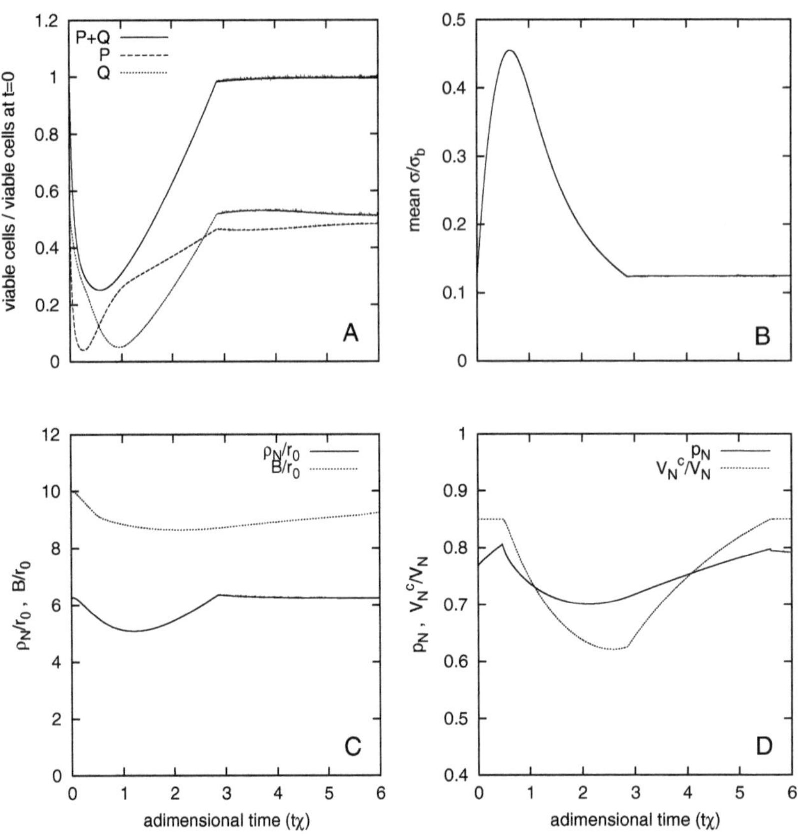

Fig. 4. a Time course of the viable cell subpopulations after a single-dose treatment; P proliferating cells, Q quiescent cells. **b** Mean oxygen concentration. **c** Cord radius ρ_N and outer boundary B. **d** Pressure and cell fraction in the necrotic region. Parameters as in Fig. 3 with $\mu_A = 1, m_P = 1, m_Q = 0.2, \tau_1 = 0.174, \tau_2 = \tau_1/20$

reports the ratio between the total volume (per unit cord length) of viable cells and its value at $t = 0$, showing the dynamics of the viable cell population following the treatment. The decrement of the amount of viable cells reduces oxygen consumption and thus causes a general reoxygenation of the cord as shown by the time course of the mean oxygen concentration (panel B). The increase in oxygen concentration induces a recruitment of quiescent cells into proliferation, so that a transient phase in which the proliferating fraction is higher than the initial one may occur. The initial value of the proliferating fraction is about 0.5 in this simulation. Thereafter, the cell populations tend to the stationary value. The radius ρ_N shows an initial shrinkage [59] followed by regrowth (panel C). The interface ρ_N quickly becomes a material boundary and remains material until, at about $t = 3$, it becomes nonmaterial again. This event is marked by a slope discontinuity. In the same panel, the time course of the boundary

B is plotted. Panel D shows the time evolution of the pressure p_N and of the cellular fraction in the necrotic region. In the initial state the constraint (66) is satisfied with the equality sign and the pressure is less than $\Psi(B)$. Due to the increased influx of liquid caused by cell death, p_N increases reaching $\Psi(B)$. At this point the regime changes, p_N is given by $\Psi(B)$ and the cellular fraction goes below $\nu^\star$. During cord regrowth, the influx of liquid decreases and the system switches again to the regime characterised by a cellular fraction equal to $\nu^\star$. We note that the time evolution of the average interstitial pressure in the viable cord closely follows the pressure in the necrotic region since, as seen in Fig. 3b, $p(r)$ exhibits a small decay between r_0 and ρ_N. As a consequence, the model predicts a transient increase in the extracellular fluid velocity v after the single-dose treatment.

The delivery of a single dose of a cycle-specific agent produces initially a preferential depletion of the proliferating subpopulation, with a reduction of the overall sensitivity of the cell population to the agent. This sensitivity, however, eventually recovers the initial level because of cell repopulation. Due to the reoxygenation of the cord, as seen in Fig. 4a, there may be a time interval in which the fraction of proliferating cells attains a value larger than the initial value (*oversensitisation*). This behaviour suggests that a second dose of the same drug, delivered in such a time window, should be more effective. The comparison between the response to a single bolus of drug (characterised by given m_P , m_Q) delivered at $t = 0$, and the response to two half doses $(m_P/2\,, m_Q/2)$ delivered at $t = 0$ and $t = T$ was performed in [20], using a slightly modified model. We used the index

$$\text{survival ratio} = \frac{\min\left[P_2(t) + Q_2(t)\right]}{\min\left[P_1(t) + Q_1(t)\right]}$$

where $P_i(t)$, $Q_i(t)$ denote the volume per unit cord length of proliferating and quiescent cells, respectively, and the subscripts refer to the single-dose response ($i = 1$) and to the split-dose response ($i = 2$). Figure 5 shows the behaviour of the survival ratio as a function of the interfraction interval T for two values of the total dose. As expected, in coincidence with the time window in which cell oversensitisation occurs this ratio is smaller than one, showing the advantage of the dose splitting, with this advantage being more marked for the higher dose. When the drug is not cycle-specific ($m_Q/m_P = 1$), this advantage obviously disappears even though the proliferating fraction of the cell population after the first dose was also augmented in this condition. We note that, even with small interfraction intervals ($T = 6$-12 h) the survival ratio is less than one, apparently in contrast with the strong depletion of the proliferating fraction that occurs at those times. This fact can be explained by considering that the drug is active for a nonnegligible time interval (about 12 h in this simulation). Therefore, even in the case of a single dose, part of the dose actually affects a cell population that has become refractory.

To represent more adequately the effect an anticancer agent in a short time horizon (1-2 days after the delivery), the pharmacodynamics of the drug and, in particular, the fact that the occurrence of cell death may extend beyond the exposition time [68] should be taken into account. A similar behaviour is also found after delivery of radiation [38]. The model in [20], used for the simulations of Fig. 5, accounts for

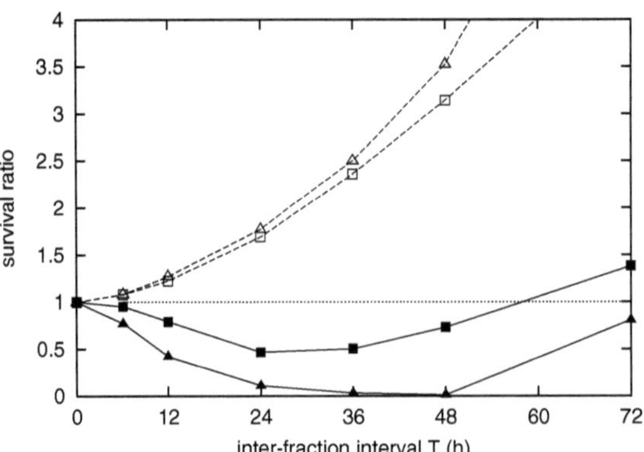

Fig. 5. Survival ratio as a function of the time interval between the two fractions. Closed symbols: $m_Q/m_P = 0.2$; $m_P = 4$ for each fraction (*squares*), $m_P = 8$ (*triangles*). Open symbols: $m_Q/m_P = 1$; $m_P = 2$ (*squares*), $m_P = 4$ (*triangles*)

this phenomenon by assuming that the exposure to the drug induces lethal damage in a fraction of cells, that undergo cell-cycle arrest and die at a subsequent time. Thus the viable tumour cells were subdivided into viable undamaged cells (proliferating and quiescent) and viable but lethally damaged cells. The transition into the compartment of lethally damaged cells occurs according to a rate related to the drug exposure. Lethally damaged cells were assumed to die following first-order kinetics. The representation of a delayed cell death was also addressed in [16, 56].

Although prescribing cell death rates proved to be useful for describing the essential features of cord response to treatment, a closer account of the response to chemotherapy would require the description of drug transport within the cord. In [16, 17], the transport of drug was modelled by assuming convection to be negligible with respect to diffusion and, as already mentioned, making the strong assumption that drug concentration is the same in the extracellular and intracellular spaces. Denoting the drug concentration by $c(r, t)$, the diffusion-consumption equation was written as:

$$\frac{\partial c}{\partial t} - D_C \Delta c = -\varphi_C(c)\nu_V - \lambda_C c,$$

$$c(r_0, t) = c_b(t), \tag{73}$$

$$\left.\frac{\partial c}{\partial r}\right|_{r=B(t)} = 0, \tag{74}$$

$$c(r, 0) = 0,$$

where D_C is the drug diffusion coefficient, $\varphi_C(c)$ represents the rate of drug consumption by the tumour cells, taken as a function of c, ν_V is the volume fraction of viable cells, and λ_C represents an additional loss, possibly related to a natural decay of the drug. The function $c_b(t)$ in (73) is the time course of drug concentration in the tumour vasculature, whereas condition (74) is consistent with the reflecting nature of the boundary B. In the model [16,17], the drug-induced death rate is a function of c, so that the diffusion-consumption equation is coupled to the equations that describe the evolution of the cell population. Because of the boundary condition (74), the calculation of drug concentration requires a model of the evolution of the necrotic region.

The results illustrated in Fig. 4 on the response of the tumour cord to a single dose of treatment can also be considered representative, to a first approximation, of the response to a single pulse of radiation. As previously mentioned, however, an important aspect of the radiation effect is the dependence of the radiosensitivity on the oxygen concentration. This point was addressed in a simplified way in [16]. A particular feature of this dependence, which must be accounted for in the modelling, is that the extent of death occurring after irradiation (even 24 h later) depends on the oxygen concentration experienced by the cell at the moment of irradiation. The advantage of a spatial model of the tumour vascularisation is evident here because it allows us to represent the oxygen distribution in the tumour tissue.

6.3 Hyperthermia treatment with geometric model of the patient

Hyperthermia treatment is a procedure which consists in irradiating the tumour with electromagnetic waves in the range of radio-frequency. The goal is to raise the tumour temperature to the range 42°C–45°C. In such a range malignant cells become much more sensitive to both radiation and drugs. Mathematics plays an essential role in the guidance of all the steps of this procedure and has provided critical help to the physicians. For this reason we decided to put some emphasis on it, even though this subject is not central in our discussion of tumour evolution.

The preliminary phase of the process is the acquisition of as good as possible information about the structure and the location of the tumour by means of magnetic resonance imaging and/or other techniques. Quite obviously a considerable amount of mathematics enters this stage too, but for our purposes we are more interested in what follows, namely: a) once the number, power and location of the emitting antennas are given, compute the electric field **E** and the power dissipated within the tumour and in the surrounding tissues; b) compute the corresponding temperature rise; c) optimise the system, so as to get good focussing on the tumour and the best possible efficiency with manageable equipment. Achieving goals a)–c) requires the construction of a *virtual patient* on which the procedure can be tested and refined. The equation assumed to govern the evolution of the temperature T is the so-called bioheat-equation, whose quasi-stationary version reads

$$k\Delta T - c_b W(T - T_b) + \frac{1}{2}\sigma|\mathbf{E}|^2 = 0, \tag{75}$$

where k is the thermal conductivity, T_b is the blood temperature, c_b is the specific heat capacity of blood and W is the mass flow rate of blood per unit volume of tissue. The last term of the equation gives the dissipated power, in which σ is the electric conductivity of the tissue. The heat source term in (75) requires the solution of Maxwell's equations. During the process, the part of the body which is crossed by radio waves is surrounded by a water container (the so-called water bolus) having the twofold purpose of favouring wave transmission and of cooling the body. Therefore a (linear) heat transmission condition is imposed on the body boundary. It must be remarked that a temperature rise tends to modify blood flow, with different effects on healthy tissues and on the tumoural tissue. Thus it makes sense to let W depend on temperature (and on space).

This problem has been the subject of a number of papers (see, e.g., the survey [12], mainly devoted to numerical methods, and [33, 49]). We can mention that the work performed by mathematicians at the Konrad-Zuse-Zentrum (Berlin) was absolutely critical to the successful hospital application of the procedure, making this a remarkable example in which mathematics, so to speak, takes the lead in a medical application.

7 Conclusions and perspectives

We reviewed classes of mathematical models of tumour growth. Roughly speaking we selected four main topics: (i) models including the analysis of stresses, (ii) analysis of morphological stability, (iii) tumour cords, (iv) treatments. These studies have different objectives and different styles. In (i) the scope is to provide a rigorous formulation in terms of the basic principles of the mechanics of mixtures, but mathematical analysis is not always pursued. In (ii) the mathematics is highly sophisticated whereas the model is so simplified that the tumour is just a support for the evolution of the nutrient concentration and of the "cell pressure". Class (iii) contains nontrivial modelling and includes mathematical results, but in idealised situations in which mechanical stresses are neglected. We also find different slants concerning treatments, from very practical to more theoretical approaches.

The subject of tumour invasion was not included for lack of space, but it is certainly not less interesting. On the contrary, it is probably the one, together with angiogenesis, that has received more attention from biologists and medical doctors. A large project on tumour invasion, coordinated by V. Quaranta (Cancer Center, Vanderbilt University, Nashville, USA), has recently been funded by NIH. The reference mathematical model [4, 5] in this project includes active cell motility, as well as the action of cell-secreted matrix degradative enzymes.

At the end of our review we can say that the complexity of tumour growth has suggested an impressive variety of ideas and also of ways of approaching the problem. Some have put a magnifying lens on a particular aspect, others have tried to consider more general situations. In any case there is not a unified view, nor any model that incorporates most of the processes going on within a real tumour. Such a goal is so difficult that one can doubt that full complexity is really a target to be pursued. It is probably better to gradually enrich some of the existing models, making them more flexible and capable of providing quantitative information on specific aspects or at least a reliable qualitative description in the framework of a broader view.

Among important open questions, we can put the formulation of 3-D invasion models, coupling them to a more accurate description of necrotic regions and of the flow of extracellular fluids, at the same time taking advantage of the knowledge acquired on modelling stresses and their action on vasculature. Therapeutical treatments should also be studied in far greater depth, encompassing the drug transport mechanisms with the distinction between extracellular and intracellular concentrations, the way drugs act on cells, the effect on the overall mechanics of the tumour produced by the massive death and degradation of cells. In general, we may say that the main challenge appears to be that of bringing together the different spatial scales at which the phenomena determining tumour growth occur (see, e.g., [1]). Moreover, classical areas of biomathematics, for instance, population dynamics, seem to provide appropriate tools to model new therapies such as immunotherapy and gene therapy by viral vectors. Thus there are many research directions which look very promising and could lead to substantial progress as well as to a deeper role of mathematicians in this crucial field of medicine.

References

[1] Alarcón, T., Byrne, H.M., Maini, P.K.: Towards whole-organ modelling of tumour growth. Prog. Biophys. Mol. Biol. **85**, 451–472 (2004)

[2] Ambrosi, D., Mollica, F.: Mechanical models in tumour growth. In: Preziosi, L. (ed.): Cancer modelling and simulation. Boca Raton, FL: Chapman & Hall/CRC 2003, pp. 121–145

[3] Ambrosi, D., Preziosi, L.: On the closure of mass balance models for tumor growth. Math. Models Methods Appl. Sci. **12**, 737–754 (2002)

[4] Anderson, A.R.: A hybrid mathematical model of solid tumour invasion: the importance of cell adhesion. Math. Med. Biol. **22**, 163–186 (2005)

[5] Anderson, A.R.A., Chaplain, M.A.J., Newman, E.L., Steele, R.J.C., Thompson, A.M.: Mathematical modelling of tumour invasion and metastasis. J. Theor. Med. **2**, 129–154 (2000)

[6] Arakelyan, L., Vainstein, V., Agur, Z.: A computer algorithm describing the process of vessel formation and maturation, and its use for predicting the effects of anti-angiogenic and anti-maturation therapy on vascular tumor growth. Angiogenesis **5**, 203–214 (2002)

[7] Araujo, R.P., McElwain, D.L.S.: A history of the study of solid tumour growth: the contribution of mathematical modelling. Bull. Math. Biol. **66**, 1039–1091 (2004)

[8] Araujo, R.P., McElwain, D.L.S.: New insights into vascular collapse and growth dynamics in solid tumors. J. Theor. Biol. **228**, 335–346 (2004)

[9] Araujo, R.P., McElwain, D.L.S.: A linear-elastic model of anisotropic tumour growth. European J. Appl. Math. **15**, 365–384 (2004)

[10] Baxter, L.T., Jain, R.K.: Transport of fluid and macromolecules in tumors. I. Role of interstitial pressure and convection. Microvasc. Res. **37**, 77–104 (1989)

[11] Bazaliy, B., Friedman, A.: Global existence and asymptotic stability for an elliptic-parabolic free boundary problem: an application to a model of tumor growth. Indiana Univ. Math. J. **52**, 1265–1303 (2003)

[12] Beck, R., Deuflhard, P., Hiptmair, R., Hoppe, R.H.W., Wohlmuth, B.: Adaptive multilevel methods for edge element discretizations of Maxwell's equations. Surveys Math. Indust. **8**, 271–312 (1999)

[13] Bellomo, N., De Angelis, E., Preziosi, L.: Multiscale modeling and mathematical problems related to tumour evolution and medical therapy. J. Theor. Med. **5**, 111–136 (2003)

[14] Bertuzzi, A., Gandolfi, A.: Cell kinetics in a tumour cord. J. Theor. Biol. **204**, 587–599 (2000)

[15] Bertuzzi, A., Fasano, A., Gandolfi, A., Marangi, D.: Cell kinetics in tumour cords studied by a model with variable cell cycle length. Math. Biosci. **177/178**, 103–125 (2002)

[16] Bertuzzi, A., d'Onofrio, A., Fasano, A., Gandolfi, A.: Regression and regrowth of tumour cords following single-dose anticancer treatment. Bull. Math. Biol. **65**, 903–931 (2003)

[17] Bertuzzi, A., Fasano, A., Gandolfi, A.: A free boundary problem with unilateral constraints describing the evolution of a tumor cord under the influence of cell killing agents. SIAM J. Math. Anal. **36**, 882–915 (2004)

[18] Bertuzzi, A., Fasano, A., Gandolfi, A.: A mathematical model for tumor cords incorporating the flow of interstitial fluid. Math. Models Methods Appl. Sci. **15**, 1735–1777 (2005)

[19] Bertuzzi, A., Fasano, A., Gandolfi, A., Sinisgalli, C.: Interstitial pressure and extracellular fluid motion in tumour cords. Math. Biosci. Engng. **2**, 445–460 (2005)

[20] Bertuzzi, A., Fasano, A., Gandolfi, A., Sinisgalli, C.: Cell resensitization after delivery of a cycle-specific anticancer drug and effect of dose splitting: learning from tumour cords. J. Theor. Biol., to appear.

[21] Bloor, M.I.G., Wilson, M.J.: The non-uniform spatial development of a micrometastasis. J. Theor. Med. **2**, 55–71 (1999)

[22] Breward, C.J.W., Byrne, H.M., Lewis, C.E.: The role of cell-cell interactions in a two-phase model for avascular tumour growth. J. Math. Biol. **45**, 125–152 (2002)

[23] Breward, C.J.W., Byrne, H.M., Lewis, C.E.: A multiphase model describing vascular tumour growth. Bull. Math. Biol. **65**, 609–640 (2003)

[24] Burton, A.C.: Rate of growth of solid tumours as a problem of diffusion. Growth **30**, 157–176 (1966)

[25] Byrne, H.M., Chaplain, M.A.J.: Growth of nonnecrotic tumors in the presence and absence of inhibitors. Math. Biosci. **130**, 151–181 (1995)

[26] Byrne, H.M., Chaplain, M.A.J.: Growth of necrotic tumors in the presence and absence of inhibitors. Math. Biosci. **135**, 187–216 (1996)

[27] Byrne, H.M., Chaplain, M.A.J.: Free boundary value problems associated with the growth and development of multicellular spheroids. European J. Appl. Math. **8**, 639–658 (1997)

[28] Byrne, H.M., King, J.R., McElwain, D.L.S., Preziosi, L.: A two-phase model of solid tumour growth. Appl. Math. Lett. **16**, 567–573 (2003)

[29] Byrne, H.M., Preziosi, L.: Modelling solid tumour growth using the theory of mixtures. Math. Med. Biol. **20**, 341–366 (2003)

[30] Cristini, V., Lowengrub, J., Nie, Q.: Nonlinear simulation of tumor growth. J. Math. Biol. **46**, 191–224 (2003)

[31] Cui, S., Friedman, A.: Analysis of a mathematical model of the effect of inhibitors on the growth of tumors. Math. Biosci. **164**, 103–137 (2000)

[32] De Angelis, E., Preziosi, L.: Advection-diffusion models for solid tumour evolution in vivo and related free boundary problems. Math. Models Methods Appl. Sci. **10**, 379–407 (2000)

[33] Deuflhard, P., Hochmuth, R.: Multiscale analysis of thermoregulation in the human microvascular system. Math. Methods Appl. Sci. **27**, 971–989 (2004)

[34] Dyson, J., Villella-Bressan, R., Webb, G.: The steady state of a maturity structured tumor cord cell population. Discrete Contin. Dyn. Syst. Ser. B **4**, 115–134 (2004)

[35] Dyson, J., Villella-Bressan, R., Webb, G.F.: The evolution of a tumor cord cell population. Commun. Pure Appl. Anal. **3**, 331–352 (2004)

[36] Folkman, J.: Tumor angiogenesis. Adv. Cancer Res. **43**, 175–203 (1985)

[37] Fontelos, M.A., Friedman, A.: Symmetry-breaking bifurcations of free boundary problems in three dimensions. Asymptot. Anal. **35**, 187–206 (2003)

[38] Forrester, H.B., Vidair, C.A., Albright, N., Ling, C.C., Dewey, W.C.: Using computerized video time lapse for quantifying cell death of X-irradiated rat embryo cells transfected with c-myc or c-Ha-ras. Cancer Res. **59**, 931–939 (1999)

[39] Fowler, J.F.: The linear-quadratic formula and progress in fractionated radiotherapy. Br. J. Radiol. **62**, 679–694 (1989)

[40] Friedman, A., Reitich, F.: Analysis of a mathematical model for the growth of tumors. J. Math. Biol. **38**, 262–284 (1999)

[41] Friedman, A., Reitich, F.: Symmetry-breaking bifurcation of analytic solutions to free boundary problems: an application to a model of tumor growth. Trans. Amer. Math. Soc. **353**, 1587–1634 (2001)

[42] Friedman, A., Hu, B., Velasquez, J.J.L.: A Stefan problem for a protocell model with symmetry-breaking bifurcations of analytic solutions. Interfaces Free Bound. **3**, 143–199 (2001)

[43] Gatenby, R.A., Gawlinski, E.T.: A reaction-diffusion model of cancer invasion. Cancer Res. **56**, 5745–5753 (1996)

[44] Greenspan, H.P.: Models for the growth of a solid tumor by diffusion. Studies Appl. Math. **52**, 317–340 (1972)

[45] Greenspan, H.P.: On the growth and stability of cell cultures and solid tumors. J. Theor. Biol. **56**, 229–242 (1976)

[46] Hahnfeldt, P., Panigrahy, D., Folkman, J., Hlatky, L.: Tumor development under angiogenic signaling: a dynamic theory of tumor growth, treatment response, and postvascular dormancy. Cancer Res. **59**, 4770–4775 (1999)

[47] Hanzawa, E.I.: Classical solutions of the Stefan problem. Tohoko Math. J. (2) **33**, 297–335 (1981)

[48] Hirst, D.G., Denekamp, J.: Tumour cell proliferation in relation to the vasculature. Cell Tissue Kinet. **12**, 31–42 (1979)

[49] Hochmuth, R., Deuflhard, P.: Multiscale analysis for the bio-heat-transfer equation – the nonisolated case. Math. Models Methods Appl. Sci. **14**, 1621–1634 (2004)

[50] Jackson, T.L.: Vascular tumor growth and treatment: consequences of polyclonality, competition and dynamic vascular support. J. Math. Biol. **44**, 201–226 (2002)

[51] Jackson, T.L.: Intracellular accumulation and mechanism of action of Doxorubicin in a spatio-temporal tumor model. J. Theor. Biol. **220**, 201–213 (2003)

[52] Jackson, T.L., Byrne, H.M.: A mathematical model to study the effects of drug resistance and vasculature on the response of solid tumors to chemotherapy. Math. Biosci. **164**, 17–38 (2000)

[53] Jain, R.K.: Normalization of tumor vasculature: an emerging concept in antiangiogenic therapy. Science **307**, 58–62 (2005)

[54] Jain, R.K., Wei, J.: Dynamics of drug transport in solid tumors: distributed parameter model. J. Bioeng. **1**, 313–329 (1977)

[55] Krogh, A.: The number and distribution of capillaries in muscles with calculations of the oxygen pressure head necessary for supplying the tissue. J. Physiol. **52**, 409–415 (1919)

[56] Lankelma, J., Fernandez Luque, R., Dekker, H., Pinedo, H.M.: Simulation model of doxorubicin activity in islets of human breast cancer cells. Biochim. Biophys. Acta **1622**, 169–178 (2003)

[57] Lubkin, S.R., Jackson, T.A.: Multiphase mechanics of capsule formation in tumors. J. Biomech. Eng. **124**, 237–243 (2002)

[58] Majno, G., Joris, I.: Apoptosis, oncosis and necrosis. An overview of cell death. Amer. J. Pathol. **146**, 3–15 (1995)

[59] Moore, J.V., Hopkins, H.A., Looney, W.B.: Dynamic histology of a rat hepatoma and the response to 5-fluorouracil. Cell Tissue Kinet. **13**, 53–63 (1980)

[60] Moore, J.V., Hopkins, H.A., Looney, W.B.: Tumour-cord parameters in two rat hepatomas that differ in their radiobiological oxygenation status. Radiat. Environ. Biophys. **23**, 213–222 (1984)

[61] Moore, J.V., Hasleton, P.S., Buckley, C.H.: Tumour cords in 52 human bronchial and cervical squamous cell carcinomas: inferences for their cellular kinetics and radiobiology. Br. J. Cancer **51**, 407–413 (1985)

[62] Mueller-Klieser, W.: Multicellular spheroids. A review on cellular aggregates in cancer research. J. Cancer Res. Clin. Oncol. **113**, 101–122 (1987)

[63] Netti, P.A., Jain, R.K.: Interstitial transport in solid tumours. In: Preziosi, L. (ed.): Cancer modelling and simulation. Boca Raton, FL: Chapman & Hall/CRC 2003, pp. 51–74

[64] Padera, T.P., Stoll, B.R., Tooredman, J.B., Capen, D., di Tomaso, E., Jain, R.K.: Cancer cells compress intratumour vessels. Nature **427**, 695 (2004)

[65] Rajagopal, K.R., Tao, L.: Mechanics of mixtures. River Edge, NJ: World Scientific 1995

[66] Rubinow, S.I.: A maturity-time representation for cell populations. Biophys. J. **8**, 1055–1073 (1968)

[67] Scalerandi, M., Capogrosso Sansone, D., Benati, C., Condat, C.A.: Competition effects in the dynamics of tumor cords. Phys. Rev. E **65**, 051918(1–10) (2002)

[68] Sena, G., Onado, C., Cappella, P., Montalenti, F., Ubezio, P.: Measuring the complexity of cell cycle arrest and killing of drugs: kinetics of phase-specific effects induced by taxol. Cytometry **37**, 113–124 (1999)

[69] Swanson, K.R., Bridge, G., Murray, J.D., Alvord, E.C.: Virtual and real brain tumors: using mathematical modeling to quantify glioma growth and invasion. J. Neurol. Sci. **216**, 1–10 (2003)

[70] Tannock, I.F.: The relation between cell proliferation and the vascular system in a transplanted mouse mammary tumour. Br. J. Cancer **22**, 258–273 (1968)

[71] Ward, J.P., King, J.R.: Mathematical modelling of drug transport in tumour multicell spheroids and monolayer cultures. Math. Biosci. **181**, 177–207 (2003)

[72] Webb, G.F.: The steady state of a tumor cord cell population. J. Evol. Equat. **2**, 425–438 (2002)

[73] Wein, L.M., Cohen, J.E., Wu, J.T.: Dynamic optimization of a linear-quadratic model with incomplete repair and volume-dependent sensitivity and repopulation. Int. J. Radiat. Oncol. Biol. Phys. **47**, 1073–1083 (2000)

Modelling the formation of capillaries

L. Preziosi, S. Astanin

Abstract. The aim of this chapter is to describe models recently developed to simulate the formation of vascular networks which mainly occurs through two different processes: vasculogenesis and angiogenesis.

The former consists in the aggregation and organisation of endothelial cells dispersed in a given environment, the latter in the formation of new vessels sprouting from an existing vessel.

The results obtained by the use of the mathematical models are compared with experimental results in vitro and in vivo. The chapter also describes the effects of the environment on network formation and investigates the possibility of governing the network structure through the use of suitably placed chemoattractants and chemorepellents.

Keywords: angiogenesis, vasculogenesis, chemotaxis, network formation.

1 Vasculogenesis and angiogenesis

Vasculogenesis and angiogenesis are two different mechanisms involved in the development of blood vessels. The former process mainly occurs when the primitive vascular network is formed. It consists of the aggregation and organisation of the *endothelial cells*, the main bricks of the capillary walls. The latter consists of the formation of new vessels which only sprout from an existing capillary or post-capillary venule. Angiogenesis already intervenes in the embryo to remodel the initial capillary network into a mature and functional vascular bed comprised of arteries, capillaries, and veins. Angiogenic remodelling co-ordinates with the establishment of blood flow and can occur through *sprouting*, i.e., by the formation of new branches from the sides of existing capillaries (see Fig. 1a) or through *intussusception*, i.e., by internal division of the vessel lumen (see Fig. 1b).

The main role of angiogenesis is, however, during adult life when it is involved in many physiological processes, for instance, the vascularization of the ovary and the uterus during the female cycle and of the mammary gland during lactation and wound healing. However, angiogenesis also plays a fundamental role in many pathological settings, e.g. tumors, chronic inflammatory diseases such as rheumatoid arthritis and psoriasis, vasculopaties such as diabetic microangiopathy, degenerative disorders such as atherosclerosis and cirrhosis, tissue damage due to ischemia. We emphasize that, though during adult life angiogenesis is the main process of capillary formation, vasculogenesis can still occur.

It is possible to divide the angiogenic process into well-differentiated stages which sometimes partially overlap. We briefly describe them below; the interested

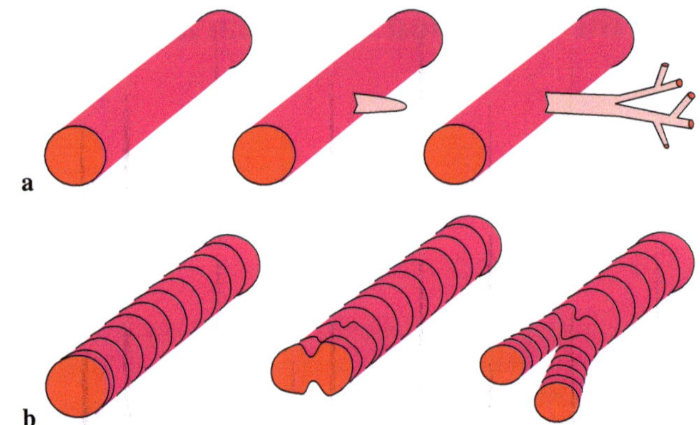

Fig. 1. Different kinds of angiogenesis: **a** sprouting and **b** intussusception

reader can find more information on the process in the recent reviews by Bussolino et al. [7] and by Mantzaris et al. [19]. These stages are regulated by precise genetic programmes and are strongly influenced by a chemical factor called Vascular Endothelial Growth Factor (VEGF).

1. The first stage is characterized by changes in the shape of the endothelial cells covering the walls of the blood vessel, by the loss of interconnection between endothelial cells, and by the reduction of vascular tonus. This in particular induces an increase in vessel permeability.
2. The stage of progression is characterized by the production of proteolitic enzymes (serine-proteins, iron-proteins) which degrade the extracellular matrix surrounding the capillary facilitating cellular movement and by the capacity of the endothelial cells to proliferate and to migrate chemotactically, i.e., up the gradient of suitable chemical factors, toward the place where it is necessary to create a new vascular network.
3. The stage of differentiation is characterized by the exit of the endothelial cells from the cellular cycle and by their capacity to survive in sub-optimal conditions and to build themselves primitive capillary structures, not yet physiologically active.
4. In the stage of maturation, the newborn vessel is completed by the formation of new extracellular matrix and by the arrival of other cells named pericytes and sometimes of flat muscle cells. During this phase a major role is played by molecules called angiopoietins leading to the development of the simple endothelial tubes into a more elaborate vascular tree composed of several cell types. In fact, they contribute to the maintenance of vessel integrity through the establishment of appropriate cell-cell and cell-matrix connections.
5. After the formation of the vascular network, a remodelling process starts. This involves the formation of anastomosis between capillaries, the loss of some physiologically useless capillaries, and remodelling of the extracellular matrix.

The inductors of angiogenesis, e.g., VEGF, cause the endothelial cell to migrate, to proliferate and to build structures which are similar to capillaries even when they are cultivated in vitro on an extracellular matrix gel. This phenomenon is called *angiogenesis in vitro*.

A further process leading to vessel formation is *arteriogenesis*, a process triggered by the occlusion of an artery. In order to overcome the problems of possible formation of ischemic tissues, the pre-existing arteriolar connections enlarge to become true collateral arteries. In this way, bypassing the site of occlusion, they have the ability to grow markedly and increase their lumen providing an enhanced perfusion to regions affected by the occlusion. We remark that the formation of collateral arteries is not simply a process of passive dilatation, but of active proliferation and remodelling.

2 In vitro vasculogenesis

Vasculogesis can be obtained in vitro using different experimental set-ups, substrata, and cell-lines, as reviewed in [31]. This is an important experiment performed not only to understand the mechanisms governing the angiogenic process, but also to test the efficacy of anti-angiogenic drugs and, in principle, to build the initial vascular network necessary to vascularize the tissues grown in vitro.

In order to understand the subsequent modelling and results we describe the experimental set-up used in [27]. A Petri dish is coated with an amount of *Matrigel*, a surface which favours cell motility and has biochemical characteristics similar to living tissues, which is $44 \pm 8\,\mu$m thick. Human Endothelial Cells from Large Veins or Adrenal Cortex Capillaries (HUVEC) are dispersed in a physiological solution which is then poured on the top of the Matrigel. The cells sediment by gravity onto the Matrigel surface and then move on the horizontal Matrigel surface giving rise to the process of aggregation and pattern formation shown in Fig. 2.

The process lasts 12–15 hours and goes through the following steps.

1. In the first couple of hours endothelial cells have a round shape. It seems that they choose a direction of motion and keep migrating with a small random component until they collide with their nearest neighbors (see Figs. 2a,b). This effect is called in biology *cell persistence* and is related to the inertia of the cell in rearranging its *cytoskeleton*, the ensemble of fibers (e.g., actin and microtubules) which drive cell motion. The direction of motion, however, is not chosen at random, but it can be shown to be correlated with the location of areas characterized by higher concentrations of cells. It is interesting to note that in this phase cells move much faster than later on when activation of focal contacts and interactions with the substratum increase. This type of motion, called *amoeboid*, can be compared with the exhibition of a gymnast with a quick sequence of jumps of the cells from handle to handle using few “arms” at a time (see Fig. 3).
2. After collision, the cells attach to their neighbors eventually forming a continuous multicellular network (Fig. 2c). The number of adhesion sites increases and the cells achieve a more elongated shape multiplying the number of adhesion sites with the substratum. The motion is much slower and resembles that of a

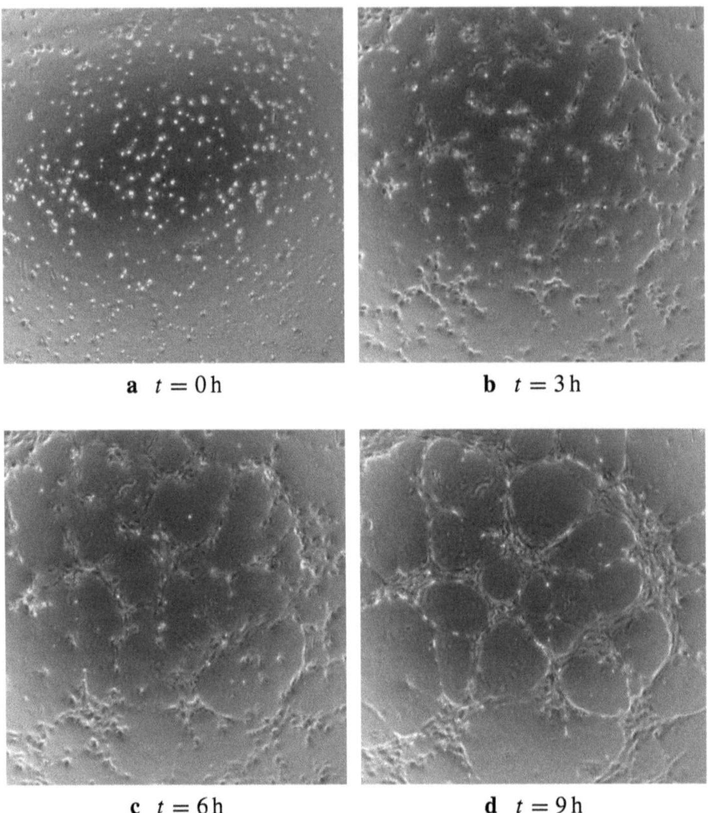

Fig. 2. The process of formation of vascular networks on a Matrigel surface. The box-side is 2 mm long

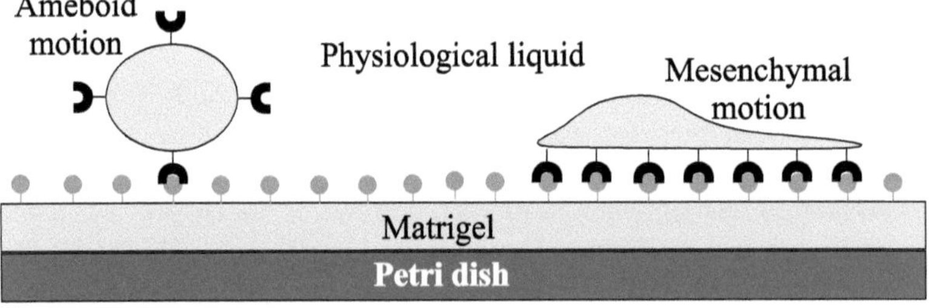

Fig. 3. Schematization of: **a** ameboid and **b** mesenchymal motion

mountain-climber who uses as many footholds as possible (see Fig. 3b) grabbing new adhesion sites and detaching from old ones one at a time. This type of motion is called *mesenchymal*.

3. The network slowly moves as a whole, undergoing a slow thinning process (Fig. 2d), which, however, leaves the network structure mainly unaltered. In

this phase the mechanical interactions among cells and between cells and the substratum become important.

4. Finally, individual cells fold up to form the lumen of the capillary, so that the formation of a capillary-like network occurs along the lines of the previously formed bidimensional structure.

The interested reader can view a film strip of an experiment (partial for easier downloads) at the web site of the EMBO journal (http://embojournal.npgjournals.com/content/vol22/issue8/index.shtml) as supplementary material to [27].

As the motion of cells in the first phase seems to be well established toward the region characterized by higher cell densities and kept until the cells encounter other cells, the main question is how the cells feel the presence of other cells. In fact, the evidence above suggests the presence of both a mechanism of persistence in cell motion and a mechanism of cross-talk among cells. As a matter of fact, endothelial cells in the process of vascular network formation exchange signals by the release and absorption of Vascular Endothelial Growth Factors-A (VEGF-A) which are also essential for their survival and growth. Moreover, autocrine/paracrine secretion of VEGF-A by endothelial cells was shown to be essential for the formation of capillary beds. This growth factor can bind to specific receptors on the cell surface and induce chemotactic motion along its concentration gradient.

In order to quantify both cell persistence and chemotactic behavior in cell motion Serini et al. [27] performed a statistical analysis of the cell trajectories. As depicted in Fig. 4, they measured the angle φ between two subsequent displacements relative to the same trajectory, which gives a measure of the persistence, and the angle θ between the velocity and the concentration gradient at the same point simulated starting from the distribution of cells and taking into account that VEGF-A, like similar soluble molecules, is degraded by the environment, mainly through oxidation processes. The angle θ then gives a measure of the chemotactic behavior.

In order to test the importance of chemotactic signalling mechanisms Serini et al. [27] also performed experiments aimed at extinguishing VEGF-A gradients spreading from individual endothelial cells plated on Matrigel by adding a saturating amount of exogenous VEGF- A. Indeed, saturation of VEGF-A gradients resulted in a strong inhibition of network formation.

The same statistical analysis as described above was repeated in saturating conditions. In this case, the diagram for φ shows that cell movement maintains a certain degree of directional persistence, while the diagram for θ shows that in saturating conditions the movement is completely de-correlated from the direction of simulated VEGF gradients showing the importance of VEGF in the process.

The final configuration achieved in the experiments is a capillary-like network which can be represented as a collection of nodes connected by cords. The amazing thing is that, over a range of values of seeded cell density extending from 100 to 200 cells/mm^2, the mean cord length measured on the experimental records is approximately constant and equal to $\ell \simeq 200 \pm 20\,\mu$m.

It is interesting to observe that capillary networks characterized by typical intercapillary distances ranging from 50 to 300 μm is instrumental for optimal metabolic

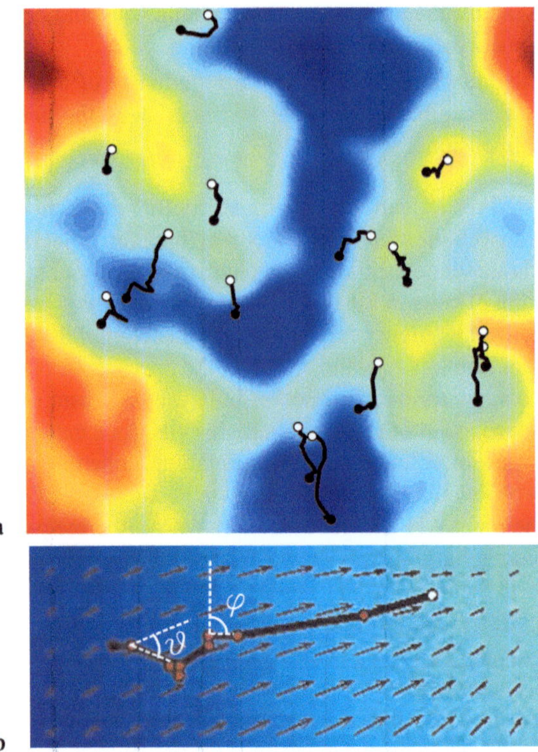

Fig. 4. Trajectories of some cells in the field of chemoattractant **a** and sample trajectory **b**. Arrows indicate concentration gradients. The angles φ and θ refer respectively to persistence and chemotaxis

exchange, so that the characteristic size of the network in vitro is biologically functional: a coarser net would cause necrosis of the tissues in the central region, a finer net would be useless.

A pathological situation in which the dimension of the capillary network changes has been described by Ruhrberg et al. [26]. In fact, they observed that mice lacking heparin-binding isoforms of VEGF-A form vascular networks with a larger mesh (see Fig. 5). This is related to the fact that the binding of some of the isoforms with lower or higher molecular weight affects the effective diffusivity of the chemical factor. Therefore VEGF plays a role in defining the mesh size and, in particular, different isoforms (with different diffusivities) can lead to different mesh sizes. As discussed in Sect. 4.1, the model proposed in [11, 27] predicts that the size of the network is related to the product of the diffusion constant and the half-life of the chemical factor, so that, if the effective diffusion increases, the typical size of the network cords increases.

If, on the one hand, the cord length is nearly independent of the density of seeded cells in a certain range, on the other hand it is observed that outside this range one

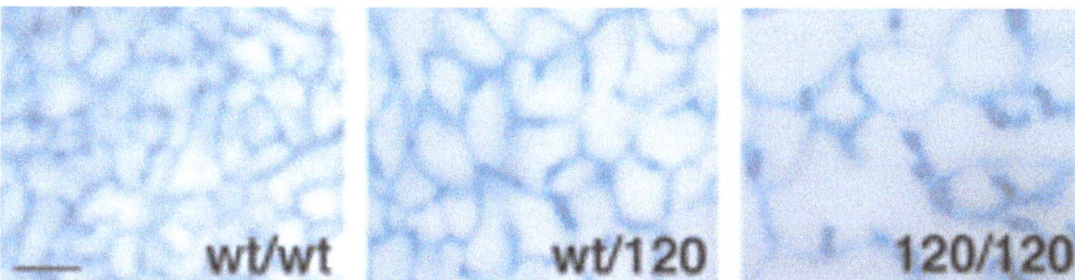

Fig. 5. Dependence of cord length on VEGF effective diffusivity (adapted from [26])

does not have a proper development of vascular networks. In fact, below a critical value of about 100 cells/mm^2, the single connected network shown in Fig. 6b breaks down in groups of disconnected structures as shown in Fig. 6a. On the other hand at higher cell densities, say above 200 cells/mm^2 (Fig. 6c) the mean cord thickness grows to accommodate an increasing number of cells. For even higher values of initial density, the network takes on the configuration of a continuous carpet of cells

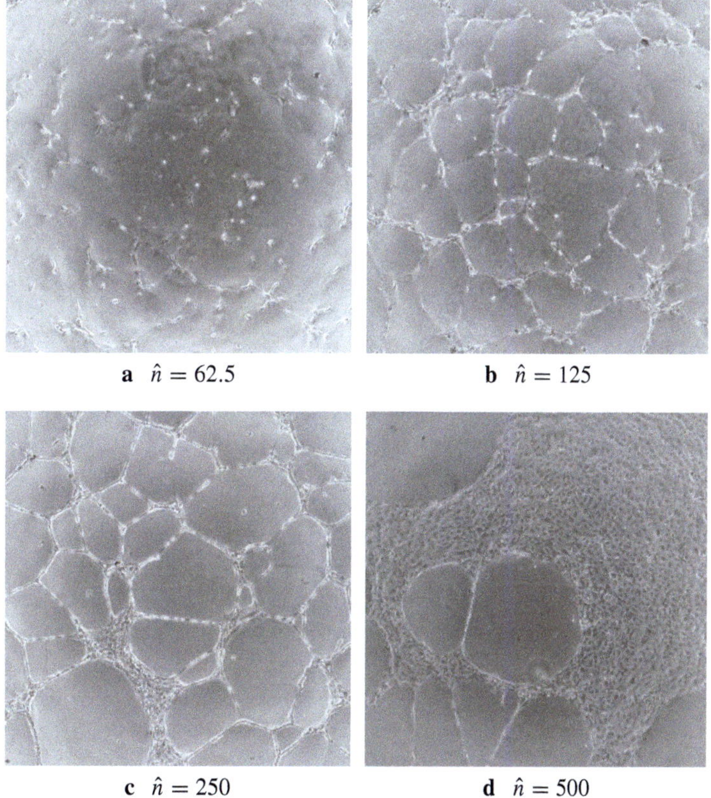

a $\hat{n} = 62.5$ **b** $\hat{n} = 125$

c $\hat{n} = 250$ **d** $\hat{n} = 500$

Fig. 6. Dependence of the types of structures formed on the density of seeded cells

with holes, called *lacunae* (Fig. 6d). This configuration is not functional. In fact, cells do not even differentiate to form the lumen in the cords.

We end this section by mentioning that the generalization of this phenomenon to the three-dimensional formation of vascular networks is not as straightforward as it might seems, because in this situation cells are surrounded by the *extracellular matrix* (the network of fibers, e.g., fibronectin, collagen, vitronectin, filling part of the extracellular space) and the possibility of ameboid motion is limited. The motion of a cell in a three-dimensional extracellular matrix can be viewed at http://www.bloodjournal.org/cgi/content/full/2002-12-3791/DC1. In fact, to move in the gel they have to cleave the extracellular matrix via the production of matrix-degrading enzymes, which alter the environment the cells move into. However, this does not exclude the possibility of the existence of a layer between two strata characterized by a reduced amount of extracellular matrix so that ameboid motion can still occur. This might be the case in the so-called *sandwich experiments* in which a second Matrigel layer is placed on top of the cells after seeding them over the layer at the bottom of the Petri dish. This gives a preferential direction of motion for the cells, the horizontal plane.

3 Modelling vasculogenesis

The objective of the mathematical model presented in this section is to simulate in silico the entire course of events occurring during vasculogenesis, i.e., a chemically dominated phase characterized by an initial ameboid motion mainly affected by gradients of endogenous chemoattractants and a subsequent mesenchymal motion in which chemotactic effects are still important but mechanics dominates because of cell anchoring to the substratum and the subsequent development of stresses.

To deduce the model, in addition to the chemical factor(s) influencing the process, we consider the following compound system:

1. the ensemble of cells, dealt with as a continuum;
2. the substratum, e.g., Matrigel;
3. the physiological liquid, which is considered as a passive constituent with negligible interactions with the others.

We develop the model in the framework of mixture theory (see, e.g., [4]), suitably adapted to the biological setting. One can then generally write for the three constituents above the following equations related to mass and momentum balance:

$$
\begin{aligned}
&\frac{\partial \rho_c}{\partial t} + \nabla \cdot (\rho_c \mathbf{v}_c) = \Gamma_c, \\
&\frac{\partial \rho_s}{\partial t} + \nabla \cdot (\rho_s \mathbf{v}_s) = \Gamma_s, \\
&\frac{\partial \rho_\ell}{\partial t} + \nabla \cdot (\rho_\ell \mathbf{v}_\ell) = \Gamma_\ell,
\end{aligned}
\tag{1}
$$

$$\frac{\partial}{\partial t}\left(\rho_c \mathbf{v}_c\right) + \nabla\cdot\left(\rho_c \mathbf{v}_c \otimes \mathbf{v}_c\right) = \nabla\cdot\mathbf{T}_c + \mathbf{F}_c + \mathbf{f}_c + \Gamma_c \mathbf{v}_c, \tag{2}$$

$$\frac{\partial}{\partial t}\left(\rho_s \mathbf{v}_s\right) + \nabla\cdot\left(\rho_s \mathbf{v}_s \otimes \mathbf{v}_s\right) = \nabla\cdot\mathbf{T}_s + \mathbf{F}_s + \mathbf{f}_s + \Gamma_s \mathbf{v}_s, \tag{3}$$

$$\frac{\partial}{\partial t}\left(\rho_\ell \mathbf{v}_\ell\right) + \nabla\cdot\left(\rho_\ell \mathbf{v}_\ell \otimes \mathbf{v}_\ell\right) = \nabla\cdot\mathbf{T}_\ell + \mathbf{F}_\ell + \mathbf{f}_\ell + \Gamma_\ell \mathbf{v}_\ell,$$

where c stands for cell, s for substratum and ℓ for liquid. For the ith constituent ($i = c, s, \ell$), Γ_i is the production rate, $\mathbf{T}_i$ is the partial stress tensor, $\mathbf{F}_i$ is the body force acting on the ith constituent, $\mathbf{f}_i$ is the momentum supply, related to the local interactions with the other constituents, and ρ_i and $\mathbf{v}_i$ are the density and the velocity of the ith constituent. In particular, we note that the density of cells, i.e., the mass of cells per unit area, can be written as $\rho_c = m_c n$, where m_c is the mass of a cell and n is the number of cells per unit area (the in vitro process is two-dimensional).

If the mixture is closed, then overall mass and momentum balance implies that

$$\Gamma_c + \Gamma_s + \Gamma_\ell = 0,$$
$$\mathbf{f}_c + \Gamma_T \mathbf{v}_c + \mathbf{f}_s + \Gamma_s \mathbf{v}_s + \mathbf{f}_\ell + \Gamma_\ell \mathbf{v}_\ell = \mathbf{0}.$$

In normal conditions endothelial cells replicate every one or two days, but this process is inhibited even further in the experimental environment. Therefore, rather than apoptosis or mithosis the right-hand side of (1) takes into account the possible change of number of cells on the substratum. This can be due to the detachment of cells from the substratum, which seems to occur during some experiments of Vailhé et al. [32], or to the sedimentation and then accumulation of cells on the substratum which may occur over a time of the order of one hour. As a consequence the last term on the right-hand side of (2) takes into account the gain/loss of momentum due to the gain/loss of mass.

We neglect this phenomenon here. We also assume that the extracellular matrix is neither produced nor degraded and therefore we can write

$$\Gamma_c = \Gamma_s = \Gamma_\ell = 0, \tag{4}$$
$$\mathbf{f}_\ell = 0, \quad \mathbf{f}_s = -\mathbf{f}_c := \mathbf{f}_n, \tag{5}$$

which shows the character of internal (interaction) force that the force $\mathbf{f}_n$ exerted by the cells on the substratum has.

3.1 Diffusion equations for chemical factors

Before studying the persistence equation (2) and the substratum equation (3) in detail we focus on the diffusion of chemotactic factors which is governed by the usual reaction-diffusion equations. We distinguish between *endogenous* chemical factors, i.e., those produced by the cell themselves, and *exogenous* chemical factors, i.e., those introduced by other components, in our case mostly from the outside. From the experimental viewpoint this can be achieved by adding to the substratum gelly sponges or gelly “spaghettis” impregnated with chemical substances able to attract endothelial cells, e.g., VEGF, or repel them, e.g., semaphorines. This is done because we envisage possible applications to tissue engineering where it is important to understand how to govern the characteristics of the network by acting from outside the system.

The diffusion of the different chemotactic factors is then governed by the equations

$$\frac{\partial c}{\partial t} = D\nabla^2 c - \frac{c}{\tau} + \alpha(\rho_c), \tag{6}$$

$$\frac{\partial c_a}{\partial t} = D_a\nabla^2 c_a - \frac{c_a}{\tau_a} + s_a(t)H_a(\mathbf{x}), \tag{7}$$

$$\frac{\partial c_r}{\partial t} = D_r\nabla^2 c_r - \frac{c_r}{\tau_r} + s_r(t)H_r(\mathbf{x}), \tag{8}$$

where c is the concentration of endogenous VEGF-A produced by endothelial cells, c_a is the concentration of exogenous chemoattractant, which might still be VEGF-A, and c_r is the density of exogenous chemorepellent.

In (6) the chemoattractant is produced by the endothelial cells at a rate α and degrades with a half life τ. In (7) and (8) the chemical factors are released at a rate $s_a(t)$ and $s_r(t)$ in certain domains identified by the indicator functions H_a and H_r, which is constantly equal to 1 in the region where the chemical factor is released and vanishes outside it. Convection is neglected because of the low fluid velocity.

The model in [11, 27] contains a production term $\alpha(\rho_c) = a\rho_c$ in the reaction-diffusion equation for the chemical factor. This implicitly means that each cell always produces a constant amount of chemoattractant independently of the environment. The model in [30] assumes a more general functional form of the production term based on the consideration: this chemoattractant is a means of communication and survival for the cell itself and its neighbors. It is known that, upon contact, cells activate mechanotransduction pathways involving cell-to-cell junctions and transmembrane proteins like cadherins. This may lead to a downregulation of the production of the chemical factor because, when cells reach an aggregate state, there is no need to communicate and recruit new cells with the release of more chemical factors. In particular, VEGF communication can be replaced by contact cadherin-cadherin signalling. At present this is a phenomenological hypothesis and we are not aware of any experimental evidence supporting or contradicting it. It would be interesting, however, to do experiments in this direction.

Specifically, the simulation below uses

$$\alpha(\rho_c) = \frac{a\rho_c}{1 + b\rho_c^2} \tag{9}$$

with $a > 0$ and $b \geq 0$, so that, for $b \neq 0$, $\alpha(\rho_c)$ has a maximum production $\frac{a}{2\sqrt{b}}$ at $\rho_c = \frac{1}{\sqrt{b}}$ and goes to zero as $\rho_c \to +\infty$.

3.2 Persistence equation for the endothelial cells

Focusing on the cell population, we use the mass balance equation (1) and the momentum balance equation for the cellular matter, which, by using (1), (4) and (5), can be simplified to

$$\frac{\partial \mathbf{v}_c}{\partial t} + \mathbf{v}_c \cdot \nabla \mathbf{v}_c = \frac{1}{\rho_c} \left(\nabla \cdot \mathbf{T}_c + \mathbf{F}_c - \mathbf{f}_n \right). \tag{10}$$

It must be observed that, in most biological phenomena, inertia is negligible. In fact, velocities are of the order of μm per second. In fact, the left-hand side of Eq. (10) should be understood as the "inertia" of the cells in changing their direction of motion, i.e., cell persistence.

The right-hand side of (10) also contains the fundamental chemotactic body force

$$\mathbf{F}_{\text{chem}} = \rho_c \beta(c) \nabla c, \tag{11}$$

where c is the concentration of a particular chemical factor and $\beta(c)$ measures the intensity of cell response which can include saturation effects, e.g.,

$$\beta(c) = \frac{\beta}{1 + \frac{c}{c_M}}, \quad \text{or} \quad \beta(c) = \beta(1 - c/c_M)_+,$$

where c_M is constant and

$$f_+ = \begin{cases} f & \text{if } f > 0 \\ 0 & \text{otherwise} \end{cases}$$

is the positive part of f. The linear dependence of the force on ρ_c corresponds to the assumption that each cell experiences a similar chemotactic action so that the momentum balance in integral form depends on the number of cells in the control volume and the related local equation on the local density. If all three chemical factors mentioned in the previous section are present, then

$$\mathbf{F}_{\text{chem}} = \rho_c (\beta \nabla c + \beta_a \nabla c_a - \beta_r \nabla c_r).$$

Lastly, the partial stress tensor gives an indication of the response of the ensemble of cells to stresses. Several constitutive equations can be formulated, but unfortunately no experimental data are available on the mechanical characteristics of ensembles of cells. It can be argued that, because the cytosol is a watery solution containing many long proteins contained in a viscoelastic membrane, the ensemble of cells might behave as a viscoelastic material. However, we can expect that the characteristic times of the viscoelastic behavior are much smaller than those related to cell motion (minutes as compared to hours), so that viscoelastic effects can be considered negligible. On the other hand, plasticity should probably be taken into account to describe the breaking of cell-to-cell adhesion bonds.

In absence of experimental evidence, in what follows the simplest constitutive equation possible,

$$\mathbf{T}_c = -p(\rho_c)\mathbf{I},$$

is considered, corresponding to an elastic fluid. This assumption implies, for instance, that the ensemble of cells cannot sustain shear, which, of course, is not true.

As we shall see, however, the presence of $\mathbf{T}_c$ is particularly important in describing the formation of lacunae.

Equation (10) then specializes to

$$\frac{\partial \mathbf{v}_c}{\partial t} + \mathbf{v}_c \cdot \nabla \mathbf{v}_c = \frac{1}{\rho_c} \nabla p(\rho_c) + \beta(c) \nabla c - \frac{1}{\rho_c} \mathbf{f}_n .$$

The exact form of $\mathbf{f}_n$, the force of interaction with the substratum, is specified in the following section.

3.3 Substratum equation

Focusing on the substratum, which is an inert intricate web of long fibers on which cells move, we can certainly state that inertial effects can be neglected. From Eq. (3), the force balance equation for the substratum then becomes

$$\nabla \cdot \mathbf{T}_s + \mathbf{f}_n + \mathbf{F}_s = \mathbf{0}, \tag{12}$$

where, in particular, $\mathbf{F}_s$ is the force of anchoring to the Petri dish, and $\mathbf{f}_n$ is the interaction force exerted by the cells on the substratum.

We observe that Eq. (12) works in the two-dimensional layer. The procedure for obtaining the two-dimensional reduction of the stress balance equation is often not clearly described in the literature. For this reason we report it in detail in an appendix to this chapter. The forces appearing in (12) derive from the interaction with the cells and with the Petri dish at the top and the bottom of the layer.

We assume that the interaction force between the substratum and the cells includes an elastic and a viscous contribution. During the ameboid motion the interaction force acting on the cells is of viscous type, which implies a weak interaction between the cells and the substratum, characterized by the rapid removal of the bonds and formation of new bonds, with weak deformation of the substratum. We can model this force as

$$\mathbf{f}_{\text{visc}} = -\gamma \rho_c (\mathbf{v}_s - \mathbf{v}_c), \tag{13}$$

where $\mathbf{v}_s = d\mathbf{u}_s/dt$ and $\mathbf{u}_s$ is the displacement of the substratum.

The elastic contribution takes into account the fact that, after some time, cells attach to the substratum with a strong bond. If the cell anchors in $\mathbf{u}_c$ and then moves to $\mathbf{u}$ we can assume that the elastic force is proportional to $\mathbf{u} - \mathbf{u}_c$. This change of behavior characterizes the transition between the chemotactic and mechanical phases. In other words, this force is absent in the initial ameboid motion and starts when the motion becomes of mesenchymal type, i.e., when cells start attaching to the adhesion molecules of the Matrigel. If we assume that there is a characteristic time t_{th} needed to anchor to the adhesion sites on the substratum and which characterizes the transition between a purely ameboid phase and a mesenchymal phase we can write

$$\mathbf{f}_{\text{elast}} = -\kappa \rho_c (\mathbf{u}_s - \mathbf{u}_c) H(t - t_{th}),$$

where κ is the anchoring rigidity and H is the Heaviside function

$$H(\tau) = \begin{cases} 1 & \text{if } \tau > 0 \\ 0 & \text{otherwise.} \end{cases}$$

Another interesting hypothesis is that ameboid motion stops when cells come in contact, so that the strongly reduced velocity allows for a better link with the adhesion molecules of the substratum. This phenomenon could be included in the previous set-up by assuming that

$$\mathbf{f}_{\text{elast}} = -\kappa(\rho_c)\rho_c\,(\mathbf{u}_s - \mathbf{u}_c)\,,$$

where, in particular, $\kappa(\rho_c)$ vanishes below a given value ρ_{th} of cell density, e.g., $\kappa(\rho_c) = \kappa H(\rho_c - \rho_{th})$.

However, if the pulling is strong enough, then some adhesion bonds could break and so the inclusion of plastic phenomena should be considered, but this is not done here and is a possible interesting development.

Referring to the two-dimensional reduction of the substratum equation described in the appendix, we need to consider the adhesion of the substratum to the Petri dish, which can be taken to be proportional to the displacement, i.e.,

$$\mathbf{f}_{\text{ext}} = -\frac{s}{h}\mathbf{u}_s,$$

where h is the substratum thickness. We emphasize the fact that this force does not act on the cellular constituent, and so it must not be considered in the momentum equation for the cells.

4 In silico vasculogenesis

According to the deduction above the mathematical model is written as follows:

$$\begin{aligned}
&\frac{\partial n}{\partial t} + \nabla\cdot(n\mathbf{v}_c) = 0,\\
&\frac{\partial \rho_s}{\partial t} + \nabla\cdot(\rho_s\mathbf{v}_s) = 0,\\
&\frac{\partial \mathbf{v}_c}{\partial t} + \mathbf{v}_c\cdot\nabla\mathbf{v}_c = \frac{1}{n}\nabla p(n) + \beta\nabla c + \beta_a\nabla c_a - \beta_r\nabla c_r +\\
&\qquad\qquad - \gamma m_c(\mathbf{v}_c - \mathbf{v}_s) - \kappa m_c(\mathbf{u}_c - \mathbf{u}_s)H(t - t_{th}),\\
&\nabla\cdot\mathbf{T}_s + \gamma m_c n(\mathbf{v}_c - \mathbf{v}_s) + \kappa m_c n(\mathbf{u}_c - \mathbf{u}_s)H(t - t_{th}) - \frac{s}{h}\mathbf{u}_s = \mathbf{0},\\
&\frac{\partial c}{\partial t} = D\nabla^2 c - \frac{c}{\tau} + \alpha(\rho_c),\\
&\frac{\partial c_a}{\partial t} = D_a\nabla^2 c_a - \frac{c_a}{\tau_a} + s_a(t)H_a(\mathbf{x}),\\
&\frac{\partial c_r}{\partial t} = D_r\nabla^2 c_r - \frac{c_r}{\tau_r} + s_r(t)H_r(\mathbf{x}),
\end{aligned}$$

where we prefer to use the number of cells per unit area, $n = \rho_c/m_c$, because this is the quantity which is given and changed in the experiments.

The experiments described in Sect. 2 start with a number of cells randomly seeded on the Matrigel. To reproduce the experimental initial conditions we always start with the following cell distribution:

$$n(\mathbf{x}, t = 0) = \frac{1}{2\pi r^2} \sum_{j=1}^{M} \exp\left[-\frac{\left(\mathbf{x} - \mathbf{x}_j(\omega)\right)^2}{2r^2} \right],$$
$$\mathbf{v}(\mathbf{x}, t = 0) = \mathbf{0}.$$

Each Gaussian bump has width of the order of the average cell radius $r \simeq 20\,\mu\text{m}$, so that from the mathematical point of view it represents a cell. Then M Gaussian bumps are centered at random locations $\mathbf{x}_j$ distributed with uniform probability on a square of size L (in the experimental set-up $L = 2$ mm). The initial velocity is null, because cells sediment from above on the horizontal surface. Unless otherwise specified, periodicity is imposed at the boundary of the domain.

4.1 Neglecting substratum interactions

As a first example we consider the formation of the vascular network in isotropic conditions, under the action of endogenous chemical factors only and neglecting mechanical interactions with the substratum.

The model then reduces to that proposed in [3,11,27]:

$$\begin{aligned}
&\frac{\partial n}{\partial t} + \nabla \cdot (n\mathbf{v}_c) = 0,\\
&\frac{\partial \mathbf{v}_c}{\partial t} + \mathbf{v}_c \cdot \nabla \mathbf{v}_c = \frac{1}{n}\nabla p(n) + \beta \nabla c - \tilde{\gamma}\mathbf{v}_c,\\
&\frac{\partial c}{\partial t} = D\nabla^2 c - \frac{c}{\tau} + \alpha(\rho_c).
\end{aligned} \tag{14}$$

The result of a simulation is shown in Fig. 7. We now consider the information encoded in the coupling of the first two equations above with the diffusion equation (8). This can be understood most simply if we neglect pressure and assume for a moment that diffusion is a faster process than pattern formation, so that the dynamics of c is "enslaved" to the dynamics of n and the derivative term $\partial c/\partial t$ can be neglected in a first approximation. Then it is possible to solve the diffusion equation for c formally and to substitute it in the persistence equation, so that one can write (for $b = 0$):

$$\frac{\partial \mathbf{v}_c}{\partial t} + \mathbf{v}_c \cdot \nabla \mathbf{v}_c = \frac{a\beta}{D} \nabla \left(\ell^{-2} - \nabla^2\right)^{-1} n, \tag{15}$$

where

$$\ell := \sqrt{D\tau}.$$

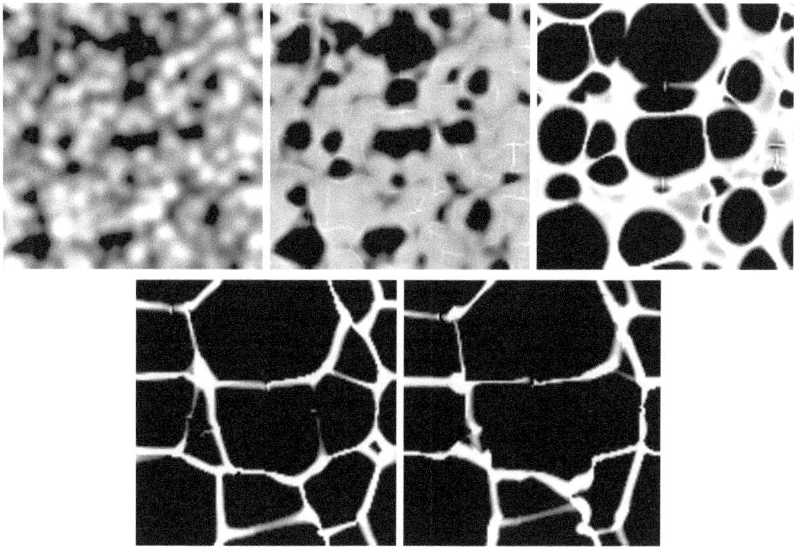

Fig. 7. Network formation in absence of substratum interaction. In particular, $\beta = 1$, $\tilde{\gamma} = 0$, $a = 1$, and $b = 0$

The appearance in the dynamical equations of the characteristic length ℓ suggests that the dynamics could favor patterns characterized by this length scale. As a matter of fact, if we rewrite the right-hand side of (15) in Fourier space as

$$\frac{a\beta}{D}\frac{\mathbf{ik}}{k^2 + \ell^{-2}} n_{\mathbf{k}},$$

we observe that the operator $\mathbf{ik}/\left(k^2 + \ell^{-2}\right)$ acts as a filter, which selects the Fourier components of n having wave numbers of order ℓ^{-2} and damps the components with higher and smaller wavenumbers.

Experimental measurements of the parameters gives $D \sim 10^{-7}\,\mathrm{cm}^2\,\mathrm{s}^{-1}$ and $\tau = 64 \pm 7$ min and therefore $\ell \sim 100\,\mu\mathrm{m}$, which is in good agreement with experimental data.

The process of network formation is then understood in the following way. Initially, non-zero velocities are built up by the chemotactic term due to the randomness in the density distribution. Density inhomogeneities are translated in a landscape of concentration of the chemoattractant factor where details of scales ℓ are averaged out. The cellular matter moves toward the ridges of the concentration landscape. A non-linear dynamical mechanism similar to that encountered in fluid dynamics sharpens the ridges and empties the valleys in the concentration landscape, eventually producing a network structure characterized by a length scale of order ℓ. In this way, the model provides a direct link between the range of intercellular interaction and the dimensions of the structure which is a physiologically relevant feature of real vascular networks.

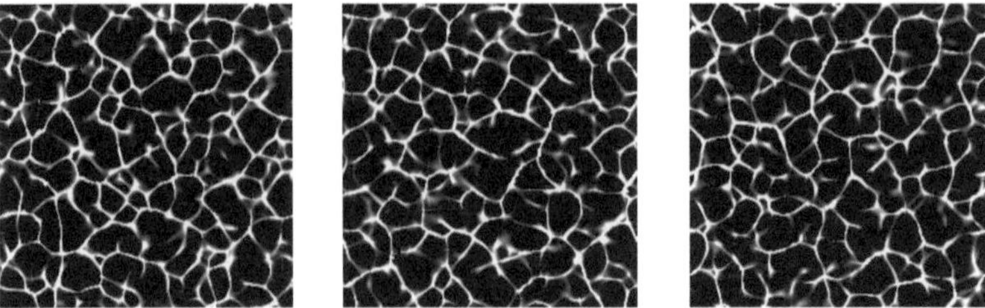

Fig. 8. Dependence of the specific network structure on the initial conditions

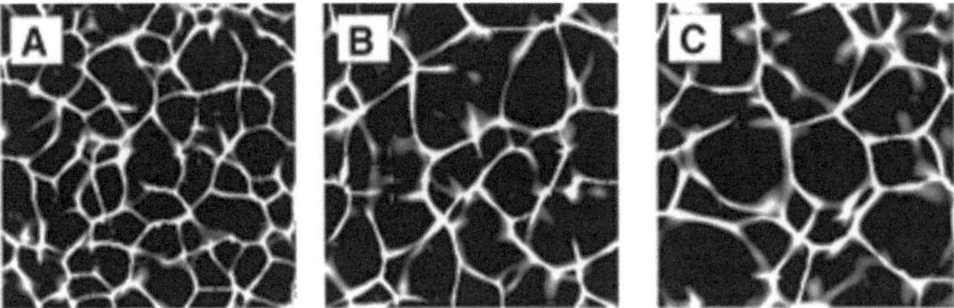

Fig. 9. Dependence of the network characteristic size on $\ell = 100, 200, 300\,\mu\text{m}$

The results of the simulations are shown in Fig. 8 which shows how the precise network structure depends on the initial conditions which are randomly set. However, at a glance the general features seem to be independent on the precise form of the initial condition and compare well with the experimental results shown in Fig. 2.

Changing the effective diffusion of the chemical factors lead to the results shown in Fig. 9, which agree with the observation that larger effective diffusivities lead to vascular networks with a larger mesh (see [26] and Fig. 5).

The model is also able to reproduce the dependence of the characteristics of the structure on the density of seeded cells. In fact, as known experimentally (see Sect. 2), on the one hand the cord length is nearly independent of the density of seeded cells in a certain range, while, on the other hand, it is observed that outside this range there is not a proper development of vascular networks. Varying the density of seeded cells one can display the presence of a percolative-like transition at small densities and a smooth transition to a "Swiss-cheese" configuration at large densities.

In fact, in the simulation, below a critical value $n_c \sim 100$ cells/mm^2 the single connected network (Fig. 10b) breaks down in groups of disconnected structures (Fig. 10a). On the other hand at higher cell densities, say above 200 cells/mm^2 (Fig. 10c), the mean cord thickness grows to accommodate an increasing number of cells. For even higher values of seeded cell density, the network takes the configuration of a continuous carpet with holes (Fig. 10d). This configuration is not functional.

Methods of statistical mechanics were used in [9, 11] to characterize the percolative transition quantitatively. They concluded that the transition occurring in the neighborhood of $n_c \sim 100$ cells/mm^2 falls in the universality class of random

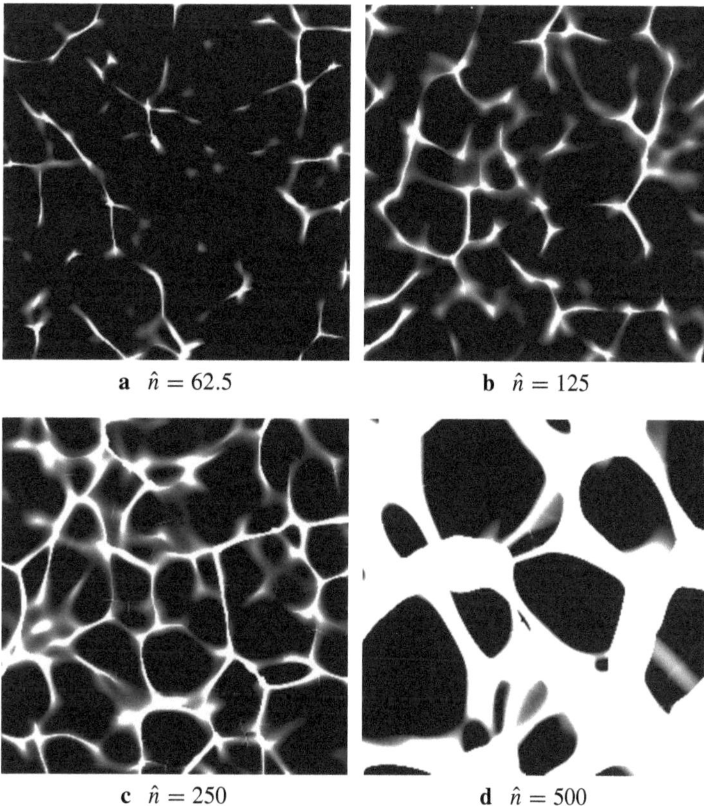

Fig. 10. Simulation of the dependence of the type of structures formed on variation of the density of cells

percolation, even in the presence of migration and dynamical aggregation. This is confirmed by the fractal dimension of the percolating cluster ($D = 1.85 \pm 0.10$). In fact, both the value obtained on the basis of the experiments and that obtained on the basis of the numerical simulations ($D = 1.87 \pm 0.03$) are close to the theoretical value expected for random percolation ($D = 1.896$). In fact, a bi-fractal behavior seems to appear at small scales, but we do not enter into detail, referring to [9] for further details.

The presence of a percolative transition in the process of formation of vascular networks is not obvious, and is linked to the average constancy of the cord length. As a matter of fact, there are at least two ways of accommodating an increasing number of cells on a vascular-type network. The first is to give priority to connectivity with respect to cord lengths as in Fig. 11a. The second corresponds to the opposite behavior. In this case, when the number of cells is too low, enforcing the constraint on the cord length makes it impossible to achieve side-to-side connectivity leading to sets of disconnected clusters as in Fig. 11b.

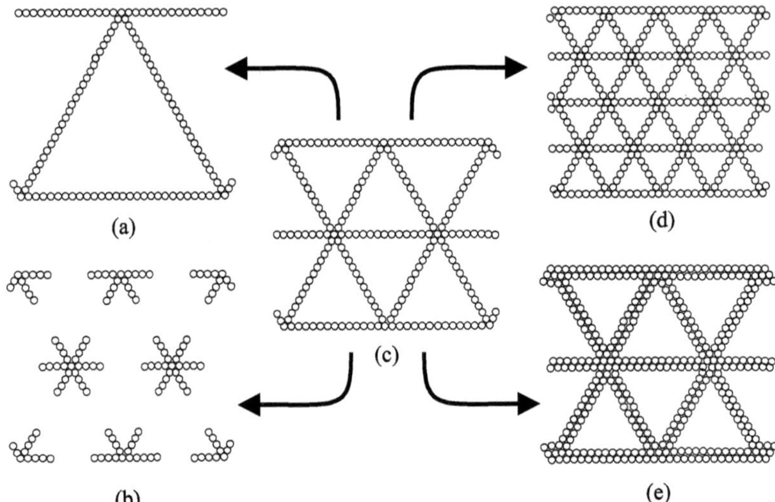

Fig. 11. Schematic representations of the percolative transition

It appears that Nature in this case chose to prioritize network size, probably because widely spaced capillary networks, like the one in Fig. 11a, would not be able to perform their main function, i.e., to supply oxygen and nutrients to the central part of the tissues.

The same mechanism might in principle explain the formation of lacunae. If the number of cells doubles, then there are two ways of accomodating the new cells. Either placing them in a more homogeneous way, forming smaller polygons, as in Fig. 11d, or adjoining the new cells to the others, as in Fig. 11e. In the first case the size of the polygons is halved, in the second it remains nearly the same, but the cords thicken. It seems that the same reasoning used in the percolative transition can be repeated here. Nature prefers to keep the size of the network as far as possible. Eventually, this leads to the formation of lacunae.

In this situation, the presence of the pressure term in the model is crucial as it avoids overcompression in the cords and enables the reproduction of the transition to the "Swiss-cheese" configuration experimentally observed for high cell densities (Fig. 6d). In addition, it avoids the blow-up of solutions characteristic of many chemotaxis models [13]. In fact, neglecting the mechanical interactions among overcrowding cells allows them to overlap in the points of maximum of the chemotactic field causing the blow-up of the solution. From the physical point of view it is easy to realize that the pressure term avoids overcrowding. In fact, among other things, Kowalczyk [13] proved that it is enough that there are $c > 0$ and $\hat{n}$ such that for all $n > \hat{n}$ $p'(n) \geq 0$ to insure the boundness of solutions in any finite time.

In order to study the formation of lacunae starting from a continuous monolayer of cells Kowalczyk et al. [14] also studied the linear stability properties of the model (14) finding that chemotaxis with the related parameters (motion, production, degradation) is the key destabilizing force while pressure is the main stabilizing force.

4.2 Substratum interactions

In this section we introduce mechanical interactions with the substratum while still assuming only endogenous chemotaxis. The model, introduced in [30], is a particular case of Eq. (14) without c_r and c_a.

The effect of mechanical stretching obtained in the simulation is compatible with what is observed in vitro, namely, in pulling on the extracellular matrix, the cells deform the substratum (see Fig. 13). However, if the substratum is too rigid or if cell adhesion is too strong, then it is very hard for the cells to form a cord. In the limit case of very stiff substrata as that in Fig. 12, then the morphogenic process leads to the formation of lacunae rather than cords. The mechanical interactions also seem to play a fundamental role in guaranteeing the stability of the network.

Figure 13a shows the contour plot of the norm of the stress tensor relative to the Matrigel,

$$\|\mathbf{T}_s\| = \sqrt{\mathbf{T}\mathbf{T}^T},$$

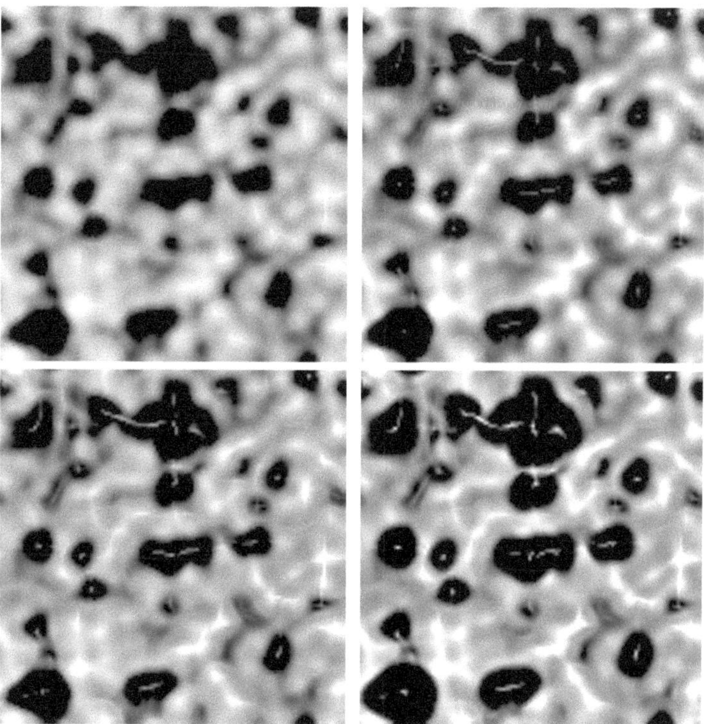

Fig. 12. Influence of the mechanical interactions on the network formation for $\kappa = 1$. The results can be compared with those in Fig. 7 which refer to no interaction with the substratum ($\kappa = 0$). The initial condition is the same as that shown in Fig. 7

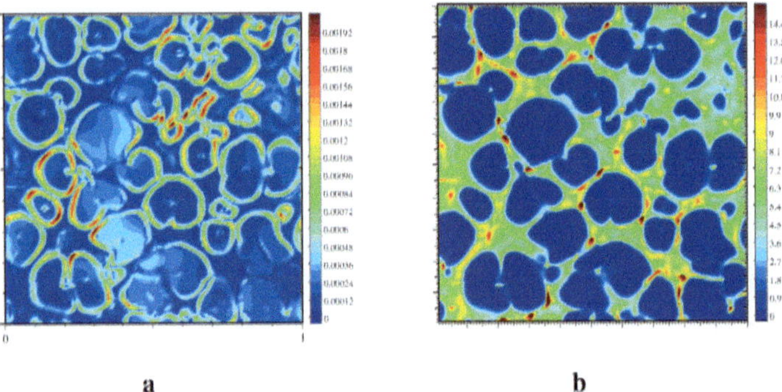

Fig. 13. Plot of the norm of the Matrigel stress tensor **a** and the corresponding density of endothelial cells **b** at the final stage of network formation. Level curves denote increasing values from blue to red

at the final stage of network formation, corresponding to the cellular density shown in Fig. 13b. It can be observed that the stress is concentrated in thin strips edging the chords and surrounding the cellular density holes. At present to our knowledge no measurements of Matrigel displacement have been made during the process of in vitro vasculogenesis, though we think that this could be done by disseminating the substratum with microspheres and monitoring their displacements.

In order to estimate the relative importance of chemotaxis versus mechanics, one can compute the L^1 norm over the domain. Figure 14 shows that, in the first instant of the simulations, the chemotactic force grows more rapidly than the elastic force, so that in the first period chemotactic effects are thus prevalent. After that, the elastic force grows, till a substantial equilibrium is reached.

In order to understand the role of cell adhesivity and substratum stiffness, Fig. 12 should be compared with Fig. 15 which presents a moderate interaction ($\kappa = 0.2$

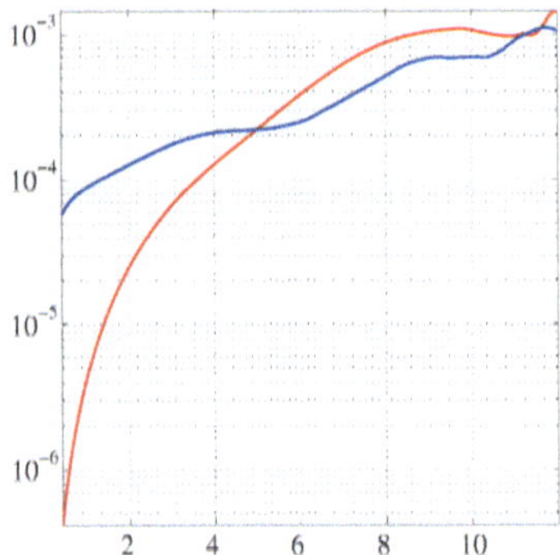

Fig. 14. Evolution of the magnitude of the chemotactic (blue line) and elastic (red line) forces averaged over the domain for the simulation reported in Fig. 15

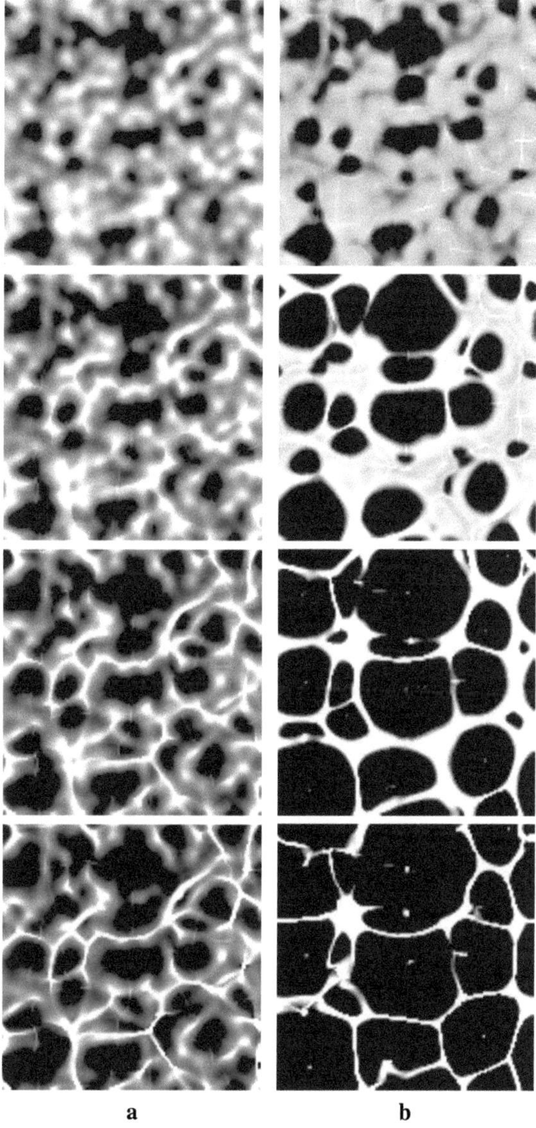

Fig. 15. Snapshots of the process of capillary network formation taken at different times as predicted by the chemomechanical model with $\kappa = 0.2$, $\gamma = 0$, and $a = 1$ and $b = 0$ in **a** and $a = 30$ and $b = 0.2$ in **b**. The initial condition is the same as that shown in Fig. 7

compared with $\kappa = 1$ in Fig. 12) and with Fig. 7 which is obtained in absence of any interaction with the substratum. We can also observe that, as expected, if the anchoring force is too weak, and therefore the chemotactic action prevalent, there is an acceleration in the formation of the cords.

We end this section by pointing out the effect of different VEGF production rates. Fig. 15a shows the formation of the vascular network using the usual production term $\alpha(n) = an$ with $a = 1$. In addition, in the interaction force $\kappa = 0.2$ and $\gamma = 0$. On the other hand, Fig. 15b is obtained for the same values as Fig. 12 but takes the effect of contact dependent production of VEGF into account. In fact, as already mentioned, cells might produce less VEGF because, upon aggregation, there is no need to recruit new cells communicating with the release of more chemical factors. Rather than the standard linear production the function $\alpha(n)$ is expressed by (9), with $a = 30$ and $b = 0.2$. It can be observed that the results of the two simulations present some differences. In particular, though the topology of the network is very similar, the saturation in the production term leads to neater structures.

Though we do not show it here, the model is still able to reproduce the transitions occurring at low and high densities, as with what is obtained in Fig. 10.

4.3 Exogenous control of vascular network formation

We now consider the case in which the formation of capillary networks is externally controlled by the use of exogenous chemoattractant (c_a) and chemorepellent (c_r), but neglecting substratum interaction, a problem studied by Lanza et al. [15].

Because diffusion is a much faster process then cell aggregation, the model can be simplified and written as:

$$\frac{\partial n}{\partial t} + \nabla \cdot (n\mathbf{v}_c) = 0,$$

$$\frac{\partial \mathbf{v}_c}{\partial t} + \mathbf{v}_c \cdot \nabla \mathbf{v}_c = \beta \nabla c + \beta_a \nabla c_a - \beta_r \nabla c_r - \gamma \mathbf{v}_c - \frac{1}{n} \nabla p(n), \tag{16}$$

$$D\nabla^2 c - \frac{c}{\tau} + \alpha n = 0, \tag{17}$$

$$D_a \nabla^2 c_a - \frac{c_a}{\tau_a} + s_a(t) H_a(\mathbf{x}) = 0, \tag{18}$$

$$D_r \nabla^2 c_r - \frac{c_r}{\tau_r} + s_r(t) H_r(\mathbf{x}) = 0. \tag{19}$$

Of course in particular cases it may be possible to integrate (18) and/or (19) so that the relative solution can be directly substituted in (16).

We have already remarked several times that the diffusion equation (17) introduces a characteristic length $\ell = \sqrt{D\tau}$ related to the size of the cords in the network structure. In the same way the other two diffusion equations (18) and (19) are characterized by two natural lengths, $\ell_a = \sqrt{D_a \tau_a}$ and $\ell_r = \sqrt{D_r \tau_r}$, related to the ranges of action of the exogenous chemoattractant and chemorepellent, respectively. We show that, within these ranges, the effect of the exogenous chemical factors strongly influence the structure of the network. On the other hand, outside these ranges endogenous chemotaxis governs the formation of a more isotropic network.

From the practical point of view this means that, having decided where to put the "spaghettis" or the "sponges" saturated with chemical factors, one can identify some strips around them where the effect of the exogenous chemical factors is felt, as shown in Fig. 16.

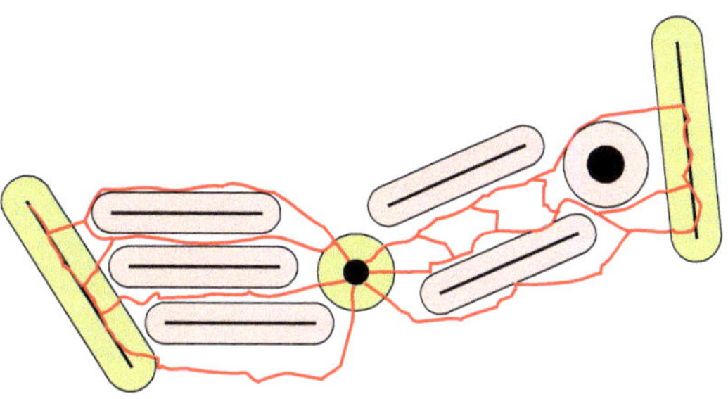

Fig. 16. Diagram illustrating the effect of the ranges of influence of the chemotactic factors. In pink the chemorepellent, in green the chemoattractant, in red a possible capillary network

As a first example consider the case in which the exogenous chemoattractant is located on two opposite sides of the domain, a situation which can be realised by putting sponges impregnated with chemoattractant on the border of the Petri dish. In this case Eq. (19) changes slightly as there is no source term and the concentration of chemoattractant in the sponges (assumed constant in time) represents the proper boundary condition for (19),

$$c_a(x = 0, y, t) = c_a(x = L, y, t) = \tilde{c}_b \quad \forall y \in [0, L], \forall t \geq 0,$$

together with periodic boundary conditions on the other two sides $y = 0, L$. In fact, in this case Eq. (18) can be solved readily so that the concentration

$$c_a = \tilde{c}_b \frac{e^{x/\ell_a} + e^{(L-x)/\ell_a}}{1 + e^{L/\ell_a}},$$

can be directly substituted in (16).

In the simulation presented in Fig. 17 the exogenous and endogenous chemoattractant were the same, so that $\ell_a = \ell = 0.196$ mm. Figure 17a then shows that, in a range ℓ from the sides $x = 0$ and $x = L$, capillaries tend to organise perpendicularly to the sides. At a distance of order ℓ they branch giving rise to a capillary network very similar to the one obtained in the endogenous case. The final structure resembles the capillary network between arteries and veins. Nonetheless, the comparison is simply qualitative as the mechanisms governing the remodelling of capillaries is probably different from that modelled here.

In Fig. 17b the chemoattractant is placed in the center x_0, i.e., $H_a(\mathbf{x}) = \delta(\mathbf{x} - \mathbf{x}_0)$. This forms a circular zone influenced by the chemical factors which is characterized by the formation of capillaries more or less arranged in the radial direction.

On the other hand, the simulation in Fig. 18 shows very clearly the action of chemorepellents. In particular, in Fig. 18a it is placed in the center of the domain. Cells then move away from the central region (more or less in a radial direction) accumulating in a moving circumference with faster cells catching up to slower ones. This process generates a circular capillary loop connected with the more isotropic structure outside it. The final size of the circumference, and therefore of the circular capillary loop, corresponds to the range of action of the chemorepellent. In fact, in

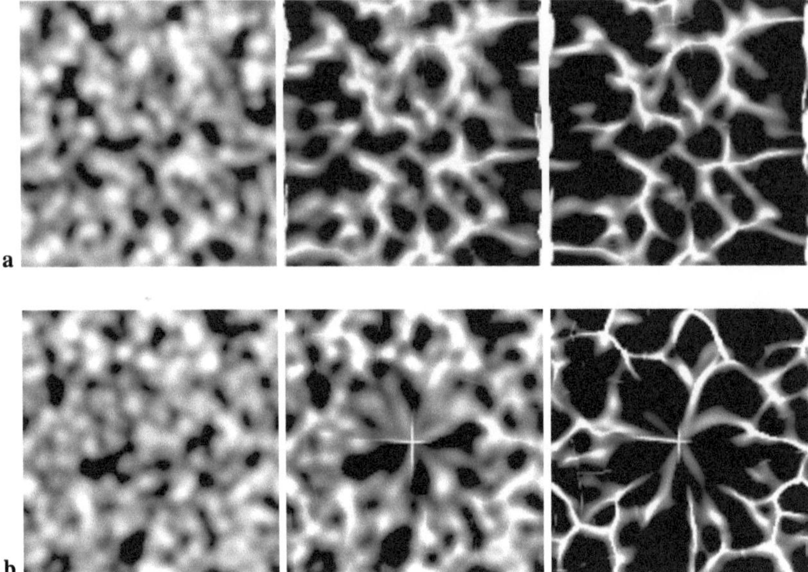

Fig. 17. Network formation influenced by an exogenous chemoattractant. In **a** it is placed on the right and on the left of the domain, in **b** in the center of the domain. Bars indicate the value of $\xi_a = 0.1$, i.e., the order of magnitude of the range of action of the exogenous chemoattractant

the simulation the values of the parameters give $\ell_r = 0.31$ mm, which is close to the theoretical value 0.316 mm.

In Fig. 18b the chemorepellent is placed in a central axis parallel to the y-axis. Also in this case cells move away from the central axis, along x, accumulating on two lines parallel to the y-axis at a distance close to the range of the chemorepellent. In this way a capillary parallel to the strip of chemorepellent is formed and connects with the outer network structure.

Again the size of the capillary-free region is nearly twice the range of action of the chemorepellent, actually a bit smaller (0.54 mm with respect to the theoretical value 0.632 mm).

In Fig. 18c three 1 mm long strips of chemorepellent are placed half a millimeter from each other. Again, cells are repelled from the strips moving in a perpendicular way and aligning in the "corridors" forming capillaries parallel to the strips. Outside the region influenced by the chemorepellent, the capillaries coalesce and connect to the external network.

In general, we can then say that chemoattractants induce, in their ranges of action, the formation of capillaries which tend to run perpendicularly to the source of chemoattractants, while chemorepellent induces the formation of capillaries which tend to run parallel to the source of chemorepellent, at a distance from the source of the order of magnitude of the range of action, as sketched in Fig. 16.

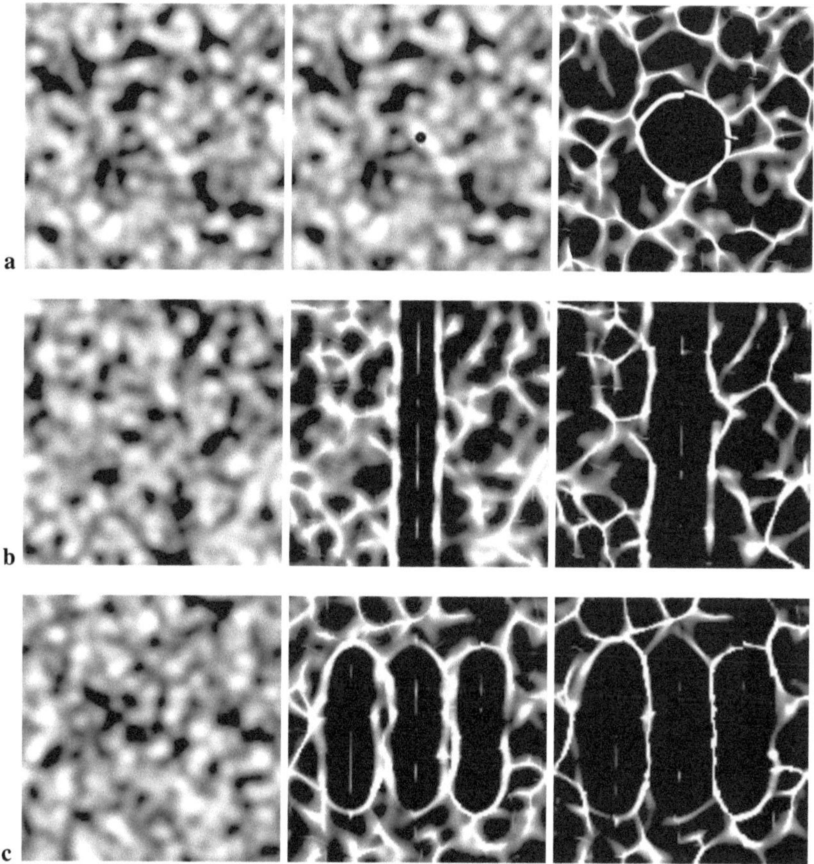

Fig. 18. Network formation influenced by an exogenous chemorepellent. In **a** the chemical factor is placed in the center, in **b** on the central axis of the domain. In **c** three $\frac{L}{2}$-long strips of chemorepellent are placed at a reciprocal distance $\frac{L}{4}$. Bars indicate the value of $\xi_r = 0.1$, i.e., the order of magnitude of the range of action of the exogenous chemorepellent

5 An angiogenesis model

As already mentioned in Sect. 1 another important process leading to the formation of vascular networks is angiogenesis, the recruitment of blood vessels from a pre-existing vasculature. Although this is a physiological process occurring, for instance, in wound healing, we focus here on tumor-induced angiogenesis, one of the most dangerous pathological aspects.

In fact, one of the crucial milestones in tumor development is the so-called *angiogenic switch*, i.e., the achieved ability of the tumor to trigger the formation of its own vascular network. In order to achieve this, the tumor cells first secrete angiogenic factors which in turn induce the endothelial cells of a neighboring blood vessel to degrade their basal lamina and begin to migrate towards the tumor. As they migrate, the endothelial cells develop sprouts which can then form loops and branches through which blood circulates. From these branches more sprouts form and the whole process repeats forming a capillary network. The biological process is described in more detail in [10].

In the literature there are several angiogenic models. Some of them are discussed in [5,16,19] where the interested reader can find more references. We focus here on a procedure introduced by Chaplain and Anderson [8] and by Sleeman and Wallis [28] which reproduces realistic capillary networks induced by a tumor.

Specifically, Chaplain and Anderson [8] study the problem by focusing on the evolution of:

- the endothelial cell density per unit area (n) at the tip of the capillary sprouts;
- the concentration c of Tumor Angiogenic Factors (TAF), e.g., VEGF;
- the concentration f of fibronectin, which, as already mentioned, is an important constituent of the extracellular matrix,

and focus on a fixed region outside the tumor.

As already seen in the previous section, the motion of the endothelial cells (at or near a capillary sprout-tip) is influenced by chemotaxis in response to TAF gradients. The chemotactic drift velocity can be taken to be of the form

$$\mathbf{v}_{chemo} = \beta(c)\nabla c, \tag{20}$$

where the receptor-kinetic law of the form

$$\beta(c) = \beta_0 \frac{c_M}{c_M + c},$$

is assumed, to reflect the fact that a cell's chemotactic sensitivity decreases with increased TAF concentration.

Also interactions between the endothelial cells and the extracellular matrix are found to be very important and to directly affect cell migration toward regions with larger amounts of extracellular matrix. The influence of fibronectin on the endothelial cells can then be modelled by the haptotactic drift velocity

$$\mathbf{v}_{hapto} = w(f)\nabla f, \tag{21}$$

where $w(f)$ is the haptotactic function. In the following, it is taken to be constant, $w(f) = w_f$.

We observe that the cell velocity obtained by the sum of (20) and (21),

$$\mathbf{v}_c = w(f)\nabla f + \beta(c)\nabla c, \tag{22}$$

might be obtained from Eq. (10) by neglecting its left-hand side and the effect due to the partial stress tensor for the cellular constituent of the mixture. In fact, considering the interaction force as given by (13) with $\mathbf{v}_s = 0$ corresponding to an undeformable substratum and

$$\mathbf{F}_c = \rho_c \tilde{w}(f)\nabla f + \rho_c \tilde{\beta}(c)\nabla c,$$

(see also (11)), one has

$$\rho_c \tilde{w}(f)\nabla f + \rho_c \tilde{\beta}(c)\nabla c - \rho_c \gamma \mathbf{v_c} = \mathbf{0},$$

which, by a suitable definition of the coefficients, leads to (22). In this way the usual chemotactic closure can be understood as a limit velocity obtained by balancing the chemotactic and haptotactic "pulling" with the "drag force" related to the difficulty of moving in the extracellular-matrix and of removing old adhesion sites while looking for new ones.

In their model Chaplain and Anderson [8] omit any birth and death terms because of the fact that they are focusing on the endothelial cells at the sprout-tips (where there is no proliferation) and that in general endothelial cells have a long half–life, on the order of months. Furthermore, they add a random motility of the endothelial cells so that the equation for the endothelial cell density n can be written as

$$\frac{\partial n}{\partial t} + \nabla \cdot \left[\left(\beta(c)\nabla c + w_f \nabla f \right) n \right] = k_n \nabla^2 n. \tag{23}$$

The evolution of the concentration of TAF and fibronectin is assumed to satisfy locally

$$\begin{aligned} \frac{\partial c}{\partial t} &= -\delta_c n c, \\ \frac{\partial f}{\partial t} &= \gamma n - \delta_f n f, \end{aligned} \tag{24}$$

where δ_c and δ_f are the uptake cofficients from the endothelial cells and γ is the production rate of fibronectin by the cells.

As initial conditions the concentration of fibronectin is taken to be constant, while the concentration of TAF, which is produced by the tumor located at the boundary of the domain, is taken to satisfy

$$k_c \nabla^2 c - \frac{c}{\tau} = 0,$$

where k_c is the diffusivity of TAF and τ its half life. The concentration of TAF at the boundary is a constant value where the tumor is located, e.g., in a central interval of one side or on the entire side, simulating a large tumor. Elsewhere no-flux boundary conditions are applied.

Therefore, the TAF secreted by the tumor diffuses into the surrounding tissue and sets up the initial concentration gradient between the tumor and any pre-existing vasculature, which is responsible for the directionality in the formation of the new capillaries. Later on, endothelial cells take up TAF. However, diffusion is neglected.

The model (23), (24) is considered to hold on a square spatial domain of side L with the parent blood vessel (e.g., limbal vessel) located along one side of the domain and the tumor located on the opposite side, either over the entire length or over part of it. Cells, and consequently the capillary sprouts, are assumed to remain within the domain of tissue under consideration and therefore no-flux boundary conditions of the form

$$\mathbf{n} \cdot \left[-k_n \nabla n + n \left(\beta(c) \nabla c + w_f \nabla f \right) \right] = 0$$

are imposed on the boundaries of the square, where $\mathbf{n}$ is the outward unit normal vector.

The aim of the technique used by Chaplain and Anderson [8] and by Sleeman and Wallis [28] which develops in the framework of a reinforced random walk, is to follow the path of the endothelial cells at the capillary sprout-tip in a discrete fashion. In order to do that they used a discretized version of the continuous model above. They then used the resulting coefficients of the five-point stencil of the standard central finite-different scheme to generate the probabilities of movement of an individual cell in response to the chemoattractant gradients and to diffusion. We briefly sketch the procedure in two dimensions, because the generalization to three dimensions is technical.

If P_0 is related to the probability of the cell of being stationary and to the probability of cells of moving away from the node $\{i, j\}$ to one of its neighbors, P_1 is related to the probability of new cells coming from the node to the right, and similarly for the others, then one can write

$$\begin{aligned} n^{i+1}_{j,k} &= P_0 n^i_{j,k} + P_1 n^i_{j+1,k} + P_2 n^i_{j-1,k} + P_3 n^i_{j,k+1} + P_4 n^i_{j,k-1}, \\ c^{i+1}_{j,k} &= \left(1 - \Delta t \delta_c n^i_{j,k}\right) c^i_{j,k}, \\ f^{i+1}_{j,k} &= \left(1 - \Delta t \delta_f n^i_{j,k}\right) f^i_{j,k} - \Delta t \gamma n^i_{j,k}, \end{aligned} \tag{25}$$

where, for instance,

$$\begin{aligned} P_0 = 1 &- \frac{4\Delta t k_n}{\Delta x^2} + \frac{\Delta t}{4\Delta x^2} \frac{\beta_0 c_M}{(c_M + c^i_{j,k})^2} \left[(c^i_{j+1,k} - c^i_{j-1,k})^2 + (c^i_{j,k+1} + c^i_{j,k-1})^2 \right] \\ &- \frac{\Delta t}{\Delta x^2} \left[\frac{\beta_0 c_M}{c_M + c^i_{j,k}} \left(c^i_{j+1,k} + c^i_{j-1,k} + c^i_{j,k+1} + c^i_{j,k-1} - 4c^i_{j,k} \right) \right. \\ &\left. + w_f \left(f^i_{j+1,k} + f^i_{j-1,k} + f^i_{j,k+1} + f^i_{j,k-1} - 4 f^i_{j,k} \right) \right], \\ P_1 = &\frac{\Delta t k_n}{\Delta x^2} - \frac{\Delta t}{4\Delta x^2} \left[\frac{\beta_0 c_M}{c_M + c^i_{j,k}} (c^i_{j+1,k} - c^i_{j-1,k}) + w_f (f^i_{j+1,k} - f^i_{j-1,k}) \right]. \end{aligned} \tag{26}$$

In particular, if there is no chemical gradient, the situation is isotropic and the probabilities $P_1, \cdots, P_4$ of moving in any direction are equal. Even in this case, the extraction of a random number decides whether the tip cell stays still or moves to a particular neighboring node rather than another. On the other hand, in presence of a chemical gradient the random walk becomes biased, because the cell has higher probabilities to move up the gradients of chemical factors.

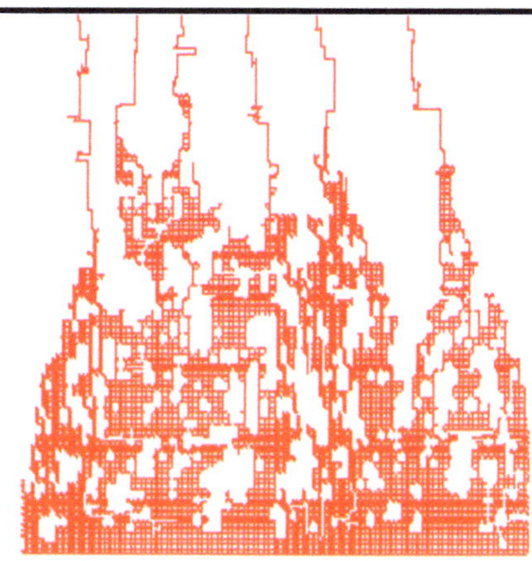

Fig. 19. Typical tumor-induced capillary network as reproduced by the procedure in (25), (26)

In addition, the discretized set-up permits the inclusion of phenomena difficult to describe using a model based on partial differential equations, e.g., capillary *branching* and *anastomosis*, i.e., the formation of capillary loops. In particular, Chaplain and Anderson [8] assumed that the density of endothelial cells necessary to allow capillary branching is inversely proportional to the distance from the tumor and proportional to the concentration of TAF. However, in order to branch a minimal distance from the previous bifurcation is needed, and of course there must be enough space in the discretized space to allow the formation of a new capillary. This assumption is consistent with the observation that the distance between successive branches along the capillaries decreases when the tumor is approached. This phenomenon is called the *brush border effect*, an effect well-described by the model and the simulation, as shown in Fig. 19.

In this approach it is even easier to describe anastomosis. When, during its motion, a capillary tip meets another capillary, then they merge to form a loop. If two sprout tips meet, then only one of the original sprouts continues to grow.

As shown in Fig. 19, the capillary networks built in this way look very realistic and compare well with what is observed experimentally. A three-dimensional animation of the angiogenic process is available at the web site http://www.maths.dundee.ac.uk/~sanderso/3d/index.html.

6 Future perspectives

This chapter is devoted to the presentation of some recent modelling approaches aimed at the description of the formation of capillary networks, with an awareness of the fact that, as reviewed in [2], there are other models of biological mechanisms that can give rise to network-like structure, starting from the seminal work by Meinhardt [21–23] and by Murray and coworkers (see, for instance, [18, 25]).

We have shown how the models presented in this chapter are able to reproduce in silico the capillary structures with correct dimensional characteristics and transitions. However, in our opinion this is only a first step.

In fact, important developments would be achieved by interfacing these models with others by considering, on the one hand, phenomena occuring at the cellular scale and, on the other hand, macroscopic effects and interactions with the surrounding environment. For instance, in the first framework, one could:

- consider that along the developed capillary network the permeability of the vessel wall changes, resulting in an increased perfusion and interstitial pressure;
- consider that new-born vessels are immature and therefore subject to mechanical collapse due, for instance, to the pressure exerted by a tumor growing around them;
- consider in more detail the remodelling process giving rules to close unnecessary branches (see, e.g. [1] and its references);
- consider adhesive properties of endothelial cells (see [24]);
- consider more closely the receptor dynamics involved, for instance, in the amplification of the chemotactic signal or in its saturation;
- consider the protein cascade linked to the VEGF-receptor to link closely the action (VEGF) to the reaction (motion and proliferation).

In the latter framework one could, for instance:

- simulate the diffusion of drugs in the capillary network as done by Stephanou et al. [29] and the diffusion of drugs and nutrients in the tissue;
- link the approach above with the one dealing with the development of tumor chords [6] and reviewed in this volume in the chapter by Fasano, Bertuzzi and Gandolfi to develop multiscale models of vascularized tumors;
- describe the oxygenation of tissues and the link between capillary distribution, hypoxic regions and remodelling.

Of course, the two directions are not mutually exclusive. For instance, in order to describe properly the perfusion of nutrients and drugs, one should take into account changes in wall permeability; also the remodelling process can give rise to the formation of temporary hypoxic regions which trigger back the formation of new vessels.

In fact, developing models which range from the sub-cellular to the tissue level is in our opinion one of the most fascinating problems in theoretical medicine to be developed in the future.

Appendix: 2D reduction of the substratum equation for the vasculogenesis model

In describing the response of the substratum to the pulling of endothelial cells in the vasculogenesis process, we can exploit the fact that the size of the Petri dish is at least one order of magnitude larger than the thickness of the layer ($\approx 50\,\mu\text{m}$). For this reason, it is convenient to reduce the force balance equation by considering the substratum as two-dimensional.

We start from the equilibrium equation for the Cauchy three-dimensional continuum, namely,

$$\hat{\nabla} \cdot \hat{\mathbf{T}} + \hat{\mathbf{f}} = \mathbf{0}, \tag{27}$$

where, e.g., $\hat{\mathbf{T}}$ is a 3×3 tensor and $\hat{\nabla}$ operates in three-dimensional space.

Integrating over the thickness of the substratum, i.e., over the interval $[0, h]$, one obtains

$$\int_0^h \hat{\nabla} \cdot \hat{\mathbf{T}} \, \mathrm{d}z + \int_0^h \hat{\mathbf{f}} \, \mathrm{d}z = \mathbf{0}. \tag{28}$$

Confining attention to the first term and using the tensorial notation, one can write

$$\int_0^h \hat{\nabla} \cdot \hat{\mathbf{T}} \, \mathrm{d}z = \left(\int_0^h \frac{\partial \hat{T}_{ij}}{\partial x_j} \mathrm{d}x_3 \right) \mathbf{e}_i, \quad i, j = 1, 2, 3, \tag{29}$$

where we have set $(x, y, z) = (x_1, x_2, x_3)$ and $(\mathbf{i}, \mathbf{j}, \mathbf{k}) = (\mathbf{e}_1, \mathbf{e}_2, \mathbf{e}_3)$ for sake of clarity and where the Einstein convention is used. In particular,

$$\begin{aligned} \int_0^h \frac{\partial \hat{T}_{ij}}{\partial x_j} \mathrm{d}x_3 &= \int_0^h \left(\frac{\partial \hat{T}_{i1}}{\partial x_1} + \frac{\partial \hat{T}_{i2}}{\partial x_2} \right) \mathrm{d}x_3 + \left[\hat{T}_{i3} \right]_0^h \\ &= \frac{\partial}{\partial x_1} \int_0^h \hat{T}_{i1} \, \mathrm{d}x_3 + \frac{\partial}{\partial x_2} \int_0^h \hat{T}_{i2} \, \mathrm{d}x_3 + \left[\hat{T}_{i3} \right]_0^h, \quad i = 1, 2, 3. \end{aligned}$$

Defining the mean stresses per unit of length as

$$T_{ij} := \int_0^h \hat{T}_{ij} \, \mathrm{d}x_3, \quad i, j = 1, 2, 3, \tag{30}$$

Eq. (29) takes the form

$$\int_0^h \nabla \cdot \hat{\mathbf{T}} \, \mathrm{d}x_3 = \left(\frac{\partial T_{i1}}{\partial x_1} + \frac{\partial T_{i2}}{\partial x_2} + \left[\hat{T}_{i3} \right]_0^h \right) \mathbf{e}_i.$$

The second term in the left-hand side of (28) can be treated analogously

$$\int_0^h \hat{\mathbf{f}} \, \mathrm{d}z = \left(\int_0^h \hat{f}_i \, \mathrm{d}x_3 \right) \mathbf{e}_i = f_i \mathbf{e}_i.$$

It is now convenient to split Eq. (28) into the system

$$\begin{aligned} &\nabla \cdot \mathbf{T} + \mathbf{f} + \left[\begin{pmatrix} \hat{T}_{13} \\ \hat{T}_{23} \end{pmatrix} \right]_0^h = \mathbf{0}, \\ &\left(\frac{\partial T_{31}}{\partial x_1} + \frac{\partial T_{32}}{\partial x_2} \right) + f_3 + \left[\hat{T}_{33} \right]_0^h = 0, \end{aligned} \tag{31}$$

where

$$\mathbf{T} := \begin{pmatrix} T_{11} & T_{12} \\ T_{21} & T_{22} \end{pmatrix} \quad \text{and} \quad \mathbf{f} := \begin{pmatrix} f_1 \\ f_2 \end{pmatrix},$$

which define the reduced stress tensor and the reduced forcing term, and where ∇ is the operator on the plane. Note that the main formal difference between (27) and (31) consists in the boundary terms which appear in the second equation.

Assuming for simplicity that the substratum behaves like an elastic material,

$$\hat{\mathbf{T}} = 2\mu\hat{\mathbf{E}} + \lambda\left(\operatorname{tr}\hat{\mathbf{E}}\right)\hat{\mathbf{I}}, \tag{32}$$

where μ, λ denote the Lamè coefficients and $\hat{\mathbf{E}} = \frac{1}{2}(\hat{\nabla}\hat{\mathbf{u}} + \hat{\nabla}\hat{\mathbf{u}}^T)$, by integrating (32) over the thickness of the substratum one has

$$\int_0^h \hat{T}_{ij}\,\mathrm{d}x_3 = 2\mu\int_0^h \hat{E}_{ij}\,\mathrm{d}x_3 + \lambda\left(\int_0^h \hat{E}_{kk}\,\mathrm{d}x_3\right)\delta_{ij}, \quad i,\ j = 1,\ 2,\ 3,$$

which, after the obvious definition of the mean strains

$$E_{ij} := \frac{1}{h}\int_0^h \hat{E}_{ij}\,\mathrm{d}x_3,$$

yields

$$T_{ij} = 2\mu h E_{ij} + \lambda h E_{kk}\delta_{ij}, \quad i,\ j = 1,\ 2,\ 3.$$

This relation can be formally rewritten by the same splitting adopted in (31):

$$\begin{aligned} \mathbf{T} &= 2\mu h\mathbf{E} + \lambda h\,(\operatorname{tr}\mathbf{E} + E_{33})\,\mathbf{I}, \\ T_{3j} &= 2\mu h E_{3j} + \lambda h E_{kk}\delta_{3j}, \quad j = 1,\ 2,\ 3, \end{aligned} \tag{33}$$

where

$$\mathbf{E} := \begin{pmatrix} E_{11} & E_{12} \\ E_{21} & E_{22} \end{pmatrix}$$

represents the reduced strain tensor and $\mathbf{I}$ denotes the 2×2 identity matrix. Thanks to the symmetry of $\hat{\mathbf{T}}$ and to Eq. (30), one has $T_{3j} = T_{j3}$, and then Eq. (33) effectively allow us to express the constitutive relation of T_{ij} for all $i,\ j = 1,\ 2,\ 3$.

Finally, integrating the usual kinematic relation over the thickness of the substratum gives

$$\begin{aligned}
\int_0^h \hat{E}_{ij}\,\mathrm{d}x_3 &= \frac{1}{2}\left\{\int_0^h \frac{\partial \hat{u}_i}{\partial x_j}\,\mathrm{d}x_3 + \int_0^h \frac{\partial \hat{u}_j}{\partial x_i}\,\mathrm{d}x_3\right\} \\
&= \frac{1}{2}\left\{\left[\frac{\partial}{\partial x_j}\left(\int_0^h \hat{u}_i\,\mathrm{d}x_3\right)(1-\delta_{j3}) + [\hat{u}_i]_0^h\,\delta_{j3}\right]\right. \\
&\quad \left. + \left[\frac{\partial}{\partial x_i}\left(\int_0^h \hat{u}_j\,\mathrm{d}x_3\right)(1-\delta_{i3}) + [\hat{u}_j]_0^h\,\delta_{i3}\right]\right\} \\
&= \frac{1}{2}\left\{\frac{\partial}{\partial x_j}\left(\int_0^h \hat{u}_i\,\mathrm{d}x_3\right)\right. \\
&\quad + \frac{\partial}{\partial x_i}\left(\int_0^h \hat{u}_j\,\mathrm{d}x_3\right) + \left([\hat{u}_i]_0^h - \frac{\partial}{\partial x_j}\int_0^h \hat{u}_i\,\mathrm{d}x_3\right)\delta_{j3} \\
&\quad \left. + \left([\hat{u}_j]_0^h - \frac{\partial}{\partial x_i}\int_0^h \hat{u}_j\,\mathrm{d}x_3\right)\delta_{i3}\right\},
\end{aligned}$$

which, on introducing the mean displacements

$$u_i := \frac{1}{h}\int_0^h \hat{u}_i\,\mathrm{d}x_3, \tag{34}$$

allows us to write

$$E_{ij} = \frac{1}{2}\left\{\frac{\partial u_i}{\partial x_j} + \frac{\partial u_j}{\partial x_i} + \left(\frac{1}{h}[\hat{u}_i]_0^h - \frac{\partial u_i}{\partial x_j}\right)\delta_{j3} + \left(\frac{1}{h}[\hat{u}_j]_0^h - \frac{\partial u_j}{\partial x_i}\right)\delta_{i3}\right\},$$

and then to split it into the system

$$\begin{aligned}
\mathbf{E} &= \frac{1}{2}\left(\nabla\mathbf{u} + \nabla\mathbf{u}^{\mathrm{T}}\right) \\
E_{3j} &= \frac{1}{2}\left[\frac{\partial u_3}{\partial x_j} + \frac{1}{h}[\hat{u}_j]_0^h + \left(\frac{1}{h}[\hat{u}_3]_0^h - \frac{\partial u_3}{\partial x_j}\right)\delta_{j3}\right], \quad j = 1,\,2,\,3,
\end{aligned} \tag{35}$$

with $\mathbf{u} := (u_1,\ u_2)^{\mathrm{T}}$; in particular

$$E_{33} = \frac{1}{h}[\hat{u}_3]_0^h.$$

Substituting (35) into (33) yields

$$\mathbf{T} = \mu h\left(\nabla\mathbf{u} + (\nabla\mathbf{u})^{\mathrm{T}}\right) + \lambda h\left(\nabla\cdot\mathbf{u} + \frac{1}{h}[\hat{u}_3]_0^h\right)\mathbf{I},$$

$$T_{3j} = \mu h\left(\frac{\partial u_3}{\partial x_j} + \frac{1}{h}[\hat{u}_j]_0^h\right) + h\left(\lambda\nabla\cdot\mathbf{u} + \frac{\mu+\lambda}{h}[\hat{u}_3]_0^h - \mu\frac{\partial u_3}{\partial x_j}\right)\delta_{3j},$$

$$j = 1,\,2,\,3;$$

with this result, Eq. (31) finally specializes to

$$h\mu\nabla^2\mathbf{u} + h(\lambda+\mu)\nabla(\nabla\cdot\mathbf{u}) + \mathbf{f} + \left[\begin{pmatrix}\hat{T}_{13}\\ \hat{T}_{23}\end{pmatrix} + \lambda\nabla\hat{u}_3\right]_0^h = \mathbf{0}, \tag{36}$$

$$h\mu\nabla^2 u_3 + f_3 + \left[\hat{T}_{33} + \mu\left(\frac{\partial\hat{u}_1}{\partial x_1} + \frac{\partial\hat{u}_2}{\partial x_2}\right)\right]_0^h = 0. \tag{37}$$

Note that Eqs. (36) and (37) are mutually independent with respect to the variables $\mathbf{u}$ and u_3, provided that the respective forcing terms $\mathbf{f}$ and f_3 do not depend on u_3 and on u_1, u_2. In this hypothesis, we can restrict our analysis to Eq. (36), which suffices by itself to obtain a two-dimensional model of the substratum, but we need to characterize the boundary term

$$\left[\begin{pmatrix}\hat{T}_{13}\\ \hat{T}_{23}\end{pmatrix} + \lambda\hat{\nabla}\hat{u}_3\right]_0^h \tag{38}$$

via the new variable $\mathbf{u}$. For this, we refer to the original three-dimensional problem and we make the following assumptions:

1. a no-slip condition of the substratum on the bottom which can be interpreted as its adhesion to the underlying Petri dish:

$$\hat{u}_1 = \hat{u}_2 = \hat{u}_3 = 0 \quad \text{for} \quad x_3 = 0; \tag{39}$$

2. because of cell motion, an imposed shear stress and a zero normal stress on the top coming from the hypothesis that cells move on the surface of the substratum without penetrating it:

$$\begin{pmatrix}\hat{T}_{13}\\ \hat{T}_{23}\end{pmatrix} = \mathbf{t}_{\text{cell}}, \quad T_{33} = 0 \quad \text{for} \quad x_3 = h. \tag{40}$$

Equation (39) then gives $\nabla\hat{u}_3 = \mathbf{0}$ for $x_3 = 0$ which, together with Eq. (40), allows us to rewrite the boundary term (38) as

$$\mathbf{t}_{\text{cell}} + \lambda\nabla\hat{u}_3\big|_{x_3=h} - \begin{pmatrix}\hat{T}_{13}\\ \hat{T}_{23}\end{pmatrix}\bigg|_{x_3=0}. \tag{41}$$

Furthermore, using the linear elastic constitutive relation (32) we obtain

$$\nabla u_3 = -\left(\frac{\partial\hat{u}_1}{\partial x_3}, \frac{\partial\hat{u}_2}{\partial x_3}\right)^{\mathrm{T}} + \frac{1}{\mu}\begin{pmatrix}\hat{T}_{13}\\ \hat{T}_{23}\end{pmatrix},$$

which yields

$$\lambda\nabla\hat{u}_3\big|_{x_3=h} = -\lambda\left(\frac{\partial\hat{u}_1}{\partial x_3}, \frac{\partial\hat{u}_2}{\partial x_3}\right)^{\mathrm{T}}\bigg|_{x_3=h} + \frac{\lambda}{\mu}\mathbf{t}_{\text{cell}},$$

so that Eq. (41) specializes to

$$\frac{\mu+\lambda}{\mu}\mathbf{t}_{\text{cell}} - \lambda\left(\frac{\partial \hat{u}_1}{\partial x_3}, \frac{\partial \hat{u}_2}{\partial x_3}\right)^{\mathrm{T}}\Bigg|_{x_3=h} - \begin{pmatrix}\hat{T}_{13}\\ \hat{T}_{23}\end{pmatrix}\Bigg|_{x_3=0},$$

and Eq. (36) becomes

$$h\mu\nabla^2\mathbf{u} + h(\lambda+\mu)\nabla\left(\nabla\cdot\mathbf{u}\right) + \mathbf{f} + \frac{\mu+\lambda}{\mu}\mathbf{t}_{\text{cell}} - \lambda\left(\frac{\partial \hat{u}_1}{\partial x_3}, \frac{\partial \hat{u}_2}{\partial x_3}\right)^{\mathrm{T}}\Bigg|_{x_3=h} - \begin{pmatrix}\hat{T}_{13}\\ \hat{T}_{23}\end{pmatrix}\Bigg|_{x_3=0} = \mathbf{0}.$$

The procedure above holds for any three-dimensional linear elastic body. The thin layer assumption has not yet been used, but its application allows us to approximate the displacements $\hat{\mathbf{u}}$ as a linear function of x_3:

$$\hat{u}_i\left(x_1, x_2, x_3\right) = \varphi_i\left(x_1, x_2\right)x_3, \quad i = 1, 2, 3,$$

where we used the no-slip condition at the interface between the substratum and the Petri dish.

According to Eq. (34),

$$u_i = \frac{1}{h}\varphi_i\int_0^h x_3\,\mathrm{d}x_3 = \frac{h}{2}\varphi_i,$$

which gives $\varphi_i = \frac{2}{h}u_i$ and then

$$\hat{u}_i\left(x_1, x_2, x_3\right) = \frac{2}{h}u_i\left(x_1, x_2\right)x_3. \tag{42}$$

Thanks to (42) we can now express

$$\begin{pmatrix}\dfrac{\partial \hat{u}_1}{\partial x_3}\\[2ex] \dfrac{\partial \hat{u}_2}{\partial x_3}\end{pmatrix}\Bigg|_{x_3=h} = \frac{2}{h}\begin{pmatrix}u_1\\ u_2\end{pmatrix},$$

$$\begin{pmatrix}\hat{T}_{13}\\ \hat{T}_{23}\end{pmatrix}\Bigg|_{x_3=0} = 2\mu\begin{pmatrix}\hat{E}_{13}\\ \hat{E}_{23}\end{pmatrix}\Bigg|_{x_3=0} = \frac{2\mu}{h}\begin{pmatrix}u_1\\ u_2\end{pmatrix},$$

and finally obtain

$$h\mu\nabla^2\mathbf{u} + h(\lambda+\mu)\nabla\left(\nabla\cdot\mathbf{u}\right) + \mathbf{f} + \mathbf{f}_{\text{cell}} - \frac{s}{h}\mathbf{u} = \mathbf{0}, \tag{43}$$

where

$$\mathbf{f}_{\text{cell}} = \frac{\mu+\lambda}{\mu}\tau_{\text{cell}}, \quad s = 2(\mu+\lambda).$$

We remark that (43) is characterized by the presence of body forces which in fact derive from the boundary conditions. In fact, the term σ_{cell} comes from the cell pulling at the top surface and the term $-\frac{s}{h}\mathbf{u}$ is a consequence of the shear stress produced at the lower boundary of the substratum by the adhesion of the Petri dish. In Sect. 4.3 $\sigma_{\text{cell}} = \mathbf{F}_{\text{visc}} + \mathbf{F}_{\text{elast}}$.

Acknowledgements

This work was funded by the EU through the Marie Curie Research Training Network Project MRTN-CT-2004-503661 "Modelling, mathematical methods and computer simulation for tumor growth and therapy". We would like to thank D. Ambrosi, A. Gamba, V. Lanza, and A. Tosin for their work and for frequent discussions.

References

[1] Alarcón, T., Byrne, H.M., Maini, P.K.: Towards whole-organ modelling of tumour growth. Prog. in Biophys. Biol. **85**, 451–472 (2004)

[2] Ambrosi, D., Bussolino, F., Preziosi, L.: A review of vasculogenesis models. J. Theor. Med. **6**, 1–19 (2005)

[3] Ambrosi, D., Gamba, A., Serini, G.: Cell directional persistence and chemotaxis in vascular morphogenesis. Bull. Math. Biol. **66**, 1851–1873 (2004)

[4] Ambrosi, D., Mollica, F.: Mechanical models in tumour growth. In: Preziosi, L. (ed.): Cancer modelling and simulation. Boca Raton, FL: Chapman & Hall/CRC 2003, pp. 121–145

[5] Bellomo, N., De Angelis, E., Preziosi, L.: Multiscale modeling and mathematical problems related to tumour evolution and medical therapy. J. Theor. Med. **5**, 111–136 (2004)

[6] Bertuzzi, A., D'Onofrio, A., Fasano, A., Gandolfi, A.: Regression and regrowth of tumour cords following single-dose anticancer treatment. Bull. Math. Biol. **65**, 903–931 (2003)

[7] Bussolino, F., Arese, M., Audero, E., Giraudo, E., Marchiò, S., Mitola, S., Primo, L., Serini, G.: Biological aspects of tumour angiogenesis. In: Preziosi, L. (ed.): Cancer modelling and simulation. Boca Raton, FL: Chapman & Hall/CRC 2003, pp. 1–22

[8] Chaplain, M.A.J., Anderson, A.R.A.: Continuous and discrete mathematical models of tumor-induced angiogenesis. Bull. Math. Biol. **60**, 857–899 (1998)

[9] Coniglio, A., de Candia, A., Di Talia, S., Gamba, A.: Percolation and Burgers' dynamics in a model of capillary formation. Phys. Rev. E **69**, 051910, 10p. (2004)

[10] Folkman, J., Haudenschild, C.: Angiogenesis in vitro. Nature **288**, 551–556 (1980)

[11] Gamba, A., Ambrosi, D., Coniglio, A., de Candia, A., Di Talia, S., Giraudo, E., Serini, G., Preziosi, L., Bussolino, F.: Percolation, morphogenesis and Burgers dynamics in blood vessels formation. Phys. Rev. Lett. **90**, 118101 (2003)

[12] Holmes, M., Sleeman, B.: A mathematical model of tumour angiogenesis incorporating cellular traction and viscoelastic effects. J. Theor. Biol. **202**, 95–112 (2000)

[13] Kowalczyk, R.: Preventing blow-up in a chemotaxis model. J. Math. Anal. Appl. **305**, 566–588 (2005)

[14] Kowalczyk, R., Gamba, A., Preziosi, L.: On the stability of homogeneous solutions to some aggregation models. Discrete Contin. Dyn. Syst. B **4**, 203–220 (2004)

[15] Lanza, V., Ambrosi, D., Preziosi, L.: Exogenous control of vascular network formation in vitro. Preprint. Turin: Dip. di Mathematica, Politecnico di Torino 2005

[16] Levine, H., Sleeman, B., Modelling tumour-induced angiogenesis. In: Preziosi, L. (ed.): Cancer modelling and simulation. Boca Raton, FL: Chapman & Hall/CRC 2003, pp. 147–184

[17] Levine, H., Sleeman, B., Nilsen-Hamilton, M.: Mathematical modeling of the onset of capillary formation initiating angiogenesis. J. Math. Biol. **42**, 195–238 (2001)

[18] Manoussaki, D., Lubkin, S.R., Vernon, R.B., Murray, J.D.: A mechanical model for the formation of vascular networks in vitro. Acta Biotheoretica **44**, 271–282 (1996)

[19] Mantzaris, N., Webb, S., Othmer, H.G.: Mathematical modeling of tumor-induced angiogenesis. J. Math. Biol. **49**, 111–187 (2004)

[20] McDougall, S.R., Anderson, A.R., Chaplain, M.A.J., Sherratt, J.A.: Mathematical modelling of flow through vascular networks: implications for tumour-induced angiogenesis and chemotherapy strategies. Bull. Math. Biol. **64**, 673–702 (2002)

[21] Meinhardt, H.: Morphogenesis of lines and nets. Differentiation **6**, 117–123 (1976)

[22] Meinhardt, H.: Models of biological pattern formation. London: Academic 1982

[23] Meinhardt, H.: Biological pattern formation as a complex dynamic phenomenon. Internat. J. Bifurcation Chaos Appl. Sci. Engrg. **7**, 1–26 (1997)

[24] Merks, R.M.H., Newman, S.A., Glazier, J.A.: Cell-oriented modeling of in vitro capillary development. In: Sloot, P. et al. (eds.): Cellular automata. (Lecture Notes in Comput. Sci. **3305**) Berlin: Springer 2004, pp. 425–434

[25] Murray, J.D., Manoussaki, D., Lubkin, S.R., Vernon, R.B.: A mechanical theory of in vitro vascular network formation. In: Little, C. et al. (eds.): Vascular morphogenesis: in vivo, in vitro, in mente. Boston: Birkhäuser Boston 1998, pp. 147–172

[26] Ruhrberg, C., Gerhardt, H., Golding, M., Watson, R., Ioannidou, S., Fujisawa, H., Betsholtz, C., Shima, D.: Spatially restricted patterning cues provided by heparin-binding VEGF-A control blood vessel branching morphogenesis. Genes and Devel. **16**, 2684–2698 (2002)

[27] Serini, G., Ambrosi, D., Giraudo, E., Gamba, A., Preziosi, L., Bussolino, F.: Modeling the early stages of vascular network assembly. EMBO J. **22**, 1771–1779 (2003)

[28] Sleeman, B., Wallis, I.P.: Tumour induced angiogenesis as a reinforced random walk: modelling capillary network formation without endothelial cell proliferation. Math. Comput. Modelling **36**, 339–358 (2002)

[29] Stephanou, A., McDougall, S.R., Anderson, A.R.A., Chaplain, M.A.J.: Mathematical modelling of flow in 2D and 3D vascular networks: applications to anti-angiogenic and chemotherapeutic drug strategies. Math. Comput. Modelling **41**, 1137–1156 (2005)

[30] Tosin, A., Ambrosi, D., Preziosi, L.: Mechanics and chemotaxis in the morphogenesis of vascular networks. Preprint Turin: Dip. di Mathematica, Politecnico di Torino 2005

[31] Vailhé, B., Vittet, D., Feige, J.-J.: In vitro models of vasculogenesis and angiogenesis. Lab. Investig. **81**, 439–452 (2001)

[32] Vailhé, B., Lecomte, M., Wiernsperger, N., Tranqui, L.: The formation of tubular structures by endothelial cells is under the control of fibrinolysis and mechanical factors. Angiogenesis **2**, 331–344 (1998)

Numerical methods for delay models in biomathematics

A. Bellen, N. Guglielmi, S. Maset

Abstract. In this chapter we direct attention to mathematical models based on delay differential equations and discuss two different approaches for their numerical approximation. Delay differential equations provide an important way of describing the time evolution of biological systems whose rate of change also depends on their configuration at previous time instances. As a significant example we review a mathematical model, due to Waltman, which describes the mechanisms by which antibodies are produced by the immune system in response to an antigen challenge. Our main goal is to emphasize the main difficulties arising in the numerical integration of such models as compared to those based on ordinary differential equations.This is done in the introduction, where a general approach is described and several aspects which are peculiar to delay differential equations are discussed. Afterwards we present two different numerical approaches, one mainly designed to solve stiff problems, as in the Waltman model, and the other for solving non-stiff problems. Numerical codes implementing the proposed approaches are also referred to.

Keywords: delay differential equations, continuous Runge-Kutta methods, Radau IIA methods, functional Runge-Kutta methods, state dependent delay, Waltman model.

1 Introduction

Models involving retarded ordinary and partial differential equations with both discrete and distributed delays are frequently encountered in mathematical biology. Introducing delays in the models has shown itself to be a powerful tool for investigating qualitative behavior of control systems and, in general, for simulating evolution phenomena in many branches of medicine and biology. Introduction of delays allowed us to improve models by taking into account important aspects previously neglected and to face more complicated phenomena based on feedback control. So, for instance, in a more realistic model for the spread of infections in large scale epidemics, a delay term may take into account the incubation period in the transmission of diseases via contacts among individuals. It was shown that episodes of periodic hematological diseases can be caused by anormalities in the feedback mechanism which regulate blood-cell numbers and, under appropriate conditions, this feedback mechanism can produce aperiodic irregular (chaotic) fluctuations. Recently, in interaction analysis between cardiovascular and respiratory function, clever models have been considered that take into account the time necessary for tissue venous blood to reach the lungs and vice versa. In the biomathematical literature there are many examples where the presence of delays makes the mathematical models much more reliable and consistent with real phenomena and laboratory observations. Indeed, the dynamics of equations including retarded arguments is much richer and this makes the models more realistic for simulation. At the same time, equations with retarded arguments become more and more complicated to analyze and the existence and uniqueness of the solution as well as important features such as oscillation and asymptotic behavior are still open problems in many cases. Finding accurate numerical solutions and sharp location of characteristic roots for establishing stability was also becoming more and more difficult and, in some cases, still represent a real challenge for numerical analysts. A careless and naive adaptation of standard numerical methods designed for ordinary and partial differential equations in the integration of equations with delays, aside from often being useless, might lead and actually did lead, as claimed and proved in Banks and Mahaffy [2], to conjectures that turned out to be erroneous and misleading for the understanding of the phenomenon studied. Algorithms for implementing delay models must be specifically designed according to the nature of the equations and the quality of the solution.

Here we confine ourselves to the case of ordinary derivatives, where the most general form of such models is given by the Retarded Functional Differential Equation (in short RFDE):

$$y'(t) = f(t, y_t), \quad t \geq t_0,$$

where the *state* $y_t(s) = y(t+s)$, $s \in [-r, 0]$, is a function belonging to the Banach space $C = C^0([-r, 0], R^d)$ of continuous functions mapping the interval $[-r, 0]$ into R^d, and $f : \Omega \longrightarrow R^d$ is a given function of the set $\Omega \subset R \times C$ into R^d. Contrary to the ordinary case, the Cauchy problem takes the form:

$$\begin{cases} y'(t) = f(t, y_t), & t \geq t_0, \\ y_{t_0}(s) = y(t_0 + s) = \varphi(s), & s \in [-r, 0], \end{cases} \tag{1}$$

where φ represents, in the Banach space C, the *initial point* or the *initial state*. Equation (1), also called the *Volterra functional differential equation*, includes both *distributed delay differential equations*, where f operates on y computed on a continuum set of past values, and *discrete delay differential equations*, where only a finite number of past values of the variable y are involved.

Models with discrete delays are characterized by the presence of the function y computed at certain deviated arguments $y(t-\tau)$, where the delay τ, which is always non-negative, may be *constant* ($\tau = \text{const}$), *time dependent* ($\tau = \tau(t)$), and *state dependent* ($\tau = \tau(t, y(t))$ or even $\tau = \tau(t, y_t)$).

A real-life example of retarded functional differential equations with both discrete and distributed delays is given by the model of Banks and Mahaffy [2] for the regulation of protein synthesis:

$$\frac{d}{dt}x^1(t) = \frac{a_1}{1 + k_1 L_1^1(z_t^n)} - b_1 x^1(t)$$

$$\frac{d}{dt}y^1(t) = \alpha_1 L_1^2(x_t^1) - \beta_1 y^1(t)$$

$$\frac{d}{dt}z^1(t) = \gamma_1 L_1^3(y_t^1) - \delta_1 z^1(t)$$

$$\left.\begin{aligned} \tfrac{d}{dt}x^i(t) &= \frac{a_i}{1+k_i L_i^1(z_t^{i-1})} - b_i x^i(t) \\ \tfrac{d}{dt}y^i(t) &= \alpha_i L_i^2(x_t^i) - \beta_i y^i(t) \\ \tfrac{d}{dt}z^i(t) &= \gamma_i L_i^3(y_t^i) - \delta_i z^i(t) \end{aligned}\right\} i = 2, \cdots, n,$$

where the operators L_i^j are given by

$$L_i^j(z_t) = \sum_{l=0}^{\nu} c_{il}^j z(t-h_l) + \int_{-r}^{0} z(t+s)\zeta_i^j(s)ds,$$

x^i is the amount of mRNA (messenger Ribo-Nucleic Acid) by the transcription of gene i, y^i is the amount of protein by the translation of x^i and z^i is the repressor produced by the protein y^i which shut down transcription in the gene $i+1$, etc.

The presence of an initial function φ, instead of an initial value y_0 in the Cauchy problem (1), entails some consequences, often unexpected, in the solution $y(t)$ for $t > t_0$. In general, there is no longer injectivity between the set of initial data φ and the set of solutions $y(t)$. Moreover, the prolongation of the initial function φ past the initial point t_0 is not smooth whenever $\varphi'(0)^- \neq y'(t_0)^+ = f(t_0, y_{t_0})$ and this lack of regularity, at t_0, propagates forward even if the ingredients f, τ and φ of the problem belong to C^∞.

For example, it is not difficult to see that, in general, the solution of the Cauchy problem:

$$\begin{cases} y'(t) = f(t, y(t-\tau(t))), & t \geq t_0, \\ y(t) = \phi(t) := \varphi(t-t_0), & t \leq t_0, \end{cases}$$

does not possess a second derivative at any point $\xi_{1,i}$ such that

$$\xi_{1,i} - \tau(\xi_{1,i}) = t_0,$$

it does not possess a third derivative at any point $\xi_{2,j}$ such that

$$\xi_{2,j} - \tau(\xi_{2,j}) = \xi_{1,i}$$

for some i, and so on for higher order derivatives. This results in a sequence of points, called *breaking points* or *primary discontinuities*, where the solution possesses only a limited number of derivatives, the *order* of the breaking point, and remains piecewise regular between two consecutive such points. Locating the breaking points and including them into the mesh is a crucial issue in the numerical integration of RFDE because it is known that any step-by-step method attains its own order of accuracy provided that the solution sought is sufficiently smooth in the current integration interval. In principle, the breaking points can be computed by recursively solving the algebraic equations above, which are trivial for constant delays and may be solved "a priori" for time dependent delays. On the contrary, in the state dependent delay case where the algebraic equation is

$$\xi_{2,j} - \tau(\xi_{2,j}, y(\xi_{2,j})) = \xi_{1,i},$$

they depend on the solution and cannot be computed in advance. In particular, their accurate computation depends on the accuracy of the approximation of y, which in turns depends on the accuracy in the computation of the breaking points themselves. This makes accurate integration of RFDEs with state dependent delay a real challenge.

Other discontinuities, called *secondary discontinuities*, may propagate along the solution caused by discontinuities in the functions f, τ or ϕ. In particular, discontinuities in the initial function ϕ may be much more harmful and cause the termination of the solution, which may reappear later and disappear again giving rise to *lacunary solutions*. It is not difficult to imagine that introduction of delays affects whether the problem is well-posed as well as the stability and asymptotic stability properties of the solution.

The general theory of RFDEs has been widely developed in the last fifty years and results in a number of, now classic, books such as those of Bellman and Cooke [8], Hale, Driver, Èl′sgol′ts and Norkin [15], Kolmanovskii and Myshkis, up to the more recent monographs of Diekmann, van Gils, Verduyn-Lunel and Walter [14] and Kuang [30], which also include many real-life examples of RFDEs and more general retarded functional differential equations.

From the numerical point of view, the presence of delays entails additional difficulties which have been tackled by different approaches. The choice of approach specifically depends on the particular kind of delay one must handle and on whether the particular aim is pursued in terms of accuracy, stability, etc. These methods range from the classical method of steps, where the RFDE is seen as a sequence of ordinary differential equations, to the use of collocation and more general continuous Runge-Kutta methods leading to piecewise polynomial approximations, or to the transformation of the delay equation into a partial differential equation with appropriate initial/boundary conditions to be integrated by direct or transverse methods of lines. All these methods, along with preliminaries and historical remarks on the numerics of delay differential equations, are described and developed in the recent book by Bellen and Zennaro [7] with particular emphasis on the class of continuous Runge-Kutta methods and their numerical stability.

Although models based on partial differential equations with delays for investigating complex biological phenomena began to be considered more than twenty years ago, they have received very little attention by numerical analysts and, to our knowledge, no public domain code is yet available for their integration. Developing algorithms for the numerical integration of partial differential equations with delays appears to be a very promising common ground of research for the numerical ODE and PDE communities and an attractive and challenging area of investigation that, in our opinion, will play a central role in research in biomathematics during the next decade. For an overview of retarded partial differential equations, see Wu [38].

2 Solving RFDEs by continuous Runge-Kutta methods

Here we present two different approaches for solving RFDEs (1) in the equivalent form

$$\begin{cases} y'(t) = f\big(t, y(t), y_t\big), & t \geq t_0, \\ y_{t_0} = y(t_0 + s) = \varphi(s), & s \in [-r, 0], \end{cases} \tag{2}$$

where we outline the possible dependence of f on the state function y_t and on the current value $y(t)$ as well. The methods are both based on the use of Continuous Runge-Kutta methods (CRK) as applied, step-by-step, to the local problems:

$$\begin{cases} w'(t) = f\big(t, w(t), w_t\big), & t_n \leq t \leq t_{n+1}, \\ w(t) = \eta(t), & t_0 \leq t \leq t_n, \\ w(t) = \phi(t), & t \leq t_0, \end{cases} \tag{3}$$

where η is the continuous approximate solution provided by the CRK method itself. The straightforward application of the CRK methods $(A, b(\theta), c)$ to the local equation (3) takes the form

$$\begin{aligned} &\eta(t_n + \theta h_{n+1}) = y_n + h_{n+1} \textstyle\sum_{i=1}^{\nu} b_i(\theta) f\big(t_{n+1}^i, Y^i, Y_{t_{n+1}^i}^i\big), \quad 0 \leq \theta \leq 1, \\ &Y^i = y_n + h_{n+1} \textstyle\sum_{j=1}^{\nu} a_{ij} f\big(t_{n+1}^j, Y^j, Y_{t_{n+1}^j}^j\big), \quad i = 1, \ldots, \nu, \end{aligned} \tag{4}$$

where $t_{n+1}^i = t_n + c_i h_{n+1}$ and the state functions $Y_{t_{n+1}^i}^i$ are suitable approximations of the $w_{t_{n+1}^i}$. For the sake of conciseness we omit the dependence of Y^i on n.

It is evident that, as long as all the state functions $Y_{t_{n+1}^i}^i(s) = Y^i(t_{n+1}^i + s)$, $s \in [-r, 0]$, have to be computed, according to the action of the functional f, only at arguments s such that $t_{n+1}^i + s \leq t_n$, we must set $Y^i(t_{n+1}^i + s) = \eta(t_{n+1}^i + s)$ and Eq. (3) is essentially an ODE. On the contrary, when for some i and for some s, $t_{n+1}^i + s > t_n$, the unknown part of the state function $Y_{t_{n+1}^i}^i(s)$ has to be provided by suitable extensions of the CRK method itself. In this case, referred to as *overlapping*, the structure of the Runge-Kutta equations changes and the problem is intrinsically different from an ODE.

A central issue in the convergence analysis of the step-by-step method for RFDEs is how its discrete and uniform global errors depend on the local discrete error and the local uniform error of the method (4) (see Bellen and Zennaro [7] for definitions). The convergence of the CRK methods for RFDEs is governed by the following theorem proved in [7].

Theorem 1. *Given the mesh $\mathcal{M}_h$, of maximum stepsize h, assume that all the breaking points of order $p+1$ are included in $\mathcal{M}_h$, so that the solution $y(t)$ is piecewise of class $C^{p+1}(t_0, t_f)$. If method (4) has discrete order p (i.e., discrete local error of order $p+1$) and uniform order q (i.e., uniform local error of order $q+1$), then the* *resulting method for RFDEs has discrete and uniform global order* $\min\{p, q+1\}$, *i.e.,*

$$\max_{t_i \in \mathcal{M}_h} \|y(t_i) - \eta(t_i)\| = \mathcal{O}(h^{\min\{p,q+1\}}).$$

$$\max_{t_0 \le t \le t_f} \|y(t) - \eta(t)\| = \mathcal{O}(h^{\min\{p,q+1\}}).$$

In other words, the local uniform error does not propagate and, in particular, the local uniform order $q = p - 1$ *is sufficient for preserving the global order* p *of the overall method.*

2.1 Continuous Runge-Kutta (standard approach) and functional continuous Runge-Kutta methods

Despite the fact that the first method which we are going to describe can be applied to the general problem (2), we will be more specific and consider, as a special case, the following RFDE with discrete state dependent delay:

$$\begin{cases} y'(t) = f\Big(t, y(t), y\big(\alpha(t, y(t))\big)\Big), & t_0 \le t \le t_f, \\ y(t) = \phi(t), \quad t \le t_0, \end{cases}$$

where, for the sake of simplicity, we have set the deviated argument $t - \tau(t, y(t))$ in the form $\alpha(t, y(t))$. One possible option in the implementation of (4), referred to as the *standard approach*, is characterized by the choice $Y^i(t_{n+1}^i + s) = \eta(t_{n+1}^i + s)$ for all i, which corresponds to setting, in (3),

$$w_t = \eta_t \quad \text{for all} \quad t \in [t_n, t_{n+1}].$$

The local problem is then:

$$\begin{cases} w'(t) = f\Big(t, w(t), \eta\big(\alpha(t, w(t))\big)\Big), & t_n \le t \le t_{n+1}, \\ w(t) = \eta(t), \quad t_0 \le t \le t_n, \\ w(t) = \phi(t), \quad t \le t_0, \end{cases}$$

and the CRK method is:

$$\eta(t_n + \theta h_{n+1}) = y_n + h_{n+1} \textstyle\sum_{i=1}^{\nu} b_i(\theta) f(t_{n+1}^i, Y^i, \tilde{Y}^i), \quad 0 \le \theta \le 1,$$
$$Y^i = y_n + h_{n+1} \textstyle\sum_{j=1}^{\nu} a_{ij} f(t_{n+1}^j, Y^j, \tilde{Y}^j), \quad i = 1, \ldots, \nu,$$
$$\tilde{Y}^i = \eta\left(\alpha(t_{n+1}^i, Y^i)\right).$$

If overlapping occurs, that is, if, for some index i,

$$t_n \le \alpha(t_{n+1}^i, Y^i) \le t_n + c_i h_{n+1},$$

then

$$\alpha(t_{n+1}^i, Y^i) = t_n + \theta_{n+1}^i h_{n+1},$$

with

$$(0 \leq)\, \theta^i_{n+1} = \frac{\alpha(t^i_{n+1}, Y^i) - t_n}{h_{n+1}} \ (\leq c_i),$$

and the *spurious stage values* $\tilde{Y}^i$ are still given by the continuous extension

$$\tilde{Y}^i = \eta(t_n + \theta^i_{n+1} h_{n+1}) = y_n + h_{n+1} \sum_{j=1}^{\nu} b_j(\theta^i_{n+1}) f(t^j_{n+1}, Y^j, \tilde{Y}^j).$$

Since the deviated function $y(\alpha(t, y(t))$ is approximated by the continuous extension $\eta(\alpha(t, y(t))$ in both current and past integration intervals, the standard approach is usually referred to simply as the continuous Runge-Kutta method.

It is worth remarking that, if overlapping occurs, the current Runge-Kutta equation turns out to be implicit even if the underlying method is explicit. However, in spite of the appearance of possible spurious stages $\tilde{Y}^i$, the dimension of the algebraic Runge-Kutta system preserves the dimension s by using the alternative K-notation:

$$\eta(t_n + \theta h_{n+1}) = y_n + h_{n+1} \sum_{i=1}^{\nu} b_i(\theta) K^i, \quad 0 \leq \theta \leq 1,$$

$$K^i = f(t^i_{n+1}, y_n + h_{n+1} \sum_{j=1}^{\nu} a_{ij} K^j,$$
$$y_n + h_{n+1} \sum_{j=1}^{\nu} b_j(\theta^i_{n+1}) K^j), \quad i = 1, \ldots, \nu,$$

where

$$\theta^i_{n+1} = \frac{\alpha(t^i_{n+1}, y_n + h_{n+1} \sum_{j=1}^{\nu} a_{ij} K^j) - t_n}{h_{n+1}}.$$

Note that overlapping takes place when the stepsize is larger than the delay as well as, independently of the stepsize, when we are integrating in a neighborhood of points where the delay vanishes. Therefore the standard approach is suitable for stiff problems where an implicit CRK method is used. However, in the presence of overlapping, the structure of the Jacobian of the Newton solver changes and this could lead to additional difficulties in the variable stepsize implementation of the overall method. The standard approach outlined above is described in detail in Sect. 3 as applied to a specific biological model leading to a system of stiff RFDEs with discrete vanishing state dependent delays, which illustrates, in its numerical integration, some of the theoretical and practical difficulties described above.

As counterpart to the standard approach we consider a second method designed for the general equation (2), including RFDEs with distributed delays, where the local problem (3) is still approximated by the CRK method (4) but, for each i, the unknown part of the state function $Y^i_{t^i_{n+1}}$ is now given by

$$Y^i(t_n + \theta h_{n+1}) = y_n + h_{n+1} \sum_{j=1}^{\nu} a_{ij}(\theta) f(t^j_{n+1}, Y^j, Y^j_{t^j_{n+1}}), \quad 0 \le \theta \le c_i. \quad (5)$$

This approach, called the *functional continuous Runge-Kutta* (FCRK) method, besides still being based on CRK methods, is quite different from the standard approach. In particular, as the RK method makes use of ν stage values Y^i, $i = 1, \dots \nu$, as approximations of $w(t^i_{n+1})$, ν different *state stage functions* $Y^i_{t^i_{n+1}}$ are defined which approximate the state functions $w_{t^i_{n+1}}$.

Remark 1. Contrary to the standard approach, the resulting method also preserves the implicit/explicit character of the underlying CRK method in the case of overlapping. This makes the FCRK method competitive for non-stiff equations with small or vanishing delays, leading to possible overlapping, as well as for functional integral equations such as

$$y'(t) = F\Big(t, y(t), \int_{t-\tau}^{t} k(t, s, y(s))ds\Big),$$

where overlapping takes place at every step.

The method, described in Sect. 4 in its general form, was presented by Cryer and Tavernini [13, 35] as a particular predictor-corrector version of polynomial collocation. The method was recently reconsidered by Maset, Torelli and Vermiglio [32] who developed it in the general form (3) and derived necessary and sufficient order conditions up to order four, along with order barriers with respect to the number of stages.

3 A threshold model for antibody production: the Waltman model

We consider a mathematical model describing the mechanism by which an antibody is formed in response to an antigen challenge. This is one of the better understood parts of the human immune system and has been widely treated in the scientific literature. The earliest models were chemical kinetics or predator-prey models (see Bell [3–5]); afterwards Hoffmann [27] and Richter [34] proposed network-based models including inhibiting and stimulating signals. We consider here a more sophisticated and widely used threshold model which makes use of an integral threshold to describe the onset of B-cell proliferation and to mark the signal which activates antibody production. Such an approach has been extensively considered in the literature (see, e.g., Gatica and Waltman [18–20], Waltman and Butz [37], Waltman [36], Cooke [12], Hoppensteadt and Waltman [28]).

These threshold models, which are still valid and topical (see, e.g., [29, 31]), are complicated to treat mathematically and have the peculiarity that they lead to functional differential equations rather than to ordinary differential equations.

The framework we consider describes a realistic, although simplified, situation describing the challenge of a chemical antigen which binds to receptor sites on the surface of lymphocytes. This determines the activation of a signal which gives rise to the lymphocyte production phase; the nature of this triggering mechanism is still not well understood by immunologists and the model aims to represent it in a very general way.

In the first phase, that preceding the onset of lymphocyte proliferation, the dynamics describing the interaction among free antigen molecules, free receptor sites and bound receptor sites is described by a *chemical reaction*-like system of *stiff* ordinary differential equations $y'(t) = f_a\,(y(t))$.

In the second phase, which is established through a first threshold effect modeled by an integral equation which depends on the concentration of bound receptor sites, the model changes form and is described by a system of delay differential equations of type $y'(t) = f_b\Big(y(t), y\big(\alpha_1(t, y(t))\big)\Big)$, where y denotes the vector of unknown concentrations of antigen molecules and receptor sites and $\alpha_1(t, y(t)) \leq t$ is a deviating argument which allows us to model the memory effect of the phenomenon. Such a deviating argument depends on the solution y itself (the so-called state of the system) and its dynamics is also described by suitable functional differential equations.

Lastly, in the third phase, which is established through a second threshold effect, the model is described by a larger system of delay differential equations which includes the proliferation of antibodies. A further memory effect is described by a second deviating argument so that the whole system is described by a system of six delay differential equations; four equations describe the dynamics of the concentrations of antigen molecules, free and bound receptor sites on the surface of lymphocytes and antibodies. They have the form $y'(t) = f_c\Big(y(t), y\big(\alpha_1(t, y(t))\big), y\big(\alpha_2(t, y(t))\big)\Big)$, where $\alpha_2(t, y(t)) \leq t$ denotes the second deviating argument modeling the memory effect in the antibody production process. The remaining two equations describe the dynamics of the deviating arguments α_1 and α_2. As time increases the memory effects tend to diminish, or even to disappear, in the model which means that $\alpha_1(t, y(t))$ and $\alpha_2(t, y(t))$ approach t.

3.1 The quantitative model

The model as a whole is an interesting system of stiff delay differential equations. To describe it mathematically, we introduce the following quantities:

(1) $y_1(t)$, the concentration of unbound antigen molecules at time t;
(2) $y_2(t)$, the concentration of unbound receptor sites at time t;
(3) $y_3(t)$, the concentration of bound receptor sites at time t;
(4) $y_4(t)$, the concentration of unbound antibodies at time t.

Initial phase. The antigen molecules and the receptor sites combine according to the mass action law:

$$\begin{aligned} y_1'(t) &= -r_1\, y_1(t)\, y_2(t) + r_2\, y_3(t) \\ y_2'(t) &= -r_1\, y_1(t)\, y_2(t) + r_2\, y_3(t) \\ y_3'(t) &= r_1\, y_1(t)\, y_2(t) - r_2\, y_3(t) \end{aligned}$$

with r_1, r_2 denoting suitable rate constants.

This model holds until time t_0 when the trigger to initiate lymphocytes proliferation acts. The model assumes that t_0 is given by the integral equation

$$\int_0^{t_0} f_1\,(y_1(s), y_2(s), y_3(s))\; ds = m_1,$$

where m_1 is an appropriate biological threshold. The previous integral models the accumulation of signals depending on the concentration of free antigen molecules, free receptor sites and receptor-antigen complexes.

Intermediate phase. In this phase new receptor sites are generated so that the system evolves according the equations:

$$\begin{aligned} y_1'(t) &= -r_1\, y_1(t)\, y_2(t) + r_2\, y_3(t) \\ y_2'(t) &= -r_1\, y_1(t)\, y_2(t) + r_2\, y_3(t) + a\, r_1\, y_1\,(\alpha_1(t))\;\, y_2\,(\alpha_1(t)) \\ y_3'(t) &= r_1\, y_1(t)\, y_2(t) - r_2\, y_3(t), \end{aligned}$$

where a is an amplification factor and $\alpha_1(t) \le t$ models a memory effect described by the integral equation

$$\int_{\alpha_1(t)}^{t} f_1\,(y_1(s), y_2(s), y_3(s))\; ds = m_1, \quad t \ge t_0. \tag{6}$$

This model holds until a certain time t_1 when the trigger to initiate antibodies production acts. The model assumes that t_1 is given by the integral equation

$$\int_0^{t_1} f_2\,(y_2(s), y_3(s))\; ds = m_2,$$

where m_2 is an appropriate biological threshold.

Final phase. In this phase antibodies (y_4) are produced by the immune system according to the equations:

$$\begin{aligned} y_1'(t) &= -r_1\, y_1(t)\, y_2(t) + r_2\, y_3(t) - s\, y_1(t)\, y_4(t) \\ y_2'(t) &= -r_1\, y_1(t)\, y_2(t) + r_2\, y_3(t) + a\, r_1\, y_1\,(\alpha_1(t))\;\, y_2\,(\alpha_1(t)) \\ y_3'(t) &= r_1\, y_1(t)\, y_2(t) - r_2\, y_3(t) \\ y_4'(t) &= -s\, y_1(t)\, y_4(t) - \gamma\, y_4(t) + b\, r_1\, y_1\,(\alpha_2(t))\;\, y_2\,(\alpha_2(t))\,, \end{aligned}$$

where s is a combination factor, b is an amplification factor related to the antibody secretion capacity of plasma cells, γ is a catabolic factor and $\alpha_2(t) \leq t$ is a second memory effect described by the integral equation

$$\int_{\alpha_2(t)}^{t} f_2\left(y_2(s), y_3(s)\right) ds = m_2, \quad t \geq t_1. \tag{7}$$

Summary. The problem as a whole consists of six equations; four to describe the interaction between the antigen and the immune system:

$$\begin{cases} y_1'(t) = -r_1\, y_1(t)\, y_2(t) + r_2\, y_3(t) - s\, y_1(t)\, y_4(t) \\ y_2'(t) = -r_1\, y_1(t)\, y_2(t) + r_2\, y_3(t) + a\, r_1\, y_1\big(\alpha_1(t)\big) y_2\big(\alpha_1(t)\big) H(t - t_0) \\ y_3'(t) = r_1\, y_1(t)\, y_2(t) - r_2\, y_3(t) \\ y_4'(t) = -s\, y_1(t)\, y_4(t) - \gamma\, y_4(t) + b\, r_1\, y_1\big(\alpha_2(t)\big)\, y_2\big(\alpha_2(t)\big) H(t - t_1), \end{cases} \tag{8}$$

and two to describe the dynamics of the deviating arguments α_1 and α_2, which are obtained by differentiating Eqs. (6) and (7):

$$\begin{cases} \alpha_1'(t) = H(t - t_0)\, \dfrac{f_1\left(y_1(t), y_2(t), y_3(t)\right)}{f_1\left(y_1\left(\alpha_1(t)\right), y_2\left(\alpha_1(t)\right), y_3\left(\alpha_1(t)\right)\right)} \\ \alpha_2'(t) = H(t - t_1)\, \dfrac{f_2\left(y_2(t), y_3(t)\right)}{f_2\left(y_2\left(\alpha_2(t)\right), y_3\left(\alpha_2(t)\right)\right)}, \end{cases} \tag{9}$$

where $H(x)$ is the Heavyside function ($H(x) = 0$ if $x < 0$ and $H(x) = 1$ if $x \geq 0$).

Numerical integration. From the numerical point of view the model presents the following difficulties (also see Figs. 1 and 2):

(i) the deviating arguments are state-dependent and hence are not known in advance;
(ii) the delays $t - \alpha_1(t, y(t))$ and $t - \alpha_2(t, y(t))$ become very small as time grows; this makes it impossible to consider the problem step-by-step as a system of ordinary differential equations;
(iii) the solution components have very different magnitudes and have very steep variations in correspondence of the triggers initiating the proliferation first of lymphocytes and later of antibodies;
(iv) the presence of discontinuities in the right-hand side, due to the threshold mechanisms, determines a certain number of breaking points, which have to be treated carefully in order to avoid a loss of accuracy;
(v) the system is *stiff* and therefore needs to be integrated numerically by an implicit method.

We have numerically integrated the problem by means of the code RADAR5, whose main features are described here. We discuss the application of stiffly accurate

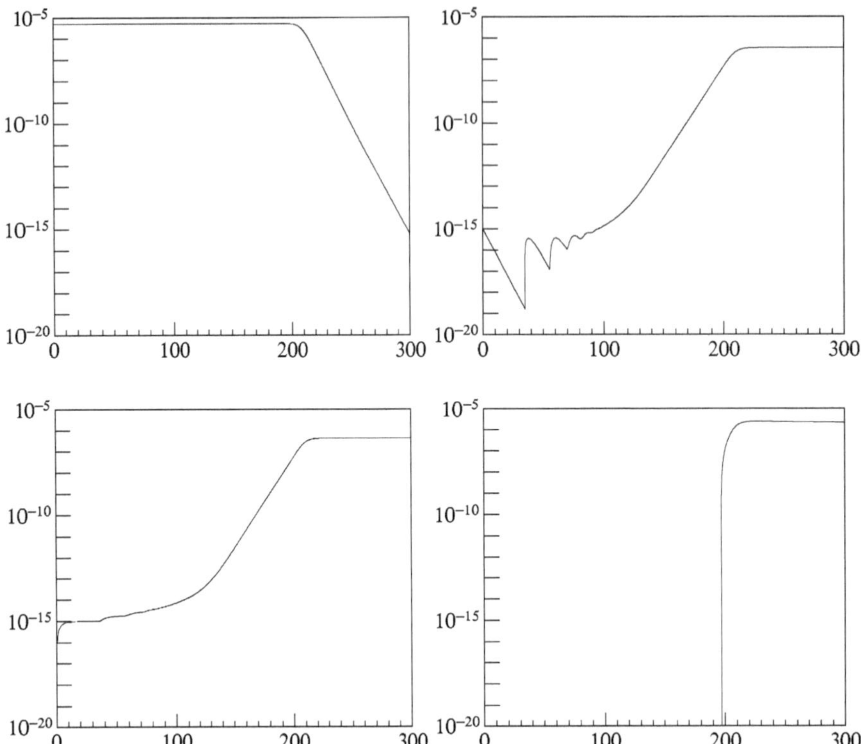

Fig. 1. Solution components of problem (8) with the choice of parameters given in Sect 3.6

collocation methods based on Radau nodes to systems of delay differential equations of the form

$$y'(t) = f\Big(t, y(t), y\big(\alpha_1(t, y(t))\big), \ldots, y\big(\alpha_p(t, y(t))\big)\Big) \tag{10}$$

with initial data

$$y(t_0) = y_0, \quad y(t) = g(t) \quad \text{for} \quad t < t_0.$$

We assume that the deviating arguments are such that $\alpha_i(t, y(t)) \leq t$ for all $t \geq t_0$ and for all i.

As we mentioned the integration of delay differential equations presents several additional difficulties with respect to ODEs.

In particular, discontinuities may occur in various orders of the derivative of the solution, independently of the regularity of the right-hand side; this could lead to a loss in the accuracy of the numerical approximation. Small delays complicate the use of explicit approximation methods and determine a structural change in the Runge–Kutta equations of implicit methods; on the other hand large delays force us to store a large amount of information (the solution in the past). Furthermore error control

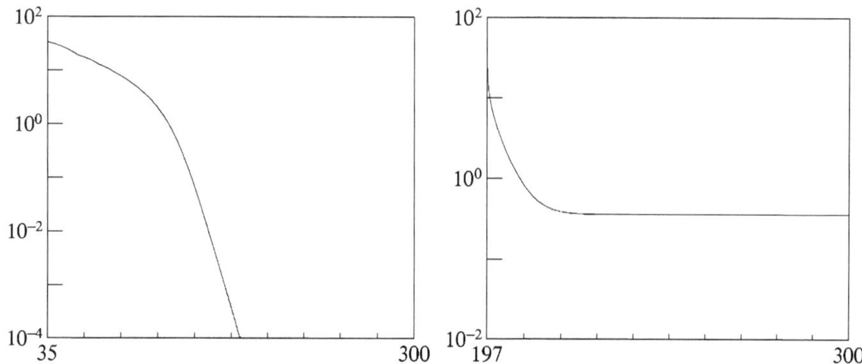

Fig. 2. Delays of problem (8) with the choice of parameters given in Sect. 3.6

strategies for ODEs may be inappropriate for delay differential equations since the continuous output also has to be controlled. For a comprehensive discussion on these issues we refer the reader to [7].

The remainder of this section is organized as follows. First we describe the integration process. Then we describe a technique to compute breaking points which is peculiar to implicit methods. Afterwards we direct our attention to the Newton process for the solution of the Runge–Kutta equations associated to each step. Lastly we deal with an error control technique which is well suited to stiff delay equations. The last section illustrates the application of the code RADAR5 to the Waltman model considered.

3.2 The integration process

We direct our attention here to stiff equations and more generally to problems where the use of an explicit method would lead to stepsize restrictions which are not due to accuracy requirements and can be overcome by using an implicit method.

The integration scheme we consider is based on the ν-stage Radau IIA collocation method (in particular the code RADAR5 uses $\nu = 3$). For a detailed description we refer the reader to [21–23]. We use the following notation:

- $f(t, y, z_1, \ldots, z_p)$ denotes the right-hand side function;
- the nodes $\{c_i\}$, the weights $\{b_i\}$ and the coefficients $\{a_{ij}\}$ are those of the Radau IIA method; in the case $\nu = 3$ its Butcher tableau is given by:

$$\begin{array}{c|ccc}
\frac{4-\sqrt{6}}{10} & \frac{88-7\sqrt{6}}{360} & \frac{296-169\sqrt{6}}{1800} & \frac{-2+3\sqrt{6}}{225} \\[4pt]
\frac{4+\sqrt{6}}{10} & \frac{296+169\sqrt{6}}{1800} & \frac{88+7\sqrt{6}}{360} & \frac{-2-3\sqrt{6}}{225} \\[4pt]
1 & \frac{16-\sqrt{6}}{36} & \frac{16+\sqrt{6}}{36} & \frac{1}{9} \\[4pt]
\hline
 & \frac{16-\sqrt{6}}{36} & \frac{16+\sqrt{6}}{36} & \frac{1}{9}
\end{array}$$

- Y^j denotes the jth stage value computed at the current step;

- $h_{n+1} = t_{n+1} - t_n$ indicates the current stepsize;
- $\alpha_{\ell j} := \alpha_\ell(t_n + c_j\, h_{n+1}, Y^j)$ yields an approximation of

$$\alpha_\ell(t_n + c_j\, h_{n+1}, y(t_n + c_j\, h_{n+1}));$$

- $\eta(t)$ is the piecewise polynomial continuous approximation to the solution. In most cases it is given step-by-step by the collocation polynomial associated to the method.

The Runge–Kutta formula applies to (10) as:

$$\begin{cases} 0 = F_i\left(Y^1, \cdots, Y^\nu, Z^{11}, \cdots, Z^{1\nu}, \ldots, Z^{p1}, \cdots, Z^{p\nu}\right), \; i = 1, \ldots, \nu, \\ y_{n+1} = Y^\nu, \end{cases} \tag{11}$$

with

$$F_i\,(\ldots) = Y^i - y_n - h_{n+1} \sum_{j=1}^{\nu} a_{ij} f\left(t_n + c_j h, Y^j, Z^{1j}, \cdots, Z^{pj}\right) \tag{12}$$

and

$$Z^{\ell j} = \begin{cases} g\left(\alpha_{\ell j}\right) & \text{if} \quad \alpha_{\ell j} \le t_0, \\ \eta\left(\alpha_{\ell j}\right) & \text{if} \quad \alpha_{\ell j} > t_0. \end{cases}$$

Here we have omitted the dependence of F_i, $\alpha_{\ell j}$ and $Z_{\ell j}$ on n. The continuous approximation η to the solution at the mth step, that is, for $t_m \le t \le t_{m+1}$ $(m \le n)$, is given by one of the polynomials:

$$u_m(t_m + \vartheta h_{m+1}) = \mathcal{L}_0(\vartheta) y_m + \sum_{i=1}^{s} \mathcal{L}_i(\vartheta) Y^i, \quad \vartheta \in [0, 1], \tag{13}$$

$$v_m(t_m + \vartheta h_{m+1}) = \sum_{i=1}^{s} \hat{\mathcal{L}}_i(\vartheta) Y^i, \quad \vartheta \in [0, 1]. \tag{14}$$

Observe that, in (13) and (14), the stage values Y^i are those relevant to the interval $[t_m, t_{m+1}]$. Here h_{m+1} is the stepsize used at the mth step, $\mathcal{L}_i(\vartheta)$ is the polynomial of degree ν satisfying $\mathcal{L}_i(c_i) = 1$ and $\mathcal{L}_i(c_j) = 0$ for $j \neq i$ (where $c_0 = 0$ and $c_1, \ldots, c_\nu$ are the nodes of the method) and $\hat{\mathcal{L}}_i$ is the polynomial of degree $\nu - 1$ satisfying $\hat{\mathcal{L}}_i(c_i) = 1$ and $\hat{\mathcal{L}}_i(c_j) = 0$ for $j \neq i$ $(j, i = 1, \ldots, \nu)$).

In most cases the first polynomial is chosen, but in those cases where t_m is a jump discontinuity for the solution (see [21]) the second polynomial provides a more accurate uniform approximation.

If $\alpha_{\ell j} \le t_n$ for all ℓ and j, then all arguments $Z^{\ell j}$ can be explicitly computed by knowledge of the continuous approximation of the solution in the past, that is, $\eta(t)$ for $t \le t_n$. This situation corresponds to the so-called method of steps (see [8]). In such a case we have to deal locally with a system of ODEs. Nevertheless, if $\alpha_{\ell j} \in (t_n, t_{n+1}]$

for some pair (ℓ, j), it means that there are delays which are smaller than the stepsize used; hence $\eta\left(\alpha_{\ell j}\right)$ is not known explicitly. In fact for $m = n$, u_m (or v_m) identifies the continuous output to be computed in the current interval $[t_n, t_{n+1}]$ (which thus depends on the unknown current stage values). As a consequence the structure of the nonlinear equations (11) would be quite different from that of the previous case.

For this reason, in order to solve the Runge–Kutta equations, we are driven to a more sophisticated scheme than that used for ODEs.

3.3 Tracking the breaking points

As we have seen in Sect. 1, a serious difficulty is the possible loss of regularity of the solution, due to breaking points, even in the presence of smooth functions $f(t, y, z)$, $g(t)$, and $\alpha_i(t, y)$ $(i = 1, \ldots, p)$ in problem (10).

In most cases, only a few breaking points are significant for numerical integration, because discontinuities in a sufficiently high derivative of the solution are not recognized by the numerical method.

In the case where α_i does not depend on $y(t)$ for all i, the breaking points can be computed in advance by solving first the scalar equations $\alpha_i(\zeta) = t_0$ for ζ, and then for every solution ζ_k the scalar equation $\alpha_i(\xi) = \zeta_k$, and so on. For an efficient integration, computed breaking points can be inserted in advance into the mesh. But in the general (the so-called state dependent) case, where, for some i, α_i depends on y, such a computation is not possible a priori.

If the breaking points are not included in the mesh and a variable stepsize integration is used, the stepsizes may be severely restricted near the low order jump discontinuities. Thus it is important to design an algorithm that allows a code to compute automatically the disturbing breaking points and to include them in the mesh of integration (for an extensive discussion we refer the reader to [22]). In this way, not only are step rejections avoided, but also the accuracy of the approximation is significantly improved.

Detection of breaking points. To compute the set B of breaking points, at the beginning we set $B = \{t_0\}$ (and possibly include irregular points of the initial function $g(t)$ of problem (10)). The problem is to find the zeros of the functions

$$d_\ell(t; \zeta) = \alpha_\ell\big(t, \eta(t)\big) - \zeta, \quad \ell = 1, \ldots, p, \tag{15}$$

where $\zeta \in B$ is a previous breaking point and $\eta(t)$ is a suitable approximation to the solution.

A very simple approach would be to test, in every accepted step, whether at least one of the functions $d_\ell(t; \zeta)$ (see (15)) changes sign. The breaking point can then be localized, computed and added to the set B. Such a strategy is often expensive because of the many step rejections that usually occur when approaching a breaking point. We consider instead the following strategy (see [22]). Suppose that the problem is integrated successfully up to t_n and a stepsize h_{n+1} is proposed for the next step. We expect a breaking point in $[t_n, t_n + h_{n+1}]$ if the following two conditions occur:

(a) the step is rejected, i.e., the iterative solver for the nonlinear system (11) fails to converge, or the local error estimate is not sufficiently small,
(b) there exists a previous breaking point ζ such that $d_\ell(t;\zeta) = \alpha_\ell(t, u_{n-1}(t)) - \zeta$ changes sign on $[t_n, t_n + h_{n+1}]$, where $u_{n-1}(t)$ is the continuous output polynomial of the preceding step.

The extrapolated use of u_{n-1} in order to approximate the solution y in the interval $[t_n, t_n + h_{n+1}]$ is safe because we assume that the solution is regular in the previous accepted step. On the other hand the use of extrapolation may lead to an approximation of the breaking point which is not accurate. For this reason we do not use the polynomial u_{n-1} for its computation.

The search is hence activated only in case of a stepsize rejection and proceeds through the following phases (we focus attention on the nth time step, where we assume a stepsize rejection).

Algorithm 2 *Assume that the step $[t_n, t_n + h_{n+1}]$ is rejected.*

1. *Look for zeros of the functions*

$$d_\ell(t;\zeta) = \alpha_\ell\Big(t, u_{n-1}(t)\Big) - \zeta, \quad \zeta \in B, \quad \ell \in \{1, \dots, p\},$$

for $t \in \big[t_n, t_n + h_{n+1}\big]$.
2. **If** *$\hat{\ell} \in \{1, \dots, p\}$ and $\hat{\zeta} \in B$ are determined such that*

$$d_{\hat{\ell}}(t_n; \hat{\zeta}) \cdot d_{\hat{\ell}}(t_n + h_{n+1}; \hat{\zeta}) < 0,$$

pass to Algorithm 3.
3. **Otherwise** *reduce the stepsize according to classical criteria.*

If Algorithm 2 actually detects a breaking point, the exact breaking point will be close to the zero of $d_{\hat{\ell}}(t;\hat{\zeta})$. We denote it by $\hat{\xi}$, that is,

$$\alpha_{\hat{\ell}}\Big(\hat{\xi}, u_{n-1}(\hat{\xi})\Big) - \hat{\zeta} = 0. \tag{16}$$

Computation of breaking points. Once a breaking point is detected the second phase of the procedure begins with the goal of computing it to the desired accuracy. We denote by $\mathbf{Y} = \big(Y^1, \cdots, Y^\nu\big)^{\mathrm{T}}$ the vectors of unknown stage values.

Algorithm 3 *Suppose that a breaking point has been detected by Algorithm 2.*

1. *We iteratively solve the augmented system*

$$Y^i = y_n + h\sum_{j=1}^{s} a_{ij} f\Big(t_n + c_j h, Y^j, Z^{1j}, \dots, Z^{pj}\Big), \quad i = 1, \dots, \nu, \tag{17}$$

$$\alpha_{\hat{\ell}}\Big(t_n + h, u_n(t_n + h)\Big) = \hat{\zeta} \tag{18}$$

with respect to the unknowns $\mathbf{Y}$ and h.

2. **If** *the iterative process converges* **then** *set* $h_{n+1} = h$ *and* **go to** *4.*
3. **Otherwise** *reduce the stepsize according to classical criteria and* **exit**.
4. **If** *the step is accepted (that is, the estimated local error is below the required error tolerance)* **then** *the new point* $\xi^* = t_n + h_{n+1}$ *is inserted into the set of computed breaking points.*
5. **Otherwise** *reduce the stepsize according to classical criteria and* **exit**.

Since we are interested in stiff problems we solve the Runge–Kutta equations by means of a suitable Newton process. In order to preserve the tensor structure of the Jacobian in the Newton process for solving (11)-(12) (see [21] and the next subsection), we alternatively solve (17) and (18) until convergence (for a convergence analysis see [22]). Experimental tests show that this strategy turns out to be very effective.

Remark 2. The need of an accurate computation of breaking points is common to all methods for integration of equations with delays. However, for non-stiff problems, the procedure above turns out to be significantly simplified by using the class of explicit methods described in Sect. 4 because, in that case, the simultaneous solution of the RK equations and the breaking point equation is implicit in the sole unknown h.

Theoretical remarks. Other authors considered alternative techniques for approximating the breaking points (see, e.g., [26] and [17]). Contrary to our approach, they do not use the continuous output of the current step, but an approximation whose error is difficult to control.

The main idea presented here is related to the fact that, in the algorithm which computes the RK-step, the stepsize is not fixed but variable; this allows for an accurate computation of the breaking point to the discrete order p of the method (the root $\hat{\xi}$ of (16) may instead be a quite inaccurate approximation of it and in any case is related to the uniform order of the method).

For the following discussion we assume that Eqs. (17) and (18) are solved exactly. Under suitable smoothness and regularity assumptions the following results hold (for a proof see [22]).

Theorem 4. *Let* $y(t)$ *be the solution of (10), and let* ζ *and* ξ *be exact breaking points of the problem such that* $\alpha_i\big(\xi, y(\xi)\big) = \zeta$ *(for some* i*). Further, let* ζ^* *be an approximation of* ζ *obtained with sufficiently small stepsizes, and let* $\xi \in (t_n, t_{n+1})$. *If*

$$\frac{d}{dt}\Big(\alpha_i\big(t, y(t)\big)\Big)\Big|_{t=\xi} \neq 0, \tag{19}$$

then the breaking point ξ^* *computed by Algorithm 3 satisfies*

$$|\xi^* - \xi| \leq C\big(\|y_{n+1} - y(t_{n+1})\| + |\zeta^* - \zeta|\big),$$

where C *is a suitable constant.*

As a consequence of this result we are able to extend Theorem 1 (also see Theorem 6.1.2 in [7]).

For problems (10) with state dependent delays it may happen that t_n is a numerically computed breaking point, and the corresponding exact breaking point is slightly different. If at this point the solution has a jump discontinuity, the global error cannot be bounded in terms of h. Nevertheless we have the following convergence result. Let H represent the maximal stepsize and $r = \max\{2\nu - 1, \nu + 1\}$, where $2\nu - 1$ is the classical order of the method.

Theorem 5. *Consider a smooth delay problem (10) on a bounded interval with well separated breaking points satisfying (19). If, instead of the exact breaking points, those obtained by Algorithm 3 are used, then*

$$\|\eta(t) - y(\vartheta)\| = \mathcal{O}(H^r),$$

where the function $\vartheta = \vartheta(t)$ satisfies $\vartheta = t + \mathcal{O}(H^r)$.

For a proof see [22].

This is equivalent to the property

$$\|\eta(t) - y(t)\| = \mathcal{O}(H^r) \quad \text{for all } t \notin J$$

with

$$J = \bigcup_{i \geq 0} \left[\xi_i, \xi_i^*\right],$$

where $\{\xi_i\}_{i\geq 0}$ and $\{\xi_i^*\}_{i\geq 0}$ denote the sets of the *exact* and corresponding *numerical* breaking points, respectively, and $\left[\xi_i, \xi_i^*\right]$ denotes the interval between them.

Continuous approximation after breaking points. Since the solution is in general not smooth corresponding to a breaking point, the continuous approximation may also be inadequate. As an example, if the solution has a jump corresponding to of a breaking point, the use of the collocation polynomial should be avoided since it forces global continuity and hence determines a loss in the uniform approximation accuracy. To obtain a more accurate approximation we prefer to consider in general the polynomial v_m (see (14)) of degree $\nu - 1$, which interpolates the values Y^i but not y_m (compare with (13)). This choice allows a globally discontinuous approximation to the solution and determines a local uniform order $q = \nu$. This choice might also be better than the use of the collocation polynomial when the solution is theoretically continuous but in practice has a jump, that is, it presents a large variation with respect to the stepsize h (see, e.g., [21]).

3.4 Solving the Runge–Kutta equations

We solve the Runge–Kutta equations by means of a suitable Newton process. For this we make use of the notation:

- $A := \{a_{ij}\}$ denotes the RK matrix;
- $U_{\ell j} := \begin{cases} u_m(\alpha_{\ell j}) & \text{if} \quad t_m \le \alpha_{\ell j} \le t_{m+1}, \\ 0 & \text{otherwise.} \end{cases}$

In order to obtain an accurate computation of the derivatives of the function

$$F_i\left(Y^1, \cdots, Y^\nu, Z^{11}, \cdots, Z^{1\nu}, \ldots, Z^{p1}, \cdots, Z^{p\nu}\right)$$

we consider the approximation

$$\frac{\partial F_i}{\partial Y^k} \approx \delta_{ik}\, I_d - h_{n+1} \sum_{\ell=1}^{p} \left(a_{ik}\, D_{\ell k} + \hat{D}_{\ell k}\right),$$

where I_d denotes the $d \times d$ identity matrix, δ_{ik} is the Kronecker delta symbol and

$$D_{\ell k} = \frac{\partial f}{\partial y} + \frac{\partial f}{\partial z_\ell}\eta'(\alpha_{\ell k})\frac{\partial \alpha_\ell}{\partial y},$$

$$\hat{D}_{\ell k} = \sum_{j=1}^{s} a_{ij}\, \frac{\partial f}{\partial z_\ell}\, \frac{\partial U_{\ell j}}{\partial Y^k},$$

with $\frac{\partial \alpha_\ell}{\partial y} = \frac{\partial \alpha_\ell}{\partial y}(t_n, y_n)$, $\frac{\partial f}{\partial y} = \frac{\partial f}{\partial y}\big(t_n, y_n, \sigma_1, \cdots, \sigma_p\big)$, and $\frac{\partial f}{\partial z_\ell} = \frac{\partial f}{\partial z_\ell}\big(t_n, y_n, \sigma_1, \cdots, \sigma_p\big)$, where $\sigma_\ell = \eta(\alpha_{\ell 0})$ and $\alpha_{\ell 0} = \alpha_\ell(t_n, y_n)$.

We note that the last term $\hat{D}_{\ell k}$ is always zero if the deviating argument falls to the left of t_n ; more precisely, we get

$$\frac{\partial U_{\ell j}}{\partial Y^k} = \mathcal{U}^{\ell}_{jk}\, I_d,$$

where

$$\mathcal{U}^{\ell}_{jk} = \begin{cases} \mathcal{L}_k(\psi_{\ell j}) & \text{if} \quad \psi_{\ell j} > 0, \\ 0 & \text{otherwise} \end{cases} \tag{20}$$

with

$$\psi_{\ell j} := \big(\alpha_\ell(t_n + c_j h_{n+1}, Y^j) - t_n\big)/h_{n+1}.$$

The approximations considered make the Newton process inexact and linearly convergent.

Structure of the Jacobian in the general case. In order to solve (11) we set $Y = \big((Y^1)^{\mathrm{T}}, (Y^2)^{\mathrm{T}}, \cdots, (Y^\nu)^{\mathrm{T}}\big)^{\mathrm{T}}$, the $\nu{\cdot}d$-dimensional vector of unknowns, and consider the Newton iteration process. In the general case the Jacobian of (11) is given by the matrix

$$J = I_\nu \otimes I_d - h_{n+1}\, A \otimes \left(\frac{\partial f}{\partial y} + \sum_{\ell=1}^{p} \frac{\partial f}{\partial z_\ell}\eta'(\alpha_{\ell 0})\frac{\partial \alpha_\ell}{\partial y}\right) - h_{n+1} \sum_{\ell=1}^{p} A \cdot \mathcal{U}^\ell \otimes \frac{\partial f}{\partial z}, \tag{21}$$

where I_ν denotes the $\nu \times \nu$ identity matrix, $\partial f/\partial y$, $\partial f/\partial z$ the matrices of partial derivatives of f with respect to the y and z variables respectively, $\partial \alpha_\ell/\partial y$ the row vector of partial derivatives of α_ℓ with respect to y and $\mathcal{U}^\ell = \left\{\mathcal{U}^\ell_{jk}\right\}_{j,k=1}^{\nu}$ (see (20)).

Although J is actually an approximation of the true Jacobian of (11), in order to distinguish it from further simplifications, we shall call the corresponding Newton iteration *quasi-exact*.

The quasi-exact iteration is always the correct one and in particular it is also substantially exact in the cases when the delay vanishes or the stepsize is larger than the delay. It allows a more efficient solution of the Runge–Kutta equations despite being quite expensive; in fact the Jacobian has a full structure (although often *sparse*), so that the cost of the LU factorization of J is $(1/3) \cdot (\nu n)^3$ ($9 \cdot n^3$ if $\nu = 3$). No general reduction to a special structure is possible in this general case. Nevertheless observe that $\mathcal{U}^\ell$ is the zero matrix if the corresponding ℓth deviating argument does not fall into the current step (see (20)). If this situation occurs for all ℓ the problem presents a so-called ODE-like structure.

The ODE-like iteration. Consider the case when $\mathcal{U}^\ell = \mathbf{0}$ for all ℓ, that is, the case where delays are larger than the stepsize or equivalently that the deviating arguments fall to the left of t_n, i.e., $\alpha_\ell(t_n + c_j h, Y^j) < t_n$ for all $j = 1, \dots, \nu$ and for all $\ell = 1, \dots, p$. Then the last term in (21) is identically zero; this means that

$$J = I_\nu \otimes I_d - h_{n+1}\, A \otimes \left(\frac{\partial f}{\partial y} + \sum_{\ell=1}^{p} \frac{\partial f}{\partial z_\ell} \eta'(\alpha_{\ell 0}) \frac{\partial \alpha_\ell}{\partial y} \right) =: J_0. \tag{22}$$

Then, on following the ideas of Bickart and Butcher, the matrix J_0 is pre-multiplied by $(h_{n+1}\, A)^{-1} \otimes I_d$. Successively, in order to exploit the structure of the system, the idea is to block-diagonalize A^{-1} (this is completely analogous to the ODE case as shown, e.g., in [25]). Denoting the transformation matrix by T, we see that $T^{-1}\, A^{-1}\, T = D$ (where D is block-diagonal); then, introducing the transformed variables $W := (T^{-1} \otimes I_d)\, Y$, we obtain an equivalent Newton iteration with Jacobian

$$\hat{J}_0 = h_{n+1}^{-1}\, D \otimes I_d - I_\nu \otimes \left(\frac{\partial f}{\partial y} + \sum_{\ell=1}^{p} \frac{\partial f}{\partial z_\ell} \eta'(\alpha_{\ell 0}) \frac{\partial \alpha_\ell}{\partial y} \right). \tag{23}$$

Since the linear system obtained has block-diagonal structure, the linear algebra is certainly more efficient than that of the original iteration (based on J_0).

But when the stepsize is larger than one or more delays, which means that, for some ℓ and for some $j \in \{1, \dots, \nu\}$, we have $\alpha_\ell(t_n + c_j h_{n+1}, Y^j) > t_n$, the situation is completely different and the previous procedure cannot fruitfully be applied to (21). In order to maintain the advantage of the tensor structure given by (23) one could proceed by considering an inexact Newton process where the correct Jacobian (that is, (21)) is only roughly approximated by (22). In this way, because of the transformation to (23), the LU factorization in the Newton process would be cheaper.

The risk lies in the fact that Newton iteration may become significantly slower or may not converge.

The stopping criteria implemented in the code RADAR5 are similar to those used for ODEs (see [21, 24]). We review them briefly. We set $\tilde{Y}^{[k]}$ as the kth iterate generated by the Newton process in order to approximate Y, the exact solution of (11), and define $\Delta^{[k]} = \tilde{Y}^{[k]} - \tilde{Y}^{[k-1]}$. If the method is linearly convergent, as we expect, then $\|\Delta^{[k]}\| \le \Theta \|\Delta^{[k-1]}\|$ with $|\Theta| < 1$. Then we get the estimate

$$\|\tilde{Y}^{[k]} - Y\| \le \frac{\Theta}{1-\Theta} \|\Delta^{[k-1]}\|. \tag{24}$$

In order to estimate Θ the ratios

$$\Theta_k = \|\Delta^{[k]}\| / \|\Delta^{[k-1]}\|$$

are computed progressively. According to (24) and with $\eta_k = \Theta_k/(1-\Theta_k)$, the iteration stops successfully if

$$\eta_k \|\Delta^{[k]}\| \le \rho \cdot \mathrm{Tol}, \tag{25}$$

where Tol is the error tolerance adopted and ρ is a suitable coefficient. We denote by $k_{\max}$ the maximum number of allowed iterations. The iteration fails if one of the following situations occur (for some $k \le k_{\max}$):

$$\Theta_k \ge 1, \tag{26}$$

$$\frac{\Theta_k^{k_{\max}-k}}{1-\Theta_k} \|\Delta^{[k]}\| > \rho \cdot \mathrm{Tol}. \tag{27}$$

Condition (26) indicates that the iteration is diverging while condition (27) indicates that (25) does not seem to be satisfied within the remaining $k_{\max} - k$ iterations.

Preserving the tensor structure of the Jacobian. As we have mentioned, using (22) as an inexact approximation of (21) may be not safe. An efficient simplification would possibly consist in approximating the matrices $\mathcal{U}^\ell$ in (21) as

$$\mathcal{U}^\ell \approx \gamma^\ell I_\nu$$

for a suitable $\gamma^\ell \in \mathbb{R}$. This would determine for the corresponding Jacobian matrix the same special tensor structure as that of the Jacobian matrix J_0 (hence allowing a transformation which is analogous to (23)).

The ODE-like iteration corresponds to choosing $\gamma^\ell = 0$ for all ℓ. But when $\mathcal{U}^\ell \ne \mathbf{0}$ we have seen (and experimentally verified) that this might be quite critical for the convergence of the Newton process. Thus a suitable choice of the parameter γ^ℓ turns out to be important for the convergence of the Newton iteration.

A structure preserving approximation. A first possibility is that of setting $\mathcal{U}^\ell = I_\nu$ if $\alpha_\ell(t_n + c_j h, Y^j) > t_n$ for some j; this strategy is implemented in the first version of the code RADAR5.

A second possibility, which is included in the second release of the code RADAR5, consists in finding the optimal coefficient γ^ℓ on the basis of a suitable optimization criterion. We propose the choice

$$\gamma^\ell \longrightarrow \min_{\gamma \in \mathbb{R}} \|\mathcal{U}^\ell - \gamma\, I_\nu\|^2, \tag{28}$$

where $\|\cdot\|$ is the Frobenius norm.

This choice is motivated both by the simple determination of the optimal coefficient, which is obtained by using such norm, and by the good results obtained in our numerical experiments.

Since the argument function in the min in (28) is a quadratic function with respect to γ, the global minimizer is computed explicitly (see [23]). We note that, in the special case where $\alpha_\ell(t, y(t)) \equiv t$, we obtain, as expected, $\gamma^\ell = 1$. This can be reasonably interpreted as the approximate situation where the stepsize is much larger than the corresponding delay.

With the previous procedure we obtain the following approximation of (21):

$$J_\gamma = I_\nu \otimes I_d - h\, A \otimes \left(\frac{\partial f}{\partial y} + \sum_{\ell=1}^{p} \left(\eta'(\alpha_{\ell 0}) \frac{\partial \alpha_\ell}{\partial y} + \gamma^\ell \right) \frac{\partial f}{\partial z_\ell} \right),$$

and, consequently, by making use of the same transformation used in order to obtain $\hat{J}_0$ (see (23)), we get

$$\hat{J}_\gamma = (h_{n+1})^{-1}\, D \otimes I_d - I_\nu \otimes \left(\frac{\partial f}{\partial y} + \sum_{\ell=1}^{p} \left(\eta'(\alpha_{\ell 0}) \frac{\partial \alpha_\ell}{\partial y} + \gamma^\ell \right) \frac{\partial f}{\partial z_\ell} \right), \tag{29}$$

which has the block-diagonal structure of $\hat{J}_0$.

Implemented algorithm in RADAR5. The inexact iterations are computationally convenient. In fact the transformation of the approximated Jacobian to (29) is very convenient; for the 3-stage Radau method the cost for the linear systems would be $(5/3) \cdot n^3$ ops.

We allow two possible Newton iterations in the method:

- a cheap *inexact* iteration which consists of setting $\mathcal{U}^\ell = \gamma^\ell\, I_\nu$ (according to (28)) and leads to a block-diagonal Jacobian matrix;
- an expensive *quasi-exact* iteration, which consists of taking $\mathcal{U}^\ell \not\equiv \mathbf{0}$ according to (20), which leads to a full-structure Jacobian matrix.

The first iteration turns out to be equivalent to the second if the stepsize is smaller than all delays, that is, all deviating arguments fall on the left-hand side of the point t_n; furthermore it is very close to the second even if the delays are much larger than the stepsize, that is, when $\alpha_\ell(s, y(s)) \approx s$ (for all ℓ) in the integration interval $[t_n, t_{n+1}]$. The strategy we have chosen is the following. We make use of an iteration indicator Iflag which drives the procedure.

Algorithm 6 *At the beginning of the step, if necessary, we compute the ODE-like Jacobian* J_0.

1. **If** *necessary we compute the optimal values* γ^{ℓ} $(\ell = 1, \ldots, p)$ *according to (28) and update the Jacobian* J_{γ} *according to (29). Then we set* Iflag $= 1$ *(inexact iteration).*
2. **Otherwise** *we set* Iflag $= 0$ *(pure ODE-like iteration).*
3. *Apply the inexact iteration.*
4. **If** *the iteration fails (see (26) and (27)) we stop it and* **go to** *6.*
5. **Otherwise** *we accept the step and* **exit**.
   Possible switch to an exact iteration.
6. **If** Iflag $= 0$ *we reduce the stepsize and* **restart** *a new step:* **go to** *3.*
7. **Otherwise** *we set* Iflag $= 2$ *(quasi-exact iteration).*
8. *Apply the quasi-exact iteration.*
9. **If** *the quasi-exact iteration fails (see (26) and (27)) we stop it and reduce the stepsize. Then we* **restart** *a new step:* **go to** *1.*
10. **Otherwise** *we accept the step and* **exit**.

3.5 Local error estimation and stepsize control

Stepsize selection strategies for stiff ordinary differential equations are usually based on error estimations at grid points. For delay equations, where the accuracy of the dense output strongly influences the performance, such an approach is not sufficient. We briefly recall the technique used in RADAU5 [24, Sect. IV.8], and we discuss a modification suitable for delay equations. Standard error estimators for ordinary differential equations are based on embedded methods. This leads to

$$\Delta y_n = h_{n+1} f(y_n, z_n) + \sum_{i=1}^{\nu} e_i \big(Y^i - y_n\big),$$

where the coefficients e_i are chosen so that $\Delta y_n = \mathcal{O}(h_{n+1}^{\nu+1})$ whenever the problem and the solution are smooth. For very stiff problems, the expression Δy_n generally overestimates the true local error; thus it is pre-multiplied by the projection matrix

$$P = \big(I_d - h_{n+1}\lambda(f_y + \ldots)\big)^{-1},$$

where λ is a real eigenvalue of the Runge–Kutta matrix A. Whenever the tensor product structure in (29) is exploited, then an LU decomposition of the matrix $I_d - h_{n+1}\lambda(f_y + \ldots)$ is already available from the simplified Newton iterations.

We consider the following norm for an arbitrary (error) vector w_n:

$$\|w_n\|^2 = \frac{1}{d} \sum_{i=1}^{d} \left(\frac{w_{n,i}}{s_i}\right)^2,$$

where $s_i = 1 + \rho\, |y_{n,i}|$ and ρ is the ratio $\mathrm{tol_r}/\mathrm{tol_a}$ between the relative ($\mathrm{tol_r}$) and absolute ($\mathrm{tol_a}$) input tolerances per step (which are used for the stepsize selection). Then we denote by ω_n the following measure of the error at grid points

$$\omega_n = \|\Delta y_n\|.$$

We call ω_n the *discrete* component of the local error. In the general case, the local order of the error-estimating method turns out to be $\nu + 1$, that is, $\omega_n = \mathcal{O}(h_{n+1}^{\nu+1})$.

Estimation of the error in the dense output. As mentioned, for delay equations, where the uniform accuracy of the numerical solution also has influence on the local error, it is necessary to control the error uniformly in time. To do this, in general we may also consider the polynomial v_m (see (14)) of degree $\nu - 1$, which interpolates the values Y^i but not y_m. It turns out that

$$\eta_n = \max_{\vartheta \in [0,1]} \|u_n(t_n + h_{n+1}\vartheta) - v_n(t_n + h_{n+1}\vartheta)\| = \|u_n(t_n) - v_n(t_n)\| = \mathcal{O}(h_{n+1}^{\nu}).$$

We use this quantity as an indicator for the uniform error and call it the *continuous* component of the local error.

The estimate used for stepsize control is then given by

$$\mathrm{err}_n = \gamma_1 \omega_n + \gamma_2 (\eta_n)^{(\nu+1)/\nu} = \mathcal{O}(h_{n+1}^{\nu+1}),$$

with the parameters $\gamma_1, \gamma_2 \geq 0$ possibly tuned by the user. This choice is the fruit of both theoretical and empirical analysis. The order of the estimation is $\nu + 1$ (that is, 4 if $\nu = 3$) when the solution is smooth, and is obtained quite cheaply. After error estimation, stepsize prediction is obtained by classical formulas (see [24]).

3.6 Numerical illustration for the Waltman problem

In this last paragraph we illustrate the behavior of the algorithms presented in this section, which are implemented in the code RADAR5 (version 2). The code applied to problem (8) behaved very well.

In particular, we focus our attention on the breaking point computation technique, on the devised Newton process, and on the error control strategy.

We consider the following choices for the model problem (8)–(9): $f_1(x, y, w) = xy + w$ and $f_2(y, w) = y + w$ are the functions modelling the accumulation effects in (6) and (7), $a = 1.8$ and $b = 20$ are the amplification factors, $\gamma = 0.002$ is a catabolic factor, $r_1 = 5 \cdot 10^4$, $r_2 = 0$ and $s = 10^5$ are combination factors. Finally, in order to simplify the problem, we fix $t_0 = 35$, $t_1 = 197$ as the activation instants.

The initial values and initial functions are given by $y_1(t) = 5 \cdot 10^{-6}$, $y_2(t) = 10^{-15}$, and $y_3(t) = y_4(t) = \alpha_1(t) = \alpha_2(t) = 0$ for $t \leq 0$.

The right-hand side of the differential equation has jump discontinuities at $t_0 = 35$ and $t_1 = 197$, but the solution is continuous and has jumps only in its derivatives. There are two state dependent delays $\alpha_1(t, y(t)) = y_5(t)$ and $\alpha_2(t, y(t)) = y_6(t)$. After activation, the first delay monotonically approaches a constant value (nearly vanishing delay); the second has an extremely steep slope and rapidly approaches 0 (see Fig. 2).

The code successfully solves this problem on the interval $[0, 300]$ for all tolerances. Breaking points for the solution are obviously t_0 and t_1 (included in the mesh). The code further computes several breaking points, some of which are indicated in Table 1. Due to the nearly vanishing delays in the problem, there are very many breaking points beyond $t = 197$, and our computation shows that only a few of them are important and need be included in the mesh. Any code that tries to compute all

Table 1. Some of the computed breaking points

breaking point	ancestor	argument
$\xi_1 = 55.21325176$	t_0	α_1
$\xi_2 = 69.26718167$	ξ_1	α_1
$\xi_3 = 79.63960593$	ξ_2	α_1
$\xi_4 = 197.0000071$	t_0	α_2
$\xi_5 = 197.0000115$	ξ_1	α_2
$\xi_6 = 197.0000125$	ξ_5	α_1
$\xi_7 = 197.0000147$	ξ_2	α_2
$\xi_8 = 197.0000173$	ξ_3	α_2

breaking points will be inefficient for this problem, because it must take excessively small steps. With relative and absolute tolerances $\text{tol}_\text{r} = 10^{-6}$, $\text{tol}_\text{a} = 10^{-6}\text{tol}_\text{r}$ at the endpoint $t = 300$ the numerically computed values of the deviating arguments are

$$\alpha_1(300) = 299.9999, \quad \alpha_2(300) = 299.6649,$$

while the stepsize $h = 16.5045$, which means that the stepsize is much larger than the delays.

Table 2 illustrates the behavior of the code for various relative error tolerances (per step), which we denote by tol_r. We write $e = -\log(\text{tol}_\text{r})$. We denote by feval the number of function evaluations and by error the computed relative error on the solution. For $\text{tol}_\text{r} = 10^{-12}$ version 1 stops soon after t_1.

Finally we look at the effects of the approximation of the Jacobian described at the end of Sect. 3.4, and implemented in version 2 of the code RADAR5 (nstep is the total number of steps).

Table 3 shows that the new version of the code is more efficient than the previous. This confirms that a better approximation of the Jacobian is achieved.

Observe that, for smaller tolerances, the advantage of the novel approach is less evident since the average stepsize decreases and overlapping occurs less frequently.

Table 2. Error behavior: comparison between versions 1 and 2 of RADAR5

	version 2		version 1	
e	feval	error	feval	error
3	2227	0.218	2800	0.778
6	3409	6.85e-4	4244	1.05e-2
9	7939	3.32e-6	8537	2.48e-4
12	22694	3.66e-8	–	–

Table 3. Behavior of Newton iteration: comparison between versions 1 and 2 of RADAR5

	version 2		version 1	
e	nstep	feval	nstep	feval
4	210	1575	393	2798
6	316	2307	443	3091
8	631	4554	798	5621
10	1183	8544	1447	10340
12	2311	16644	2799	20021

3.7 Software

Release 2.1 of the code RADAR5 is presently being distributed at the web-sites:
http://univaq.it/~guglielm
http://www.unige.ch/~hairer/software.html
with several examples, including the Waltman model considered here.

4 The functional continuous Runge-Kutta method

In this section we describe the functional continuous Runge-Kutta method introduced in Sect. 2 for the general retarded functional differential equation in the form:

$$\begin{cases} y'(t) = f(t, y_t), & t \geq t_0, \\ y(t) = \phi(t), & t \leq t_0. \end{cases} \tag{30}$$

According to (4) with the option (5) for the stages functions $Y^i_{t^i_{n+1}}$, the FCRK method takes the form

$$\eta(t_n + \theta h_{n+1}) = \eta(t_n) + h_{n+1} \sum_{i=1}^{\nu} b_i(\theta) K^i, \quad \theta \in [0, 1], \tag{31}$$

where the derivatives K^i are given by

$$K^i = f\left(t^i_{n+1}, Y^i_{t^i_{n+1}}\right) \tag{32}$$

and $Y^i_{t^i_{n+1}}$ is a stage function given by:

$$\begin{aligned} &Y^i(t_n + \theta h_{n+1}) = \eta(t_n) + h_{n+1} \sum_{j=1}^{\nu} a_{ij}(\theta) K^j, \quad \theta \in [0, c_i], \\ &Y(t) = \eta(t), \quad t \leq t_n. \end{aligned} \tag{33}$$

Note that the coefficients a_{ij}, $i, j = 1, \ldots, \nu$, are polynomial functions of the parameter $\theta \in [0, 1]$ and this feature renders these schemes different from Continuous Runge-Kutta (CRK) methods where only the weights b_i, $i = 1, \ldots, \nu$, are polynomial functions of θ.

The method is called explicit if $a_{ij}(\cdot)$ is the zero function for $j \geq i$. In the case of explicit methods the derivatives K^i, $i = 1, \dots, \nu$, can be successively computed as:

- $K^1 = f\left(t^1_{n+1}, Y^1_{t^1_{n+1}}\right)$, where:

$$Y^1(t_n + \theta h_{n+1}) = \eta(t_n), \quad \theta \in [0, c_1],$$
$$Y^1(t) = \eta(t), \quad t \leq t_n,$$

(note that $K^1 = f(t_n, \eta_{t_n})$ when $c_1 = 0$);

- for $i = 2, \dots, \nu$, $K^i = f\left(t^i_{n+1}, Y^i_{t^i_{n+1}}\right)$, where:

$$Y^i(t_n + \theta h) = \eta(t_n) + h_{n+1} \sum_{j=1}^{i-1} a_{ij}(\theta) K^j, \quad \theta \in [0, c_i],$$
$$Y^i(t) = \eta(t), \quad t \leq t_n.$$

On the contrary, in the general implicit case described by (32) and (33), the derivatives vector $K = (K^1, \dots, K^\nu) \in \mathbb{R}^{\nu d}$ is the solution of an νd-dimensional algebraic system.

Remark 3. We note that, if $a_{ij}(\cdot) = b_j(\cdot)$, then $Y^i = \eta$ for $i = 1, \dots, \nu$ and the FCRK method coincides with the standard approach. On the contrary a CRK method $(A, b(\cdot), c)$ coincides with a FCRK method when it is natural, i.e., $a_{ij} = b_j(c_i)$, $i, j = 1, \dots, \nu$, and the coefficients are given by $a_{ij}(\cdot) = b_j(\cdot)$.

When the FCRK method is applied to ODEs, only the value $\eta(t_n + h_{n+1})$ is needed and (31–33) become:

$$\eta(t_n + h_{n+1}) = \eta(t_n) + h_{n+1} \sum_{i=1}^{\nu} b_i(1) K^i,$$

$$K^i = f\left(t^i_{n+1}, Y^i\left(t^i_{n+1}\right)\right)$$

and

$$Y^i\left(t^i_{n+1}\right) = \eta(t_n) + h_{n+1} \sum_{j=1}^{\nu} a_{ij}(c_i) K^j,$$

respectively. Thus it is the same as the ν-stage RK method for ODEs (A, b, c) where A is $\nu \times \nu$ matrix of elements

$$a_{ij} = a_{ij}(c_i), \; i, j = 1, \dots, \nu,$$

and b is the ν-vector of components

$$b_i = b_i(1), \; i = 1, \dots, \nu.$$

We denote the FCRK method by the usual Butcher tableau:

$$\begin{array}{c|ccc} c_1 & a_{11}(\theta) & \dots & a_{1\nu}(\theta) \\ \cdot & \cdot & & \cdot \\ \cdot & \cdot & & \cdot \\ \cdot & \cdot & & \cdot \\ c_\nu & a_{\nu 1}(\theta) & \dots & a_{\nu\nu}(\theta) \\ \hline & b_1(\theta) & \dots & b_\nu(\theta) \end{array}$$

Here are some elementary examples of FCRK method.

- One-stage FCRK method:

$$\begin{array}{c|c} c & \theta \\ \hline & \theta \end{array} \tag{34}$$

i.e.,

$$\eta(t_n + \theta h_{n+1}) = \eta(t_n) + h_{n+1}\theta K^1, \quad \theta \in [0, 1],$$

where $K^1 = f\left(t_n + ch_{n+1}, Y^1_{t_n + ch_{n+1}}\right)$ and

$$Y^1(t_n + \theta h_{n+1}) = \eta(t_n) + h_{n+1}\theta K^1, \quad \theta \in [0, c],$$
$$Y^1(t) = \eta(t), \ t \leq t_n.$$

In particular, for $c = 0, 1/2, 1$ we get the explicit Euler, midpoint and implicit Euler FCRK methods, respectively.

- Trapezoidal FCRK method:

$$\begin{array}{c|cc} 0 & 0 & 0 \\ 1 & \frac{1}{2}\theta & \frac{1}{2}\theta \\ \hline & \frac{1}{2}\theta & \frac{1}{2}\theta \end{array} \tag{35}$$

i.e.,

$$\eta(t_n + \theta h_{n+1}) = \eta(t_n) + h_{n+1}\frac{\theta}{2}\left(K^1 + K^2\right), \quad \theta \in [0, 1],$$

where $K^1 = f\left(t_n, \eta_{t_n}\right)$, $K^2 = f\left(t_{n+1}, Y^2_{t_{n+1}}\right)$ and

$$Y^2(t_n + \theta h_{n+1}) = \eta(t_n) + h_{n+1}\tfrac{\theta}{2}\left(K^1 + K^2\right), \quad \theta \in [0, 1],$$
$$Y^2(t) = \eta(t), \quad t \leq t_n.$$

According to Remark 3, the foregoing examples are of the standard approach. The next two are not.

- Another version of the trapeziodal FCRK method:

$$\begin{array}{c|cc} 0 & 0 & 0 \\ 1 & \frac{1}{2}\theta & \frac{1}{2}\theta \\ \hline & \theta-\frac{1}{2}\theta^2 & \frac{1}{2}\theta^2 \end{array} \qquad (36)$$

i.e.,

$$\eta\left(t_n+\theta h_{n+1}\right)=\eta\left(t_n\right)+h_{n+1}\left[\left(\theta-\frac{\theta^2}{2}\right)K^1+\frac{\theta^2}{2}K^2\right], \quad \theta\in[0,1],$$

where $K^1=f\left(t_n,\eta_{t_n}\right)$, $K^2=f\left(t_{n+1},Y^2_{t_{n+1}}\right)$ and

$$\begin{aligned} &Y^2\left(t_n+\theta h_{n+1}\right)=\eta\left(t_n\right)+h_{n+1}\tfrac{\theta}{2}\left(K^1+K^2\right), \quad \theta\in[0,1],\\ &Y^2(t)=\eta(t), \quad t\le t_n. \end{aligned}$$

This version differs from the previous in that it has uniform order two instead of one.

- Heun method:

$$\begin{array}{c|cc} 0 & 0 & 0 \\ 1 & \theta & 0 \\ \hline & \theta-\frac{1}{2}\theta^2 & \frac{1}{2}\theta^2 \end{array} \qquad (37)$$

i.e.,

$$\eta\left(t_n+\theta h_{n+1}\right)=\eta\left(t_n\right)+h_{n+1}\left[\left(\theta-\frac{\theta^2}{2}\right)K^1+\frac{\theta^2}{2}K^2\right], \quad \theta\in[0,1],$$

where $K^1=f\left(t_n,\eta_{t_n}\right)$, $K^2=f\left(t_{n+1},Y_{t_{n+1}}\right)$ and

$$\begin{aligned} &Y\left(t_n+\theta h_{n+1}\right)=\eta\left(t_n\right)+h_{n+1}\theta K^2, \quad \theta\in[0,1],\\ &Y(t)=\eta(t), \quad t\le t_n. \end{aligned}$$

The explicit Euler and Heun methods for the general RFDE (30) were first presented by Cryer and Tavernini in [13]. In the subsequent paper [35], Tavernini considered particular implicit FCRK methods derived from collocation and particular explicit FCRK methods derived from predictor-corrector versions of the earlier methods. In particular, he obtained the four-stage explicit method:

$$\begin{array}{c|cccc} 0 & 0 & 0 & 0 & 0 \\ 1 & \theta & 0 & 0 & 0 \\ \frac{1}{2} & \theta-\frac{\theta^2}{2} & \frac{\theta^2}{2} & 0 & 0 \\ 1 & \theta-\frac{\theta^2}{2} & \frac{\theta^2}{2} & 0 & 0 \\ \hline & \theta-\frac{3\theta^2}{2}+\frac{2\theta^3}{3} & 0 & 2\theta^2-\frac{4\theta^3}{3} & -\frac{\theta^2}{2}+\frac{2\theta^3}{3} \end{array}$$

and the seven-stage explicit method:

$$
\begin{array}{c|ccccccc}
0 & 0 & 0 & 0 & 0 & 0 & 0 & 0 \\
1 & \theta & 0 & 0 & 0 & 0 & 0 & 0 \\
\frac{1}{2} & \theta - \frac{\theta^2}{2} & \frac{\theta^2}{2} & 0 & 0 & 0 & 0 & 0 \\
1 & \theta - \frac{\theta^2}{2} & \frac{\theta^2}{2} & 0 & 0 & 0 & 0 & 0 \\
\frac{1}{3} & \theta - \frac{3\theta^2}{2} + \frac{2\theta^3}{3} & 0 & 2\theta^2 - \frac{4\theta^3}{3} & -\frac{\theta^2}{2} + \frac{2\theta^3}{3} & 0 & 0 & 0 \\
\frac{2}{3} & \theta - \frac{3\theta^2}{2} + \frac{2\theta^3}{3} & 0 & 2\theta^2 - \frac{4\theta^3}{3} & -\frac{\theta^2}{2} + \frac{2\theta^3}{3} & 0 & 0 & 0 \\
1 & \theta - \frac{3\theta^2}{2} + \frac{2\theta^3}{3} & 0 & 2\theta^2 - \frac{4\theta^3}{3} & -\frac{\theta^2}{2} + \frac{2\theta^3}{3} & 0 & 0 & 0 \\
\hline
 & b_1(\theta) & 0 & 0 & 0 & b_5(\theta) & b_6(\theta) & b_7(\theta)
\end{array}
\qquad (38)
$$

where:

$$
\begin{aligned}
b_1(\theta) &= \theta - \frac{11\theta^2}{4} + 3\theta^3 - \frac{9\theta^4}{8} \\
b_5(\theta) &= \frac{9\theta^2}{2} - \frac{15\theta^3}{2} + \frac{27\theta^4}{8} \\
b_6(\theta) &= -\frac{9\theta^2}{4} + 6\theta^3 - \frac{27\theta^4}{8} \\
b_7(\theta) &= \frac{\theta^2}{2} - \frac{3\theta^3}{2} + \frac{9\theta^4}{8}.
\end{aligned}
$$

More recently Maset, Torelli and Vermiglio in [32] provided, for the FCRK method, uniform and discrete order conditions up to order four and found the minimum number of stages necessary in the explicit case.

In the remainder of this section we provide order conditions for FCRK methods and construct explicit methods of uniform global order two, three and four. Lastly, we analyze the effect of perturbations due to approximations in the evaluation of the right-hand side function f in (30).

4.1 Order conditions

Henceforth we assume the following simplifying conditions for the ν-stage FCRK method $(A(\cdot), b(\cdot), c)$

$$
\begin{aligned}
&\sum_{i=1}^{\nu} b_i(\theta) = \theta, \quad \theta \in [0,1], \\
&\sum_{j=1}^{\nu} a_{ij}(\theta) = \theta, \quad \theta \in [0, c_i], \quad i = 1, \ldots, \nu,
\end{aligned}
\qquad (39)
$$

which guarantee uniform order one. Moreover, we set $b_i = b_i(1)$, $i = 1, \ldots, \nu$, and denote the distinct c_i's by $c_1^*, \ldots, c_{\nu^*}^*$.

The (necessary and sufficient) condition for uniform order two is

$$
\sum_{i=1}^{\nu} b_i(\theta) c_i = \frac{\theta^2}{2}, \quad \theta \in [0,1], \qquad (40)
$$

whereas the condition for discrete order two is the same but with $\theta = 1$.

Table 4. Uniform order conditions for FCRK methods

Order	Conditions
3	$\sum_{i=1}^{\nu} b_i(\theta) c_i^2 = \frac{\theta^3}{3}, \quad \theta \in [0,1]$
	for $m = 1, \ldots, \nu^*$: $\sum_{\substack{i=1 \\ c_i = c_m^*}}^{\nu} b_i(\theta) \left(\sum_{j=1}^{\nu} a_{ij}(\beta) c_j - \frac{\beta^2}{2} \right) = 0, \quad \theta \in [0,1], \quad \beta \in [0, c_m^*]$
4	$\sum_{i=1}^{\nu} b_i(\theta) c_i^3 = \frac{\theta^4}{4}, \quad \theta \in [0,1]$
	for $m = 1, \ldots, \nu^*$: $\sum_{\substack{i=1 \\ c_i = c_m^*}}^{\nu} b_i(\theta) \left(\sum_{j=1}^{\nu} a_{ij}(\beta) c_j^2 - \frac{\beta^3}{3} \right) = 0, \quad \theta \in [0,1], \quad \beta \in [0, c_m^*]$
	for $l, m = 1, \ldots, \nu^*$: $\sum_{\substack{i=1 \\ c_i = c_l^*}}^{\nu} \sum_{\substack{j=1 \\ c_j = c_m^*}}^{\nu} b_i(\theta) a_{ij}(\beta) \left(\sum_{k=1}^{\nu} a_{jk}(\gamma) c_k - \frac{\gamma^2}{2} \right) = 0, \quad \theta \in [0,1], \quad \beta \in [0, c_l^*], \quad \gamma \in [0, c_m^*]$

Therefore the one-stage methods (34), which have uniform order one, have discrete order two if and only if $c = \frac{1}{2}$. On the basis of Theorem 1, method (34) with $c = \frac{1}{2}$ has global order two.

Moreover the two versions (35) and (36) of the trapezoidal rule have discrete order two and uniform order one and two, respectively. Thus, both of them have global order two.

In Table 4 we show the (necessary and sufficient) additional conditions for uniform orders three and four. The conditions for discrete orders three and four are obtained by setting $\theta = 1$. Note that the conditions are quite different from the order conditions for the RK methods. The most striking difference lies in the sums

$$\sum_{\substack{i=1 \\ c_i = c_m^*}}^{\nu}$$

where we sum not over all nodes but only over those that are equal to c_m^*.

4.2 Explicit methods

In this section we specialize the previous order conditions to explicit methods by setting $a_{ij}(\cdot)$ equal to the zero function for $j \geq i$ and then construct methods up to global order four. Explicit methods satisfying (39) must have $c_1 = 0$. Moreover we assume, without loss of generality, that $c_i \neq 0$ for $i = 2, \ldots, \nu$ and set $c_1^* = 0$.

Two-stage explicit methods satisfying (39) take the form:

$$\begin{array}{c|cc} 0 & 0 & 0 \\ c_2 & \theta & 0 \\ \hline & \theta - b_2(\theta) & b_2(\theta) \end{array} \tag{41}$$

with $c_2 \neq 0$. By (40), we see that the methods of uniform order two are given by

$$b_2(\theta) = \frac{\theta^2}{2c_2}, \qquad \theta \in [0,1].$$

For example, for $c_2 = 1$ we obtain the Heun method (37) and, for $c_2 = \frac{1}{2}$, we obtain the method:

$$\begin{array}{c|cc} 0 & 0 & 0 \\ \frac{1}{2} & \theta & 0 \\ \hline & \theta - \theta^2 & \theta^2 \end{array}$$

that we might call the Runge method since it reduces, in the ODE case, to the classical Runge method:

$$\begin{array}{c|cc} 0 & 0 & 0 \\ \frac{1}{2} & \frac{1}{2} & 0 \\ \hline & 0 & 1 \end{array}.$$

Since methods (41) have uniform order one, discrete order two suffices for global order two and, by (40) with $\theta = 1$, this is obtained when

$$b_2 = \frac{1}{2c_2}.$$

Orders three and four. Consider three-stage explicit methods satisfying the simplifying assumptions (39):

$$\begin{array}{c|ccc} 0 & 0 & 0 & 0 \\ c_2 & \theta & 0 & 0 \\ c_3 & \theta - a_{32}(\theta) & a_{32}(\theta) & 0 \\ \hline & \theta - b_2(\theta) - b_3(\theta) & b_2(\theta) & b_3(\theta) \end{array} \tag{42}$$

where $c_2, c_3 \neq 0$. By (40) the condition for uniform order two is

$$b_2(\theta)\, c_2 + b_3(\theta)\, c_3 = \frac{\theta^2}{2}, \qquad \theta \in [0,1]. \tag{43}$$

The first condition in Table 4 for uniform order three is

$$b_2(\theta)\, c_2^2 + b_3(\theta)\, c_3^2 = \frac{\theta^3}{3}, \qquad \theta \in [0,1]. \tag{44}$$

Thus there are polynomials b_2, b_3 satisfying (43) and (44) only if $c_2 \neq c_3$. In this case one of the two remaining conditions for uniform order three reads:

$$b_2(\theta)\left(-\frac{\beta^2}{2}\right) = 0, \qquad \theta \in [0,1], \quad \beta \in [0, c_2].$$

Since $b_2(\cdot) \neq 0$ (this follows by (43) and (44)) explicit FCRK methods (42) of uniform order three do not exist. Note that the same order barrier holds for explicit CRK methods where four stages are necessary (and sufficient) for uniform order three (see [7]).

Now we look for three-stage explicit methods (42) of uniform order two and discrete order three, and then of global order three. First, we consider the case $c_2 \neq c_3$. By (40) and Table 4 with $\theta = 1$, necessary and sufficient conditions for uniform order two and discrete order three are:

$$\begin{cases} b_2(\theta)\, c_2 + b_3(\theta)\, c_3 = \frac{\theta^2}{2}, \quad \theta \in [0,1], \\ b_2 c_2^2 + b_3 c_3^2 = \frac{1}{3} \\ b_2 \left(-\frac{\beta^2}{2}\right) = 0, \quad \beta \in [0, c_2], \\ b_3 \left(a_{32}(\beta)\, c_2 - \frac{\beta^2}{2}\right) = 0, \quad \beta \in [0, c_3]. \end{cases}$$

The third condition yields $b_2 = 0$ and then the first two conditions are satisfied when $c_3 = \frac{2}{3}$ and $b_3 = \frac{3}{4}$. Thus a method (42) is of uniform order two and discrete order three if and only if:

$$\begin{aligned} & c_3 = \tfrac{2}{3} \\ & b_2 = 0 \\ & b_3(\theta) = \tfrac{3\theta^2}{4} - \tfrac{3}{2} b_2(\theta)\, c_2, \quad \theta \in [0,1], \\ & a_{32}(\theta) = \tfrac{\theta^2}{2c_2}, \quad \theta \in [0, c_3]. \end{aligned}$$

An example of such a method (obtained with $b_2(\cdot) = 0$ and $c_3 = \frac{1}{3}$) is:

$$\begin{array}{c|ccc} 0 & 0 & 0 & 0 \\ \frac{1}{3} & \theta & 0 & 0 \\ \frac{2}{3} & \theta - \frac{3}{2}\theta^2 & \frac{3}{2}\theta^2 & 0 \\ \hline & \theta - \frac{3}{4}\theta^2 & 0 & \frac{3}{4}\theta^2 \end{array}$$

which reduces to the three-stage Heun method:

$$\begin{array}{c|ccc} 0 & 0 & 0 & 0 \\ \frac{1}{3} & \frac{1}{3} & 0 & 0 \\ \frac{2}{3} & 0 & \frac{2}{3} & 0 \\ \hline & \frac{1}{4} & 0 & \frac{3}{4} \end{array}$$

in the ODE case.

Other methods (42) of uniform order two and discrete order three can be obtained when $c_2 = c_3$. In this case (42) is of uniform order two and discrete order three if and only if:

$$\begin{aligned} & c_2 = c_3 = \tfrac{2}{3} \\ & b_3 \neq 0 \\ & b_2(\theta) = \tfrac{3\theta^2}{4} - b_3(\theta), \quad \theta \in [0,1], \\ & a_{32}(\theta) = \tfrac{9\theta^2}{16 b_3}, \quad \theta \in [0, c_3]. \end{aligned}$$

An example (obtained with $b_3(\theta) = \frac{1}{2}\theta$) is given by:

$$
\begin{array}{c|ccc}
0 & 0 & 0 & 0 \\
\frac{2}{3} & \theta & 0 & 0 \\
\frac{2}{3} & \theta - \frac{9}{8}\theta^2 & \frac{9}{8}\theta^2 & 0 \\
\hline
 & -\frac{3}{4}\theta^2 + \theta & \frac{3}{4}\theta^2 - \frac{1}{2}\theta & \frac{1}{2}\theta
\end{array}
$$

which reduces to:

$$
\begin{array}{c|ccc}
0 & 0 & 0 & 0 \\
\frac{2}{3} & \frac{2}{3} & 0 & 0 \\
\frac{2}{3} & \frac{1}{6} & \frac{1}{2} & 0 \\
\hline
 & \frac{1}{4} & \frac{1}{4} & \frac{1}{2}
\end{array}
$$

in the ODE case.

We can conclude that three-stage FCRK methods of global order three do in fact exist as in the ordinary case.

When we pass to consider FCRK methods of global order four, i.e., uniform order three and discrete order four, six stages turn out to be necessary and sufficient. Note that, for explicit CRK methods, uniform order three and discrete order four is achieved with only four stages (see [7]).

Consider then a six-stage explicit method satisfying (39):

$$
\begin{array}{c|cccccc}
0 & & & & & & \\
c_2 & \theta & & & & & \\
c_3 & \theta - a_{32}(\theta) & a_{32}(\theta) & & & & \\
c_4 & \theta - \sum_{j=2}^{3} a_{4j}(\theta) & a_{42}(\theta) & a_{43}(\theta) & & & \\
c_5 & \theta - \sum_{j=2}^{4} a_{5j}(\theta) & a_{52}(\theta) & a_{53}(\theta) & a_{54}(\theta) & & \\
c_6 & \theta - \sum_{j=2}^{5} a_{6j}(\theta) & a_{62}(\theta) & a_{63}(\theta) & a_{64}(\theta) & a_{65}(\theta) & \\
\hline
 & \theta - \sum_{i=2}^{6} b_i(\theta) & b_2(\theta) & b_3(\theta) & b_4(\theta) & b_5(\theta) & b_6(\theta)
\end{array}
\qquad (45)
$$

where $c_2, c_3, c_4, c_5, c_6 \neq 0$.

By using the conditions for uniform order three and the conditions for discrete order four ($\theta = 1$) in Table 4, we can show that a six-stage explicit FCRK method (45) is of uniform order three and discrete order four if:

$$\begin{aligned}
&c_3 \neq c_4\\
&\tfrac{c_5+c_6}{3} - \tfrac{c_5 c_6}{2} = \tfrac{1}{4}\\
&b_2(\cdot) = 0\\
&b_3, b_4 = 0\\
&b_3(\theta)\, c_3 + b_4(\theta)\, c_4 + b_5(\theta)\, c_5 + b_6(\theta)\, c_6 = \tfrac{\theta^2}{2}, \ \theta \in [0, 1],\\
&b_3(\theta)\, c_3^2 + b_4(\theta)\, c_4^2 + b_5(\theta)\, c_5^2 + b_6(\theta)\, c_6^2 = \tfrac{\theta^3}{3}, \ \theta \in [0, 1],\\
&a_{32}(\theta) = \tfrac{\theta^2}{2c_2}, \quad \theta \in [0, c_2],\\
&a_{42}(\theta)\, c_2 + a_{43}(\theta)\, c_3 = \tfrac{\theta^2}{2}, \quad \theta \in [0, c_4],\\
&a_{52}(\cdot) = 0\\
&a_{53}(\theta) = \tfrac{\theta^2 c_4}{2c_3(c_4-c_3)} - \tfrac{\theta^3}{3c_3(c_4-c_3)}, \quad \theta \in [0, c_5],\\
&a_{54}(\theta) = -\tfrac{\theta^2 c_3}{2c_4(c_4-c_3)} + \tfrac{\theta^3}{3c_4(c_4-c_3)}, \quad \theta \in [0, c_5],\\
&a_{62}(\cdot) = 0\\
&a_{63}(\theta)\, c_3 + a_{64}(\theta)\, c_4 + a_{65}(\theta)\, c_5 = \tfrac{\theta^2}{2}, \quad \theta \in [0, c_6],\\
&a_{63}(\theta)\, c_3^2 + a_{64}(\theta)\, c_4^2 + a_{65}(\theta)\, c_4^2 = \tfrac{\theta^3}{3}, \quad \theta \in [0, c_6].
\end{aligned} \tag{46}$$

So, by taking the abscissae c_3, c_4, c_5, c_6 such that $c_3 \neq c_4,\ c_4 \neq c_5$ and

$$\frac{c_5 + c_6}{3} - \frac{c_5 c_6}{2} = \frac{1}{4}, \tag{47}$$

and weights and coefficients such that:

$$\begin{aligned}
&b_2(\cdot) = b_3(\cdot) = b_4(\cdot) = 0\\
&a_{42}(\cdot) = 0\\
&a_{52}(\cdot) = 0\\
&a_{62}(\cdot) = a_{63}(\cdot) = 0,
\end{aligned}$$

we obtain the tableau:

$$\begin{array}{c|cccccc}
0 & & & & & & \\
c_2 & \theta & & & & & \\
c_3 & \theta - \frac{\theta^2}{2c_2} & \frac{\theta^2}{2c_2} & & & & \\
c_4 & \theta - \frac{\theta^2}{2c_3} & 0 & \frac{\theta^2}{2c_3} & & & \\
c_5 & \theta - a_{53}(\theta) - a_{54}(\theta) & 0 & a_{53}(\theta) & a_{54}(\theta) & & \\
c_6 & \theta - a_{64}(\theta) - a_{65}(\theta) & 0 & 0 & a_{64}(\theta) & a_{65}(\theta) & \\
\hline
 & \theta - b_5(\theta) - b_6(\theta) & 0 & 0 & 0 & b_5(\theta) & b_6(\theta)
\end{array}$$

where:

$$\begin{aligned}
&a_{53}(\theta) = \tfrac{\theta^2 c_4}{2c_3(c_4-c_3)} - \tfrac{\theta^3}{3c_3(c_4-c_3)}, \quad \theta \in [0, c_5],\\
&a_{54}(\theta) = -\tfrac{\theta^2 c_3}{2c_4(c_4-c_3)} + \tfrac{\theta^3}{3c_4(c_4-c_3)}, \quad \theta \in [0, c_5],\\
&a_{64}(\theta) = \tfrac{\theta^2 c_5}{2c_4(c_5-c_4)} - \tfrac{\theta^3}{3c_4(c_5-c_4)}, \quad \theta \in [0, c_6],\\
&a_{65}(\theta) = -\tfrac{\theta^2 c_4}{2c_5(c_5-c_4)} + \tfrac{\theta^3}{3c_5(c_5-c_4)}, \quad \theta \in [0, c_6],\\
&b_5(\theta) = \tfrac{\theta^2 c_6}{2c_5(c_6-c_5)} - \tfrac{\theta^3}{3c_5(c_6-c_5)}, \quad \theta \in [0, 1],\\
&b_6(\theta) = -\tfrac{\theta^2 c_5}{2c_6(c_6-c_5)} + \tfrac{\theta^3}{3c_6(c_6-c_5)}, \quad \theta \in [0, 1].
\end{aligned}$$

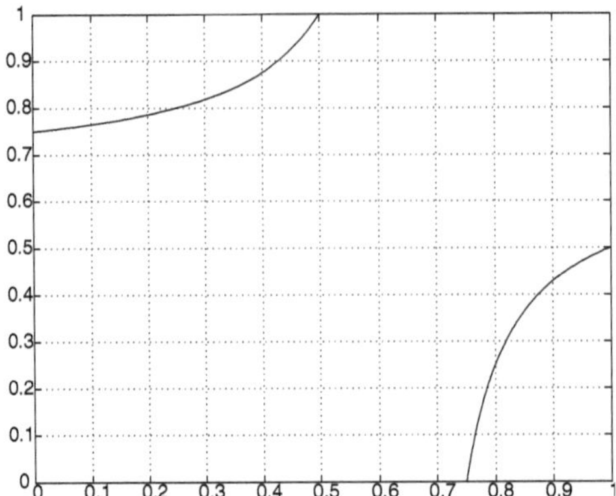

Fig. 3. The curves are the set of couples $(c_5, c_6) \in [0, 1]^2$ satisfying condition (47)

Figure 3 displays the couples (c_5, c_6) satisfying the relation (47). Among them we note $c_5 = 1/2$ and $c_6 = 1$ and, symmetrically, $c_5 = 1$ and $c_6 = 1/2$.

It is worth remarking that conditions (46), which are sufficient for uniform order three and discrete order four, are also necessary when the abscissae c_2, c_3, c_4, c_5, c_6 are distinct.

The construction of higher order FCRK methods is in progress. So far it is proven that seven stages are sufficient for uniform order four and are necessary in the case of distinct abscissae. An example of a seven-stage method of uniform order four is given by (38).

4.3 The quadrature problem

For RFDEs with distributed delay:

$$\begin{cases} y'(t) = F\left(t, y(t), \int\limits_{t-\tau}^{t} k(t, s, y(t+s))\, ds\right), & t \geq t_0, \\ y(t) = \phi(t), \ t \leq t_0, \end{cases} \tag{48}$$

the function f in (30) is given by

$$f(t, \varphi) = F\left(t, \varphi(0), \int\limits_{-\tau}^{0} k(t, s, \varphi(s))\, ds\right)$$

and involves an integral. So, in general, we can provide only approximated values of f by a quadrature rule. In other words, we use an approximation $\widetilde{f}$.

Another situation where an approximation of f is required is the RFDE:

$$\begin{cases} y'(t) = F\left(t, y(t), \sum\limits_{m=0}^{\infty} k(t, m, y(t-\tau_m))\right), & t \geq t_0, \\ y(t) = \phi(t) \ t \leq t_0 \end{cases}$$

where the function f is given by

$$f(t,\varphi)=F\left(t,\varphi(0),\sum_{m=0}^{\infty}k(t,m,\varphi(-\tau_m))\right).$$

In this subsection the effect of using an approximation $\widetilde{f}$ instead of f in a FCRK method is considered. We report only the main result; the details can be found in [32].

We denote by $\widetilde{f}(t,\varphi;\lambda)$ the approximation of $f(t,\varphi)$, where the parameter λ takes into account the approximation procedure adopted in the computation of $f(t,\varphi)$ such as, for example, the quadrature rule selected for the integral in (48).

We introduce the errors

$$\epsilon_{n+1}^{i}=\widetilde{f}\left(t_{n+1}^{i},y_{t_{n+1}^{i}};\lambda_{n+1}^{i}\right)-f\left(t_{n+1}^{i},y_{t_{n+1}^{i}}\right),\quad i=1,\ldots,\nu,$$

where λ_{n+1}^{i} is the parameter relevant to the procedure used for the approximation of $f\left(t_{n+1}^{i},Y_{t_{n+1}^{i}}^{i}\right)$ in (32). It can be easily proved that, if

$$\epsilon_{n+1}^{i}=O\left((h_{n+1})^{\min\{q+1,p\}}\right)$$

for all n and i, then the global order remains $\min\{q+1,p\}$ even if the values $f\left(t_{n+1}^{i},Y_{t_{n+1}^{i}}^{i}\right)$ are replaced by their approximations $\widetilde{f}\left(t_{n+1}^{i},Y_{t_{n+1}^{i}}^{i};\lambda_{n+1}^{i}\right)$.

For instance, in the case of (48), replacing the integrals in

$$F\left(t_{n+1}^{i},Y^{i}\left(t_{n+1}^{i}\right),\int_{t_{n+1}^{i}-\tau}^{t_{n+1}^{i}}k\left(t_{n+1}^{i},s,Y^{i}\left(t_{n+1}^{i}+s\right)\right)ds\right),\quad i=1,\ldots,\nu,$$

by a composite l-point Gaussian quadrature rule, with

$$l=\left\lceil\frac{\min\{q+1,p\}}{2}\right\rceil,$$

across the intervals $\left[t_{n+1}^{i}-\tau,t_m\right]\subset[t_{m-1},t_m]$, $\left[t_k,t_{k+1}\right]$, $k=m,\ldots,n-1$, and $\left[t_n,t_{n+1}^{i}\right]$, the global order $\min\{q+1,p\}$ is preserved.

We end this section by remarking that (48) can also be integrated by the numerical methods, specifically designed for integro-differential equations, described in [9,11] or [10], where the quadrature rule for the integrals is a part of the method.

References

[1] Baker, C.T.H., Paul, C.A.H., Willè, D.R.: Issues in the numerical solution of evolutionary delay differential equations. Adv. Comput. Math. **3**, 171–196 (1995)

[2] Banks, H.T., Mahaffy, J.M.: Stability of cyclic gene models for systems involving repression. J. Theoret. Biol. **74**, 323–334 (1978)
[3] Bell, G.I.: Mathematical model of clonal selection and antibody production. J. Theor. Biol. **29**, 191–232 (1970)
[4] Bell, G.I.: Mathematical model of clonal selection and antibody production II. J. Theor. Biol. **33**, 339–378 (1971)
[5] Bell, G.I.: Mathematical model of clonal selection and antibody production III. The cellular basis of immunological paralysis. J. Theor. Biol. **33**, 378–398 (1971)
[6] Bell, G.I.: Predator-prey equations simulating an immune response. Math. Biosci. **16**, 291–314 (1973)
[7] Bellen, A., Zennaro, M.: Numerical methods for delay differential equations. Oxford: Oxford University Press 2003
[8] Bellman, R., Cooke, K.L.: Differential-difference equations. New York: Academic Press 1963
[9] Brunner, H.: The numerical analysis of functional integral and integro-differential equations of Volterra type. Acta Numer. **13**, 55–145 (2004)
[10] Brunner, H.: Collocation methods for Volterra integral and related functional differential equations. (Cambridge Monographs on Appl. Comput. Math. **15**). Cambridge: Cambridge University Press 2004
[11] Brunner, H., van der Houwen, P.J.: The numerical solution of Volterra equations. (CWI Monographs **3**). Amsderdam: North-Holland 1986
[12] Cooke, K.L.: Functional-differential equations: Some models and perturbation problems. In: Hale, J.K., LaSalle, J.P. (eds.): Differential equations and dynamical systems. New York: Academic Press 1967, pp. 167–183
[13] Cryer, C., Tavernini, L.: The numerical solution of Volterra functional differential equations by Euler's method. SIAM J. Numer. Anal. **9**, 105–129 (1972)
[14] Diekmann, O., van Gils, S.A., Verduyn Lunel, S.M., Walter, H.O.: Delay equations. Functional, complex, and nonlinear analysis. (Appl. Math. Sci.) **110**. Berlin: Springer 1995
[15] Èl′sgol′ts, L.E., Norkin, S.B.: Introduction to the theory and application of differential equations with deviating arguments. New York: Academic Press 1973
[16] Enright, W.H., Hayashi, H.: A delay differential equation solver based on a continuous Runge-Kutta method with defect control. Numer. Algorithms **16**, 349–364 (1998)
[17] Feldstein, A., Neves, K.W.: High order methods for state-dependent delay differential equations with nonsmooth solutions. SIAM J. Numer. Anal. **21**, 844–863 (1984)
[18] Gatica, J.A., Waltman, P.: A threshold model of antigen antibody dynamics with fading memory. In: Lakshmikantham, V. (ed.): Nonlinear phenomena in mathematical sciences. New York: Academic Press 1982, pp. 425–439
[19] Gatica, J.A., Waltman, P.: Existence and uniqueness of solutions of a functional-differential equation modeling thresholds. Nonlinear Anal. **8**, 1215–1222 (1984)
[20] Gatica, J.A., Waltman, P.: A system of functional-differential equations modeling threshold phenomena. In: Lakshmikantham, V. (ed.): Nonlinear analysis and applications. (Lecture Notes in Pure and Appl. Math. **109**) New York: Dekker 1987, pp. 185–188
[21] Guglielmi, N., Hairer, E.: Implementing Radau IIA methods for stiff delay differential equations. Computing **67**, 1–12 (2001)
[22] Guglielmi, N., Hairer, E.: Automatic computation of breaking points in implicit delay differential equations. Submitted. (2005)
[23] Guglielmi, N.: On the Newton iteration in the application of collocation methods to implicit delay equations. Appl. Numer. Math. **53**, 281–297 (2005)

[24] Hairer, E., Wanner, G.: Solving ordinary differential equations. II. Stiff and differential-algebraic problems. 2nd. ed. Berlin: Springer 1996

[25] Hairer, E., Wanner, G.: Stiff differential equations solved by Radau methods. J. Comput. Appl. Math. **111**, 93–111 (1999)

[26] Hauber, R.: Numerical treatment of retarded differential-algebraic equations by collocation methods. Adv. Comput. Math. **7**, 573–592 (1997)

[27] Hoffmann, G.W.: A theory of regulation and self-nonself discrimination in an immune network. Europ. J. Immunol. **5**, 638–647 (1975)

[28] Hoppensteadt, F., Waltman, P.: A problem in the theory of epidemics. Math. Biosci. **9**, 71–91 (1970)

[29] Hoppensteadt, F., Waltman, P.: Did something change? Thresholds in population models. In: Kirkilionis, M. et al. (eds.): Trends in nonlinear analysis. Berlin: Springer 2003, pp. 341–374

[30] Kuang, Y.: Delay differential equations with applications in population dynamics. Boston: Academic Press 1993

[31] Mahaffy, J.M., Bélair, J., Mackey, M.C.: Hematopoietic model with moving boundary condition and state dependent delay: applications in erythropoiesis. J. Theor. Biol. **190**, 135–146 (1998)

[32] Maset, S., Torelli, L., Vermiglio, R.: Runge-Kutta methods for retarded functional differential equations. Math. Models Methods Appl. Sci. **15**, 1203–1251 (2005)

[33] Paul, C.A.H.: A test set of functional differential equations. Numerical Analysis Report 243. Manchester: Univ. Manchester UMIST 1994

[34] Richter, P.H.: A network theory of the immune system. Europ. J. Immunol. **5**, 350–354 (1975)

[35] Tavernini, L.: One-step method for the numerical solution of Volterra functional differential equations. SIAM J. Numer. Anal. **8**, 786–795 (1971)

[36] Waltman, P.: A threshold model of antigen-stimulated antibody production. In: Bell, G.I. et al. (eds.): Theoretical immunology. (Immunology Ser. **8**) New York: Dekker 1978, pp. 437–453

[37] Waltman, P., Butz, E.: A threshold model of antigen-antibody dynamics. J. Theoret. Biol. **65**, 499–512 (1977)

[38] Wu, J.: Theory and applications of partial functional-differential equations. (Applied Mathematical Sciences **119**) Berlin: Springer 1996

Computational electrocardiology: mathematical and numerical modeling*

P. Colli Franzone, L.F. Pavarino, G. Savaré

Abstract. This paper deals with mathematical models of cardiac bioelectric activity at both the cell and tissue levels, their integration in coupled models and their numerical simulation. The macroscopic bidomain model of the cardiac tissue is derived by the two-scale homogenization method. Existence and uniqueness results for the cellular and bidomain models are reviewed. A rigorous derivation of the bidomain model is presented in the framework of Γ-convergence theory, and approximation results concerning its time and space discretization are given.The bidomain model of the myocardium is coupled with the extracardiac medium and extracardiac potentials, computed from given cardiac sources by means of differential or integral representations in order to obtain body surface maps and electrograms. Various approximate models of the bidomain model are examined and discussed such as the monodomain model, the eikonal equations and a relaxed monodomain model. These continuous cardiac models are then numerically approximated by isoparametric finite elements in space and adaptive finitie difference methods in time. Numerical simulations of the monodomain and bidomain models are discussed and examples of large-scale parallel computations are reported; these simulate excitation and repolarization processes in three-dimensional anisotropic domains.

Keywords: Computational electrocardiology, reaction-diffusion systems, bidomain and monodomain models, ionic membrane models, eikonal equations, numerical approximations, parallel simulations, anisotropic cardiac excitation and repolarization.

1 Introduction

Electrocardiology deals with the description of both intracardiac bioelectric phenomena and the extracardiac electric field generated in the animal or human body. The practice of modern medicine relies on noninvasive imaging technologies, such as CT, MRI and PET, for diagnostic purposes and to drive therapeutic procedures. Even though cardiac arrhythmias are among the major causes of death and disability, a noninvasive imaging technique yielding an accurate and reliable diagnosis of the electrophysiological state of the heart is not yet available. Clinic electrocardiography deals with the detection and interpretation of noninvasive potential measurements collected from the time course of the usual Electrocardiograms (ECG) at a few points on the body surface or from the evolution of body surface maps, i.e., potential distribution maps on the body surface reconstructed from measurements at numerous

* This work was partially supported by grants from M.I.U.R (PRIN 2003011441), from the Istituto di Matematica Applicata e Tecnologie Informatiche, Pavia, Italy and from Progetto Intergruppo Istituto Nazionale di Alta Matematica, Roma, Italy

electrodes (100 or more; see the recent surveys [146, 147]). Since the electrode location of the ECG is centimeters away from the heart surface and the current conduction from heart to thorax results in strong signal attenuation and smoothing, the information content of ECGs and body maps is limited and it is a difficult task to extract from these signals detailed information on pathological heart states associated with ischemia or sudden death. Indeed, the origin of arrhythmogenic activity or the existence of abnormal electrophysiological substrates may not be easily inferred from the sequence of cardiac excitation in many cases.

The scientific basis of electrocardiology is the so-called *forward problem of electrocardiology*, i.e., modeling the bioelectric cardiac sources and the conducting media in order to derive the potential field. Of considerable interest for applications are the so-called *inverse problems of electrocardiography* in terms of potentials (see, e.g., the reviews [55, 125] and [16, 17, 116]) or in terms of the cardiac sources (see, e.g., [22, 124]). In this paper, we focus on the Forward Problem alone.

The formulation of models at both cellular and tissue levels provide essential tools for integrating the increasing knowledge of bioelectrochemical phenomena occurring through cardiac cellular membranes. Detailed cellular phenomena are described in microscopic membrane models and the latter are then inserted in macroscopic tissue models in order to investigate their effects at tissue level. Ultimately, the coupled models are validated by comparing simulated results with experimental *in vitro* and *in vivo* data.

From a macroscopic point of view, the forward problem of electrocardiology is described by the so-called bidomain model for the evolution of the intracellular, extracellular and extracardiac potential fields. The two main components of the bidomain model are: a) the dynamics of the ionic current flow through the cardiac cellular membrane, modeled by a system of ordinary differential equations; b) a macroscopic representation of the cardiac tissue modeled as a bidomain superposition of the intra- and extracellular media characterized by anisotropic conductivity tensors associated with the fiber architecture of the myocardium.

In this survey, we investigate aspects related to the formulation of mathematical models of cardiac bioelectric activity, the numerical discretization of these models and their computer simulation. For other important aspects of heart modeling, such as cardiac mechanics, blood flow, electro-mechanical and fluid-mechanical coupling, see, e.g., [38, 67, 101, 126, 137].

The work is organized as follows. In Sect. 2, we survey the main mathematical models of bioelectric activity at a cellular level: the ionic current membrane models for ventricular cells in Sect. 2.1, and models of cellular aggregates of interconnected cells in Sect. 2.2, formally deriving a homogenized model at a macroscopic level. In Sect. 3, we introduce an interpretation of the macroscopic model as a bidomain model and its equivalent formulation. In Sect. 4, we discuss various approximate models: eikonal models, linear and nonlinear monodomain models. In Sect. 5, numerical approximations of the monodomain and bidomain models are discussed. Lastly, in Sect. 6, we display parallel simulations of the processes of excitation, repolarization and re-entry in an idealized geometry of the left ventricular wall.

2 Mathematical models of the bioelectric activity at cellular level

The bioelectric activity of the heart during a heartbeat is a fairly complex phenomenon: here we give only a brief description of major features related to the ventricular myocardium.

The cardiac structure is composed of a collection of elongated cardiac cells having roughly a cylindrical form with a diameter $d_c \approx 10\,\mu\text{m}$ and length $l_c \approx 100\,\mu\text{m}$. The cells are coupled together mainly in end-to-end but also in side-to-side apposition by gap-junctions [65, 127]. These specialized membrane regions of densely packed channels provide direct intercellular communication between the cytoplasmatic compartments of two adjacent cells; they are large at the longitudinal cell ends and small along the lateral borders. The end-to-end contacts form the long fiber structure of the cardiac muscle whereas the presence of lateral junctions establishes a connection between the elongated fibers.

The potential jump v across the membrane is called the transmembrane potential. The whole process of potential generation is quite complicated and is essentially due to current flows of sodium, potassium and calcium ions through the cellular membrane separating the intra- (i) and extracellular (e) media and to their diffusion in these two conducting media.

Starting from the sino-atrial node, which acts as a pacemaker, a front-like variation of the transmembrane potential v spreads first in the atria and then through the myocardium with a very fast transition from the resting value v_r to the plateau value v_p. The values of v_r, v_p for cardiac cells are about $-90\,\text{mV}$ and $10\,\text{mV}$. This phase constitutes the excitation or depolarization phase; it is followed by an interval of almost constant potential (refractory period) and a subsequent less rapid return to the initial state (repolarization). The time profile of the transmembrane potential v may depend in general on the position $\mathbf{x}$ and on the local state of the heart; the whole bioelectric cycle lasts about 300 msec in the human heart.

The fiber structure strongly affects both the excitation and repolarization processes and, in particular, is the main factor of the anisotropic conductivity in cardiac tissue; see [80, 139].

2.1 Ionic current membrane models

We provide a brief account of the structure of the models describing the ionic current across the cellular membrane. The electrical behavior of the membrane is represented by a circuit consisting of a capacitor, modeling the phospholipidic double-layer structure of the membrane, connected in parallel with a resistor, modeling the various ionic channels regulating the selective and independent ionic fluxes through the membrane. The total transmembrane current is the sum of the ionic current I_{ion} and the capacitive current $I_C = C_m \dfrac{dv}{dt}$, where v is the transmembrane potential and C_m the membrane capacity per unit area. By conservation of current, this total current must equal the applied current I_{app},

$$C_m \frac{dv}{dt} + I_{\mathrm{ion}} = I_{\mathrm{app}}.$$

Models for the ionic current I_{ion} are based on the channel gating formalism.

Channel gating. In the simplest models of a given type of ionic channel, the total number $[T]$ of such channels embedded in the cellular membrane is given by $[T] = [O] + [C]$, the sum of the numbers $[O]$ and $[C]$ of channels in the open and closed states, respectively. Denoting by α the rate of channel opening and by β the rate of channel closing, by the law of mass action we have

$$\frac{d[O]}{dt} = \alpha(v)[C] - \beta(v)[O];$$

written in terms of the fraction of open channels $g = [O]/[T]$, this becomes

$$\frac{dg}{dt} = \alpha(v)(1-g) - \beta(v)g. \tag{1}$$

The dependence of the rates $\alpha(v)$ and $\beta(v)$ on the potential v is modeled by different functions in different ionic models, obtained by fitting experimental data. The differential equation (1) can be rewritten as

$$\frac{dg}{dt} = \frac{g_\infty - g}{g_\tau}, \tag{2}$$

where $g_\infty = \alpha/(\alpha+\beta)$ can be interpreted as the steady-state value of $g(t)$ and $g_\tau = 1/(\alpha+\beta)$ as the time constant for the transient state of $g(t)$, since, if g_∞ and g_τ were independent of v, the exact solution of (2) for an initial state g_0 would be $g(t) = g_\infty + (g_0 - g_\infty)e^{-t/g_\tau}$.

Hodgkin-Huxley formalism. Most mathematical models of the ionic currents through the cellular membrane generating the cardiac action potential are based on appropriate extensions of the formalism introduced by A. Hodgkin and A. Huxley in [61] (Nobel Prize in Medicine in 1963) for the quantitative description of the nerve action potential. In this model, the ionic current is the sum of three currents,

$$I_{\mathrm{ion}} = I_{\mathrm{Na}} + I_{\mathrm{K}} + I_{\mathrm{L}},$$

a sodium current I_{Na}, a potassium current I_{K} and a leakage current I_{L}. A linear current-voltage relationship is assumed for all three currents, i.e., we have $I_{\mathrm{Na}} = g_{\mathrm{Na}}(v - v_{\mathrm{Na}})$, $I_{\mathrm{K}} = g_{\mathrm{K}}(v - v_{\mathrm{K}})$, $I_{\mathrm{L}} = g_{\mathrm{L}}(v - v_{\mathrm{L}})$, where $v_{\mathrm{Na}}, v_{\mathrm{K}}, v_{\mathrm{L}}$ are the respective equilibrium potentials and $g_{\mathrm{Na}}, g_{\mathrm{K}}, g_{\mathrm{L}}$ are the conductance coefficients. Here g_{L} is assumed constant, while $g_{\mathrm{Na}} = \overline{g}_{\mathrm{Na}} m^3 h$ and $g_{\mathrm{K}} = \overline{g}_{\mathrm{K}} n^4$ are modeled by gating variables h, m, n satisfying channel-gating first-order kinetic equations such as (1). We recall that the Nernst equilibrium potential for the kth ion is given by $v_k = \frac{RT}{zF} \log \frac{c_k^e}{c_k^i}$, where F is the Faraday constant, R the ideal gas constant, T the absolute temperature, z the valence of the ion and $c_k^{i,e}$ are the intra- and extracellular ion concentrations.

More detailed ionic models. Current progress in molecular biology continues to produce more detailed data on and understanding of the dynamic of the ionic fluxes through the cellular membrane; see, e.g., [41] and the recent review [96]. These new and more accurate experimental data are progressively incorporated in more complex membrane models by parameter identification and data fitting. The ionic current is generally given by

$$I_{\text{ion}}(v, w, c) = \sum_{k=1}^{P} g_k(c) \prod_{j=1}^{M} w_j^{p_{jk}} (v - v_k(c)) + I_0(v, c), \tag{3}$$

where $g_k(c)$ and $v_k(c)$ are the conductance coefficient and Nernst equilibrium potential for the kth ion and p_{j_k} are integers. In (3), we have split the ionic current as the sum of a term related to ionic fluxes modulated by the gating dynamics and a time-independent term $I_0(v, c)$. The gating variables $w := (w_1, \dots, w_M)$ regulate the conductances of the various ionic fluxes and $c := (c_1, \dots, c_Q)$ are variables regulating the intracellular concentrations of the various ions. The dynamics of the gating variables w is given by differential equations such as (1), while the ionic concentrations c satisfy specific differential equations:

$$\begin{aligned}
&\frac{dw_j}{dt} = R_j(v, w_j) = \alpha_j(v)(1 - w_j) - \beta_j(v)(w_j), && \alpha_j,\ \beta_j > 0,\\
&w_j(\mathbf{x}, 0) = w_{j,0}(\mathbf{x}), \qquad 0 \le w_j \le 1, && j = 1, \dots, M,\\
&\frac{dc_k}{dt} - S_k(v, w, c) = 0, \quad c_k(\mathbf{x}, 0) = c_{k,0}(\mathbf{x}), && k = 1, \dots, Q.
\end{aligned}$$

Among the most used models are:

- Beeler-Reuter (1977) [11]: mammal ventricular cells, M = 6, Q=1
- Di Francesco-Noble (1985) [41]: mammal Purkinje fibers
- Luo-Rudy 1 (LR1, 1991) [84]: mammal ventricular cells, M = 6, Q=1
- Noble et al. (1991) [95]: ventricular cells
- Luo-Rudy 2 (LR2, 1994) [85]: guinea pig ventricular cells, M= 6, Q= 5
- Winslow et al. (1999) [159]: canine ventricular cells: M = 25, Q= 6
- ten Tusscher et al. (2004) [148]: human ventricular cells
- Gima-Rudy-Hund LRd (2002, 2004) [53, 66]: ventricular cells, M = 12, Q=6.

We refer to the original references for the complete specifications of the models. Figure 1 shows the time evolution of the LR1 action potential, gating and ion concentration variables. In particular, the morphology of the action potential presents an excitation (depolarization) phase lasting about 1 msec, followed by an early repolarization and then a plateau phase of about 200 msec and a final repolarization phase lasting about 50 msec, with a return to the resting value.

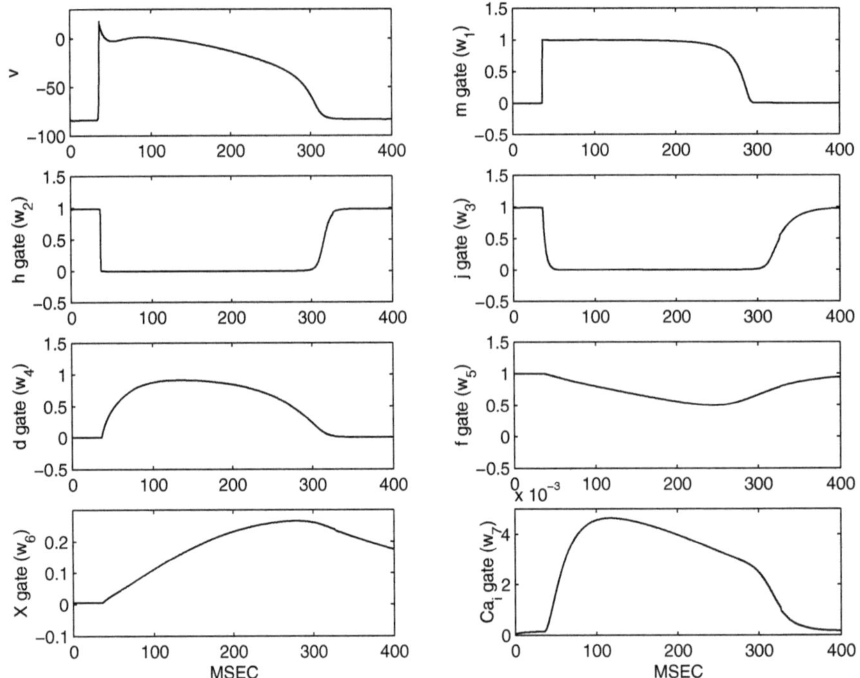

Fig. 1. LR1 membrane model: action potential v, gating variables $w_1, \cdots, w_6$, calcium concentration w_7 at a given point as a function of time

Reduced models. Simplified models of lower complexity, with only one or two gating variables and no ionic concentrations, were proposed and employed in many numerical simulations. The simplest and most used is the FitzHugh Nagumo (FHN) model (M = 1). Assuming that at rest the potential v is zero, in this model the ionic current is described by using only one gating variable w:

$$\begin{cases} I_{\text{ion}}(v, w) := g(v) + \beta w \\ R(v, w) := \eta v - \gamma w, \end{cases} \tag{4}$$

where g is a cubic-like function and $\beta, \eta, \gamma > 0$.

The FHN gating model yields only a coarse approximation of a typical cardiac action potential, particularly in the plateau and repolarization phases. A more recent simplified ionic model with two gating variables (M = 2) was extensively investigated in [42, 43] in simulations of re-entry phenomena.

2.2 Mathematical models of cardiac cell arrangements

At a cellular level the structure of the cardiac tissue can be viewed as composed of two ohmic conducting media: the intracellular space Ω_i (inside the cells) and the extracellular space Ω_e (outside) separated by the active membrane Γ_m. Due to the presence of gap junctions connecting the cardiac cells end-to-end and side-to-side, Ω_i and Ω_e are regarded as two simply-connected open sets of $\mathbb{R}^3$. The effects of the microstructure on current flow are also included in the conductivity tensors $\Sigma_i(\mathbf{x})$, $\Sigma_e(\mathbf{x})$ as inhomogeneous functions of space that reflect the local variations of conductances because of the presence of structural intra- and extracellular inhomogeneities of resistance associated with, e.g., gap junctions, connective tissue, collagen, blood vessel. Let u_i, u_e be the intra- and extracellular potentials and $\mathbf{J}_{i.e} = -\Sigma_{i,e}\nabla u_{i,e}$ their current densities. Let ν_i, ν_e denote the unit exterior normals to the boundaries of Ω_i and Ω_e respectively, satisfying $\nu_i = -\nu_e$ on Γ_m. Under quasi-stationary conditions (see [108]), due to the current conservation law, the normal current flux through the membrane is continuous, and so $\mathbf{J_i} \cdot \nu_i = \mathbf{J_e} \cdot \nu_i$, i.e, in terms of potentials,

$$\nu_i^T \; \Sigma_i \; \nabla u_i + \nu_e^T \Sigma_e \; \nabla u_e = 0 \quad \text{on } \Gamma_m,.$$

On the other hand, since the only active source elements lie on the membrane Γ_m, each flux is equal to the membrane current per unit area I_m which consists of a capacitive and an ionic term (see [68, 86]):

$$-\nu_i^T \; \Sigma_i \; \nabla u_i = \nu_e^T \; \Sigma_e \; \nabla u_e = I_m = C_m \frac{\partial v}{\partial t} + I_{\text{ion}}. \tag{5}$$

In this expression C_m is the surface capacitance of the membrane per unit area and $v := u_i|_{\Gamma_m} - u_e|_{\Gamma_m}$ is the transmembrane potential which reflects the fact that Γ_m is a discontinuity surface for the potential (in the following, we simply write $v = u_i - u_e$).

Denoting by I_i^s, I_e^s the stimulation currents applied to the intra- and extracellular spaces, we have

$$-\operatorname{div}(\Sigma_i \nabla u_i) = I_i^s \quad \text{in } \Omega_i, \qquad -\operatorname{div}(\Sigma_e \nabla u_e) = I_e^s \quad \text{in } \Omega_e. \tag{6}$$

Assuming that $\Omega := \Omega_i \cup \Omega_e \cup \Gamma_m$ is embedded into an insulating media, then we must assign homogeneous Neumann boundary conditions for u_i, u_e on the remaining part of the boundaries $\Gamma_{i,e} = \partial\Omega_{i,e} \setminus \Gamma_m$, namely,

$$\nu_{i,e}^T \Sigma_{i,e} \nabla u_{i,e} = 0.$$

Finally, the system (5), (6) must be supplemented by the initial conditions

$$v(\cdot, 0) = v_0, \quad w(\cdot, 0) = w_0 \quad \text{on } \Gamma_m.$$

For the electric potentials u_i, u_e we can consider two characteristic length scales: a *micro* scale related to typical cell dimensions $\{d_c, l_c\}$ and a *macro* scale determined by a length constant appropriate to the tissue. At the latter scale, i.e., at a macroscopic level, in spite of the discrete cellular structure, the cardiac tissue can be represented by a continuous model. To identify this macroscopic scale, following [94], we consider a suitable nondimensional form of the cellular mathematical model.

The cellular conductivity matrices $\Sigma_i(\mathbf{x})$ and $\Sigma_e(\mathbf{x})$ are symmetric positive definite matrices; setting $\bar{\mu} = \bar{\mu}_i + \bar{\mu}_e$, with $\bar{\mu}_i$, $\bar{\mu}_e$ the average eigenvalues on a cell element, we consider the dimensionless conductivity matrices

$$\sigma_{i,e} = \Sigma_{i,e}/\bar{\mu}.$$

Let R_m be an estimate of the passive membrane resistance near the equilibrium point v_r, i.e., the resting transmembrane potential. Multiplying both sides of Eq. (5) by $R_m/\bar{\mu}$, we obtain

$$-\sigma_i R_m \mathbf{n}_i^T \nabla u_i = \frac{C_m R_m}{\bar{\mu}} \frac{\partial v}{\partial t} + \frac{R_m}{\bar{\mu}} I_{\text{ion}}. \tag{7}$$

We introduce the membrane time constant $\tau_m = R_m C_m$, the length scale unit $\Lambda = \sqrt{l_c \bar{\mu} R_m}$ and we consider the following scaling of the space and time variables:

$$\widehat{\mathbf{x}} = \mathbf{x}/\Lambda, \quad \widehat{t} = t/\tau_m.$$

Disregarding the presence of applied current terms and rescaling Eqs. (6), (7) in the intra- and extracellular media, we obtain:

$$\begin{aligned} &\operatorname{div}_{\widehat{\mathbf{x}}}(\sigma_i \nabla_{\widehat{\mathbf{x}}} u_i) = 0 \quad \text{in } \Omega_i, \qquad \operatorname{div}_{\widehat{\mathbf{x}}}(\sigma_e \nabla_{\widehat{\mathbf{x}}} u_e) = 0 \quad \text{in } \Omega_e \\ &-\nu_i^T \sigma_i \nabla_{\widehat{\mathbf{x}}} u_i = \nu_e^T \sigma_e \nabla_{\widehat{\mathbf{x}}} u_e = \varepsilon \frac{\partial v}{\partial \widehat{t}} + I_{\text{ion}}(v, w, c)) \quad \text{on } \Gamma_m, \end{aligned}$$

where the dimensionless parameter is the ratio $\varepsilon = l_c/\Lambda$ between the micro and the macro length constants.

The two-scale method of homogenization can be applied to the previous current conservation equations. The microscopic space variable measured in the unit cell is defined by $\boldsymbol{\xi} := \mathbf{x}/l_c$; then, for the dimensionless macroscopic coordinates, the micro- and macro scales are related to each other by the scaling parameter ε

$$\boldsymbol{\xi} := \widehat{\mathbf{x}}/\varepsilon.$$

For convenience, in the following we omit the superscripts $\widehat{}$ of the dimensionless variables.

Following the standard approach of homogenization theory, we are assuming that the cells are distributed according to an ideal periodic organization similar to a regular lattice of interconnected cylinders. Due to the longitudinal and transverse intercellular interconnections, in the modeled periodic cellular aggregate the intra- and extracellular media are connected regions. If $\{\boldsymbol{e}_1, \boldsymbol{e}_2, \boldsymbol{e}_3\}$ is an orthogonal basis of $\mathbb{R}^3$, we let

$$E_i, \quad E_e := \mathbb{R}^3 \setminus \overline{E}_i, \quad \text{with common boundary } \Gamma_m := \partial E_i \cap \partial E_e,$$

denote two reference open, connected and periodic subsets of $\mathbb{R}^3$ with Lipschitz boundary, i.e., satisfying

$$E_{i,e} + \boldsymbol{e}_k = E_{i,e}, \quad k = 1, 2, 3.$$

The elementary periodicity region

$$Y := \Big\{ \sum_{k=1}^{3} \alpha_k \boldsymbol{e}_k : 0 \le \alpha_k < 1, \quad k = 1, 2, 3 \Big\}$$

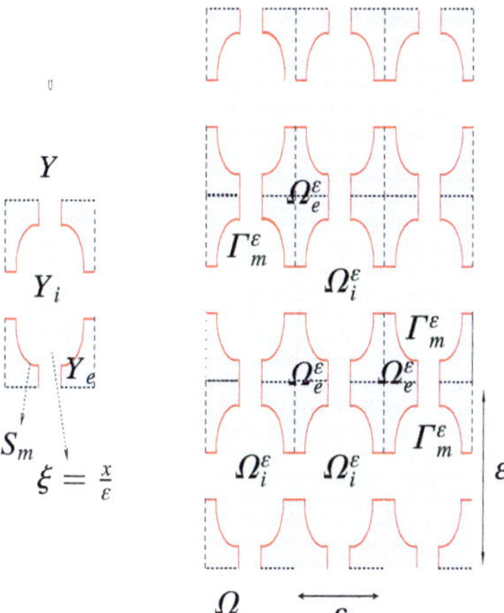

Fig. 2. *Right*: The ideal periodic geometry in a bidimensional section of the simplified 3-D periodic network of interconnected cells. *Left*: Unit cell in the microscopic variable $\xi = x/\varepsilon$

is composed of the intra- and extracellular volumes $Y_{i,e} = Y \cap E_{i,e}$ and represents a reference volume box containing a single cell Y_i with cell membrane surface $S_m = \Gamma_m \cap Y_i$.

The main geometrical assumption is that the physical intra- or extracellular regions are the ε-dilations of the reference lattices $E_{i,e}$, defined as

$$\varepsilon E_{i,e} = \{\varepsilon\xi : \xi \in E_{i,e}\} \quad \text{with} \quad \varepsilon\Gamma_m := \{\varepsilon\xi : \xi \in \Gamma_m\}.$$

Therefore, the decomposition of the physical region Ω, occupied by the cardiac tissue, into intra- and extracellular domains $\Omega_{i,e}^\varepsilon$ (see Fig. 2) can be obtained simply by intersecting Ω with $\varepsilon E_{i,e}$, i.e.,

$$\Omega_i^\varepsilon = \Omega \cap \varepsilon E_i, \quad \Omega_e^\varepsilon = \Omega \cap \varepsilon E_e, \quad \Gamma_m^\varepsilon = \partial\Omega_i^\varepsilon \cap \partial\Omega_e^\varepsilon = \Omega \cap \varepsilon\Gamma_m.$$

The common boundary Γ_m^ε models the cellular membrane.

Since cardiac tissue exhibits a number of significant inhomogeneities, such as those related to cell-to-cell communications, the conductivity tensors are considered dependent on both the slow and the fast variables, i.e., $\sigma_{i,e}(x, \frac{x}{\varepsilon})$. The dependence of σ_i on $\frac{x}{\varepsilon}$ models the inclusion of gap-junction effects. We then define the rescaled symmetric conductivity matrices

$$\sigma_{i,e}^\varepsilon(x) = \sigma_{i,e}(x, \frac{x}{\varepsilon}),$$

where $\sigma_{i,e}(x,\xi) : \Omega \times E_{i,e} \to \mathbb{M}^{3\times 3}$ are continuous functions satisfying the usual uniform ellipticity and periodicity conditions:

$$\left.\begin{aligned} \sigma|y|^2 \le \sigma_{i,e}(x,\xi)y \cdot y \le \sigma^{-1}|y|^2 \\ \sigma_{i,e}(x, \xi + \boldsymbol{e}_k) = \sigma_{i,e}(x,\xi) \end{aligned}\right\} \quad \forall\,(x,\xi) \in \Omega \times E_{i,e},\ y \in \mathbb{R}^3,$$

for a given constant $\sigma > 0$.

The dimensionless cellular model P^ε. Summarizing, we formulate the full reaction-diffusion system associated with the cellular model P^ε as follows:

let $\Omega := \Omega_i^\varepsilon \cup \Omega_e^\varepsilon \cup \Gamma_m^\varepsilon$ be the cardiac tissue volume,
$\Omega_i^\varepsilon :=$ the intracellular space, with dimensionless conductivity $= \sigma_i^\varepsilon$,
$\Omega_e^\varepsilon :=$ the extracellular space, with intracellular conductivity $= \sigma_e^\varepsilon$,
$\nu_i, \nu_e :=$ exterior unit normals to $\partial\Omega_i^\varepsilon, \partial\Omega_e^\varepsilon$,
$\Gamma_m^\varepsilon :=$ surface cellular membrane,
$\boldsymbol{n} = \nu_i = -\nu_e$ normal to Γ_m^ε pointing towards Ω_e^ε.

Then the vector $(u_i^\varepsilon, u_e^\varepsilon, w^\varepsilon, c^\varepsilon)$, with $v^\varepsilon = u_i^\varepsilon - u_e^\varepsilon$, satisfies the problem:

$$\begin{cases} -\operatorname{div} \sigma_i^\varepsilon(x)\nabla u_i^\varepsilon = 0 & \text{in } \Omega_i^\varepsilon \\ -\operatorname{div} \sigma_e^\varepsilon(x)\nabla u_e^\varepsilon = 0 & \text{in } \Omega_e^\varepsilon \end{cases} \tag{8}$$

$$I_m^\varepsilon = \begin{cases} -\sigma_i^\varepsilon(x)\,\boldsymbol{n}^T\nabla u_i^\varepsilon = \varepsilon \left[\dfrac{\partial v^\varepsilon}{\partial t} + I_{\text{ion}}(v^\varepsilon, w^\varepsilon, c^\varepsilon) \right] \\ -\sigma_e^\varepsilon(x)\,\boldsymbol{n}^T\nabla u_e^\varepsilon = \varepsilon \left[\dfrac{\partial v^\varepsilon}{\partial t} + I_{\text{ion}}(v^\varepsilon, w^\varepsilon, c^\varepsilon) \right] \end{cases} \tag{9}$$

$$\frac{\partial w^\varepsilon}{\partial t} - R(v^\varepsilon, w^\varepsilon) = 0 \quad \text{on } \Gamma_m^\varepsilon, \qquad \frac{\partial c^\varepsilon}{\partial t} - S(v^\varepsilon, w^\varepsilon, c^\varepsilon) = 0,$$

supplemented by the following boundary conditions of Neumann type (assuming, for instance, that the cellular aggregate is embedded in an insulated medium):

$$\mathbf{n}^T\nabla u_i^\varepsilon = 0 \quad \text{on} \quad \partial\Omega_i^\varepsilon/\Gamma_m^\varepsilon \quad \text{and} \quad \mathbf{n}^T\nabla u_e^\varepsilon = 0 \quad \text{on} \quad \partial\Omega_e^\varepsilon/\Gamma_m^\varepsilon,$$

and the following degenerate initial conditions on $v^\varepsilon, w^\varepsilon, c^\varepsilon$:

$$v^\varepsilon(\mathbf{x}, 0) = v_0^\varepsilon(\mathbf{x}), \quad w^\varepsilon(\mathbf{x}, 0) = w_0^\varepsilon(\mathbf{x}), \quad c^\varepsilon(\mathbf{x}, 0) = c_0^\varepsilon(\mathbf{x}) \quad \text{on } \Gamma_m^\varepsilon.$$

The variables $v^\varepsilon, w^\varepsilon, c^\varepsilon$ and I_m^ε are defined only on the surface of the cellular membrane Γ_m^ε.

2.3 Formal two-scale homogenization

We briefly indicate how to use the two-scale method (see [7,14,70,98,129]) and formal asymptotic expansions to convert the microscopic model of the periodic cellular aggregate into an averaged continuum representation of the cardiac tissue, neglecting the presence of stimulation currents. We seek a solution $(u_i^\varepsilon, u_e^\varepsilon, w^\varepsilon, c^\varepsilon)$, where each component has an asymptotic form in powers of ε of the form

$$u = u_0(\mathbf{x}, \boldsymbol{\xi}, t) + \varepsilon u_1(\mathbf{x}, \boldsymbol{\xi}, t) + \varepsilon^2 u_2(\mathbf{x}, \boldsymbol{\xi}, t) + \cdots$$

with coefficients u_k Y–periodic functions of $\boldsymbol{\xi}$. Considering the full derivative operators

$$\nabla u = \varepsilon^{-1}\nabla_\xi u + \nabla_x u,$$

$$\operatorname{div} \sigma \nabla u = \varepsilon^{-2} \operatorname{div}_{\xi} \sigma \nabla_{\xi} u + \varepsilon^{-1} \operatorname{div}_{\xi} \sigma \nabla_{x} u + \varepsilon^{-1} \operatorname{div}_{x} \sigma \nabla_{\xi} u + \operatorname{div}_{x} \sigma \nabla_{x} u,$$

substituting the asymptotic forms into the first equations of (8), (9) and equating the coefficients of the powers -1, 0, 1, of ε to zero, we obtain the following equations for the functions $u_k(\mathbf{x}, \boldsymbol{\xi}, t)$, $k = 0, 1, 2$, associated with $u = u_i^{\varepsilon}$:

$$\begin{cases} \text{find } u_k \ \ Y\text{–periodic in } \xi \text{ such that:} \\ -\operatorname{div}_{\xi} \sigma_i(\mathbf{x}, \boldsymbol{\xi}) \nabla_{\xi} u_k = f_k(\mathbf{x}, \boldsymbol{\xi}) - \operatorname{div}_{\xi} \sigma_i \mathbf{F}_k(\mathbf{x}, \boldsymbol{\xi}) & \text{in} \quad E_i \\ \mathbf{n}_{\xi}^T \sigma_i(\mathbf{x}, \boldsymbol{\xi}) \nabla_{\xi} u_k = g_k(\mathbf{x}, \boldsymbol{\xi}) + \mathbf{n}_{\xi}^T \sigma_i \mathbf{F}_k(\mathbf{x}, \boldsymbol{\xi}) & \text{on} \quad \Gamma_m \end{cases} \tag{10}$$

with

$$\begin{cases} f_0 = 0, \ \mathbf{F}_0 = \mathbf{0} & \text{in} \quad E_i \\ g_0 = 0 & \text{on} \quad \Gamma_m, \end{cases} \tag{11}$$

$$\begin{cases} f_1 = 2 \operatorname{div}_x \sigma_i \nabla_{\xi} u_0, \quad \mathbf{F}_1 = \mathbf{0} & \text{in} \quad E_i \\ g_1 = \mathbf{n}_{\xi}^T \sigma_i \nabla_x u_0 & \text{on} \quad \Gamma_m, \end{cases} \tag{12}$$

$$\begin{cases} f_2 = \operatorname{div}_x \sigma_i \nabla_{\xi} u_1 + \operatorname{div}_x \sigma_i \nabla_x u_0, \quad \mathbf{F}_2 = -\sigma_i \nabla_x u_1 & \text{in} \quad E_i \\ g_2 = -(\dfrac{\partial v_0}{\partial t} + I_{\text{ion}}(v_0, w_0, c_0)) & \text{on} \quad \Gamma_m \\ v_0 = u_0^i - u_0^e, \quad \partial_t w_0 - R(v_0, w_0) = 0, \quad \partial_t c_0 - S(v_0, w_0, c_0) = 0. \end{cases} \tag{13}$$

In problem (10), the variable $\mathbf{x}$ appears as a parameter. Let $f_k(\mathbf{x}, \boldsymbol{\xi}), \mathbf{F}_k(\mathbf{x}, \boldsymbol{\xi}), g_k(\mathbf{x}, \boldsymbol{\xi})$ be Y–periodic functions in ξ. Then the problems for $k = 0, 1, 2$ admit a unique solution u_k, apart from an additive constant (a consequence of an easy extension of the result [7, Thm. 2] or [98, Thm. 6.1]) if and only if

$$\int_{Y_i} f_k d\xi + \int_{S_m} g_k ds_{\xi} = 0, \quad k = 0, 1, 2.$$

From the first cellular problem (11) for $k = 0$, it follows that the Y–periodic solution u_0 is independent of $\boldsymbol{\xi}$ and that $u_0(\mathbf{x})$, depending only on the macroscopic variable $\mathbf{x}$, represents a potential average over Y_i if the subsequent terms $u_k(\mathbf{x}, \boldsymbol{\xi}, t)$ are determined with zero mean value on Y_i. Since $f_1 = 0$ and $\int_{S_m} \sigma_i ds_{\xi} = 0$, the solvability of problem (11) related to the data (12) is assured and it is easy to check that the solution with zero mean on Y_i, i.e., $\int_{Y_i} u_1 d\xi = 0$, can be represented as

$$u_1(\mathbf{x}, \boldsymbol{\xi}, t) = -\mathbf{w}(\mathbf{x}, \boldsymbol{\xi})^T \nabla_x u_0, \tag{14}$$

where $\mathbf{w}(\mathbf{x}, \boldsymbol{\xi}) = (w_1, w_2, w_3)^T$ is the unique zero mean value solution on Y_i satisfying:

$$\begin{cases} \operatorname{div}_{\xi} \sigma_i(\mathbf{x}, \boldsymbol{\xi}) \nabla_{\xi} w_k = 0 & \text{in } Y_i \\ \mathbf{n}_{\xi}^T \sigma_i(\mathbf{x}, \boldsymbol{\xi}) \nabla_{\xi} w_k = n_{\xi_k} & \text{on } S, \quad k = 1, 2, 3. \end{cases}$$

Then problem (10) related to the data (13) becomes:

$$\begin{cases} -\operatorname{div}_\xi \sigma_i \nabla_\xi u_2 = -\operatorname{div}_\xi \sigma_i \nabla_x (\mathbf{w}^T \nabla_x u_0) + \\ \qquad + \operatorname{div}_x \sigma_i \nabla_x u_0 - \operatorname{div}_x \sigma_i \nabla_\xi (\mathbf{w}^T \nabla_x u_0) \quad \text{in} \quad Y_i \\ n_\xi^T \sigma_i \nabla_\xi u_2 = -(\dfrac{\partial v_0}{\partial t} + I_{\text{ion}}(v_0, w_0)) + n_\xi^T \sigma_i (\mathbf{w}^T \nabla_x u_0) \quad \text{on} \quad S_m. \end{cases}$$

In order to assure its solvability, the following compatibility relation must be satisfied:

$$-\int_{Y_i} \operatorname{div}_x \sigma_i \nabla_\xi (\mathbf{w}^T \nabla_x u_0)\, d\xi + \int_{Y_i} \operatorname{div}_x \sigma_i \nabla_x u_0\, d\xi \\ - \int_{S_m} (\frac{\partial v_0}{\partial t} + I_{\text{ion}}(v_0, w_0))\, ds_\xi = 0.$$

As u_0 is independent of $\boldsymbol{\xi}$ and by (14) it follows that

$$\operatorname{div}_x \left[\int_{Y_i} \sigma_i(\mathbf{x}, \boldsymbol{\xi}) \left\{ I - \nabla_\xi \mathbf{w}(\mathbf{x}, \boldsymbol{\xi})^T \right\} d\xi \right] \nabla_x u_0 = |S|(\frac{\partial v_0}{\partial t} + I_{\text{ion}}(v_0, w_0)),$$

where I is the identity matrix, $\nabla_\xi \mathbf{w}^T = [\nabla_\xi w_1, \nabla_\xi w_2, \nabla_\xi w_3]$ and $|Y_i|$, $|S_m|$ denote the volume and the area of Y_i and S_m, respectively.

Let $\beta = |S_m|/|Y|$ be the ratio between the membrane surface area and the volume of the reference cell and $\beta_{i,e} = |Y_{i,e}|/|Y|$. With reference to medium (i), $u_0 = u_0^i$, $\mathbf{w} = \mathbf{w}^i$, we set

$$D_i(\mathbf{x}) = \frac{1}{|Y|} \int_{Y_i} \sigma_i(\mathbf{x}, \boldsymbol{\xi}) \left\{ I - \nabla_\xi (\mathbf{w}^i)^T \right\} d\xi.$$

Hence, we obtain the following *"averaged equation"* for the intracellular potential:

$$\operatorname{div} D_i(\mathbf{x}) \nabla_x u_0^i = \beta \left(\frac{\partial v_0}{\partial t} + I_{\text{ion}}(v_0, w_0) \right).$$

Following [14], we easily check that the macroscopic conductivity tensor of the intracellular D_i is symmetric and positive definite.
Proceeding as for $u = u_e$ and taking into account the fact that the unit normal $\mathbf{n}$ points inside Ω_e, we obtain the following averaged equations for the extracellular potential:

$$\operatorname{div} D_e(\mathbf{x}) \nabla_x u_0^e = -\beta \left(\frac{\partial v_0}{\partial t} + I_{\text{ion}}(v_0, w_0) \right).$$

The dimensionless averaged model *P*. In summary: for a periodic network of interconnected cells, the governing dimensionless equations of the macroscopic intra- and extracellular potentials at zero order in ε are given by:

$$\begin{cases} \operatorname{div} D_i(\mathbf{x}) \nabla_x u_i = \beta \left(\dfrac{\partial v}{\partial t} + I_{\text{ion}}(v, w) \right) \\ \operatorname{div} D_e(\mathbf{x}) \nabla_x u_e = -\beta \left(\dfrac{\partial v}{\partial t} + I_{\text{ion}}(v, w) \right) \\ v = u_i - u_e, \quad \dfrac{\partial w}{\partial t} - R(v, w) = 0, \quad \dfrac{\partial c}{\partial t} - S(v, w, c) = 0. \end{cases} \tag{15}$$

Here the effective conductivity tensors are given by

$$D_{i,e}(\mathbf{x}) = \frac{1}{|Y|}\int_{Y_{i,e}} \sigma_{i,e}(\mathbf{x},\boldsymbol{\xi})\left\{I - \nabla_{\xi}(\mathbf{w}^{i,e})^T\right\} d\xi$$

and $\mathbf{w}^{i,e} = (w_1^{i,e}, w_2^{i,e}, w_3^{i,e})^T$ are solutions of:

$$\begin{cases} \operatorname{div}_{\xi}\, \sigma_{i,e}(\mathbf{x},\boldsymbol{\xi})\nabla_{\xi} w_k^e = 0 & \text{in } Y_{i,e} \\ \mathbf{n}_{\xi}^T \sigma_{i,e}(\mathbf{x},\boldsymbol{\xi})\nabla_{\xi} w_k^{i,e} = n_{\xi_k} & \text{on } S, \quad k = 1,2,3. \end{cases}$$

The previous derivation based on the two-scale method is only formal but the averaged model can be rigorously justified in the framework of Γ-convergence theory as a limit problem of the cellular model for $\varepsilon \to 0$; see [5].

2.4 Theoretical results for the cellular and averaged models

We introduce the functional space

$$\mathbb{V}^{\varepsilon} = \left\{H^1(\Omega_i^{\varepsilon}) \times H^1(\Omega_e^{\varepsilon})\right\} / \{(\gamma,\gamma) : \gamma \in \mathbf{R}\} \times L^2(\Gamma_m^{\varepsilon})^M \times L^2(\Gamma_m^{\varepsilon})^Q, \quad (16)$$

the vector variables $\boldsymbol{U} := (u_i, u_e, w, c) \in \mathbb{V}^{\varepsilon}$, $\hat{\boldsymbol{U}} := (\hat{u}_i, \hat{u}_e, \hat{w}, \hat{c}) \in \mathbb{V}^{\varepsilon}$, the vector time derivative $\partial_t \boldsymbol{U} := (\frac{\partial u_i}{\partial t}, \frac{\partial u_e}{\partial t}, \frac{\partial w}{\partial t}, \frac{\partial c}{\partial t})$ and we set $v = u_i - u_e$, $\hat{v} = \hat{u}_i - \hat{u}_e$. Then we introduce the forms:

$$\begin{aligned} b^{\varepsilon}(\boldsymbol{U},\hat{\boldsymbol{U}}) &:= \varepsilon \int_{\Gamma_m^{\varepsilon}} [\, v\,\hat{v} + w\,\hat{w} + c\hat{c}\,]\, d\gamma \\ a^{\varepsilon}(\boldsymbol{U},\hat{\boldsymbol{U}}) &:= \sum_{i,e} \int_{\Omega_{i,e}^{\varepsilon}} \sigma_{i,e}^{\varepsilon}\, \nabla u_{i,e} \cdot \nabla \hat{u}_{i,e}\, d\mathbf{x} \\ F^{\varepsilon}(\boldsymbol{U},\hat{\boldsymbol{U}}) &:= \varepsilon \int_{\Gamma_m^{\varepsilon}} [I_{\text{ion}}(v,w,c)\,\hat{v} - R(v,w)\,\hat{w} - S(v,w,c)\,\hat{c}\,]\, d\gamma, \end{aligned}$$

and we consider the variational formulation of the differential problem $\boldsymbol{P}^{\varepsilon}$:

$$\text{find } \boldsymbol{U}^{\varepsilon} : [0,T] \to \mathbb{V}^{\varepsilon} \;:\; b^{\varepsilon}(\partial_t \boldsymbol{U}^{\varepsilon}, \hat{\boldsymbol{U}}) + a^{\varepsilon}(\boldsymbol{U}^{\varepsilon}, \hat{\boldsymbol{U}}) + F^{\varepsilon}(\boldsymbol{U}^{\varepsilon}, \hat{\boldsymbol{U}}) = 0 \quad \forall \hat{\boldsymbol{U}} \in \mathbb{V}^{\varepsilon}, \quad (17)$$

where the parabolic b^{ε} and elliptic a^{ε} forms are degenerate, but their sum is coercive on $\mathbb{V}^{\varepsilon}$. The variational formulation is supplemented by the initial conditions:

$$v^{\varepsilon}(\cdot,0) = v_0^{\varepsilon}, \quad w^{\varepsilon}(\cdot,0) = w_0^{\varepsilon}, \quad c^{\varepsilon}(\cdot,0) = c_0^{\varepsilon} \quad \text{on } {\Gamma_m}^{\varepsilon}.$$

We consider the functional space

$$\mathbb{V} := \left\{H^1(\Omega) \times H^1(\Omega)\right\} / \{(\gamma,\gamma) : \gamma \in \mathbf{R}\} \times L^2(\Omega)^M \times L^2(\Omega)^Q$$

and we introduce the following bilinear and nonlinear forms associated with the averaged model (15): for $\boldsymbol{U} = (u_i, u_e, w, c) \in \mathbb{V}$, $\hat{\boldsymbol{U}} = (\hat{u}_i, \hat{u}_e, \hat{w}, \hat{c}) \in \mathbb{V}$, and $v = u_i - u_e$, $\hat{v} = \hat{u}_i - \hat{u}_e$, we set:

$$b(\boldsymbol{U}, \hat{\boldsymbol{U}}) := \beta \int_{\Omega} [\partial_t v\, \hat{v} + \partial_t w\, \hat{w} + \partial_t c\, \hat{c}\,]\, dx$$

$$a(\boldsymbol{U}, \hat{\boldsymbol{U}}) := \sum_{i,e} \int_{\Omega_{i,e}} Di, e\, \nabla u_{i,e} \cdot \nabla u_{i,e}\, d\mathbf{x}$$

$$F(\boldsymbol{U}, \hat{\boldsymbol{U}}) := \beta \int_{\Omega} [I_{\text{ion}}(v, w, c)\, \hat{v} - R(v, w)\, \hat{w} - S(v, w, c)\, \hat{c}\,]\, dx.$$

The homogenized conductivity tensors $D_{i,e}$ can be characterized by solving the following variational cellular problems for every $y \in \mathbb{R}^3$:

$$y^T D_{i,e}(x)\, y := \min \Big\{ \frac{1}{|Y|} \int_{Y_{i,e}} \big(\nabla u(\xi) + y\big)^T \sigma_{i,e}(x, \xi) \big(\nabla u(\xi) + y\big)\, d\xi : \; u \in H^1_{\text{loc}}(\mathbb{R}^d),\ u\ Y\text{-periodic} \Big\}. \tag{18}$$

These tensors are symmetric and positive definite matrices, and the bilinear forms $b(\cdot, \cdot)$, $a(\cdot, \cdot)$ are coercive on $\mathbb{V}$.

The variational formulation of the averaged problem $\boldsymbol{P}$, related to the macroscopic model (15), is given as:

$$\text{find } \boldsymbol{U} : [0, T] \to \mathbb{V} : b(\partial_t \boldsymbol{U}, \hat{\boldsymbol{U}}) + a(\boldsymbol{U}, \hat{\boldsymbol{U}}) + \mathcal{F}(\boldsymbol{U}, \hat{\boldsymbol{U}}) = 0 \quad \forall \hat{\boldsymbol{U}} \in \mathbb{V}, \tag{19}$$

supplemented with the initial conditions

$$v(\cdot, 0) = v_0, \quad w(\cdot, 0) = w_0, \quad c(\cdot, 0) = c_0 \quad \text{in } \Omega.$$

We now focus on the FitzHugh-Nagumo membrane model ([46, 47]): the ionic current is a cubic-like function in v and is linear in the recovery variable w. In this simplified model, the unknown is the vector $(u_i^\varepsilon, u_e^\varepsilon, w^\varepsilon)$ and it was shown in [37] that both the cellular and averaged models share the same variational structure and yield well-posed problems. More precisely, introducing as in (16) the quotient space

$$\mathbb{V}^\varepsilon = \Big\{ H^1(\Omega_i^\varepsilon) \times H^1(\Omega_e^\varepsilon) \Big\} / \{(\gamma, \gamma) : \gamma \in \mathbf{R}\} \times L^2(\Gamma_m^\varepsilon),$$

we have the following result for problem $\boldsymbol{P}^\varepsilon$.

Theorem 1. *Assuming that Γ^ε is regular and that the initial data satisfy*

$$(v_0^\varepsilon, w_0^\varepsilon) \in L^2(\Gamma_m^\varepsilon) \times L^2(\Gamma_m^\varepsilon),$$

then there exists a unique solution $\boldsymbol{U}^\varepsilon = (u_i^\varepsilon, u_e^\varepsilon, w^\varepsilon) \in C^0(]0, T]; \mathbb{V}^\varepsilon)$ of the variational formulation (17) of Problem $\boldsymbol{P}^\varepsilon$ with

$$\frac{\partial v^\varepsilon}{\partial t}, \ \frac{\partial w^\varepsilon}{\partial t} \in L^2(0, T; L^2(\Gamma_m^\varepsilon));$$

here $\underline{u}^\varepsilon := (u_i^\varepsilon, u_e^\varepsilon)$ solves the differential equations $\boldsymbol{P}^\varepsilon$ in the standard distributional sense.

Introducing the quotient space

$$\mathbb{V} := \left\{H^1(\Omega) \times H^1(\Omega)\right\} / \{(\gamma, \gamma) : \gamma \in \mathbf{R}\} \times L^2(\Omega)$$

we have the following result for problem $\boldsymbol{P}$.

Theorem 2. *Assuming that the initial data satisfy*

$$(v_0, w_0) \in L^2(\Omega) \times L^2(\Omega),$$

then there exists a unique solution $\boldsymbol{U} = (u_i, u_e, w) \in C^0(]0, T]; \mathbb{V})$ *of the variational formulation (19) of the averaged Problem* $\boldsymbol{P}$ *with*

$$\frac{\partial v}{\partial t}, \ \frac{\partial w}{\partial t} \in L^2(0, T; L^2(\Omega));$$

here $\underline{u} := (u_i, u_e)$ *solves the differential equations* $\boldsymbol{P}$ *in the standard distributional sense.*

We remark that the above abstract variational framework of the cellular (17) and averaged (19) models in terms of the forms $a^\varepsilon, b^\varepsilon, F^\varepsilon$ and a, b, F respectively, share the same structural properties; see [37].

For the cellular and the averaged models with ionic current membrane dynamics, described by the classical Hodgkin-Huxley model [61] or by the Luo-Rudy Phase I model [84], results an when they are well-posed can be found in [153, 154].

2.5 Γ-convergence result for the averaged model with FHN dynamics

We now present a convergence result for the homogenization process related to the bidomain model with Nagumo membrane model (i.e., FHN without the recovery variable); see [107] for details of the general FHN case.

The problem $\boldsymbol{P}^\varepsilon$ is not a standard parabolic homogenization problem and its main difficulties are associated with the fact that b^ε depends explicitly on ε, it is *degenerate* and the boundaries of $\Omega^\varepsilon_{i,e}$ could be quite irregular. For $z^\varepsilon \in L^2(\Gamma^\varepsilon)$, $\underline{\boldsymbol{u}}^\varepsilon = (u_i^\varepsilon, u_e^\varepsilon) \in H^1(\Omega_i^\varepsilon) \times H^1(\Omega_e^\varepsilon)$, and $z \in L^2(\Omega)$, $\underline{\boldsymbol{u}} = (u_i, u_e) \in (H^1(\Omega))^2$, we define the energy-like functionals:

$$b^\varepsilon(z^\varepsilon) := \varepsilon \int_{\Gamma_m^\varepsilon} |z^\varepsilon|^2 d\gamma, \quad a^\varepsilon(\underline{\boldsymbol{u}}^\varepsilon) := \sum_{i,e} \int_{\Omega^\varepsilon_{i,e}} \sigma^\varepsilon_{i,e} \nabla u^\varepsilon_{i,e} \cdot \nabla u^\varepsilon_{i,e} dx,$$

$$b(z) := \beta \int_\Omega |z|^2 dx, \quad a(\underline{\boldsymbol{u}}) := \sum_{i,e} \int_\Omega D_{i,e} \nabla u_{i,e} \cdot \nabla u_{i,e}\, dx,$$

$$\mathcal{G}^\varepsilon(z^\varepsilon) := \varepsilon \int_{\Gamma_m^\varepsilon} G(z^\varepsilon) d\gamma, \quad \mathcal{G}(z) := \beta \int_\Omega G(z) dx, \tag{20}$$

where G is a positive primitive of g in the FHN model (4).

Theorem 3. *Assume that* $v_0^\varepsilon = u_{i,0}^\varepsilon - u_{e,0}^\varepsilon$, w_0^ε *converge to* $v_0 = u_{i,0} - u_{e,0}$, w_0 *in the following "distributional" sense:*

$$\lim_{\varepsilon\downarrow 0} \varepsilon \int_{\Gamma_m^\varepsilon} v_0^\varepsilon(x)\zeta(x)\,d\gamma = \beta \int_\Omega v_0(x)\zeta(x)\,dx \quad \forall\zeta \in C_0^\infty(\Omega)$$

$$\lim_{\varepsilon\downarrow 0} \varepsilon \int_{\Gamma_m^\varepsilon} w_0^\varepsilon(x)\zeta(x)\,d\gamma = \beta \int_\Omega w_0(x)\zeta(x)\,dx \quad \forall\zeta \in C_0^\infty(\Omega)$$

$$\lim_{\varepsilon\downarrow 0} \int_{\Omega_{i,e}^\varepsilon} u_{i,e}^\varepsilon(x)\zeta(x)\,dx = \beta_{i,e} \int_\Omega u_{i,e}(x)\zeta(x)\,dx \quad \forall\zeta \in C_0^\infty(\Omega)$$

and that the related energies satisfy

$$\lim_{\varepsilon\downarrow 0} b^\varepsilon(v_0^\varepsilon) = b(v_0), \ \lim_{\varepsilon\downarrow 0} b^\varepsilon(w_0^\varepsilon) = b(w_0), \quad \limsup_{\varepsilon\downarrow 0} \left(a^\varepsilon(u_0^\varepsilon) + \mathcal{G}^\varepsilon(v_0^\varepsilon)\right) < +\infty.$$

Let $\Omega_0 \subset\subset \Omega$ *be a reference open subdomain with positive measure,* $\boldsymbol{u}^\varepsilon = (u_i^\varepsilon, u_e^\varepsilon)$, $v^\varepsilon = u_i^\varepsilon - u_e^\varepsilon$, w^ε *and* $\underline{\boldsymbol{u}} = (u_i, u_e)$, $v = u_i - u_e$, w *be the solutions of the cellular and averaged models with* $\int_{\Omega_0\cap\Omega^\varepsilon} u_e^\varepsilon dx = 0$ *and* $\int_{\Omega_0\cap\Omega} u_e dx = 0$*, respectively. Then, for every time* $t \in [0, T]$,

$$(u_{i,e}^\varepsilon, v^\varepsilon, w^\varepsilon) \to (u_{i,e}, v, w) \quad \textit{as } \varepsilon \downarrow 0, \quad \textit{in the distributional sense,}$$

with

$$a^\varepsilon(\boldsymbol{u}^\varepsilon) \to a(\boldsymbol{u}), \quad b^\varepsilon(v^\varepsilon) \to b(v), \quad b^\varepsilon(w^\varepsilon) \to b(w).$$

Moreover, there exist extensions $\mathcal{T}_i^\varepsilon u_i^\varepsilon, \mathcal{T}_e^\varepsilon u_e^\varepsilon$ *of* $u_i^\varepsilon, u_e^\varepsilon$ *in the whole domain* Ω*, solutions of the cellular problem* $\boldsymbol{P}^\varepsilon$*, which converge in* $L^2(0, T; H^1_{\mathrm{loc}}(\Omega))$ *to the unique solution* $(u_i, u_e) \in \mathbb{V}$ *of the averaged model* $\boldsymbol{P}$.

The variational approach for the convergence of the evolution problem is based on the introduction of the time-semidiscrete approximation by the implicit Euler method (see Sect. 5), which reduces the evolution system to discrete families of stationary problems. More precisely, given $\boldsymbol{U}_0^\varepsilon$, we introduce the sequence of variational problems:

find $\boldsymbol{U}_{\tau,n}^\varepsilon \in \mathbb{V}$, $n = 1, \ldots, N$

$$b^\varepsilon\Big(\frac{\boldsymbol{U}_{\tau,n}^\varepsilon - \boldsymbol{U}_{\tau,n-1}^\varepsilon}{\tau}, \hat{\boldsymbol{U}}\Big) + a^\varepsilon(\boldsymbol{U}_{\tau,n}^\varepsilon, \hat{\boldsymbol{U}}) + F^\varepsilon(\boldsymbol{U}_{\tau,n}^\varepsilon, \hat{\boldsymbol{U}})) = 0 \quad \forall\hat{\boldsymbol{U}} \in \mathbb{V}.$$

For $\tau = T/N$ sufficiently small, the coercivity of $a + b$ and the one-sided Lipschitz condition on F^ε guarantee the recursive solvability of these equations.

The previous theorem follows by combining Γ-convergence and uniform error estimates for the Euler discretization; assuming, for simplicity, an instantaneous ionic current without recovery, i.e., $I_{\mathrm{ion}}(v) = g(v)$, the discrete solution $\boldsymbol{U}_{\tau,1}^\varepsilon, \ldots, \boldsymbol{U}_{\tau,N}^\varepsilon$ of the Euler scheme solves the iterated (convex) minimization problem

$$\boldsymbol{U}_{\tau,n}^\varepsilon = \operatorname*{arg\,min}_{\boldsymbol{V}\in\mathbb{V}} \left\{\frac{1}{2\tau} b^\varepsilon(\boldsymbol{V} - \boldsymbol{U}_{\tau,n-1}^\varepsilon) + \Phi^\varepsilon(\boldsymbol{V})\right\}, \tag{21}$$

where Φ^{ε} is the Lyapunov functional

$$\Phi^{\varepsilon}(\boldsymbol{U}^{\varepsilon}) := a^{\varepsilon}(\underline{\boldsymbol{u}}^{\varepsilon}) + \mathcal{G}^{\varepsilon}(v^{\varepsilon})$$

and a^{ε}, b^{ε}, $\mathcal{G}^{\varepsilon}$ were defined in (20). Analogously, we can define the same quantities for the limit case $\varepsilon = 0$, obtaining a discrete solution $\boldsymbol{U}_{\tau,n}$ that solves an iterated convex minimization problem (21) without ε involving the Lyapunov functional

$$\Phi(\boldsymbol{U}) := a(\underline{\boldsymbol{u}}) + \mathcal{G}(v)$$

with a, b, $\mathcal{G}$ again defined in (20).

The general strategy for passing to the limit (see [107]) can be summarized in the following diagram:

Evolution Problem $\boldsymbol{P}^{\varepsilon}\ :\ \boldsymbol{U}^{\varepsilon}$	$\rightarrow$	**Stationary Problem** $\boldsymbol{U}_{\tau}^{\varepsilon,n+1}$: $\min_{\boldsymbol{W}} \frac{1}{2\tau} b^{\varepsilon}(\boldsymbol{W} - \boldsymbol{U}_{\tau}^{\varepsilon,n}) + \Phi^{\varepsilon}(\boldsymbol{W})$
		$\varepsilon \downarrow 0$
Limit of the Evolution Problem $\boldsymbol{P}\ :\ \boldsymbol{U}$	$0 \leftarrow \tau$	**Limit of the Stationary Problem** $\boldsymbol{U}_{\tau}^{n+1} : \min_{\boldsymbol{W}} \frac{1}{2\tau} b(\boldsymbol{W} - \boldsymbol{U}_{\tau}^{n}) + \Phi(\boldsymbol{W})$

2.6 Semidiscrete approximation of the bidomain model with FHN dynamics

We conclude this section by mentioning examples of approximation results for the averaged model $\boldsymbol{P}$; for other approximation results in the context of reaction-diffusion problems (see, e.g., [62, 69, 88, 89]).

First, we consider a semidiscrete scheme in space obtained by using conforming linear finite elements (see, e.g., [115] for a general introduction to the finite element method). Assuming that Ω is a polygonal convex domain in $\mathbb{R}^3$ and $\mathcal{T}_h$ is a family of triangulations associated with a reference polyhedron $\widehat{E}$ by an invertible affine map T_E for every $E \in \mathcal{T}_h$; we denote by V_h the finite-dimensional space of continuous functions whose restrictions to each element of $\mathcal{T}_h$ are polynomial of degree one. A semidiscrete problem is obtained by applying a standard Galerkin procedure on $\mathbb{V}_h$ to the averaged model (15). In [130] various stability results are given as well as the following error estimate.

Theorem 4. *For regular and quasi-uniform mesh and a regular initial datum $w_0 \in H^1(\Omega)$, and with $\boldsymbol{U}_h(t)$ denoting the finite element approximation of the semidiscrete approximation of (19) with FitzHugh-Nagumo dynamics, then the following optimal a priori error estimate holds:*

$$\mathrm{e}_h^2 := \max_{t\in(0,T)} b\big(\boldsymbol{U}(t) - \boldsymbol{U}_h(t)\big) + \int_0^T a\big(\boldsymbol{U}(t) - \boldsymbol{U}_h(t)\big)\, dt \le Ch^2.$$

We now consider a semidiscrete approximation in time of the averaged model by applying the implicit Euler scheme. More precisely, we choose a partition of the time interval $[0, T]$ into N subintervals

$$\mathcal{P} = \{0 = t_0 < t_1 < t_2 < \dots t_{N-1} < t_N = T\} \quad \text{with variable steps } \tau_n = t_n - t_{n-1}$$

and we set $\tau = \max_{1 \le n \le N} \tau_n$.

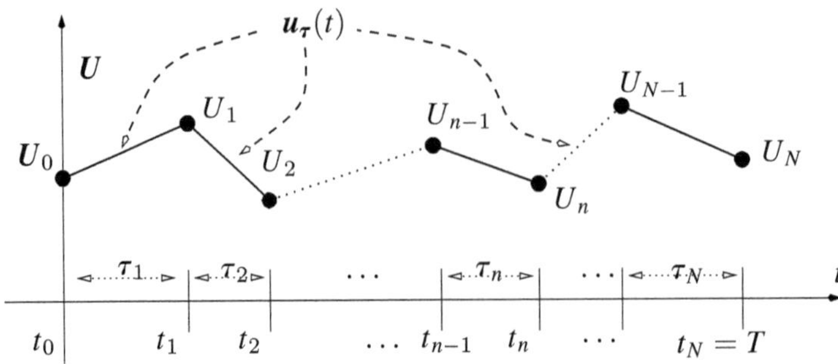

Given U_0, we introduce the sequence of variational problems: find $U_n \in \mathbb{V}$, $n = 1, \ldots, N$, such that

$$b(\frac{U_n - U_{n-1}}{\tau_n}, \hat{U}) + a(U_n, \hat{U}) + F(U_n, \hat{U})) = 0 \quad \forall\ \hat{U} \in \mathbb{V}.$$

Considering the discrete solution $U_\tau(t)$ given by the continuous piecewise linear function interpolating the values $\{U_n\}_{n=0}^N$ on the grid $\mathcal{P}$, we have the following error estimate.

Theorem 5. *For sufficiently regular initial data, the following a priori error estimate between u and U_τ, measured by the natural variational (semi)norms, holds:*

$$\mathrm{e}_\tau^2 := \max_{t \in (0,T)} b\big(U(t) - U_\tau(t)\big) + \int_0^T a\big(U(t) - U_\tau(t)\big)\, dt \le C\tau^2,$$

or, equivalently,

$$\|\sqrt{b(U - U_\tau)}\|_{L^\infty(0,T)} + \|U - U_\tau\|_{L^2(0,T;\mathbb{V})} \ \le\ C\,\tau$$

with C independent of τ.

We conclude by presenting a result related to a posteriori error estimates. In [10], by resorting to the theory developed in [97] for evolution variational problem, a posteriori error estimates were derived for general degenerate evolution equations.

Theorem 6. *For sufficiently regular initial data, let e_τ be the error between U and U_τ measured by the natural variational (semi)norms*

$$\mathrm{e}_\tau^2 := \max\Big[\max_{t \in (0,T)} e^{-2\lambda_g t} b\big(U(t) - U_\tau(t)\big), \int_0^T e^{-2\lambda_g t} a\big(U(t) - U_\tau(t)\big)\, dt\Big]$$

with $\lambda_g = \Big(\inf_{v \in R} g'(v)\Big)^-$.

Then, by applying the theory of [10] to the bidomain model, the error e_τ can be estimated a posteriori by

$$e_\tau^2 \le \sum_n \tau_n^2 \mathcal{D}_n + \frac{\lambda_g^2}{2} \sum_n \tau_n^2 \sqrt{b(\delta \boldsymbol{U}^n)},$$

where $\mathcal{D}_n = a(\boldsymbol{U}_n - \boldsymbol{U}_{n-1}) + F(\boldsymbol{U}_n, \boldsymbol{U}_n - \boldsymbol{U}_{n-1}) - F(\boldsymbol{U}_{n-1}, \boldsymbol{U}_n - \boldsymbol{U}_{n-1})$ *and*

$$\delta \boldsymbol{U}^n = \frac{\boldsymbol{U}_n - \boldsymbol{U}_{n-1}}{\tau_n}.$$

3 The anisotropic bidomain model

The macroscopic variational model, derived rigorously from the asymptotic behavior of the cellular model in the periodic case, has the following interpretation in differential terms: the macroscopic cardiac tissue can be represented as the superposition of two anisotropic continuous media, the intra- (i) and extra- (e) cellular media, coexisting at every point of the tissue and separated by a distributed continuous cellular membrane; see, e.g., [24, 59, 76, 121].

The cardiac ventricular tissue is modeled as an arrangement of cardiac fibers organized in toroidal layers nested within the ventricular wall and rotating counterclockwise from epi- to endocardium (see, e.g., [139]). More recently, [80, 81] have shown that this fiber structure has an additional laminar organization modeled as a set of muscle sheets running radially from epi- to endocardium. Therefore, at any point $\mathbf{x}$, it is possible to identify a triplet of orthonormal principal axes $\mathbf{a}_l(\mathbf{x}),\ \mathbf{a}_t(\mathbf{x}),\ \mathbf{a}_n(\mathbf{x})$, with $\mathbf{a}_l(\mathbf{x})$ parallel to the local fiber direction, $\mathbf{a}_t(\mathbf{x})$ and $\mathbf{a}_n(\mathbf{x})$ tangent and orthogonal to the radial laminae, respectively, and both transversal to the fiber axis [39, 81]. Denoting by $\sigma_l^{i,e},\ \sigma_t^{i,e},\ \sigma_n^{i,e}$ the conductivity coefficients in the intra- and extracellular media measured along the corresponding directions $\mathbf{a}_l, \mathbf{a}_t, \mathbf{a}_n$, then the anisotropic conductivity tensors $M_i(\mathbf{x})$ and $M_e(\mathbf{x})$ related to the *orthotropic anisotropy* of the media are given by

$$M_{i,e}(\mathbf{x}) = \sigma_l^{i,e}\, \mathbf{a}_l(\mathbf{x})\mathbf{a}_l^T(\mathbf{x}) + \sigma_t^{i,e}\, \mathbf{a}_t(\mathbf{x})\mathbf{a}_t^T(\mathbf{x}) + \sigma_n^{i,e}\, \mathbf{a}_n(\mathbf{x})\mathbf{a}_n^T(\mathbf{x}). \tag{22}$$

For *axisymmetric* anisotropic media, $\sigma_n^{i,e} = \sigma_t^{i,e}$ and

$$M_{i,e}(\mathbf{x}) = {\sigma_t}^{i,e} I + ({\sigma_l}^{i,e} - {\sigma_t}^{i,e})\mathbf{a}_l(\mathbf{x})\mathbf{a}_l^T(\mathbf{x}),$$

where I denotes the identity matrix.

The intra- and extracellular electric potentials u_i, u_e in the anisotropic bidomain model are described by a reaction-diffusion system coupled with a system of ODEs for ionic gating variables w and for the ion concentrations c. We denote by $v = u_i - u_e$ the transmembrane potential and by

$$J_m = \chi I_m = c_m \frac{\partial v}{\partial t} + i_{\text{ion}}(v, w, c)$$

the membrane current per unit volume, where $c_m = \chi C_m$, $i_{\text{ion}} = \chi I_{\text{ion}}$, with χ the ratio of membrane area per tissue volume, C_m the surface capacitance and I_{ion}

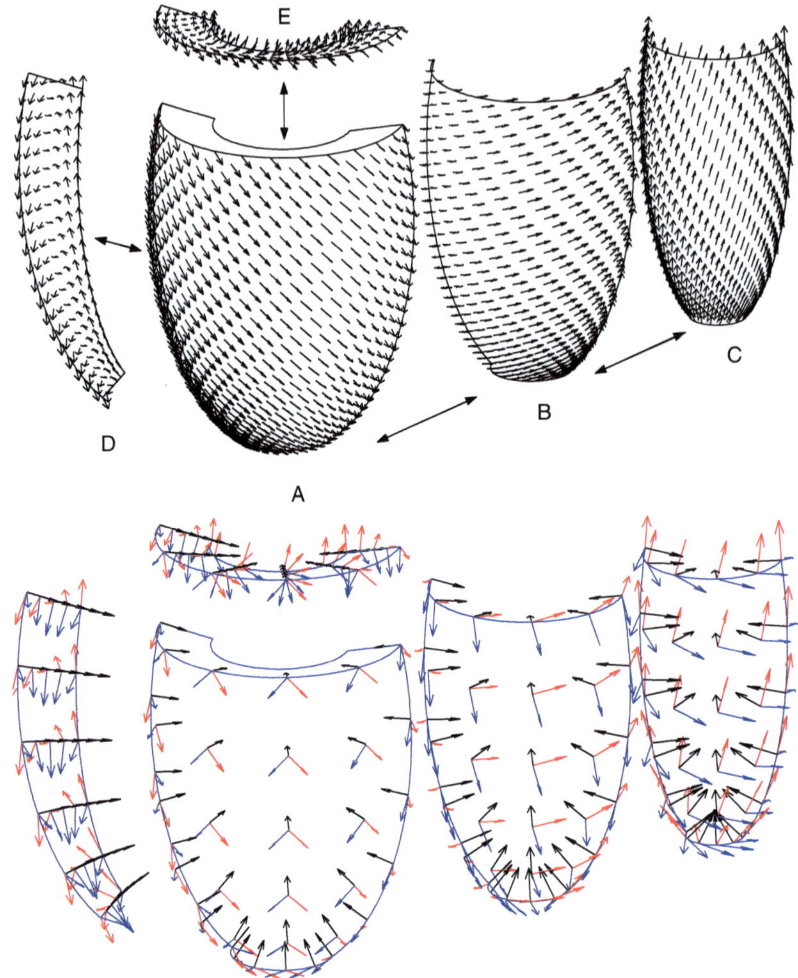

Fig. 3. Fiber direction (*top*) and orthonormal triplet $\mathbf{a}_l$, $\mathbf{a}_t$, $\mathbf{a}_n$ (*bottom*) on the epicardium (A), midwall (B), endocardium (C), intramural sections (D, E)

the ionic current of the membrane per unit area. Let I^e_{app} be an applied extracellular current per unit volume, satisfying the compatibility condition $\int_\Omega I^e_{\text{app}}\, dx = 0$, and $\mathbf{j}_{i,e} = -M_{i,e}\nabla u_{i,e}$ the intra- and extracellular current density. Due to the current conservation law, we have

$$\operatorname{div} \mathbf{j_i} = -J_m + I^i_{\text{app}}, \quad \operatorname{div} \mathbf{j_e} = J_m + I^e_{\text{app}}.$$

Let Ω_H, $\Gamma_H = \partial\Omega_H$ denote the volume and the heart surface, respectively. Then the anisotropic bidomain model in the unknown potentials $(u_i(\mathbf{x}, t), u_e(\mathbf{x}, t))$, $v(\mathbf{x}, t) = u_i(\mathbf{x}, t) - u_e(\mathbf{x}, t)$, gating variables $w(\mathbf{x}, t)$ and ion concentrations $c(\mathbf{x}, t)$ can be written as:

$$\begin{cases} c_m \dfrac{\partial v}{\partial t} - \mathrm{div}(M_i \nabla u_i) + i_{\mathrm{ion}}(v, w, c) = I^i_{\mathrm{app}} & \text{in } \Omega_H \times (0, T) \\ -c_m \dfrac{\partial v}{\partial t} - \mathrm{div}(M_e \nabla u_e) - i_{\mathrm{ion}}(v, w, c) = I^e_{\mathrm{app}} & \text{in } \Omega_H \times (0, T) \\ \dfrac{\partial w}{\partial t} - R(v, w) = 0, \qquad \dfrac{\partial c}{\partial t} - S(v, w, c) = 0 & \text{in } \Omega_H \times (0, T) \\ \mathbf{n}^T D_{i,e} \nabla u_{i,e} = 0 & \text{in } \Gamma_H \times (0, T) \\ v(\mathbf{x}, 0) = v_0(\mathbf{x}), \quad w(\mathbf{x}, 0) = w_0(\mathbf{x}), \quad c(\mathbf{x}, 0) = c_0(\mathbf{x}) & \text{in } \Omega_H, \end{cases} \tag{23}$$

where we have imposed insulated boundary conditions. The reaction-diffusion (R-D) system (23) uniquely determines v, while the potentials u_i and u_e are defined only up to the same additive time-dependent constant relating to the reference potential. Until now the bidomain model was formulated in terms of the potential fields u_i and u_e but it can be equivalently expressed in terms of the transmembrane and extracellular potentials $v(\mathbf{x}, t)$ and $u_e(\mathbf{x}, t)$; in fact, adding the two evolution equations of the system (23) and substituting $u_i = v - u_e$, we obtain an elliptic equation in the unknown (v, u_e) which coupled with one of the evolution equations gives the following equivalent formulation of the anisotropic bidomain model:

$$\begin{cases} c_m \dfrac{\partial v}{\partial t} + i_{\mathrm{ion}}(v, w, c) - \mathrm{div}(M_i \nabla v) - \mathrm{div}(M_i \nabla u_e) = I^i_{\mathrm{app}} & \text{in } \Omega_H \times (0, T) \\ -\mathrm{div}((M_i + M_e) \nabla u_e) - \mathrm{div}(M_i \nabla v) = I^i_{\mathrm{app}} + I^e_{\mathrm{app}} & \text{in } \Omega_H \times (0, T) \end{cases} \tag{24}$$

or

$$\begin{cases} -\mathrm{div}((M_i + M_e) \nabla u_e) - \mathrm{div}(M_i \nabla v) = I^i_{\mathrm{app}} + I^e_{\mathrm{app}} & \text{in } \Omega_H \times (0, T) \\ c_m \dfrac{\partial v}{\partial t} + i_{\mathrm{ion}}(v, w, c) + \mathrm{div}(M_e \nabla u_e) = I^e_{\mathrm{app}} & \text{in } \Omega_H \times (0, T). \end{cases}$$

In order to establish a connection between the noninvasive potential measurements on the body surface and the bioelectric cardiac source currents we must couple the macroscopic bidomain model of the cardiac tissue with the description of the current conduction in the extracardiac medium. We denote by Ω_0, M_0, $\mathbf{j_0} = -M_0 \nabla u_0$, u_0, the extracardiac volume, the conductivity tensor, the current density and the extracardiac potential respectively, and the body surface by $\Gamma_0 = \partial\Omega_0 - \Gamma_H$. Disregarding, e.g., the presence of external applied currents, we may assume that no current sources lie outside the heart and, since the body is embedded in the air, which is an insulating medium, the current vector is tangent to the body surface. Current conservation at the interface Γ_H requires that $\mathbf{n}^T(\mathbf{j_i} + \mathbf{j_e})) = \mathbf{n}^T \mathbf{j_0}$, where $\mathbf{n}$ denotes the exterior normal with respect to Ω_H; then, in the bidomain representation: we have:

$$\begin{cases} \mathrm{div}\,(\mathbf{j_i} + \mathbf{j_e}) = 0 \quad \text{in } \Omega_H, \quad \mathrm{div}\,\mathbf{j_0} = 0 \quad \text{in } \Omega_0 \\ \mathbf{n}^T(\mathbf{j_i} + \mathbf{j_e}) = \mathbf{n}^T \mathbf{j_0} \quad \text{on } \Gamma_H, \quad \mathbf{n}^T \mathbf{j_0} = 0 \quad \text{on } \Gamma_0, \end{cases}$$

and, with

$$\widehat{M} = \widehat{M}(\mathbf{x}) = \begin{cases} M_i(\mathbf{x}) + M_e(\mathbf{x}), & \mathbf{x} \in \Omega_H \\ M_0(\mathbf{x}), & \mathbf{x} \in \Omega_0, \end{cases} \qquad u(\mathbf{x}, t) = \begin{cases} u_e(\mathbf{x}, t), & \mathbf{x} \in \Omega_H \\ u_0(\mathbf{x}, t), & \mathbf{x} \in \Omega_0, \end{cases}$$

then the extracellular and extracardiac potential field u satisfies the following elliptic problem:

$$\begin{cases} \operatorname{div} \widehat{M} \nabla u(\mathbf{x}, t) = \begin{cases} \operatorname{div} \mathbf{J}_v(\mathbf{x}, t) & \text{in } \Omega_H \\ 0 & \text{in } \Omega_0 \end{cases} \\ [\![\, u(\mathbf{x}, t) \,]\!]_{\Gamma_H} = 0, \quad [\![\, \mathbf{n}^T \widehat{M} \nabla u(\mathbf{x}, t) \,]\!]_{\Gamma_H} = \mathbf{n}^T \mathbf{J}_v(\mathbf{x}, t) \\ \mathbf{n}^T M_0 \nabla u(\mathbf{x}, t) = 0 \qquad \text{on } \Gamma_0, \end{cases} \tag{25}$$

where

$$\mathbf{J}_v(\mathbf{x}, t) = -M_i \nabla v(\mathbf{x}, t)$$

and $[\![\, \Phi \,]\!]_S = \Phi_{S^+} - \Phi_{S^-}$ denotes the jump through a surface S, with $\Phi|_{S^\pm}$ equal to the trace taken on the positive and negative sides of Σ with respect to the oriented normal. We remark that the right-hand sides div $\mathbf{J}_v(\mathbf{x}, t)$ and $\mathbf{n}^T \mathbf{J}_v(\mathbf{x}, t)$ act as *impressed* current or *current source density*. Thus, if we assume the transmembrane potential distribution $v(\mathbf{x}, t)$ known, the above elliptic problem fully characterizes the extracellular and extracardiac field $u(\mathbf{x}, t)$ apart from an additive constant.

The interface condition $[\![\, \mathbf{n}^T \widehat{M} \nabla u(\mathbf{x}, t) \,]\!]_{\Gamma_H} = \mathbf{n}^T \mathbf{J}_v(\mathbf{x}, t)$ is equivalent to the conservation of current relationship $\mathbf{n}^T \mathbf{j_i} + \mathbf{n}^T \mathbf{j_e} = \mathbf{n}^T \mathbf{j}_0$. Moreover, another interface condition should be added in order to fully define the reaction-diffusion system in the media (i)+(e). While there is no general agreement on this additional interface condition (see, e.g., [25, 58]), the three most used in practice are $\mathbf{n}^T \mathbf{j_i} = 0$ [58, 78, 119, 151], $\mathbf{n}^T \mathbf{J}_v = 0$ [23–25, 109, 122] and $\mathbf{n}^T \mathbf{j_i} = C_m \frac{\partial v}{\partial t} + I_{\text{ion}}$ [110]. At present, a rigorous derivation of homogenized interface conditions at the cardiac tissue boundary in contact with a conducting medium is still missing.

3.1 Boundary integral formulation for ECG simulations

One of the major tasks in computational electrocardiology is to explain the genesis of the electrocardiograms, i.e., the interpretation of the morphology of the ECGs in terms of the underlying bioelectric cardiac events which have generated them. In order to reduce the complexity related to both the geometrical description and the conduction properties of the various structures embedded in the extracardiac medium, in large scale simulations associated with the full body, the conductivity tensor M_0 is usually assumed isotropic with constant or piecewise constant conductivity.

Under this simplifying assumption, the simulation of the evolution of the extracardiac potential at a limited number of locations distributed on the body surface, i.e., the usual derivation of ECGs, can be afforded with reduced computational load by considering a boundary integral formulation instead of a differential representation.

We briefly describe this widely used integral approach (see [50, 111], [29, 36], [45, 152]), assuming for simplicity that $M = \sigma_0 I$ with σ_0 constant.

The previous elliptic problem (25) with boundary conditions of Neumann type selects, as expected, the distribution of the potential field apart from a time-dependent constant determined by choosing a reference potential. The usual reference potential in electrocardiography is the so-called *Wilson central terminal* which is approximately equal to the potential average on the thorax body surface or on the the epicardial surface as shown in experiments on animals (see [143]). With $\Sigma \subseteq \Gamma_0$ denoting a part of the body surface and choosing as a reference the potential average on Σ, then the potential w measured with respect to this reference is given by

$$w(\mathbf{x}, t) = u(\mathbf{x}, t) - \frac{1}{|\Sigma|} \int_{\Sigma} u(\boldsymbol{\xi}, t)\, d\sigma_{\xi}$$

with $|\Sigma|$ denoting the area of the surface Σ.

For an observation point $\mathbf{x}$ on the body surface Γ_0 we introduce the Green function ψ, also called the *lead field*, solution of the following elliptic problem with Neumann boundary conditions:

$$\begin{cases} -\mathrm{div}_{\xi} \widehat{M} \nabla_{\xi} \psi = 0 \quad \text{in } \Omega_H \cup \Omega_0 \\ [\![\ \psi \ ]\!]_{\Gamma_H} = 0, \quad [\![\ \mathbf{n}^T \widehat{M} \nabla_{\xi} \psi \ ]\!]_{\Gamma_H} = 0 \\ \mathbf{n}^T \widehat{M} \nabla_{\xi} \psi = -\dfrac{1}{|\Sigma|} \, \chi_{\Sigma}(\boldsymbol{\xi}) \, + \, \delta(\boldsymbol{\xi} - \mathbf{x}) \quad \text{on } \Gamma_0, \end{cases} \tag{26}$$

where $\chi_{\Sigma}(\boldsymbol{\xi})$ denotes the characteristic function of the surface Σ. Proceeding formally, we apply the second Green formula to the couple (u, ψ) in Ω_0 and in Ω_H, where ψ is a solution of (26). Adding these two relations we obtain

$$\begin{aligned} &\int_{\Omega_0 \cup \Omega_H} \left[u \ \mathrm{div} \ \widehat{M} \nabla \psi - \psi \ \mathrm{div} \ \widehat{M} \nabla u \right] d\xi \\ &= \int_{\Gamma_0} u \ \mathbf{n}^T M_0 \nabla \psi \ d\gamma_{\xi} - \int_{\Gamma_0} \psi \mathbf{n}^T M_0 \nabla u \ d\gamma_{\xi} \\ &+ \int_{\Gamma_H} [\![\ \mathbf{n}^T \widehat{M} \nabla \psi \ ]\!]_{\Gamma_H} \ u \ d\gamma_{\xi} - \int_{\Gamma_H} [\![\ \mathbf{n}^T \widehat{M} \nabla u \ ]\!]_{\Gamma_H} \ \psi \ d\gamma_{\xi}, \end{aligned}$$

where the unit normal $\mathbf{n}$ is outward to Ω_H or to $\bar{\Omega} = \bar{\Omega}_H \cup \bar{\Omega}_0$ according to whether the boundary condition is given on Γ_H or on Γ_0. For (26) we then have

$$\begin{aligned} - \int_{\Omega_0 \cup \Omega_H} \psi \ \mathrm{div} \ \widehat{M} \nabla u \ d\xi &= -u_{ref} + u(\mathbf{x}, t) - \int_{\Gamma_0} \psi \mathbf{n}^T \widehat{M} \nabla u \ d\gamma_{\xi} \\ &\quad - \int_{\Gamma_H} [\![\ \mathbf{n}^T \widehat{M} \nabla u \ ]\!]_{\Gamma_H} \ \psi \ d\gamma_{\xi}. \end{aligned}$$

Hence, taking (25) into account, it follows that

$$w(\mathbf{x}, t) = - \int_{\Omega_H} \psi \ \mathrm{div} \ \mathbf{J}_v \ d\xi + \int_{\Gamma_H} \psi \ \mathbf{n}^T \mathbf{J}_v \ d\gamma_{\xi},$$

and, applying the first Green formula, we obtain the following boundary integral representation (see [29, 36]):

$$w(\mathbf{x},t) = \int_{\Omega_H} \mathbf{J}_v^T \nabla_\xi \psi(\boldsymbol{\xi},\mathbf{x}) d\xi = -\int_{\Omega_H} (\nabla v(\boldsymbol{\xi},t))^T M_i(\boldsymbol{\xi}) \nabla_\xi \psi(\boldsymbol{\xi},\mathbf{x}) d\xi \quad (27)$$

$$= -\int_{\Gamma_H} v(\boldsymbol{\xi},t) \left\{ \mathbf{n}^T M_i(\boldsymbol{\xi}) \nabla_\xi \psi(\boldsymbol{\xi},\mathbf{x}) \right\} d\gamma_\xi$$

$$+ \int_{\Omega_H} v(\boldsymbol{\xi},t)) \left\{ \operatorname{div} M_i(\boldsymbol{\xi}) \nabla_\xi \psi(\boldsymbol{\xi},\mathbf{x}) \right\} d\xi .$$

The Neumann boundary condition on Σ for the lead field ψ models the chosen reference potential, i.e., guarantees that $w(\mathbf{x},t)$ has zero mean value on Σ. In the case of equal anisotropic ratio $M = \text{const } M_i$, the volume integral in the right-hand side of the last line of (27) disappears and the extracardiac potentials are generated by impressed currents concentrated only on the heart surface. This approximate model coincides with the so-called *heart surface model* [50, 51] widely used in spite of its approximate value (see [28, 29]). A rigorous derivation of the integral representation of w is given in [36] when the reference potential is the potential at a fixed point $\mathbf{x}_0$ of Ω; a similar integral representation can be found in [161].

The numerical computation of the electrograms (EGs) based on the integral representation (27) requires special care because of the singularity of ψ at the observation point $\mathbf{x}$ and of the presence of the moving steep wave front characterizing $v(\boldsymbol{\xi},t)$ during the depolarization phase when simulating epicardial and intramural EGs. An adaptive numerical procedure was proposed and implemented in [36], where it is shown how to compute, with limited computational effort, epicardial and intramural EGs free of numerical artifacts.

Factors influencing the shape of the EGs are: the intramural fiber rotation, the anisotropy ratio of the intra- and extracellular conductivity coefficients, the reference potential, and the proximity of the observation site to the epi- or endocardial surface. The EG QRS waveform, as recorded directly from the heart, can vary from monophasic to multiphasic due to the appearance of humps and spikes. Simulation studies of EGs incorporating the tissue anisotropy [26, 29, 49, 54, 60, 82, 120, 145] have confirmed the appearance of R–waves and Q–waves in the EGs recorded from regions through which excitation spread along and across fibers, respectively, for sites near a pacing site. The effect of the reference potential on the unipolar EGs in a large parallelepipedal slab of cardiac tissue was studied in [145] using a source splitting technique (also see [28, 29] for details). In [29] we extended the investigation to a more realistic geometry of the myocardial wall showing that anisotropy of the cardiac sources plays an essential role in producing the multiple humps and spikes that appear in the polymorphic EG wave forms.

A qualitative comparison between the simulated EGs and potentials generated by epicardial pacing and those recorded from isolated or exposed dog hearts can be found in [29, 92, 93].

4 Approximate modeling of cardiac bioelectric activity by reduced models

In the bidomain model (23), the transmembrane potential v during the excitation phase of the heartbeat exhibits a steep propagating layer spreading throughout the myocardium with a thickness of about 0.5 mm. At every point of the cardiac domain, this upstroke phase lasts about 1 msec. Therefore, the simulation of the excitation process requires the numerical solution of problems with small space and time steps (of the order of 0.1 mm and 0.01 msec). This fact constraints 3-D simulations to limited blocks with dimensions of a few cm; see [26, 31, 60]. For large scale simulations involving the whole ventricles, computer memory and time requirements become excessive and less demanding approximations have been developed, such as monodomain and eikonal models.

4.1 Linear anisotropic monodomain model

In order to reduce the computational load further, many large scale simulations have been performed using the so-called monodomain model; it is well-known that, if the two media have the same anisotropy ratio, then the bidomain system reduces to the monodomain model. We remark that this is not the physiological case, as clearly follows from well established experimental evidence. We present an interesting derivation of a reduced bidomain model which does not make such a priori assumptions (also see [20, 72]) and that we will still call it the monodomain model.

Denoting by $J_{\text{tot}} = \mathbf{j_i} + \mathbf{j_e}$ the total current flowing in the two media and by $M = M_i + M_e$ the conductivity of the bulk medium, since $J_{\text{tot}} = -M_i \nabla u_i - M_e \nabla u_e$, substituting $u_i = v + u_e$, we obtain

$$\nabla u_e = -M^{-1} M_i \nabla v - M^{-1} J_{\text{tot}}. \tag{28}$$

Therefore, the second equation in the bidomain system (23) can be written as

$$-c_m \frac{\partial v}{\partial t} + \operatorname{div}(M_e M^{-1} M_i \nabla v) + \operatorname{div}(M_e M^{-1} J_{\text{tot}}) - i_{\text{ion}}(v, w, c) = I^e_{\text{app}}. \tag{29}$$

Since the conductivity tensors are given by (22), then

$$M_e M^{-1} = \mu^e_l I + (\mu^e_t - \mu^e_l)\mathbf{a}_t(\mathbf{x})\mathbf{a}_t^T(\mathbf{x}) + (\mu^e_n - \mu^e_l)\mathbf{a}_n(\mathbf{x})\mathbf{a}_n^T(\mathbf{x}), \tag{30}$$

with $\mu^e_{l,t,n} = \sigma^e_{l,t,n}/(\sigma^e_{l,t,n} + \sigma^i_{l,t,n})$. Assuming constant conductivity coefficients and taking into account the fact that $\operatorname{div} J_{\text{tot}} = I^i_{\text{app}} + I^e_{\text{app}}$, we have

$$\begin{aligned}\operatorname{div}(M_e M^{-1} J_{\text{tot}}) &= \mu^e_l \operatorname{div} J_{\text{tot}} \\ &\quad + (\mu^e_t - \mu^e_l)\operatorname{div}[\mathbf{a}_t(\mathbf{x})\mathbf{a}_t^T(\mathbf{x}) J_{\text{tot}}] + (\mu^e_n - \mu^e_l)\operatorname{div}[\mathbf{a}_n(\mathbf{x})\mathbf{a}_n^T(\mathbf{x}) J_{\text{tot}}] \\ &= \mu^e_l (I^i_{\text{app}} + I^e_{\text{app}}) + (\mu^e_t - \mu^e_l)\operatorname{div}[\mathbf{a}_t(\mathbf{x})\mathbf{a}_t^T(\mathbf{x}) J_{\text{tot}}] \\ &\quad + (\mu^e_n - \mu^e_l)\operatorname{div}[\mathbf{a}_n(\mathbf{x})\mathbf{a}_n^T(\mathbf{x}) J_{\text{tot}}].\end{aligned} \tag{31}$$

From (28) it follows that $-M_eM^{-1}M_i\nabla v = M_eM^{-1}J_{tot} + M_e\nabla u_e$, so that we have the flux relationship

$$-\mathbf{n}^T M_eM^{-1}M_i\nabla v = \mathbf{n}^T M_eM^{-1}J_{tot} + \mathbf{n}^T M_e\nabla u_e.$$

By using the split form (30), the first term on the right-hand side can be written as

$$\begin{aligned}&\mathbf{n}^T(M_eM^{-1}J_{tot})\\ &= \mu_l^e\mathbf{n}^T J_{tot} + (\mu_t^e - \mu_l^e)(\mathbf{n}^T\mathbf{a}_t)(\mathbf{a}_t^T J_{tot}) + (\mu_n^e - \mu_l^e)(\mathbf{n}^T\mathbf{a}_n)(\mathbf{a}_n^T J_{tot}).\end{aligned} \quad (32)$$

The insulating conditions $\mathbf{n}^T\mathbf{j_i} = \mathbf{n}^T\mathbf{j_e} = 0$ imply $\mathbf{n}^T J_{tot} = 0$, i.e., J_{tot} is tangent to Γ_H, and, assuming that fibers are also tangent to Γ_H, we have $\mathbf{n}^T\mathbf{a}_n = 0$ and $\mathbf{a}_t^T J_{tot} = 0$; substituting these conditions in (32), we obtain

$$\mathbf{n}^T M_eM^{-1}M_i\nabla v = 0. \quad (33)$$

We remark that, for media having equal anisotropic ratio, i.e., $\sigma_l^e/\sigma_l^i = \sigma_t^e/\sigma_t^i = \sigma_n^e/\sigma_n^i$, we have $\mu_l^e = \mu_t^e = \mu_n^e$. Thus, two additional terms in (31) related to the projections of J_{tot} on the directions across the fibers disappear. Disregarding these two additional source terms $\mathbf{a}_t^T J_{tot}$ and $\mathbf{a}_n^T J_{tot}$, we have $\mathrm{div}(M_eM^{-1}J_{tot}) \approx \mu_l^e(I_{app}^i + I_{app}^e)$. Substituting this approximation in (29) and considering the boundary condition (33), we obtain an approximate model consisting in a single parabolic reaction-diffusion equation for v with the conductivity tensor $M_m = M_eM^{-1}M_i$, $I_{app}^m = \left(I_{app}^i\sigma_l^e - I_{app}^e\sigma_l^i\right)/(\sigma_l^e + \sigma_l^i)$ and coupled with the same gating system:

$$\begin{cases} c_m\dfrac{\partial v}{\partial t} - \mathrm{div}(M_m\nabla v) + i_{ion}(v, w, c) = I_{app}^m & \text{in } \Omega_H \times (0, T)\\ \dfrac{\partial w}{\partial t} - R(v, w) = 0, \quad \dfrac{\partial c}{\partial t} - S(v, w, c) = 0 & \text{in } \Omega_H \times (0, T)\\ \mathbf{n}^T M_m\nabla v = 0 & \text{in } \Gamma_H \times (0, T)\\ v(\mathbf{x}, 0) = v_0(\mathbf{x}), \quad w(\mathbf{x}, 0) = w_0(\mathbf{x}), \quad c(\mathbf{x}, 0) = c_0(\mathbf{x}) & \text{in } \Omega_H. \end{cases} \quad (34)$$

The evolution equation determines the distribution of $v(\mathbf{x}, t)$ and then the extracellular potential distribution u_e is derived by solving the elliptic boundary value problem:

$$\begin{cases} -\mathrm{div}(M\nabla u_e) = \mathrm{div}(M_i\nabla v) + I_{app}^i + I_{app}^e & \text{in } \Omega_H\\ -\mathbf{n}^T M\nabla u_e = \mathbf{n}^T M_i\nabla v & \text{on } \Gamma_H. \end{cases} \quad (35)$$

We refer to the system consisting of Eqs. (34) and (35) as the anisotropic monodomain model. We remark that the bidomain and the monodomain models are described by systems of a parabolic equation coupled with an elliptic equation, but in the latter the evolution equation is fully uncoupled with the elliptic one in the case of an insulated domain Ω_H.

4.2 Eikonal models

Another way of avoiding the high computational costs of the full bidomain model is based on the use of *eikonal − curvature models* for the evolution of the excitation wavefront surface.

With these models the simulation of the activation sequence in large volumes of cardiac tissue is computationally practical but at the price of a loss of fine details concerning the thin layer where the upstroke of the action potential occurs. These numerical simulations are based on evolution geometric laws describing the macroscopic kinetic mechanism of the spreading of the excitation wavefronts, and do not require fine spatial and temporal resolution.

We now outline the derivation of such approximated models. The Fitz Hugh–Nagumo approximation of the membrane kinetic is very useful for a qualitative analysis of the nonlinear dynamics of the R–D system. As we focus only on the excitation phase, we can neglect the recovery variable w and hence $i_{\mathrm{ion}} = g(v)$. The resulting simplified ionic model is widely used for gaining general insight into wave propagation in the cardiac excitable media. Although this model is not suitable in a quantitative detailed study, at a fine scale, of the upstroke of the action potential v through the excitation wavefront, it is nevertheless appropriate if we are interested in the large scale behavior of the front-like solution. We note that, during the excitation phase of the heart beat, the main feature, at a macroscopic level, is the excitation wavefront configuration and its motion. In order to investigate the propagation of this wavefront we must analyze more deeply the internal layer of v which affects the spreading.

By proceeding as in [24] with a suitable scaling, the macroscopic dimensionless form of the bidomain model can be written as the following singularly perturbed R–D system:

$$\begin{cases} \dfrac{\partial v^{\varepsilon}}{\partial t} + \dfrac{1}{\varepsilon}\, g(v) - \varepsilon \,\mathrm{div}\,(M_i \nabla u_i^{\varepsilon}) \;= 0 & \text{in } \Omega_H \\[2ex] -\dfrac{\partial v^{\varepsilon}}{\partial t} - \dfrac{1}{\varepsilon} g(v) - \varepsilon \,\mathrm{div}\,(M_e \nabla u_e^{\varepsilon}) = 0 & \text{in } \Omega_H, \end{cases} \tag{36}$$

where $v^{\varepsilon} = u_i^{\varepsilon} - u_e^{\varepsilon}$, the dimensionless parameter ε is of the order $10^{-3} - 10^{-2}$ and g is a scaled form of a cubic–like ionic current $i_{\mathrm{ion}}(v)$. We denote by $v_r < v_{\mathrm{th}} < v_p$ the three zeros of g representing the resting, threshold and excited transmembrane values, respectively and we assume that $\int_{v_r}^{v_p} g(v)dv < 0$.

The R-D systems with excitable dynamics are studied by mathematical tools from singular perturbation theory; see, e.g., [44,76]. Because of the previous singular perturbation structure, $u_{\varepsilon}^{i}, u_{\varepsilon}^{e}$ diffuse quite slowly, while the reaction takes place much faster; hence, the development of a moving layer associated with a traveling wavefront solution is to be expected. Exploiting the singular perturbation approach, we can derive anisotropic geometric evolution laws capturing the asymptotic behavior of traveling wavefront solutions of the R–D system (36) (see [12,13,23,24,71,73,74,76] and for isotropic media [138]).

Assuming that the excitation propagates in fully recovered tissue, a monotonic temporal behavior of v is expected; then the excitation wavefront $S_\varepsilon(t)$ can be represented by the level surface of the transmembrane potential of value $(v_r + v_p)/2$, that is:

$$S_\varepsilon(t) = \{\mathbf{x} \in \Omega_H, \;\; v_\varepsilon(\mathbf{x}, t) = (v_r + v_p)/2\}. \tag{37}$$

We define the *activation time* $\psi(\mathbf{x})$ as the time instant at which the potential v at $\mathbf{x}$ reaches the value $(v_r + v_p)/2$. Then the excitation wavefront $S_\varepsilon(t)$ is also represented by the level surface of the activation time at the time instant t, that is:

$$S_\varepsilon(t) = \{\mathbf{x} \in \Omega_H, \;\; \psi(\mathbf{x}) = t\}.$$

Let

$$q_{i,e}(\mathbf{x}, \boldsymbol{\xi}) = \boldsymbol{\xi}^T M_{i,e}(\mathbf{x})\boldsymbol{\xi}$$

be the conductivity coefficients at a point $\mathbf{x}$ in the intra- and extracellular media measured along the direction of the unit vector $\boldsymbol{\nu}$. We define the harmonic mean of the quadratic forms associated with the conductivity tensors $M_{i,e}$ by

$$q(\mathbf{x}, \boldsymbol{\xi}) = (q_i(\mathbf{x}, \boldsymbol{\xi})^{-1} + q_i(\mathbf{x}, \boldsymbol{\xi})^{-1})^{-1}.$$

The nonlinear form $q(\mathbf{x}, \boldsymbol{\xi})$ admits the representation

$$q(\mathbf{x}, \boldsymbol{\xi}) = \boldsymbol{\xi}^T Q(\mathbf{x}, \boldsymbol{\xi})\boldsymbol{\xi} \;\text{ with }\; Q(\mathbf{x}, \boldsymbol{\xi}) = \frac{q_i(\mathbf{x}, \boldsymbol{\xi})^2 M_e(\mathbf{x}) + q_e(\mathbf{x}, \boldsymbol{\xi})^2 M_i(\mathbf{x})}{[q_i(\mathbf{x}, \boldsymbol{\xi}) + q_i(\mathbf{x}, \boldsymbol{\xi})]^2}, \tag{38}$$

which gives the conductivity measured along the direction $\boldsymbol{\xi}$ of the bulk medium composed by coupling in series the media (i) and (e). Then we introduce the indicatrix function

$$\Phi(\mathbf{x}, \boldsymbol{\xi}) = \sqrt{q(\mathbf{x}, \boldsymbol{\xi})}. \tag{39}$$

Let (c,a) be the unique bounded solution of the eigenvalue problem:

$$\begin{cases} a'' + c\, a' - g(a) = 0 \\ a(-\infty) = v_p\,, \quad a(\infty) = v_r, \quad a(0) = (v_p + v_r)/2. \end{cases} \tag{40}$$

A formal inner asymptotic expansion in powers of ε of $(u_i^\varepsilon, u_e^\varepsilon)$, a solution of (36), and $v^\varepsilon = u_i^\varepsilon - u_e^\varepsilon$ can be performed by using two different stretchings of variables. Let $\boldsymbol{\nu}$ be the Euclidean unit vector normal to the wavefront $S_\varepsilon(t)$ oriented toward the resting tissue and, for $\mathbf{s}(t) \in S_\varepsilon(t)$, we define the vector $\mathbf{n}_\Phi(\mathbf{s}) = \Phi_{\boldsymbol{\xi}}(\mathbf{s}, \boldsymbol{\nu})$. As in [12] we take a Lagrangian point of view and we consider the moving reference $(\mathbf{s}, y, \tau)$ defined by

$$y = \frac{\eta}{\varepsilon}, \qquad \mathbf{x} = \mathbf{s}(t) + \eta\, \mathbf{n}_\Phi(\mathbf{s}(t)), \qquad \tau = t, \quad \forall\; \mathbf{s}(t) \in S_\varepsilon(t), \tag{41}$$

i.e. stretching the space coordinate along the $\mathbf{n}_\Phi$ direction.

By using the moving frame (41) the asymptotic expansion for the bidomain model (36) shows that (see [12, Appendix B]), at least formally, the front associated with (37) moves along the relative normal vector $\mathbf{n}_{\Phi}$ with velocity $\theta_{\epsilon}(\mathbf{n}_{\Phi})$ given at any $\mathbf{s}(t) \in S_{\varepsilon}(t)$ by

$$\theta(\mathbf{n}_{\Phi}) = c - \varepsilon \operatorname{div} \mathbf{n}_{\Phi} + \mathcal{O}(\varepsilon^2),$$

where c is related to the traveling wave solution a of (40). Since $\mathbf{n}_{\Phi} \cdot \boldsymbol{\nu} = \Phi(\mathbf{s}, \boldsymbol{\nu})$, the velocity $\theta_{\epsilon}(\boldsymbol{\nu})$ in the Euclidean normal direction $\boldsymbol{\nu}$ of $S_{\varepsilon}(t)$ is given by $\Phi(\mathbf{s}, \boldsymbol{\nu})\theta_{\epsilon}(\mathbf{n}_{\Phi})$; then,

$$\theta_{\epsilon}(\boldsymbol{\nu}) = \Phi(\mathbf{s}, \boldsymbol{\nu})(c - \varepsilon \operatorname{div} \Phi_{\boldsymbol{\xi}}(\mathbf{s}, \boldsymbol{\nu})) + \mathcal{O}(\varepsilon^2). \tag{42}$$

Therefore, on dropping $\mathcal{O}(\varepsilon^2)$ terms, the front behaves as a hypersurface $S(t)$, propagating according to the anisotropic geometric law with normal velocity $\theta(\boldsymbol{\nu})$ given by

$$\theta(\boldsymbol{\nu}) = \Phi(\mathbf{s}, \boldsymbol{\nu})(c - \varepsilon \operatorname{div} \Phi_{\boldsymbol{\xi}}(\mathbf{s}, \boldsymbol{\nu})). \tag{43}$$

Equations of this type are also called *eikonal-curvature models* since $K_{\Phi} = \operatorname{div} \mathbf{n}_{\Phi} = \operatorname{div} \Phi_{\boldsymbol{\xi}}(\mathbf{s}, \boldsymbol{\nu})$ is the *anisotropic* mean curvature with respect to a suitable Finsler metric; see [12, 13] for definitions and details.

A formal derivation based on an Eulerian point of view was developed in [23,24]; in this approach the new frame $(\boldsymbol{\chi}, \tau)$ is defined by stretching the time variable with respect to the activation time, that is:

$$\boldsymbol{\chi} = \mathbf{x}, \quad \tau = \frac{t - \varphi(\mathbf{x})}{\varepsilon}. \tag{44}$$

By using the fixed Eulerian frame (44), the asymptotic expansion for the bidomain model (36) yields

$$\frac{\theta_{\varepsilon}(\boldsymbol{\nu})}{\Phi(\mathbf{x}, \boldsymbol{\nu})} \left(1 + \varepsilon \operatorname{div} \left(\frac{\Phi(\mathbf{x}, \boldsymbol{\nu})\Phi_{\boldsymbol{\xi}}(\mathbf{x}, \boldsymbol{\nu})}{\theta_{\varepsilon}(\boldsymbol{\nu})}\right)\right) = c + \mathcal{O}(\varepsilon^2), \tag{45}$$

or, equivalently,

$$\frac{\theta_{\varepsilon}(\boldsymbol{\nu})}{\Phi(\mathbf{x}, \boldsymbol{\nu})} = c - \varepsilon \operatorname{div} \Phi_{\boldsymbol{\xi}}(\mathbf{x}, \boldsymbol{\nu}) + \varepsilon \frac{\Phi(\mathbf{x}, \boldsymbol{\nu})}{\theta_{\varepsilon}(\boldsymbol{\nu})} \nabla \left(\frac{\theta_{\varepsilon}(\boldsymbol{\nu})}{\Phi(\mathbf{x}, \boldsymbol{\nu})}\right) \cdot \Phi_{\boldsymbol{\xi}}(\mathbf{x}, \boldsymbol{\nu}) + \mathcal{O}(\varepsilon^2). \tag{46}$$

Since both Eqs. (42), (46) imply that $\dfrac{\theta_{\varepsilon}(\boldsymbol{\nu})}{\Phi(\mathbf{x}, \boldsymbol{\nu})} = c + \mathcal{O}(\varepsilon)$, the two eikonal equations (42), (46) are equivalent up to second-order terms. Equations (43) and (46), on disregarding the $\mathcal{O}(\varepsilon^2)$ term, are called, respectively, the *eikonal–curvature* and *eikonal–diffusion* equations [30, 149].

The rigorous justification of the connection between the evolution of a suitable level set of v and the surface flowing according to geometric evolution law, remains to our knowledge an open problem. A partial rigorous characterization in the Γ-convergence framework was obtained for the stationary bidomain model [5]. We introduce the family of vectorial integral Lyapunov functionals dependent on the pair $\boldsymbol{u} := (u_i, u_e)$:

$$\begin{aligned} \mathcal{F}^\varepsilon(\boldsymbol{u}) := \varepsilon \int_\Omega \Big(M_i \nabla u_i \cdot \nabla u_i + M_e \nabla u_e \cdot \nabla u_e \Big)\, dx \\ + \frac{1}{\varepsilon} \int_\Omega G(u_i - u_e)\, dx, \end{aligned} \tag{47}$$

where $G' = g$. The degenerate reaction-diffusion system associated with (36) in the pair of unknowns $\boldsymbol{u}^\varepsilon := (u_i^\varepsilon, u_e^\varepsilon)$ can be obtained by taking the gradient flow of the Lyapunov functional with respect to the positive but degenerate bilinear form in $L^2(\Omega; \mathbb{R}^2)$:

$$b(\boldsymbol{u}, \boldsymbol{w}) := \int_\Omega (u_i - u_e)\,(w_i - w_e)\, dx, \quad \boldsymbol{u} = (u_i, u_e), \quad \boldsymbol{w} = (w_i, w_e).$$

This yields the following system of variational evolution equations:

$$b(\partial_t \boldsymbol{u}^\varepsilon, \boldsymbol{v}) + \delta\mathcal{F}^\varepsilon(\boldsymbol{u}^\varepsilon, \boldsymbol{v}) = 0 \quad \forall\ \boldsymbol{v} \in H^1(\Omega; \mathbb{R}^2),$$

where $\partial_t \boldsymbol{u}^\varepsilon = (\frac{\partial u_i^\varepsilon}{\partial t}, \frac{\partial u_e^\varepsilon}{\partial t})$ and $\delta\mathcal{F}^\varepsilon(\boldsymbol{u}, \cdot)$ is the Euler-Lagrange first variation of the functional $\mathcal{F}^\varepsilon$ [52], which is the variational formulation of (36). Since the anisotropic curvature K_Φ corresponds to the first variation of the anisotropic surface energy integral functional associated with Φ, in order to justify the form of the anisotropic curvature term, in [5] we studied the asymptotic limit, as $\varepsilon \downarrow 0$, of the stationary problem associated with the singular reaction–diffusion system (36) for a function $g = G'$, where now G is a potential having wells of equal depth. More precisely, the Γ-limit of Lyapunov functionals associated with the family (47) is a surface integral functional whose energy density is a continuous family of norms $\Phi^*(\mathbf{x}, \cdot)$ characterized by solving a localized minimization problem; see [5] for details.

Formal asymptotic results in the bidomain case (see [12, 23, 24]) suggest that $\Phi^*(\mathbf{x}, \nu) = \Phi(\mathbf{x}, \nu) = c\sqrt{q(x, \nu)}$. On the other hand, in certain pathophysiological settings, such as regional ischemia and a healed infarction, the corresponding conductivity tensors M_i, M_e yield a nonconvex Φ; since Φ^* is always convex, in this case the previous equality does not hold. It would be interesting to check whether the convex envelope of Φ is a good substitute in this case.

4.3 Relaxed nonlinear anisotropic monodomain model

We can easily see that, by rescaling as in (36) the reaction-diffusion equation related to the monodomain model (34) and by using formal asymptotic expansions as before, the anisotropic evolution law of the front does not coincide with that derived from the bidomain model. In fact, although the eikonal-curvature equation up to terms of order $O(\varepsilon^2)$ presents the same structure

$$\theta(\boldsymbol{\nu}) = \Phi(\mathbf{s}, \boldsymbol{\nu})(c - \varepsilon \operatorname{div} \Phi_{\boldsymbol{\xi}}(\mathbf{s}, \boldsymbol{\nu})),$$

the nonlinear function $\Phi(\mathbf{x}, \boldsymbol{\xi}) = \sqrt{q(\mathbf{x}, \boldsymbol{\xi})}$ for the monodomain model is defined by

$$q(\mathbf{x}, \boldsymbol{\xi}) = \boldsymbol{\xi}^T M_m(\mathbf{x}, \boldsymbol{\xi})\boldsymbol{\xi}, \qquad \text{where} \quad M_m(\mathbf{x}) = M_e(M_i + M_e)^{-1} M_i,$$

i.e., M_m is the harmonic tensor associated with $M_{i,e}$.

We now consider the following monodomain model with nonlinear diffusion term, which we call the *relaxed monodomain model*

$$\frac{\partial v}{\partial t} + \frac{1}{\varepsilon} f(v) - \varepsilon \operatorname{div}(Q(\mathbf{x}, \nabla v)\nabla v) = I_{\text{app}}$$

with $Q(\mathbf{x}, \boldsymbol{\xi})$ defined in (38) and written explicitly in terms of the conductivity tensors as

$$Q(\mathbf{x}, \boldsymbol{\xi}) := \left(\frac{\xi^T M_i \xi}{\xi^T M \xi}\right)^2 M_e(\mathbf{x}) + \left(\frac{\xi^T M^e \xi}{\xi^T M \xi}\right)^2 M_i(\mathbf{x}), \quad M := M_i + M_e.$$

The diffusion term is nonlinear except when $M_e = \lambda M_i$, with a constant $\lambda \in \mathbb{R}$, i.e., for equal anisotropic ratio of the two media, $Q(\mathbf{x}, \boldsymbol{\xi}) = \frac{\lambda}{1+\lambda} M_i(\mathbf{x})$.

Proceeding by formal asymptotic expansions (see [12, Appendix A]), we see that the *relaxed monodomain model* admits the same eikonal-curvature equation as that associated with the bidomain model. The nonlinear conductivity tensor of the medium $Q(\mathbf{x}, \nabla v)$, which is homogeneous of degree zero in the second variable, is a function of the local direction of propagation of the front-like solution given by the unit vector $\nabla v/|\nabla v|$. In [12], we show that the eikonal-curvature equation as an approximate model for describing the evolution of the relaxed transmembrane potential v can be justified rigorously by estimating the distance between a suitable level set of the relaxed evolution v and the surface flowing according to the geometric law. We observe here that a suitable convexity property of Φ is crucial for this rigorous result (see [12] for details). Such a property is true in a wide range of physiological choices but is not guaranteed for generic choices of matrices M^i, M^e. Pathological anisotropies, e.g., modeling ischemic tissue, can lead to a nonconvex Φ, hence to a relaxed model which is not well-posed. This issue requires further study since it could be related to mechanisms of re-entry phenomena associated with cardiac arrhythmias in the presence of ischemic substrates (see [16]). While we have focused so far on the use of reduced models for the excitation phase, we note that the relaxed monodomain model could also be used as a reduced model in all phases of the heartbeat by solving the following problem in dimensional form:

$$\begin{cases} c_m \dfrac{\partial v}{\partial t} - \operatorname{div}(Q(\mathbf{x}, \nabla v)\nabla v) + i_{\text{ion}}(v, w, c) = I_{\text{app}} & \text{in } \Omega_H \times (0, T) \\ \dfrac{\partial w}{\partial t} - R(v, w) = 0, \qquad \dfrac{\partial c}{\partial t} - S(v, w, c) = 0 & \text{in } \Omega_H \times (0, T) \\ \mathbf{n}^T (Q(\mathbf{x}, \nabla v)\nabla v = 0 & \text{in } \Gamma_H \times (0, T) \\ v(\mathbf{x}, 0) = v_0(\mathbf{x}), \quad w(\mathbf{x}, 0) = w_0(\mathbf{x}), \quad c(\mathbf{x}, 0) = c_0(\mathbf{x}) & \text{in } \Omega_H \\ -\operatorname{div}(M\nabla u_e) = \operatorname{div}(M_i \nabla v) + I^i_{\text{app}} + I^e_{\text{app}} \quad \text{in } \Omega_H & \\ -\mathbf{n}^T M \nabla u_e = \mathbf{n}^T M_i \nabla v \quad \text{on } \Gamma_H. & \end{cases}$$

5 Discretization and numerical methods

The anisotropic cardiac models discussed in the previous sections are discretized by the finite element method in space and a semi-implicit finite difference method in time; see [115] for an introduction to these methods. Various numerical techniques have been used in order to discretize the bidomain and monodomain models; for example, we mention here finite differences [18, 63, 74, 75, 90, 100, 117, 120, 128, 131, 156], finite elements [21, 26, 31, 32, 64, 118, 132, 133, 155], finite volumes [40, 56, 57, 104], hybrid finite element difference methods [15], and the interconnected cable method [156].

We focus on the numerical approximation of the eikonal, monodomain (34) and bidomain (23) models in three-dimensional domains representing a portion of the ventricular wall.

5.1 Numerical approximation of the Eikonal–Diffusion equation

Numerical methods for finding the evolving surface $S(t)$ based on a direct discretization of (43) encounters many difficulties, e.g., front–tracking techniques, when high curvature and topological changes of the wavefronts occur. One way to overcome the singularities due to collisions, merging and extinction is the level set approach of (43) (see, e.g., [99, 134]) which represents $S(t)$ as the zero level surface of a function $w(\mathbf{x}, t)$ formally solving

$$\frac{\partial w}{\partial t} = \Phi(\mathbf{x}, \nabla w)(c + \varepsilon \operatorname{div} \Phi_\xi(\mathbf{x}, \nabla w)).$$

As observed before, during the excitation sequence in a fully recovered tissue the wavefront surface $S_\varepsilon(t)$ admits a Cartesian representation. Therefore $\theta_\varepsilon(\boldsymbol{\nu}) = 1/|\nabla\psi|$ and hence, dropping $\mathcal{O}(\varepsilon^2)$, the eikonal-curvature equation (42) reduces to

$$\Phi(\mathbf{x}, \nabla\psi)(c - \varepsilon \operatorname{div} \Phi_\xi(\mathbf{x}, \nabla\psi)) = 1,$$

and the eikonal-diffusion equation (45) to

$$-\varepsilon \operatorname{div} \Phi(\mathbf{x}, \nabla\psi)\Phi_\xi(\mathbf{x}, \nabla\psi) + c\, \Phi(\mathbf{x}, \nabla\psi) = 1. \tag{48}$$

In both equations, the term of order ε is related to the influence of the wavefront curvature on the propagation in an anisotropic medium.

Since, in a fully recovered tissue, the propagation front admits a Cartesian representation, the eikonal–diffusion equation is more convenient. In fact, at the collision points, $\nabla\psi = 0$ and only a discontinuity appears in the divergence term. On the other hand, the level set approach [99, 134] of the eikonal-curvature model, under the same circumstances, exhibits singularities requiring a regularization of Φ. For this reason, in [26, 28, 30] we chose to work with the eikonal–diffusion model using an equivalent formulation of Eq. (48). From (38), (39) it is easy to verify that

$\Phi(\mathbf{x}, \mathbf{p}) = \sqrt{\boldsymbol{\xi}^T Q(\mathbf{x}, \boldsymbol{\xi})\boldsymbol{\xi}}$ and $\Phi_{\boldsymbol{\xi}}(\mathbf{x}, \mathbf{p}) = \dfrac{Q(\mathbf{x}, \boldsymbol{\xi})\boldsymbol{\xi}}{\Phi(\mathbf{x}, \boldsymbol{\xi})}$. Therefore, setting $\mathbf{p} = \nabla\psi$, the eikonal–diffusion equation (48) can be written as:

$$\begin{cases} -\varepsilon \operatorname{div} Q(\mathbf{x}, \mathbf{p})\mathbf{p} + c\sqrt{\mathbf{p}^T Q(\mathbf{x}, \mathbf{p})\mathbf{p}} = 1 \quad \text{in } \Omega_H \\ \mathbf{n}^T\ Q\ \mathbf{p} = 0 \quad \text{on } \Gamma, \quad \psi(\mathbf{x}) = t_a(\mathbf{x}) \quad \text{on } S_a, \end{cases} \tag{49}$$

where S_a is the boundary of the initial activated region and $t_a(\mathbf{x})$ the corresponding activation instants. The solution $\psi(\mathbf{x})$ of this nonlinear problem can be viewed as the steady–state solution of the following parabolic problem associated with (49):

$$\begin{cases} \dfrac{\partial w}{\partial t} - \varepsilon \operatorname{div} Q(\mathbf{x}, \nabla w)\nabla w + c\sqrt{\nabla w^T Q(\mathbf{x}, \nabla w)\nabla w} = 1 \quad \text{in } \Omega_H \\ \mathbf{n}^T\ Q\ \nabla w = 0 \quad \text{on } \Gamma \\ w(\mathbf{x}, t) = w_a(\mathbf{x}) \quad \text{on } S_a, \quad w(\mathbf{x}, 0) = w_0. \end{cases} \tag{50}$$

Equation (50) belongs to a broad class of equations known as "Hamilton–Jacobi" equations, with the Hamiltonian term given by $c\,\Phi(\mathbf{x}, \nabla w)$ and a second-order nonlinear diffusion term $-\varepsilon \operatorname{div} Q(\mathbf{x}, \nabla w)\nabla w$. We considered the following discretization in time obtained by applying a semi–implicit approximation for the diffusive term and explicit for the transport term

$$\frac{w_{n+1} - w_n}{\tau_n} - \varepsilon \operatorname{div} Q(\mathbf{x}, \nabla w_n)\nabla w_{n+1} + c\sqrt{(\nabla w_n)^T Q(\mathbf{x}, \nabla w_n)\nabla w_n} = 1,$$

with w_n the approximate solution at time t_n and $\tau_n = t_{n+1} - t_n$. The space discretization was carried out by the usual Galerkin finite element method. We remark that the solution of the discrete problem can exhibit spurious secondary fronts originating at the domain boundary (see [26]). This is due to the fact that in Eq. (49) the transport term is dominant with respect to the nonlinear diffusion term. To overcome this problem the term $\Phi(\mathbf{x}, \nabla w)$ required special treatment. In order to avoid these numerical artifacts, we adapted the upwind scheme proposed by Osher-Sethian for propagating fronts with curvature dependent speed. This hybrid upwind scheme proved to be quite efficient and allowed us to solve our equation in every case and using a mesh-size h of the order of 1 mm (see [26, 28] and also [149]). Interested readers can find many results of numerical simulations with the eikonal approach in [26–30] and [71, 73, 74, 101].

Lastly, we remark that the nonlinear parabolic equation shares the same nonlinear diffusion term that of the relaxed monodomain model. A similar implicit–explicit time discretization could be applied for the numerical solution of the relaxed monodomain model where the reaction term, modeling the ionic current membrane, is treated explicitly instead of the Hamiltonian term.

5.2 Numerical approximations of the monodomain and bidomain models

We recall that, in the bidomain (monodomain) model, we have an evolution R–D system (equation) coupled with a system of ordinary differential equations. In order to perform a time discretization, it seems natural to first advance the solution v by solving the R–D system (equation) and subsequently to update the gating and ionic concentration variables by solving the membrane system. The time advancement of the solution of the R–D system (equation) can be obtained by using either explicit, semi-implicit or implicit schemes, requiring accordingly vector updates or the solution of a linear or nonlinear system.

Another popular technique is based on operator splitting, i.e., on separating the diffusion operator, related to conduction in the media, from the reaction operator, related to the ionic current, gating and ionic concentration dynamics. In the monodomain case, the time discretization is realized, for given v_n, w_n, C_n, by first solving in a time step τ_n explicitly or semi-implicitly a linear parabolic equation $\partial_t v - \operatorname{div} M_m(\mathbf{x})\nabla v = I^m_{\text{app}}(t_n) - i_{\text{ion}}(v_n, w_n, c_n)$ and subsequently solving for a step τ_n a system of ordinary differential equations $\partial_t w - R(v_n, w) = 0, \quad \partial_t c - S(v_n, w, c) = 0$. See, e.g., [113,150] for a second-order operator splitting.

Different operator splitting schemes were also applied to the bidomain model based on the formulation (24) in terms of (v, u_e). For instance, a semi-discrete operator splitting of the system of an elliptic equation coupled with an evolution equation (see [75,83,140,150,155]) can be written as:

$$\begin{cases} \text{given } (v_n, w_n, c_n): \\ -\operatorname{div}((M_i + M_e)\nabla u_e^n) = \operatorname{div}(M_i \nabla v^n) + I^i_{\text{app}}(t_n) + I^e_{\text{app}}(t_n) \\ c_m \dfrac{v^{n+1} - v^n}{\tau_n} - \operatorname{div}(M_i \nabla v^{n+1}) = -i_{\text{ion}}(v^n, w^n, c^n) + \operatorname{div}(M_i \nabla u_e^n) + I^i_{\text{app}}(t_{n+1}) \\ \dfrac{w^{n+1} - w^n}{\tau_n} - R(v^{n+1}, w^{n+1}) = 0, \qquad \dfrac{c^{n+1} - c^n}{\tau_n} - S(v^{n+1}, w^{n+1}, c^n) = 0. \end{cases}$$

We remark that the depolarization and repolarization phases show different time and space constants, particularly the small thickness (1–2 mm) of the activation layer with respect to the much larger size of the cardiac tissue. Thus, in order to reduce the computational load, especially during the excitation process, adaptive techniques [19,21,90,100,150] and domain decomposition methods [106,107,114,123,162,163] have been developed. Adaptive techniques for nonlinear parabolic PDEs can be found in [79].

Here we briefly sketch a finite element space approximation coupled with a semi-implicit method for the time discretization of the bidomain model (23) formulated in terms of $\boldsymbol{u} = (u_i, u_e)$ and of the monodomain model.

Finite element space discretization. The computational domain Ω in our numerical simulations is either a Cartesian slab or a curved slab modeled as a truncated ellipsoidal volume with parametric equations:

$$\begin{cases} x = a(r)\cos\theta\cos\varphi, & \varphi_{\min} \le \varphi \le \varphi_{\max}, \\ y = b(r)\cos\theta\sin g, & \theta_{\min} \le \theta \le \theta_{\max}, \\ z = c(r)\sin\theta, & 0 \le r \le 1, \end{cases} \tag{51}$$

where $a(r) = a_1 + r(a_2 - a_1)$, $b(r) = b_1 + r(b_2 - b_1)$, $c(r) = c_1 + r(c_2 - c_1)$, and a_i, b_i, c_i, $i = 1, 2$, are given coefficients determining the main axes of the ellipsoid. The fibers rotate intramurally linearly with the depth for a total amount of 120° and when the point of view is from the cavity side, the rotation is CounterClockWise (CCW) proceeding from the endocardium to epicardium. More precisely, in a local ellipsoidal reference system $(\boldsymbol{e}_\varphi, \boldsymbol{e}_\theta, \boldsymbol{e}_r)$, the fiber direction $\mathbf{a}_l(\mathbf{x})$ at a point $\mathbf{x}$ is given by

$$\mathbf{a}_l(\mathbf{x}) = \boldsymbol{e}_\varphi \cos\alpha(r) + \boldsymbol{e}_\theta \sin\alpha(r) \text{ with } \alpha(r) = \frac{2}{3}\pi(1-r) - \frac{\pi}{4}, \quad 0 \le r \le 1. \tag{52}$$

We discretized the ellipsoidal slab with a structured grid of $n_i \times n_j \times n_k$ hexahedral isoparametric Q_1 finite elements. Introducing the associated finite element space V_h, we obtain a semidiscrete problem by applying a standard Galerkin procedure. Choosing a finite element basis $\{\varphi_i\}$ for V_h we denote by

$$\mathrm{M} = \{\mathrm{m}_{rs} = \int_\Omega \varphi_r \, \varphi_s \mathrm{dx}\}, \qquad \mathrm{A}_{m,i,e} = \{\mathrm{a}_{rs}^{m,i,e} = \int_\Omega (\nabla\varphi_r)^T D_{m,i,e} \nabla\varphi_s \mathrm{dx}\}$$

the symmetric mass and stiffness matrices, respectively, and by $\mathrm{i}_{\mathrm{ion}}^h$, $\mathrm{I}_{\mathrm{app}}^{m,h}$, $\mathrm{I}_{\mathrm{app}}^{(i,e),h}$ the finite element interpolants of i_{ion}, I_{app}^m and $I_{\mathrm{app}}^{m,e}$, respectively. Integrals are computed with a 3D trapezoidal quadrature rule, so the mass matrix M is lumped to a diagonal form. In our implementation, we reordered the unknowns by writing, for every node, the $\mathbf{u}_i$ and $\mathbf{u}_e$ components consecutively, so as to minimize the bandwidth of the stiffness matrix.

Semi-implicit time discretization. The time discretization is performed by an implicit-explicit method using the implicit Euler method for the diffusion term, while the nonlinear reaction term $\mathrm{i}_{\mathrm{ion}}$ is treated explicitly. The use of an implicit treatment of the diffusion terms appearing in the monodomain or bidomain models is essential to allow an adaptive change of the time step according to the stiffness of the various phases of the heartbeat. The ODE system for the gating variables is discretized by the semi-implicit Euler method and the explicit Euler method is applied for solving the ODE system for the ion concentrations. As a consequence, the full evolution system is decoupled by first solving the gating and ion concentrations system (given the potential $\mathbf{v}^n$ at the previous time-step):

$$\mathbf{w}^{n+1} - \Delta t\, R(\mathbf{v}^n, \mathbf{w}^{n+1}) = \mathbf{w}^n$$

$$\mathbf{c}^{n+1} = \mathbf{c}^n + \Delta t\, S(\mathbf{v}^n, \mathbf{w}^{n+1}, \mathbf{c}^n),$$

and then solving for $\mathbf{u}_i^{n+1}$, $\mathbf{u}_e^{n+1}$ in the *bidomain case*:

$$\left(\frac{c_m}{\Delta t}\begin{bmatrix} \mathrm{M} & -\mathrm{M} \\ -\mathrm{M} & \mathrm{M} \end{bmatrix} + \begin{bmatrix} \mathrm{A}_i & 0 \\ 0 & \mathrm{A}_e \end{bmatrix}\right)\begin{pmatrix} \mathbf{u}_i^{n+1} \\ \mathbf{u}_e^{n+1} \end{pmatrix}$$

$$= \frac{c_m}{\Delta t}\begin{pmatrix} \mathrm{M}(\ \mathbf{u}_i^n - \mathbf{u}_e^n) \\ \mathrm{M}[-\mathbf{u}_i^n + \mathbf{u}_e^n] \end{pmatrix} + \begin{pmatrix} \mathrm{M}[-\mathrm{i}_{\mathrm{ion}}^h(\mathbf{v}^n, \mathbf{w}^{n+1}, \mathbf{c}^{n+1}) + \mathrm{I}_{\mathrm{app}}^{i,h}] \\ \mathrm{M}[\ \mathrm{i}_{\mathrm{ion}}^h(\mathbf{v}^n, \mathbf{w}^{n+1}, \mathbf{c}^{n+1}) + \mathrm{I}_{\mathrm{app}}^{e,h}] \end{pmatrix},$$

where $\mathbf{v}^n = \mathbf{u}_i^n - \mathbf{u}_e^n$. As in the continuous model, $\mathbf{v}^n$ is uniquely determined, while $\mathbf{u}_i^n$ and $\mathbf{u}_e^n$ are determined only up to the same additive time-dependent constant chosen by imposing the condition $\mathbf{1}^T \mathbf{M} \mathbf{u}_e^n = 0$.

In the *monodomain case*, we have to solve the following equation for $\mathbf{v}^{n+1}$:

$$\left(\frac{c_m}{\Delta t}\mathrm{M} + \mathrm{A}_m\right)\mathbf{v}^{n+1} = \frac{c_m}{\Delta t}\mathrm{M}\mathbf{v}^n - \mathrm{M}\,\mathrm{i}_{\text{ion}}^h(\mathbf{v}^n, \mathbf{w}^{n+1}, \mathbf{c}^{n+1}) + \mathrm{M}\mathrm{I}_{\text{app}}^{m,h}.$$

We employ an adaptive time-stepping strategy based on controlling the transmembrane potential variation $\Delta v = \max(\mathbf{v}^{n+1} - \mathbf{v}^n)$ at each time-step (see [84]):

- if $\Delta v < \Delta v_{\min} = 0.05$ then we select $\Delta t = (\Delta v_{\max}/\Delta v)\Delta t$ (if smaller than $\Delta t_{\max} = 6$ msec);
- if $\Delta v > \Delta v_{\max} = 0.5$ then we select $\Delta t = (\Delta v_{\min}/\Delta v)\Delta t$ (if greater than $\Delta t_{\min} = 0.005$ msec).

The dynamics of the gating variables are described by equations of the form (1). In order to guarantee control of the variation of $\mathbf{w}_j$ as well, given $\mathbf{v}^n$, each equation is integrated exactly due to the linearity in $\mathbf{w}_j$.

Parallel linear solver. Parallelization and portability are realized using the PETSc parallel library [8,9], while visualization of the results is based on MATLAB. Among other works using parallel tools in cardiac simulations, see [48, 87, 91, 112, 114, 128, 140, 155, 158]. We partition the computational domain into subdomains of approximately the same volume and assign them to different processors. All vectors and matrices are partitioned accordingly and never assembled globally. The computational core of our code is the linear system that must be solved at each time step. The symmetric coefficient matrices of this system in the discrete bidomain and monodomain models are

$$\frac{c_m}{\Delta t}\begin{bmatrix} \mathrm{M} & -\mathrm{M} \\ -\mathrm{M} & \mathrm{M} \end{bmatrix} + \begin{bmatrix} \mathrm{A}_i & 0 \\ 0 & \mathrm{A}_e \end{bmatrix} \qquad \text{and} \qquad \frac{c_m}{\Delta t}\mathrm{M} + \mathrm{A}_m,$$

respectively, in which the first is positive semidefinite and the second positive definite. The associated systems are solved iteratively by the Preconditioned Conjugate Gradient (PCG) method, using as initial guess the solution at the previous time step. The parallel PCG provided by the PETSc library is then preconditioned by a block Jacobi preconditioner with blocks built from the local stiffness and mass matrices on each subdomain. On each block, we use an incomplete LU factorization ILU(0) solver; see, e.g., [115]. The numerical experiments reported in our previous works [31,33,102] on the parallel solver validation, show that the resulting one-level iterative solver performs well in the monodomain case, but not in the bidomain case. Therefore, more research is needed in order to build better bidomain preconditioners, particularly with two or more levels; see [136].

The parallel machines available to us were: an IBM SP4 with 512 processors (Power 4–1300 MHz) of the Cineca Consortium (www.cineca.it) and a Linux Cluster with 72 processors (Xeon - 2.4 GHz) of the University of Milan (cluster.mat.unimi.it). Due to different costs, user loads and availability, we ran different simulations on different machines.

6 Numerical simulations

We report in this final section the results of numerical simulations in three dimensions with the anisotropic bidomain and monodomain models, coupled with the LR1 ionic model. We assume homogeneous cellular membrane properties, i.e., all individual cells have the same intrinsic transmembrane action potential. The LR1 parameters are as in the original model except G_{si} which was reduced by a factor of $2/3$ in order to obtain an Action Potential Duration (APD) of about 265 msec. The time evolution at a given point of the transmembrane potential v and of the LR1 variables is shown in Fig. 1. The orthotropic bidomain and monodomain models have parameters

$$\chi = 10^3\,\mathrm{cm}^{-1},\ C_m = 10^{-3}\,\mathrm{mF/cm}^2,$$

$$\sigma_l = 1.2\cdot 10^{-3},\ \sigma_t = 3.46\cdot 10^{-4},\ \sigma_n = 4.35\cdot 10^{-5},\ \text{ all in }\ \Omega^{-1}\,\mathrm{cm}^{-1}.$$

The macroscopic features of the excitation and subsequent repolarization process are described by extracting from the spatio-temporal transmembrane potential the sequence of the propagating excitation and repolarization wave fronts. In particular, we define the excitation time $t_e(\mathbf{x})$ at a given point $\mathbf{x}$ as the unique time when $v(\mathbf{x}, t_e(\mathbf{x})) = -60\,\mathrm{mV}$ during the upstroke of the excitation phase. Analogously, during the downstroke of the repolarization phase, there is a unique time instant $t_r(\mathbf{x})$ when $v(\mathbf{x}, t_r(\mathbf{x})) = -60\,\mathrm{mV}$. We denote the action potential duration by APD $= t_r - t_e$.

Endocardial stimulation with the bidomain - LR1 model. We simulate the excitation and repolarization processes elicited by a stimulus applied at the center of the endocardial face of an orthotropic slab of dimensions $2 \times 2 \times 0.5\,\mathrm{cm}^3$, using the bidomain - LR1 model. We recall that, as viewed from the epicardium, proceeding from the endocardial to the epicardial surface the fibers rotate CW from 75° to −45° for a total amount of 120°. The simulation uses $200 \times 200 \times 100$ finite elements.

It is well-known that, in the presence of rotational anisotropy, intramural excitation starting from an endocardial stimulation site, first proceeds toward the epicardium but subsequently, due to CW fiber rotation, comes back pointing toward the endocardial plane (see, e.g., [28, 141, 144]). Due to faster propagation in the upper layers, where the fiber rotates CW, the spread of excitation and recovery on the endocardium undergoes an acceleration, in particular, in areas where the excitation moves mainly across fibers, causing the appearance of dimple-like inflections in the isochrone profiles; see Fig. 4. Proceeding from endo- to epicardium, on the intramural planes parallel to the endocardium the spacing between excitation (recovery) isochrones increases, the wave front shapes become rounder and we observe a transmural twisting of the isochrones, i.e., the major axis of the oblong isochrones progressively rotates CW. On the epicardial plane both the excitation and recovery front-boundary collision first occur at the center of the face and the large spacing between successive isochrones indicates a fast excitation and repolarization progression with a maximum apparent speed at the breakthrough point, where a sudden change of the wave front curvature occurs. With regard to repolarization, the recovery

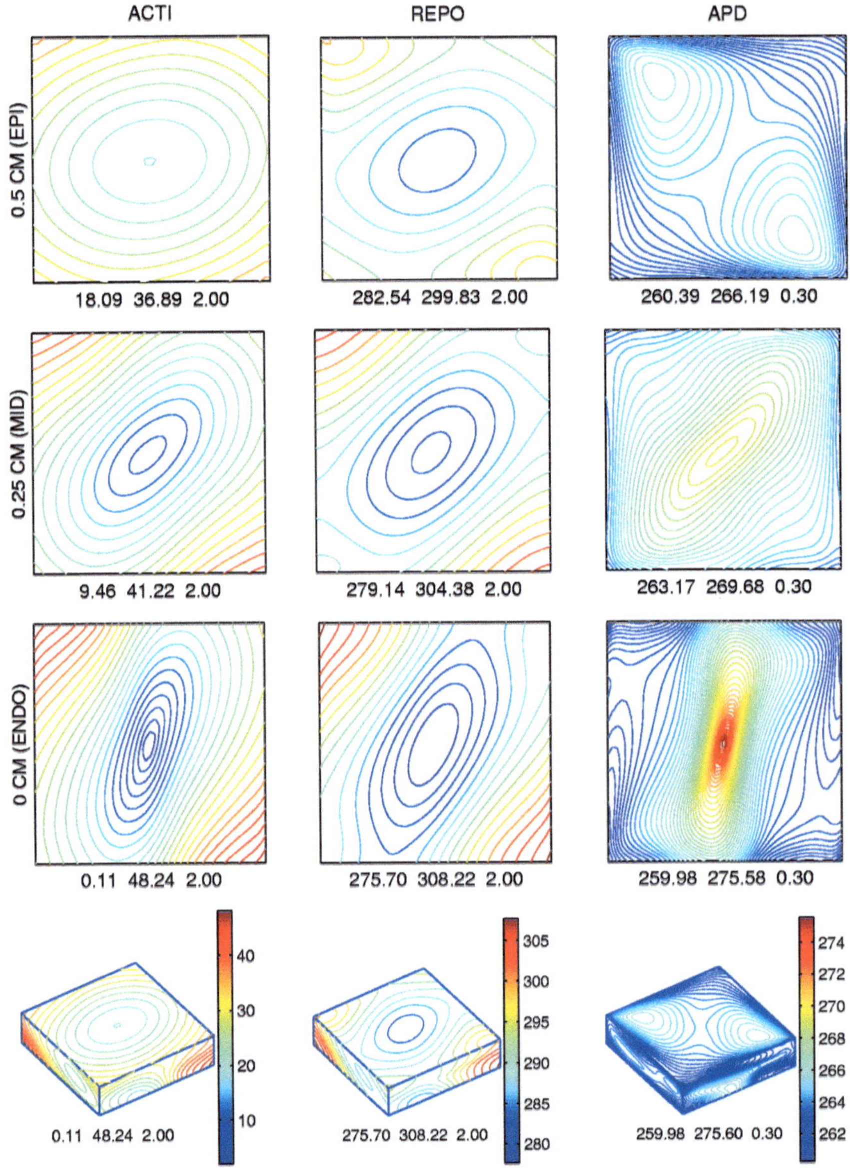

Fig. 4. Endocardial stimulation with the bidomain - LR1 model on a ventricular slab $2 \times 2 \times 0.5\,\mathrm{cm}^3$, mesh $201 \times 201 \times 51$. Top panels report the isochrone lines of activation (*first column*, ACTI), repolarization (*second column*, REPO), APD (*third column*) on the epicardium (*first row*), midwall (*second row*), endocardium (*third row*), 3D slab with intramural sections (*fourth row*). The viewpoint in each image is from the epicardial side and below each panel are the minimum, maximum and step in msec of the displayed map

isochrones, on the endocardial, midwall and epicardial planes, exhibit a somewhat smoother shape and faster propagation compared with the excitation sequence, since the spacing between the recovery isochrones is greater than the corresponding excitation spacing. In particular, the endocardial repolarization isochrones propagate across fibers faster than the excitation wave, yielding a progressive APD shortening in the cross-fiber direction of propagation as shown in Fig. 4. The APD distributions on the endocardial and midwall planes exhibit a central maximum surrounded by level lines elongated along the local fiber direction, indicating that the APD decreases more rapidly when moving from the center of the face in the cross-fiber direction than along fibers. The APD distribution on the epicardial plane displays a saddle point at the epicardial breakthrough and the APD increases reaching a maximum when moving away from the saddle point in a direction parallel to the epicardial fibers of $-45°$. These results show that APD patterns present anisotropic features and a definite spatial dispersion in spite of the assigned homogeneity of the individual cellular membrane properties. The extracellular waveforms on endocardial selected points, displayed in Fig. 5, show the various morphologies of the QRS complex (related to the excitation phase) and of the T-wave (related to the recovery phase) when moving away from the stimulation site.

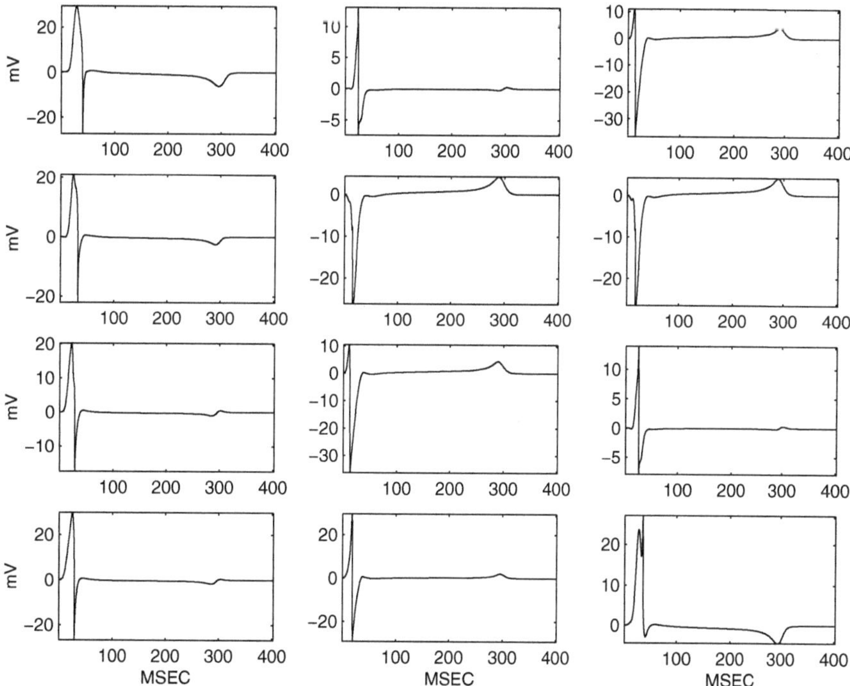

Fig. 5. Same simulation as in Fig. 4. Time course of the extracellular potentials u_e at selected locations on the endocardial plane at the 3×4 mesh points, starting from the lower-left corner and proceeding with horizontal and vertical space steps equal to 0.66 and 0.25 cm, respectively

Endocardial stimulation with the monodomain - LR1 model. Next, we simulate a normal heartbeat generated by an idealized Purkinje stimulation. The domain is an idealized half ventricle described by the ellipsoidal coordinates (51) with parameters:

$$\varphi_{\min} = -\pi/2,\ \varphi_{\max} = \pi/2,\ \theta_{\min} = -3\pi/8,\ \theta_{\max} = \pi/8,$$
$$a_1 = b_1 = 1.5 \text{ cm},\ a_2 = b_2 = 2.7 \text{ cm},\ c_1 = 4.4 \text{ cm},\ c_2 = 5 \text{ cm}.$$

We recall that, proceeding from the endocardial to the epicardial surface, the fibers rotate CCW for a total amount of 120°. The simulation uses $500 \times 500 \times 100$ finite elements and the excitation process is started by applying an idealized Purkinje network on the endocardium (modeling the Purkinje Ventricular Junctions (PVJ)), consisting in slightly delayed stimuli of 200 μA/cm^3, lasting 1 msec, at 13 small areas (of $4 \times 4 \times 2$ elements each) around the center of the endocardium; see Fig. 6. The adaptive time stepping strategy selects time step sizes of 0.05 msec in the activation phase, about 0.74 msec in the plateau phase and about 0.19 msec in the repolarization phase, for a total of 3481 time steps. This simulation of the whole cardiac cycle on 52 processors of our Linux cluster takes about 11.8 hours.

In Fig. 6, we report the isochrone lines of the activation time t_e (first column, ACTI), repolarization time t_r (second column, REPO), APD = $t_r - t_e$ (third column) on the epicardium (first row), midwall (second row), endocardium (third row) and cardiac volume (fourth row). The viewpoint in each image is located inside the ventricle and below each panel are the minimum, maximum and step in msec of the displayed map. On the endocardial surface, collisions and subsequent merging of the multiple excitation and recovery fronts originating from the PVJ are observed; hence the intramural wavefront surfaces display an undulating shape. Similar patterns, but more smoothed, with isochronal profiles with bulges are displayed on the midwall intramural section. On the epicardial surface, excitation and recovery isochrones show multiple bulges and minima, the latter related to breakthrough sites. In general, repolarization isochrone profiles exhibit a somewhat smoother shape compared to the excitation sequence. In spite of the homogeneous cellular membrane properties which we assumed (i.e., all individual cells elicited the same intrinsic transmembrane action potential), the rotational anisotropy and wave front propagation produce a spatial variation of the APD throughout the slab, with a total dispersion on the order of 15 msec.

The role of the myocardial architecture on the excitation process has been intensively investigated both experimentally (see, e.g., [141] and its references) and by simulations [26, 28, 30, 60, 93]. The influence of the rotational anisotropy on the recovery phase is less studied, especially in *in vivo* experiments [142] and simulation results could help in the interpretation of the experimental data for the recovery phase (see, e.g., [18, 33, 34, 103]).

Our current work is extending these simulations to include inhomogeneities of the tissue and heterogeneity of the cellular membrane properties (see, e.g., [33–35, 64, 103]), in order to investigate their influence in both normal and diseased tissues.

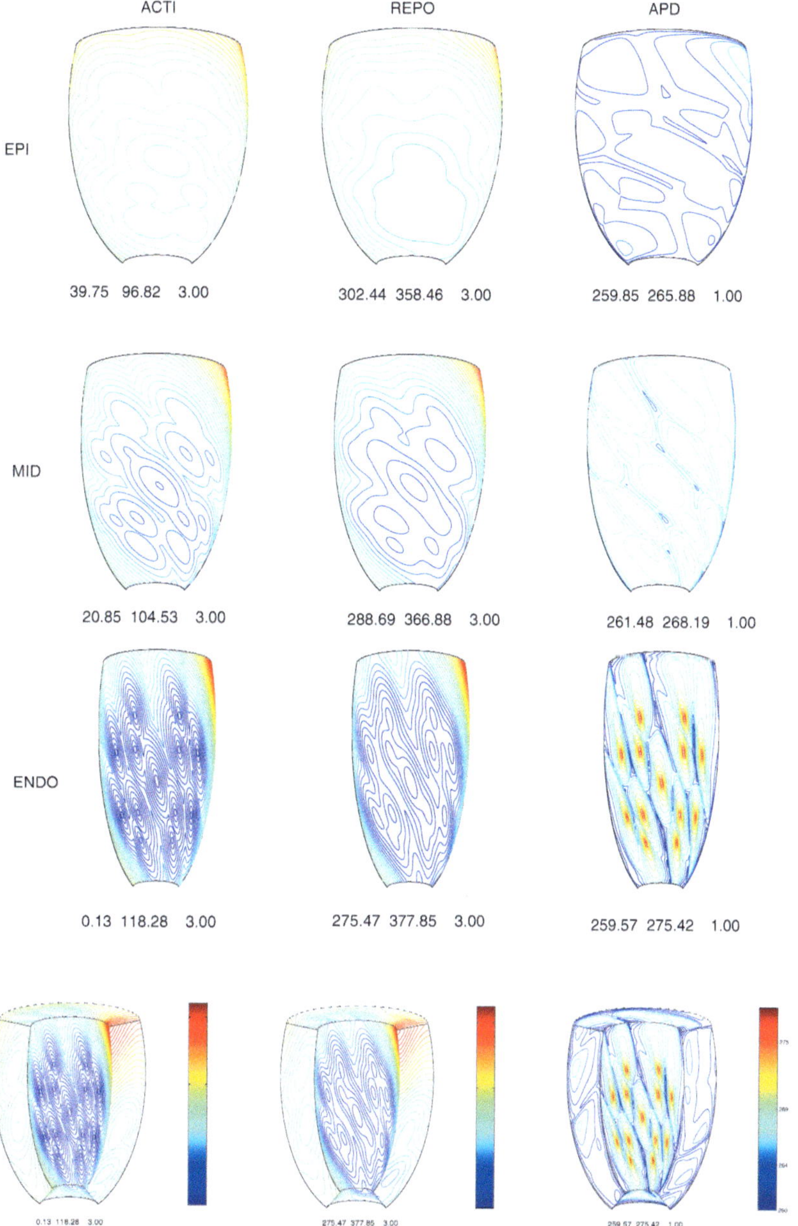

Fig. 6. Normal heartbeat with the monodomain - LR1 model on an idealized half ventricle, mesh $501 \times 501 \times 101$. Isochrone lines of activation (*first column*, ACTI), repolarization (*second column*, REPO), APD (*third column*) on the epicardium (*first row*), midwall (*second row*), endocardium (*third row*), intramural sections with endocardium (*fourth row*). The viewpoint in each image is located inside the ventricle and below each panel are the minimum, maximum and step in msec of the displayed map

Numerical simulations of re-entry phenomena. We now consider re-entry phenomena, see, e.g., [77], [164, Part V and VII], simulated in three dimensions, where the possible configurations of re-entrant fronts are much more complex and less understood than in two dimensions. Due to the high computational complexity of large scale simulations, virtually all works in the vast existing literature on cardiac re-entry simulations employ model simplifications in order to obtain a tractable discrete problem [1–4, 101].

Bidomain - LR1 stable scroll waves. We start with the simulation of a stable scroll wave using the anisotropic bidomain - LR1 model; for more detailed results see [32]. The conductivity tensors are assumed axisymmetric with values

$$\sigma_l^i = 3\cdot 10^{-3},\ \sigma_t^i = 3.1525\cdot 10^{-4},\ \sigma_l^e = 2\cdot 10^{-3},\ \sigma_t^e = 1.3514\cdot 10^{-3}\ (\Omega^{-1}\,\mathrm{cm}^{-1}).$$

The original LR1 model is modified in order to shorten the APD according to [48], by setting $G_{\mathrm{Na}} = 16$, $G_{\mathrm{K}} = 0.432$, $G_{\mathrm{si}} = 0$. We assume homogeneous intrinsic cellular membrane properties throughout the slab. Due to the high computational costs of the bidomain model, we limit our simulation to a Cartesian slab of dimensions $2\times 2\times 0.5\,\mathrm{cm}^3$, discretized with $200\times 200\times 50$ finite elements. The intramural fibers rotate linearly with depth for a total amount of 90°, i.e., 18°/mm, starting from −90° (0°) on the lower-endocardial (upper-epicardial) surface of the slab with respect to a side on the slab.

Re-entry is initiated with a cross-gradient procedure, applying first an impulse of $200\,\mu\mathrm{A/cm}^3$ for 1 msec along one of the main intramural sides of the slab, generating a plane wave initially, and then eliciting an orthogonal front by applying a second impulse at an appropriate time (in this case $t = 68.4$ msec) in the bottom-left quarter volume of the domain. This cross-gradient stimulation elicits a vortex-like pattern, usually called a scroll wave, rotating around a tube-like filament which is a 3D analog of the core of a spiral wave.

The contour plots of the resulting scroll waves are shown in Fig. 7, from $t = 100$ to $t = 200$ msec, every 20 msec. The colormap of the transmembrane potential distribution ranges from blue (resting values around −84 mV) to red (excitation front around 10 mV). The effect of the anisotropy is shown by the elongated spirals on the epicardium, by the twisted scroll waves in the intramural sections and by the meandering of the epicardial spiral tip.

Monodomain - LR1 scroll waves in ellipsoidal geometry. We now consider monodomain - LR1 simulations on an idealized half ventricle modeled by a truncated half ellipsoid (with parameters described at the beginning of this section), discretized with $500\times 500\times 80$ isoparametric finite elements. The conductivity coefficients are

$$\sigma_l = 1.2\cdot 10^{-3},\ \sigma_t = 2.5\cdot 10^{-4}\ (\Omega^{-1}\mathrm{cm}^{-1}),$$

and the original LR1 model is modified as described before in order to shorten the APD ($G_{\mathrm{Na}} = 16$, $G_{\mathrm{K}} = 0.432$, $G_{\mathrm{si}} = 0$); in addition, we set $G_{\mathrm{K1}} = 0.6047$ and

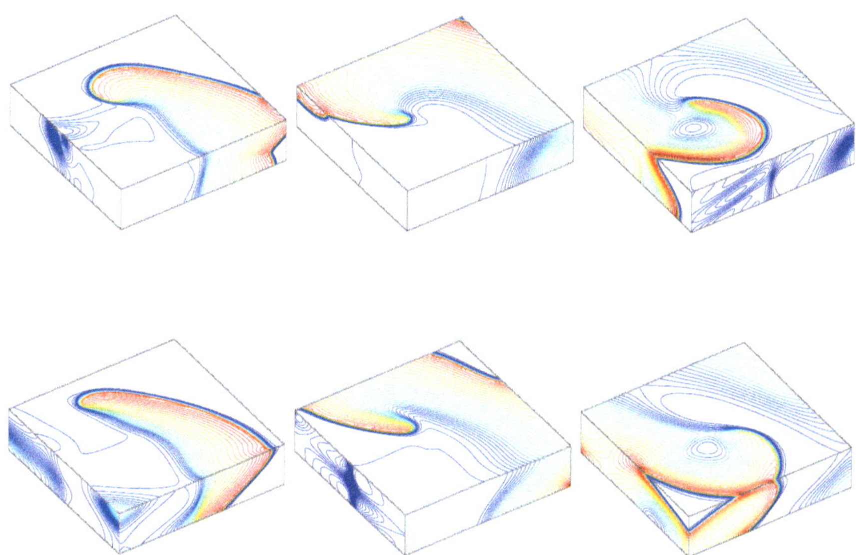

Fig. 7. Stable scroll wave with bidomain - LR1 model on a slab $2 \times 2 \times 0.5$ cm^3, mesh $201 \times 201 \times 51$, $G_{si} = 0$. Panels show the transmembrane potential v at times t = 100, 120, 140, 160, 180, 200 msec. The colormap ranges from blue (resting value) to red (excitation front)

scale the time constants τ_d and τ_f by a factor of 10. In order to see more clearly the effect of the curved geometry, we considered parallel fibers in this case, i.e., $\alpha(r) = 0$ in Eq. (52). Re-entry is initiated with a broken wave procedure, where at time $t = 0$ we set $v = 10\,\text{mV}$ on a vertical intramural section running from epi- to endocardium and from the bottom to about 3/4 of the ventricle height (see the first panel of Fig. 8); moreover, on another vertical section to the right of the previous, we set at $t = 0$ the gating variables w to their steady state corresponding to the fixed value $v = -10\,\text{mV}$, which is in the refractory phase of the action potential. In this way, an excitation front starts from the vertical section where $v = 10\,\text{mV}$, but, on the right-hand side, the front is blocked by the other section where the gating variables are inhibiting propagation because they are in their refractory phase. Therefore, the front curls around the upper end of the sections and initiates a scroll wave. Unexpectedly, the front also curls around the bottom end of the sections, possibly because of the high curvature of the domain geometry there, resulting in a second counter-rotating scroll wave. After an initial adjustment, the two scroll waves seem to reach a stable counter-rotating configuration shown in Fig. 8 at $t = 500$ msec, together with the time course of the transmembrane potential at a point. This configuration is also known as *figure-8* or *double loop* re-entry; see, e.g., [160].

Monodomain - LR1 scroll wave breakup. Finally, we consider scroll wave breakup in three dimensions. The computational domain is a slab of size $6 \times 6 \times 0.6\,\mathrm{cm}^3$, with a linear intramural fiber rotation of 120°. The simulation employs $400 \times 400 \times 40$ finite elements and is run up to 1300 msec. A scroll wave is started with the same cross-gradient stimulation as before. In Fig. 9 is shown a head-to-tail collision, leading to a subsequent breakup (first row, $t = 705, 715, 725$ msec) generating additional scroll waves and spiral tips with many broken fronts (last row, $t = 1250, 1275, 1300$ msec) displaying a spatio-temporal chaotic configuration of multiple wave re-entry.

7 Conclusions

We presented the main mathematical models used in computational electrocardiology to describe the complex multiscale structure of the bioelectrical activity of the heart, from the microscopic activity of ion channels of the cellular membrane

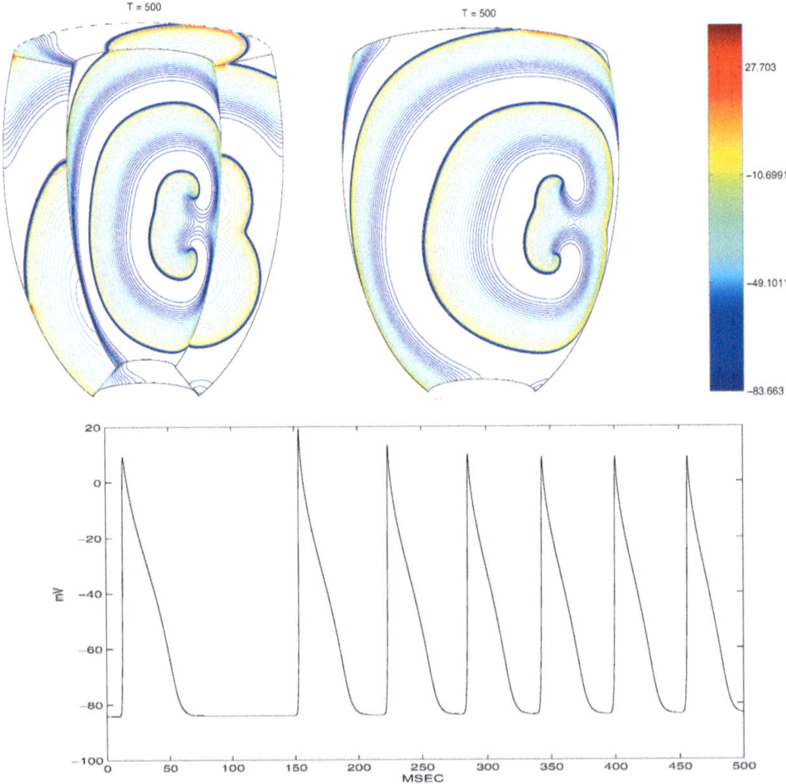

Fig. 8. Counter-rotating double scroll wave with the monodomain - LR1 model on idealized half ventricle, mesh $501 \times 501 \times 81$. Distribution of v on the endocardium and intramural sections (*top left*), epicardium (*top right*) at time t = 500 msec (viewpoint is inside the ventricle); time course of the transmembrane potential v at a given point (*bottom*)

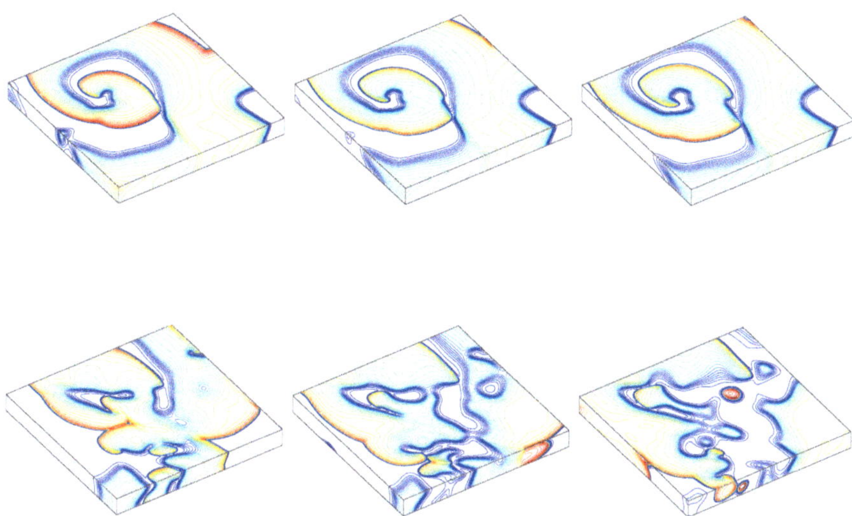

Fig. 9. Scroll wave breakup with monodomain – LR1 model, $G_{si} = 0.056$, domain: $6 \times 6 \times 0.6\ \mathrm{cm}^3$. *First row*: t = 705, 715, 725; *second row*: t = 1250, 1275, 1300 msec. The colormap ranges from blue (resting values around -84 mV) to red (excitation front around 10 mV)

to the macroscopic properties of the anisotropic propagation of excitation and recovery fronts in the whole heart. We described how reaction-diffusion systems can be rigorously derived from microscopic models of cellular aggregates by homogenization methods and asymptotic expansions. Models of cardiac tissue include the anisotropic bidomain and monodomain models, as well as the eikonal and various relaxed approximations, while the ionic cellular models include Luo-Rudy type models as well as simpler FitzHugh-Nagumo variants. We also presented advanced numerical methods for discretizing and numerically solving these complex models on three-dimensional domains, using adaptive and parallel techniques. The resulting solvers are able to reproduce accurately a complete normal heartbeat phenomena in large ventricular volumes, simulating, e.g., various potential waveforms, activation and recovery fronts, and action potential dispersion. The solvers can also simulate re-entry phenomena such as spiral and scroll waves, their breakup and the transition to electrical turbolence. Current work is investigating the role of inhomogeneities of the tissue and heterogeneity of the cellular membrane properties, due, e.g., to ischemia, and the coupling of electrocardiological models with mechanical and fluid dynamic models, with the future goal of their integration with cardiovascular and circulatory models.

Appendix: list of symbols

$u_{i,e}$ = intra- and extracellular potentials
v_i = $u_i - u_e$ transmembrane potential
C_m = membrane capacity per unit surface area
I_C = $C_m \frac{dv}{dt}$ capacitive current
I_{app} = applied current
I_m = membrane current per unit surface area
J_m = membrane current per unit tissue volume
$I^s_{i,e}$ = intra- and extracellular stimulation currents
I_{ion} = ionic current
g_k = membrane conductance for the kth ion
v_k = Nernst potential for the kth ion
w = $(w_1, \ldots, w_M)$ gating variables
c = $(c_1, \ldots, c_R)$ ion concentrations
$\Omega_{i,e}$ = intra- and extracellular space
Γ_m = cellular membrane
Ω = $\Omega_i \cup \Omega_e \cup \Gamma_m$ cardiac domain
$\Sigma_{i,e}$ = intra- and extracellular conductivity tensors
$\overline{\mu}_{i,e}$ = average eigenvalues of $\Sigma_{i,e}$ on a cell element
$\overline{\mu}$ = $\overline{\mu}_i + \overline{\mu}_e$
$\sigma_{i,e}$ = $\Sigma_{i,e}/\overline{\mu}$ dimensionless intra- and extracellular conductivity tensors
$\mathbf{J}_{i,e}$ = $-\Sigma_{i,e}\nabla u_{i,e}$ intra- and extracellular current densities
R_m = passive membrane resistance
τ_m = $R_m C_m$ membrane time constant
Λ = $\sqrt{l_c \overline{\mu} R_m}$ length scale unit
$\widehat{\mathbf{x}}$ = $\mathbf{x}/\Lambda$, $\widehat{t} = t/\tau_m$ scaled space and time variables
ε = l_c/Λ ratio between micro and macro length constants
$\boldsymbol{\xi}$ = $\mathbf{x}/l_c = \widehat{\mathbf{x}}/\varepsilon$ microscopic space variable
$E_{i,e}$ = intra- and extracellular reference periodic lattices
$\varepsilon E_{i,e}$ = ε-dilation of $E_{i,e}$
Y = elementary periodicity region
$Y_{i,e}$ = $Y \cap E_{i,e}$
S_m = $\Gamma_m \cap Y_i$
$\Omega^\varepsilon_{i,e}$ = $\Omega \cap \varepsilon E_{i,e}$
Γ^ϵ_m = $\Omega \cap \epsilon \Gamma_m$
$\sigma^\varepsilon_{i,e}(x), \sigma_{i,e}(x, \frac{c}{\varepsilon})$ rescaled conductivity matrices
$\nu_{i,e}$ = exterior unit normals to $\partial\Omega^\varepsilon_{i,e}$
$\boldsymbol{P}^\varepsilon$ = dimensionless cellular model
β = $|S_m|/|Y|$ membrane surface area per reference cell volume
$\beta_{i,e}$ = $|Y_{i,e}|/|Y|$ intra- and extra cellular volume per reference cell volume
$\boldsymbol{P}$ = dimensionless averaged model
$\mathbb{V}^\varepsilon$ = $\{H^1(\Omega^\varepsilon_i) \times H^1(\Omega^\varepsilon_e)\} / \{(\gamma, \gamma) : \gamma \in \mathbf{R}\} \times L^2(\Gamma^\varepsilon_m)^M \times L^2(\Gamma^\varepsilon_m)^Q$
$\boldsymbol{U}^\varepsilon$ = $(u^\varepsilon_i, u^\varepsilon_e, w^\varepsilon, c^\varepsilon)$, $\boldsymbol{U} = (u_i, u_e, w, c)$, $\hat{\boldsymbol{U}} = (\hat{u}_i, \hat{u}_e, \hat{w}, \hat{c})$

$b^\varepsilon(\boldsymbol{U},\hat{\boldsymbol{U}}),\ a^\varepsilon(\boldsymbol{U},\hat{\boldsymbol{U}}),\ F^\varepsilon(\boldsymbol{U},\hat{\boldsymbol{U}})$ forms defining problem $\boldsymbol{P}^\varepsilon$
$\mathbb{V} \ = \{H^1(\Omega)\times H^1(\Omega)\}/\{(\gamma,\gamma):\gamma\in\mathbf{R}\}\times L^2(\Omega)^M\times L^2(\Omega)^Q$
$b(\boldsymbol{U},\hat{\boldsymbol{U}}),\ a(\boldsymbol{U},\hat{\boldsymbol{U}}),\ F(\boldsymbol{U},\hat{\boldsymbol{U}})$ forms defining problem $\boldsymbol{P}$
$\mathbf{a}_l,\ \mathbf{a}_t,\ \mathbf{a}_n$ orthonormal axes along fiber, tangent and normal to radial laminae
$\sigma_l^{i,e}, \sigma_t^{i,e} \sigma_n^{i,e}$ bidomain conductivity coefficients along $\mathbf{a}_l, \mathbf{a}_t, \mathbf{a}_n$
$M_{i,e}$ = bidomain intra- and extracellular conductivity tensors
χ = membrane surface area per tissue volume
$c_m \ = \chi C_m$
$I_{\text{app}}^{i,e}$ = bidomain intra- and extracellular applied currents
Ω_H = heart volume, $\Gamma_H = \partial\Omega_H$ heart surface
$\Omega_0,\ M_0,\ u_0$ extracardiac volume, conductivity tensor, potential
$\Gamma_0 \ = \partial\Omega_0 - \Gamma_H$ body surface
$M_m \ = M_e M^{-1} M_i$ monodomain conductivity tensor, $M = M_i + M_e$
$I_{\text{app}}^m = (I_{\text{app}}^i \sigma_l^e - I_{\text{app}}^e \sigma_l^i)/(\sigma_l^e + \sigma_l^i)$ monodomain applied current
$S_\varepsilon(t),\ S(t)$ excitation wavefronts
$\theta_\epsilon(\mathbf{n}_\Phi),\ \theta(\boldsymbol{v})$ wavefront velocities along suitable directions $\mathbf{n}_\Phi$ and $\boldsymbol{v}$
$q_{i,e}(\mathbf{x},\boldsymbol{\xi}) = \boldsymbol{\xi}^T M_{i,e}(\mathbf{x})\boldsymbol{\xi}$ conductivity coefficients of the media (i) and (e) along the direction $\boldsymbol{\xi}$
$q(\mathbf{x},\boldsymbol{\xi}) = (q_i(\mathbf{x},\boldsymbol{\xi})^{-1} + q_i(\mathbf{x},\boldsymbol{\xi})^{-1})^{-1}$ conductivity coefficient of the bulk medium along the direction $\boldsymbol{\xi}$
$\Phi(\mathbf{x},\boldsymbol{\xi}) \ = \ \sqrt{q(\mathbf{x},\boldsymbol{\xi})}$ indicatrix function of the media

Acknowledgments

The authors would like to thank Bruno Taccardi for introducing them to the field of mathematical physiology and for many stimulating discussions.

References

[1] Special issue on "Fibrillation in normal ventricular myocardium". Chaos **8**(1) (1998)
[2] Special issue on "Mapping and control of complex cardiac arrhythmias". Chaos **12**(3) (2002)
[3] Special issue on "From excitable media to virtual cardiac tissue". Chaos Solititons Fractals **13**(8) (2002)
[4] Special issue on "Virtual tissue engineering of the heart". Internat J. Bifurcat. Chaos Appl. Sci. Eng. **13**(12) (2003)
[5] Ambrosio, L., Colli Franzone, P., Savaré, G.: On the asymptotic behaviour of anisotropic energies arising in the cardiac bidomain model. Interfaces Free Bound. **2**, 213–266 (2000)
[6] Antzelevitch, C., Fish, J.: Electrical heterogeneity within the ventricular wall. Basic Res. Cardiol. **96**, 517–527 (2001)
[7] Bakhvalov, N., Panasenko, G.: Homogenisation: averaging processes in periodic media. Dordrecht: Kluwer 1989

[8] Balay, S., Buschelman, K., Eijkhout, V., Gropp, W.D., Kaushik, D., Knepley, M., Curfman McInnes, L., Smith, B.F., Zhang, H.: PETSc Users Manual. Tech. Rep. ANL-95/11 – Revision 2.1.5. Argonne, IL: Argonne National Laboratory 2004

[9] Balay, S., Buschelman, K., Gropp, W.D., Kaushik, D., Knepley, M., Curfman McInnes, L., Smith, B.F., Zhang, H.: PETSc web page. 2001 http://www.mcs.anl.gov/petsc

[10] Bassetti, F.: Variable time-step discretization of degenerate evolution equations in Banach spaces. Numer. Funct. Anal. Optim. **24**, 391–426 (2003)

[11] Beeler, G.W., Reuter, H.T.: Reconstruction of the action potential of ventricular myocardial fibers. J. Physiol. **268**, 177–210 (1977)

[12] Bellettini, G., Colli Franzone, P., Paolini, M.: Convergence of front propagation for anisotropic bistable reaction-diffusion equations. Asymptot. Anal. **15**, 325–358 (1997)

[13] Bellettini, G., Paolini, M.: Anisotropic motion by mean curvature in the context of Finsler geometry. Hokkaido Math. J. **25**, 537–566 (1996)

[14] Bensoussan, A., Lions, J.-L., Papanicolaou, G.: Asymptotic analysis for periodic structures. Amsterdam: North-Holland 1978

[15] Buist, M., Sands, G., Hunter, P., Pullan, A.: A deformable finite element derived finite difference method for cardiac activation problems. Ann. Biomed. Eng. **31**, 577–588 (2003)

[16] Burnes, J.E., Taccardi, B., Ershler, P.R., Rudy, Y.: Noninvasive electrocardiogram imaging of substrate and intramural ventricular tachycardia in infarcted hearts. J. Am. Coll. Cardiol. **38**, 2071–2078 (2001)

[17] Burnes, J.E., Taccardi, B., Rudy, Y.: A noninvasive imaging modality for cardiac arrhythmias. Circulation **102**, 2152–2158 (2000)

[18] Cates, A.W., Pollard, A.E.: A model study of intramural dispersion of action potential duration in the canine pulmonary conus. Ann. Biomed. Eng. **26**, 567–576 (1998)

[19] Cherry, E.M., Greenside, H.S., Henriquez, C.S.: A space-time adaptive method for simulating complex cardiac dynamics. Phys. Rev. Lett. **84**, 1343–1346 (2000)

[20] Clements, J.C., Nenonen, J., Li, P.K., Horacek, B.M.: Activation dynamics in anisotropic cardiac tissue via decoupling. Ann. Biomed. Eng. **32**, 984–990 (2004)

[21] Colli Franzone, P., Deuflhard, P., Erdmann, B., Lang, J., Pavarino, L.F.: Adaptivity in space and time for reaction-diffusion systems in electrocardiology. Report 05-30. Berlin: Konrad-Zuse-Zentrum für Informationstechnik 2005

[22] Colli Franzone, P., Guerri, L., Magenes, E.: Oblique double layer potential for the direct and inverse problems of electrocardiology. Math. Biosci. **68**, 23–55 (1984)

[23] Colli Franzone, P., Guerri, L., Rovida, S.: Wavefront propagation in an activation model of the anisotropic cardiac tissue: asymptotic analysis and numerical simulations. J. Math. Biol. **28**, 121–176 (1990)

[24] Colli Franzone, P., Guerri, L., Tentoni, S.: Mathematical modeling of the excitation process in myocardial tissue: Influence of fiber rotation on wavefront propagation and potential field. Math. Biosci. **101**, 155–235 (1990)

[25] Colli Franzone, P., Guerri, L.: Models of the spreading of excitation in myocardial tissue. Crit. Rev. Biomed. Eng. **20**, 211–253 (1992) and in: Pilkington, T.C. et al. (eds.): High-performance computing in biomedical research. Boca Raton, FL: CRC Press 1993, pp. 359–401

[26] Colli Franzone, P., Guerri, L.: Spreading of excitation in 3-D models of the anisotropic cardiac tissue. I: Validation of the eikonal approach. Math. Biosci. **113**, 145–209 (1993)

[27] Colli Franzone, P., Guerri, L., Taccardi, B.: Spread of excitation in a myocardial volume: simulation studies in a model of anisotropic ventricular muscle activated by point stimulation. J. Cardiovasc. Electrophysiol. **4**, 144–160 (1993)

[28] Colli Franzone, P., Guerri, L., Pennacchio, M., Taccardi, B.: Spread of excitation in 3-D models of the anisotropic cardiac tissue. II: Effects of fiber architecture and ventricular geometry. III: Effects of ventricular geometry and fiber structure on the potential distribution. Math. Biosci. **147**, 131–171; **151**, 51–98 (1998)

[29] Colli Franzone, P., Guerri, L., Pennacchio, M., Taccardi, B.: Anisotropic mechanisms for multiphasic unipolar electrograms: simulation studies and experimental recordings. Ann. Biomed. Eng. **28**, 1326–1342 (2000)

[30] Colli Franzone, P., Guerri, L., Taccardi, B.: Modeling ventricular excitation: axial and orthotropic anisotropy effects on wavefronts and potentials. Math. Biosci. **188**, 191–205 (2004)

[31] Colli Franzone, P., Pavarino, L.F.: A parallel solver for reaction-diffusion systems in computational electrocardiology. Math. Models Methods Appl. Sci. **14**, 883–911 (2004)

[32] Colli Franzone, P., Pavarino, L.F.: Numerical simulation of cardiac reaction-diffusion models: normal and reentry dynamics. Istit. Lombardo Accad. Sci. Lett. Rend. A., to appear

[33] Colli Franzone, P., Pavarino, L.F., Taccardi, B.: A parallel solver for anisotropic cardiac models. Computers in Cardiology 2003. IEEE Conf. Proc. **30**, 781–784 (2003)

[34] Colli Franzone, P., Pavarino, L.F., Taccardi, B.: Simulating patterns of excitation, repolarization and action potential duration with cardiac bidomain and monodomain models. Math. Biosci. **197**, 35–66 (2005)

[35] Colli Franzone, P., Pavarino, L.F., Taccardi, B.: Monodomain simulations of excitation and recovery in cardiac blocks with intramural heterogeneity. In: Frangi, A.F. et al. (eds.): Functional imaging and modeling of the heart. (Lecture Notes Comput. Sci. **3504**) Berlin: Springer 2005, pp. 267–277

[36] Colli Franzone, P., Pennacchio, M., Guerri, L.: Accurate computation of electrograms in the left ventricular wall. Math. Models Methods Appl. Sci. **10**, 507–538 (2000)

[37] Colli Franzone, P., Savaré, G.: Degenerate evolution systems modeling the cardiac electric field at micro- and macroscopic level. In: Lorenzi, A., Ruf, B. (eds.): Evolution equations, semigroups and functional analysis. Basel: Birkhäuser 2002, pp. 49–78

[38] Costa, K.D., Holmes, J.W., McCulloch, A.D.: Modelling cardiac mechanical properties in three dimensions. Philos. Trans. Roy. Soc. London Ser. A **359**, 1233–1250 (2001)

[39] Costa, K.D., May-Newman, K., Farr, D., O'Dell, W.G., McCulloch, A.D., Omens, J.H.: Three-dimensional residual strain in midanterior canine left ventricle. Am. J. Physiol. **273**, H1968–H1976 (1997)

[40] Coudière, Y., Pierre, C.: Stability and convergence of a finite volume method for two systems of reaction-diffusion equations in electro-cardiology. Nonlinear Anal. Real World Appl. to appear. DOI: 10.1016/j.nonrwa.2005.02.006

[41] Di Francesco, D., Noble, D.: A model of cardiac electrical activity incorporating ionic pumps and concentration changes. Philos Trans. Roy. Soc. London Ser. **307**, 353–398 (1985)

[42] Fenton, F.H., Cherry, E.M., Hastings, H.M., Evans, S.J.: Multiple mechanisms of spiral wave breakup in a model of cardiac electrical activity. Chaos **12**, 852–892 (2002)

[43] Fenton, F.H., Karma, A.: Vortex dynamics in three-dimensional continuous myocardium with fiber rotation: filament instability and fibrillation. Chaos **8**, 20–47 (1998)

[44] Fife, P.C.: Dynamics of internal layers and diffusive interfaces. (CBMS-NSF Regional Conf. Ser. Appl. Math. **53**) Philadelphia: SIAM 1988

[45] Fischer, G., Tilg, B., Modre, R., Huiskamp, G.J.M., Fetzer, J., Rucker, W., Wach, P.: A bidomain model based BEM-FEM coupling formulation for anisotropic cardiac tissue. Ann. Biomed. Eng. **28**, 1229–1243 (2000)

[46] FitzHugh, R.: Impulses and physiological states in theoretical models of nerve membrane. Biophys. J. **1**, 445–466 (1961)

[47] FitzHugh, R.: Mathematical models of excitation and propagation in nerve. In: Schwan, H.P. (ed.): Biological Engineering. New York: McGraw-Hill 1969, pp. 1–85

[48] Garfinkel, A., Kim, Y-H., Voroshilovsky, O., Qu, Z., Kil, J.R., Lee, M.-H., Karagueuzian, H.S., Weiss, J.N., Chen, P.-S.: Preventing ventricular fibrillation by flattening cardiac restitution. Proc. Nat. Acad. Sci. USA **97**, 6061–6066 (2000)

[49] Geselowitz, D.B., Barr, R.C., Spach, M.S., Miller III, W.T.: The impact of adjacent isotropic fluids on electrocardiograms from anisotropic cardiac muscle. A modeling study Circ. Res. **51**, 602–613 (1982)

[50] Geselowitz, D.B.: On the theory of the electrocardiogram. Proc. IEEE **77**, 857–876 (1989)

[51] Geselowitz, D.B.: Description of cardiac sources in anisotropic cardiac muscle. Application of bidomain model. J. Electrocardiol. **25**(Suppl.), 65–67 (1992)

[52] Giaquinta, M., Hildebrandt, S.: Calculus of Variations. I. The Lagrangian formalism. (Grundlehren der mathematischen Wissenschaften **310**, Berlin: Springer 1996

[53] Gima, K., Rudy, Y.: Ionic current basis of electrocardiographic waveforms: a model study. Circ. Res. **90**, 889–896 (2002)

[54] Gulrajani, R.M.: Models of the electrical activity of the heart and computer simulation of the electrocardiogram. Crit. Rev. Biomed. Eng. **16**, 1–66 (1988)

[55] Gulrajani, R.M., Roberge, F.A., Savard, P.: The inverse problem of electrocardiography. In: MacFarlane, P.W., Lawrie, T.T.V. (eds.): Comprehensive electrocardiology. I. Oxford: Pergamon 1989, pp. 237–288

[56] Harrild, D.M., Henriquez, C.S.: A finite volume model of cardiac propagation. Ann. Biomed. Eng. **25**, 315–334 (1997)

[57] Harrild, D.M., Penland, R., Henriquez, C.: A flexible method for simulating cardiac conduction in three-dimensional complex geometries. J. Electrocardiol. **33**, 241–251 (2000)

[58] Henriquez, C.S., Plonsey, R.: Simulation of propagation along a cylindrical bundle of cardiac tissue. I: Mathematical formulation. IEEE Trans. Biomed. Eng. **37**, 850–860 (1990)

[59] Henriquez, C.S.: Simulating the electrical behavior of cardiac tissue using the bidomain model. Crit. Rev. Biomed. Eng. **21**, 1–77 (1993)

[60] Henriquez, C.S., Muzikant, A.L., Smoak, C.K.: Anisotropy, fiber curvature, and bath loading effects on activation in thin and thick cardiac tissue preparations: simulations in a three-dimensional bidomain model. J. Cardiovasc. Electrophysiol. **7**, 424–444 (1996)

[61] Hodgkin, A., Huxley, A.: A quantitative description of membrane current and its application to conduction and excitation in nerve. J. Physiol. **117**, 500–544 (1952)

[62] Hoff, D.: Stability and convergence of finite difference methods for systems of nonlinear reaction-diffusion equations. SIAM J. Numer. Anal. **15**, 1161–1177 (1978)

[63] Hooke, N.: Efficient simulation of action potential propagation in a bidomain. Ph.D. thesis. Durham, NC: Duke Univ. 1992

[64] Hooks, D.A., Tomlinson, K.A., Marsolen, S.G., LeGrice, I.J., Smaill, B.H., Pullan, A.J., Hunter, P.J.: Cardiac microstructure: implications for electrical propagation and defibrillation in the heart. Circ. Res. **91**, 331–338 (2002)

[65] Hoyt, R.H., Cohen, M.L., Saffitz, J.E.: Distribution and three-dimensional structure of intercellular junctions in canine myocardium. Circ. Res. **64**, 563–574 (1989)

[66] Hund, T.J., Rudy, Y.: Rate dependence and regulation of action potential and calcium transient in a canine cardiac ventricular cell model. Circulation **110**, 3168–3174 (2004)

[67] Hunter, P.J., McCulloch, A.D., ter Keurs, H.: Modelling the mechanical properties of cardiac muscle. Prog. Biophys. Mol. Biol. **69**, 289–331 (1998)

[68] Jack, J.J.B., Noble, D., Tsien, R.W.: Electric current flow in excitable cells. Oxford: Clarendon 1983

[69] Jerome, J.W.: Convergence of successive iterative semidiscretizations for FitzHugh-Nagumo reaction diffusion systems. SIAM J. Numer. Anal. **17**, 192–206 (1980)

[70] Jikov, V.V., Kozlov, S.M., Oleĭnik, O.A.: Homogenization of differential operators and integral functionals. Berlin: Springer 1994

[71] Keener, J.P.: An eikonal-curvature equation for action potential propagation in myocardium. J. Math. Biol. **29**, 629-651 (1991)

[72] Keener, J.P.: Direct activation and defibrillation of cardiac tissue. J. Theor. Biol. **178**, 313–324 (1996)

[73] Keener, J.P., Panfilov, A.V.: Three–dimensional propagation in the heart: the effects of geometry and fiber orientation on propagation in myocardium. In: Zipes, D.P., Jalife, J. (eds.): Cardiac electrophysiology: from cell to bedside. 2nd ed. Philadelphia: Saunders 1995, pp. 335-347

[74] Keener, J.P., Panfilov, A.V.: The effects of geometry and fibre orientation on propagation and extracellular potentials in myocardium. In: Panfilov, A.V., Holden, A.V. (eds.): Computational biology of the heart. Chichester: Wiley 1997, pp. 235–258

[75] Keener, J.P., Bogar, K.: A numerical method for the solution of the bidomain equations in cardiac tissue. Chaos **8**(1), 234–241 (1998)

[76] Keener, J.P., Sneyd, J.: Mathematical physiology. New York: Springer 1998

[77] Kleber, A.G., Rudy, Y.: Basic mechanisms of cardiac impulse propagation and associated arrhythmias. Physiol. Rev. **84**, 431–488 (2004)

[78] Krassowska, W., Neu, J.C.: Effective boundary conditions for syncytial tissue. IEEE Trans. Biomed. Eng. **41**, 143–150 (1994)

[79] Lang, J.: Adaptive multilevel solution of nonlinear parabolic PDE systems. Theory, algorithm, and applications. (Lecture Notes Comput. Sci. Eng. **16**) Berlin: Springer 2000

[80] LeGrice, I.J., Smaill, B.H., Chai, L.Z., Edgar, S.G., Gavin, J.B., Hunter, P.J.: Laminar structure of the heart: ventricular myocyte arrangement and connective tissue architecture in the dog. Am. J. Physiol. Heart Circ. Physiol. **269**, H571–H582 (1995)

[81] LeGrice, I.J., Smaill, B.H., Hunter, P.J.: Laminar structure of the heart: a mathematical model. Am. J. Physiol. Heart Circ. Physiol. **272**, H2466–H2476 (1997)

[82] Leon, L.J., Horacek, B.M.: Computer model of excitation and recovery in the anisotropic myocardium. I: Rectangular and cubic arrays of excitable elements. II: Excitation in the simplified left ventricle. III: Arrhythmogenic conditions in the simplified left ventricle. J. Electrocardiol. **24**, 1–15, 17–31, 33–41 (1991

[83] Lines, G.T., Grøttum, P., Tweito, A.: Modeling the electric activity of the heart: a bidomain model of the ventricles embedded in a torso. Comput. Vis. Sci. **5**, 195–213 (2003)

[84] Luo, C., Rudy, Y.: A model of the ventricular cardiac action potential: depolarization, repolarization, and their interaction. Circ. Res. **68**, 1501–1526 (1991)

[85] Luo, C., Rudy, Y.: A dynamic model of the cardiac ventricular action potential. I. Simulations of ionic currents and concentration changes. II. After depolarizations, triggered activity, and potentiation. Circ. Res. **74**, 1071–1096, 1097–1113 (1994)

[86] Malmivuo, J., Plonsey, R.: Bioelectromagnetism. New York: Oxford University Press 1995

[87] Mardal, K.-A., Sundnes, J., Langtangen, H.P., Tveito, A.: Systems of PDEs and block preconditioning. In: Langtangen, H.P., Tveito, A. (eds.): Advanced topics in computational partial differential equations. (Lecture Notes Comput. Sci. Eng. **33**) Berlin: Springer 2003, pp. 199–236

[88] Mascagni, M.: The backward Euler method for numerical solution of the Hodgkin–Huxley equations of nerve conduction. SIAM J. Numer. Anal. **27**, 941–962 (1990)

[89] Miura, R.M.: Accurate computation of the stable solitary wave for the FitzHugh–Nagumo equations. J. Math. Biol. **13**, 247–269 (1982)

[90] Moore, P.K.: An adaptive finite element method for parabolic differential systems: some algorithmic considerations in solving in three space dimensions. SIAM J. Sci. Comput. **21**, 1567–1586 (2000)

[91] Murillo, M., Cai, X.-C.: A fully implicit parallel algorithm for simulating the non-linear electrical activity of the heart. Numer. Linear. Algebra Appl. **11**, 261–277 (2004)

[92] Muzikant, A.L., Henriquez, C.S.: Validation of three-dimensional conduction models using experimental mapping: are we getting closer? Prog. Biophys. Mol. Biol. **69**, 205–223 (1998)

[93] Muzikant, A.L., Hsu, E.W., Wolf, P.D., Henriquez, C.S.: Region specific modeling of cardiac muscle: comparison of simulated and experimental potentials. Ann. Biomed. Eng. **30**, 867–883 (2002)

[94] Neu, J.S., Krassowska, W.: Homogenization of syncytial tissues. Crit. Rev. Biomed. Eng. **21**, 137–199 (1993)

[95] Noble, D., Noble, S.J., Bett, G.C., Earm, Y.E., Ko, W.K., So, I.K.: The role of sodium-calcium exchange during the cardiac action potential. Ann. NY Acad. Sci. **639**, 334–353 (1991)

[96] Noble, D., Rudy, Y.: Models of cardiac ventricular action potentials: iterative interaction between experiment and simulation. Philos. Trans. Roy. Soc. London Ser. A **359**, 1127–1142 (2001)

[97] Nochetto, R.H., Savaré, G., Verdi, C.: A posteriori error estimates for variable time-step discretizations of nonlinear evolution equations. Comm. Pure Appl. Math. **53**, 525–589 (2000)

[98] Oleĭnik, O.A., Shamaev, A.S., Yosifian, G.A.: Mathematical problems in elasticity and homogenization. Amsterdam: North-Holland 1992

[99] Osher, S., Fedkiw, R.: Level set methods and dynamic implicit surfaces. (Applied Mathematical Sciences **153**) New York: Springer 2003

[100] Otani, N.F.: Computer modeling in cardiac electrophysiology. J. Comput. Phys. **161**, 21–34 (2000)

[101] Panfilov, A.V., Holden, A.V.: Computational biology of the heart. Chichester: Wiley 1997

[102] Pavarino, L.F., Colli Franzone, P.: Parallel solution of cardiac reaction-diffusion models. In: Kornhuber, R. et al. (eds.): Domain decomposition methods in science and engineering. (Lecture Notes Comput. Sci. Eng. **40**) Berlin: Springer 2005, pp. 669–676

[103] Penland, R.C., Henriquez, C.S.: Impact of transmural structural and ionic heterogeneity on paced beats in the ventricle. In: Schalij, M.J., Janse, M.J., van Oosterom, A., van der Wal, E., Wellens, H.J. (eds.): Einthoven 2002: 100 Years of Electrocardiography. Leiden: Einthoven Foundation 2002, pp. 35-44

[104] Penland, R., Harrild, O., Henriquez, C.: Modeling impulse propagation and extracellular potential distributions in anisotropic cardiac tissue using a finite volume element discretization. Comput. Vis. Sci. **4**, 215–226 (2002)

[105] Pennacchio, M.: The mortar finite element method for the cardiac "bidomain" model of extracellular potential. J. Sci. Comput. **20**, 191–210 (2004)

[106] Pennacchio, M.: A nonconforming domain decomposition method for the cardiac potential problem. Computers in Cardiology 2001. IEEE Conf. Proc. **28**, 537–540 (2001)
[107] Pennacchio, M., Savaré, G., Colli Franzone, P.: Multiscale modelling for the bioelectric activity of the heart. SIAM J. Math. Anal. **37**(4), 1333–1370 (2006)
[108] Plonsey, R., Heppner, D.: Considerations of quasi-stationarity in electrophysiological systems. Bull. Math. Biophys **29**, 657–664 (1967)
[109] Plonsey, R., Barr, R.C.: Interstitial potentials and their change with depth into cardiac tissue. Biophys. J. **51**, 547–555 (1987)
[110] Plonsey, R.: Bioelectric sources arising in excitable fibers (ALZA lecture). Ann. Biomed. Eng. **16**, 519–546 (1988)
[111] Plonsey, R., Barr, R.C.: Bioelectricity: a quantitative approach. New York: Plenum 1988
[112] Pormann, J.: A modular simulation system for the bidomain equations. Ph.D. thesis. Durham, NC: Duke Univ. 1999
[113] Qu, Z., Garfinkel, A.: An advanced algorithm for solving partial differential equation in cardiac conduction. IEEE Trans. Biomed. Eng. **46**, 1166–1168 (1999)
[114] Quan, W., Evans, S.J., Hastings, H.M.: Efficient integration of a realistic two-dimensional cardiac tissue model by domain decomposition. IEEE Trans. Biomed. Eng. **45**, 372–385 (1998)
[115] Quarteroni, A., Valli, A.: Numerical approximation of partial differential equations. Berlin: Springer 1994
[116] Ramanathan, C., Ghanem, R.N., Jia, P., Ryu, K., Rudy, Y.: Noninvasive electrocardiographic imaging for cardiac electrophysiology and arrhythmia. Nat. Med. **10**, 422–428 (2004)
[117] Rappel W.J.: Filament instability and rotational tissue anisotropy: a numerical study using detailed cardiac models. Chaos **11**, 71–80 (2001)
[118] Rogers, J.M., McCulloch, A.D.: A collocation-Galerkin finite element model of cardiac action potential propagation. IEEE Trans. Biomed. Eng. **41**, 743–757 (1994)
[119] Roth, B.J.: A comparison of two boundary conditions used with the bidomain model of cardiac tssue. Ann. Biomed. Eng. **19**, 669–678 (1991)
[120] Roth, B.J.: Action potential propagation in a thick strand of cardiac muscle. Circ. Res. **68**, 162–173 (1991)
[121] Roth, B.J.: How the anisotropy of the intracellular and extracellular conductivities influences stimulation of cardiac muscle. J. Math. Biol. **30**, 633–646 (1992)
[122] Roth, B.J., Wikswo, J.P.: A bidomain model for the extracellular potential and magnetic field of cardiac tissue. IEEE Trans. Biomed. Eng. **33**, 467–469 (1986)
[123] Rousseau, G., Kapral, R.: Asynchronous algorithm for integration of reaction-diffusion equations for inhomogeneous excitable media. Chaos **10**, 812–825 (2000)
[124] Rudy, Y.: The electrocardiogram and its relationship to excitation of the heart. In: Sperelakis, N. (ed.): Physiology and pathophysiology of the heart. 3rd ed. Boston: Kluwer 1995, pp. 201–239
[125] Rudy, Y., Oster, H.S.: The electrocardiographic inverse problem. Crit. Rev. Biomed. Eng. **20**, 25–45 (1992)
[126] Sachse, F.B.: Computational cardiology. Modeling of anatomy, electrophysiology, and mechanics. (Lecture Notes Comput. Sci. **2966**) Berlin: Springer 2004
[127] Saffitz, J.E., Kanter, H.L., Green, K.G., Tolley, T.K., Beyer, E.C.: Tissue-specific determinants of anisotropic conduction velocity in canine atrial and ventricular myocardium. Circ. Res. **74**, 1065–1070 (1994)
[128] Saleheen, H.I., Ng, K.T.: A new three-dimensional finite-difference bidomain formulation for inhomogeneous anisotropic cardiac tissues. IEEE Trans. Biomed. Eng. **45**, 15–25 (1998)

[129] Sánchez-Palencia, E., Zaoui, A. (eds.): Homogenization techniques for composite media. (Lectures Notes Phys. **272**) Berlin: Springer 1987

[130] Sanfelici, S.: Convergence of the Galerkin approximation of a degenerate evolution problem in electrocardiology. Numer. Methods Partial Differential Equations **18**, 218–240 (2002)

[131] Sanfelici, S.: Numerical simulations of fractioned electrograms and pathological cardiac action potential. J. Theor. Med. **4**, 167–181 (2002)

[132] Seemann, G., Sachse, F.B., Chiasaowong, K., Weiss, D.: Quantitative reconstruction of cardiac electromechanics in human myocardium: assembly of electrophysiologic and tension generation models. J. Cardiovasc. Electrophysical **14**, S210–S218 (2003)

[133] Seemann, G., Sachse, F.B., Weiss, D.L., Dossel, O.: Quantitative reconstruction of cardiac electromechanics in human myocardium: regional heterogeneity. J. Cardiovasc. Electrophysical **14**, S219–S228 (2003)

[134] Sethian, J.A.: Level set methods and fast marching methods: evolving interfaces in computational geometry, fluid mechanics, computer vision, and materials science. 2nd ed. Cambridge: Cambridge University Press 1999

[135] Simms, H.D., Geselowitz, D.B.: Computation of heart surface potentials using the surface source model. J. Cardiovasc. Electrophysiol. **6**, 522–531 (1995)

[136] Smith, B.F., Bjørstad, P., Gropp, W.D.: Domain decomposition: parallel multilevel methods for elliptic partial differential equations. Cambridge: Cambridge University Press 1996

[137] Smith, N.P., Nickerson, D.P., Crampin, E.J., Hunter, P.J.: Multiscale computational modelling of the heart. Acta Numer. **13**, 371–431 (2004)

[138] Soravia, J.P., Souganidis, P.E.: Phase-field theory for Fitzhugh-Nagumo type systems. SIAM J. Math. Anal. **27**, 1341–1359 (1996)

[139] Streeter, D.: Gross morphology and fiber geometry in the heart. In: Berne, R.M. ed al. (eds.): Handbook of physiology. Sect. 2: The cardiovascular system. Vol. 1: The heart. Bethesda, MD: Amer. Physiolog. Soc. 1979, pp. 61–112

[140] Sundnes, J., Lines, G.T., Grøttum, P., Tveito, A.: Electrical activity in the human heart. In: Langtangen, H.P., Tveito, A. (eds): Advanced topics in computational partial differential equations. (Lecture Notes Comput. Sci. Eng. **33**) Berlin: Springer 2003, pp. 401–449

[141] Taccardi, B., Macchi, E., Lux, R.L., Ershler, P.R., Spaggiari, S., Baruffi, S., Vyhmeister, Y.: Effect of myocardial fiber direction on epicardial potentials. Circulation **90**, 3076–3090 (1994)

[142] Taccardi, B., Punske, B., Helie, F., MacLeod, R., Lux, R., Ershler, P., Dustman, T., Vyhmeister, Y.: Epicardial recovery sequences and excitation recovery intervals during paced beats. Role of myocardial architecture. Pacing Clin. Electrophysiol. **22**(4) part II: 833 (1999)

[143] Taccardi, B., Lux, R.L., MacLeod, R.S., Ershler, P.R., Dustman, T.J., Scott, M., Vyhmeister, Y., Ingebrigtsen, N.: ECG waveforms and cardiac electric sources. J. Electrocardiol. **29**(Suppl.), 98–100 (1996)

[144] Taccardi, B., Lux, R.L., Ershler, P.R., MacLeod, R.S., Dustman, T.J., Ingebrigtsen, N.: Anatomical architecture and electrical activity of the heart. Acta Cardiol. **52**, 91–105 (1997)

[145] Taccardi, B., Veronese, S., Colli Franzone, P., Guerri, L.: Multiple components in the unipolar electrocardiogram: a simulation study in a three-dimensional model of ventricular myocardium. J. Cardiovasc. Electrophysiol. **9**, 1062–1084 (1998)

[146] Taccardi, B., Punske, B., Lux, R., MacLeod, R., Ershler, P., Dustman, T., Vyhmeister, Y.: Useful lessons from body surface mapping. J. Cardiovasc. Electrophysiol. **9**, 773–786 (1998)

[147] Taccardi, B., Punske, B.B.: Body surface potential mapping. In: Zipes, D. Jalife, J. (eds.): Cardiac electrophysiology: from cell to bedside. 4th ed. Philadelphia: Saunders 2004, pp. 803–811

[148] ten Tusscher, K., Noble, D., Noble, P.J., Panfilov, A.V.: A model for human ventricular tissue. Am. J. Physiol. Heart Circ. Physiol. **286**, H1573–H1589 (2004)

[149] Tomlinson, K.A., Hunter, P.J., Pullan, A.J.: A finite element method for an eikonal equation model of myocardial excitation wavefront propagation. SIAM J. Appl. Math. **63**, 324–350 (2002)

[150] Trangenstein, J.A., Kim, C.: Operator splitting and adaptive mesh refinement for the Luo-Rudy I model. J. Comput. Physics **196**, 645–679 (2004)

[151] Tung, L.: A bidomain model for describing ischemic myocardial DC potentials. Ph.D. thesis. Cambridge, MA: M.I.T. 1978

[152] van Oosterom, A.: Forward and inverse problems in electrocardiography. In: Panfilov, A.V., Holden, A.V. (eds.): Computational biology of the heart. Chichester: Wiley 1997, pp. 295–343

[153] Veneroni, M.: Reaction-diffusion systems for the microscopic cellular model of the cardiac action potential. In preparation.

[154] Veneroni, M.: Reaction-diffusion systems for the macroscopic Bidomain model of the cardiac action potential. In preparation.

[155] Vigmond, E.J., Aguel, F., Trayanova, N.A.: Computational techniques for solving the bidomain equations in three dimensions. IEEE Trans. Biomed. Eng. **49**, 1260–1269 (2002)

[156] Vigmond, E.J., Leon, L.J.: Computationally efficient model for simulating electrical activity in cardiac tissue with fiber rotation. Ann. Biomed. Eng. **27**, 160–170 (1999)

[157] Viswanathan, P.C., Shaw, R.M., Rudy, Y.: Effects of I_{Kr} and I_{Ks} heterogeneity on action potential duration and its rate dependence: a simulation study. Circulation **99**, 2466–2474 (1999)

[158] Weber dos Santos, R., Plank, G., Bauer, S., Vigmond, E.: Parallel multigrid preconditioner for the cardiac bidomain model. IEEE Trans. Biomed. Eng. **51**, 1960–1968 (2004)

[159] Winslow, R.L., Rice, J., Jafri, S., Marban, E., O'Rourke, B.: Mechanisms of altered excitation-contraction coupling in canine tachycardia-induced heart failure. II: Model studies. Circ. Res. **84**, 571–586 (1999)

[160] Wit, A.L., Janse, M.J.: The ventricular arrhythmias of ischemia and infarction: electrophysiological mechanisms. Mt. Kisco, NY: Futura 1993

[161] Yamashita, Y., Geselowitz, D.B.: Source–field relationships for cardiac generators on the heart surface based on their transfer coefficients. IEEE Trans. Biomed. Eng. **32**, 964–970 (1985)

[162] Yu, H.: Solving parabolic problems with different time steps in different regions in space based on domain decomposition methods. Appl. Numer. Math. **30**, 475–491 (1999)

[163] Yu, H.: A local space-time adaptive scheme in solving two-dimensional parabolic problems based on domain decomposition methods. SIAM J. Sci. Comput. **23**, 304–322 (2001)

[164] Zipes, D., Jalife, J.: Cardiac electrophysiology: from cell to bedside. 4th ed. Philadelphia: Saunders 2004

The circulatory system: from case studies to mathematical modeling

L. Formaggia, A. Quarteroni, A. Veneziani

Abstract. In this work we illustrate actual case studies in vascular medicine and surgery that we have recently investigated with the support of mathematical models and numerical simulations. We present six examples, where our investigation has different purposes, ranging from a better understanding of phenomena of clinical interest to the optimization of surgical procedures. For each case, the description of the problem is followed by an illustration of the mathematical model and the numerical technique used for its investigation, including the discussion of numerical results. Each example thus provides the conceptual framework to introduce mathematical models and numerical methods whose applicability, however, goes beyond the specific case that is addressed.

Keywords: Blood flow modeling, computational haemodynamics, vascular diseases and surgical planning.

1 An overview of vascular dynamics and its mathematical features

The use of mathematical models, originally applied mainly in sectors with a strong technical content (such as, e.g., automotive and aerospace engineering), is now widespread in many fields of the life sciences as well, where human factors often prevail. Bioinformatics, mathematical analysis and scientific computing support investigations in different fields of biology (such as genetics or physiology) and medicine. Mathematical models and numerical simulations can, for example, establish a bond between molecular structures and clinically evident behavioral patterns. Providing quantitative data on the behavior of organs, systems, or even the entire body, in terms of subcellular functions, they can also contribute, through the interpretation of medical images and maps of electric potentials, to the definition of therapies and the design of medical devices.

One subject that has caught the attention of important mathematicians and scientists in the past (from Aristotle to Bernoulli, Euler, Poiseuille and Young) is the functioning of the cardiocirculatory system. Recently, the socio-economic impact of cardiovascular pathologies has further motivated this research, which presents challenging mathematical difficulties. Up to the 1970s, in vitro and animal experiments were the main means of investigation in this field. However, progress in computational fluid dynamics as well as the increase in computer power has added numerical experimentation to the tools at the disposal of medical researchers, biologists and bioengineers. Quantities such as shear stresses on the endothelium surface,

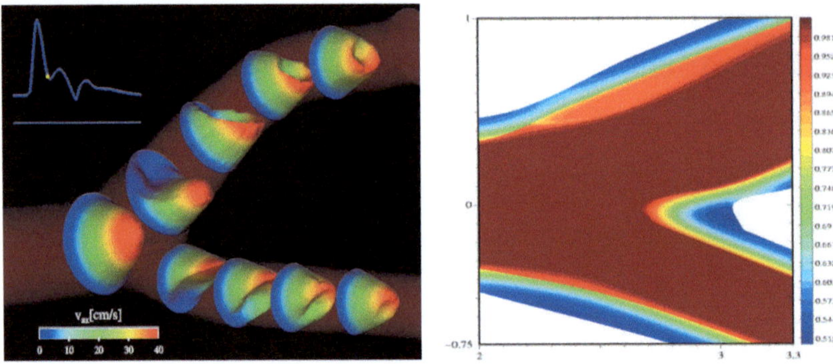

Fig. 1. *On the left*: Velocity profiles computed in a carotid bifurcation at the end of systole (courtesy of M. Prosi). *On the right*: Simulations of the concentration of oxygen in the lumen and the wall of a carotid bifurcation (courtesy of P. Zunino)

which are quite hard, if not impossible, to measure in vitro, can now be calculated from simulations carried out on real geometries obtained with three-dimensional reconstruction algorithms. In this respect, a decisive thrust has been provided by the development of modern non-invasive data collection technologies such as nuclear magnetic resonance, digital angiography, CT scans and Doppler anemometry (see, e.g., [56]).

The separation of the blood flow and the generation of a secondary motion are today recognized as potential factors for the development of arterial pathologies (such as the atherosclerotic plaques formation). They may be induced by a particular vascular morpohology, e.g., a bifurcation; an example related to the carotid artery is illustrated on the left-hand side of Fig. 1. A detailed understanding of the local haemodynamic patterns and of their effects on the vessel wall is today a possibility thanks to accurate computer simulations.

In large arteries, blood flow interacts mechanically with the vessel wall, giving rise to a complex fluid-structure interaction mechanism with a continuous transfer of energy between the blood and the vessel structure. Moreover, a thorough investigation of the role of heamodynamics in vascular pathologies needs to monitor the concentration of relevant chemical components (such as oxygen, lipids or, possibly, drugs). The blood flow problem must therefore be coupled with models describing the transport, diffusion and absorption of chemicals in all the various layers that form the arterial wall (intima, media and adventitia). The complexity (and non-linearity) of the coupling is increased by the fact that wall shear stresses influence the orientation and deformation (or even damaging) of the endothelial cells. Consequently, wall permeability typically depends on wall shear stress. Numerical simulations of this type, such as that on the right-hand side of Fig. 1, can explain the biochemical modifications produced by blood flow alterations caused, for example, by the presence of a stenosis.

Basic mathematical features of the problem. We next indicate basic features of the phenomena previously illustrated that drive the choice of the mathematical models and the numerical methods used for their approximation. The first, and perhaps most relevant, is *unsteadiness.* To quote [37]: "The most obvious thing of blood flow in arteries is that it is pulsatile." The arterial pulsatility induced by the action of the heart strongly influences haemodynamics. The basic time scale in this context is given by the heart beat (about 0.8 s). We may recognize an initial phase called the *systole* (about 0.3 s), when the aortic valve is open and blood is thrust into the arterial system, followed by the *diastole* (about 0.5 s), initiated by the closure of the aortic valve. Fast transients are therefore a relevant feature of blood flow and specific numerical techniques for their reliable simulations are required, particularly when one is interested in blood flow in large arteries (those whose diameter is above 0.4 mm).

Heterogeneity is another key feature. In fact, haemodynamics entails different phenomena interacting at different levels. At the mathematical level, this implies the coupling of different models acting either in the same computational domain or in adjacent subdomains and related by appropriate interface or matching conditions. Their numerical treatment may require ad-hoc methods developed on split regions (*domain decomposition techniques*, see [45]).

A further important feature of haemodynamics problems is the presence of *multiple scales* in both time and space. An illustrative instance is represented by the (either active or passive) regulation of blood flow distribution. A stenosis in a carotid or a cerebral artery, even when yielding significant lumen reduction, does not necessarily cause a relevant reduction of blood supply to the downstream compartments. In fact, blood flow is redistributed through secondary vessels and continues to ensure an almost physiological blood flow. These morphological changes are activated by biochemical mechanisms which govern vessel dilation and may even drive the oxygen exchange between blood and tissues. Here, we face different time scales (blood flow and regulation mechanisms) and spatial scales (local heamodynamics and global circulation). Similar mechanisms are present, for instance, in the Willis circle, making the vascular network in the brain a robust system.

Another example is the occurence and growth of cerebral aneurysms, a major pathology with many aspects still to be clarified. Here, complex interactions involving systemic factors, such as hypertension or high cholesterol levels, and local blood flow features associated to particular vascular morphologies can induce the occurence of the pathology. Again, different time scales are involved.

This multiscale nature requires us to devise suitable numerical techniques for coupling the different models, capable of reproducing the interaction between small scale phenomena as well as at the macroscopic level. In this respect, the term *geometrical multiscale* has been coined for techniques that take account of different space scales involving local and systemic dynamics. Several examples and applications may be found in [15, 16, 46] and in [18].

The tasks of computational haemodynamics. All the features previously listed must be adequately accounted for when developing mathematical models and numerical methods for the circulatory system. The rigorous mathematical derivation of these models is, however, beyond the scope of these notes; the interested reader can refer, e.g., to [42]. Here, we wish to take a different route. We present a set of problems arising from realistic clinical cases, and illustrate their specific characteristics that drive the choice of mathematical and numerical models.

The selection of our case studies was motivated by the different objectives that one wants to achieve with numerical simulations. One goal of computational haemodynamics is to help the *understanding of the occurence and development of (individual) physiopathologies.*

In general, in the medical field, carrying out in vivo and in vitro experiments has clear practical and ethical limitations, in particular for a deep understanding of the features of a single patient that could be responsible for a pathological behavior. Nowadays, for instance, geometrical reconstructions of an individual carotid morphology starting from angiographs, CT scans or MR images can be extensively used for evaluating the impact of the vessel shape on the wall shear stress and consequently on the possible development of atherosclerotic plaques. Other examples are considered in the present overview. In particular, numerical simulations of the different pulmonary artery banding in neonates affected by left ventriculum hypoplasia, which has clarified the impact of the banded vessel profile on the shear stress and provided a quantitative explanation of the observed follow-up in the patients (see Sect. 2.1). Another example addressed here refers to the modifications of the blood flow and metabolic dynamics induced by exercise (e.g., in sport), or by ageing (Sect. 2.2).

Another issue is *prediction and design*. In some engineering fields numerical simulations represent a consolidated tool for supporting design and the setting up of a new prototype, with the aim of reducing more expensive experimental assessment. To quote [32]: "Since the late 1950s, CFD (Computational Fluid Dynamics) has played a major role in the development of more versatile and efficient aircraft. It has now become a crucial enabling technology for the design and development of flight vehicles. No serious aeronautical engineer today would consider advancing a new aircraft design without extensive computational testing and optimization. The potential of CFD to play a similar role in cardio-vascular intervention is very high." With a similar perspective, in this work we address the design of drug-eluting stents. The role of numerical simulations in setting up a coating film ensuring correct drug delivery is essential (Sect. 2.3). Another example is given by numerical simulations for comparing different possibilities of a surgical intervention in pediatric heart diseases, providing practical indications for the surgeon (Sect. 2.4).

Finally, a third - and perhaps the most ambitious - task is *identification and optimization*. Scientific computing is nowadays used to solve not only direct, but also inverse problems, i.e., to help devise a solution which fulfills prescribed optimality criteria. The task is therefore not only to simulate the fluid dynamics in a given vascular district or, more generally, in a *compartment* (i.e., a set of organs and tissues). Rather, the desired dynamics inside the compartment is specified (or given by measures in identification problems), and the computations have the role of identifying the "parameters" of the problem, ensuring that these features are best satisfied (e.g., the "optimal" shape of a prosthetic implant). The major difficulty in solving optimization problems in general (and for the life sciences in particular) is represented by the severe computational costs. Optimization solvers are usually based on iterative procedures and this could be prohibitely expensive if each iteration requires the solution of a system of non-linear time-dependent partial differential equations. For this reason, specific techniques are under development, aiming at reducing computational costs. Here, we address two cases. In the first, optimization has been applied to setting up a procedure for the so-called *peritoneal dialysis* (Sect. 2.5). The optimal solution computed by numerical means was implemented in an electronic device with the aim of regulating the process individually for each patient. In the second case that we present, shape optimization techniques, together with specific methods for the reduction of computational costs, are applied to coronary by-pass anastomosis in order to find the "best" post-surgery configuration, which reduces the risk of operation failure (Sect. 2.6).

In Sect. 3, we give a synthesis of the different examples considered, highlighting open problems and perspectives of this interesting and rapidly growing research field.

2 Case studies

We present here a set of examples in the field of computational haemodynamics emerging from actual clinical cases which we have studied in cooperation with medical doctors, surgeons and bioengineers. Each problem is first described in medical terms; then we introduce its mathematical formulation and the numerical tools that were used for its solution.

2.1 Numerical investigation of arterial pulmonary banding

The problem

Artificial regulation of the pulmonary blood flow is sometimes necessary to deal with serious heart congenital defects; see [7]. A surgical procedure to achieve this goal consists of banding the pulmonary artery, so that the vessel lumen is suitably modified and the blood flow rate adjusted as needed. In fact, this technique modifies the pulmonary artery *resistance*[1]. This represents a palliative for pathologies such as functional univentricular hearts, multiple ventricular and atrio-ventricular septal defects. More recently, it has been used in hypoplastic left heart malformations, either as a rescue procedure for critical neonates or as a preparatory measure for subsequent surgical operations (such as the Norwood procedure; see Sect. 2.4 or the literature on heart transplants).

Conventional pulmonary artery banding in neonates and infants, however, has drawbacks which have limited its use so far. In particular:

[1] *Vascular resistance* is related to the vessel lumen and the flow viscosity. It is introduced in Sect. 2.2.

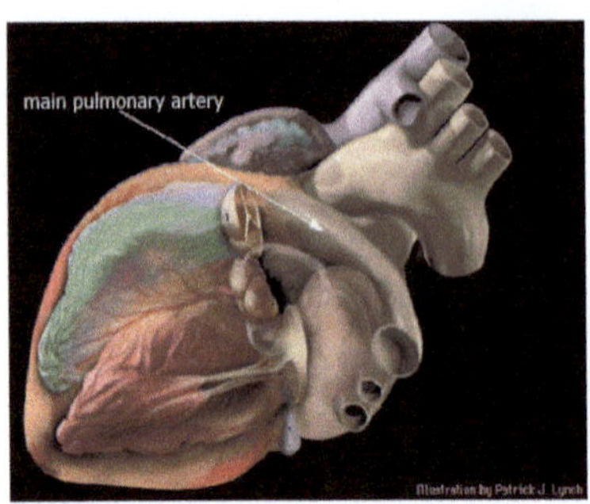

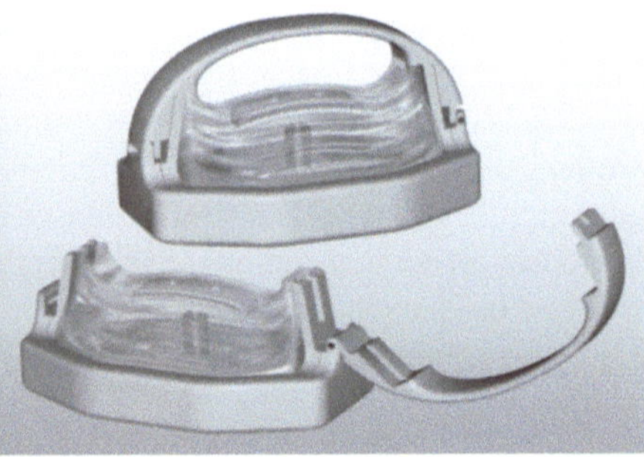

Fig. 2. *On the left*: The heart with the indication of the pulmonary artery where the banding is applied. (Modified from P. J. Lynch, see patricklynch.net). *On the right*: FloWatch device in open and closed positions. Notice the banana-shape of the clipping (photo from [5] with permission)

1. the banding has to be adapted in time since blood flow demand changes with growth;
2. the pulmonary arterial wall is damaged by the procedure to such an extent that frequently a surgical repair procedure is needed after de-banding.

Recently, a new device for arterial banding, called *FloWatch©-PAB* (designed by EndoArt, Lausanne, Switzerland), was developed, originally to overcome the first limitation. It was later found that it helps to overcome the second problem as well, as we show next.

The FloWatch system comprises an implantable device and an external control unit. The former features a clip which is placed around the pulmonary artery in a fashion similar to a watch band. The area of the vessel lumen can be adjusted by means of a piston which acts on the clip and is driven by a micro-engine. The engine is electrically activated and controlled by telemetry by means of the external unit. This system, called *adjustable pulmonary artery banding*, allows us to adapt the banding in time without further surgical intervention on the patient.

Clinical data have shown that, with FloWatch banding, the second drawback also seems to be avoided. To quote [7]: "...we didn't see any lesion in the pulmonary artery in the experimental study...the histology showed in all the cases almost normal pulmonary artery with very pliable tissue."

A quantitative analysis of the functioning of the device was carried out by means of numerical simulations as reported in [6] and it has highlighted the relations between the perimeter and the area of the banding, the banding pressure gradient distribution and the induced stresses, comparing them with the traditional approach. The research was carried out in cooperation with Dr. A. Corno of the Royal Liverpool Children's Hospital, Alder Hey.

Numerical models and simulations

We next introduce mathematical models that have proved to be well suited for the simulation of pulmonary arterial banding and we briefly discuss the numerical results.

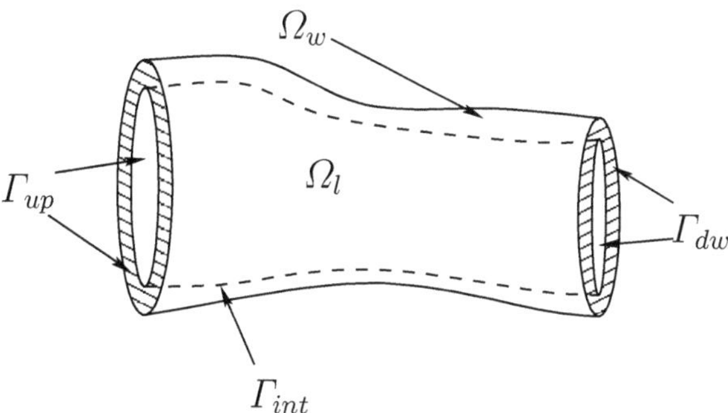

Fig. 3. A possible domain in a haemodynamic simulation

Models and methods. We denote by $\boldsymbol{u}(\mathbf{x}, t)$ and $P(\mathbf{x}, t)$ the blood velocity and pressure respectively in the domain Ω_l, the vascular lumen. By Ω_w we denote the vessel wall (see Fig. 3). The two domains are separated by the so-called *endothelial membrane* Γ_{int} and are delimited by the boundaries Γ_{up} (upstream or proximal sections) and Γ_{dw} (downstream or distal sections). Lastly, the external boundary of the vascular wall is denoted by Γ_{ext}.

The application of the physical principles of momentum and mass conservation for an incompressible fluid leads to the equations:

$$\begin{cases} \dfrac{\partial \boldsymbol{u}}{\partial t} + (\boldsymbol{u} \cdot \nabla)\boldsymbol{u} - \nabla \cdot \mathbf{S} + \nabla p = \mathbf{f} \\ \nabla \cdot \boldsymbol{u} = 0 & \mathbf{x} \in \Omega,\ t \in (0, T] \\ \boldsymbol{u}|_{t=0} = \boldsymbol{u}_0 \end{cases} \tag{1}$$

For simplicity the momentum equation has been divided by the density ρ; therefore $p = P/\rho$ here and $\mathbf{f}$ is a generic field of forces per unit of mass.

The quantity $\mathbf{S}$ is the so-called *deviatoric* stress tensor and is a function of the velocity. The actual dependence of $\mathbf{S}$ on $\boldsymbol{u}$ is a matter of *blood rheology*, i.e., the mathematical description in terms of constitutive laws of the complex interaction between the suspended particles and the behavior of the fluid as a continuum. The adoption of a specific rheological law depends on the features of the particles (in particular, the red cells), and on the characteristics of the vascular district at hand. For the derivation of Eqs. (1) and the rheological laws, see, e.g., [42] and its bibliography. In the case of the pulmonary artery, which has a diameter of about 2 cm, Newtonian rheology is accurate enough to describe the blood behavior. Here, the deviatoric tensor is proportional to the symmetric part of the velocity gradient, namely,

$$\mathbf{S} \equiv 2\nu\frac{1}{2}\left(\nabla\boldsymbol{u} + \nabla^T\boldsymbol{u}\right), \tag{2}$$

where ν is the (constant) *kinematic viscosity* of blood.

Equations (1) must be completed with boundary conditions on $\partial\Omega_l$. In particular, those on $\Gamma_{up} \cup \Gamma_{down}$ should account for the presence of the remaining part of the circulatory system. For the purpose of the present study, in which we want to compare the flow conditions induced by different types of banding, it is sufficient to assume that, for $t > 0$,

$$u = \mathbf{g}, \quad \mathbf{x} \in \Gamma_{\text{up}}, \quad \text{and} \quad p\mathbf{n} - \mathbf{S} \cdot \mathbf{n} = \mathbf{0}, \quad \mathbf{x} \in \Gamma_{\text{down}}. \tag{3}$$

In particular, $\mathbf{g}$ has been chosen as a Poiseuille parabolic fully developed profile [42].

On Γ_{int} we need to prescribe conditions associated to the interaction between the blood and the vascular wall. In general, the setting-up of these conditions is a challenging problem at both the mathematical and the numerical level, because of the compliance of the vascular wall. However, for simplicity, in the present case we assume that the wall is rigid, which corresponds to taking

$$u = \mathbf{0}, \quad \mathbf{x} \in \Gamma_{\text{int}}, \quad t > 0. \tag{4}$$

Equations (1), (2) are the so-called *Navier-Stokes equations* for an incompressible Newtonian fluid. They have been extensively investigated from the mathematical viewpoint, e.g., in [58]. Together with (3) and (4), they provide the mathematical model for our banding problem.

The description of the computational domain Ω_1 was obtained in our case by a geometrical reconstruction from images taken with a camera, effected by using tools provided by the software package *Mathematica* [64]. Camera data cover the arterial segment from the pulmonary valve to the section upstream of the pulmonary bifurcation. However, since the bifurcation strongly affects the local haemodynamics, we have included it by extending the reconstructed domain with a T-bifurcation (see Fig. 4 left). The diameter of the vascular lumen is here $D = 18$ mm.

For the numerical solution of (1) we used finite differences for the time derivative and finite elements for the space derivatives [42]. In Fig. 4 (right) we show a detail of the finite element grid around the FloWatch device.

Results. Several numerical simulations were carried out for both the traditional banding and the one produced by the FloWatch. Three levels of occlusion of the

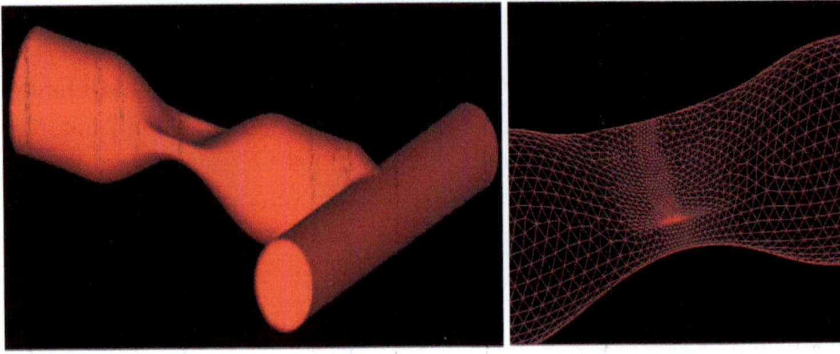

Fig. 4. *On the left*: Computational domain including the pulmonary banding and the pulmonary bifurcation. *On the right*: Detail of the mesh around the banding (courtesy of M. Prosi)

lumen were considered, namely, 26%, 50% and 62%, as well as two flow rates, 1.5 and 2.0 lit/min. The latter were set by choosing the Poiseuille inflow datum **g** appropriately.

The banding controls the flow by adjusting the vascular resistance. Indeed, a vasoconstriction induces a higher pressure gradient for a given flow rate and, conversely, a reduced flow for a given pressure difference at the ends of the vessel. In a perfect cylindrical domain the ratio between pressure gradient and flow rate is inversely proportional to the section area (see [63]). The first goal of numerical simulation was therefore to investigate the behavior of this quantity in the case of the more complex shape produced by the FloWatch banding (see the picture on the right-hand side of Fig. 4).

In Fig. 5 (right) we report the pressure differences obtained with the two bandings for an occlusion of 50% and 26% of the section area, respectively, and for two input flow rates. The conclusion is that the efficiency is almost independent of the type of banding and is strongly related to the section area. In fact, although the shape induced by the two bandings is very different, the pressure drop for a given flow and occlusion level is much the same. This is also confirmed by the clinical data reported in [6]. In Fig. 6 we illustrate the computed wall shear stress maps induced by blood in the two cases. It is possible to observe that the conventional circular banding is associated to a lower shear stress field and this is frequently recognized as a possible pathogenic factor in wall tissue degeneration or even atherosclerosis (see, e.g., [36]).

Another task of the numerical simulations was to investigate the situation after debanding. It was observed that, with FloWatch banding, the perimeter of the constricted section conforms to the banana-shape section of the FloWatch banding as shown in the right-hand picture in Fig. 2. Consequently, it is practically independent of the luminal section area, as shown in the left-hand graph of Fig. 5, where we have plotted the cross-sectional area versus the external perimeter for both the conventional and the FloWatch banding, for the normal range of constriction level.

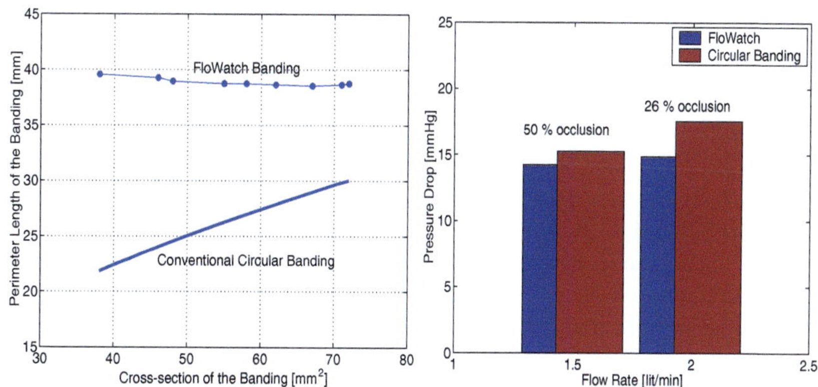

Fig. 5. *On the left*: Section area vs. section perimeter for the circular and the FloWatch banding, respectively. *On the right*: Computed pressure difference vs. flow rate for the two types of banding and different occlusion levels

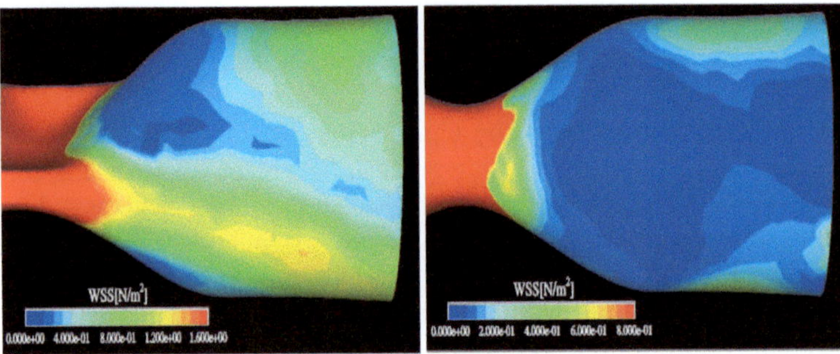

Fig. 6. Wall shear stress fields in the FloWatch (*left*) and circular (*right*) bandings (courtesy of M. Prosi)

A realistic hypothesis is then that, after the removal of the constriction, the pulmonary artery reopens to a circular section having the same perimeter as in the banded configuration. Therefore, after conventional banding the vessel is unable to reopen fully, and this is verified by clinical evidence. Furthermore, it is reasonable to assume that the perimeter reduction induced by conventional banding yields local stresses in the vessel structure which responds with unwanted morphological changes. Indeed, after removing the conventional band, the arterial wall is often found to be severely damaged, with increased thickness, dense fibrosis and consequent loss of elasticity and pliability, so that surgical resection and reconstruction is necessary. The reconstruction procedure is an additional intrusive operation and often induces residual undesired resistances in the pulmonary artery.

Using the FloWatch device it was found that, after de-banding, the perimeter of the luminal section is only slightly reduced and therefore the residual constriction is not significant. For a 50% occlusion, corresponding to 39 mm^2 cross-section area, the perimeter after de-banding is about 39.5 mm which corresponds to a diameter of 12.6 mm, not far from the physiological diameter (18 mm). Numerical simulations on the de-banded configuration confirm that the residual pressure drop is in the range of 2.5 mm Hg for a flow rate of 1.5 lit/min, which is reasonably small.

Geometrical and numerical analysis have provided a possible interpretation for the clinical evidence, showing that the new FloWatch banding:

1. is effective at all realistic constriction levels, showing the same flow control characteristics as conventional banding;
2. achieves a given area reduction with a smaller contraction of the vessel perimeter; this implies a reduced stress on the banded arterial wall and better behavior after de-banding, to such an extent that a surgical reconstruction of the pulmonary artery is in general no longer needed.

2.2 Numerical investigation of systemic dynamics

The problem

What are the effects of aging on the vascular system? How does heavy exercise influence the transport by the blood stream of oxygen or other metabolites and their consumption by tissues? Which self-regulatory process governs the dynamics of blood solutes? Why does the implant of an endovascular prothesis induce an overload on the heart?

To answer all these questions requires a quite different viewpoint than the one adopted in the previous example. There, the goal was to investigate a *local* process. Here, the answer needs a systemic perspective, able to correlate actions and reactions in different cardiovascular compartments. The "Digital Astronaut" programme launched in October 2004 by the National Space Biomedical Research Institute of NASA can be regarded as an ambitious example in this perspective. The long-term goal of this program is to develop a model accounting for the mutual interaction of the different parts of the body and able to evaluate the response of the cardiovascular system to different external conditions (see [49]). The final aim is to find countermeasures to the syndromes and pathologies affecting an astronaut living in a low gravity environment for a long period, and to speedup the return to normal conditions at the end of a mission.

Using appropriate mathematical models we can simulate the regulating processes which the human body activates to adapt to changes in outside conditions. For instance, the elasticity of a blood vessel changes in the absence of gravity as the smooth muscles that surround it are controlled by the nervous system, which in turn reacts to biochemical or mechanical variations. Indeed, adjustment and regulating processes are characteristics common to all biological systems: thousands of feedback mechanisms influence the conditions of cells and organs, and are eventually the foundation of life. Such processes are encoded by complex enzymic reactions and are particularly hard to describe in a purely phenomenological and experimental manner, especially in complex organisms like human beings.

The mathematical tools for this simulations cannot, in general, be the same as those used in the previous section. Even if we just focus on blood flow dynamics, carrying out a simulation of a large part of the circulatory system by solving the three-dimensional Navier-Stokes equations (1) would require the availability of a large set of morphological data (quite difficult to obtain). Not to mention the high computational costs. Furthermore, in certain vascular compartments the hypothesis of Newtonian rheology would be questionable.

However, the level of detail given by a 3D model is unnecessary when one is primarily interested in the global response. We need therefore to find a reasonable hierarchy of models, with different levels of detail, but capable of answering our questions. Moving from reasonable simplifying assumptions, we can basically derive two kinds of model: networks of 1D models and lumped parameter models. In this section, we take a brief glance at their basic features, highlighting to what extent they are able to represent the behavior of the circulatory system. On this basis, we try to give an answer to some of the questions suggested at the beginning of the section.

The studies we present in the following paragraphs were motivated by the "Sport and Rehabilitation Engineering Programme 2004" at EPFL, and carried out in cooperation with M. Tuveri, vascular surgeon at the Policlinico S. Elena, Cagliari.

Numerical models and simulations

1D models. If we exploit the fact that an artery is a quasi-cylindrical vessel and that blood flows mainly in the axial direction, we may build a simplified model that ignores the transversal components of the velocity. Moreover, we could assume that the wall displaces only along the radial direction and describe the fluid-structure interaction blood flow problem in terms of the measure $A(z,t)$ of a generic axial section $\mathcal{A}(z)$ of the vessel (see Fig. 7 (left)) and the mean flux

$$Q(z,t) = \int_{\mathcal{A}(z)} \boldsymbol{u}_z d\sigma .$$

Here, z indicate the axial coordinate. Under simplifying yet realistic hypotheses, the following one dimensional (1D) model is obtained [42]:

$$\begin{aligned} &\frac{\partial A}{\partial t} + \frac{\partial Q}{\partial z} = 0, \\ &\frac{\partial Q}{\partial t} + \frac{A}{\rho}\frac{\partial p}{\partial A} - \alpha \overline{\boldsymbol{u}}_z^2 \frac{\partial A}{\partial z} + 2\alpha \overline{\boldsymbol{u}}_z \frac{\partial Q}{\partial z} + K_R \left(\frac{Q}{A}\right) Q = 0 \end{aligned} \tag{5}$$

for $z \in (0, L)$, and $t > 0$, which describes the flow of a Newtonian fluid in a compliant straight cylindrical pipe of length L. Here, $\overline{\boldsymbol{u}}_z \equiv A^{-1} \int_{\mathcal{A}} \boldsymbol{u}_z d\sigma$, and the parameter α, called the *momentum correction* and also the *Coriolis coefficient*, is defined as $\alpha = (A\overline{\boldsymbol{u}}_z^2)^{-1} \int_{\mathcal{A}} \boldsymbol{u}_z^2 d\sigma$. The pressure is assumed to be a function of A according to a constitutive law that specifies the mechanical behavior of the vascular tissue. Different models can be obtained by choosing different pressure-area laws. Lastly, K_R is a parameter accounting for the viscosity of the fluid, whose expression depends on the simplifying assumptions made (see [42] and [4]).

The hyperbolic system (5) can be used to describe blood flowing in a vascular segment. Since the arterial system can in fact be assimilated to an hydraulic network, it may be modelled as a network of 1D hyperbolic PDE's as long as suitable matching conditions are found at the branching points. If we denote by Ω_1 a proximal segment, and by Ω_2 and Ω_3 the two branches (see Fig. 7 (right)), a possible set of conditions that ensure mass and energy conservation is [42]:

$$\begin{aligned} &Q_1 + Q_2 + Q_3 = 0, \\ &\left(p + \frac{1}{2}|\overline{\boldsymbol{u}}_z|^2\right)_1 = \left(p + \frac{1}{2}|\overline{\boldsymbol{u}}_z|^2\right)_2 = \left(p + \frac{1}{2}|\overline{\boldsymbol{u}}_z|^2\right)_3, \end{aligned} \tag{6}$$

where all quantities are computed at the bifurcation point. More sophisticated bifurcation conditions may also consider the effect of the angles among the branches (see, e.g., [14]).

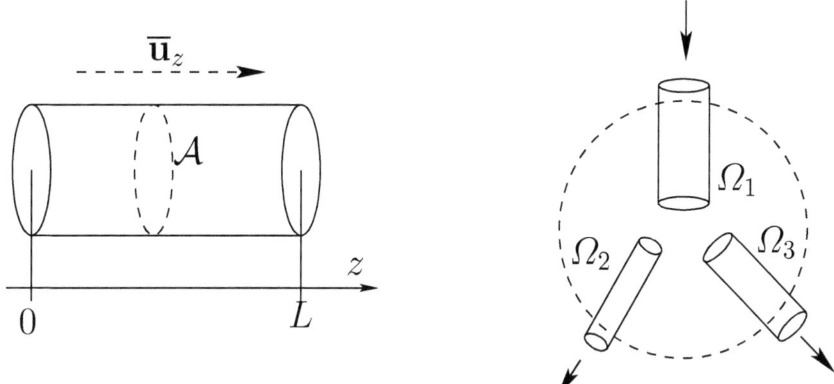

Fig. 7. *On the left*: Representation of an arterial cylindrical segment. *On the right*: Sketch of a bifurcation

The system of equations formed by (5) and (6) is stable [14] and is used in the next paragraph to model circulation. We warn the reader that research is still active in finding different 1D models that could improve the computation of wall shear stress [52] or the description of curved segments [30]. Furthermore, the estimation of the physical parameters of the model needed to obtain realistic computations is a rather complex task, an account is given in [31].

Lumped parameter models. A further simplification in the mathematical description of circulation relies on the subdivision of the vascular system into *compartments*, according to criteria suited to the problem at hand. Blood flow, as well as the other quantities of interest, is described in each compartment by a set of parameters depending only on time. For blood flow, these parameters are the average flux and the pressure in the compartment. The mathematical model is typically given by a system of ordinary differential equations in time that govern the dynamics of each compartment, and their mutual coupling. Often, these models are called *lumped parameter* models and also (with a little abuse of notation) 0D models.

In this way, large parts of the circulation (if not all) can be modeled. The level of detail can be varied according to the needs of problem. For instance, if the objective is the study of the regulatory mechanism in the circle of Willis and its interaction with global circulation, we would adopt a more detailed description of the former, while we might describe the latter with just a few compartments.

A useful way of representing lumped parameter models of the circulation is based on the analogy with electric networks. In this analogy the flow rate is represented by the electric current and the pressure by the voltage. The equations coupling the different compartments are given by the *Kirchhoff balance laws*, which assert the continuity of mass and pressure. The effects on blood dynamics due to the vascular compliance is here represented by means of capacitances. Similarly, inductances and

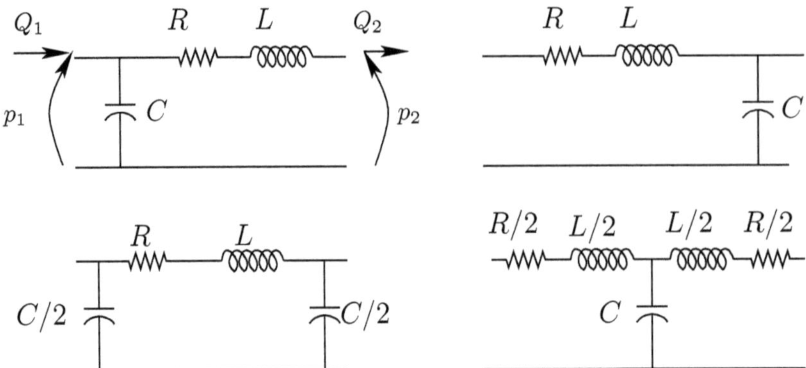

Fig. 8. Four possible lumped parameter representations of a compliant vessel in terms of electrical circuits. The four cases differ in the state variables and the upstream/downstream data to be prescribed

resistances represent the inertial terms and the effect of blood viscosity, respectively (see, e.g., [17]).

By exploiting the same analogy, it is also possible to devise a lumped parameter representation of the heart. In particular, the electric analog of each ventricle is given in Fig. 9 where the presence of heart valves is taken into account by *diodes* which allow the current to flow in one direction only. For more details about this model, see [17].

Figure 8 illustrates different electrical schemes that may be used to describe blood flow in a passive compartment. By coupling together these schemes and the model of the heart it is possible to derive a lumped parameter model of the whole circulatory system.

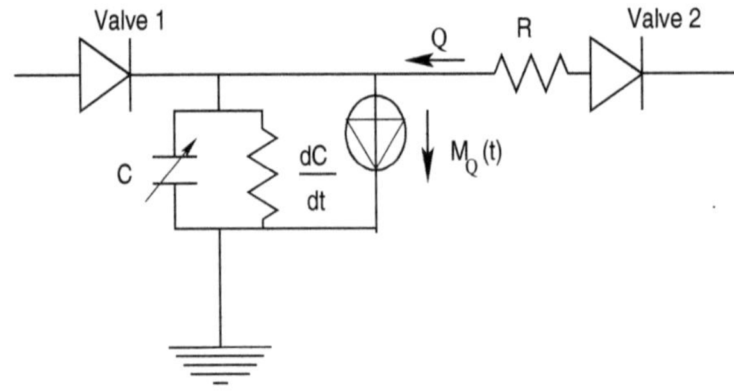

Fig. 9. Network for the lumped parameter modeling of a ventricle

From the mathematical viewpoint, a general representation of a lumped parameter model is a Differential Algebraic Equation (DAE) system in the form:

$$\begin{cases} \dfrac{d\mathbf{y}}{dt} = B(\mathbf{y}, \mathbf{z}, t), & t \in (0, T], \\ G(\mathbf{y}, \mathbf{z}) = 0, \end{cases}$$

supplemented with the *initial condition vector* $\mathbf{y}|_{t=t_0} = \mathbf{y}_0$. Here, $\mathbf{y}$ is the vector of *state variables*, associated to the flux in an inductance and the pressure in a capacitance, while $\mathbf{z}$ are the variables of the network which do not appear under time derivatives. Lastly, G represents the set of algebraic equations that come from the Kirchhoff laws. If we assume that the Jacobian matrix $\partial G/\partial \mathbf{z}$ is non-singular, by the implicit function theorem we can express $\mathbf{z}$ as a function of $\mathbf{y}$ and, with additional algebraic manipulations, resort to the following reduced non-linear Cauchy problem

$$\frac{d\mathbf{y}}{dt} = \boldsymbol{\Phi}(\mathbf{y}, t) = A(\mathbf{y}, t)\mathbf{y} + \mathbf{r}(t), \quad t \in (0, T],$$
$$\mathbf{y} = \mathbf{y}_0 \quad \text{at } t = t_0.$$

An example of a systemic model. The arterial system can be considered as a transmission line where the pressure wave generated by the heart propagates to the periphery. The propagation (velocity, reflections, etc.) clearly depends on the line characteristics. This is a rather schematic picture of the basic mechanism of "pulse wave" propagation which can aid our understanding of why a peripheral occlusion or an endovascular prosthesis could induce, for instance, an overload on the heart. In fact, an occlusion induces a wave reflection that might back-propagate along the transmission line and reach the heart.

A similar effect may be induced by a vascular prosthesis. Indeed, the replacement of part of a diseased artery by a prosthesis corresponds to replacing a portion of the transmission line by one with different physical characteristics. This introduces a discontinuity that stimulates reflections back-propagating to the heart. One-dimensional hyperbolic models of the type described here are very well suited to describe these propagation phenomena. In Fig. 10 we reproduce snapshots of the numerical solution obtained by simulating with 1D models the implant of a prosthesis at the abdominal bifurcation to cure an aneurysm. The pictures in the top row represent the case of an endo-prosthesis made with material softer that the vascular tissue. In the bottom, we illustrate the case where the prosthesis is stiffer. The presence of a strong back-reflection in the latter case is evident. When the reflected wave reaches the heart it may induce a pressure overload. These results may guide the design of better prostheses.

A more complete 1D network, such as that including the largest 55 arteries shown in Fig. 11 (left), may be adopted for a more thorough numerical investigation of the systemic dynamics. Since "left ventricle and arterial circulation represent two mechanical units that are joined together to form a coupled biological system" [37, Chap. 13], we need to couple the 1D model with a model of the heart (or at least of

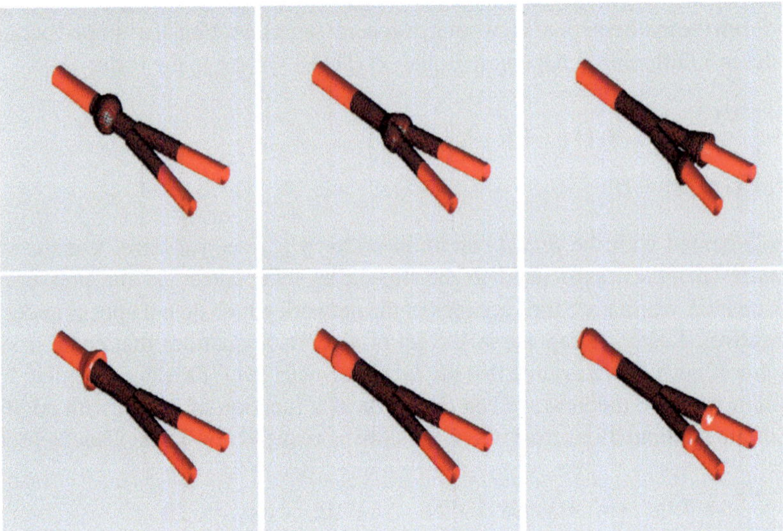

Fig. 10. Snapshots of the simulation of a vascular bifurcation with a prosthesis, carried out with a 1D model. The three pictures in the top row illustrate the case of a prosthesis softer than the arterial wall. The most relevant reflection is at the distal interface between the prosthesis and the vessels (*right*). In the bottom row the results are obtained using the same boundary data but with a prosthesis stiffer than the vascular wall. The most relevant reflection is at the proximal interface between the vessel and the prosthesis (*left*) and it back-propagates to the heart (simulations carried out by D. Lamponi)

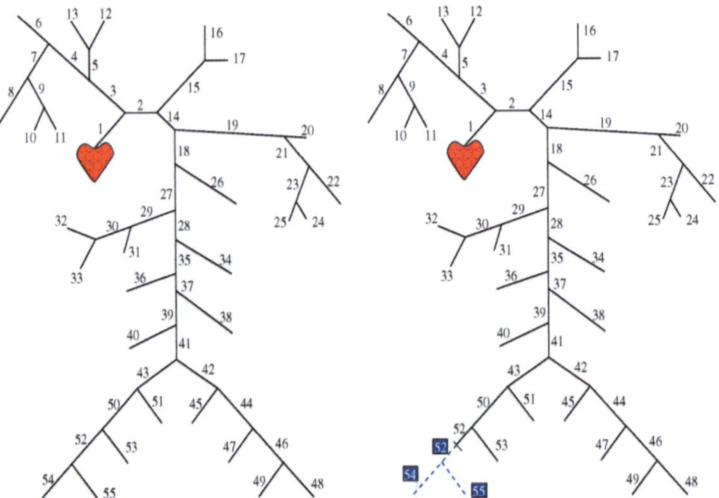

Fig. 11. Arterial tree composed of a set of 55 straight vessels, described by 1D models (see [61]). *On the right*: A pathological case, in which some of the vessels are assumed to be completely occluded

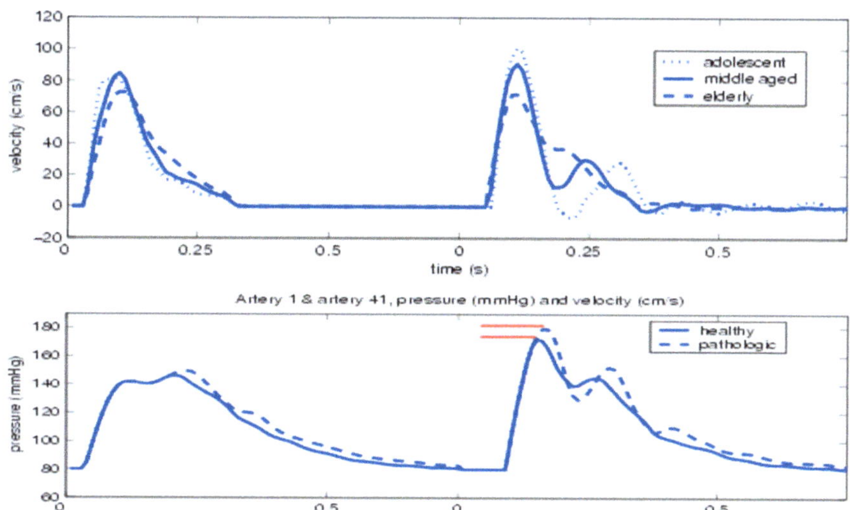

Fig. 12. *Top*: Time history of blood velocity in the thoracic aorta for different aged individuals. *Bottom*: Time history of the pressure in the ascending aorta for a healthy individual (*solid line*) and one suffering a complete occlusion of the right femoral artery (*dashed*)

the left ventricle), e.g., a lumped parameter model. In the numerical results presented here, the opening of the aortic valve is driven by the difference between the ventricular and the aortic pressures, P_v and P_a, while the closing is governed by the flux. Peripheral circulation in smaller arteries and capillaries may also be accounted for by lumped parameter models of the kind represented in Fig. 8.

Figure 12 (top) illustrates the behavior of the arterial pressure and flow in arteries in subjects of different age. Ageing is indirectly simulated by changing the physical characteristics of the arterial walls in the 1D model. More precisely, the stiffness of the arterial walls has been increased with age, in accordance with clinical evidence. The effects are evident.

Finally, Fig. 12 (bottom) compares the results of a physiological and a pathological case. More precisely, we report the different behaviour of the pressure in the abdominal aorta when a femoral artery is occluded, for instance, by a thrombus.

Mathematical description of cardiovascular self-regulation. So far, we have implicitly assumed that the parameters that govern our model, such as resistances and compliances, are given values, obtained from measurements or by other means. This is not true, as is well-known from daily experience: the duration of the heart beat is different at rest or after a long run! The circulatory system is extremely robust, in the sense that it ensures the correct blood supply to organs and tissues in very diverse situations.

This is possible thanks to self-regulating mechanisms. One such mechanism ensures that the arterial pressure is maintained within a physiological range (about 90–100 mm Hg). Indeed, if pressure falls below this range, the oxygenation of the

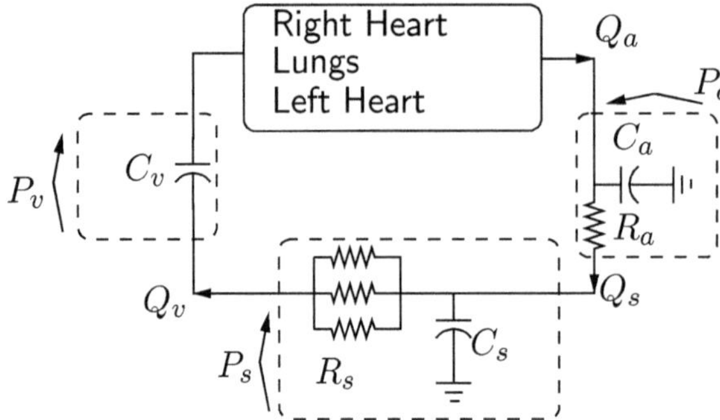

Fig. 13. A four-compartment description of the vascular system with self-regulating controls

peripheral tissues would be gravely reduced; on the other hand, a high arterial pressure would induce vascular diseases and heart overload. This regulatory system is called the *baroreflex effect* and is described, e.g., in [25] and [27]. The elements of the feedback baroreceptor loop are: (i) the sets of *baroreceptors* located in the carotid arteries and the aortic arc (they transmit impulses to the brain at a rate increasing with the arterial pressure); (ii) the *parasympathetic nervous system*, which is excited by the activity of baroreceptors and can slow the heart rate down; (iii) the *sympathetic nervous system*, which is inhibited by the baroreceptors and can increase the heart rate (it also controls venous pressure and the systemic resistance). Another ingredient of the self-regulating capabilities of the arterial system is the so-called *chemoreflex effect*, a mechanism able to induce capillary dilation and opening when an increment of oxygen supply is required by the organs, e.g., during heavy exercise.

The models presented so far do not include these feedback mechanisms. Moving from the simple four-compartment scheme depicted in Fig. 13 (comprising heart and lungs, an arterial compartment, a systemic compartment and a venous compartment) a possible model including the baroreflex and chemoreflex effects is described in the following paragraphs (more details are found in [10]).

The model is formed by different systems, mutually interacting.

1. *Haemodynamics:*

$$\begin{cases} C_{\mathrm{a}}\dfrac{dP_{\mathrm{a}}}{dt} = Q_{\mathrm{a}} - \dfrac{P_{\mathrm{a}} - P_{\mathrm{s}}}{R_{\mathrm{a}}} \\ C_{\mathrm{s}}\dfrac{dP_{\mathrm{s}}}{dt} = \dfrac{P_{\mathrm{a}} - P_{\mathrm{s}}}{R_{\mathrm{a}}} - \dfrac{P_{\mathrm{s}} - P_{\mathrm{v}}}{R_{\mathrm{s}}} \\ C_{\mathrm{v}}\dfrac{dP_{\mathrm{v}}}{dt} = Q_{\mathrm{a}} - C_{\mathrm{a}}\dfrac{dP_{\mathrm{a}}}{dt} - C_{\mathrm{s}}\dfrac{dP_{\mathrm{s}}}{dt} \\ Q_{\mathrm{a}} = \dfrac{1}{T}\left(\Delta V(P_{\mathrm{v}}) - \dfrac{P_{\mathrm{a}}}{E}\right), \end{cases}$$

where C_{a}, C_{s} and C_{v} are the arterial, systemic and venous compliances, respectively, while R_{a}, R_{s} and R_{v} are the corresponding resistances. In particular, R_{s} is given by the contribution of the skeletal muscle (sm), the splanchic compartment (sp) and other organs (o) so that $R_s = \left(R_{sm}^{-1} + R_{sp}^{-1} + R_o^{-1}\right)^{-1}$. Finally, T is the heart beat duration, E the cardiac elastance and ΔV the ventricular volume variation during the heart beat, which is a function of the venous pressure.

2. *Baroreceptor control:*

$$\begin{cases} \dfrac{dT}{dt} = f_T(T, f_B(P_a)) \\ \dfrac{dE}{dt} = f_E(E, f_B(P_a)), \end{cases}$$

where f_T, f_E and f_B are suitably defined non-linear functions. In particular f_B describes the action of the baroreceptor loop.

3. *Chemoreflex control:*

$$\begin{cases} \dfrac{d\tilde{R}_i}{dt} = f_i(\tilde{R}_i, f_B(P_a)) \quad i \in (\text{sm}, \text{p}, \text{o}) \\ \dfrac{d\tilde{x}}{dt} = f_{x_i}(x_i, f_C(c_{\text{oxy},i})) \\ R_{sm} = \dfrac{\tilde{R}_{sm}}{1 + x_{sm}}, \quad R_{sp} = \tilde{R}_{sp}(1 + x_{sp}), \quad R_o = \dfrac{\tilde{R}_o}{1 + x_o}, \end{cases}$$

where $c_{\text{oxy},i}$ is the oxygen concentration in compartment i (i = sm, sp, o).

4. *Metabolism:* The oxygen concentration is in turn given by a model for the tissue metabolism, possibly including the reaction with other chemicals. If $\mathbf{c}_i$ denotes the vector of chemical concentrations in compartment i, a general formulation of this model reads

$$\frac{d\mathbf{c}_i}{dt} = \mathbf{A}_i\mathbf{c}_i + Q_i(\mathbf{c}_a - \mathbf{c}_i),$$

where $\mathbf{A}_i$ is the *stochiometric matrix* describing the interactions among chemicals in the compartment, and the second term on the right-hand side is a *transport term*.

An important concern in devising such a model is the identification and tuning of all the parameters appearing in the equations [10]. Often, they have to be inferred by indirect measurements and observations.

Despite of its apparent simplicity, this model is able to simulate different realistic situations, which are the virtual counterparts of actual protocols in sport medicine and physiology. For instance, in Fig. 14 we illustrate the time evolution of the haemodynamic variables during an incremental exercise with a linear increasing workload and a total duration of 10 minutes. The extensive validation of such models is the subject of current research.

2.3 The design of drug-eluting stents

The problem

The treatment of coronary pathologies in an advanced stage or the cure of stenosis-caused atherosclerotic plaques may be carried out by the implant of a *stent*. The stent

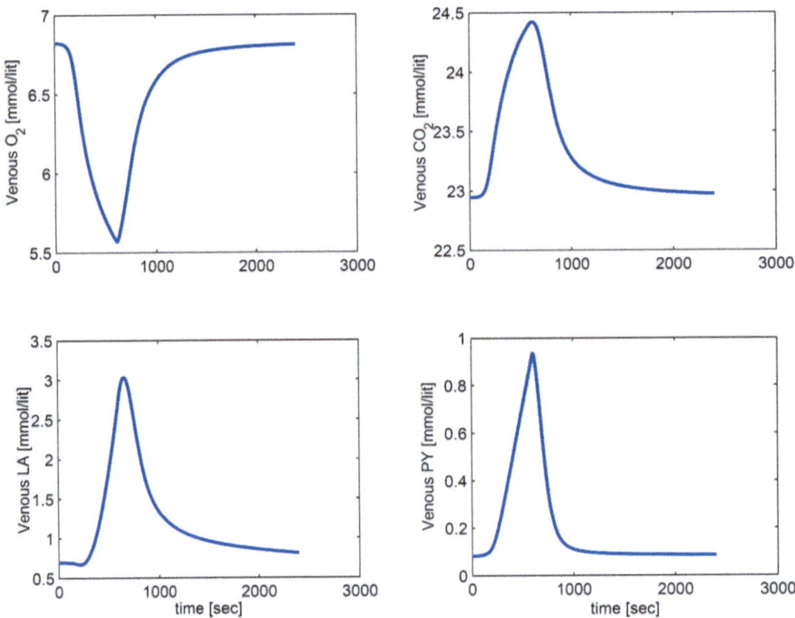

Fig. 14. Time history of haemodynamics variables in an individual under an incremental 10 minute exercise with a linearly increasing workload (courtesy of C. D'Angelo)

is a microstructure, which is formed by interwoven and appropriately shaped metal filaments. It is driven inside the arterial system until it is near the artherosclerotic plaque. Then it is expanded to bring the arterial lumen to its original diameter and restore an adequate blood flow. Generally, these medical devices are left permanently in the implantation. The implant of a stent is a less invasive procedure than bypass surgery and therefore its adoption by vascular surgeons is increasing. Data extracted from the American Heart Association's *Heart Diseases and Stroke Statistics 2004* confirm that, from 1979 to 2001, the number of stent implants in the United States has tripled, and about 1,208,000 such operations were performed just in 2001.

Cardiovascular stents have to meet many requirements, which are at times in conflict with each other. For example, they must be extremely flexible along their longitudinal axis in order to be able to find their way through contorted arteries and reduced diameters. They must be adequately visible with X-ray techniques, since the implant is guided from the outside. When they are driven through the arterial system they are in a compressed state, and their radial dimension is minimal. Yet they must expand easily to their original size once the final position is reached. Furthermore, they should have enough stiffness to maintain the final expanded shape under the mechanical strain exercised by the atherosclerotic plaque and the vessel wall. Last, but not least, they must be biocompatible to minimize thrombogenesis.

The study of the impact of a stent implant on the blood flow, both locally and in the whole cardiovascular system, is an extremely complex problem and mathematical models can be of assistance. As we have pointed put in the previous section, an implant that increases the rigidity of the vascular wall reflects back the part of the energy of the flow and in some cases could generate an increase in the peak pressure in the proximal region and possibly an overload of the heart. However, in the case of a stent there is a second, perhaps more important, source of disturbance, namely the interaction with the cells of the vessel wall in the contact region. Metals like iron and nickel, which are used to manufacture certain families of stents, can interact with the cells of the endothelium, the tunica intima and media, causing an inflammatory reaction which can lead to the uncontrolled proliferation of smooth muscle cells, yielding a narrowing of the vessel lumen. To counteract this, biomedical researchers have developed stents coated with a microlayer of material designed to store an anti-inflammatory drug which is slowly released into the vessel wall tissue. Here, the most relevant points are the choice of the drug and the design of a suitable matrix that can store and release the anti-inflammatory agent at the correct rate. The latter point calls for the development of new nanostructure materials and technologies to store the drug into them.

A numerical simulation of the release of drug into the arterial tissue requires the development of pharmacokinetic models, to be coupled with transport-diffusion equations. Numerical computations will enable us to test several design configurations of the stent, and help us to select the most appropriate. For more details, see [65].

This research is currently developed in cooperation with the Laboratory for Biological Structure Mechanics, Politecnico di Milano, the Department of Structural, Environmental and Biological Chemistry, University of Bologna, the CMCS-EPFL, Lausanne and the Service de Chirurgie Cardio-Vasculaire, Centre Hospitalier Universitaire Vaudois (CHUV) in Lausanne.

Numerical models and simulations

Following [65], we assume that the tissues constituting the arterial walls, as well as the stent coating, behave as porous media with respect to the filtration of plasma and the transfer of molecules [19, 50]. For simplicity, we consider here only the interaction of the stent with the media, which is the thickest tissue layer constituting the arterial wall, and we assume that the stent is completely embedded into the wall (see Fig. 15). The domain of the problem is therefore given by two subdomains, the media and the stent coating, where the filtration of plasma and the diffusion, transport and chemical binding of the drug have to be modelled.

Models and methods. We denote by subscript 1 the quantities related to the media and by subscript 2 the quantities related to the coating of the stent. The portion of the wall is therefore denoted by Ω_1, while Ω_2 represents the portion of the stent under consideration, as shown in Fig. 15. The interface Γ between Ω_1 and Ω_2 can be seen as a boundary for the governing equations on Ω_1 and Ω_2, respectively. Further, let

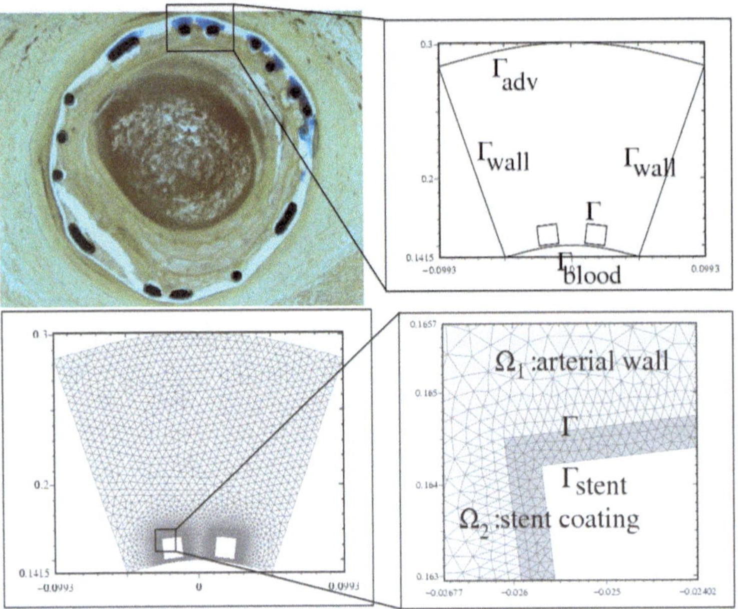

Fig. 15. Detail of a stented artery, with the representation of the domain used for the numerical simulation of the stent design problem, and the associated computational mesh (courtesy of P. Zunino)

Γ_{blood} be the boundary separating the arterial wall from the arterial lumen and Γ_{adv} that separating the wall from the outer tissues, the latter corresponding to the surface of the adventitia. We denote by $\boldsymbol{u}_1$ the volume-averaged filtration velocity of the plasma in the media. The velocity $\boldsymbol{u}_1$ is governed by Darcy's equation:

$$\begin{cases} \boldsymbol{u}_1 = -\dfrac{K_1}{\mu_1}\nabla p_1 \text{ with } \quad \nabla \cdot \boldsymbol{u}_1 = 0 \text{ in } \Omega_1, \\ p_1 = p_{blood} \text{ on } \Gamma_{blood}, \quad p_1 = p_{adv} \text{ on } \Gamma_{adv}, \quad \boldsymbol{u}_1 \cdot \mathbf{n}_1 = 0 \text{ on } \Gamma \cup \Gamma_{wall}, \end{cases} \tag{7}$$

where $\mathbf{n}_1$ and $\mathbf{n}_2$ are the outward normal vectors with respect to $\partial\Omega_1$ and $\partial\Omega_2$, respectively. Moreover, in Eq. (7), K_1 is the Darcy permeability of the media, while μ_1 is the viscosity of plasma.

As for the chemical dynamics, we are interested in the volume-averaged concentration in each domain, c_i $(i = 1, 2)$, including the amount of drug present in the fluid and that bound with the tissue. These concentrations satisfy the equations:

$$\begin{cases} \dfrac{\partial c_1}{\partial t} + \nabla \cdot (-D_1 \nabla c_1) + \dfrac{\gamma \boldsymbol{u}_1 c_1}{k_1 \epsilon_1} = 0 \qquad \text{in } \Omega_1, \\ c_1 = 0 \text{ on } \Gamma_{blood}, \\ -D_1 \nabla c_1 \cdot \mathbf{n}_1 = 0 \text{ on } \Gamma_{adv} \cup \Gamma_{wall} \end{cases} \tag{8a}$$

and

$$\begin{cases} \dfrac{\partial c_2}{\partial t} + \nabla \cdot (-D_2 \nabla c_2) = 0 & \text{in } \Omega_2 \\ -D_2 \nabla c_2 \cdot \mathbf{n}_2 = 0 \text{ on } \Gamma_{\text{stent}} \end{cases} \tag{8b}$$

Here, the parameters D_1 and D_2 represent the diffusivities of the drug in the media and in the stent coating, respectively, k_i is the so-called partition coefficient of the drug, and ϵ_i is the porosity of the media. The coefficient γ accounts for possible frictional effects between the molecules of drug and the pores of the arterial walls. The boundary condition on Γ_{blood} reflects the assumption that the concentration of the drug in the blood is negligible, while that on Γ_{adv} states that the diffusive flux on this boundary is vanishing. The same condition is applied to Γ_{wall} by virtue of the axial symmetry of the domain Ω_1.

Equations (8a) and (8b) should be paired with a suitable set of matching conditions. This is a delicate issue: in fact, the physical properties of the arterial walls with respect to mass transfer are very different from those of the stent coating, where the drug is initially stored, and this can induce strong variations in drug concentration when passing from one medium to the other. The region at the interface is a permeable membrane that accounts for possible jumps of concentration on its sides. Matching conditions can be derived from the so-called Kedem-Katchalsky model (see [26]), which enforces the mass conservation across the subdomain interface:

$$\begin{cases} -D_1 \nabla c_1 \cdot \mathbf{n}_1 = D_2 \nabla c_2 \cdot \mathbf{n}_2 \\ -D_1 \nabla c_1 \cdot \mathbf{n}_1 = \sigma(c_1/k_1\epsilon_1 - c_2/k_2\epsilon_2) \end{cases} \quad \text{on } \Gamma,$$

where σ is the membrane permeability.

In this model the drug dynamics does not affect the plasma flow, so we can solve (7) independently of the concentration problem, yielding a simplification of the numerical scheme as well. For a discussion and analysis of the previous model, we refer to [48] and [47] where finite elements were used for the space discretization and an implicit Euler scheme for the time discretization. The application of this technique yields at every time step a system of linear algebraic equations featuring a block structure that reflects the multidomain nature of the problem. The strongly heterogeneous nature of the two subdomains is reflected in unfavorable conditioning properties of such a matrix, so that special techniques have to be adopted for its numerical solution. A possible approach is to resort to the domain decomposition method applied to the multidomain structure of the problem (see [45]). In [47] this technique was used to precondition a GMRes iterative method, obtaining optimal convergence properties. In particular, it is shown that the preconditioner has the same spectral properties as those the matrix governing the problem.

Results. The drug release by the stent coating is influenced by many factors, namely, the shape of the fibers and the coating, the properties of the drug and those of the arterial wall. Numerical simulations enable the evaluation of the behavior of various configurations highlighting the most effective technological solutions (see

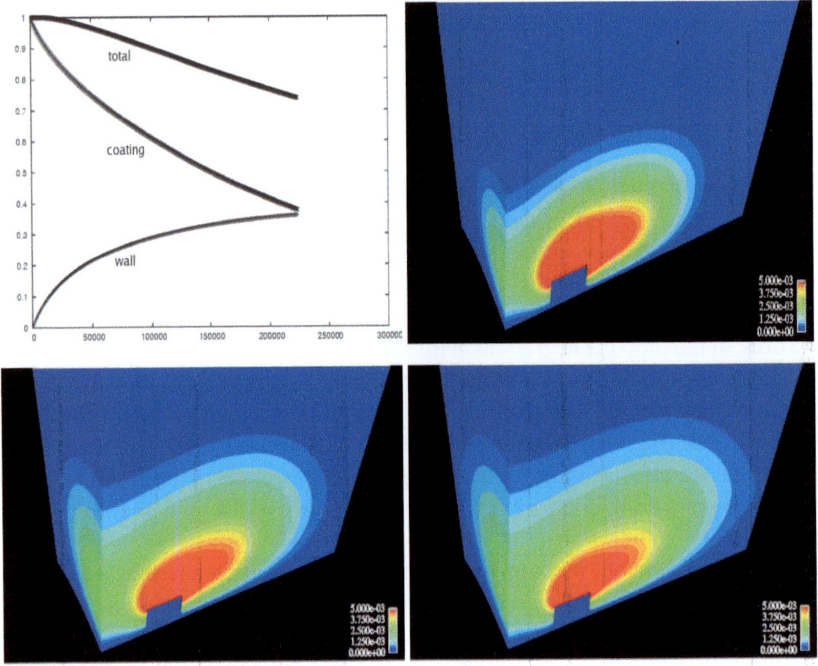

Fig. 16. Programmable stent simulation. *Top left*: Drug (heparin) concentration evolution in time in the coating and in the wall. *Top right*: Concentration iso-lines after one day. *Bottom left*: After two days. *Bottom right*: After three days (courtesy of M. Prosi)

[65]). In Fig. 16 we illustrate numerical results obtained from the simulation of a *programmable stent*. In particular, in the top-left picture we illustrate the evolution in time of the drug concentration (normalized with respect to the concentration at the initial instant) in the stent coating and in the vascular wall and as the sum of the two curves. Part of the drug is lost in the blood flow and this explains the reduction in time of the total amount of drug. In the other pictures we show the iso-lines of concentration after one (top-right), two (bottom-left) and three (bottom-right) days respectively.

The profile of the drug release rate is a key factor for the design of a drug-eluting stent. Numerical simulations of this kind may help to choose the most appropriate drug components or the best coating matrix characteristics. In particular, from Fig. 16 we conclude that the programmable stent considered here ensures a slow drug release over three days.

2.4 Pulmonary and systemic circulation in individuals with congenital heart defects

The problem

Some serious congenital heart anomalies feature a marked hypoplasia of the left heart, including the aorta, aortic valve, left ventricle and mitral valve. These pathologies are indicated with the name Hypoplastic Left Heart Syndrome (HLHS) and need to be treated surgically. A possible approach is based on a staged reconstruction involving three operative procedures (see [3, 38]). At the first stage, called the *Norwood procedure*, the main pulmonary trunk is attached to the augmented aorta to establish an unobstructed systemic circulation. An interposition graft called a *systemic-to-pulmonary artery shunt* (diameter 3, 3.5 or 4 mm) is placed so as to provide pulmonary perfusion and gas exchange. Two possible options for this stage are the Central Shunt (CS) and the right Modified Blalock-Taussig Shunt (MBTS). In the former, a bypass is placed directly between the aorta and the pulmonary artery, while in the MBTS the conduit is interposed between the innominate artery and the pulmonary artery (see Fig. 17). At the subsequent stages, pulmonary perfusion is achieved by connecting the superior vena cava (second stage) and the inferior vena cava (third stage) directly to the pulmonary artery. The systemic-to-pulmonary shunt is removed at the second stage. In some cases, this intervention has induced coronary insufficiency, for reasons that are not completely understood, and, in particular, the influence of shunt position and diameter on systemic haemodynamics is not clear.

Another surgical approach was recently proposed by Sano in [54] as a replacement of the Norwood procedure, and is referred to here as the Sano Operation (SO). It consists mainly in the connection of the systemic circulation to the pulmonary one by means of a synthetic vessel, whose diameter typically ranges from 4 to 6 mm, connecting the right ventricle to the pulmonary arteries (see Fig. 17). This alternative seems to have many potential advantages. In particular, diastolic coronary perfusion may be more stable. However, many questions are still open concerning, for instance, the optimal shunt size and shunt material, the growth and distortion of the pulmonary arteries, possible ventricular volume overload due to shunt backward flow, and potential risk of arrhythmia.

Mathematical models and numerical simulations can help us to understand the problem, provide quantitative data for comparing different options, and eventually support the decisions of the surgeon. A crucial breakthrough for the correct simulation of this problem has been the *geometrical multiscale approach*. This is motivated by the following considerations. On the one hand, we need an accurate simulation of the local haemodynamics in order to investigate the influence of the different possible shunt options and the possible presence of backward flow. On the other hand, it is crucial to analyze the mutual influence of the local haemodynamics on the systemic dynamics and, for instance, to assess to what extent coronary perfusion is affected by the presence of the shunt. As we pointed out in Sect. 1, numerical models of this type demand specifical techniques that we briefly illustrate in the next sections. For more details see [17, 18, 29]. Numerical results on the Norwood procedure and the Sano operation are taken from [28] and [33].

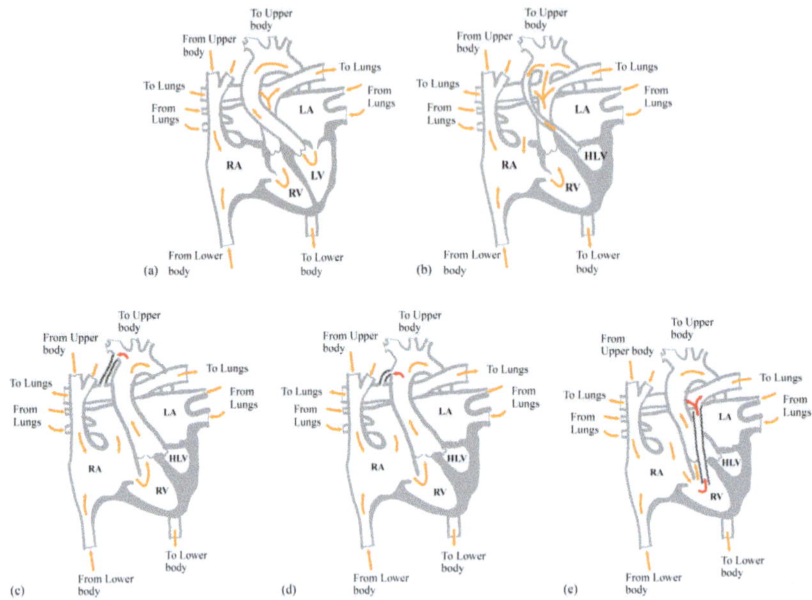

Fig. 17. *Top left*: Normal heart. *Top right*: Hypoplastic left ventricle. *Bottom left*: Modified Blalock-Taussig shunt. *Bottom middle*: Central shunt. *Bottom right*: Sano operation (courtesy of F. Migliavacca)

This research was developed in cooperation with the Laboratory for Biological Structure Mechanics, Politecnico di Milano with the partnership of the Cardiac Unit, Great Ormond Street Hospital for Chidren, London and the Section of Cardiac Surgery, University of Michigan Health System, Ann Arbor.

Numerical models and simulations

A possible approach to account for both local and systemic dynamics is to couple the Navier-Stokes equations in the domain of interest, the shunt and its neighborhood, with simplified models such as those introduced in Sect. 2.2 for the description of the remainder of the circulatory system. A diagram of the model obtained in this way in the case of the Sano operation shunt is shown in the left-hand picture of Fig. 19. At the mathematical level, this model implies a coupling between partial and ordinary differential equations. For their numerical solution, it is then natural to resort to an iterative approach based on the splitting of the whole problem into its basic components. A schematic representation of a possible numerical approach is given in Fig. 18. An explicit scheme is used for the lumped parameter model to advance time from level t^n to t^{n+1}. The computed pressures on the interface are then imposed as boundary conditions for solving the Navier-Stokes problem, advanced in time by an implicit scheme.

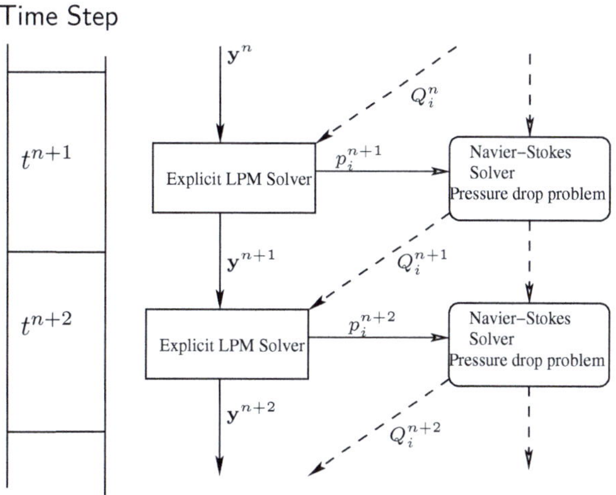

Fig. 18. Possible numerical scheme for the coupling of a Lumped Parameter Model (LPM) and the Navier-Stokes problem

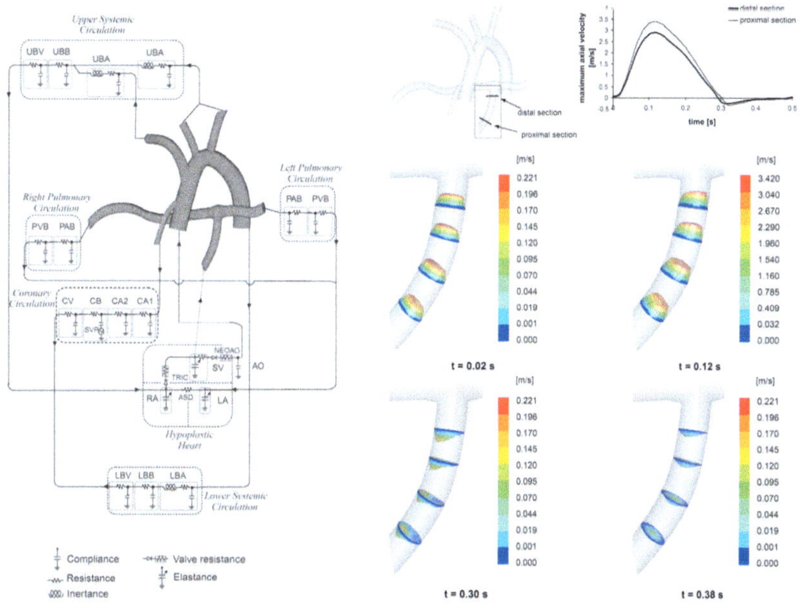

Fig. 19. *Left*: Geometrical multiscale model of the Sano operation which couples a detailed model of the shunt and a lumped parameter representation of the circulation. *Right*: Flow profiles in the shunt of the Sano operation (courtesy of F. Migliavacca)

We may observe that the pressures provided by the lumped parameter models at the interface are in fact average quantities. Therefore, the Navier-Stokes subproblem requires the solution of Eqs. (1) with the interface boundary conditions

$$\frac{1}{\text{meas}(\Gamma_i)} \int_{\Gamma_i} p\, ds = P_i(t), \quad i = 1, 2, \ldots, m\,, \tag{9}$$

where m is the number of interfaces between the local and systemic subproblems. In Fig. 19, for instance, we have $m = 8$. This initial-boundary value problem is not complete, since the Navier-Stokes problem would require pointwise boundary data. This mismatch can be overcome by completing the defective data given by the systemic submodel. One task of the geometrical multiscale approach is to minimize the numerical artifacts introduced by this completion.

In [24] a particular *weak or variational formulation of the boundary problem* has been devised which allows us to satisfy conditions (9), giving rise to a well-posed problem. In fact, this formulation implicitly forces natural (Neumann-like) boundary conditions which select one particular solution among all the possible ones. A well posedness analysis of the multiscale model obtained in this way can be found in [46]. Depending on the specific choice of the hydraulic network representing the circulatory system in the lumped model, different "defective" boundary conditions could be prescribed for the Navier-Stokes problem on the interfaces. In particular, we may have *flow rate boundary conditions*, corresponding to imposing the conditions

$$\int_{\Gamma_i} \boldsymbol{u}(t) \cdot \mathbf{n}\, ds = Q_i(t) \text{ for } i = 0, \ldots, n.$$

In principle, for these conditions we may also find a suitable variational formulation that ensures the existence of a solution, but it requires the use of non-standard functional spaces which make the finite elements discretization problematic. For this reason, a reformulation of the problem was proposed in [13] which is more suited for the numerical approximation.

Results

Extensive numerical simulations and comparisons with available clinical data were carried out in [28, 33].

Clinical evidence and numerical results agree in showing that the cardiac output is higher in the Norwood procedure (with both CS and MBTS approaches) than with the SO when the size of the shunt is the same (see Fig. 20 (top-left)). It is therefore worth comparing the different techniques for a similar value of the cardiac output, e.g., MBTS or CS with a 3 mm shunt versus SO with a 4 mm shunt, and so on. By doing so the numerical results show that SO features a lower pulmonary flow and higher coronary perfusion and pressure with respect to the corresponding Norwood procedures (see Fig. 20 (top-right, bottom-left and bottom-right)). This is consistent with clinical evidence. Also the minimal, clinically irrelevant, presence of backward fluxes in the SO shunt highlighted in Fig. 19 is in agreement with the available data.

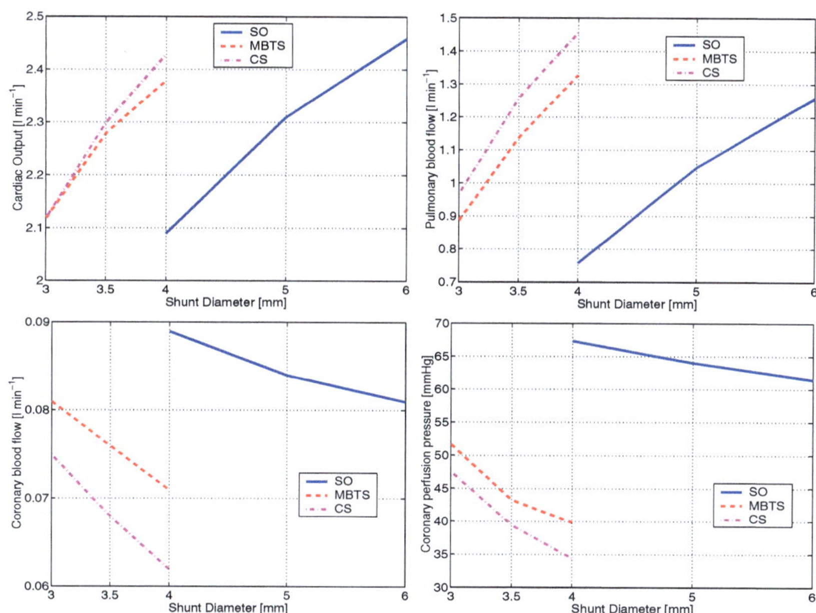

Fig. 20. Comparison among MBTS, CS, and SO with shunt sizes ensuring comparable cardiac output (*top left*): pulmonary blood flow (*top right*), coronary flow (*bottom left*) and pressure (*bottom right*)

It is worth pointing out that this favorable agreement between numerical and clinical results was only obtained thanks to the development of the multiscale techniques. To quote [33]: "the use of simpler, stand-alone 3-D or lumped parameter models would not yield results as meaningful as those obtained here. Indeed, the adopted approach allows one to evaluate quantitatively the post-operative situation, thus suggesting its use as a tool for preoperative planning".

2.5 Peritoneal dialysis optimization

The problem

Chronic Kidney Disease (CKD) affects approximately 17 million Americans and about 400,000 of them are either on dialysis or require kidney transplant ([2]). To support End Stage Renal Disease (ESRD) patients amounts to an approximate cost of 13.82 billion dollars annually. Peritoneal Dialysis (PD) occupies a well established place among the therapeutic options for patients with ESRD [35, 39]. With this technique, blood purification is obtained by the exchange of chemicals between blood and a solution injected in the peritoneal cavity. The solution is periodically replaced by injections or extractions from the patient, through an external pump. The exchange of chemicals takes place across the net of capillaries permeating the peritoneum (see Fig. 21).

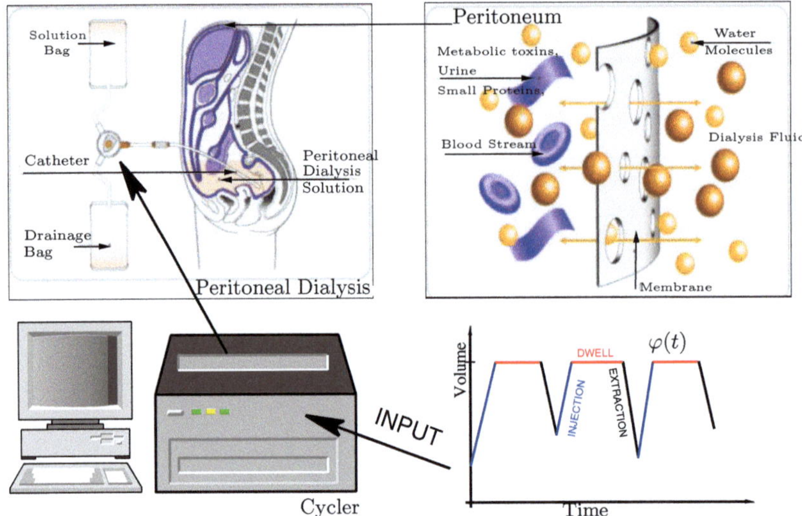

Fig. 21. Schematic representation of the peritoneal dyalisis process: the cycler drives the injection-dwell-extraction profile φ of the solution injected into the peritoneal cavity

However, this technique sometimes fails, essentially because of alterations in the peritoneal membrane transport characteristics leading to an inefficient small solute and/or water removal (see [21]). The effectiveness of the therapy is directly related to the dynamics of the injection/extraction of the solution, as well as to the individual characteristics of the patient. However, *standard therapies*, i.e., injection/extraction profiles, such as CAPD (Continuous Ambulatory Peritoneal Dialysis), CCPD (Continuous Cycling Peritoneal Dialysis), TPD (Tidal Peritoneal Dialysis), and NPD (Nocturnal Peritoneal Dialysis), are often adopted with only an incomplete characterization of the patient. The Dynamic Peritoneal Dialysis (DPD) which we describe represents an alternative that customizes the therapy to a specific patient by making it more efficient and/or more biocompatible. The main difficulty in the development of APD is the complexity of its prescriptions and set-up. Mathematical models allow us to optimize PD therapies and facilitate the use of more elaborated therapeutic options. The final goal is to develop a procedure able to find for each patient the injection/extraction patterns that ensure the best blood purification and water removal. These tools are based on classical models proposed in the literature (see, e.g., [60] and [51]), derived from equations describing exchanges of chemical species across a membrane separating two solutions with different concentrations. This mathematical framework has been validated and tuned for different patient categories, in particular, for different characteristics of the peritoneal membrane [66].

This simulation environment is the result of a fruitful multi-disciplinary collaboration between a med-tech company (Debiotech S.A., Lausanne) and clinical partners (Inselspital, Bern, University Hospital, Ghent and Azienda Sanitaria Ospedaliera Molinette, Turin).

Numerical models and simulations

The model. During PD therapy, the exchange of chemicals takes place through the net of capillaries within the folded peritoneal membrane. The geometrical modeling of the domain to account for spatial variations would be extremely difficult and computationally expensive. Moreover, the exchanges are very rapid, due to the high concentration difference between the two sides of the membrane. Furthermore, we are mainly interested in the variation of global quantities and not in the local details. Consequently, a *lumped parameter model* which describes the kinetics of chemicals during the therapy looks most suitable to this specific study. Two compartments are considered, one accounting for the body (denoted by the index b), and one for the peritoneal cavity of the patient (denoted by d). The latter compartment is filled by a solution of N chemicals, denoted by the indexes $i = 1, 2, \dots, N$. Apart from the boundary layer near the membrane, which is accounted for by the interface condition, the concentrations may be assumed to be uniform in each compartment. The physical quantities of interest are then the *volume of the solution* and the *total amount of each solute* in the two compartments, namely V_{b}, V_{d}, $V_{\mathrm{b}}c_{\mathrm{b},i}$, $V_{\mathrm{d}}c_{\mathrm{b},i}$, where $c_{\mathrm{b},i}$, $c_{\mathrm{d},i}$ for $i = 1, 2, \dots, N$ are the concentrations (mass of solute per volume of solution). The interaction between the two compartments is governed by the equations prescribing the flux of solvent J_{v} and of each solute $J_{\mathrm{s},i}$ across the membrane. In the Kedem-Katchalsky model (see [26]) introduced in Sect. 2.3, the membrane is characterized by a set of pores that allow the exchange of the solvent and the solutes between the two compartments. The pores are subdivided in different classes, denoted by the index $j = 1, \dots, M$, depending on their size. We denote by $L_{\mathrm{p}}S_j$ and $PS_{i,j}$, respectively, the hydraulic conductivity and permeability of the membrane relative to the ith molecule through the set of pores indexed by j. Moreover $\sigma_{i,j}$ denotes the *reflection coefficient* of the solute i with respect to the pore of class j. From the *Starling law of filtration* [27] we have that

$$J_{\mathrm{v},j} = L_{\mathrm{p}}S_j \left(\Delta p - \sum_{i=1,\dots,N} \sigma_{i,j} \Delta\pi_i \right),$$

where Δp and $\Delta\pi_i$ $(i = 1, \dots, N)$ are the static and osmotic pressure differences between the two compartments respectively. In particular, $\Delta\pi_i$ depends on the solute concentration on the two sides of the membrane, according to the *Van't Hoff law*

$$\Delta\pi_i = RT \left(c_{\mathrm{b},i} - c_{\mathrm{d},i} \right),$$

where R is the gas constant and T is the absolute temperature. The volumes of solute in the two compartments are therefore governed by the following system of ordinary differential equations:

$$\frac{dV_{\mathrm{d}}}{dt} = Q + \sum_{j=1}^{M} J_{\mathrm{v},j} + J_{\mathrm{v,lymph}}, \qquad \frac{dV_{\mathrm{b}}}{dt} = -\sum_{j=1}^{M} J_{\mathrm{v},j} + J_{\mathrm{v,lymph}} \tag{10}$$

where $J_{\mathrm{v,lymph}}$ is the flux of the lymphatic liquid and Q is the injection-extraction profile executed by the pump (corresponding to the time derivative of the time-volume curve depicted in Fig. 21). The condition $Q(t) > 0$ corresponds to the *injection phase* while $Q(t) < 0$ corresponds to the *extraction phase*. Lastly, $Q(t) = 0$ corresponds to the *dwell phase*, when the liquid is left inside the peritoneal cavity. From now on we set $\varphi(t) = \int_{t_0}^{t} Q(\tau)d\tau$. In practice, $\varphi(t)$ is characterized by a suitably small number of parameters.

The solute flux $J_{s,i}$ (see [9, 22]) is the sum of a diffusive term, depending on the jump of concentration across the membrane, and a transport term, defined as the product of effective solvent flux and the average concentration within the membrane,

$$J_{s,i} = \sum_{j=1,\ldots,M} \left[PS_{i,j}\left(c_{b,i} - c_{d,i}\right) + J_{v,j}(1-\sigma_{i,j})\left(w_i c_{b,i} + (1-w_i)c_{d,i}\right)\right]$$

so that

$$\begin{cases} \dfrac{d(V_b c_{b,i})}{dt} = g_i - J_{s,i} \\ \dfrac{d(V_d c_{d,i})}{dt} = J_{s,i} + Q c_{d,i}. \end{cases} \tag{11}$$

Here g_i represents the metabolic contribution to the ith solute. Equations (10) and (11) provide a system of $2N+2$ non-linear equations that describe the rate of change of the unknowns V_b, V_d, $V_b c_{b,i}$, and $V_d c_{b,i}$.

It is worth pointing out that this model can be applied to a large number of chemical species, with very weak limitations. In particular, it takes into account the basic chemicals considered in dialysis, e.g., *urea, glucose* and *creatinine*. Furthermore, it can also be applied to *sodium*, in order to study its removal, or to *large polymers*, which are nowadays becoming an alternative to glucose for driving water out of patients with more severe kidney inefficiency.

Moreover, the model can account for different models of the peritoneal membrane, in particular, the *iso-pore model* ($M = 1$) and the *three-pore model* ($M = 3$), where medium-sized and large pores account for large molecule (e.g., proteins) dynamics, and ultra-small pores account for the exchange of water.

From the numerical viewpoint, we have to solve an ordinary differential system. This can be done in different ways (for a general introduction see [44]). However, because of the succession of injection-dwell-extraction phases, the dynamics of the process is more critical during the periods of injection and extraction, requiring in these phases a more accurate time discretization than during the dwell phase. For this reason, *adaptive time discretization methods* have been studied to balance the need for accuracy and low computational costs.

Results. In order to determine the efficiency of a therapy, clinicians mostly focus on two molecules, urea and creatinine. and on the net amount of fluid extracted during a therapy, the so-called *ultra-filtration*. Consequently, an effective therapy is characterized by a suitable balance of the following indicators:

1. the normalized extracted urea over a week KT/V_{urea};
2. the creatinine clearance Cl_{creat};
3. the ultrafiltration UF.

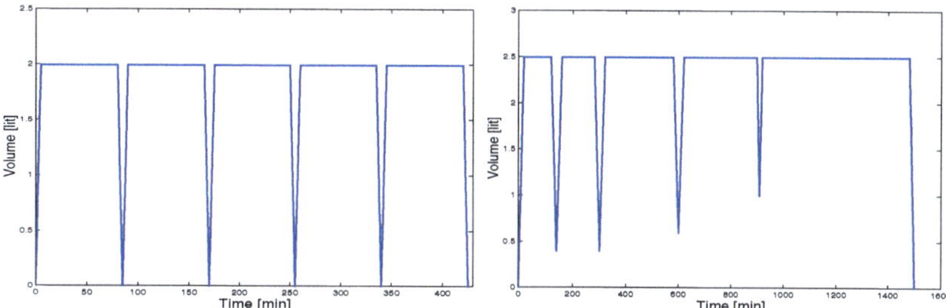

Fig. 22. Different possible patterns φ of the injection-dwell-extraction dynamics. *On the left*: The case of a standard therapy. *On the right*: A profile computed after optimization (courtesy of D. Mastalli)

We define the *efficacy parameter* η as a weighted combination of the previous indicators,

$$\eta = w_1 KT/V_{\text{urea}} + w_2 Cl_{\text{creat}} + w_3 UF,$$

where w_1, w_2 and w_3 are suitable weighting coefficients satisfying $w_1+w_2+w_3 = 1$. Additional factors, which can be taken to evaluate the adequacy of PD, such as the amount of glucose and sodium that are exchanged during therapy, can be introduced as well.

Each therapy features a particular injection/extraction pattern $\varphi(t)$. In Fig. 22 we illustrate two possible therapies. Typically, the following constraints on the injection-dwell-extraction pattern must be satisfied:

1. the total duration T_{tot} of the therapy must not be exceeded; a typical range is 4-10 hours;
2. the total amount of dialysate V_{tot} must be fully exploited; a realistic range is 4-16 liters;
3. the peritoneal cavity should be empty at the end of the therapy.

By means of the numerical simulations, the efficacy of a therapy can be computed for several values of the inputs. The data resulting from the numerical simulations are summarized in Fig. 23, which describes the trend of KT/V_{urea} and sodium removal for a specific patient. In particular, the picture shows that, as expected, an increase of the therapy duration or of the volume causes an increase of the KT/V_{urea} index, and that a therapy of just 5 hours cannot achieve a sodium removal greater than about 1.5g. These results on therapy performance on a specific patient basis can help us to set up an appropriate PD treatment. The final goal is the automatic determination of the injection-extraction pattern $\varphi(t)$ able to satisfy the given requirements for each patient, and its prescription by setting the cycler that executes the therapy. This new framework is called DPD (Dynamic Peritoneal Dialysis) and is made possible by the development of new, fully programmable, cycler pumps. Since DPD enjoys a larger number of degrees of freedom in specifying the pattern of φ with respect to the more classical CCPD or APD, its prescription has to be provided directly by a numerical optimization process. To this aim we need to set up a *multi-objective optimization*

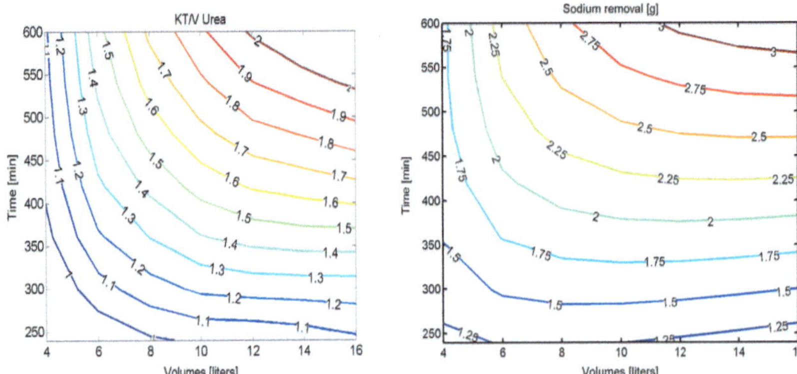

Fig. 23. Simulation of the peritoneal dialysis for a given patient: KT/V_{urea} (*left*) and sodium removal (*right*) as functions of the therapy time and the dialysate volume (courtesy of D. Mastalli)

strategy that allows us to take into account the several factors coming into play with different weights. This requires us to identify a global efficiency parameter η whose maximum within the therapy constraints would provide the optimal therapy.

Let $\eta = \eta(\varphi)$ describe the relationship between the efficiency and the input profile. Then a control problem can be formulated as:

$$\text{find } \varphi \text{ such that } \eta\left(\varphi_{\text{opt}}\right) = \max_{\varphi} \eta(\varphi)\,.$$

To solve this problem the optimization algorithm used is based on Pontryagin's maximum principle (see [62]), and it consists of an iterative process that starts at an initial guess φ_0, corresponding to a standard therapy, e.g., an APD. By solving the PD problem and a related problem, the *adjoint problem*, it is possible to find a sequence of iterates φ_n $(n = 1, 2, \ldots)$ that converges to the optimal one. The numerical algorithm terminates when the optimum has been approximated with sufficient accuracy. The algorithm requires us to run a numerical simulation of the PD problem for different values of φ at each iteration. For instance, by assuming $w_1 = 1$, $w_2 = w_3 = 0$ (that means that we optimize KT/V_{urea}) starting with a standard therapy profile (Fig. 22 (left)), KT/V_{urea} is improved as indicated in Tab. 1. The associated "optimal" profile is depicted in Fig. 22 (right). Since the optimization procedure is rather efficient it may be directly coded in the software that drives the cycler.

Table 1. Improvement in KT/V_{urea} yielded by the optimization algorithm

Iterations	KT/V_{urea}	% improvement
0	1.3299	0
10	1.3354	0.4
30	1.3554	1.9

This new simulation environment is opening new perspectives for a PD treatment carried out at the patient's home. By means of mathematical tools, the therapy is actually tuned in the "best" way for the individual patient who may choose the shortest therapy for a given efficiency target, or the more efficient therapy for a given period of dialysis. Furthermore, sensors measuring the concentrations in the patient may provide a feedback control. The mathematical model can then be improved by a "data assimilation" procedure, that adapts the therapy constantly.

2.6 Anastomosis shape optimization

The problem

When a coronary artery is (partially or totally) occluded, blood is unable to oxygenate the heart muscle properly. The oxygen supply can be restored surgically by means of a bypass from the aorta to the coronary artery downstream of the occlusion. Various implant procedures and bypass types are currently available (see Fig. 24). A bypass can be composed either of organic material (e.g., the saphena vein taken from one of the patient's legs, or the mammary artery) or of prosthetic material. They may feature very different shapes such as, e.g., cuffed arteriovenous access grafts.

Current statistics [23] show that unfortunately 18% of patients who undergo surgery for a bypass implant risk re-occlusion and the 80% of bypasses implanted must be replaced after 10 years. The repetition of surgical procedures involves a high risk of complications. This is why it is worthwhile to investigate the aspects that may cause complications and post-operative failures, such as recirculation, abnormal disturbed flows, re-stenoses, hyperplasia, etc., with the aim of finding a strategy for their reduction.

In Fig. 25 we illustrate simulations in two simplified anastomoses, highlighting the impact of the angle between the stenosed branch and the graft on the downstream haemodynamics during the diastole. In fact, mathematical shape optimization tools can be applied for suggesting optimized configurations at various levels, from the local geometry (especially in the implant area) to the quantities that form the entire

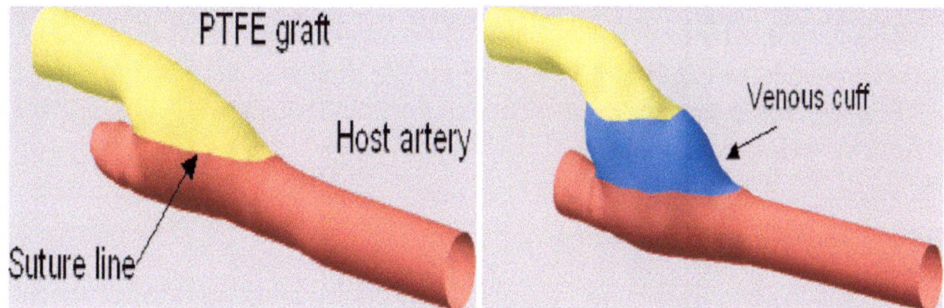

Fig. 24. Two different possible morphological variants of a coronary bypass (from the web site www.numerik.math.tugraz.at/biomech/cfd/selected_studies/flow.html, courtesy of K. Perktold, M. Prosi.). *On the left*: A conventional procedure. *On the right*: The so-called: Miller cuff

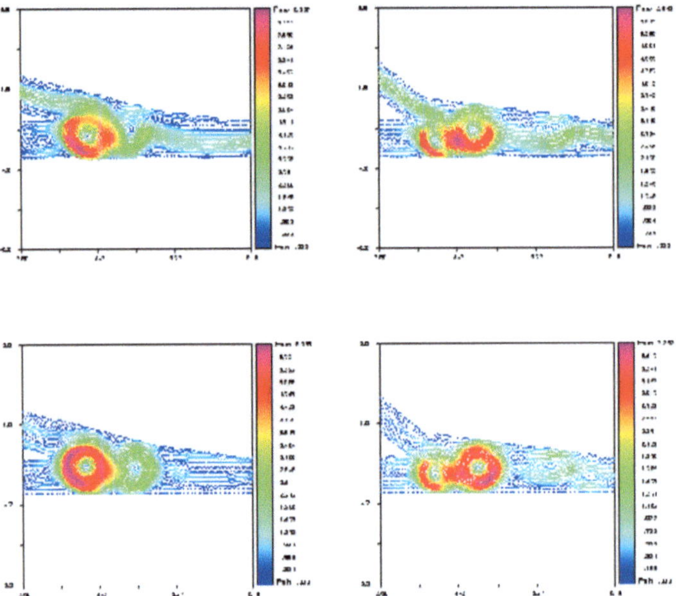

Fig. 25. Velocity field during diastole in two different models of a bypass

bypass structure (implant angle, ratio between bypass diameter and the artery in the implant area, distance between the new implant and the stenoses, etc.).

Shape optimization is currently applied in many fields of engineering (see, e.g., [41]) and, in principle, can be extended to many biomedical applications. Here, the domain of interest is not given, as it is the outcome of the computation. The goal is to find the domain that allow us to satisfy requirements for the velocity and pressure fields, which are normally specified as the minimum of a suitable cost functional, subject to constraints. This procedure implies high computational costs that have so far prevented its practical application in surgical planning. Here we present nevertheless some basic results.

We note that, aside from the problem we are considering here, shape optimization techniques could be applied as well to the FloWatch device discussed in Sect. 2.1.

This research is carried out in cooperation with A. Patera (MIT, Cambridge) and V. Agoshkov (Russian Academy of Sciences, Moscow).

Numerical models and simulations

Approaches to control and shape design. Our goal is to find a geometrical configuration of the anastomosis which could reduce the downstream (or distal) occlusion risk, which in turn is related to the local haemodynamics generated by the bypass. To this aim, we need to: (i) find an appropriate cost function, related to the local haemodynamics, which measures how the latter affects the distal re-stenosis risk and

(ii) devise a procedure to find the bypass shape which, among all admissible shapes, reduces the risk.

Certain physical quantities, often called *indices*, have been proposed for measuring the re-occlusion risk (e.g., the Oscillatory Shear Index, the Mean Wall Shear Stress Gradient, the Oscillatory Flow Index). They are all derived from the velocity and pressure fields that can be computed by solving the Navier-Stokes equations (1) around the anastomosis region. In the optimization process, these equations play the role of *state equations*, establishing the relations between the control variables (the shape of the domain) and the function to be controlled. As a cost functional, here we have chosen a measure of the *fluid vorticity* in the domain since it was found that all the above indices have a favorable value in a flow with low vorticity. So it seems resonable to take it as a possible indicator for anomalous downstream conditions induced by the anastomosis (see Fig. 25). More precisely, if Ω_{d} denotes the anastomosis downstream region, a reliable cost functional is (see [43]):

$$\mathcal{J}(\boldsymbol{u}) \equiv \int_{\Omega_{\mathrm{d}}} |\nabla \times \boldsymbol{u}|^2 d\mathbf{x}. \tag{12}$$

The second issue can be faced with (at least) two different approaches. In the first, the domain is locally deformed by moving each point on the boundary following the indication of the optimization algorithm. In practice, this means that the Navier-Stokes equations are discretized (e.g., by finite elements) on a computational mesh; then the optimization algorithm computes the displacement of the boundary nodes. The mesh is deformed or recomputed correpondingly (see Fig. 26 (top-left)).

A different approach is based on perturbation theory. Suppose that the boundary to be adjusted is described by a function $f(x, \varepsilon)$ (see Fig. 26 (bottom-left)), and in particular that the dependence on the parameter ε is expressed as

$$f(x, \varepsilon) = f_0(x) + \varepsilon f_1(x) + \varepsilon^2 f_2(x) + \dots,$$

where $f_0(x)$ corresponds to the unperturbed shape. The Navier-Stokes solutions $\boldsymbol{u}$ and p are assumed to be regular functions of the parameter ε, so that the expansions $\boldsymbol{u} = \boldsymbol{u}_0 + \varepsilon \boldsymbol{u}_1 + \varepsilon^2 \boldsymbol{u}_2 + \dots$ and $p = p_0 + \varepsilon p_1 + \varepsilon^2 p_2 + \dots$ make sense. Then optimal control theory can be used for computing f_i by solving the problem for the first perturbations $\boldsymbol{u}_i, p_i, i = 1, 2, \dots$.

In the sequel, we briefly address the first approach. The reader interested in the second approach is referred to [1].

We note that it is possible to effect the optimization at a different, more global, geometrical level, by considering the configuration of the whole bypass parametrized by a few geometrical quantities (see Fig. 26 (right)), e.g., the length and the angle of the graft, the distance between the anastomosis and the stenosis, etc. The two optimization levels (local and global) can be suitably combined. In fact, the global optimization can be attacked by using the so-called *reduced basis techniques*, which can obtain "optimal" parameter estimates with a low computational cost. The results of this step can be used as the initial configuration of the local shape optimization (see [53]).

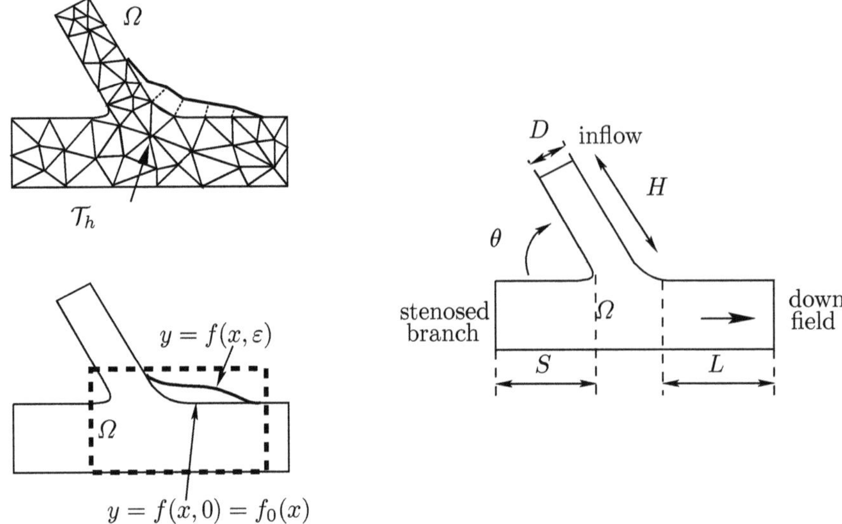

Fig. 26. Shape control approaches: local optimization based on the local control of the nodes (*top left*), local control based on perturbation theory (*bottom left*), global control based on a few geometrical parameters (*right*)

Results. The basic mathematical ingredients for an optimal control problem are:

1. a *control variable* (possibly a vector of scalar variables), belonging to a functional space $\mathcal{U}$ of admissible controls;
2. the *state equations*, that represent the physical system to be optimized; for the sake of simplicity, in the numerical simulations presented here, we refer to the steady linear Stokes problem;
3. the *cost functional* $\mathcal{J}$; in our case, we choose (12).

The general statement of the problem then reads:

$$\text{find} \quad w \in \mathcal{U} \quad \text{such that} \quad \mathcal{J}(w) \leq \mathcal{J}(v) \quad \forall v \in \mathcal{U}.$$

The minimization can be obtained iteratively. Starting from a given configuration, the state problem, suitably discretized, is solved to estimate the cost functional. The boundary deformation is suggested by a descent gradient-type method (see [41]). This step requires the evaluation of the cost functional gradient $\mathcal{J}'$ which is computed by solving another partial differential system, the *adjoint problem*. When the displacement is computed, the domain is moved and the mesh is deformed accordingly. The loop continues until a given stopping criterion is satisfied.

Figure 27 illustrates the results obtained by this algorithm: at the top-left we have the initial unperturbed configuration and then, in clockwise order, we have configurations featuring 22%, 38% and 45% (optimal shape) vorticity reduction. The full loop was carried out in 25 iterations. It is interesting to note that the optimal

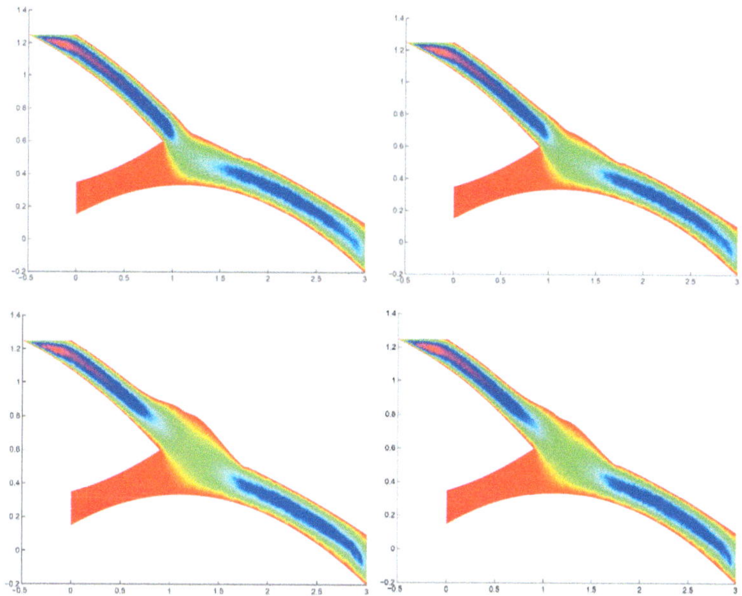

Fig. 27. Different solutions found during the shape optimization process: initial shape is at the top-left and in clockwise order we have shapes featuring 22%, 38% and 45% reduction in vorticity (courtesy of G. Rozza)

shape resembles an intervention in use in surgical practice, the so-called Taylor patch (see [43]). Therefore, control theory furnishes in this case a possible rigorous "explanation" of a practice so far based only on intuition and experience.

3 A wider perspective

In this work we present just a few practical examples where the mathematical and numerical modeling of the cardiovascular system has made relevant contributions, and probably will make even more important ones in the future. Many other problems of a relevant medical interest have been investigated by developing mathematical models and numerical methods. We cite, for instance, vessel tissue dynamics and the mechanics of the heart wall. These problems stimulate the development of accurate and computationally affordable models for biological tissues. The description of the mechanics of the walls is quite often based on the definition of a *strain energy density function*, whose derivatives yield the components of the stress-deformation tensor. An overview of recent contributions in this field can be found in [57]. In particular, as regards heart mechanics and functionality, recent investigations show that the ventricular myocardium can be unwrapped by blunt dissection into *a single continuous muscle band* (see [59]): this anatomical evidence could be included in mathematical models for heart mechanics. Correspondingly, numerical methods for the simulation of the fluid-structure interaction in arteries and in the heart have been extensively investigated. The *immersed boundary method* proposed by C. Peskin for heart dynamics in 1970's is receiving growing attention in different fields of computational biology (see, e.g., [34, 40]). Other methods based on the iterative solution of fluid and structure problems and the Arbitrary Lagrangian Eulerian (ALE) approach for managing mesh motion have been investigated in recent years ([11, 12, 20]). In following this approach, the fluid-structure interaction problem is split at each time level into the separate computation of the fluid velocity and pressure fields and of the structure displacement, together with the computation of the stress on the wall (which typically is used as a boundary term for the structure solver) and the mesh displacement ALE computation. This research has yielded a 10-fold reduction in CPU time for a typical haemodynamic simulation, even if, for stability reasons, about 10^4 time steps are needed for one second of time simulated (about one heart beat). This stimulates active research aimed at achieving more effective methods.

In Fig. 28 we reproduce snapshots (taken from [11]) of a fluid-structure simulation in a carotid artery obtained with this method, implemented in the code LifeV (see www.lifev.org).

Another issue that we have not considered in our overview is *temperature* and its role in physiology and also in such medical treatments as hyperthermia therapy in oncology [55].

In this context, and in general in the field of microcirculation, modeling requires mathematical techniques for the correct description of the behavior of heterogeneous media featuring small scales with respect to the scale of interest.

Besides the problems just mentioned, which belong to the *core* of computational haemodynamics, other relevant topics strictly related to numerical computing are

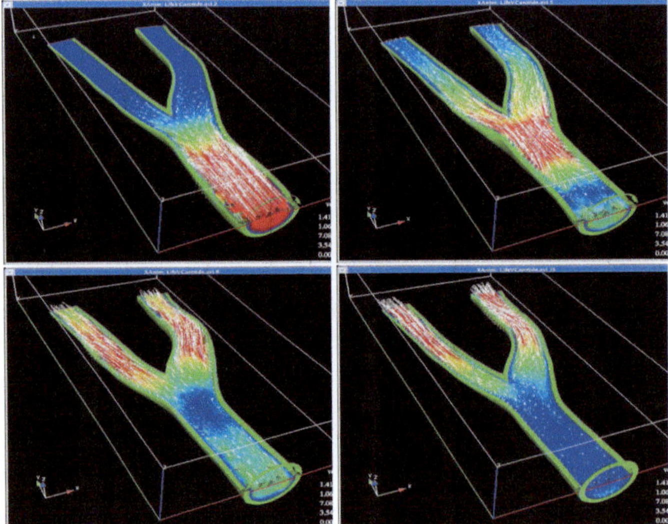

Fig. 28. Snapshots of a fluid-structure interaction simulation in a compliant carotid artery (courtesy of G. Fourestay)

to be addressed which concern pre- and post-processing in blood flow simulations. *Accurate geometrical reconstruction* from medical images is an important and active field of research (see, e.g., [56]); in particular, the impact of geometrical modeling on the accuracy and reliability of numerical simulations is a crucial aspect yet to be extensively investigated. On the other hand, an *effective synthesis of the large amounts of data* obtained by numerical simulations is of paramount relevance for medical purposes. In fact, the definition and accurate computation of a few quantities or indices, able to summarize the relevance of a pathology, e.g., the residence time inside an aneurysm, are decisive steps in translating numerical results into practical indications for medical doctors.

The case studies presented and these final comments give evidence of the great development of mathematical and numerical models for the cardiovascular system in recent years. Basic aspects of the problems at hand have been understood and in some cases this has yielded practical answers. Future challenges will concern the numerical *integration* of the basic components developed so far. To quote [8]: this goal can be regarded in the framework of: "*in silico* organs, organ systems and, ultimately, organisms. In silico models will be crucial tools for biomedical research and development in the new millennium, extracting knowledge from the vast amount of increasingly detailed data, and integrating this into a comprehensive analytical description of biological functions with predictive power: the Physiome." The present overview pinpoints the need for modeling complex, heterogeneous and interacting dynamics, ranging from single cell dynamics up to complex network analysis. An instance of the most interesting tasks of such "in silico" vision is the coupled electrical and mechanical simulation of the heart.

The complexity of the subject is also reflected in the necessity of integrating different mathematical and numerical tools in the same solution environment. For example, statistical and numerical tools could be integrated in establishing correlations between clinical data, understanding their driving mechanisms and defining precise decision trees, which are common tools in clinical practice. The effective integration of mathematical/numerical methods can therefore be targeted as a key answer to the new (or even old) challenges of vascular medicine and surgery.

Acknowledgments

The authors wish to thank everyone involved in the various projects addressed in this work: in no specific order, we mention A. Corno (Liverpool), C. D'Angelo, D. Mastalli, G. Rozza (CMCS-EPFL, Lausanne), F. Nobile, M. Prosi, P. Zunino (MOX, Politecnico di Milano), S. Deparis (MIT, Cambridge), G. Dubini, F. Migliavacca, G. Pennati, R. Balossino, K. Laganà (LABS, Milan), M.A. Fernandez, J.F. Gerbeau (INRIA, Paris). This work was made possible by INDAM support through the Project: "Integrazione di Sistemi Complessi in Biomedicina: modelli, simulazioni, rappresentazioni" and the EU Project HPRN-CT-2002-00270 "HaeModel".

References

[1] Agoshkov, V., Quarteroni, A., Rozza, G.: Shape design in aorto-coronaric bypass using perturbation theory. SIAM J. Num. Anal., to appear (2006)

[2] American Society Of Nephrology (2005) www.asn-online.org

[3] Bove, E.: Current status of staged reconstruction for hypoplastic left heart syndrome. Pediatr. Cardiol. **19**, 308–315 (1998)

[4] Čanić, S., Kim, E.H.: Mathematical analysis of the quasilinear effects in a hyperbolic model for blood flow through compliant axi-symmetric vessels. Math. Methods Appl. Sci. **26**,1161–1186 (2003)

[5] Corno, A., Fridez, P., von Segesser, L., Stergiopulos, N.: A new implantable device for telemetric control of pulmonary blood flow. Interact. Cardiov. Thorac. Surg. **1**, 46–49 (2002)

[6] Corno, A., Prosi, M., Fridez, P., Zunino, P., Quarteroni, A., von Segesser, L.: No more pulmonary artery reconstruction after banding. Submitted. J. Thorac. Cardiovasc. Surg. (2005)

[7] Corno, A., Sekarski, N., Bernath, M., Payot, M., Tozzi, P., von Segesser, L.: Pulmonary artery banding: long-term telemetric adjustment. Eur. J. Cardiothorac. Surg. **23**, 317–322 (2003)

[8] Crampin, E., Halstead, M., Hunter, P., Nielsen, P., Noble, D., Smith, N., Tawhai, M.: Computational physiology and the Physiome Project. Exp. Physiol. **89**, 1–26 (2004)

[9] Curry, F.: Mechanics and thermodynamics of transcapillary exchange. In: Renkin, E.M., Michel, C.C. (eds.): Handbook of physiology. Sect. 2: The cardiovascular system. Vol. IV. Microcirculation. Bethesda, MD: Amer. Physiol. Soc. 1984, pp. 309–374

[10] D'Angelo, C., Papelier, Y.: Mathematical modelling of the cardiovascular system and skeletal muscle interaction during exercise. In: Cancès, E., Gerbeau J.-F. (eds.): CEMRACS 2004 – Mathematics and applications to biology and medicine. (ESAIM: Proc. **14**) LesUlis: EDP Sciences 2005, pp. 72–88

[11] Deparis, S., Discacciati, M., Fourestey, G., Quarteroni, A.: Fluid-structure algorithms based on Steklov-Poincaré operators. Comput. Methods Appl. Mech. Engrg., to appear (2006)

[12] Deparis, S., Fernández, M.A., Formaggia, L.: Acceleration of a fixed point algorithm for fluid-structure interaction using transpiration conditions. M2AN Math. Model. Numer. Anal. **37**, 601–616 (2003)

[13] Formaggia, L., Gerbeau, J.F., Nobile, F., Quarteroni, A.: Numerical treatment of defective boundary conditions for the Navier-Stokes equations. SIAM J. Numer. Anal. **40**, 376–401 (2002)

[14] Formaggia, L., Lamponi, D., Quarteroni, A.: One-dimensional models for blood flow in arteries. J. Engrg. Math. **47**, 251–276 (2003)

[15] Formaggia, L., Nobile, F., Quarteroni, A., Veneziani, A.: Multiscale modelling of the circulatory system: a preliminary analysis. Comput. Visual. Sci. **2**, 75–83 (1999)

[16] Formaggia, L., Veneziani, A.: One dimensional models for blood flow in the human vascular system. In: Biological Fluid Dynamics. Rhode-St. Genèse, Belgium: von Karman Institute 2003

[17] Formaggia, L., Veneziani, A.: Geometrical multiscale models of the cardiovascular system: from lumped parameters to 3D simulations. In: Biological Fluid Dynamics. Belgium: von Karman Institute 2003

[18] Formaggia, L., Veneziani, A.: Geometrical multiscale models for the cardiovascular system. In: Kowalewski, T.A. (ed.): Blood flow modelling and diagnostics. Warsaw: Abiomed, Polish Academy of Sciences 2005

[19] Fry, D.: Mass transport, atherogenesis and risk. Atheroscl. **7**, 88–100 (1987)
[20] Gerbeau, J.-F., Vidrascu, M., Frey, P.: Fluid-structure interaction in blood flows on geometries based on medical imaging. Comput. Structures **83**, 155–165 (2005)
[21] Gokal, R., Blake, P., Passlick-Deetjen, J., Schaub, T., Prichard, S., Butkart, J.: What is the evidence that peritoneal dialysis is underutilized as an ESRD therapy? Semin. Dial. **15**, 149–161 (2002)
[22] Gokal, R., Nolph, K.: The textbook of peritoneal dialysis. Dordrecht: Kluwer 1994
[23] Health Centers On Line (2005) heart.healthcentersonline.com
[24] Heywood, J., Rannacher, R., Turek, S.: Artificial boundaries and flux and pressure conditions for the incompressible Navier-Stokes equations. Internat. J. Numer. Methods Fluids **22**, 325–352 (1996)
[25] Hoppensteadt, F., Peskin, C.: Mathematics in medicine and the life sciences. New York: Springer 1992
[26] Kedem, O., Katchalsky, A.: Thermodynamic analysis of the permeability of biological membranes to non-electrolytes. Biochim. Biophys. Acta. **27**, 229–246 (1958)
[27] Keener, J., Sneyd, J.: Mathematical physiology. New York: Springer 1998
[28] Lagana, K., Balossino, R., Migliavacca, F., Pennati, G., Bove, E., de Leval, M., Dubini, G.: Multiscale modeling of the cardiovascular system: application to the study of pulmonary and coronary perfusions in the univentricular circulation. J. Biomech. **38**, 1129–1141 (2005)
[29] Laganà, K., Dubini, G., Migliavacca, F., Pietrabissa, R., Pennati, G., Veneziani, A., Quarteroni, A.: Multiscale modelling as a tool to prescribe realistic boundary conditions for the study of surgical procedures. Biorheol. **39**, 359–364 (2002)
[30] Lamponi, D.: One dimensional and multiscale models for blood flow circulation. Thèsis. Lausanne: EPFL 2004
[31] Martin, V., Clément, F., Decoene, A., Gerbeau, J.-F.: Parameter identification for a one-dimensional blood flow model. In: Cancès, E., Gerbeau, J.-F. (eds.): CEMRACS 2004 – Mathematics and applications to biology and medicine. (ESAIM: Proc. **14**) LesUlis: EDP Sciences 2005, pp. 174–200
[32] Metcalfe, R.: The promise of computational fluid dynamics as a tool for delineating therapeutic options in the treatment of aneurysms. Am. J. Neurorad. **24**, 553–554 (2003)
[33] Migliavacca, F., Balossino, R., Pennati, G., Dubini, G., Hsia, T.-Y., de Leval, M.R., Bove, E.L.: Multiscale modelling in biofluiddynamics: Application to reconstructive paediatric cardiac surgery. J. Biomech. **39**, 1010–1020 (2006)
[34] Mittal, R., Iaccarino, G.: Immersed boundary methods. In: Lumley, J.L. et al. (eds.): Annual Review of Fluid Mechanics. Vol. 37, Palo Alto, CA: Annual Reviews 2005, pp. 239–261
[35] Moncrief, J., Popovich, R.: Continuous ambulatory peritoneal dialysis. Contrib. Nephrol. **17**, 139–145 (1979)
[36] Nerem, R., Levesque, M.J.: Fluid mechanics in atherosclerosis. In: Skalak, R., Chien, S. (eds.): Handbook of bioengineering. New York: McGraw-Hill 1987, pp. 21.1–21.22
[37] Nichols, W., O'Rourke, M.: McDonald's Blood flow in arteries. 3rd ed. London: Arnold 1990
[38] Norwood, W.: Hypoplastic left heart syndrome. Ann. Thorac. Surg. **52**, 688–695 (1991)
[39] Oreopoulos, D., Lobbedez, T., Gupta, S.: Peritoneal dialysis: where is it now and where is it going? Int. J. Artif. Organs **27**, 88–94 (2004)
[40] Peskin, C., McQueen, D.: A three-dimensional computational method for blood flow in the heart. I. Immersed elastic fibers in a viscous incompressible fluid. J. Comput. Phys. **81**, 372–405 (1989)

[41] Pironneau, O.: Optimal shape design for elliptic systems. (Springer Series in Computational Physics) Berlin: Springer 1984
[42] Quarteroni, A., Formaggia, L.: Mathematical modelling and numerical simulation of the cardiovascular system. In: Ayache, N. (ed.): Handbook of numerical analysis, Computational models for the human body. Vol. XII. Amsterdam: North-Holland 2004, pp. 3–127
[43] Quarteroni, A., Rozza, G.: Optimal control and shape optimization of aorto-coronaric bypass anastomoses. Math. Models Methods Appl. Sci. **13**, 1801–1823 (2003)
[44] Quarteroni, A., Sacco, R., Saleri, F.: Numerical mathematics. Berlin: Springer 2000
[45] Quarteroni, A., Valli, A.: Domain decomposition methods for partial differential equations. Oxford: Clarendon Press 1999
[46] Quarteroni, A., Veneziani, A.: Analysis of a geometrical multiscale model based on the coupling of ODEs and PDEs for blood flow simulations. Multiscale Models. Simul. **1**, 173–195 (2003)
[47] Quarteroni, A., Veneziani, A., Zunino, P.: A domain decomposition method for advection-diffusion processes with application to blood solutes. SIAM J. Sci. Comput. **23**, 1959–1980 (2002)
[48] Quarteroni, A., Veneziani, A., Zunino, P.: Mathematical and numerical modeling of solute dynamics in blood flow and arterial walls. SIAM J. Numer. Anal. **39**, 1488–1511 (2002)
[49] Raker, B.: Researchers announce advances in medical models and tools for space exploration. Northwest Science and Technology Winter 2005 https://depts.washington.edu/nwst
[50] Rappitsch, G., Perktold, K.: Pulsatile albumin transport in large arteries: a numerical simulation study. J. Biomech. Eng. **118**, 511–519 (1996)
[51] Rippe, B.: A three-pore model of peritoneal transport. Perit. Dial. Int. **13** (Suppl. 2), S35–S38 (1993)
[52] Robertson, A., Sequeira, A.: A director theory approach for modelling blood flow in the arterial system: an alternative to classical ID models. Math. Models Methods Appl. Sci. **15**, 871–906 (2005)
[53] Rozza, G.: On optimization, control and shape design for an arterial bypass. Internat. J. Numer. Methods Fluids **47**, 1411–1419 (2005)
[54] Sano, S., Ishino, K., Kawada, M., Arai, S., Kasahara, S., Asai, T., Masuda, Z., Takeuchi, M., Ohtsuki, S.: Right ventricle-pulmonary artery shunt in first-stage palliation of hypoplastic left heart syndrome. J. Thorac. Cardiovasc. Surg. **126**, 504–509 (2003)
[55] Sreenivasa, G., Gellermann, J., Rau, B., Nadobny, J., Schlag, P., Deuflhard, P., Felix, R., Wust, P.: Clinical use of the hyperthermia treatment planning system HyperPlan to predict effectiveness and toxicity. Int. J. Radiat. Oncol. Biol. Phys. **55**, 407–419 (2003)
[56] Steinman, D.: Image-based computational fluid dynamics modeling in realistic arterial geometries. Ann. Biomed. Eng. **30**, 483–497 (2002)
[57] Telega, J.J., Stańcyzk, M.: Modelling of soft tissues beahaviour. In: Telega, J.J.: (ed.): Modelling in biomechanics (AMAS **19**) Lecture Notes. Warsaw: Abiomed, Polish Academy of Sciences 2005
[58] Temam, R.: Navier-Stokes equations. Theory and numerical analysis. 3rd ed. Amsterdam: North-Holland 1984
[59] Torrent-Guasp, F.: The cardiac muscle. Madrid: Foundacioń Juan March 1973
[60] Vonesh, E., Lysaght, M., Moran, J., Farrell, P.: Kinetic modeling as a prescription aid in peritoneal dialysis. Blood Purif. **9**, 246–270 (1991)
[61] Wang, J.J., Parker, K.: Wave propagation in a model of the arterial circulation. J. Biomech. **37**, 457–470 (2004)

[62] Weisstein, E.: Pontryagin maximum principle. mathworld.wolfram.com
[63] White, F.: Viscous fluid flow. 2nd ed. New York: McGraw-Hill 1991
[64] Wolfram: www.mathematica.org
[65] Zunino, P.: Multidimensional pharmacokinetic models applied to the design of drug-eluting stents. Cardiovasc. Eng. **4**, 181–191 (2004)
[66] Zunino, P., Mastalli, D., Quarteroni, A., VanBiesen, W., Vecten, D., Pacitti, A., Lameire, N., Neftel, F., Wauters, J.: Development of a new mathematical approach to optimize peritoneal dialysis. In preparation. 2005

Index